AF753268

Classic Works in RF Engineering

Combiners, Couplers, Transformers, and Magnetic Materials

For a listing of recent titles in the *Artech House Microwave Library,* turn to the back of this book.

Classic Works in RF Engineering

Combiners, Couplers, Transformers, and Magnetic Materials

John L. B. Walker
Daniel P. Myer
Frederick H. Raab
Chris Trask
Editors

ARTECH
HOUSE
BOSTON | LONDON
artechhouse.com

Library of Congress Cataloging-in-Publication Data
A catalog record for this book is available from the U.S. Library of Congress.

British Library Cataloging in Publication Data

Classic works in RF engineering : combiners, couplers, transformers, and magnetic materials. — (Artech House microwave library)
1. Radio—Equipment and supplies 2. Radio frequency
I. Walker, John L.B.
621.3′84

ISBN-10: 1-59693-376-3

Cover design by Eleanor Roe Studio, Inc.

International Standard Book Number: 1-59693-376-3

10 9 8 7 6 5 4 3 2 1

CONTENTS

Preface . . . ix

Acknowledgments . . . xi

Chapter 1 Introduction . . . 1
John L.B. Walker

1.1 Background . . . 1
1.2 Aims and Organization of This Book . . . 1
Reference . . . 2

Chapter 2 Magnetic Materials . . . 3
Chris Trask

2.1 Introduction . . . 3
2.2 Powdered Iron Materials . . . 3
2.3 Ferrite Materials . . . 3
2.4 Complex Permeability . . . 4
2.4.1 Temperature Variations . . . 5
2.4.2 Effects of Machining . . . 5
2.5 Dimensional Resonance . . . 5
2.6 Losses and the Wideband Transformer Model . . . 5
2.6.1 Eddy Current Losses . . . 6
2.6.2 Hysteresis Losses and the Hysteresis Curve . . . 6
2.7 Barkhausen Noise . . . 7
2.8 Ferromagnetic Distortion . . . 7
2.9 Conclusion . . . 7
References . . . 7
Reprints
Gary A. Breed, "An Introduction to Iron Powder and Ferrite Cores," *RF Design*, March 1992, pp. 121–124. . . . 9
Jim Cox, "Iron Powder Cores for RF Power Applications," *RF Design*, February 1987, pp. 39, 41, and 43 (correction in March 1987). . . . 13
M.F. "Doug" DeMaw, "Magnetic Cores in RF Circuits," *RF Design*, Vol. 3, No. 4, April 1980, pp. 12–16. . . . 17
D. Polder, "Ferrite Materials," *Proceedings of the IEE*, Part 2, Vol. 97, April 1950, pp. 12–16. . . . 21
Hideo Saotome and Yo Sakaki, "Complex Permeability of Polycrystalline Mn-Zn and Ni-Zn Ferrites," *Electrical Engineering in Japan*, Vol. 123, No. 2, February 1998, pp. 1–7. . . . 33
V. Cagan and M. Guyot, "Fast and Convenient Technique for Broadband Measurements of the Complex Permeability of Ferrimagnets," *IEEE Transactions on Magnetics*, Vol. MAG-20, No. 5, September 1984, pp. 1732–1734. . . . 41
J.J. Suh, B.M. Song, and Y.H. Han, "Temperature Dependence of Power Loss of Mn-Zn Ferrites at High Frequency," *IEEE Transactions on Magnetics*, Vol. 36, No. 5, September 2000, pp. 3402–3404. . . . 45
Glenn R. Skutt and Fred C. Lee, "Characterization of Dimensional Effects in Ferrite-Core Magnetic Devices," *1996 IEEE Power Electronics Specialists Conference (PESC '96)*, pp. 1435–1440. . . . 49
Toshiro Sato and Yo Sakaki, "Physical Meaning of Equivalent Loss Resistance of Magnetic Cores," *IEEE Transactions on Magnetics*, Vol. 26, No. 5, September 1990, pp. 2894–2897. . . . 55

M. Barloti, A. Reatti, and M.K. Kazimierczuk, "Modeling Iron-Powder Inductors at High Frequencies," *1994 IEEE Industry Applications Meeting (IAS '94)*, Vol. 2, pp. 1225–1232. 59
Marian K. Kazimierczuk, Giuseppe Sancineto, Gabriele Grandi, Ugo Reggiani, and Antonio Massarini, "High-Frequency Small-Signal Model of Ferrite Core Inductors," *IEEE Transactions on Magnetics*, Vol. 35, No. 5, September 1999, pp. 4185–4191. 67
K. Hollaus and O. Bíró, "Derivation of a Complex Permeability From the Preisach Model," *IEEE Transactions on Magnetics*, Vol. 38, No. 2, March 2002, pp. 905–908. 75
Heinz Bittel, "Noise of Ferromagnetic Materials," *IEEE Transactions on Magnetics*, Vol. MAG-5, No. 3, September 1969, pp. 359–365. 79
Shane Collins and Kieran Flynn, "Intermodulation Characteristics of Ferrite-Based Directional Couplers," *Microwave Journal*, November 1999, pp. 122–130. 87

Chapter 3 Baluns/Hybrids/Anti-Phase Power Splitters 91
John L.B. Walker
3.1 Introduction 91
3.2 Lossless Two- or Three-Port Baluns 92
3.2.1 Bridge (Lattice) Baluns 92
3.2.2 Marchand Balun 93
3.2.3 Guanella Balun 95
3.2.4 Anti-Phase Power Splitters Derived from Symmetrical 4-Port Networks 98
3.3 Lossy Three-Port and Lossless Four-Port Baluns 98
3.4 Conclusions 99
References 99
Reprints
Winfried Bakalski, Werner Simbürger, Herbert Knapp, Hans-Dieter Wohlmuth, and Arpad L. Scholtz, "Lumped and Distributed Lattice-Type LC-Baluns," *IEEE International Microwave Symposium*, Seattle, WA, June 3–7, 2002, pp. 209–212. 101
Nathan Marchand, "Transmission-Line Conversion Transformers," *Electronics*, Vol. 17, December 1944, pp. 142–145. 105
J.H. Cloete, "Exact Design of the Marchand Balun," *Microwave Journal*, Vol. 23, No. 5, May 1980, pp. 99–102. 109
J.H. Cloete, "Graphs of Circuit Elements for the Marchand Balun," *Microwave Journal*, Vol. 24, No. 5, May 1981, pp. 125–128. 111
Charles L. Goldsmith, Anthony Kikel, and Nancy L. Wilkens, "Synthesis of Marchand Baluns Using Multilayer Microstrip Structures," *International Journal of Microwave and Millimeter-Wave Computer-Aided Engineering*, Vol. 2, No. 3, 1992, pp. 179–188. 115
G. Guanella, "New Method of Impedance Matching in Radio-Frequency Circuits," *The Brown Boveri Review*, September 1944, pp. 327–329. 125
Yoke Choy Leong, Kian Sen Ang, and Chee How Lee, "A Derivation of a Class of 3-Port Baluns from Symmetrical 4-Port Networks," *IEEE International Microwave Symposium*, Seattle, WA, June 3–7, 2002, pp. 1165–1168. 129
Chi-Yang Chang and Chu-Chen Yang, "A Novel Broad-Band Chebyshev-Response Rat-Race Ring Coupler," *IEEE Transactions on Microwave Theory and Techniques*, Vol. 47, No. 4, April 1999, pp. 455–462. 133
James S. McLean, "Balancing Networks for Symmetric Antennas—I: Classification and Fundamental Operation," *IEEE Transactions on Electromagnetic Compatibility*, Vol. 44, No. 4, November 2002, pp. 503–514. 141

Chapter 4 Transmission Line Transformers 153
Daniel P. Myer
4.1 Introduction 153
4.2 Philosophy/Organization 154
4.3 Historical Aspects 154
4.4 Transformer Network Synthesis 155
4.5 Transformer Network Analysis 155
4.6 Transformer Design, Construction, and Application 156

4.7 Conclusions and Commentary on Next Generation Transmission Line Transformers 156
References 156
Reprints
C.L. Ruthroff, "Some Broad-Band Transformers," *Proceedings of the IRE*, Vol. 47, August 1959, pp. 1337–1342. 159
S.E. London and S.V. Tomashevich, "Line Transformers with Fractional Transformation Factor," *Telecommunications and Radio Engineering*, Vol. 28/29, April 1974, pp. 129–130. 165
Thomas J. Gluszczak and John D. Harmer, "Transmission Line Transformers with Integer-Ratio Voltage Transformations," *Proceedings of the IEEE International Symposium on Circuits and Systems*, 1974, pp. 368–370. 167
Daniel Myer, "Synthesis of Equal Delay Transmission Line Transformer Networks," *Microwave Journal*, Vol. 35, No. 3, March 1992, pp. 106–114. 171
Donald A. McClure, "Broadband Transmission Line Transformer Family Matches a Wide Range of Impedances," *RF Design*, Vol. 17, No. 2, February 1994, pp. 62–66. 175
Donald A. McClure, "Broadband Transmission Line Transformer Family Matches a Wide Range of Impedances—Part 2," *RF Design*, Vol. 18, No. 5, May 1995, pp. 40–49. 179
Ersch Rotholz, "Transmission-Line Transformers," *IEEE Transactions on Microwave Theory and Techniques*, Vol. MTT-29, No. 4, April 1981, pp. 327–331. 185
G. Soundra Pandian and R. Arockiasamy, "Equivalent Representation of Broad-Band Transmission-Line Transformer with Vector Impedances," *Proceedings of the IEEE*, Vol. 69, No. 10, October 1981, pp. 1368–1369. 191
K.B. Niclas, R.R. Pereira, and A.P. Chang, "Transmission Lines Accurately Model Autotranformers," *Microwaves & RF*, Vol. 31, No. 11, November 1992, pp. 67–76. 193
Enzo Carpentieri, "Model Characterizes Transmission-Line Transformers," *Microwaves & RF*, Vol. 35, No. 11, November 1996, pp. 73–80. 199
Enzo Carpentieri, "Equations Model Transmission-Line Transformers," *Microwaves & RF*, Vol. 36, No. 6, June 1997, pp. 94–98. 205
Wade Blocker, "The Behavior of the Wide-Band Transmission-Line Transformer for Nonoptimum Line Impedance," *Proceedings of the IEEE*, Vol. 66, No. 4, April 1978, pp. 518–519. 207
Suhash C. Dutta Roy, "A Transmission Line Transformer Having Frequency Independent Matching Properties," *International Journal of Circuit Theory and Applications*, Vol. 8, 1980, pp. 55–64. 209
Peter Lefferson, "Twisted Magnet Wire Transmission Line," *IEEE Transactions on Parts, Hybrids, and Packaging*, Vol. PHP-7, No. 4, December 1971, pp. 148–154. 219
O. Pitzalis, Jr., "Practical Design Information for Broadband Transmission Line Transformers," *Proceedings of the IEEE*, Vol. 56, No. 4, April 1968, pp. 738–739. 227
Octavius Pitzalis, Jr., and Thomas P.M. Couse, "Broadband Transformer Design for RF Transistor Power Amplifiers," *1968 Electronic Components Conference*, pp. 207–216. 229
H.O. Granberg, "New MOSFETs Simplify High Power RF Amplifier Design," *RF Design*, Vol. 9, No. 10, October 1986, pp. 43–52. 241

Chapter 5 Hybrid Power Combiners and Splitters 247
Frederick H. "Fritz" Raab

5.1 Terminology and Classification 247
5.2 Basic Theory of Hybrid Splitter 248
5.3 Basic Theory of Hybrid Combiner 249
5.4 Literature 250
References 251
Reprints
Roderick K. Blocksome, "Practical Wideband RF Power Transformers, Combiners, and Splitters," *Proceedings of RF Technical Expo West*, 1986. 253
Helge O. Granberg, "A Two-Stage 1 kW Solid-State Linear Amplifier," *Application Note AN-758*, Motorola Semiconductor Products, Phoenix, AZ, 1976. 295
Ernest J. Wilkinson, "An *N*-Way Hybrid Power Divider," *IRE Transactions on Microwave Theory and Techniques*, Vol. MTT-8, No. 1, January 1960, pp. 116–118. 311
Ken Dufour, "60 Watt VHF Amplifier Uses Splitting/Combining Techniques," *Engineering Bulletin EB93*, Motorola (Freescale) Semiconductor, Phoenix, AZ, circa 1981. 315

Roderick K. Blocksome, "Predicting RF Output Power From Combined Power Amplifier Modules," *RF Design*, Vol. 11, No. 2, February 1988, pp. 40–49. 321
Adel A.M. Saleh, "Improving the Graceful-Degradation Performance of Combined Power Amplifiers," *IEEE Transactions on Microwave Theory and Techniques*, Vol. MTT-28, No. 10, October 1980, pp. 1068–1070. 327

Chapter 6 Quadrature Combiners and Splitters 331
Frederick H. "Fritz" Raab

6.1 Basic Theory of Quadrature Splitter 331
6.2 Quadrature Combiner with Nominal Load 332
6.3 Quadrature Combiner with Arbitrary Load 334
6.4 Cancellation of Harmonics 338
6.5 Cancellation of Backward IMD 339
6.6 Literature 339
References 340
Reprints
R.E. Fisher, "Broad-Band Twisted-Wire Quadrature Hybrids," *IEEE Transactions on Microwave Theory and Techniques*, Vol. MTT-21, No. 5, May 1973, pp. 355–357. 341
Chen Y. Ho and Ge-Lih Chen, "Design of Wideband Quadrature Couplers for UHF/VHF: Part I," *RF Design*, Vol. 12, No. 11, November 1989, pp. 58–61. 343
Chen Y. Ho and Ge-Lih Chen, "Design of Wideband Quadrature Couplers for UHF/VHF: Part II," *RF Design*, Vol. 12, No. 12, December 1989, pp. 51–55. 347
D.P. Andrews and C.S. Aitchison, "Synthesis of Symmetric Lumped Element Quadrature Hybrids," *IEE Proceedings—Microwaves, Antennas & Propagation*, Vol. 148, No. 5, October 2001, pp. 269–274. 351
D.P. Andrews and C.S. Aitchison, "Design of Cascaded Lumped Element Quadrature Hybrids," *IEE Proceedings—Microwaves, Antennas & Propagation*, Vol. 148, No. 5, October 2001, pp. 275–279. 357
Lee B. Max, "FM Broadcast 88 to 108 MHz 4,000 Watt Amplifier Subsystem," *RF Design*, Vol. 9, No. 2, February 1986, pp. 36–38. 363
Jonathan Shumaker, Raymond Basset, and Alex Skuratov, "High-Power GaAs FET Amplifiers: Push-Pull Versus Balanced Configurations," *Applied Microwave & Wireless*, Vol. 14, No. 5, May 2002, pp. 26–32. 367

About the Editors 371

PREFACE

Subcommittee MTT-17 of the Technical Coordinating Committee of the IEEE Microwave Theory and Techniques Society was created in 1997 to provide a focal point for the 26,000 RF engineers whose work involved frequencies below 1 GHz. Prior to the creation of MTT-17, there was no focal point within IEEE for work at HF, VHF, and UHF. The RF Expo Conference provided a focal point during the late 1980s and early 1990s, but was discontinued. Publications in this area were scattered among a variety of journals such as *Ham Radio* and *RF Design*, as well as the RF Expo proceedings.

Since its inception, MTT-17 has served the RF engineering community through HF/VHF/UHF sessions and workshops at the annual International Microwave Symposium. RF transformers are used in many HF/VHF/UHF designs. However, many of the important publications are no longer readily accessible. This led MTT-17 to propose a book that collected in one place the most significant papers from various sources. MTT-17 assigned members Dan Myer, Fritz Raab, and John Walker to the project, and Chris Trask was added because of his expertise on materials.

As noted, the aim of this book is to collect in one place important reprints related to RF transformers. Chapter 2, prepared by Chris Trask, discusses the characteristics of the ferrite and powdered-iron materials used in these transformers. Baluns are discussed in Chapter 3, prepared by John Walker. Dan Myer then discusses the more general forms of transformer in Chapter 4. Finally, in Chapters 5 and 6 Fritz Raab describes power splitters and combiners of both conventional-hybrid and quadrature-hybrid types. The important reprints associated with each of these topics are included at the end of each chapter. Altogether, 53 classic papers are reprinted in this book for the convenience of the RF engineer.

Frederick H. Raab
Colchester, Vermont
January 2006

ACKNOWLEDGMENTS

John L.B. Walker
I gratefully acknowledge the tremendous support provided by my fellow editors without whose efforts this book would not have been possible. This book has had a tortuous history and I would like to acknowledge what a pleasure it has been to work with Artech House who instantly appreciated its potential. Finally, I would like to acknowledge the very helpful comments from the anonymous reviewer.

Daniel P. Myer
I would like to acknowledge the important contributions of the following outstanding individuals:

- Coeditors Dr. Frederick Raab (and his wife Becky Raab), Chris Trask, and especially Dr. John Walker whose personal tenacity and persistence made this valuable work a reality;
- Tiina Ruonamaa, commissioning editor at Artech House Books, for her high professional caliber in managing this project;
- Robert Vitkovich, Thomas J. Dowling, Dr. Jerry Sevick, D.C Youla, Dr. Fred Winter, Dr. Richard Myer, Dr. Veljko Radeka, Clyde L. Ruthroff, Rowland Lee, and the late Helge O. Granberg for their valuable insight into the history, theory, and operation of transmission line transformers;
- All the authors I have had the pleasure of contacting for reprint authorization: Simon Y. London, Thomas J. Gluszczak, Donald A. McClure, Ersch Rotholz, Dr. G. Soundra Pandian, Wade Blocker, Suhash C. Dutta Roy, and Peter Lefferson;
- Finally, I dedicate this work to my mother and father Mary and Clyde Myer (the quintessential engineer), my family, Eugene, Paul, Richard, and Eleanor Myer, and my wife Catherine and son Justin.

Frederick (Fritz) H. Raab
Preparation of Chapters 5 and 6 was due in no small part to the efforts of Becky Raab. Becky cleverly tracked-down the authors and publishers of candidate reprints and then diligently obtained their permissions. She also assembled and proofread the manuscript, and made copies of the reprints for these chapters. Hans Raab prepared the figures for Chapter 5.

ACKNOWLEDGMENTS

Chapter 1

INTRODUCTION

1.1 BACKGROUND

While there is no formal definition of what constitutes radio frequency (RF), it is generally accepted that it covers at least the frequency range of 1 to 1,000 MHz, including the sub-bands of high frequency (HF) of 2 to 30 MHz, very high frequency (VHF) of 30 to 300 MHz, and ultrahigh frequency (UHF) of 300 to 1,000 MHz. However, personal mobile phones operating around 2 GHz, as well as the plethora of different wireless connection technologies based on the IEEE 802.11 standard at 2.4 GHz, all use RF construction techniques rather than traditional microwave methods. Therefore, it is generally accepted that RF extends above 1 GHz to at least 2.4 GHz. This conclusion is partly the result of the introduction of monolithic microwave integrated circuits (MMICs) that enable lumped elements to be used well above UHF.

The RF frequency range is used for a diverse range of applications including AM, FM, and digital radio broadcast, TV broadcast, over-the-horizon (OTH) radar, various communication systems (HF, troposcatter, meteor-burst, and mobile radio), magnetic resonance imaging (MRI), and RF heating. This list is by no means exhaustive, and the interested reader can find further information about RF systems in the article by Raab et al. [1]. Some of these applications require very high power levels, up to the MW level in the case of broadcast and OTH radar, but most of the other applications need at least kW power levels. Another feature of the RF frequency range is the extraordinarily large bandwidth required by some of the systems. As an example, HF radio operates over more than a decade bandwidth, and a minimum of 10 W is needed over the 2 to 30 MHz frequency range, which further extends the design challenge. UHF TV requires much higher power levels of up to 50 kW, but over the reduced frequency range of almost an octave (470–860 MHz). The combination of high power and broad bandwidth has spurred some very innovative designs for the components that are used at RF. This is remarkable considering that some of these truly creative designs originated in the 1940s, long before the advent of powerful computer-aided design programs, and were based on a deep understanding of fundamental physics.

Another feature of the RF frequency range is the transitional nature between the lower frequency audio range, where traditional lumped elements are the norm, and the higher frequency microwave range, where designs are based around lengths of transmission line. At RF lumped components cease to behave as ideal components, but instead have unwanted parasitic elements associated with them. As an example, an inductor has a parasitic shunt capacitance between the windings that creates a resonance frequency above which the inductor actually appears as a capacitor rather than an inductor. On the other hand, a quarter-wave length of transmission line is uncomfortably long at RF unless a very high dielectric constant substrate material is used, and this generally precludes the use at RF of many of the sophisticated microwave transmission line components that use multiple lengths of transmission line in the design. Consequently, RF components commonly require very innovative approaches to solving the often-conflicting requirements of small size, high power, large bandwidth, and high performance. Good examples of this are the Guanella anti-phase power splitter, (described in Chapter 3) which uses a piece of transmission line whose length is not critical, but which has ferrite wrapped around the outside of the transmission line to extend the low frequency response, and the equal-delay transformer (described in Chapter 4) which, once again, uses lengths of transmission configured in such a way that the length is not critical and so can be applied at RF.

1.2 AIMS AND ORGANIZATION OF THIS BOOK

As mentioned earlier, much of the groundbreaking development of RF components occurred in the 1940s, and many of these components still form the backbone of designs that are undertaken today. Today's designs may incorporate additional refinements, and may perhaps use construction techniques that were unheard of in the 1940s, such as MMICs, but they are nevertheless fundamentally based on decade-old concepts. Unfortunately, much of the original work is hard to find as it was published in journals that have long since ceased to be published or had limited circulation. The primary purpose of this book is to make available to today's engineers this wealth of information in a readily accessible form.

This book focuses on some passive signal processing components commonly used at RF, namely impedance conversion and power splitting/combining. Impedance conversion covers both impedance level transformation as well as balanced to unbalanced transformation. As many of these passive components require the use of magnetic materials for optimum performance, the book continues with a chapter devoted to this topic. Chapter 2 begins by looking at the history of powdered iron materials before considering ferrite materials that use the oxides of magnetic materials as their basic constituent. The chapter then continues with a discussion on losses in magnetic materials caused by eddy currents and hysteresis before concluding with noise and distortion considerations.

Chapter 3 on baluns/hybrids/anti-phase power splitters initially discusses the distinction between these terms before detailing the fundamental performance limitations associated with this type of component. The chapter then continues with a description of lossless two- and three-port baluns including the lattice (bridge), Marchand and Guanella baluns, and concludes with lossy three-port and lossless four-port baluns.

Chapter 4 is concerned with impedance transformation, and it is shown that at RF this is best undertaken by using a transmission-line transformer. The chapter then continues with a historical perspective focused on the Guanella and Ruthroff designs before considering the synthesis and analysis of these and related structures. The chapter concludes with a discussion on design, construction and application.

Finally, Chapters 5 and 6 are devoted to hybrid and quadrature power combiners and splitters, respectively. Chapter 5 begins by discussing the basic terminology and then develops the theory of hybrid splitters and hybrid combiners. This is followed by a survey of relevant papers in this area. Chapter 6 describes quadrature combining of power amplifiers, a system otherwise known as a balanced amplifier. However, this chapter also considers aspects that are not normally addressed, such as the effect of arbitrary loads on the overall performance as well as their effect upon the constituent amplifiers in terms of the impedances that they see, the supply voltages required, and so on. Harmonic and backward IMD performance is also evaluated. This material is based on hitherto unpublished research by Fritz Raab. Finally, the chapter concludes with a survey of relevant literature in this field.

REFERENCE

[1] Raab, F.H., R. Caverly, R. Campbell, M. Eron, J.B. Hecht, A. Mediano, D.P. Myer, and J.L.B.Walker, "HF, VHF, and UHF Systems and Technology," *IEEE Trans. Microwave Theory and Techniques,* Vol. MTT-50, No. 9, March 2002, pp. 653–658.

Chapter 2

MAGNETIC MATERIALS

2.1 INTRODUCTION

Magnetic materials, which encompass both powdered iron and ferrites, are an essential part in the design of wideband transformers at HF, VHF, and UHF frequencies, and, as such, the designer needs to be aware of the various properties associated with these materials. These materials provide opportunities to realize transformers of relatively small size, but, at the same time, they will have impacts in the form of added losses, noise, and in some instances, such as in association with power amplifiers, an increase in nonlinearities in the form of harmonics and intermodulation distortion (IMD).

2.2 POWDERED IRON MATERIALS

The usage of small particles of iron for magnetic applications extends back to the late 19th century. At that time, iron flakes made by way of the electrolytic and milling process were used to lower the eddy current losses in transformers and loading coils in telephone exchanges and undersea cables. This process was later used to make audio and telecommunications core materials from iron, nickel, and molybdenum alloys, such as Permalloy. The process is still in use today, as it is an economical choice for those applications, as well as for high efficiency switching power supplies.

For even low RF frequencies, the large grain size of electrolytic iron produces excessive eddy current losses. In the late 1920s, the production of fine iron powder by way of carbonyl vapormetallurgy was reduced to practice [1] primarily for making fine metal castings, and was later developed for the production of magnetic materials for the telecommunications and radio industry [2]. Since then, the process has been refined to produce iron powders of uniform physical and magnetic properties. With diameters of 5 μm (carbonyl W) and less being available, powdered iron magnetic materials can be readily used well into VHF frequencies, as shown by the included papers by Breed [3], Cox [4], and DeMaw [5]. An additional process by which iron powder is produced by hydrogen reduction is still in practice, but the larger size of the particles limits it's practical usage to frequencies below 1 MHz. A partial listing of the types of powdered iron materials that are commercially available is shown in Table 2.1.

2.3 FERRITE MATERIALS

Ferrites are defined as magnetic materials composed of oxides containing ferric (iron) ions as the main constituent. The development of modern ferrite magnetic materials began in the Netherlands in 1936 by E.C. Snoek and simultaneously in Japan by T. Takei after they realized that by combining the high electrical resistivity of iron oxides with their useful magnetic properties the resulting materials could be useful in transformers and inductors at higher frequencies than were possible with iron powders. The addition of manganese and zinc introduced an air gap which reduced the losses while keeping the permeability high. This led to the development of manganese-zinc-ferrous ferrites, and, by 1945, Snoek had established the basis of physics and technology for their practical production [6, 7]. As shown in the included paper by Polder [8], the introduction of ferrite materials with their favorable properties was met with considerable interest by the electronics industry, where they found im-

Table 2.1 *Commercial Carbonyl Iron Powder Materials*

Designation	*Diameter (mm)*	*Density (g/cm³)*	*Composition (%)* Fe	C	O	N	*Permeability* (u_i)	*Stability (ppm/°C)*	*Range (MHz)*
E	8	7.8	98.0	0.8	0.3	0.8	10.0	95	0.25–10
TH	5	7.8	98.0	0.8	0.3	0.8	9.0	30	1–25
SF	3	7.8	98.0	0.8	0.3	0.8	8.5	35	3–40
J	10	7.3	90.0	0.8	1.1	8.0	9.0	280	3–40
W	5	7.3	90.0	0.8	1.1	8.0	6.0	150	15–100
C	10	7.8	99.5	0.2	0.3	0.1	20.0	280	0.15–3
HP	10	7.8	99.5	0.05	0.3	0.05	35.0	370	0.02–1
GQ4	8	7.8	98.0	0.05	0.3	0.05	35.0	255	0.02–1
GS6	5	7.8	98.0	0.05	0.03	0.05	25.0	190	0.15–3

mediate application at higher frequencies where the losses of powdered iron were undesirable [9–11] and are widely used in the design of wideband transformers [3, 12].

There are essentially three types of ferrite materials that find usage in HF, VHF, and UHF frequencies: Manganese-Zinc (MnZn), Nickel-Zinc (NiZn), and Cobalt-Nickel-Zinc (CoNiZn). The first of these, MnZn is suitable for applications at VLF, LF, MF, and low HF frequencies, where the higher permeabilities and lower losses make very compact wideband transformers practical. The NiZn ferrites have the widest range of use, and can be found in wideband transformers throughout the HF spectrum well into the lower range of the VHF region. The last of these, CoNiZn, has higher resonance frequencies and is capable of higher flux densities than NiZn, but can suffer permanent damage if subjected to excessively high flux densities. A partial listing of the types of ferrite materials that are commercially available and which are suitable for the design of HF, VHF, and UHF transformers is shown in Table 2.2.

2.4 COMPLEX PERMEABILITY

While we tend to think of magnetic materials in terms of the initial permeability (μ_i), the permeability is complex, having both real (μ') and imaginary (μ'') components, which represent reactive and loss terms, respectively [6, 13]. The chart of Figure 2.1 shows a typical pair of real and imaginary permeabilities in series form. The point at which μ'' is maximum is known as the ferroresonance frequency (f_{res}), which tends to vary inversely with the initial permeability [7]. The ferroresonance is an important factor to consider in the design of wideband transformers as it represents the point at which losses become significant. The

Table 2.2 *Commercial Ferrite Materials*

Manganese-Zinc (MnZn) Ferrites

Manufacturer	*Type*	*Permeability (u_i)*	*Saturation Flux (Gauss)*	*Stability ppm/°C*	*Loss Factor*	*Usable Frequency*	*Resonance Frequency*
Fair-Rite	77	2000	4900		15	1MHz	2MHz
Fair-Rite	33	800	2500	1000	35	2MHz	10MHz
Ferroxcube	3D3	750±20%	300		30	3MHz	8MHz
Neosid	F58	750±20%	450	0.5–2.3	20	4MHz	8MHz

Nickle-Zinc (NiZn) Ferrites

Manufacturer	*Type*	*Permeability (u_i)*	*Saturation Flux (Gauss)*	*Stability ppm/°C*	*Loss Factor*	*Usable Frequency*	*Resonance Frequency*
Ceramic Magnetics	CMD5005	1300	3200			800kHz	
Ceramic Magnetics	CN20	800	3800			2MHz	
Ceramic Magnetics	C2025	175	3500			15MHz	
Ceramic Magnetics	C2050	100	3400			25MHz	
Ceramic Magnetics	C2075	35	2700			75MHz	
Ceramic Magnetics	N40	15	1800			150MHz	
Fair-Rite	43	850	2750	10,000	85	5MHz	10MHz
Fair-Rite	61	125	2350	1500	32	25MHz	50MHz
Fair-Rite	65	100	2150	500	80	25MHz	35MHz
Fair-Rite	67	40	3000	1300	150	100MHz	150MHz
Fair-Rite	68	20	2000	600	400	150MHz	280MHz
Ferroxcube	4C65	125±25%	300		130	20MHz	55MHz
Neosid	F14	220±20%	350	12–30	50	15MHz	25MHz
Neosid	F16	125±20%	340	20–50	100	30MHz	50MHz
Neosid	F25	50±20%		10–15	300	50MHz	60MHz
Neosid	F28	30±20%			250	100MHz	120MHz
Neosid	F29	12±20%			1000	200MHz	250MHz
Steward	28	850	3250				
Steward	25	125	3600			15MHz	25MHz
Steward	21	16	2500			200MHz	250MHz

Cobalt-Nickle-Zinc (CoNiZn) Ferrites

Manufacturer	*Type*	*Permeability (u_i)*	*Saturation Flux (Gauss)*	*Stability ppm/°C*	*Loss Factor*	*Usable Frequency*	*Resonance Frequency*
Ferronics	K	125	3200	1000		50MHz	60MHz
Ferronics	P	40	2150	1000		80MHz	100MHz

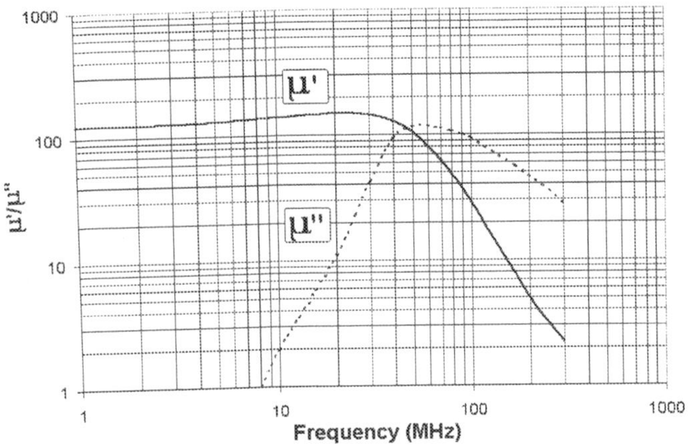

Figure 2.1 *Typical permeability curves.*

measurement of the complex permeability of the magnetic material is fairly straightforward for low frequencies, one such method being described in the included paper by Cagan and Guyot [14], which also desribes a method that is suitable for UHF and low microwave (L-band) frequencies.

2.4.1 Temperature Variations

Both the complex permeability and ferroresonance vary over temperature, the degree of which is dependent on the material being used and which is not necessarily a linear function of temperature. The designer should always refer to the manufacturer's data sheet for information concerning temperature variations, which should then be taken into account in the design of wideband transformers, just as with inductors [15]. The included paper by Suh et al. [16] illustrates the temperature variation of complex permeability and losses for a class of MnZn ferrites.

2.4.2 Effects of Machining

In cases where some machining must be performed, it should be understood that grinding induces stresses within the material which, in turn, will permanently affect the permeability in the immediate area, and higher permeability materials are more sensitive to the effects of grinding [17]. The degree to which the overall permeability is affected is dependent upon the amount of machining performed, the mechanical pressure applied, and the orientation of the machined surface to the magnetic flux. The machining of magnetic materials by way of chemical etching has a much lesser effect on the permeability [17]. However, if grinding is necessary, then some consideration should be given to annealing the material after grinding so as to restore the permeability to its initial state [17, 18]. In either case, magnetic materials that are machined should be thoroughly tested before committing to a design, in order to ensure that the machining has not compromised the magnetic properties of the material.

2.5 DIMENSIONAL RESONANCE

In addition to the ferroresonance of the permeability of magnetic materials, an improperly designed magnetic component (meaning both transformers and inductors) may experience a phenomenon known as dimensional resonance. Dimensional resonance will result if the smallest cross-sectional dimension of the core perpendicular to the magnetic field is a half wavelength, at which point a fundamental standing wave will occur across the section. Under this condition, the flux at the surface will be cancelled by the flux at the center of the core, reducing both the net flux and the permeability to zero [6]. Determining the cross-sectional dimension of the core is no simple matter, and the designer is best advised to obtain a copy of IEC Publication 205 [19] for guidance.

Similarly, a corresponding standing wave in the electric flux density will occur when the smallest dimension perpendicular to the electric field is a half wavelength, the result of which is a quadrature standing wave and a corresponding increase in the magnetic loss [6]. Since the matter of dimensional resonance is function of the physical dimensions of the core, the problem is easily avoided in practice [20–23], and is most likely to occur in applications using large cores covering many octaves as the core size is generally chosen to accommodate the lower frequencies, such that the core may be sufficiently large as to produce the magnetic and/or electric field standing waves at the higher frequencies of operation. The paper by Skutt and Lee [23] that is included in the reprints gives an adequate treatment of the problem as it relates to the design of high frequency switching power supplies.

2.6 LOSSES AND THE WIDEBAND TRANSFORMER MODEL

The schematic of Figure 2.2 illustrates the complete wideband transformer model [6], which includes an ideal transformer as well as the parasitic reactive and loss elements, which are generally understood to be distributed. Since it is inconvenient to model transformers by way of distributed inductances, capacitances, and resistances per se, lumped elements are used.

In the complete wideband transformer model of Figure 2.2, the inductors L_{l1} and L_{l2} represent the leakage inductances of the primary and secondary windings, respectively, and are a result of the fact that not all of the flux produced in one winding links to the second winding, a deficiency known as *leakage* [24], and are the primary limiting factor in the high frequency performance of the transformer. Since the leakage flux paths are primarily in air, these leakage inductances are practically constant [25, 26]. The mutual inductance L_M is a result of the flux in the transformer core that links the two windings [27], and is the primary

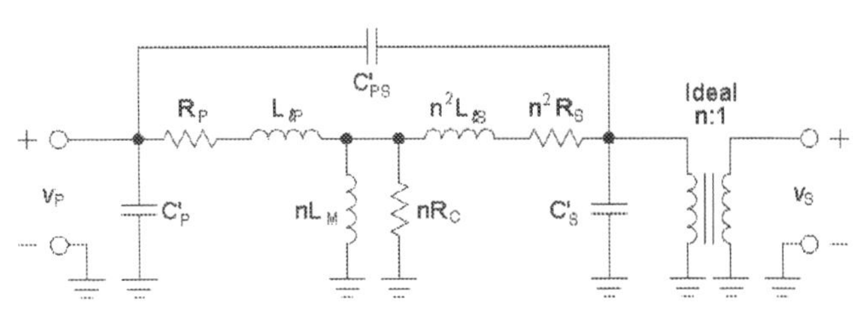

Figure 2.2 *Complete wideband transformer model.*

limiting factor in the low frequency performance of the transformer.

Capacitor C'_P represents the distributed primary capacitance which is a result of the shunt capacitance of the primary winding. Likewise, C'_S represents the shunt capacitance of the secondary winding. Capacitor C'_{PS} is referred to as the interwinding capacitance [28], which in transformers having a significant amount of conductors can interact with the inductances of the transformer and create a transmission zero.

The losses associated with wideband transformers are generally attributed to the conductors themselves and the magnetic material. The conductor losses consist, first of all,with the bulk, or copper losses and, secondly, with the skin effect. Copper losses are those associated with the energy dissipated by resistance in the conductors, and skin effect losses are a result of the current flowing in a conductor being increasingly concentrated in the outer surface as the frequency increases. At low frequencies, skin effect losses are negligible and the current is generally distributed evenly through the conductor. Increasing the frequency reduces the depth to which the current penetrates the conductor, thereby resulting in an increase in the apparent resistance of the transformer conductors. Additional losses may be experienced by way of the dielectric properties of any insulating, coating, or encapsulation materials as well as the dielectric properties of the magnetic material itself [6].

These losses are linear and are distributed throughout the length of the conductors, but in wideband transformer models it is convenient to depict them as individual lumped elements, just as with the reactive elements. In the complete wideband transformer model of Figure 2.2, the resistor R_P describes the conductor losses associated with the primary winding and the resistor n^2R_S describes the conductor losses associated with the secondary winding as referred to the primary side of the transformer.

The third resistor R_C in the complete wideband transformer model of Figure 2.2 represents the losses associated with the magnetic material, and these losses are generally comprised of, but not limited to, eddy current losses and hysteresis losses [29], the latter of which increase with ω^2 or even ω^3, and is significant in transformers that are operated near the ferroresonance of the core material [30]. Losses are also dependent upon the level of magnetization, such as bias flux as discussed in the included paper by Sato and Sakaki [31]. In cases in which the transformer is designed with sufficient performance margins, the losses can be ignored and the lossless wideband transformer model of Figure 2.3 may be used [32]. Generally, it is best to consider these losses in the initial design stages so as not to compromise the design.

The included papers by Bartoli et al. [33] and Kazimierczuk et al. [34] provide suitable guidelines for derivations of the losses for inductors utilizing powdered iron and ferrite magnetic materials, respectively (as well as the losses due to the skin effect of the conductors), which can also be applied to the modeling of wideband transformers.

2.6.1 Eddy Current Losses

Eddy current losses are directly related to the electrical conductivity of the magnetic materials, and since they contain iron and other metals, the problem becomes fairly obvious. The magnitude of the eddy current losses depends greatly on the size and shape of the core material and may be reduced by dividing the material into electrically insulated laminations, such as with power and audio transformers, or grains, as with powdered iron.

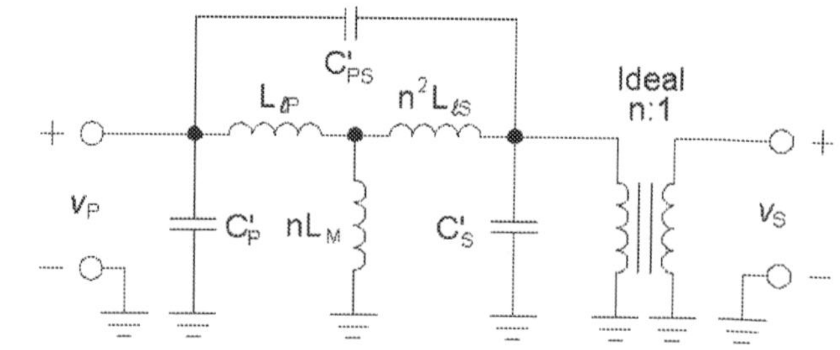

Figure 2.3 *Lossless wideband transformer model.*

In ferrites, the material is regarded as being homogeneous and the eddy losses are usually included with the hysteresis losses [6]. To avoid the eddy current losses in ferrites due to the low resistance of the ferrous ions in the bulk material, a reoxidation process is often employed to enclose the grains in a layer of high resistivity [35]. For powdered iron materials, the fine particles of iron are insulated by exposing them to a mild acid or a variety of insulating salts to which the encapsulating material can adhere to [36].

Eddy current losses are also frequency dependent, as the conductivity of the magnetic material reduces the penetration of alternating electromagnetic fields [37], a phenomenon known as magnetic skin effect [30] or eddy current shielding [38]. The reduced penetration of the magnetic field also results in a lower effective permeability of the material [37].

2.6.2 Hysteresis Losses and the Hysteresis Curve

To understand the losses associated with the hysteresis of the magnetic material, it is first necessary to comprehend the nature of hysteresis itself. A typical hysteresis, or B-H curve, is provided in Figure 2.4 for the purpose of a generalized discussion. Initially, there is no magnetic flux or field in the material. As a magnetic field is applied, a flux develops and the B-H relationship proceeds towards the point of saturation. As the magnetic field decreases, the path proceeds along the upper portion of the

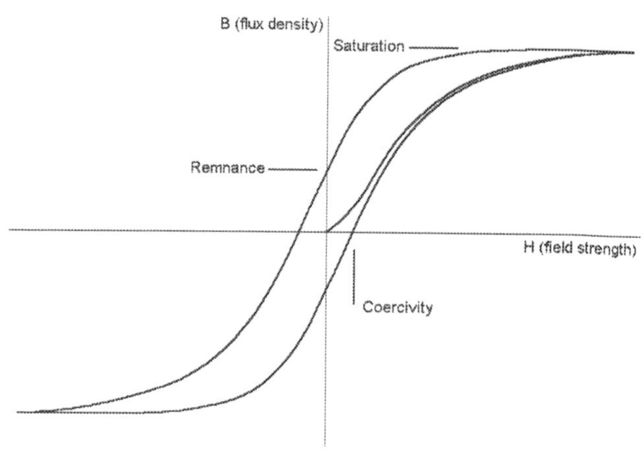

Figure 2.4 *Typical hysteresis (B-H) loop.*

curve arriving at a point where there is no magnetic field (H=0), but some flux remains within the material. This point is referred to as remnance (or remanence), and it is properly defined as the flux density that remains in a magnetic material when the applied field is removed.

The path now proceeds further along the upper edge of the curve, arriving at the point where there is no flux (B=0). This is referred to as the coercivity (or coercive force) of the material, and it is properly defined as the field strength that must be applied in order to reduce (or coerce) the remnant flux to zero. Magnetic materials with a high measure of coercivity are referred to as being hard, and are generally found in permanent magnets and digital recording media. On the other hand, materials with a low measure of coercivity are referred to as being soft and are used to produce linear components such as transformers and inductors.

Hysteresis losses are a result of creating the remnant flux from energy from the applied magnetic field. It is, therefore, easily understood that magnetic materials with low coercivity will have lower hysteresis losses. The included paper by Hollaus and Biró [39] shows that the hysteresis curve, the hysteresis losses, and the complex permeability may be derived by way of the Preisach model, which may be of interest to those involved in the intensive computer modelling of RF components and systems.

2.7 BARKHAUSEN NOISE

It should be obvious that these losses equate to an increase in noise in a circuit, and that a poorly designed wideband transformer used in the input stages of a receiver or in antenna matching circuitry can have serious effects on the noise figure (NF) of a receiving system. An additional source of noise in magnetic materials that should be of concern in small signal systems results from the fact that magnetization is not a continuous process, but is, instead, a series of extremely small, discontinuous steps known as the Barkhausen Effect [20, 36, 37, 40–42]. The noise attributed to this phenomenon is called Barkhausen Noise, and it has a power spectrum that is proportional to $1/f^2$ [43]. The subject of sources of additional noise is a serious consideration in the design of low-noise amplifiers (LNAs), mixers, and oscillators. The paper by Bittel [42] that is included in the reprints gives a very good overview of the essential aspects of the main types of noise in ferromagnetic materials.

2.8 FERROMAGNETIC DISTORTION

As can be seen from the typical hysteresis curve of Figure 2.4, the relationship between the flux density B and the applied field strength H is not linear, and this nonlinear relationship results in the generation of both harmonics and intermodulation distortion products [6, 30, 44–46]. In most cases, the symmetry of the hysteresis curve will yield odd-ordered harmonics, but in cases where bias current is passed through the windings, even-ordered harmonics may result if the bias current results in any net magnetic flux in the material. Although a minor consideration in the design of wideband transformers for small signal applications, distortion becomes a significant factor in the design of power amplifier components where high signal levels may create sufficient flux to approach saturation of the material, and where large bias currents may cause the flux to be assymetrical in a poorly designed transformer, leading to undesirable second-order harmonic products. The included paper by Collins and Flynn [46] provides an adequate introduction to the otherwise poorly documented subject of ferromagnetic distortion.

2.9 CONCLUSION

Most designers are aware of the losses associated with the use transformers and inductors in circuit design. This brief overview of various characteristics of powdered iron and ferrite materials reveals that these losses are attributed to more than just the resistive losses of the conductors themselves, and that even the aspects of eddy current and hysteresis losses due to the core material may not be sufficient.

For instance, the subject of dimensional resonance is an aspect of transformer design that receives little attention, but which should be considered in the design process in order to avoid compromising high frequency performance. This can be a problem in the design of a transformer intended to cover many octaves as a magnetic core chosen to give adequate coupling at lower frequencies may be large enough to cause standing waves in the core at the higher operating frequencies.

For those whose interests lie with the design of components intended for small signal amplifiers and frequency sources, the losses incurred by magnetic components is insufficient when considering the sources of noise degradation. In such cases, the matter of Barkhausen Noise should be of immediate interest as all sources of noise play a decisive rôle in the overall noise performance of high performance communications equipment. This is especially true in the design of equipment intended for use with high-density digital modulation formats such as 256 QAM.

Conversely, those involved in the design of magnetic components for high power applications need to be equally aware of the consequences of hysteresis as an insufficiently large core, as well as the generation of bias flux from assymetrical bias current in the conductors, can give rise to the generation of both harmonics and intermodulation distortion. Thus, when bringing together the need for a large core to accommodate high power along with the possibility of excessive losses due to dimensional resonance at high frequencies, the design of wideband transformers for high-power applications can be viewed as a very serious undertaking that demands close attention to details.

References

[1] Mittash, A., C. Muller, and W. Schubadt, *Manufacture of Iron Powder,* U.S. Patent 2,772,956, December 4, 1930.

[2] Polgreen, G.R., "The Production and Application of Metal Powders," *G.E.C. Journal,* July 1952, pp. 152–169.

[3] Breed, Gary, "An Introduction to Iron Powder and Ferrite Cores," *RF Design,* March 1992, pp. 121–124.

[4] Cox, Jim, "Iron Powder Cores for RF Power Applications," *RF Design*, February 1987, pp. 39, 41 and 43 (correction in March 1987).

[5] DeMaw, Doug, "Magnetic Cores in RF Circuits," *RF Design*, Vol. 3, No. 4, April 1980, pp. 12–16.

[6] Snelling, E.C., *Soft Ferrites: Properties and Applications*, 2nd edition, London: Butterworths, 1988.

[7] Snoek, J.L., *New Developments in Ferromagnetic Materials*, 2nd edition, Amsterdam: Elsevier, 1949.

[8] Polder, D., "Ferrite Materials," *Proceedings of the IEE*, Part 2, Vol. 97, April 1950, pp. 246–256.

[9] Fairweather, A., F.F. Roberts, and J.E. Welch, "Ferrites," *Reports on Progress in Physics*, Vol. 15, 1952, pp. 142–172.

[10] Latimer, K.E., and H.B. Macdonald, "A Survey of the Possible Applications of Ferrites," *Proceedings of the IEE*, Part II, Vol. 97, 1949, pp. 257–267.

[11] Snelling, E.C., "Ferrites for Linear Applications," *IEEE Spectrum*, January 1972, pp. 42–51 (Part I—Properties), February 1972, pp. 26–32 (Part II—Performance Requirements).

[12] "Use of Ferrites for Wideband Transformers," Fair-Rite Products Corporation Application Note.

[13] Saotome, H., and Y. Sakaki, "Complex Permeability of Polycrystalline Mn-Zn and Ni-Zn Ferrites," *Electrical Engineering in Japan*, Vol. 123, No. 2, Feb 1998, pp. 1–7.

[14] Cagan, V., and M. Guyot, "Fast and Convenient Technique for Broadband Measurements of the Complex Permeability of Ferrimagnets," *IEEE Trans. Magnetics*, Vol. 20, No. 5, September 1984, pp. 1732–1734.

[15] Takanori T., M. Ueshima, and T. Tokunaga, "Frequency Dispersion and Temperature Variation of Complex Permeability of Ni-Zn Ferrite Composite Materials," *Journal of Applied Physics*, Vol. 78, No. 6, 15 September 1995, pp. 3983–3991.

[16] Suh, J.J., B.M. Song, and Y.H. Han, "Temperature Dependence of Power Loss of Mn-Zn Ferrites at High Frequency," *IEEE Trans. Magnetics*, Vol. 36, No. 5, September 2000, pp. 3402–3404.

[17] Knowles, J.E., "The Effect of Surface Grinding Upon the Permeability of Manganese-Zinc Ferrites," *Journal of Physics, D, Applied Physics*, Vol. 3, No. 9, September 1970, pp. 1346–1351.

[18] Stern, E., and D. Temme, "Magnetostriction Effects in Remanence Phase Shifters," *IEEE Trans. Microwave Theory and Techniques*, Vol. 13, No. 6, November 1965, pp. 873–874.

[19] "Calculation of the Effective Parameters of Magnetic Piece Parts," International Electrotechnical Commission (IEC) Publication 205, 1966 (amended December 1976 and June 1981).

[20] Tebble, R.S., and D.J. Craik, *Magnetic Materials*, New York: Wiley, 1969.

[21] Smit, J., and H.P.J. Wijn, *Ferrites: Physical Properties of Ferrimagnetic Oxides in Relation to Their Technical Applications*, Eindhoven, the Netherlands: Wiley, 1959.

[22] Brockman, F.G., P.H. Dowling, and W.G. Steneck, "Dimensional Effects Resulting from a High Dielectric Constant Found in a Ferromagnetic Ferrite," *Physical Review*, Vol. 77, No. 1, January 1950, pp. 85–93.

[23] Skutt, G.R., and F. C. Lee, "Characterization and Dimensional Effects in Ferrite-Core Magnetic Devices," *1996 IEEE Power Electronics Specialists Conference (PESC '96)*, pp. 1435–1440.

[24] O'Meara, T.R., "Analysis and Synthesis with the 'Complete' Equivalent Circuit for the Wide-Band Transformer," *Transactions of the AIEE*, Part I, Mar 1962, pp. 55–62.

[25] Lord, H.W., "The Design of Broad-Band Transformers for Linear Electronic Circuits," *Transactions of the AIEE*, Part 1, Vol. 69, 1950, pp. 1005–1010.

[26] Clarke, K.K., and D.T. Hess, *Communication Circuits: Analysis and Design*, Reading, MA: Addison-Wesley, 1971.

[27] Gillette, P.R., K. Oshima, and R.M. Rowe, "Measurement of Parameters Controlling Pulse Front Response of Transformers," *IRE Trans. Component Parts*, Vol. 3, No. 1, March 1956, pp. 20–25.

[28] Nordenberg, H.M., *Electronic Transformers*, New York: Reinhold, 1964.

[29] Saotome, H., and Y. Sakaki, "Iron Loss Analysis of Mn-Zn Ferrite Cores," *IEEE Trans. on Magnetics*, Vol. 33, No. 1, January 1997, pp. 728–734.

[30] Grossner, N.R., *Transformers for Electronic Circuits*, New York: McGraw-Hill, 1967.

[31] Sato, T., and Y. Sakaki, "Physical Meaning of Equivalent Loss Resistance of Magnetic Cores," *IEEE Trans. on Magnetics*, Vol. 6, No. 5, September 1990, pp. 2894–2897.

[32] Trask, C., "Wideband Transformers: An Intuitive Approach to Models, Characterization and Design," *Applied Microwave & Wireless*, November 2001, pp. 30–41.

[33] Barloti, M., A. Reatti, and M.K. Kazimierczuk, "Modeling Iron-Powder Inductors at High Frequencies," *1994 IEE Industry Applications Meeting* (IAS '94), Vol. 2, pp. 1225–1232.

[34] Kazimierczuk, M.K., G. Sancineto, G. Grandi, U. Reggiani, and A. Massarini, "High-Frequency Small-Signal Model of Ferrite Core Inductors," *IEEE Trans. Magnetics*, Vol. 35, No. 5, September 1999, pp. 4185–4191.

[35] Smit, Jan, (ed.), *Magnetic Properties of Materials*, New York: McGraw-Hill, 1971.

[36] Heck, C., *Magnetic Materials and Their Applications*, New York: Crane, Russak & Co., 1974.

[37] Bradley, F.N., *Materials for Magnetic Functions*, New York: Hayden, 1971.

[38] Welsby, V.G., *The Theory and Design of Inductance Coils*, London: Macdonald & Co. Ltd., 1950.

[39] Hollaus, K., and O. Biró, "Derivation of a Complex Permeability from the Preisach Model," *IEEE Trans. Magnetics*, Vol. 38, No. 2, March 2002, pp. 905–908.

[40] Barkhausen, H., "Zwei mit Hilfe der neuen Verstärker entdeckte Erscheinungen," *Physikalische Zeitshrift*, Vol. 20, 1919, pp. 401–407.

[41] Warren, K.G., "Barkhausen Noise in Transformer Cores," *Electronic Technology*, March 1961, pp. 89–94.

[42] Bittel, H., "Noise of Ferromagnetic Materials," *IEEE Trans. Magnetics*, Vol. 5, No. 3, September 1969, pp. 359–365.

[43] Bertotti, Giorgio, *Hysteresis in Magnetism for Physicists, Materials Scientists, and Engineers*, San Diego, CA: Academic Press, 1998.

[44] Macfadyen, K.A., *Small Transformers and Inductors*, London: Chapman & Hall, 1953.

[45] Kalb, R.M., and W.R. Bennett, "Ferromagnetic Distortion of a Two-Frequency Wave," *Bell System Technical Journal*, Vol. 14, 1935, pp. 322–359.

[46] Collins, S., and K. Flynn, "Intermodulation Characteristics of Ferrite-Based Directional Couplers," *Microwave Journal*, November 1999, pp. 122–130.

An Introduction to Iron Powder and Ferrite Cores

By Gary A. Breed
Editor

The subject of magnetic materials for inductors and transformers cannot be adequately covered in a single article. To provide an introduction to these materials, this tutorial describes some very basic principles, and offers a few practical rules for common applications.

In grade school science class, we observed that a simple electromagnet became much stronger when a bolt or nail was placed in its core. In RF circuits, where inductors are defined by their electromagnetic behavior, iron-based materials can serve the same purpose; concentrating the magnetic lines of force in the core of an inductor. The result is an increase in the inductance over the same coil of wire in air. Useful materials for this purpose have "soft" magnetic properties, which means they do not become permanently magnetized like "hard" materials.

The degree of enhancement of the magnetic field is a function of the permeability (μ) of the core material; a numerical factor that defines the increase in concentration of the magnetic flux due to the presence of the material. The materials discussed here may have permeabilities from 3, only a little more than that of air, to 2500 or more. The ratio of two permeabilities is linear with regard to magnetic flux, but only when both are measured under exactly the same conditions. Magnetic materials are not linear with regard to magnetic flux versus current through the winding (and hence, with increasing flux density). Refer to your college Electromagnetic Fields textbook for a better explanation of permeability, flux density, coercivity, and other terminology (1). Also read the technical explanations contained in manufacturers' catalogs and applications notes to determine how μ is measured for their products.

Rather than describe the physics of these materials in detail, the rest of this article will describe the properties that affect their performance as RF inductors, transformers and chokes.

Iron Powders

This low permeability class of materials is made from iron compounds, mainly oxides. Carbonyl iron materials will have μ values from about 3 to 35, hydrogen reduced iron compounds can have μ values in the 25 to 125 range. Other compounds have been developed as well, with considerable recent effort to create materials with greater temperature stability.

The most common applications of iron powder materials are in toroidal inductor cores and adjustable slug-tuned coils. The doughnut shape of a torus has the advantage of creating a magnetic field that is contained within the core. Without the open ends of a linear winding, a toroidal coil has relatively little radiation. This is true even for air coils, although a core of magnetic material greatly increases the containment of the field around the circle defined by the core. As a result, inductors wound on toroid cores have low interaction with surrounding components and rarely require shielding.

Slug-tuned inductors are coils wound on a tubular forms of plastic, ceramic or other material, with a piece of magnetic material that can move in an out of the coil. The movement may be accomplished either with threaded core material that screws into the coil form, or the material may be a short rod supported by an adjustable mechanism such as a lead screw to move it in relation to the coil. Adjustable inductors using these methods are among the most common RF components.

Besides the self-shielding nature of the toroid coil and the adjustable capabilities of a slug coil, the use of iron

Core Size	26-mix yel/wht u=75 0-1. MHz	3-mix gray u=35 .05-.5 MHz	15-mix red/wht u=25 .1-2. MHz	1-mix blue u=20 .5-5. MHz	2-mix red u=10 1.-30 MHz	6-mix yellow u=8 2-50. MHz	10-mix black u=6 10-100 MHz	12-mix grnwht u=3 20-200 MHz	0-mix tan u=1 50-300 MHz
T-12-—	NA	60	59	43	20	16	12	6,5	3.0
T-16-—	NA	61	55	44	22	19	13	8.0	3.0
T-20-—	NA	90	65	52	27	22	16	10.0	3.5
T-25-—	NA	100	85	70	34	27	19	12.0	4.5
T-30-—	325	140	93	85	43	36	25	16.0	6.0
T-37-—	275	120	90	80	40	30	25	15.0	4.9
T-44-—	360	180	160	105	52	42	33	18.5	6.5
T-50-—	320	175	135	100	49	40	31	18.0	6.4
T-68-—	420	195	180	115	57	47	32	21.0	7.5
T-80-—	450	180	170	115	55	45	33	22.0	8.5
T-94-—	590	248	200	160	84	70	58	32.0	10.6
T-106-	900	450	345	325	135	116	NA	NA	19.0

Table 1. Some typical core materials and A_L values.

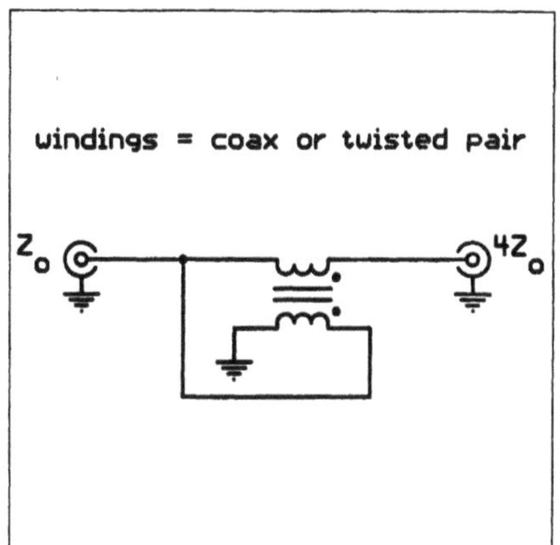

Figure 1. A unbalanced 4:1 transmission line transformer.

powders allows greater inductance with fewer turns of wire. To compute the required number of turns to achieve a desired inductance, manufacturers of these products have established a standard reference, A_L values. A_L is the inductance in microhenries for a 100-turn winding on that core. Each core of different materials and different sizes will have a unique A_L. Table 1 is an abbreviated list of core materials, sizes and A_L values (2,3).

To compute the required number of turns for a given inductance in microhenries, use the following formula:

$$\text{Turns} = 100 \sqrt{\frac{\text{desired L (uH)}}{A_L}}$$

Also included in most catalogs is a summary of wire sizes versus possible number of turns on each size toroid core. Table 2 lists this data for just one common core size, the T-37, with 0.375 inch outside and 0.205 inch inside diameter. A rule of thumb is that best results are achieved with a winding that covers approximately 3/4 of the core.

For example, if we want a 1.0 uH inductor on a T-37 core, we might choose 6-mix material, with an A_L of 30 uH/100t. Using the above formula, we would need an 18-turn winding. From the wire size chart, we see that 23 turns of #24 wire will fill a T-37 toroid. The required 18 turns would fill about 78 percent of the core, which is close to our 3/4 rule of thumb. If some allowance for compressing or spreading the windings is desired, smaller size wire should be used.

Ferrites

The class of materials generally called ferrites covers many combinations of iron oxide and other metal alloys. The alloys in RF ferrites typically are nickel-zinc, manganese-zinc, and manganese (4,5). Ferrites can have very high μ, which means that they greatly increase the inductance of a given winding.

Wire size	No. turns	Wire size	No. turns
#10	1	#26	31
12	3	28	41
14	5	30	53
16	7	32	67
18	9	34	87
20	12	36	110
22	17	38	140
24	23	40	177

Table 2. Number of turns that fit on a 0.375 inch toroid core.

Only the lowest permeability ferrites are suitable for simple inductors at RF, although they are often used as cores for low frequency inductors. Often, high-μ ferrites are used where inductance values are not critical, such as RF chokes and EMI suppression.

Ferrite's most common use at RF is for transformers and other coupled functions (6,7). The large increase in inductance provided by these materials means that only a few turns are required to get significant inductive reactance. This reactance can be used to block common-mode currents in transmission line transformers. Related transmission line applications include directional couplers, quadrature hybrids, power dividers and combiners. Figure 1 shows a single-ended 4:1 transformer, a typical application for a ferrite-loaded transmission line. Conventional isolated-winding transformers can also be constructed with relatively few turns.

High inductance per turn has two main advantages: minimum conductor length provides better high frequency response, and fewer turns allows larger conductors and higher currents. With proper selection of the ferrite material, the size and type of conductor, and system impedance, components using ferrites can be extremely broadband; several decades in some applications.

Ferrite cores for transformer applications are often supplied in a two-hole arrangement. This is usually referred to as a balun core because of its use in balanced to unbalanced transformers, the most common of which is 75 ohm coaxial to 300 ohm balanced transformers for television. This physical arrangement makes easy layout possible, and doubles the inductance per turn since the wire passes through the core in each direction (Figure 2).

Figure 2. A balun core can simplify transformer construction.

Limitations

Iron powder and ferrite materials have limitations on frequency of operation, inductor Q, losses, temperature stability, power handling and DC current handling.

Operating frequency and maximum Q are related characteristics which correspond to μ. Magnetic materials offer resistance to the change in magnetic fields as the AC voltage rises and falls. This effect follows the permeability value, so low μ materials will have less effect and be better for higher frequencies. High permeability materials have practical limits on high frequency performance for the same reason.

These limits will be different for inductors, conventional transformers, and transmission line applications since the mechanism of operation is different for each of these. Inductors need high Q and, normally, the best choice is the lowest permeability core that fits wire size and inductance requirements. Conventional transformers can be low Q, and will usually benefit from smaller number of turns. In this case, the user would select a high permeability material that doesn't exhibit excessive loss at the operating frequency. Transmission line transformers rely on the differential currents between conductors, typically either a twisted pair or a coaxial cable. The function of the magnetic material is to provide enough reactance to impede the flow of common mode currents. High permeability is an advantage, but with considerations for power handling, flux saturation and high frequency losses.

Magnetic core materials all are rated by their manufacturers for performance over temperature and for a range of flux densities. Excessive temperature rise will change the characteristics of the core (or even damage it). Magnetic performance is not linear at high flux densities, and materials must be operated within their recommended ranges (8). These limitations on power handling are primarily a function of core material and core size.

As a simple example, a broadband transformer might be constructed on high permeability ferrite to achieve maximum bandwidth. As a low-level device it may be acceptable to have 0.2 dB loss or more, but the same configuration in a high power application could cause excessive temperature rise in the core. Attempting to increase power handling by using a larger core could result in too much inductance in the windings, further degrading performance.

Other characteristics to note include performance well above normal operating frequencies, where these materials tend to be very lossy. This can be used to advantage for suppression of unwanted VHF or UHF responses. It is common to use ferrite beads for suppression of VHF oscillations, but be careful that the added inductance does not affect operation at lower frequencies!

Temperature stability is another area to understand. For critical circuits such as oscillators or very narrow filters, be certain that the inductance value will not have excessive drift with temperature. Some materials have better temperature performance than others. When in doubt, check the manufacturers' temperature curves, or call them for a recommendation regarding your application.

Conclusion

Iron powder and ferrite cores have extremely valuable properties for RF applications. But the proper selection of a material for a specific application is not always simple. Like many areas of RF engineering, there are some common rules of thumb that will help in most cases. But, there are also many additional performance characteristics that must be understood and included in the process of selecting the right magnetic core material for a specific application. If you work with these materials regularly, it is essential to learn how all these characteristics affect performance. ***RF***

References

1. R.K. Wangsness, *Electromagnetic Fields*, John Wiley & Sons, New York, 1979.
2. *RF Applications*, Catalog 3, Micrometals.
3. Terminology as used by Micrometals.
4. *Fair-Rite Soft Ferrites*, Fair-Rite Products Corp. catalog, 11th Edition, 1991.
5. *Linear Ferrite Materials and Components*, Ferroxcube catalog, Seventh Edition.
6. J. Sevick, *Transmission Line Transformers*, Second Edition, ARRL, 1990.
7. H. Granberg, *Broadband Transformers and Power Combining Techniques for RF*, Application Note AN-749, Motorola, Inc.
8. J. Cox, "Iron Powder Cores for RF Power Applications," *RF Design*, February 1987.

Iron Powder Cores for RF Power Applications

By Jim Cox
Micrometals, Inc.

As RF engineers strive for higher power, they need to have the right components, capable of handling that power. The purpose of this article is to present new information which will allow the RF design engineer to select the proper iron powder core for inductors and transformers handling power in the 250 kHz to 5 MHz frequency range. Much of the data presented can be used for applications throughout the HF range.

Iron powder cores are commonly used to produce high Q inductors and transformers for selective circuits. Iron powder cores used in RF applications are composed of extremely small particles of highly pure carbonyl iron. the distributed air-gap of iron powder cores contributes to their rather low permeability and very good stability. In applications involving low level signals, the choice of core size, material and winding is normally based on required Q and/or packaging requirements. A listing of the general magnetic properties of carbonyl iron powders is shown in Figure 1.

Low level broadband transformers and RF chokes are commonly built on high permeability ferrite cores. Ferrites are a non-metallic, ceramic ferromagnetic compound with a spinel crystalline structure. Ferrite cores have higher permeability than iron powder cores, but are less stable. While much of the information presented here will be applicable to ferrite cores, the data presented is primarily intended for iron powder cores.

Inductor and transformer applications involving higher power signals require additional considerations. A common misconception is that core saturation is the primary limiting factor in selecting a core for RF power applications. While it needs to be determined how much voltage drop or current flow a given inductor or transformer can support before a limit is reached, this limit will be either magnetic saturation *or* excessive temperature rise resulting from both winding (copper) and core material losses.

General Magnetic Properties

Mix #	Basic Iron Powder	Material Permeability (μo)	Temperature Stability (+)	Resonant Circuit Frequency Range[1] (MHz)	Color Code
1	Carbonyl C	20	280 ppm/°C	.15-2.0	Blue
2	Carbonyl E	10	95	.25-10.	Red
3	Carbonyl HP	35	370	.02-1.0	Gray
6	Carbonyl SF	8.5	35	2.0-30.	Yellow
7	Carbonyl TH	9.0	30	1.0-20.	White
8	Carbonyl GQ4	35	255	.02-1.0	Orange
10	Carbonyl W	6.0	150	10-100	Black
12	Synthetic Oxide	4.0	170*	20-200	Green/White
15	Carbonyl GS6	25	190	.10-3.0	Red/White
17	Carbonyl	4.0	50	20-200	Blue/Yellow
22	Synthetic Oxide	4.0	410*	20-200	Green/Orange
0	Phenolic	1	0	50-250	Tan

[1]Frequency range indicated is for maximum Q. For wide-band applications where high Q is not required, the useful frequency range will typically extend 10 to 100 times higher.
*Non-linear

Figure 1. Table of iron powder core materials.

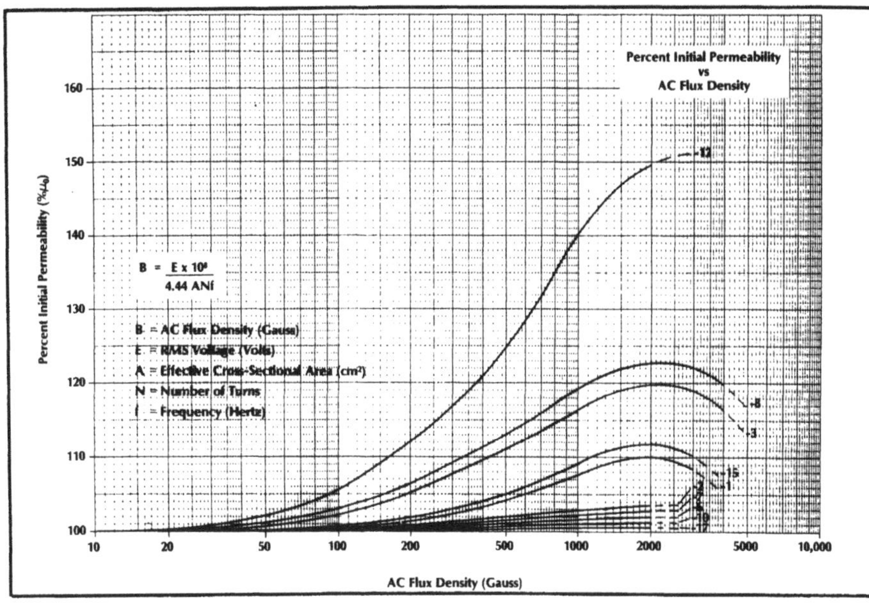

Figure 2. Core saturation determination.

Magnetic saturation is the point at which an increase in magnetization force does not result in any further increase in flux density. This implies that there will be a loss of permeability. An inductor will show a decrease in inductance and a transformer will show both a decrease in impedance and will not transform the additional signal. Carbonyl iron powders typically reach their maximum permeability at about 3,000 gauss and then begin to saturate. They reach full saturation at ap-

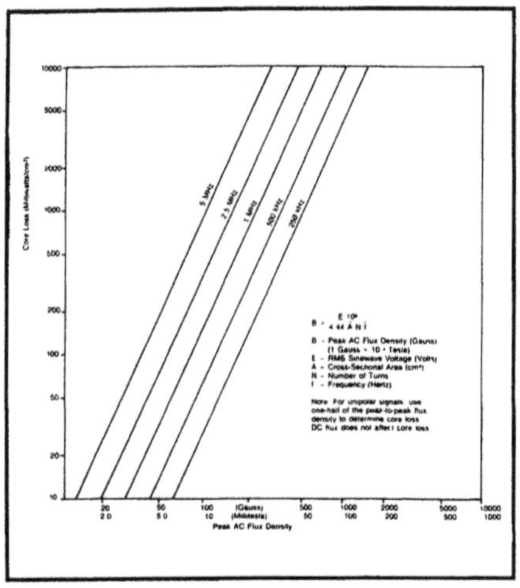

Figure 3. Core loss vs. AC flux density for −6 material (u = 8.5).

proximately 10,000 gauss. This characteristic is illustrated in Figure 2.

With continuous sinewave signals, iron powder cores are limited by temperature rise resulting from losses rather than magnetic saturation. Temperature rise results from losses in both the winding and the core material. The fundamental winding losses are equal to I^2R where I is the RMS current flowing in amperes and R is the effective AC resistance. At high frequency the current carried by a conductor tends to be concentrated near the surface (skin effect). The skin depth of the AC current in a copper conductor at room temperature is described by:

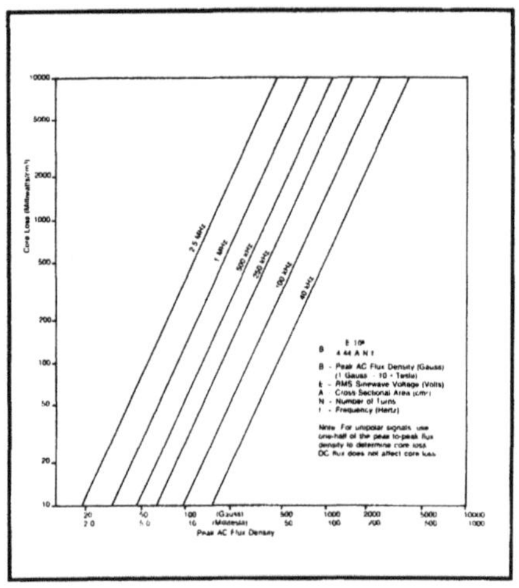

Figure 4. Core loss vs. AC flux density for −2 material (u = 10).

$$\text{Skin Depth (cm)} = \frac{6.62}{\sqrt{f}}$$

where f is frequency in hertz.

Wire size should be chosen for a skin depth equal to the conductor radius. For example, at 1 MHz a wire larger than #35 AWG will not be fully utilized and will thus show an increased AC resistance. Due to this, the use of a large number of strands of fine wire that are insulated from each other and interwoven can be useful in reducing the AC resistance of conductors at high frequency. Such a conductor is known as litz wire. Practical litz conductors are very effective at frequencies below 500 kHz, but begin to lose effect above 3 MHz.

Core Losses

Figures 3 and 4 show core loss graphs for two iron powder core materials (Micrometals, Inc. materials 6 and 2) with permeabilities of 8.5 and 10. The losses generated by the core materials were determined experimentally. The total loss of a wound coil was measured and the losses due to the test set-up, the winding, and the core material were separated. The information presented is for the core material alone.

	Power dissipation (mW/cm³)		
Core	10 C	25 C	40 C
T30	400	1148	2026
T37	412	1170	2065
T44	310	884	1556
T50	307	874	1535
T68	234	664	1167
T80	212	602	1056
T94	160	454	802
T106	114	322	566
T130	117	331	582
T157	94	266	468
T200	87	260	436
T300	62	186	327
T400	43	130	228

Figure 5. Chart of power dissipation for various core sizes.

als were determined experimentally. The total loss of a wound coil was measured and the losses due to the test set-up, the winding, and the core material were separated. The information presented is for the core material alone.

The core loss is expressed in milliwatts per cubic centimeter as a function of peak AC flux density (DC flux does not generate core loss) for a number of frequencies from 40 kHz to 5.0 MHz. This power dissipated by a core will generate temperature rise. The formula used which describes this relationship is:

$$\text{Temp. Rise (°C)} = \left[\frac{\text{Power Dissipation (Milliwatts)}}{\text{Surface Area (cm}^2)}\right]^{.833}$$

This formula provides a reasonable approximation for the temperature rise of a core in free air. In other environments, such as moving air or an enclosed case,

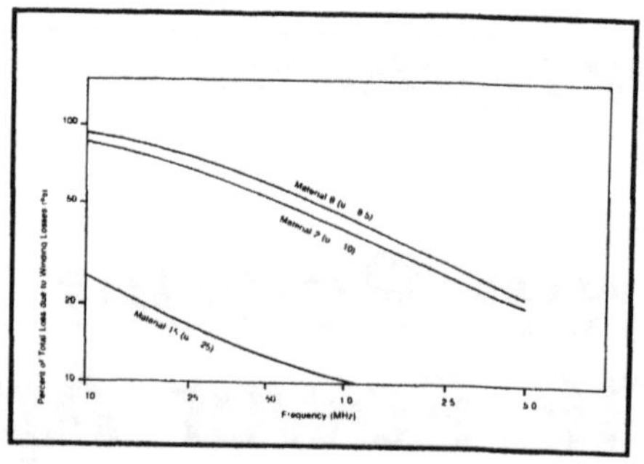

Figure 6. Percent winding loss for single-layer coils.

other relationships will need to be used. With constant power dissipation, it typically takes a core about 2 hours to reach its final temperature. Applications involving low-duty or intermittent operation can time average the losses.

It is important to limit the operating temperature of inductors and transformers using iron powder cores. Long term operation of iron powder cores above 100°C can cause a permanent reduction in both Q and inductance.

A listing of the power dissipation in milliwatts per cubic centimeter for temperature rises of 10, 25 and 40 degrees C for most of the common iron powder core sizes is shown in Figure 5. For those not familiar with the part numbering system: A T30 is a toroidal core with an outside diameter of about .30 inches, while a T400 is a 4.00 inch outside diameter toroid. It can be seen that the physically small parts can dissipate more power per unit volume than the physically large parts.

The peak AC flux density (Figures 3 and 4) for sinewaves is described by Faraday's Law:

$$B = \frac{E\ 10^8}{4.44\ A\ N\ f}$$

Where:
B = Peak AC Flux Density (Gauss) (1 Gauss = 10^{-4} Tesla)
E = RMS Sinewave Voltage (Volts)
A = Core Cross-Sectional Area (cm^2)
N = Number of Turns
f = Frequency (Hertz)

This form of Faraday's Law is generally more useful for transformer applications where the applied voltage is normally known. For an inductor,

$$E = 2\pi f L I$$

Where:
E = RMS Sinewave Voltage (Volts)
f = Frequency (Hertz)
L = Inductance (Henries)
I = RMS Current (Amperes)

Comparing the core loss characteristics of the three materials shown in Figures 4, 5 and 6 at a particular frequency does not indicate a significant difference in the power dissipated in milliwatts per cubic centimeter at a given AC flux density. However, with further investigation it becomes evident that for a given inductance and voltage, the lower permeability materials generate less AC flux density and, thus, lower core loss, than the higher permeability materials.

Coils wound on lower permeability materials, however, require more turns than higher permeability materials to produce the same inductance, and will, therefore, produce greater winding loss. In general, the best compromise in performance will occur when the winding and core losses are approximately equal.

While the distribution of total loss between the core and winding for any given frequency is dependent on the winding details, as well as the physical size of the coil, it is useful to have a rough approximation of this distribution. Figure 6 illustrates this distribution for three iron powder core materials over the frequency range of 250 kHz to 5.0 MHz with a typical single layer winding. This graph shows that at a frequency of 1 MHz, of the total loss generated in a typical coil wound on material 2, about 40 percent of that loss is due to the winding. In the higher permeability material 15, at 1 MHz, about 10 percent of the total loss generated is due to the winding. With this guideline, it is then possible to estimate how much core loss can be tolerated without exceeding reasonable temperature rise limits.

In an effort to take the basic core loss data, which is expressed in magnetic terms and present it in more common engineering terms, application curves have been generated by Micrometals, Inc. The set of curves shown in Figure 7 shows the maximum voltage as a function of inductance (and inductive reactance) for various core sizes in Micrometals material 2 at a frequency of 1 MHz for a temperature rise of 25°C due to core loss. These curves are based on continuous sinewave signals. A similar set of curves in terms of current as a function of inductance appears in Figure 8. Similar graphs are available for other frequencies and core materials from Micrometals, Inc.

The use of these curves can be illustrated with the following examples:

1) If an inductor application at 1 MHz requires 10 uH (microhenries) and the AC RMS current level is 1 ampere, Figure 10 shows that a T68 size core will be required (Part number T68-2). It can also be seen in Figure 9 that this condition corresponds to a voltage drop across this coil of about 70 volts.

2) To select a core for a transformer requiring 100 ohms of primary inductive reactance and capable of supporting 200 volts RMS, Figure 9 should be used. This shows that 100 ohms of reactance is equal to 15.9 uH and that a T157 size core will be required to support the 200 volts RMS.

Inductance ratings in microhenries for 100 turns are commonly provided by iron powder core manufacturers. The number of turns required to produce a desired inductance for a core with a known inductance rating is given by:

Required Turns =

$$100\left[\frac{\text{desired L (uH)}}{\text{L rating uH/100 turns}}\right]^{1/2}$$

In example (1) the T68-2 has an inductance rating of 57 uH for 100 turns and will thus require 42 turns for 10 uH. The T157-2 in example (2) has an inductance rating of 140 uH for 100 turns. The 15.9 uH will require 34 turns. In general, for inductors, the largest wire which will fit in a full single layer will produce the best results. A winding table with this information is available.

It is further possible to define a "power rating" for the various core sizes. This "power rating" will be defined to be the product of the current flowing through a coil times the voltage being dropped across that coil. For a given temperature rise due to core loss, this product is independent of the number of turns wound on the core. Figure 9 shows this information for material 2 at 1 MHz.

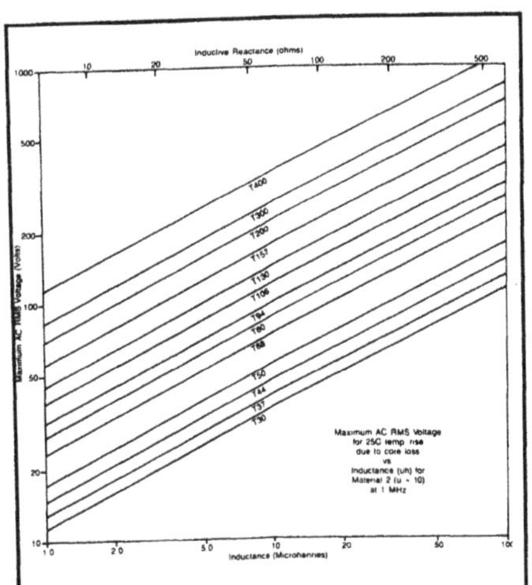

Figure 7. Maximum AC RMS voltage characteristic.

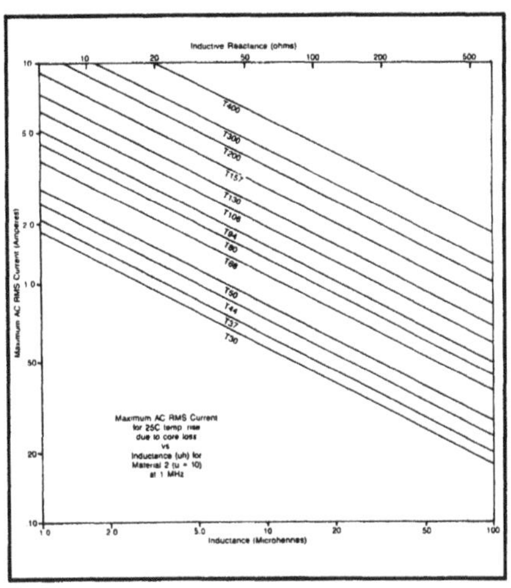

Figure 8. Maximum AC RMS current characteristic.

Material 2 — 1 MHz

Core	Watts
T30	21
T37	26
T44	37
T50	49
T68	88
T80	125
T94	160
T106	236
T130	331
T157	515
T200	794
T300	1127
T400	2108

Figure 9. "Power Rating" for 25C temperature rise due to core loss.

Summary

The selection of the core size and material for RF power inductors and transformers has typically relied on the "cut and try" method. New data and application information is now available which allows the design engineer to more easily and accurately select the optimum iron powder core size and material for these applications.

About the Author

Jim Cox is the Chief Applications Engineer at Micrometals, Inc., 1190 North Hawk Circle, Anaheim, CA 92807-1788. He received his BSEE from the University of California at Irvine and has been with Micrometals for 13 years. Jim can be reached at (714) 630-7420.

rf letters

Letters should be addressed to: Editor, *RF Design*, 6300 S. Syracuse Way, Suite 650, Englewood, CO 80111.

Hybrid Splitter/Mixer IMD Comments

Editor:

With regards to Daniel Marz article "Hybrid Splitter Improves Mixer IMD Performance" (Jan. 87) I wonder how much of the improvement in IMD is due to the 3 dB padding effect of the hybrid. Virtually the same performance can be achieved by terminating the mixer wiith a 3 dB resistive attenuator. This would properly terminate the filter as well. The resultant circuit would be less costly and easier to implement.

Better performance can be obtained by using an LC diplexer at the output of the mixer to terminate the LO, but this would add to the circuit complexity and therefore its cost.

Stephan E. Sykes
Harris/RF Communications
Rochester, NY

Editor:

The author of the article "Hybrid Splitter Improves Mixer IMD Performance," January 1987, p. 62, implies that all energy reflected from the filter is absorbed by the isolating resistor in the power divider, thus accounting for the mixer IM improvement. This is not true; the reflected energy is equally divided between the isolating resistor and the mixer. A 3 dB pad would have the same effect.

David E. Norton
Adams-Russell, Anzac Division
Burlington, MA

The Author's Response

Editor:

Both the 3 dB pad and the splitter will give similar results, although at high frequencies, where a splitter can be easily constructed using microstrip techniques, the splitter will need only two resistors against the three required by the 3 dB pad.

Daniel Marz
General Instrument, Jerrold Div.
Hatboro, PA

Errata

Editor:

I especially enjoyed my old colleague Dick Bain's article "The Engineer's Toolkit," but unfortunately it suffered from having received less than your usual grade of editing. I suggest a contest: a prize for finding the most errors on page 134!

William B. Lurie
Boca Raton, FL

Figure 1, page 134 of the February 1987 RF Design *should have rho (ϱ) in place of p. — Editor*

Editor: Feb. issue, page 41 — center — part of paragraph one appears below Figure 5. Was something of importance left out? Where is Fig. 10? I hope this is not standard procedure.

C.B. Shuttle
Belmont, CA

No, it's not standard procedure! Disregard the repeated text just below Fig. 5. On page 43, the references to Figures 10 and 9 in the center paragraphs should be changed to Figures 8 and 7. — Editor

Magnetic Cores in RF Circuits

Understanding core parameters helps the RF Engineer in designing inductors and transformers.

M.F. "Doug" DeMaw
Technical Department Manager, ARRL, Inc.
Newington, Conn.

Getting the proper ferrite or powdered-iron toroid, sleeve or pot-core into your RF circuit is not as casual an exercise as some designers believe! We must consider the core performance versus operating frequency (Q_L and bandwidth), the power-handling ability of the chosen core and the physical size of the completed inductor or transformer. A factor of equal importance in many circuits is the core-material stability in the presence of heat (ambient and current-induced heating). The design approach has been confusing to some RF engineers and technicians who have not had the occasion to develop circuits in which magnetic-core devices were required. This paper will, therefore, address the matter of fundamental design procedures for narrow and broadband RF inductors and transformers which contain magnetic core materials. For the most part, the discussion deals with the practical aspects of the design exercise.

Powdered Iron or Ferrite?

A notable advantage of powdered-iron over ferrite-core material is that the powdered iron B_{op} in gauss (operational flux density) is much greater for a given cross-sectional area than that of ferrite. Owing to this virtue the size of an inductor or transformer can be substantially smaller if iron is used rather than ferrite. In essence, powdered-iron cores yield higher values of B_{max} (maximum flux density in gauss) than the ferrite cores of equivalent physical dimensions.

Another advantage in the use of powdered-iron cores is *core stability*. Generally, ferrite materials are more prone to permeability changes with temperature than is the case with iron. Furthermore, if the B_{max} of a ferrite core is exceeded in an RF power circuit the permeability of the material will undergo a permanent change. This is not true of powdered-iron core material.

Why then should the designer consider using ferrite core material when iron seems to be so much better? Paramount among the decisions to employ ferrite cores is the very high value of μ_i (initial permeability) which is available: Ferrite cores can be purchased with permeability factors in excess of 5000. The common upper limit for powdered-iron cores is approximately 75 in terms of μ_i. In view of the foregoing it becomes apparent that ferrite materials lend themselves well to the fabrication of low and medium-frequency inductors and transformers. This is especially significant with respect to broadband RF transformers. The principle is illustrated dramatically in

Figure 1 where two inductors are wound for use at 455 kHz on 0.5-inch OD toroid cores, using powdered iron and ferrite materials for the frequency of operation versus good Q. A substantially greater number of turns is needed for L1 than is required for L2. The advantage of the ferrite over the powdered iron is clear: In the case of L2 we can use fewer turns of heavier gauge wire. The net result is higher tuned-circuit Q and reduced fabrication time. Both inductors have been calculated for single-layer coil windings.

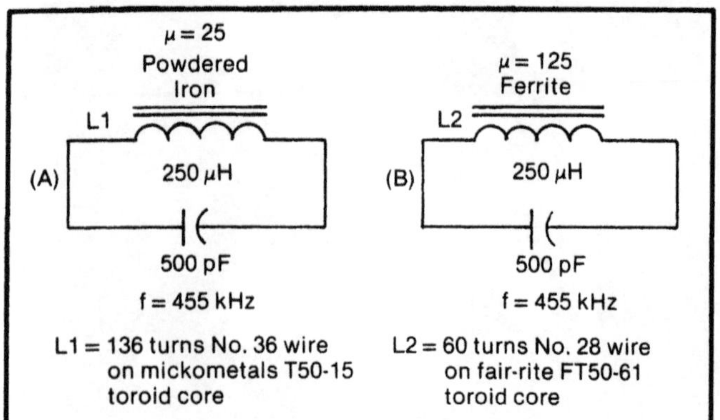

Figure 1.

Powdered-iron core materials have been characterized mainly as being best suited to applications at medium frequency and higher. Even though the irons have been used successfully for many years as low as a few hundred Hertz, their greatest application has been at RF. This is because it is possible to obtain low values of μ_i with powdered iron (as low as 1, for use at VHF). This permits the designer to obtain suitable values of Q_u through the VHF spectrum, which is not possible with ferrite core materials. For example, the Micrometals 0-mix iron is recommended for use between 50 and 300 MHz. It has a μ_i of 1. The next frequency bracket (20 to 250 MHz) lists 12-mix as the optimum core material. It has a μ_i of 3. It can be seen from this that the higher the μ_i the lower the recommended operating frequency in terms of maximum attainable Q_u. At the opposite end of the scale we find the Micrometals mix-41 iron, which has a μ_i of 75. The suggested frequency range for this material is 1 to 10 kHz.

Fair Rite Corporation produces a material-68 ferrite core which has a μ_i of 20 and is specified for use between 80 and 180 MHz. Most designers seem to limit their use of ferrites to a maximum frequency of approximately 25 MHz when designing narrow-band circuits. A suitable ferrite material for this range is mix 63. It has a μ_i of 40. The manufacturer suggests using mix 63 from 15-25 MHz.

Why Use Magnetic Cores at RF?

There are a number of advantages we can realize by using magnetic core material in RF inductors and transformers. When we consider the alternative — air-wound solenoidal inductors — the advantage becomes obvious. The magnetic-core component allows the mass of the circuit to be reduced markedly.

Still another advantage to be realized is seen in the case of toroidal inductors. As opposed to conventional inductors they are *self-shielding*. This eliminates the need for metal shield enclosures around the coil or transformer. The same is true of pot-core inductors and transformers. This advantage is lost when rods are used as the core material.

A more subtle advantage attendant to magnetic cores is improved Q_u when large amounts of inductance are required. This is because an equivalent air-core inductance would require many turns of light-gauge wire, whereas the magnetic-core equivalent inductance could be realized with substantially fewer turns of heavier-gauge wire. This would reduce the series R component to enhance the Q.

Broadband Transformers

Perhaps the greatest use of magnetic cores in RF circuits is found in broadband-transformer applications. These take the form of *conventional* and *transmission-line transformers*. They are commonly used in solid-state transmitters and receivers. They also lend themselves to viable use in antenna systems as baluns, combiners and matching transformers. Toroids, balun cores and rods are suitable for use in antenna systems. Pot cores are used as well as the foregoing for small-signal broadband transformers.

Figure 2 illustrates the basic difference in the winding format of conventional and transmission-line transformers. Illustration A shows a conventional transformer schemat-

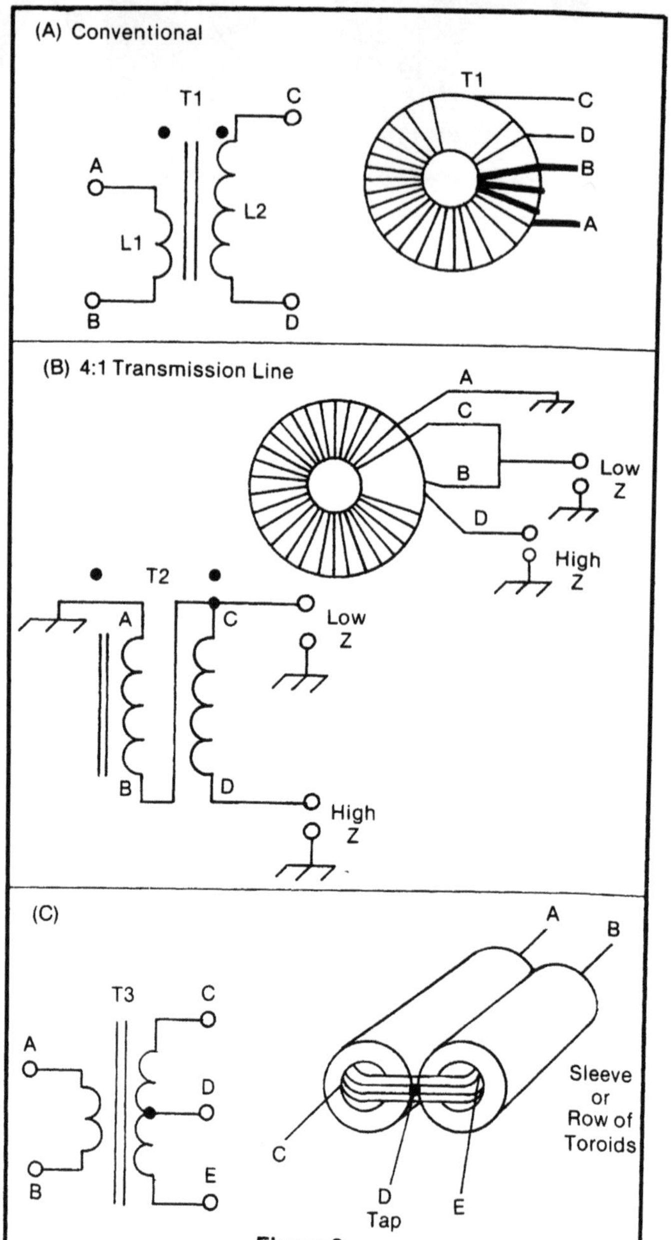

Figure 2.

ically and pictorially. The major advantage of this style of broadband transformer is that a wide latitude of transformation ratios are possible, since the primary and secondary windings are separate. The consensus is that conventional transformers are not as efficient as transmission-line equivalents are. Figure 2C illustrates another common form of conventional transformer. It uses two rows of toroid cores (stacked) or long ferrite sleeves (2 each). This type of broadband transformer is used frequently in solid-state RF power amplifiers for input and output matching.

A simple illustration of the transmission-line transformer is seen at B of Figure 2. The example shows an unbalanced-to-unbalanced 4:1 transformer which contains a bifilar winding. The wires can be laid on the core in parallel or they can be twisted an appropriate number of times per inch. Alternatively, lengths of subminiature coaxial cable are sometimes used. Microdot 25-ohm line is frequently the designer's choice. The disadvantage in using this style of transformer is that only fixed transformation ratios are possible — 1:1, 4:1, 9:1, 16:1, etc. It is recommended that a winding gap of approximately 30° be left on the toroid core as shown in Figure 2 at A and B. This technique aids in reduction of unwanted distributed capacitance across the winding and between the coil turns. The benefits from maintaining a gap in the winding are most pronounced when large numbers of turns are laid on a core. For example, a 900-turn winding might exhibit 70 pF of distributed capacitance with the coil ends directly adjacent to one another. The same inductor, but with a 30° winding gap, might show only 40 pF of distributed capacitance. The undesirable effects of large distributed capacitances are high "effective inductance" and degraded Q. In this context, the closer the coil's self-resonance is to the operating frequency, the poorer the Q.

Winding Characteristics

We must ensure that the individual windings of a broadband transformer exhibit sufficient reactance (X_L) at the lowest operating frequency to be effective as transformers or chokes. A viable rule of thumb is to make the smallest transformer winding approximately 4 to 5 times higher in X_L than the load resistance connected to the winding. Hence, if we were to match a 50-ohm source to a 10-ohm load with a 5:1 broadband transformer, the 10-ohm winding should have an X_L of approximately 40 to 50 ohms at the lowest operating frequency. This is demonstrated in Figure 3. L2 of T1 looks into a 10-ohm load at the base of Q1. Using the X4 rule, the winding reactance must be 40 ohms. In order to ensure ample winding inductance at 1.5 MHz we will need 4.2 μH. Once this requirement is satisfied, L1 will also have sufficient X_L for the 50-ohm source impedance of the generator. Too many transformer turns could lead to excessive X_c, as discussed earlier. This might result if we followed the X10 X_L rule that has been suggested by some designers.

Ferrite core material is used almost exclusively in broadband transformers. Generally, the materials are of fairly high μ_i, such as 950 for use between 1.5 and 30 MHz. This is done to ensure ample inductance in the windings at the lower end of the spectrum being covered. As the operating frequency is elevated the core effects begin to diminish, thereby giving rise to excellent transformer characteristics at the upper end of the desired frequency range. The actual permeability chosen will depend largely upon the frequency spread desired and the operating-frequency limits.

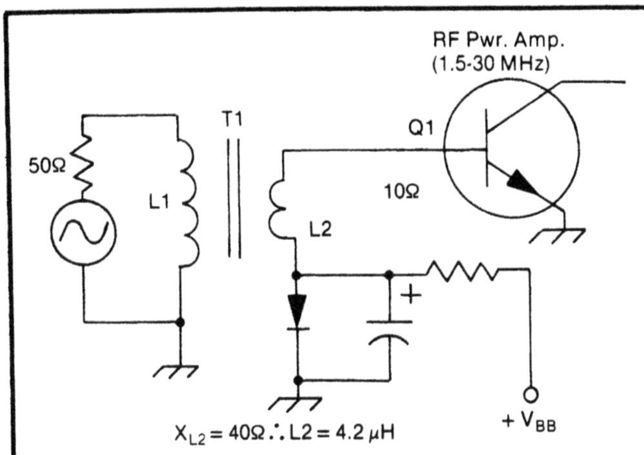

Figure 3.

The Proper Size Core for the Job

The undesirable effects of using too small a core are manifested in several ways. First, saturation will take place at or below peak-power level. Examination of the hysterisis profile of a saturating core will reveal a perfect condition for the enhancement of harmonic currents. The output waveform of an RF transformer which is saturating will show a square-wave characteristic when viewed on a scope. This is definitely an unwanted characteristic in most RF circuits.

Saturation changes the μ_e (effective permeability), which can lead to instability and losses. In a worst-case situation the core can become permanently damaged or might even shatter. Ideally, the core should be operated in the linear portion of its B_{max} curve. This data is available from the manufacturer.

The first consideration in determining the required core size is the B_{op} that will prevail. This can be found from

$$B_{op(ac)} = \frac{E_{rms} \times 10^8}{4.44\, f\, N\, A_e} \text{ gauss} \qquad (1)$$

where B_{op} is the operational flux density, E_{rms} is the voltage across the winding, N is the number of turns, A_e is the equivalent area of the magnetic path in cm^2 and f is in Hz.

The situation changes somewhat when both AC and DC currents flow in the winding. $B_{op(total)}$ is obtained from

$$B_{op(total)} = \frac{E_{rms} \times 10^8}{4.44\, f\, n\, A_e} + \frac{N\, I_{DC}\, A_L}{10\, A_e} \text{ gauss} \qquad (2)$$

where I_{DC} is the direct current through the winding and A_L is the manufacturer's inductance index for the specified core.

This requires some knowledge of the operating parameters which will prevail before a trial core is chosen for the equations. As cumbersome an approach as this is, it becomes part of the design procedure. If a given core will not yield the required safe margin between B_{op} and B_{max}, then the next larger size core is selected for the equations.

Some designers prefer to use E_{pk} in equations 1 and 2. This allows a safety factor beyond that which the rms value of the voltage will allow. However, the writer prefers to use the largest core that will fit into the available space on a circuit board or equipment module. This usually permits the designer to constrict B_{op} to the linear portion of the core's curve. The price differential between an electrically suitable core and the next larger size is minimal.

The Correct Number of Turns

Regardless of the physical structure of the magnetic core material being used, each core type and size has a specific A_L factor which can be used to determine how many turns are needed for a given inductance. A_L information is published by most manufacturers of magnetic core material. If the A_L factor for a given core can't be obtained, it can be determined easily by placing a few trial turns of wire on the unknown core and measuring the inductance of the winding. A_L can then be obtained from

$$A_L = L_{\mu H} \times 10^4/N^2 \qquad (3)$$

where $L_{\mu H}$ is the measured inductance and N is the number of turns. Thus, if we wound 6 turns of wire on an unknown toroid core and the measured inductance was 1.0 μH, the A_L would equal

$$1.0 \times 10^4/36 = 277.7$$

Once the A_L factor is known it is an easy matter to calculate the turns necessary for a given inductance. For example, assume that we used the core for which we obtained the A_L factor by means of Equation 3. We desire an inductance of, say, 112 μH. The winding size can be determined from

$$N = 100\sqrt{L_{\mu H}/A_L}\ turns$$

hence

$$N = 100\sqrt{112/277.7} = 63.5\ turns \qquad (4)$$

When dealing with large amounts of inductance on cores with relatively high permeability factors, it is common to change Equation 4 to

$$N = 3162\sqrt{L_{mH}/A_L}\ turns \qquad (5)$$

These equations apply to any magnetic core material — toroids, pot cores and rods.

Summary

Additional applications information and design data can be obtained from the publications listed in the bibliography. Also included are the names and addresses of some leading U.S. manufacturers of ferrite and powdered-iron core materials. Their literature contains detailed information which will be of use to the designer. □

Bibliography

Application Notes:

1. Granberg, *Broadband Transformers and Combining Techniques for RF,* Motorola AN-749.
2. *Applying Ferroxcube Ferrite Cores to the Design of Power Magnetics,* Ferroxcube Corp. of America, Bulletin 330.

Technical Papers:

1. DeMaw, "On Ferrite-Rod Inductors," *QST* Magazine for March 1979, p. 38.
2. DeMaw, "The Practical Side of Toroids," *QST* Magazine for June 1979, p. 29.
3. Granberg, "One KW — Solid-State Style," parts 1 and 2, *QST* Magazine for April and May 1976.
4. Pitzalis and Couse, "Broadband Transformer Design for RF Power Amplifiers," *ECOM-2989,* U.S. Army Electronics Command, Ft. Monmouth, N.J.
5. Pitzalis, Horn and Baranello, "Broadband 60-W HF Linear Amplifier," *IEEE Journal of Solid-State Circuits,* vol. SC-6, No. 3, June 1971.
6. Pitzalis, Horn and Baranello, "400-Watt HF Linear Amplifier," *ECOM-3338,* U.S. Army Electronics Command, Ft. Monmouth, N.J.
7. Ruthroff, "Some Broadband Transformers," *Proc. IEEE,* Aug. 1959, pp. 1337-1342.

Books:

DeMaw, *Practical RF Communications for Engineers and Technicians,* No. 21557, Howard W. Sams & Co., Inc., Chapter 1.

Hayward and DeMaw, *Solid-State Design for the Radio Amateur,* ARRL, Inc., Newington, CT 06111.

Hilberts, *Amperex High-Power Transistor Book,* "On the Design of HF Wideband Power Transformers," *ECOM* report 6907.

Polydoroff, *High-Frequency Magnetic Materials,* John Wiley and Sons, N.Y.

Practical Circuit Information:

The ARRL Radio Amateur's Handbook, 1980 Edition, ARRL, Inc. Newington, CT 06111. Contains numerous applications of magnetic cores in HF and VHF transmitting and receiving equipment.

Manufacturers and Distributors:

Amidon Associates, 12033 Otsego Street, N. Hollywood, CA 91607. Single lot sales by mail plus catalog and data sheets.

Arnold Engineering Co., Box G, Morengo, IL 60152 (mfgr.)

Elna Ferrite Labs, Inc., 9 Pine Grove, Woodstock, NY 12498 (small-quantity, by-mail distributor).

Fair-Rite Products Corp., 1 Commercial Row, Walkill, NY 12589 (mfgr.)

Ferroxcube Corp., 5083 Kings Hwy., Saugerties, NY 12477 (mfgr.).

Indiana General Ferrite Products, Crows Mill Rd., Keasby, NJ 08832 (mfgr.).

Magnetics, Div. of Spang Industries, Box 391, Butler, PA 16001 (mfgr.)

Micrometals, Inc., 1190 N. Hawk Circle, Anaheim, CA 92807 (mfgr.).

Palomar Engineers, Box 455, Escondido, CA 92025. (small-quantity, by-mail distributor).

621.318.22 : 621.317.4

Paper No. 883
MEASUREMENTS SECTION

FERRITE MATERIALS

By Dr. D. POLDER.

(*The paper was first received* 18*th May, and in revised form* 29*th July,* 1949. *It was read before the* MEASUREMENTS SECTION 8*th November,* 1949.)

SUMMARY

In Section 1.1 it is argued that ferromagnetic materials with a high resistivity are of considerable practical importance. In Section 1.2 the electric and magnetic properties of ferrite materials at various frequencies are discussed. In Section 1.3 an account is given of the physicochemical background of the properties of ferrites. In Section 1.4 the manufacture is dealt with and the important possibilities for application of different ferrite materials are briefly reviewed. In the following Sections 2.1–2.5 the experimental methods are described which are used for the measurements of the electric and magnetic properties of the materials. Section 2.2 deals with a bridge used at frequencies between 1 and 100 kc/s. Sections 2.3 and 2.4 contain a description of the methods used for measurements at frequencies up to 500 Mc/s, while Section 2.5 deals with experiments in the microwave region.

(1) GENERAL PROPERTIES

(1.1) Introduction

It is well known that the use of a solid or even laminated magnetic metal as a core material is restricted to comparatively low frequencies because of eddy-current losses. The frequency range can be considerably extended by the use of metal-powder cores. The chief disadvantage of dust cores, however, is their considerable loss of permeability, since the insulating layers enveloping the grains behave as air-gaps. The ideal magnetic core material, therefore, is a homogeneous material with a high permeability and such a low conductivity that eddy-current losses may be completely ignored. Such a material has several advantages over dust-core material. The effective magnetic permeability can be varied at will by the introduction of an air-gap and fixed at a relatively high value. Furthermore, a better magnetic shielding can be obtained than with the inhomogeneous powder material. Also, the hysteresis losses will be favourably affected by the homogeneity of the material, since the magnetic flux will be more uniformly distributed than in the grains of a powder core.

It is clear that a sufficiently low conductivity can be found only in non-metallic ferromagnetic substances. It happens that the substance in which ferromagnetism was first observed is a non-metallic one. It is the mineral magnetite Fe_3O_4. It contains one divalent iron ion or ferrous ion for every two trivalent or ferric ions, so we may write $Fe^{II}Fe_2^{III}O_4$ and call it ferrous ferrite. Its d.c. resistivity is about 10^{-2} ohm-cm, i.e. a factor 10^3 higher than that of iron. For our purpose the conductivity of magnetite is still too high. It is found that the replacement of the ferrous ions by different divalent ions such as Cu, Ni, Zn or Pb leads to ferrites (copper ferrite, nickel ferrite, etc.) with a d.c. resistivity of the order of 10^6 ohm-cm, i.e. a factor 10^{11} higher than for iron.

Several of these ferrites were prepared by Hilpert[1] in 1909. From a magnetic point of view his results were unsatisfactory. The ferrites showed low permeabilities and high total losses, even though the eddy-current losses were negligible. Moreover, the properties were not reproducible. Since 1933 the ferrites have been the subject of extensive investigations in the Philips laboratory.[2] By a sintering process mixed crystals of different ferrites can be prepared, and Snoek[2] has shown that several of the resulting non-metallic magnetic substances show excellent properties from various points of view. The products have in common a high resistivity (10–10^8 ohm-cm) and a high initial permeability (10–3 000). Examples are NiZn-ferrite and MnZn-ferrite. It is desired to emphasize that different samples of these and other ferrites have been prepared which show considerably different properties. The properties strongly depend on the precise chemical composition and the heat treatment. A fuller discussion of these points will be found in the following Sections.

(1.2) Electric and Magnetic Properties of Ferrite Materials

A quantity of principal interest is the magnetic susceptibility. This quantity, defined as the ratio of the magnetic induction and magnetic field, strongly depends on the value of the magnetic field and on the field to which the material has been exposed previously. For practical applications one is often interested in the initial permeability μ, which is shown by the material in its demagnetized state when it is exposed to a small magnetic field. Since the considerations in this paper mainly concern alternating fields of small amplitude, the well-known convention of writing the field quantities as complex quantities proportional to $\exp(j\omega t)$ will be adopted. The magnetic permeability μ defined by $B = \mu H$ is then a complex quantity itself. If we write $\mu = \mu' - j\mu''$, μ' refers to the part of the magnetic induction B that is in phase with the field H, while μ'' refers to the part of B that shows a phase-lag of $\frac{1}{2}\pi$ with H. The magnetic loss-angle θ_m is defined by $\tan\theta_m = \mu''/\mu'$. In a similar way a complex permittivity $\kappa = \kappa' - j\kappa''$ and a dielectric loss angle can be introduced. Obviously, μ' and μ'' as well as κ' and κ'' depend on frequency. Data for several types of ferrites will be given in this Section. In the following the values of κ and μ relative to κ_0 and μ_0 of free space will always be used, thus $\mu_{rel} = \mu/\mu_0$ and $\kappa_{rel} = \kappa/\kappa_0$. Furthermore, the subscript (*rel*) will be omitted so that from now onwards κ and μ indicate the values relative to vacuum.

A further important quantity is the conductivity σ of a ferrite. For experiments in which alternating currents are involved it appears to be rather artificial to maintain the strict distinction between conductivity σ and the imaginary part of the permittivity. This may be most easily seen from Maxwell's equation

$$\text{curl}\, H = \sigma\mathscr{E} + \partial D/\partial t$$

which reads in the a.c. case, using $D = \kappa\kappa_0\mathscr{E}$

$$\text{curl}\, H = j\omega\left[\kappa' - j\left(\kappa'' + \frac{\sigma}{\kappa_0\omega}\right)\right]\kappa_0\mathscr{E}$$

It follows that a conductivity σ is equivalent to an imaginary part of κ given by

$$\kappa'' = \frac{\sigma}{\kappa_0\omega} \quad . \quad . \quad . \quad . \quad . \quad . \quad (1)$$

This is an "integrating" paper. Members are invited to submit papers in this category, giving the full perspective of the developments leading to the present practice in a particular part of one of the branches of electrical science.

Dr. Polder is with Philips Research Laboratories, N. V. Philips' Gloeilampenfabrieken Eindhoven, Netherlands.

In this paper, therefore, we shall sometimes interpret the experimental results on the dielectric behaviour of ferrites in terms of a (frequency-dependent) conductivity only and sometimes in terms of κ'' only.

For the practical application of ferrites it is of considerable importance to know the losses which may be expected in a coil with a ferrite core. Several factors contribute to the losses: hysteresis losses, which vanish for small amplitudes of the alternating magnetic field, eddy-current losses or, more generally, losses that disappear for samples with sufficiently small dimensions and residual losses which are due to the presence of a residual magnetic loss angle θ_m. Though in the majority of practical cases the last-mentioned losses are predominant in ferrites, the discussion will deal first with the losses due to dimensional effects and the hysteresis losses.

It is well known that eddy-current losses are considerable as soon as the skin depth δ becomes of the order of or smaller than the smallest dimensions of the specimen. For an ordinary material in which the real part of the permittivity as well as the imaginary part of the permeability are negligible, the skin depth is given by

$$\delta = \sqrt{\left(\frac{2}{\omega\mu'\mu_0\sigma}\right)} = \frac{\lambda}{2\pi}\sqrt{\left(\frac{2}{\mu'\kappa''}\right)} \quad . \quad . \quad . \quad (2)$$

The second equality is obtained by rewriting the conductivity in terms of κ'' with the aid of (1). λ is the wavelength of the electromagnetic wave in vacuum. Let us now assume for a moment that the formula (2), which is valid for metals, is also applicable to ferrite materials. Let us further assume that we may use the d.c. conductivity and the d.c. magnetic permeability in order to obtain an estimate of the value of δ at a given frequency. In that case we find that for a ferrite with a d.c. resistivity ρ of, say, 10^5 ohm-cm, and $\mu' = 1\,000$, δ is of the order of 30 cm at a frequency of about 2·5 Mc/s.

This figure reveals at once the advantage of poorly-conducting magnetic materials over magnetic metals. It must be emphasized, however, that an estimate of the cross-section of a core which may be tolerated at a given frequency may be considerably in error if the calculation is carried out in the way we have just described. The reason is twofold. Firstly it has been found by Brockman, Dowling and Steneck[3] that the high-frequency conductivity of a ferrite may be considerably larger than the d.c. conductivity. In Fig. 1 it is seen that the resistivity of a sample of MnZn-ferrite at 10 Mc/s was about 10 ohm-cm (corresponding to $\kappa'' = 18\,000$), to be compared with a d.c. resistivity of 130 ohm-cm. Secondly, this MnZn-ferrite appears to have an extremely high permittivity κ', which may be as large as 10^5 (cf. Fig. 1).[3] This is a particularly interesting phenomenon, but it is considered that a fuller discussion of its physical basis is beyond the scope of the present paper. It is, however, desirable to point out that a high permittivity (though in several cases only high at lower frequencies) seems to be a common property of ferrites and also of many non-magnetic semi-conductors.[4] For our considerations the high permittivity is of importance, since the simple formula (2) for the skin effect is no longer applicable in cases where $\kappa' \gg \kappa''$. As a matter of fact there is no proper skin effect in substances like the MnZn-ferrites mentioned. There is a change of the pattern of lines of force inside the core which can be described by the term "dimensional resonance," a phenomenon that can also give rise to considerable losses.[5] It can be proved by a reasoning similar to that used for the derivation of (2), that, generally speaking, dimensional effects will be of importance at a frequency for which the quantity δ', defined by

$$\delta' = \frac{\lambda}{2\pi}\sqrt{\left(\frac{2}{|\mu||\kappa|}\right)} \quad . \quad . \quad . \quad . \quad . \quad (3)$$

is of the order of the smallest dimensions of the specimen. For a core of 1×1 cm cross-section made from MnZn-ferrite of the type described above (κ' about 5×10^4) this frequency is somewhat below 1 Mc/s.

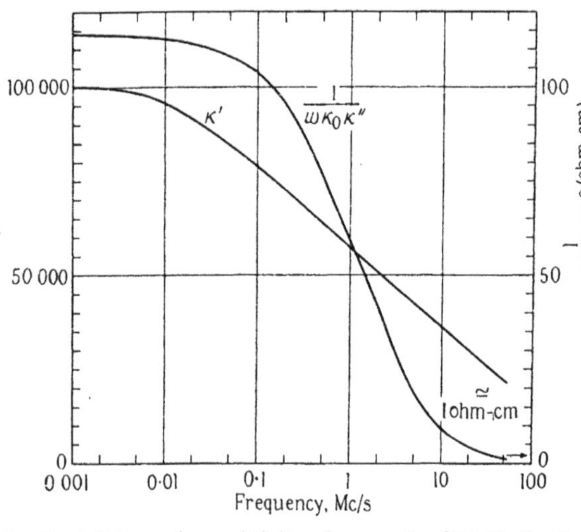

Fig. 1.—Resistivity and permittivity of a sample of MnZn-ferrite as a function of frequency.

Since δ' computed from (3) may be considerably smaller than δ obtained from (2), the question arises whether the observed high permittivity seriously restricts the use of ferrite materials above 1 Mc/s. It must be emphasized that a high κ' which persists at frequencies above 1 Mc/s has been found only in the better-conducting ferrites like the sample of MnZn-ferrite discussed above. This type of MnZn-ferrite was developed for application below 1 Mc/s, where it shows excellent properties from several points of view. Its magnetic losses are too large at higher frequencies. In Section 3 it will be explained why this MnZn-ferrite has been given a comparatively low resistivity. Most ferrites, however, e.g. NiZn-ferrite or different types of MnZn-ferrite, have a higher resistivity. It appears that ferrites with a greater resistivity have a much more favourable high-frequency dielectric behaviour. NiZn-ferrite, for instance, can easily be prepared with a d.c. resistivity of $10^5 - 10^6$ ohm-cm, while its high-frequency resistivity is smaller by a factor 3 or so only. At low frequencies its permittivity is quite high, about 2 000 at 1 kc/s, but at higher frequencies it decreases rapidly to the value 15, being practically constant from 100 kc/s onwards (cf. Fig. 2). A core of 1×1 cm cross-section from this material for which $\mu' \simeq 700$, will not give rise to dimensional effects up to about 40 Mc/s.

Let us now turn to the hysteresis losses. In ferrite materials these can be minimized by a suitable choice of the chemical composition and heat treatment. If the amplitude of the magnetic induction in a core is B_{max} gauss, the heat developed by hysteresis per cycle and per cubic cm of the core is, for small amplitudes,

$$C_h B^3_{max}/8\pi \text{ erg/cm}^3*$$

C_h is the hysteresis coefficient. For materials for which the so-called Raleigh relation holds (cf. Ref. 2) C_h does not depend on $\hat{B}$ at small amplitudes. The hysteresis coefficient of ferrite materials does depend on $\hat{B}$, however, and usually the figures are given for $\hat{B} = 7{\cdot}5$ gauss and a frequency of 2 kc/s. The experimental figures show that ferrite materials compete favourably with core materials such as Sendust. MnZn-ferrite has

* C_h, used in this paper, is equivalent to h used in other published work.

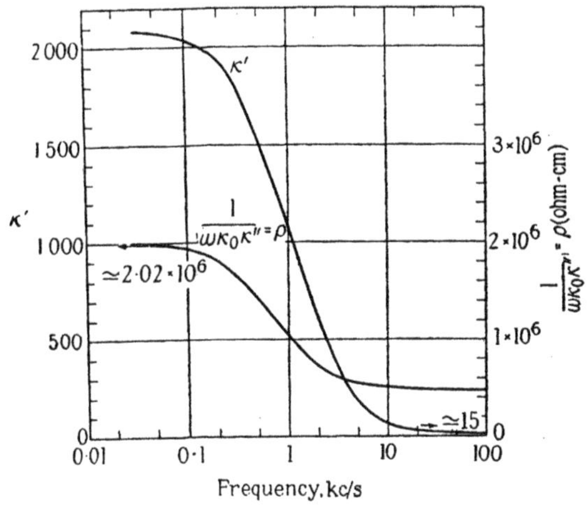

Fig. 2.—Resistivity and permittivity of a sample of NiZn-ferrite as a function of frequency.

been prepared having $\mu' = 2\,000$ and $C_h = 1{\cdot}6 \times 10^{-6}$, to be compared, for instance, with $C_h = 5 \times 10^{-6}$ for a Sendust core with $\mu' = 65$.

The most important part of the losses in ferrite materials at low inductions is accounted for by the presence of an imaginary part μ'' of the permeability. For small values of μ'', $(\tan \theta_m)/\mu'$ is the appropriate quantity to characterize the quality of the material with respect to the magnetic losses. The reason is that for a fixed value of B_{max} the energy dissipation in ergs per cycle and per cubic cm of the material is equal to $\frac{1}{4}B_{max}^2$ $(\tan \theta_m)/\mu'$. This is the case in a solid core as well as in a core in which a small air-gap has been introduced. Sometimes the residual loss factor $C_r = 2\pi\,(\tan \theta_m)/\mu'$ is used. Experimentally it is found that C_r depends strongly on frequency. With increasing frequency the losses at first increase slowly, but later more rapidly, so that a fairly well-defined frequency can be indicated at which the material ceases to be useful for most practical purposes. As a general rule it has been found that this frequency limit is the lower, the higher the initial permeability of the material. For instance, MnZn-ferrite having $\mu' = 1\,000$ may show $(\tan \theta_m)/\mu'$ of the order 10^{-6} for frequencies below 10 kc/s. At 1 Mc/s $(\tan \theta_m)/\mu$ is of the order 10^{-4} $(\tan \theta_m \simeq 0{\cdot}1)$ and a rapid increase is found at frequencies above 1 Mc/s. On the other hand NiZn-ferrite has been prepared having $\mu' = 42$, for which $(\tan \theta_m)/\mu'$ is still of the order 10^{-4} at 25 Mc/s. Curves for $(\tan \theta_m)/\mu'$ of various samples of MnZn- and NiZn-ferrites are shown in Fig. 3.

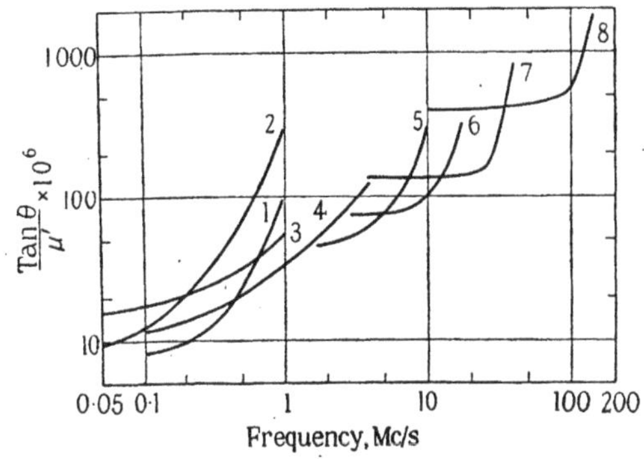

Fig. 3.—$(\tan \theta_m)/\mu'$ as a function of frequency for different ferrite materials. Curves 1, 2 and 3 refer to samples of MnZn-ferrite, curves 4–8 to samples of NiZn-ferrite.

The deterioration of the magnetic behaviour at high frequencies appears to be a general phenomenon in ferrites. Its physical basis will be briefly discussed in the following Section. For all ferrites a rapid increase of μ'' is found somewhere in the frequency region between 100 kc/s and 200 Mc/s (cf. Fig. 4).

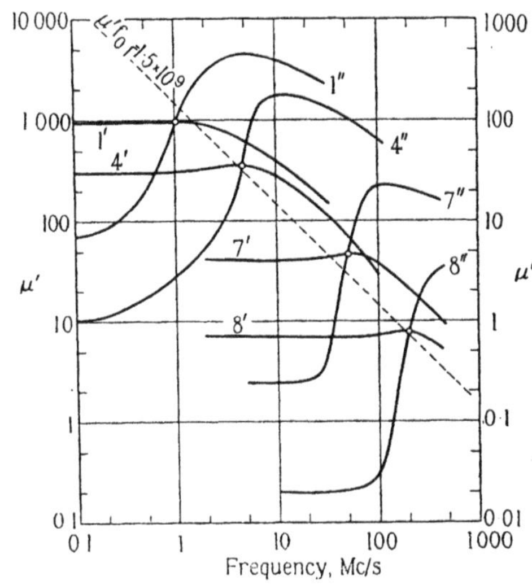

Fig. 4.—Curves 1′, 4′, 7′, 8′ give μ' of the samples 1, 4, 7, and 8 in Fig. 3. Curves 1″, 4″, 7″, 8″ give μ''. Note the different scale for μ' and μ''.

The increase of μ'' is accompanied by a decrease of μ' at a somewhat higher frequency. At still higher frequencies μ'' decreases as well as μ'. In this region μ'' may be considerably larger than μ'. It is interesting to note that an approximate relation exists between μ' and the frequency $f_{0\cdot1}$ for which $\tan \theta_m \simeq \mu''/\mu' = 0{\cdot}1$. In Fig. 4, curves for μ' and μ'' of several ferrites are shown. A different logarithmic scale is used for μ' and μ'', so that the intersection of the μ' curve and the μ'' curve for the same sample indicates the point where $\mu''/\mu' = 0{\cdot}1$. It is seen that the relation $\mu' f_{0\cdot1} = 1{\cdot}5 \times 10^9$ holds approximately for different ferrites. This experimental relation may be of some use for practical purposes.

The saturation magnetization $(B - \mu_0 H)_{sat} = M_{sat}$ of the ferrites is comparatively low and practically never exceeds 5 000 gauss at room temperature. Therefore the material cannot be used in power transformers, etc. Another important point is the relatively low Curie-temperature of the ferrites with a high permeability, so that the ambient temperature must not be allowed to approach the Curie point closely. In Fig. 5, curves are given for M as a function of H at room temperature, for M_{sat} as a function of temperature, and for μ' as a function of temperature. The curves refer to a MnZn-ferrite with a Curie point at 105° C.

(1.3) The Properties of Ferrites in Connection with Chemical Composition and Heat Treatment

The ferrites in which the divalent ions have a Goldschmidt radius between approximately 0·75 and 1·00 Å show a cubic structure. Mixed crystals between these ferrites can be formed in all proportions as far as known. These ferrites are $MgFe_2O_4$, $NiFe_2O_4$, $CoFe_2O_4$, $Fe^{II}Fe_2O_4$, $MnFe_2O_4$, $ZnFe_2O_4$ and $CdFe_2O_4$; $PbFe_2O_4$ and $CaFe_2O_4$ have hexagonal and tetragonal structures, $CuFe_2O_4$ is cubic only at elevated temperatures and is tetragonal at room temperature.

The cubic structure is very essential for the magnetic properties of the ferrites. We shall see that temperatures between

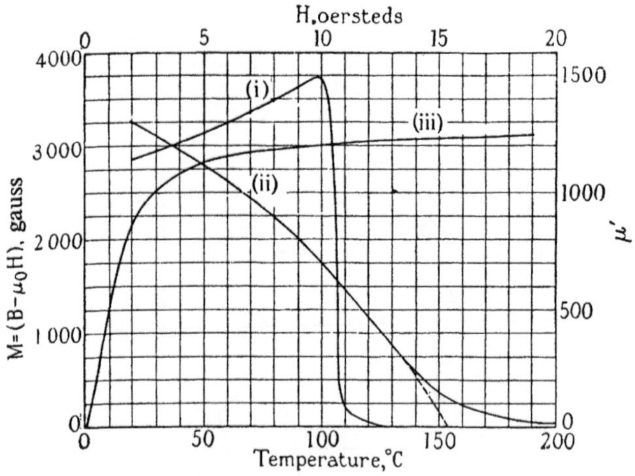

Fig. 5.—(i) μ' as a function of T; (ii) M_{sat} as a function of T; (iii) $M = B - \mu_0 H$ as a function of H. The data refer to a sample of MnZn-ferrite.

1 000 and 1 400° C are used for the preparation of solid objects of ferrite material. The cubic structure is the only structure in which stresses due to unequal shrinkage in different directions of the crystallites during cooling can be avoided to a large extent. It is important to keep internal stresses as low as possible, since stresses give rise to internal anisotropy of the material and it is well known that this reduces the initial permeability and enhances the hysteresis losses.

A cubic ferrite has the same crystal structure as the mineral spinel ($MgAl_2O_4$) (Fig. 6). The spinel structure consists of a

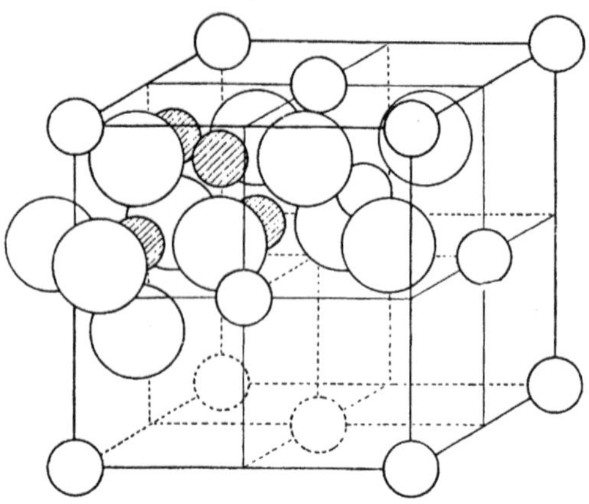

Fig. 6.—Unit cell of spinel structure (large spheres: oxygen; small spheres, hatched: ions in octahedral positions; unhatched: ions in tetrahedral positions. The ions are drawn in only 2 of the 8 octants.

very nearly close-packed arrangement of oxygen ions, between which there are interstices in which the divalent and trivalent ions are found. A closer inspection of the structure shows that there are two types of interstices, called octahedral and tetrahedral positions respectively, according to the way in which they are surrounded by oxygen ions. For every 32 oxygen ions in the unit cell of the spinel structure there are 16 octahedral and 8 tetrahedral positions occupied by metal ions (cf. Fig. 6).

It appears that the electric and magnetic properties of ferrites are closely connected with the way in which the various metal ions are distributed over the available positions. So far as the resistivity is concerned, a general rule given by Verwey and co-workers[6] predicts that the conductivity is increased by the presence of ions of the same metal, but of different valency on equivalent crystallographic positions. Thus it can be understood that ferrites of the type $Me^{II}Fe_2O_4$, where Me^{II} is any metal ion different from Fe^{II}, show low conductivity, while the introduction of a small amount of Fe^{II} ions by incorporation of Fe_3O_4 into the mixed crystal increases the conductivity by several factors of 10. For reasons to be discussed later, such an incorporation has been proved useful for the preparation of, for instance, MnZn-ferrite with high permeability. The d.c. conductivity of this material is about 100 ohm-cm, i.e. a factor 10^4 higher than for a ferrite without ferrous ions.

In order to understand the relation between the magnetic properties and the distribution of the metal ions over the available positions, some further peculiarities of the spinel structure will be briefly recalled. According to Barth and Posnjak,[7] we can distinguish between two types of spinel structure: the normal structure, in which the divalent ions occupy the tetrahedral positions, and the inverted structure in which they are statistically distributed over the octahedral positions. Verwey and co-workers[8] have shown that the non-(ferro-)magnetic ferrites ($ZnFe_2O_4$ and $CdFe_2O_4$) are of the normal type, whereas the ferromagnetic single ferrites (e.g. Mg-ferrite and, according to indirect evidence also Cu-ferrite, Mn-ferrite, ferrous ferrite, Ni-ferrite and Co-ferrite) show the structure of the inverted type. In the inverted structure one half of the ferric ions occupy tetrahedral positions and the other half octahedral positions. Thus we can write $Fe(NiFe)O_4$, in which formula the symbols between parentheses indicate ions occupying octahedral positions. A mixed ferrite of which the two constituent simple ferrites show a normal and an inverted structure respectively, e.g. NiZn-ferrite, is represented by $Zn_aFe_{1-a}(Ni_{1-a}Fe_{1+a})O_4$, where a and $1 - a$ are the molecular fractions of Zn-ferrite and Ni-ferrite in this mixed ferrite.

Let us now consider the saturation magnetization that would be expected in ferrites if all magnetic ions fully contributed to the magnetization. For this we assume that the magnetic moments of the ions are identical with the paramagnetic moments as shown, for instance, in hydrated salts, i.e. 5 Bohr magnetons for a ferric ion, 2·3 Bohr magnetons for nickel and so on. Thus for NiZn-ferrite one would expect at least 10 magnetons per molecule. The experimental data for the saturation magnetization at low temperature show much lower values (cf. Table 1). Zn-ferrite, as well as Cd-ferrite, do not even show ferromagnetism.

Table 1

SATURATION MAGNETIZATION IN BOHR MAGNETONS PER MOLECULE

	Sum of ionic moments	M_s		Δ
		Experimental	Calculated	
$Fe(NiFe)O_4$	12·3	2·3	2·3	0
$Zn_{0·1}Fe_{0·9}(Ni_{0·9}Fe_{1·1})O_4$ ..	12·1	2·9	3·1	0·2
$Zn_{0·3}Fe_{0·7}(Ni_{0·7}Fe_{1·3})O_4$ ..	11·6	4·2	4·7	0·5
$Zn_{0·5}Fe_{0·5}(Ni_{0·5}Fe_{1·5})O_4$ ..	11·1	5·0	6·1	1·1
$Zn_{0·7}Fe_{0·3}(Ni_{0·3}Fe_{1·7})O_4$ ..	10·7	3·9	7·7	3·8
$Zn(Fe_2)O_4$	10	0	10	10

A qualitative understanding of these data can be obtained by assuming with Neel[9] that in the magnetic spinels a magnetic dipole in a tetrahedral position has a strong tendency to an anti-parallel orientation relative to the dipoles of ions in adjacent octahedral positions. Therefore, the total moment should be the difference of the sum of the moments of all dipoles in octahedral positions, and the sum of the dipoles in tetrahedral positions (cf. col. 3 in Table 1). Gorter[10] has found that for low percentage of Zn there is approximate agreement between this expectation and the experimental figures. The deviations for ferrites with a higher percentage of Zn, and in particular the vanishing moment for Zn-ferrite, can be understood in the following way. The strong interaction between dipoles in tetrahedral and octahedral positions indirectly favours the parallel orientation of all dipoles in octahedral positions. Apart from this mechanism there is a weak direct interaction between dipoles on octahedral positions, which favours a mutually anti-parallel orientation. When the number of magnetic ions in tetrahedral positions decreases, it becomes more and more difficult for the first mechanism to compete with the second. The result is a decrease of saturation magnetization for ferrites with a higher percentage of Zn, and absence of ferromagnetism in Zn-ferrite.

It is seen that the saturation magnetization, considered as a function of the molecular fraction of Zn-ferrite, passes through a maximum. A similar behaviour of the saturation magnetization is found in other mixed ferrites, e.g. MnZn-ferrite, MgZn-ferrite, etc. Further details on the subject of the saturation magnetization are given in a paper by Gorter.[10]

In order to obtain a ferrite with a high initial permeability the material must show a sufficiently large saturation magnetization. There are several more factors that affect the initial permeability. First of all the density; it is well known that the permeability is the higher, the less porous the material. By an appropriate milling and prefiring treatment and by sintering the pressed material at a high temperature for a sufficiently long time, the density can be increased up to a value that differs only little from the maximum possible density. In commercial MnZn-ferrite the deviation is of the order of 5%.

The further factors determining the initial permeability have been discussed extensively by Snoek[2] and can be described by the term "magnetic anisotropy." This is said to be large if there are large forces impeding the process of magnetization of the specimen in the direction of the applied magnetic field. A large anisotropy means low initial permeability and high hysteresis losses. It may be that the movement of Bloch boundaries between the Weiss domains is hindered by inhomogeneities (inclusions, etc.) or internal stresses, and also that there are forces trying to keep the direction of magnetization in the individual Weiss domains along one or more preferred directions. By making the material as homogeneous and as dense as possible with the aid of an appropriate milling and heating technique, magnetic anisotropy can be reduced to a large extent, but nevertheless we still have to consider the effect of residual stresses (stress anisotropy) and of crystal anisotropy.

Crystal anisotropy is the preference of the magnetization for one of the crystal axes, e.g. the [100] or [111] axis in a cubic crystal. It is known that the crystal anisotropy decreases with increasing temperature and reaches a very low value already well below the Curie temperature. By preparing mixed crystals with the appropriate amount of Zn-ferrite, the Curie point of every ferrite may be lowered to any desired value. By choosing the Curie temperature as low as compatible with the highest ambient temperature to be expected, Snoek has been able to obtain the optimum conditions for a high permeability.

Internal stresses give rise to a preferred orientation either parallel or perpendicular to the direction of the stress, depending on the sign of the magnetostriction. Since internal stresses introduced during cooling can never be completely avoided, it is important to have a material with low magnetostriction in order to obtain a high permeability.

Usually the magnetostriction decreases if the temperature increases towards the Curie temperature, so that a low Curie point is also favourable with respect to magnetostriction. Went[2] has found that a further improvement can be obtained by making use of the fact that the magnetostriction is negative for all ferrites except ferrous ferrite, which shows a large positive magnetostriction. Therefore the introduction of only a few per cent of Fe_3O_4 in mixed crystals which have already low magnetostriction suffices to reduce magnetostriction to a negligible value. A large amount of Fe_3O_4 cannot be tolerated, of course, because of its effect on the conductivity.

For the purpose of obtaining a high permeability the magnetic anisotropy must be minimized. For a good high-frequency behaviour (i.e. a high frequency-limit), however, the magnetic anisotropy is essential, for the following reason. The increase of μ'' and the corresponding decrease of μ' of ferrite material between 1 and 200 Mc/s is most probably due to so-called ferromagnetic resonance absorption. It is well known that the individual magnetic dipoles of a ferromagnetic material are spinning magnetic tops. When there are forces which keep the tops in a preferred direction, the tops can perform a precession movement around this direction. Ferromagnetic resonance absorption occurs if the frequency of the alternating magnetic field is equal to the precession frequency. This frequency is the higher, the stronger the forces acting upon the dipole tops.

If the force is a constant external magnetic field, the resonance absorption takes place at one well-defined frequency. The corresponding experiments are described in Section 2.5. In a demagnetized ferrite the forces are due to the magnetic anisotropy,[11] and these may vary considerably inside the material. Therefore, there will be a broad region of absorption. In accordance with experiment, the absorption is expected in a frequency region that is the higher, the larger the magnetic anisotropy. This explains the incompatibility of a high initial permeability and a high frequency-limit.

(1.4) Manufacture; Varieties; Applications

Objects are prepared using a sintering technique such as is common in ceramic industry. The oxides—or other compounds which give oxides on heating—are ground together to a homogeneous mass of small particle size and usually heated at a comparatively low temperature in order to reduce shrinkage during the final sintering process. The powder is then ground again and pressed in a steel die to the desired shape, allowing for a shrinkage of about 20% during firing, a value which depends on the raw materials employed and the temperature at which the mixture has been heated previously.

In this way objects are obtained with a tolerance of $\simeq 2\%$ in dimensions. In order to obtain tubes or rods the powder is mixed with an organic binder, extruded and fired.

The final firing is carried out in a carefully chosen gas atmosphere at a temperature of 1 000–1 400° C, depending on the type of ferrite and on the raw materials involved. The final product is hard but may be ground to a very close tolerance, e.g. to give extremely flat surfaces. This is important for the design of magnetic cores in which air-gaps of a very precisely known width have to be introduced.

A brief survey of the important varieties of ferrite materials ("Ferroxcube") will now be given.

Ferroxcube I(*a*) (Copper-zinc ferrite 0·20 CuO, 0·30 ZnO, 0·50 Fe_2O_3). This ferrite has high μ' ($\mu' > 1\,000$) and has been used for h.f. transformers. Since it proved possible to reach equal μ' and residual losses, lower hysteresis losses and a higher Curie temperature with Manganese-zinc ferrite, the use of CuZn-ferrite has been discontinued.

Ferroxcube II (Magnesium-zinc ferrite 0·25 MgO, 0·25 ZnO, 0·50 Fe_2O_3, $\mu' \leqslant 360$) has high residual losses and is now considered obsolete.

Ferroxcube III (Manganese-zinc ferrite, 0·25 MnO, 0·25 ZnO, 0·50 Fe_2O_3, usually with a slight excess of Fe_2O_3) is being produced in increasing quantities. It has high initial permeability, low hysteresis losses, low residual and negligible eddy-current losses up to 500 kc/s.

Ferroxcube V is a new variety of Ferroxcube for applications where a high permeability at high flux densities is important rather than μ'. Its saturation value is higher than for all other ferrites, and decreases less when the temperature is raised.

Ferroxcube IV (Nickel-zinc ferrite). In the mixed crystal series $NiFe_2O_4$-$ZnFe_2O_4$ the whole range of magnetic properties as they occur in ferrites is traversed. The new development in this field is to be found in a paper by Gorter.[12]

Nickel ferrite is very anisotropic, but by incorporating increasing amounts of Zn-ferrite it is made increasingly isotropic. Thus Ni-ferrite has low permeability and a high frequency-limit because of a large anisotropy field, whereas, for example, 15 NiO, 35 ZnO, 50 Fe_2O_3 has large μ' (1 000–4 000) and a correspondingly low frequency-limit. Furthermore, the increase of the isotropy is accompanied by a decrease of the hysteresis losses and the coercive force.

It is outside the scope of the present general survey of ferrite materials to give detailed information on the different combinations of the important properties such as resistivity, initial permeability and its temperature-coefficient, high-frequency behaviour, hysteresis losses, Curie temperature and saturation magnetization, which can be obtained. From the preceding paragraphs it will be clear that, in order to obtain a material for which some of these properties are very favourable, one will have to be satisfied with somewhat less favourable properties in other aspects (cf. the incompatibility of a high initial permeability and a high frequency-limit). The research carried out, however, has revealed to a large extent the laws that govern the dependence of the important properties on the composition and the method of preparation of ferrite materials. From this general knowledge it is possible to judge which type of ferrite material is most suitable for the special purpose for which it is wanted.

Because of its favourable properties ferrite material has a wide field of application. It is eminently suitable for use in carrier telephony, e.g. in band-pass filters, transformers and equalizers. Here the application of MnZn-ferrite made it possible to gain a factor 2·5 in the quality of the filter and a reduction of the volume by a factor 5, compared with the old filter consisting of a core of 40μ nickel-iron wires and a jacket of powder-core material, the low permeability of which necessitated the use of a screening box at a considerable distance from the magnetic material.

Also, in band-pass filters for the intermediate frequency in superheterodyne wireless sets, ferrite material can give important economy. In television it may be used for the time-base and pulse transformers. Further application are for tuning oscillator circuits by a movable core or by adjusting the effective permeability by means of d.c. magnetization (so-called permeability tuning).

For input transformers the material is very suitable; for output transformers or power transformers at low frequencies its saturation value is too low.

(2) METHODS OF MEASUREMENT

(2.1) Introduction

In this Section of the paper some experimental arrangements that have been used in the Philips laboratory for the determination of the electric and magnetic properties of ferrite materials will be described. In general it is possible to distinguish between two experimental methods: measurements with the aid of a bridge circuit and measurements with the aid of a resonance circuit. The former method has been used up to a frequency of about 100 kc/s. In the latter method the resonance circuit may consist either of a separated capacitance and inductance (*LC* circuit, frequency range from about 10 kc/s up to 50 or even 100 Mc/s in some cases), or of a coaxial line (frequency range from 1 to 500 Mc/s), or of a resonance cavity (in the centimetre-wave region). The coaxial line has been used only for determination of the magnetic properties.

The sample of the ferrite material may be a thin plate or a small cube provided with electrodes being used as a condenser in the case of dielectric measurements with the aid of a bridge circuit or an *LC*-resonance circuit. For the magnetic measurements a thin toroid or ring is used, which is inserted into a coaxial line or which is provided with a winding and used as a self-inductance in an *LC* circuit or bridge circuit. In the resonance cavity either cylindrical or spherical samples of small dimensions are used.

For these measurements it may be of great importance to have the dimensions, or in any case the smallest dimension of the sample, sufficiently small, for the following reason. In all experimental arrangements mentioned, the calculation of the complex permittivity or the complex magnetic permeability of the material from the actual experimental observations is relatively simple, provided that inside the sample there is a homogeneous flux-density (i.e. the dielectric displacement and current in the case of a dielectric measurement and the magnetic induction in the magnetic case). If, however, the dimensions of the specimen are too large, dimensional effects may occur, so that the simple calculation referred to above would give an effective value of κ or μ rather than the actual values for the material. The discrepancy may be already quite large, even when the change in the flux pattern is still relatively small. The danger of an erroneous interpretation of the experimental observation is especially great in the case of ferrites, since it has been shown that they may exhibit a high value of $|\kappa|$ and $|\mu|$, which, of course, favours the occurrence of dimensional effects already at a relatively low frequency.

These statements can be illustrated by a brief discussion of the dimensional effect which may have to be taken into account if the magnetic permeability is measured on a toroid sample either of a rectangular or of a circular cross-section. It is assumed that the experimental arrangement is such that the external alternating magnetic field H is tangential to the surface of the ferrite sample. Further it is assumed that the cross-sectional dimensions are so small that the value of H at the surface is equal to H_s everywhere at the surface. The assumption implies that these dimensions are very small compared with the wavelength λ in free space. Under these assumptions the effective complex permeability of the material may be defined as the total flux of the complex magnetic induction B through the cross-section divided by H_s times the surface of the cross-section. For a circular cross-section with radius r, application of the Maxwell theory gives

$$\mu_{eff} = 2\mu J_1(kr)/krJ_0(kr)$$

For a thin flat ring with a height $2a$ we find

$$\mu_{eff} = \mu \tan(ka)/(ka)$$

Here $$k = 2\pi\sqrt{(\mu\kappa)}/\lambda$$

In order to demonstrate the character and the importance of the corrections which have to be considered, the explicit expressions for μ'_{eff} and μ''_{eff} for a thin ring are written down to a first approximation ($|ka| \ll 1$)

$$\mu'_{eff} = \mu' + \tfrac{1}{3}(k_0 a)^2[\kappa'(\mu'^2 - \mu''^2) - 2\kappa''\mu'\mu'']$$

$$\mu''_{eff} = \mu'' + \tfrac{1}{3}(k_0 a)^2[\kappa''(\mu'^2 - \mu''^2) + 2\kappa'\mu'\mu'']$$

where $k_0 = 2\pi/\lambda$.

For a sample of MnZn-ferrite (sample 1 in Section 1.2) at 24 Mc/s we find for a ring with $2a = 0{\cdot}16$ cm ($\kappa' = 2{\cdot}8 \times 10^4$, $\kappa'' = 1{\cdot}9 \times 10^4$, $\mu' = 210$, $\mu'' = 270$)

$$\mu'_{eff} - \mu' \simeq -175$$

$$\mu''_{eff} - \mu'' \simeq +7$$

These values have been calculated with the aid of the exact formula. It may be pointed out that under certain conditions the dimensional effect may be so important that a negative μ'_{eff} is measured while the actual μ' is definitely positive. In order to avoid dimensional effects as much as possible, thin rings with $2a = 0{\cdot}06$ cm have been used for the measurements of the magnetic properties at high frequencies.

For the same reason the measurements of the dielectric properties have been carried out on a cubic sample of 2 × 2 × 2mm. Moreover, at very high frequencies an additional constant external magnetic field has been applied which quenches the magnetic permeability of the material, thus shifting the occurrence of dimensional effects to a higher frequency.

(2.2) Measurements with the Aid of a Bridge

A bridge method has been used for the measurements of the residual magnetic losses of ferrites, for the measurements of the dielectric properties of ferrites and for the measurements of hysteresis losses. The frequency range used was 1–100 kc/s.

Residual magnetic losses have been determined with the resonance bridge shown in Fig. 7. Here L_x, R_x denotes the

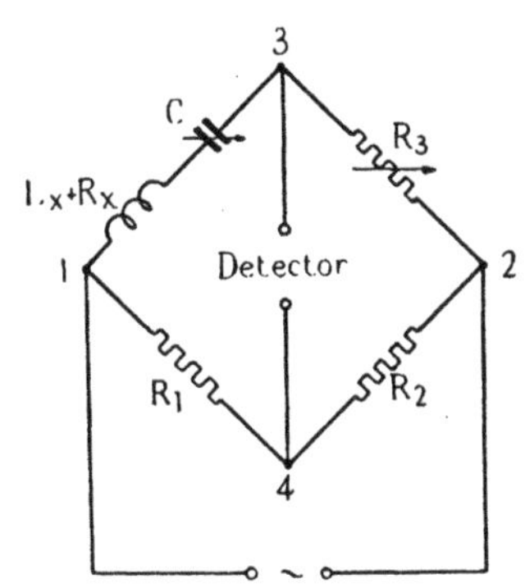

Fig. 7.—Bridge circuit used for measurements of electric and magnetic properties.

sample under test in the form of a toroidal core with a number of turns providing a self-inductance of about 10 mH. It is switched in series with the variable condenser C to form the branch 1–3. R_1 and R_2 are two equal fixed resistors and R_3 is a resistance box. When balance has been obtained L_x and C are in resonance, whence $\omega^2 L_x C = 1$ and R_3 equals the sum of R_x and the series resistance of C due to dielectric losses. In most cases the latter may be ignored, but as its value is known a correction can be made if necessary.

All samples were measured at 2, 16, 40 and 100 kc/s, the current through the coil being kept so low as to make hysteresis losses negligible. At 100 kc/s C could be an air condenser; for the lower frequencies a mica capacitance-box was used. The whole measuring apparatus had been built for precision measurements and has been described elsewhere.[13]

The measurements of the dielectric properties of ferrites were carried out on rectangular blocks, discs and sometimes rings. Electrodes were prepared by evaporation of silver in high vacuum.

The precision bridge just mentioned was also used for these measurements. The sample under test was placed in branch 1–3, whilst a variable condenser C′ was switched in parallel with resistor R_3. The resistances R_1 and R_2 were chosen to have a convenient ratio. From the readings of R_3 and the condenser C′ the parallel capacitance and resistance of the sample are immediately calculated.

Two sources of error must be considered in carrying out these measurements. First the ratio between R_1 and R_2 must be so chosen as to make the phase angle of R_3 small compared with the phase angle of the sample. (Most resistance boxes have the lowest phase angles for resistance values between 100 and 1 000 ohms.) Secondly, special care must be taken when samples of low parallel resistance are measured. A stray inductance in branch 1–3 may then easily cause a large error in the parallel capacitance and could even make the whole branch 1–3 inductive instead of capacitive.

Measurements of the hysteresis factor have been carried out with the aid of a Maxwell bridge, shown in Fig. 8.

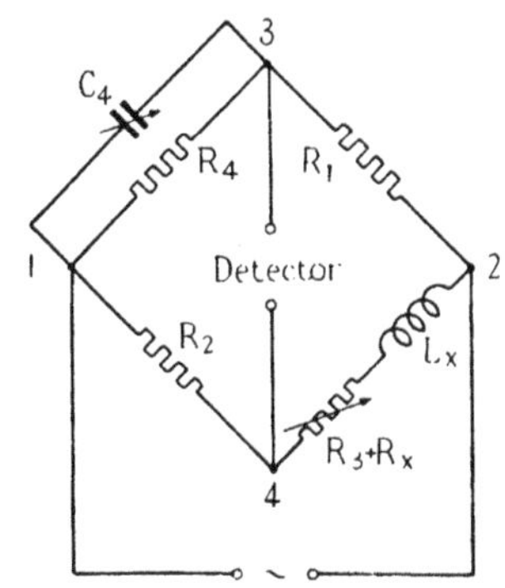

Fig. 8.—Bridge circuit used for determination of hysteresis losses.

Here R_1, R_2 and R_4 denote three fixed resistors, whilst in branch 2–4 the coil under test, with inductance L_x and resistance R_x, is switched in series with a variable resistor R_3. C_4 denotes a variable condenser. In the balanced condition we have

$$(R_3 + R_x)R_4 = R_1R_2 \text{ and}$$

$$L_x = R_1R_2C_4$$

L_x and R_x can easily be calculated from the values of C_4 and R_3. Now L_x and R_x are measured as a function of the current through the coil at a frequency of 2 000 c/s. The current is measured in the following way: a voltmeter is switched across the branch 1–3 where it disturbs the balance only slightly, its input capacitance being very small compared with the capacitance C_4. The voltage across 1–3 equals that across 1–4, provided the bridge has been balanced, but the current flowing through 1–4 equals that flowing through the branch 2–4 containing the coil. The voltage observed is a simple measure of the current through 2–4. A current of 0·1 mA could be measured conveniently. As neither L_x nor R_x is a linear function of the current, a number of values of the current were chosen, covering a wide range of magnetic force. The number of turns on each sample was chosen so as to give an inductance of about 0·3 mH, and the values of the current were 0·1, 0·2, 0·5, 1, 2, and 5 mA.

The non-linearity mentioned above proves that the B/H curve of the ferrites examined cannot be in accordance with the well-known Rayleigh relation. As a consequence the measurement of the distortion as a function of current is also of interest. The measurement of the third harmonic could easily be performed simultaneously with the measurements described above. For this purpose a band-pass filter around 6 000 c/s was switched in parallel with the bridge detector. The filtered-out third-harmonic voltage was then amplified and measured. Presuming that the current through the coil is of a practically pure waveform, the incidence of a third harmonic in the magnetic induction is equivalent to a generator of third-harmonic e.m.f. switched in series with L_x in Fig. 8. The relation between this voltage and the voltage measured across the detector was calculated, the result being that with $R_1 = 10$ ohms and $R_2 = 1\,000$ ohms the measured voltage differed only about 1% from the e.m.f. for values of L_x below 1 mH.

(2.3) Measurements with an LC Circuit

(2.3.1) Measurements of the Magnetic Properties.

For the measurements of the magnetic properties in the frequency range from 10 kc/s to 100 Mc/s the following circuit has been used (cf. Fig. 9). A very accurately calibrated oscillator is

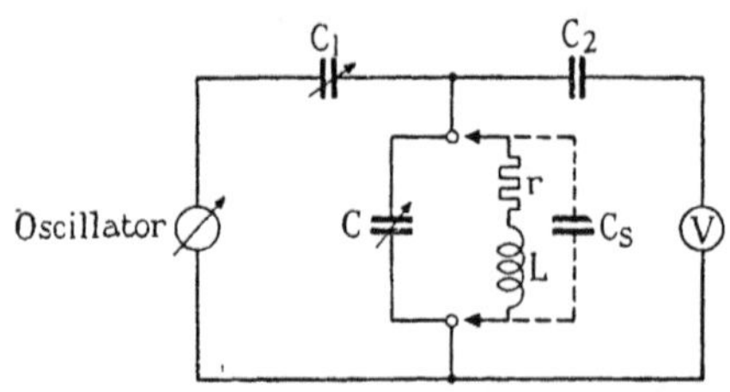

Fig. 9.—Circuit used for measurements of magnetic properties.

coupled to a resonance circuit by a small variable condenser C_1. The voltage on the resonance circuit is measured by an electronic voltmeter which is coupled very loosely to the circuit. We assume C_1 to be so small that the voltage across the circuit is proportional to its impedance when the resonance curve of the circuit is delineated either by the variation of the tuning capacitance C or by the variation of the frequency.

The measurement consists in the determination of the self-inductance L and the resistance r of a toroidal coil with a ferrite core. In all cases where this method has been applied r was smaller than $0{\cdot}1\omega L$, so that r^2 could be neglected with respect to $(\omega L)^2$. The principle of the measurement is simple. The value of C_r of C is determined for which a maximum voltage is observed at the voltmeter, the frequency being kept constant. Then $L = (C_r + C_s)(r^2 + \omega^2 L^2)$ or $L = 1/\omega^2(C_r + C_s)$. Here C_s is the self-capacitance of the coil, which should be known with some accuracy, especially in the case of a measurement at a high frequency. The effective permeability of the material follows immediately from

$$\mu' \simeq |\mu| \simeq \frac{1}{\mu_0 NnA}\,\frac{1}{\omega^2(C_r + C_s)}$$

where N is the total number of turns, n the number of turns per unit length of the core, and A the cross-section of the core.

The difference ΔC between the two values of C for which the impedance of the circuit is $2^{-\frac{1}{2}}$ times its maximum value, gives the resistance of the coil, since

$$r/\omega L = \Delta C/2(C_r + C_s)$$

From this the magnetic loss angle θ_m would follow at once, if losses in the copper wires could be neglected. The copper losses are determined separately with the aid of a dummy coil, which contains a core of dielectric material with negligible loss angle. The wire used for the dummy coil is identical with that of the coil with the ferrite core. The total length l_0 of the wire is made so long that the self-inductance of the dummy coil equals L. The ΔC value for the dummy coil (ΔC_0) is determined. The copper losses of the original coil will be smaller than that of the dummy by a factor l/l_0, if l is the length of the wire of the coil with the ferrite core. It follows that the magnetic loss angle of the ferrite material is given by

$$\tan\theta_m = \frac{\mu''}{\mu'} = \left(\Delta C - \frac{l}{l_0}\Delta C_0\right) \div 2(C_r + C)$$

As mentioned already the formula holds if $(\mu'')^2$ may be neglected compared with $(\mu')^2$. Furthermore, effective values of μ rather than the actual values are obtained. Dimensional effects must be taken into account separately. Finally it should be mentioned that the magnetic field in the coil is kept so small that hysteresis losses are negligible. The actual performance of the measurement is somewhat simpler if the resonance curve is delineated by the variation of the frequency rather than the capacitance C. In this case the interpretation of the experimental observations in terms of the permeability of the material is somewhat more complicated, however, because of the dependence of the permeability on frequency. The explicit expressions involved will not be discussed here.

(2.3.2) Measurements of the Dielectric Properties.

The dielectric properties have been measured with the aid of the circuit shown in Fig. 10. The method can be used between

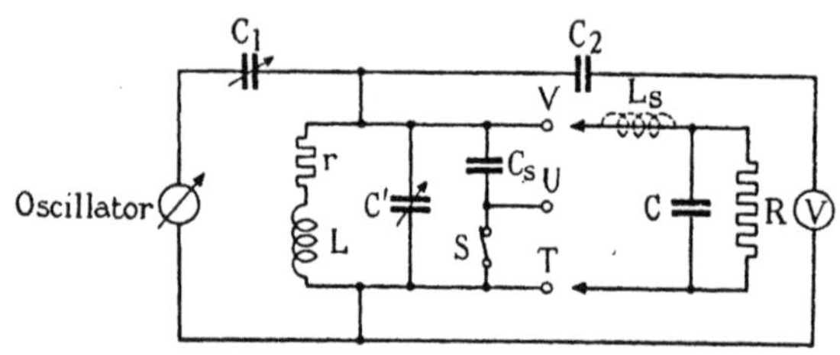

Fig. 10.—Circuit used for measurements of electric properties.

100 kc/s and 50 Mc/s. It consists of the determination of the capacitance C and the parallel resistance R of a ferrite condenser in the form of a thin plate or a small cube provided with silver electrodes. In Fig. 10 L_s is the self-inductance of the connecting leads. The procedure is the following. First the ferrite condenser is not connected with the resonance circuit consisting of the elements L, r and $C' + C_s$, the circuit being closed at S. The value of C' is determined for which the voltmeter V shows the maximum amplitude V_0. The impedance of the circuit at resonance is $2/(\omega\Delta C')$, where $\Delta C'$ measures the width of the resonance curve as described in the preceding Section. Then the ferrite condenser is connected to the terminals T and V, and the variation $\delta C'$ of C' is measured which brings the new circuit into resonance. In this Section it is always assumed that Q^{-2} can be neglected with respect to unity, Q being the quality of a resonance circuit either loaded or unloaded. Provided that $\omega^2 L_s C$ and L_s/CR^2 can be neglected with respect to unity, which is usually the case, we have $|\delta C'| = C$ and the effective dielectric constant κ' of the material follows from

$$\kappa' = \frac{d}{\kappa_0 A}|\delta C'|$$

where d is the distance between the electrodes and A the area of an electrode. The impedance of the new circuit is most easily found by observing the voltage V_1 across the circuit at resonance and comparing it with the voltage V_0 for the unloaded circuit. The value of R follows immediately and we have

$$\kappa'' = \frac{d}{\kappa_0 A}\frac{1}{\omega R} = \frac{d}{\kappa_0 A}\frac{V_0 - V_1}{V_1}\frac{\Delta C'}{2}$$

The method has its disadvantages, however, if R is too small, so that the impedances of the loaded and unloaded circuit are of a different order of magnitude. In this case the ferrite condenser is connected to the terminals U and T and the circuit is disconnected at S. As compared with the original unloaded circuit, a variation $\delta C'$ of the condenser C′ is required to restore resonance. The voltage V_1' is observed at the voltmeter under resonance conditions. With the aid of V_1' the equivalent parallel resistance R_p of the circuit between V and T may be computed from

$$R_p = \frac{V_1'}{V_0 - V_1'}\frac{2}{\omega \Delta C'}$$

A straightforward calculation shows that the unknown quantities R and C follow from

$$R = R_p\left[\alpha^2 + \left(\frac{1}{\omega C_s R_p}\right)^2\right]$$

$$1 + C/C_s = \alpha R_p/R$$

where $$\alpha = \delta C'/C_s + \omega^2 L_s C_s$$

In these calculations it is assumed that $\omega^2 L_s C_s$ can be neglected with respect to unity. In view of the small values of R which have been measured by this second method, it may be mentioned that it is not necessary to assume that $L_s/CR^2 \ll 1$, but only $\omega^2 L_s^2/R^2 \ll C/C_s$, which is a weaker condition. The second term in the equation for α is of importance as soon as L_s/CR^2 can no longer be neglected with respect to unity. In view of the high values of the permittivity which have been measured by this second method, it may be noted that the formulae given above are still correct if $\omega^2 L_s C$ is of the order of unity.

The effective permittivity $\kappa = \kappa' - j\kappa''$ can easily be computed from C and R. At frequencies between 10 and 50 Mc/s dimensional effects may occur even in a cube of 2 × 2 × 2 mm if the permittivity of the material is very high, e.g. in the sample of MnZn-ferrite with a d.c. resistivity of 100 ohm-cm having $\kappa' > 10^4$. In such cases a reliable measurement of the dielectric properties can be carried out only if a strong constant external magnetic field is applied which quenches the magnetic permeability. At somewhat lower frequencies it was checked separately that the external field does not affect the dielectric properties of the material.

(2.4) Measurements with a Coaxial Line

The methods described in this Section have been used in the frequency range between 1 and 500 Mc/s and at a frequency of about 3 000 Mc/s.

A coaxial line with a length l is coupled to an accurately calibrated oscillator. The adjustable couplings 1 and 2 (cf. Fig. 11) are checked to be so loose that the properties of the coaxial line do not affect the frequency and the amplitude of the oscillator when the resonance curve of the circuit is delineated either by variation of an element of the resonance circuit itself (the length l or the top-capacitance C_T) or by variation of the frequency of the oscillator. The coaxial line can be used in two different resonance circuits.

(*a*) $l = \frac{1}{4}\lambda$. *No top-condenser* (*cf. Fig.* 11).

This method is used at frequencies where l has a convenient value, i.e. from 75 Mc/s up to 500 Mc/s and higher. The

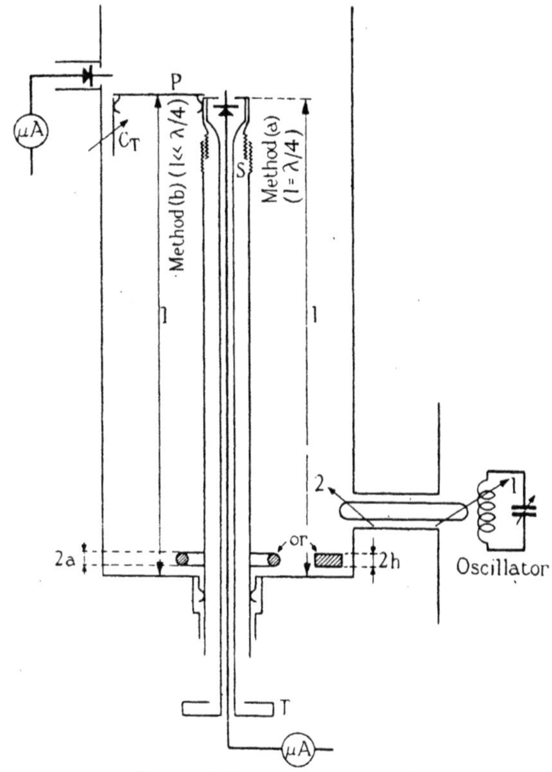

Fig. 11.—Coaxial line used for measurements of magnetic properties.

voltage at the open end is measured by means of a crystal detector inside the upper end of the inner conductor. The line may be tuned coarsely by adjusting the length of the inner conductor. A ring of spring contacts at the short-circuited end assures low series resistance for the main current. Accurate tuning and delineation of the resonance curve is achieved by varying the frequency or by raising and lowering the upper end of the inner conductor (variation of l) by means of a fine screw S near the top where the current is almost zero and hence fluctuations of the series resistance do not affect the measurements.

(*b*) $l \ll \frac{1}{4}\lambda$. *Lumped top-capacitance* C_T (*on the left of Fig.* 11).

This method is suitable for measurements at frequencies between 1 and 100 Mc/s. The voltage across the low-loss top-condenser C_T is measured by means of a crystal detector on the outer conductor. The circuit may be tuned coarsely by adjusting the top-condenser. Accurate tuning and delineation of the resonance curve is achieved by varying the frequency or by variation of the top-capacitance. For this purpose C_T includes both a large adjustable condenser of concentric plates with an air-gap and a small precisely calibrated condenser. A piston P with two rings of spring contacts may be removed and readjusted without detuning the circuit and without altering the series resistance. Hence a ring or toroid may be placed on the bottom of the coaxial line without disturbing effects.

The measurement consists of the determination of the amount of detuning and the variation of the quality of the circuit when a ferromagnetic ring is inserted. In method (*a*) the amount of detuning is found from the variation δl of the length l; in method (*b*) it is found from the variation δC of the top-capacitance C_T which is required to restore resonance. In either method a variation of the frequency of the oscillator can also be used for determining the amount of detuning, but, just as in the preceding Section, this may be disadvantageous in view of the frequency dependence of the permeability μ. The quality of the coaxial line is determined by the width of the resonance curve expressed by the quantity Δl (cf. Section 2.3) or ΔC respectively. The computation of the effective μ' and μ'' from δC and ΔC is considerably simplified if the values δC_c and ΔC_c for a copper dummy ring of identical dimensions as the ferrite ring are measured first. Provided that the quality of the loaded circuit as well as the unloaded circuit is sufficiently high ($Q > 10$) and that $\delta C \ll C_T$, we have

$$\mu' - 1 = -\frac{\delta C}{\delta C_c}$$

The relation is obvious, if it is kept in mind that at these frequencies the copper dummy ring acts as a substance with permeability $\mu' = 0$. For method (*a*) the similar formula $\mu' - 1 = -\delta l/\delta l_c$ holds.

The measurement of μ'' may be slightly more difficult if the quality of the loaded circuit does not differ very much from that of the unloaded circuit. In that case it is not correct to neglect δl with respect to l in the first stage of the calculations, and we have for method (*a*)

$$\mu'' = \frac{\Delta l - \Delta l_c(1 + \delta l/l)^2}{2\delta l_c}$$

while for method (*b*) we find

$$\mu'' = \frac{\Delta C[C_T/(C_T + \delta C)]^2 - \Delta C_c}{2\delta C_c}$$

It has already been mentioned that these simple formulae are correct only if the power factor of the loaded circuit is sufficiently small. This does not mean that the loss angle of the material should also be small, since the ferrite ring occupies only a small part of the total volume of the coaxial line. In contradistinction to the measurements with an LC circuit as described in Section 2.3, the coaxial line can easily be used for the determination of the magnetic properties of ferrite samples having a large loss angle at the frequency of measurement.

It need not be emphasized again that corrections for dimensional effects must be taken into account separately, and that the magnetic permeability measured in the way described above is an effective one rather than the actual permeability.

Finally it should be mentioned that, though the use of a copper dummy ring is the simplest procedure, the permeability can also be calculated directly from the observed δC and ΔC values together with the dimensions of the ring. Good agreement has been found between the results of direct calculations and those obtained from comparison with the dummy rings.

(2.5) Ferromagnetic Resonance Absorption and other Measurements at 3·2 cm Wavelength

Experiments have been carried out on the absorption of microwaves in small ferrite spheres placed in an external magnetizing field H_0. It has already been pointed out in Section 1.3 that strong absorption may be expected at a frequency that is roughly equal to the frequency of precession of the spinning magnetic tops in the ferromagnetic material around the external magnetic field H_0. If H_0 is sufficiently large, so that saturation of the sample is obtained, the frequency ω_0 of maximum absorption in a spherical specimen is given by a relation given by Kittel[14]

$$\omega_0 = |\gamma| H_0$$

Here $|\gamma|$ is the "gyromagnetic ratio" of the magnetic dipoles responsible for ferromagnetism in the ferrite material. The following relation holds

$$|\gamma| = \left|\frac{e\mu_0}{2m_e}\right| g = 1\cdot105\ 10^5 g \text{ m/(ampere-turns)-sec}$$

where μ_0 is the permeability of free space, e the charge of an electron, m_e its mass and g a g-factor whose value is 2 for free spins. The g-factor for most ferromagnetic materials differs only slightly from 2.

If the external field is 3 000 oersteds, ω_0 is about 10^{10} Mc/s. At this frequency the precessing magnetic tops resonate with the microwave magnetic field, provided that the latter field is perpendicular to H_0. The resonance character of the phenomenon is most clearly demonstrated when we measure at various frequencies the magnetic dipole moment m of the spherical sample that is induced by the microwave in the direction of its magnetic field vector h_0.

In the actual experiments, however, the frequency is kept constant and measurements are carried out for different values of the external magnetic field H_0. The real part m' of the total induced alternating magnetic moment m, as well as the imaginary part m'', are determined.

For the measurements[15] a cylindrical cavity is used which resonates at 3·2 cm in the mode TM_{020} (Fig. 12). Small ferrite

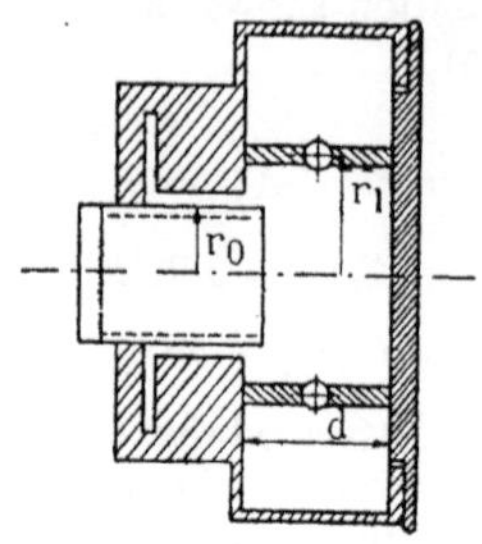

Fig. 12.—Cavity used for microwave experiments.

spheres (diameter 2 or 3 mm) are inserted into the cavity at an electric nodal circle with radius r_1 fixed by the first root of the equation $J_0(2\pi r/\lambda) = 0$. The external field H_0 points in the direction of the axis of the cylindrical cavity, and the microwave field h_0 is tangential to the electric nodal circle and perpendicular to H_0. The shift δf of the resonance frequency f of the cavity by loading with ferrite spheres can be compensated with the aid of a concentric tuning rod incorporating a micrometer scale. The tuning rod is calibrated in terms of a variation of the resonance frequency of the cavity. Therefore the value of m'/h follows immediately from the readings of the micrometer.

A calculation gives

$$\frac{\delta f}{f} = -\frac{J_1^2(2\pi r_1/\lambda)}{J_1^2(2\pi R/\lambda)} \frac{m'}{2h_0\mu_0 V}$$

where V is the volume and R the radius of the cavity. Experimental results for a sample of NiZn-ferrite are shown in Fig. 13. The tuning rod can also be used for determining the quality of the resonance cavity loaded with ferrite spheres. In this way m'' can be obtained for the given frequency and magnetic field H_0. The corresponding Δf is also shown in Fig. 13. The curves in this Figure show the characteristic features of dispersion curves: m' has a positive sign if H_0 is so large that the corresponding frequency $\omega_0 = |\gamma| H_0$ is larger than the frequency used in the experiment, while it has a negative sign if smaller. Maximum losses are observed in the region where the sign of m' changes.

It is interesting to observe that the g-factors for ferrite materials computed from the experimental data with the aid of the first two formulae in this Section show deviations from the spin-only value 2. In a series of NiZn-ferrites with varying Ni : Zn ratio, g-factors have been found varying from 1·95 to 2·35. A further discussion of this point will be published elsewhere.[16]

A cylindrical cavity can also be used for the measurement of the permittivity at microwaves. A cylindrical specimen of the

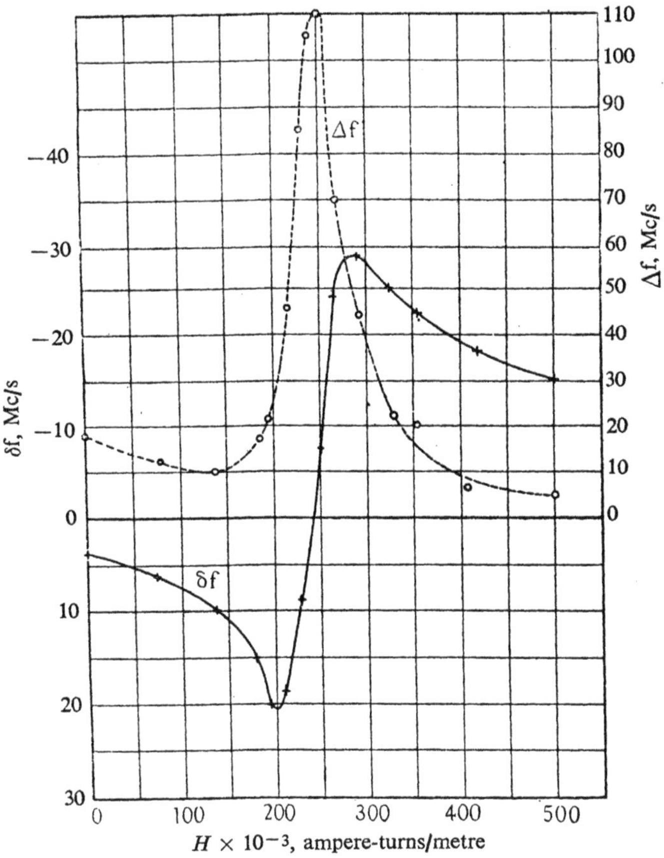

Fig. 13.—Frequency shift δf and change of damping Δf of the resonating cavity due to ferrite spheres, as a function of external magnetizing field.

material with radius a is placed in the centre of the cavity. The modes employed were TM_{010} or TM_{020}.

To a first approximation ($\kappa^{\frac{1}{2}}a/R \ll 1$) the observed frequency shift δf, preferably compensated by a calibrated tuning rod, and the width Δf of the resonance curve, as defined in the preceding Sections, are proportional to ($\kappa - 1$) according to

$$\frac{\delta f + j\frac{\Delta f}{2}}{f} = -\frac{(\kappa - 1)}{2}\frac{\mathscr{E}_1^2 \Delta v}{\int \mathscr{E}^2 dv}$$

where $\mathscr{E}_1$ is the local field in the centre of the cavity and Δv the volume of the cylindrical specimen.

For several samples of NiZn-ferrite κ' has been found to be approximately 9 or 10 and $\kappa'' \ll 1$ in the microwave region.

(3) ACKNOWLEDGMENTS

This paper is a summary of the work on ferrite materials carried out in Philips' Research Laboratories during the last decade. Since most results have been obtained by a close co-operation of several investigators it is impossible to indicate precisely the contributions of the individual workers to the contents of this paper. Much of the extensive earlier work is due to J. L. Snoek. An account of his work and that of his co-workers is found in his book "New Developments in Ferromagnetic Materials."[2] The present paper contains, in addition to some of the earlier work, a number of new results, many of which will eventually form the subject of independent publications. Broadly speaking, the new contributions to the physical chemistry of ferrite materials are due to E. W. Gorter, many of whose results are presented in Sections 1.3 and 1.4. The investigations on the dielectric properties by Brockman, Dowling and co-workers of Philips Laboratories Inc., Irvington on Hudson, New York, gave the impulse to the measurements of these properties reported in Section 1.2 of the paper. In Section 2.2, C. G. Koops discusses the bridge method he has developed in order to measure the electric and magnetic properties. In Sections 2.3.1, 2.3.2 and 2.4 the methods of measurements with the aid of resonance circuits designed by M. Gevers and C. M. v.d. Burgt are presented. Many results obtained by the last three investigators are found in Section 1.2. The work on resonance absorption of microwaves reported in Section 2.5 is by H. G. Beljers. The author of the present paper has contributed to its theoretical aspects. Numerous discussions with H. B. G. Casimir and E. J. W. Verwey, directors of Philips' Research Laboratories, have proved invaluable for the accomplishment of the work on ferrites.

(4) BIBLIOGRAPHY

(1) HILPERT, S.: "Genetische und konstitutive Zusammenhänge in den magnetischen Eigenschaften bei Ferriten und Eisenoxyden," *Berichte der Deutschen Chemischen Gesellschaft*, 1909, **42**, p. 2248.

(2) SNOEK, J. L.: "New Developments in Ferromagnetic Materials" (Elsevier Publishing Company, Amsterdam-New York, 1947).

SNOEK, J. L.: "Non-metallic Magnetic Materials for High Frequencies," *Philips Technical Review*, 1946, **8**, p. 353.

(3) BROCKMAN, F. G., DOWLING, P. H., and STENECK, W. G.: "Magnetic Properties of Ferromagnetic Ferrite," *Physical Review*, 1949, **75**, p. 1298.

(4) To be published.

(5) DOWLING, P. H., and BROCKMAN, F. G.: "Magnetic Properties of Ferromagnetic Ferrite," *Physical Review*, 1949, **75**, p. 1298.

(6) VERWEY, E. J. W., HAAYMAN, P. W., and ROMEYN, F. C.: "Physical Properties and Cation Arrangement of Oxides with Spinel Structures," *Journal of Chemical Physics*, 1947, **15**, p. 181.

(7) BARTH, T. F. W., and POSNJAK, E.: "Spinel Structures: with and without Variate Atom Equipoints," *Zeitschrift für Kristallographie*, 1932, **82**, p. 325.

(8) VERWEY, E. J. W., and HEILMAN, E. L.: "Physical Properties and Cation Arrangement of Oxides with Spinel Structures," *Journal of Chemical Physics*, 1947, **15**, p. 174.

(9) NEEL, M. L.: "Propriétés magnétiques des ferrites; ferrimagnétisme et antiferromagnétisme," *Annales de Physique*, 1948, **3**, p. 137.

(10) GORTER, E. W.: To be published in *Nature*.

(11) SNOEK, J. L.: "Dispersion and Absorption in Magnetic Ferrites at Frequencies above one Mc/s," *Physica*, 1948, **14**, p. 207.

(12) GORTER, E. W.: To be published in *Nature*.

(13) KÖHLER, J. W. L., and KOOPS, C. G.: "A New Alternating-Current Bridge for Precision Measurements," *Philips Research Reports*, 1946, **1**, p. 419.

(14) KITTEL, C.: "On the Theory of Ferromagnetic Resonance Absorption," *Physical Review*, 1948, **73**, p. 155.

(15) BELJERS, H. G.: "Measurements on Gyromagnetic Resonance of a Ferrite using Cavity Resonators," *Physica*, 1949, **14**, p. 629.

(16) BELJERS, H. G., and POLDER, D.: "Ferromagnetic Resonance Absorption and g-factors in Ferrites." To be published.

Electrical Engineering in Japan, Vol. 123, No. 2, 1998
Translated from Denki Gakkai Ronbunshi, Vol. 117-A, No. 8, August 1997, pp. 813–817

Complex Permeability of Polycrystalline Mn-Zn and Ni-Zn Ferrites

HIDEO SAOTOME and YO SAKAKI
Chiba University, Japan

SUMMARY

The complex permeability of ferrites is frequency dependent. The real part of the complex permeability deteriorates in a high frequency range and the imaginary part has a peak after starting the deterioration. This paper examines the possibility that the frequency characteristics for some ferrites can be approximately derived from a first-order linear differential equation for the magnetic field intensity and magnetic flux density. The first-order differential equation is expressed by the reciprocal of the complex permeability and provides first-order magnetic and electric circuits for ferrite cores. In contrast with the commonly used series R_s-L_s circuit for the cores, obtained from $B = (\mu' - j\mu'')H$, the first-order electric circuit derived consists of an inductance (L) and resistance connected in parallel. In this paper, it is demonstrated that the inductance L remains constant, whereas Ls decrease with the increase in frequency. In other words, the real part of the reciprocal of the complex permeability remains constant for an increase in frequency. In addition, it is found that the imaginary part of the reciprocal of the complex permeability is approximately proportional to the frequency. © 1998 Scripta Technica, Electr Eng Jpn, 123(2): 1–7, 1998

Key words: Ferrites; complex permeability; Snoek limit; dynamic magnetic loss; iron loss.

1. Introduction

For toroidal cores made of the same material but with different Mn-Zn and Ni-Zn ferrite dimensions, the authors have carried out measurements and analyses of iron loss up to MHz range. From the comparison of these results, it is clear that the loss which has previously been described as a residual loss is in fact a dynamic magnetic loss inside the cores [1, 2].

On the other hand, the complex permeability $\dot{\mu}$, which has been previously used to express the ferrite loss, is generally expressed as ($\mu'' > 0$)

$$\dot{\mu} = \mu' - j\mu'' \tag{1}$$

with $j = \sqrt{-1}$ (in this paper, an upper dot will be attached to complex numbers other than j). Taking into account the temporal phase lag with respect to the magnetic field $\dot{H}$, the magnetic flux density $\dot{B}$ becomes

$$\dot{B} = \dot{\mu}\dot{H} = (\mu' - j\mu'')\dot{H} = \mu'\dot{H} - j\mu''\dot{H} \tag{2}$$

where μ' and μ'' express the inherent frequency characteristics of the material. μ' starts to decrease with an increase in the frequency from the vicinity of the frequency at which μ'' becomes maximum. It is known that this decrease in μ' occurs at a relatively low frequency in materials for which μ' is large at low frequency. In the measurements of the frequency characteristics of μ' and μ'' performed for the Ni-Zn ferrites with plural different compositions, the line connecting the points at which μ' is one half of that at low frequency is called the limit line of Snoek, and this is generally explained as due to the natural resonance [3–5]. Recently, on the other hand, in addition to a paper by the authors [7], a report was presented [6] that shows that this drop of μ' in the ferrites is not a resonance phenomenon of a second-order system due to a magnetic characteristic such as natural resonance or magnetic wall resonance. Thus, this issue needs to be studied.

In this paper, we will show for several ferrite materials that the decrease in μ' can be explained by a dynamic magnetic loss parameter introduced to express the dynamic magnetic loss. We show that this differs from the characteristic based on a second-order system with an unusual resonance phenomenon.

CCC0424-7760/98/020001-07

2. Complex Permeability and Dynamic Magnetic Loss Parameter

2.1 Measurement of complex permeability

When Eq. (2) is multiplied by the cross-sectional area A of the magnetic path of the core and the number of turns N of the exciting winding provided in the core and then differentiated with respect to time (multiplying by jw where w is the exciting angular frequency), the left-hand side becomes the induced voltage $\dot{E}$ of the exciting winding. On the other hand, assuming the length of the magnetic path of the core to be l, the exciting current becomes $\dot{I} = \dot{H}l/N$. Moreover, if the $L_s - R_s$ series equivalent circuit shown in Fig. 1 is assumed for this core, we have, from $\dot{E} = (R_s + jwL_s)\dot{I}$,

$$L_s = \mu' N^2 A/l \tag{3}$$
$$R_s = \omega\mu'' N^2 A/l \tag{4}$$

Therefore, μ' and μ'' can be determined from the measurement of L_s and R_s. The catalog data on the frequency characteristics of the initial permeability μ_i presented by the ferrite manufacturers usually use μ' of Eq. (3).

Measured examples of the frequency characteristics of μ' and μ'' are shown by ● and ▲, respectively, in Fig. 2. Figure 2 shows the measured values of the frequency characteristics of μ' and μ'' for Mn-Zn ferrite core No. 1 whose dimension and medium parameters are shown in Table 1 and will be discussed below. An impedance/gain phasor analyzer (HP made 4194A) was used in the measurement of L_s and R_s.

2.2 Magnetic field equation for expressing magnetic loss

In Eq. (2), the magnetic flux density $\dot{B}$ is decomposed into components that are in phase with, and are lagging the magnetic field $\dot{H}$ by $\pi/2$, i.e., $\mu'\dot{H}$ and $\mu''\dot{H}$. This can be expressed by saying $\dot{B}$ is expanded with $\dot{H}$. On the other

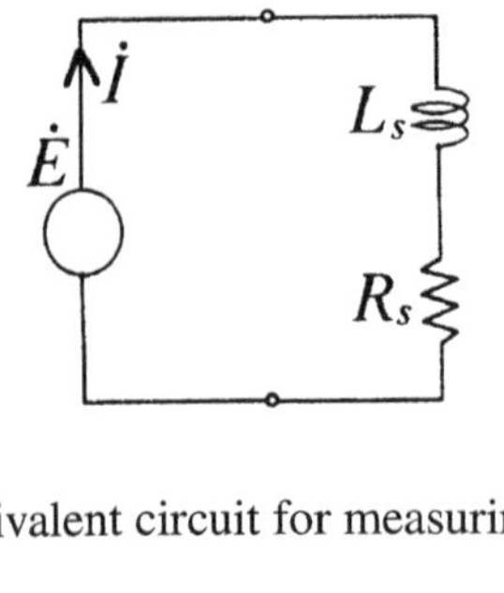

Fig. 1. Equivalent circuit for measuring μ' and μ''.

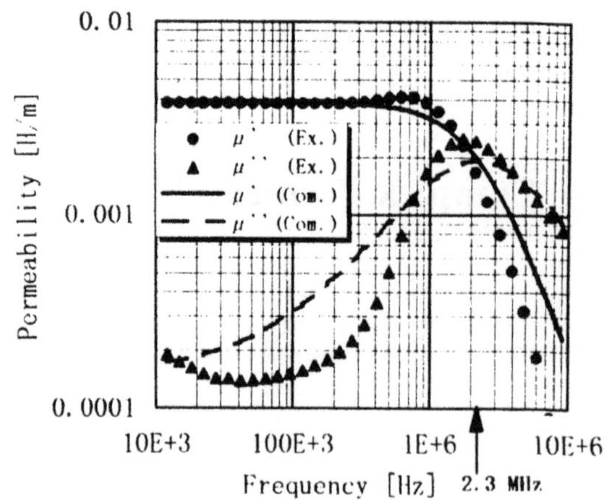

Fig. 2. Frequency characteristics of the permeabilities for core No. 1.

hand, if $\dot{H}$ is expanded into components that are in phase with and leading $\dot{B}$ by $\pi/2$, it can be written as [1, 2]

$$\dot{H} = \frac{1}{\mu_0\mu_r}\dot{B} + j\frac{\omega_0}{\lambda_{h0}}\dot{B} + j\frac{\omega}{\lambda_f}\dot{B} \tag{5}$$

Here, μ_0 is the permeability of vacuum, μ_r is the relative permeability, λ_{h0} [Ω/m] is a hysteresis loss parameter, w_0 [rad/s] is the angular frequency on measurement of λ_{h0}, and λ_f [Ω/m] is a dynamic magnetic loss parameter introduced to explain the residual loss as a dynamic magnetic loss. Figure 3 shows the B–H loops. The inner and outer loops are the B–H loops at low frequency and high frequency, respectively. Here, low frequency means the frequency range in which the hysteresis loss of the ferrite is dominant with respect to the iron loss. The respective terms on the right-hand side of Eq. (5) are explained below by using the

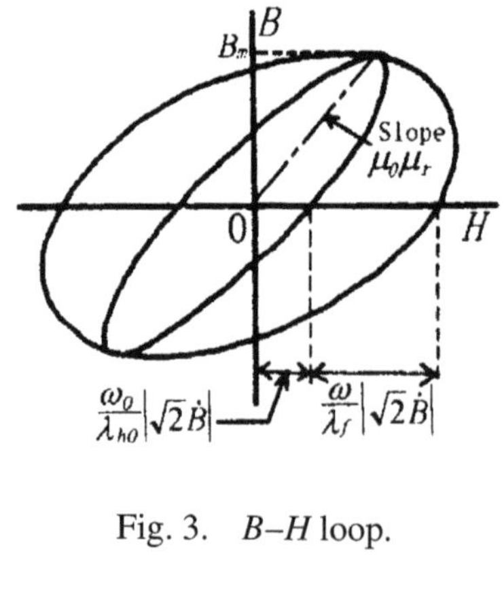

Fig. 3. B–H loop.

B–H loops of Fig. 3. Since the first term on the right-hand side of Eq. (5) is the in-phase component of $\dot{H}$ with $\dot{B}$, $\mu_0\mu_r$ expresses the slope of a straight line connecting the origin and the point at which B becomes maximum in the B–H loops of Fig. 3 (in this case, dB/dt = 0 and H is the only component in phase with B). Therefore, μ_r is the relative permeability (real number) defined by the slope of the B–H loop (μ_r is a real number that is different from $\dot{\mu}_r = \dot{\mu}/\mu_0 = \mu_r' - j\mu_r''$, and also different from μ_r' to be described in (1) of section 3). Moreover, the second term on the right-hand side of Eq. (5) expresses the magnetic field component that is leading $\dot{B}$ by phase angle $\pi/2$, but does not depend on the frequency. The first and second terms on the right-hand side express the inner B–H loop of Fig. 3, i.e., the hysteresis loop. From the complex permeability at low frequency (angular frequency w_0) where the iron loss can be regarded as a hysteresis loss alone, λ_{h0} can be determined from

$$\lambda_{h0} = \omega_0(\mu'' + \mu'^2/\mu'') \tag{6}$$

Moreover, if the magnitude of $\dot{B}$ is the effective value, $(w_0/\lambda_{h0})|\sqrt{2}\dot{B}|$ becomes the coercive force. The third term on the right-hand side of Eq. (5) expresses the magnetic field component that increases in proportion to the excitation frequency and is leading $\dot{B}$ by a $\pi/2$ phase angle. The product of this magnetic field component $(w/\lambda_f)|\dot{B}|$ and the time variation rate $w|B|$ of the magnetic flux density gives the dynamic magnetic loss density

$$p_f = (\omega|\dot{B}|)^2/\lambda_f \ [\text{W/m}^3] \tag{7}$$

which increases in proportion to the square of the excitation frequency. The dynamic magnetic loss [J/m^3] per period corresponds to the portion that forms the area of the outer loop of Fig. 3 after subtracting the area of the inner loop. As an example of the medium parameter values of Eq. (5), the values of μ_r, λ_{h0}($w_0 = 2\pi \times 1000$ [rad/s]) and λ_f (not related to the frequency) of core No. 1 are shown in Table 1 [1]. Here, λ_{h0} depends on the magnitude of the hysteresis loop; however, μ_r and λ_f do not fluctuate widely in the unsaturated range of the magnetic flux density except for extremely small excitation conditions [1, 2].

If Eq. (5) is multiplied by the magnetic path length l, to give the magnetomotive force $\dot{\boldsymbol{F}}$, then the magnetic circuit of Fig. 4 is derived. However, $\dot{\phi}$ is the magnetic flux (= $\dot{B}A$),

$$\mathscr{R} = \frac{l}{\mu_0\mu_r A} \tag{8}$$

$$\Lambda_h = \frac{l}{\lambda_{h0}(\omega/\omega_0)A} \tag{9}$$

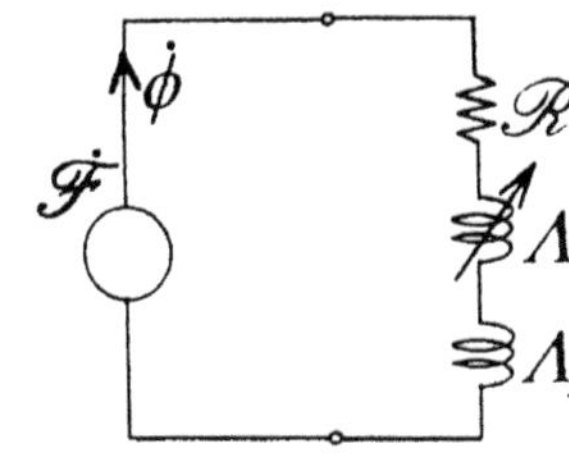

Fig. 4. Magnetic circuit taking account of the magnetic loss.

$$\Lambda_f = \frac{l}{\lambda_f A} \tag{10}$$

and Λ_h decreases with an increase in the frequency. The magnetic circuit of Fig. 4 gives the electrical equivalent circuit for an inductor provided with an exciting winding (number of turns N) in the core. If the applied voltage and winding current are $\dot{E}$ and $\dot{I}$, respectively, when the winding resistance is neglected, it becomes as shown in Fig. 5. The respective components of the current $\dot{I}$ that produce the magnetomotive forces imposed on $\mathfrak{R}$, Λ_h, and Λ_f, respectively, flow in branches to L, R_h, and R_f of Fig. 5. The relations between the respective circuit elements of Figs. 4 and 5 become

$$L = N^2/\mathscr{R} \tag{11}$$

$$R_h = N^2/\Lambda_h \tag{12}$$

$$R_f = N^2/\Lambda_f \tag{13}$$

Here, R_h increases with an increase in the frequency. From Fig. 5, the hysteresis loss P_h [W] and the dynamic magnetic loss P_f [W] of the inductor, with the magnitude of the voltage $\dot{E}$ as the effective value, become, respectively,

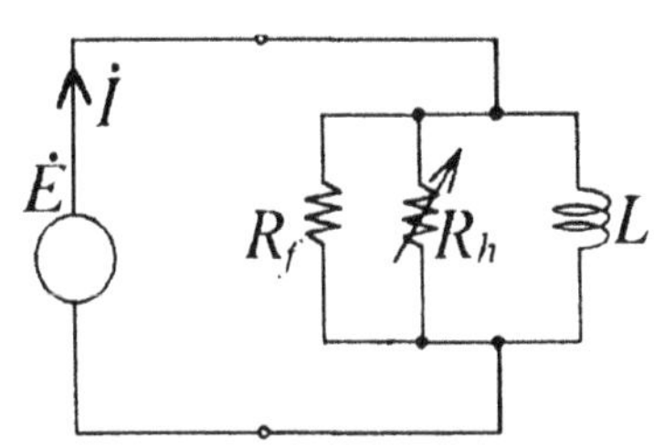

Fig. 5. Equivalent circuit derived from Fig. 4.

$$P_h = |\dot{E}|^2 / R_h \tag{14}$$

$$P_f = |\dot{E}|^2 / R_f \tag{15}$$

It can be seen from the above that Eq. (2) gives the inductance-resistance series circuit of Fig. 1 [4]. On the other hand, Eq. (5) derives the parallel circuit in Fig. 5.

2.3 Expressions by phasor diagrams

Equations (2) and (5) express essentially the same thing since they are mutually reciprocal. The frequency characteristics of μ' and μ'' can be expressed as follows, using coefficients μ_0, μ_r, w_0/λ_{h0}, and λ_f of Eq. (5) that are not related to the excitation frequency.

$$\mu' = \frac{\mu_0 \mu_r \lambda^2}{\omega^2 (\mu_0 \mu_r)^2 + \lambda^2} \tag{16}$$

$$\mu'' = \frac{\omega (\mu_0 \mu_r)^2 \lambda}{\omega^2 (\mu_0 \mu_r)^2 + \lambda^2} \tag{17}$$

where

$$\lambda = \frac{\lambda_h \lambda_f}{\lambda_h + \lambda_f} \tag{18}$$

$$\lambda_h = \lambda_{h0} \frac{\omega}{\omega_0} \tag{19}$$

Using the values μ_r, λ_{h0}, and λ_f for core No. 1 shown in Table 1, the frequency characteristics of μ' and μ'' determined from Eqs. (16) and (17) are indicated by the solid and broken lines of Fig. 2, respectively. Thus they become roughly the same characteristics as the measured values. That is, the decrease in μ' from the vicinity of the frequency at which μ'' is maximum can be expressed by Eq. (5). On the other hand, concerning the frequency characteristics of μ'', although there is a large deviation around 100 kHz between the measured values and the calculated values, both agree roughly near the maximum value of μ''. Using the magnitudes of the solid and broken lines of Fig. 2 as the vertical axis and horizontal axis, respectively, Fig. 6 shows the frequency locus of the complex permeability $\dot{\mu}$ when $\dot{H}$ is used a reference phasor [3]. What can be seen from Fig. 6 is that the phase angle of $\dot{\mu}$, i.e., the phase angle of $\dot{B} = \dot{\mu}\dot{H}$, opens wider with an increase in the frequency; moreover, the magnitudes of $\dot{B}$ and $\dot{\mu}$ decrease. On the other hand, for the same core and with $\dot{B}$ as the reference phasor, Fig. 7 shows the drawing of the frequency locus of

$$\frac{1}{\dot{\mu}} = \frac{1}{\mu_0 \mu_r} + j\frac{\omega}{\lambda} \tag{20}$$

It is seen that the magnitude of $\dot{H} = (1/\dot{\mu})\,\dot{B}$ increases with an increase in the frequency. However, the real part of $1/\dot{\mu}$ is constant regardless of the frequency. This means that from Eqs. (8) and (11), the inductance L of Fig. 5 is constant regardless of the frequency. Thus, the core will exhibit a transformer action even in the high-frequency range. In the iron loss measurement of core No. 1 at 10 MHz excitation, this agrees with the experimental fact that the search coil (secondary winding) voltage occurs according to the turn ratio with respect to the excitation voltage.

3. Analysis

(1) $\mu'(=\mu_0\mu_r')$ and $\mu_0\mu_r$ roughly agree in the low-frequency range. However, as shown in Fig. 2, μ' exhibits the frequency characteristics and μ_r can be treated as a constant value as described above. Accordingly, consider the difference between μ' and $\mu_0\mu_r$. Figures 8(a) and 8(b) show, by providing a proper reference phasor, the phasor diagram with magnetic field $\dot{H} = H_R + jH_I$ and magnetic flux density

Table 1. Properties of cores No. 1, No. 2, and No. 3

Core	Inner diameter [mm]	Outer diameter [mm]	Height [mm]	μ_r*	λ_{h0} [Ω/m]* at 1 kHz	λ_f [Ω/m]
No. 1	5.0	10.0	2.5	3080	615	57,000
No. 2	7.5	12.8	2.6	1900	240	29,000
No. 3	15.0	25.0	4.5	2400	240	37,000

* Measured value at crest value $B_m = 20$ [mT] of exciting magnetic flux density.

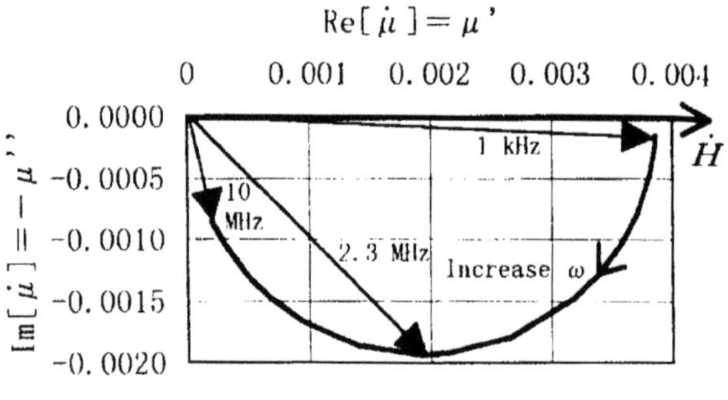

Fig. 6. Phasor diagram of $\dot{\mu}$.

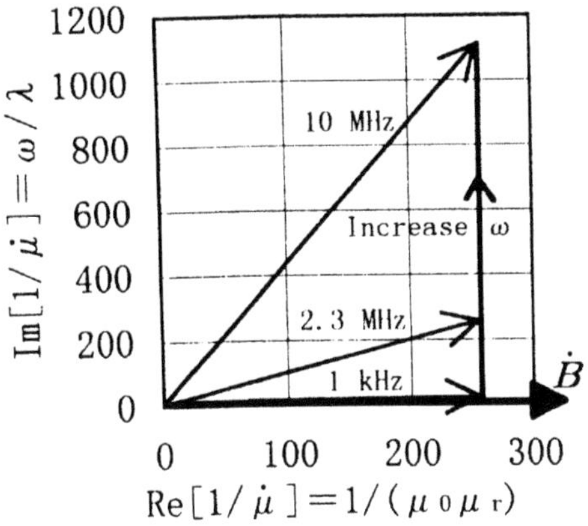

Fig. 7. Phasor diagram of $1/\dot{\mu}$.

$\dot{B} = B_R + jB_I$. $\mu_0\mu_r$ is defined as the ratio of the magnetic flux density to the magnetic field in phase with it, and from Fig. 8(a), becomes

$$\mu_0\mu_r = |B|/H' \qquad (21)$$

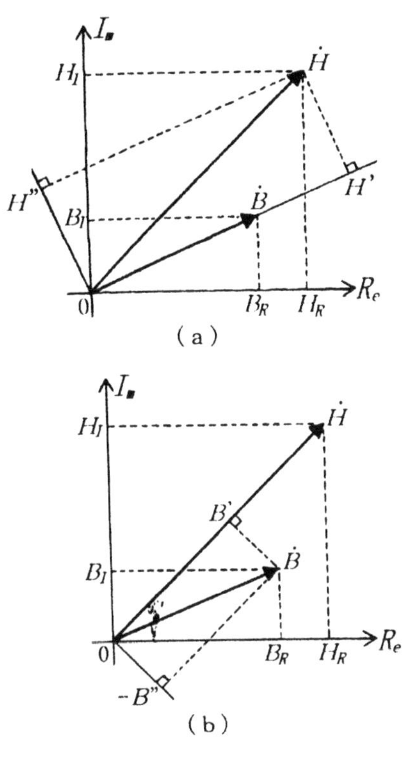

Fig. 8. Phasor diagram of $\dot{H}$ and $\dot{B}$.

On the other hand, μ' is defined as the reciprocal of the ratio of the magnetic field to the magnetic flux density in phase with it, and from Fig. 8(b), is

$$\mu' = B'/|H| \qquad (22)$$

It is seen from Figs. 8(a) and 8(b) that $\mu' \cong \mu_0\mu_r$ (constant value) in the low-frequency range where the phase difference between $\dot{H}$ and $\dot{B}$ is small. Moreover, the phase difference between $\dot{H}$ and $\dot{B}$ increases when the excitation frequency is increased. However, this can be expressed by either phasor diagram of Fig. 8. The experimental fact that μ_r can be treated as constant regardless of the frequency means that H' of Fig. 8(a) is constant regardless of the frequency under the condition of constant maximum exciting magnetic flux density B_m (refer to Fig. 3), and the component that increases with the frequency is only H'' of the same figure, that is, only the imaginary component of Eq. (5) (to be exact, the third term on the right-hand side).

(2) From the circle diagram of Fig. 6, the value of μ' becomes half that at low frequency at the frequency where μ'' becomes maximum. Let this frequency be f_h [8]. We have $\mu' = \mu''$ in this case and, under this condition, f_h determined by the following equation.

$$f_h = \frac{1}{2\pi}\left(\frac{1}{\mu_0\mu_r} - \frac{\omega_0}{\lambda_{h0}}\right)\lambda_f \qquad (23)$$

As an example, $f_h = 2.3$ [MHz] in the case of core No. 1. Equation (23) can also be determined from $\partial\mu''/\partial w = 0$ by using Eq. (19). It is seen from Eq. (23) that f_h is proportional to the dynamic magnetic loss parameter λ_f. Moreover, since the second term on the right-hand side of this equation is smaller than the first term (for example, about one order of magnitude in the case of core No. 1), f_h is roughly in inverse proportion to the relative permeability μ_r. This fact agrees with an experimental observation that has been known qualitatively for many ferrites previously.

(3) When the excitation frequency is increased, R_f of Fig. 5 becomes relatively small with respect to wL. Under the condition of B_m constant ($|\dot{E}| \propto f$), since the dynamic magnetic loss increases in proportion to the square of $|E|$, i.e., the square of the frequency (refer to Eqs. (15) and (7)), it is necessary to use the inductor (or the transformer provided with secondary winding) for pulse applications whose duty is lower or to cool the core sufficiently. On the other hand, under the condition of constant voltage $\dot{E}$, since R_f is constant regardless of the frequency, the dynamic magnetic loss P_f of Eq. (15) is constant regardless of the frequency.

(4) Figures 9 and 10 show the analogous measurement and calculation results as Fig. 2 for Ni-Zn ferrite cores No. 2 and No. 3 shown in Table 1; and the respective f_h are also shown. In these cases, it is also seen that the frequency

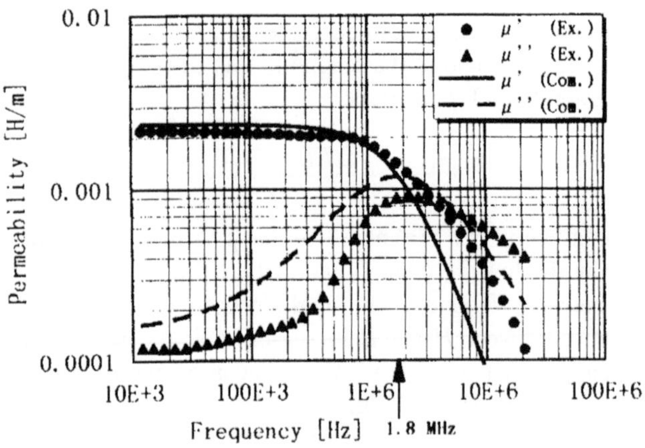

Fig. 9. Frequency characteristics of the permeabilities for core No. 2.

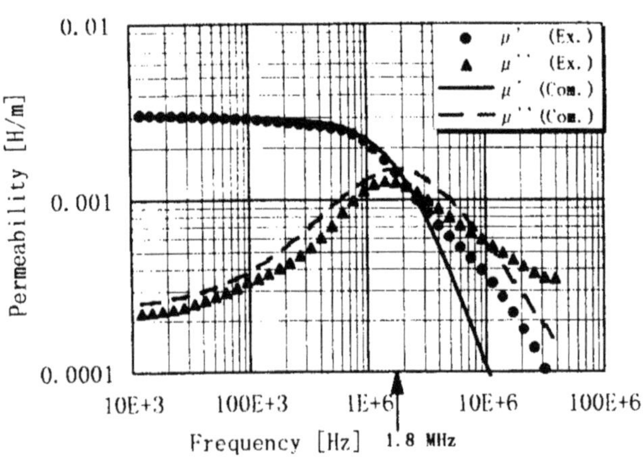

Fig. 10. Frequency characteristics of the permeabilities for core No. 3.

characteristics of μ' and μ'' can generally be expressed by substituting by substituting μ_r, λ_{h0}, and λ_f shown in Table 1 into Eq. (5).

(5) In the high-frequency range where the dynamic magnetic loss becomes dominant, the second term on the right-hand side of Eq. (5) can be neglected and the relation between $\dot{B}$ and $\dot{H}$ can be expressed as a first-order linear differential equation with constant coefficients. Hence, the frequency characteristics of μ' and μ'' (Eqs. (16) and (17)) derived from Eq. (5) are not a second-order system (resonance type) but are a model of a first-order system (relaxation type).

4. Conclusions

The complex permeability of ferrites generally exhibits frequency characteristics and the magnitude of the real part drops with an increase in the frequency from the vicinity of the maximum value of the imaginary part. In this paper, we have measured the frequency characteristics of the complex permeability for three kinds of ferrite materials and shown that the same frequency characteristics can be obtained by a first-order linear differential equation for the magnetic flux density using the dynamic magnetic loss parameter which the authors have proposed. Moreover, this differential equation gives the first-order magnetic circuit, from which a parallel electric circuit of resistance and inductance is derived, and it was shown that this inductance is constant, independent of the frequency. This is in contrast to the fact that $\dot{B} = (\mu' - j\mu'')\dot{H}$ derives the resistance-inductance series circuit and the inductance in this case exhibits frequency characteristics. In other words, it has been shown that the complex permeability has the frequency characteristics, however, it can be treated as if the real part of its reciprocal is constant and the imaginary part is approximately proportional to the frequency.

REFERENCES

1. H. Saotome and Y. Sakaki. Iron loss analysis of Mn-Zn ferrite cores. IEEE Trans. Magn., Vol. 33, pp. 728–734 (1997).
2. H. Saotome, T. Kawai, and Y. Sakaki. Iron loss evaluation of ferrites by dynamic magnetic loss parameter. Trans. I.E.E. Japan, Vol. 116-A, No. 6, pp. 559–565 (1996).
3. J. Smith and H.P.J. Wijn. Ferrites, 269, Phillips Research Laboratories, The Netherlands (1959).
4. K. Ohta. Fundamentals of Magnetic Engineering II: Applications of Magnetism. Kyoritsu-sha, p. 363 (1973).
5. S. Chikazumi. Physics of Ferromagnetic Materials (Second Volume)—Magnetic Characteristics and Applications. Shokabo, p. 323 (1984).
6. S. Goto et al. Frequency characteristics of permeabilities of Mn-Zn ferrites. Records of the 19th Scientific Lecturer of the Japan Society of Applied Magnetism, No. 263 (1995).
7. H. Saotome and Y. Sakaki. Is Snoek limit caused by resonance phenomena? Ibid., No. 272 (1995).
8. H. Saotome and Y. Sakaki. Iron loss estimation of ferrites with the frequency characteristics of the real part of the complex permeability. Trans. I.E.E. Japan, Vol. 116-A, No. 11, pp. 1027–1028 (1996).

AUTHORS (from left to right)

Hideo Saotome (member) completed the first half of his Doctor's course at the Engineering Research Division of the Graduate School of Hosei University in March 1983. He joined Fuji Electric Co. in April of that year and worked at Fuji Electric Corporate R & D Center. In March 1994, he completed the second half of his Doctor's course at the Engineering Research Division of the Graduate School of Hosei University. He became an assistant at Chiba University in June of that year and has been an associate professor since September 1995. He is engaged in research mainly on inverse problem of electromagnetic field system, iron loss analysis of magnetic materials, reactive power compensation control, and new magnetic actuators. He has a Doctor (Engineering) degree. He is a member of the IEEE and Japan Society of Applied Magnetism.

Yo Sakaki (member) graduated from the Department of Electrical Engineering of the Faculty of Engineering of Ibaraki University in March 1954. After working for Shindengen Electric Mfg. Co., etc., he completed his graduate study in electrical and communication engineering at the Engineering Research Division of Tohoku University in March 1965. After serving as an assistant at Tohoku University, he is currently a professor at Chiba University. He is engaged in research mainly on analysis of magnetizing mechanism and clarification of high-frequency power-loss generating mechanism of magnetic materials, magnetism applied devices, as well as verification of magnetic compass view of excursion fishes. He has a Doctor of Engineering degree. He is a member of the Institute of Electronics, Information and Communication Engineers, Japan Society of Applied Magnetism, IEEE, and Japan Society of Fisheries Science.

IEEE TRANSACTIONS ON MAGNETICS, VOL. MAG-20, NO. 5, SEPTEMBER 1984

FAST AND CONVENIENT TECHNIQUE FOR BROADBRAND MEASUREMENTS OF THE COMPLEX INITIAL PERMEABILITY OF FERRIMAGNETS

V. Cagan and M. Guyot, C.N.R.S. Laboratoire de Magnetisme

ABSTRACT

A new technique is proposed for the determination of spectra of the complex initial permeability of ferrimagnets at variable temperature over a broad frequency range.
Different from the conventional techniques commonly used, the present method is based on a direct determination of the flux through the cross section of the sample. This is achieved by measuring the complex voltage induced in a pick-up coil, the reference voltage needed to drive the vector-voltmeter being taken at the right place, depending on the frequency range. Two measurement cells are used : a high frequency cell covering the range from 1 or 2 GHz down to 0.1 MHz and a low frequency cell below 0.1 MHz down to about 1 Hz.
A technical description of the cells is given, mainly for the high frequency one which is the most sophisticated. Details are also provided on the calculation methods used for the permeability determination and examples of applications are shown. This technique is very suitable for automatization with a microcomputer.

INTRODUCTION

Spectra of the complex initial permeability at various temperatures are of a great interest when investigating ferrimagnetic insulators such as ferrites, both for fundamental research and applications. The measurement frequency should range from a few hertz to a few GHz while the thermal capability must vary from liquid helium or liquid nitrogen temperatures up to the Curie temperature (860 K for a nickel ferrite).
The conventional techniques (1) used for such measurements are based on the determination of the complex inpedance of a circuit loaded with a toroid shaped sample, in such a way that the magnetizing field is along the toroid circumference, which results in a zero demagnetizing field. Covering the whole frequency range requires several different techniques : at high frequencies, either vector reflectometers or microwave bridges, at lower frequencies, different types of impedance bridges. At room temperature, plotting a whole frequency spectrum is then time consuming and can be very tedious. Moreover, when the sample is placed inside a temperature controlled cell, additional difficulties or even impossibilities occur :
1) when using standard bridges because of the influence of the connection cable length, typically above 1 MHz.
2) in the microwave region because of the problems arising for precise impedance determination over a wide temperature range (for example : moving the sample when using a coaxial admittance-meter; thermal variations of the empty coaxial line behavior for any technique, etc.).

V. Cagan and M. Guyot

C.N.R.S. Laboratoire de Magnétisme

92195 Meudon/Bellevue, France

To overcome such difficulties, we perfected a different technique very suitable for microwave frequencies up to 2 GHz, which has been also extended to the lower frequency range. The present technique can be achieved owing to high frequency vector-volmeters which are now available.

PRINCIPLE

It is based on the direct determination of the flux through the cross section of the sample by measuring the complex voltage induced in a pick-up coil, the reference voltage needed to drive the vector-voltmeter being taken at the right place in order to be in-phase with the magnetizing field.

Experimentally it is quite different according to the frequency range : in the upper part of the frequency range, say roughly 0.5 MHz up to 1 or 2 GHz, one uses a truly coaxial measurement cell which is driven by reflection ; at lower frequencies (down to a few Hz), the sample is still magnetized by a single wire placed in a transmission network, but standard windings are used for picking up the induced voltage.

EXPERIMENTAL

I- High frequency set-up : The schematic diagram of our present set-up, covering 0.1 MHz up to 1 GHz, is shown in fig.1.

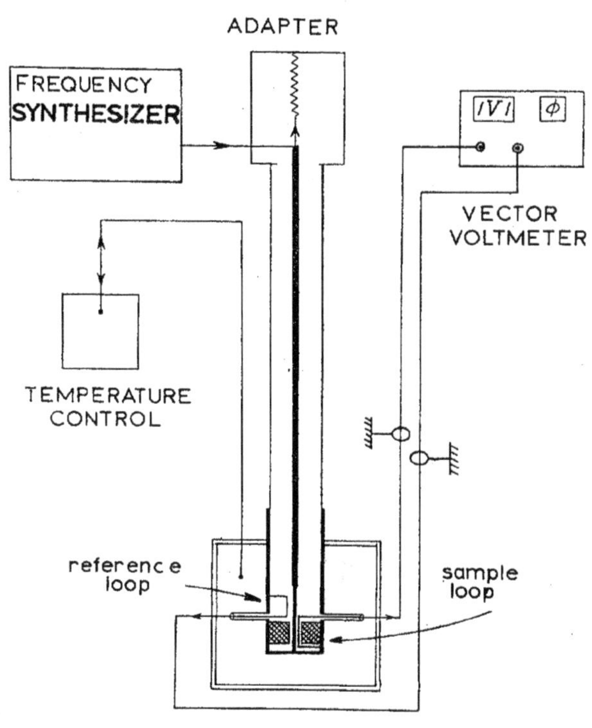

Fig.1 : High frequency set-up

The sample is located at the short-circuited end of a 50 Ω rigid coaxial line (inner conductor 6.2 mm OD). In such a position, the sample is

0018-9464/84/0900-1732$01.00©1984 IEEE

always at a maximum of r.f. field, whatever the frequency. A SWR bridge is used as an adapter between the frequency synthesizer and the rigid coaxial line. The length of the toroidal sample must be chosen short enough in order to insure a good field homogeneity : for all ferrimagnets, the standard values of permeability and permittivity at a given frequency result in a wavelength inside the material which permits a sample length of a few mm.
As shown in the figure, the sample is surrounded by a single wire loop which reads the voltage induced by the flux variations.
A second single wire loop, located against the sample, yields a voltage used as a phase reference of the magnetizing field. This reference voltage locks-in the vector-voltmeter which measures the sample loop voltage.
Both the frequency synthesizer and vector-voltmeter are monitored by a micro-computer, which allows measurements to be automatically performed over the whole frequency range at a given temperature.

At a given frequency, the complex voltage measured from the sample loop is proportional to the flux variations over the cross section of the toroid, hence to the permeability value, providing the reference signal is in-phase with the magnetizing field. Assuming that the reference loop is close enough against the sample, this can be optimized at high frequencies by matching the two coaxial lines which feed the vector-voltmeter in order to minimize the phase mismatch due to propagation effects.

At frequencies lower than 10 MHz, the reference signal produced by a single-wire loop is not large enough to drive the vector-voltmeter. Since the propagation effects become negligible at such frequencies, we then use a multi-wire loop located at the input of the rigid coaxial line. Two different ways can be used for the calculation of the complex permeability value. The first one in an exact evaluation of the total flux variations through the toroid cross-section, while the second one consists of a comparison method.

We prefer using the comparison method at high frequencies (0.1 MHz up to 1000 MHz) since it better takes into account the possible phase mismatches due to the cable length.
The empty cell is first calibrated, the sample loop being positionned at the right place by a non-magnetic toroid. The ratio A /B (reference loop voltage/sample loop voltage) and the phase difference φ_0 between A_0 and B_0 are recorded over the whole frequency range, converted into simple laws versus frequency and then stored in the computer memory to be used in the final calculation.

Several tests carried out at different temperatures have shown that the room temperature calibration remains valid for the whole temperature range.

Let A_1, B_1 and φ_1 be the corresponding parameters measured at a given temperature for the cell loaded with a magnetic sample, and K the ratio of sample loop length/toroid length ; the complex permeability value is given by the relations.

$$\mu' = K \frac{A_0}{B_0} \frac{B_1}{A_1} \cos(\varphi_1 - \varphi_0)$$

$$\mu'' = K \frac{A_0}{B_0} \frac{B_1}{A_1} \sin(\varphi_1 - \varphi_0)$$

Introducing the factor K permits the same loop to be used for various sample lengths but may result in a small error in the absolute value of (because of the inaccuracy of the loop length determination).
A convenient mounting of the sample has been obtained by designing the cell composed of several dismountable parts : around the loops, the outer conductor is made of two semi-cyclindrical brass shells which include the temperature control elements (heater, temperature sensors) ; the inner conductor, which has a reduced diameter to leave space for the loop inside the sample, can be removed when positioning the sample, as well as the loop and the short-circuit cap.
Special care has been taken for insuring good contact between metallic parts over the whole temperature range (low temperatures up to 600K at the present) by using clamping springs. The sample loop is made of 0.5 mm diameter platinium wire and fits on the same sockets as those used in SMB female connectors.
The loops are connected to the upper flange of the dewar by high temperature coaxial cables (maximum temperature 620 K).

II - Lower frequencies set-up : Fig.2 shows the schematic diagram of the configuration used in the frequency range from 0.1 MHz down to a few Hz. The sample is placed inside a small brass furnace which is located in the dewar.
The magnetizing field is still produced by a single wire, but the reference voltage is taken from a resistive impedance which loads the magnetizing circuit.

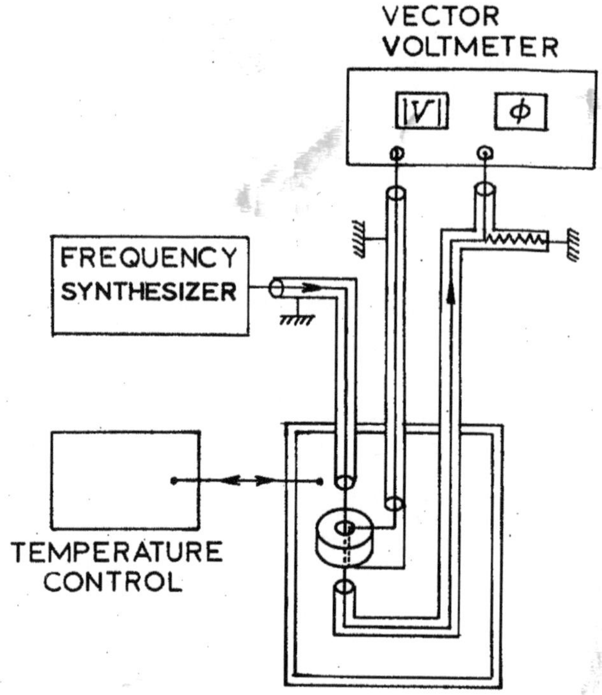

Fig.2 : Low frequency set-up

Since the propagation effects are negligible, the load can be placed far from the sample. To accurately measure the induced voltage, we use two different windings : the first one for frequencies higher than 1 k Hz, the second one, including more turns, below 1 k Hz. The number of turns is chosen according to the permeability value expected.

The permeability value determination is carried out through the exact calculation of the flux variations.
A_1, B_1 and φ_1 having the same meaning as above, the permeability is given by the relations :

$$\mu' = \frac{C}{F} \frac{B_1}{A_1} \cos\varphi_1$$

$$\mu'' = \frac{C}{F} \frac{B_1}{A_1} \sin\varphi_1$$

Where F = Frequency in MHz and

$$C = \frac{10^3 R}{4\ N_2\ h\ \mathrm{Log}\ D/d}$$

Where R = Load impedance in Ω

N_2= number of turns of the winding

h, D, d = length in cm, outer and inner diameters of the sample, respectively.

As above, a micro computer is used to perform automatic measurements.

TYPICAL RESULTS

Fig.3 shows a typical initial permeability spectrum of a slightly iron-deficient YIG sample recorded from 4 Hz to 1 GHz. From the left hand side, one can see 1) a partial relaxation due to the presence of diffusants, 2) the main domain wall relaxation, and 3) the natural spin resonance. The insert shows an enlargement of the spin resonance and its thermal shift from 290 K to 77K.

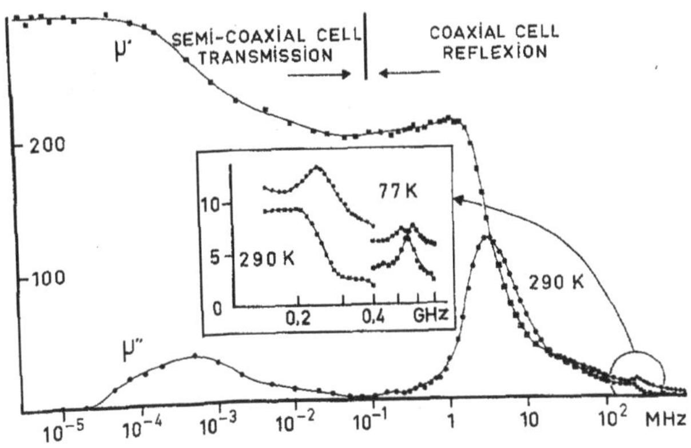

Fig.3 : Initial permeability spectrum of a YIG sample.

From this typical result, one must point out the significance of low frequency measurements for an accurate determination of the quasi-static value of the initial permeability of certain samples.

One can see indeed that the permeability value at 1 k Hz, which is the frequency commonly used for quasi-static measurements, is not at all a good figure for the static value because of the existence of a partial relaxation.

More generally, carrying out broadhand frequency measurements over a large temperature range allows us to follow the evolution, the nature, and the characteristics of the various dispersions which are observed.

Fig.4 shows typically a series of spectra of a YIG sample. From such a set of curves, one can deduce that the conditions for the motion and the relaxation of the domain wall are very different, according to whether the temperature is above or below room temperature, for this particular sample.

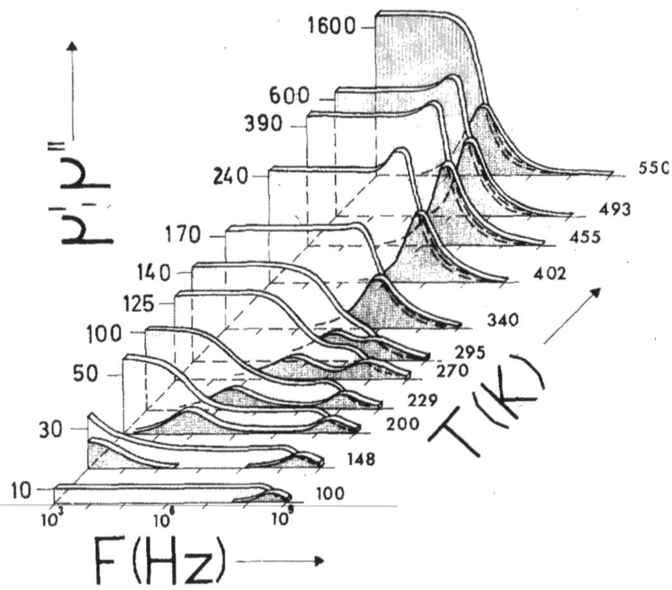

Fig.4 : Tri-dimensional representation of the complex permeability as a function of frequency and temperature of a YIG sample. Light curves are for μ', dark curves for μ''.
The vertical scale is compressed with increasing temperatures.

CONCLUSION

The technique which has been described presents several advantages compared with the standard methods :

- a very broad frequency range can be covered by using only two different measurement cells.
- at high frequencies and microwaves, measurements are easily performed at variable temperature.
- at lower frequencies, measurements are made possible down to a few hertz, even 1 Hz with adequate care, resulting in a better approach to the static initial permeability value.
- in both cases measurements are very rapid and easy, in addition, the technique is very suitable for automatization.

REFERENCE

1) - P.A. MILES, W.B. WESTPHAL and A. VON HIPPEL
Rev. Mod. Phys. 29, 279, (1957)

Temperature Dependence of Power Loss of Mn-Zn Ferrites at High Frequency

J. J. Suh, B. M. Song, and Y. H. Han

Abstract—**The power loss characteristics of Mn-Zn ferrites were observed up to 1 MHz at 25 mT as a function of temperature. At 100 kHz the temperature at which the power loss minimum occurred was near 70°C, at 500 kHz 50°C and at 1 MHz even lower. The contribution of eddy current loss to the temperature dependence of the total power loss was negligibly small at 1 MHz because the resistivity is independent of temperature. The contribution of the residual loss due to magnetic rotation was significant and strongly temperature dependent at 1 MHz. The frequency where the residual loss starts to increase was in good agreement with the onset frequency of dissipation of the complex permeability.**

Index Terms—**Mn-Zn ferrites, hysteresis loss, eddy current loss, residual loss.**

I. Introduction

Mn-Zn ferrites are widely used as transformer core materials for high frequency switched mode power supplies. For these applications, the temperature (T_m) at which the power loss minimum occurs as well as the power loss reduction should be controlled. In practical cases, the combination of core losses and winding losses would generate heat and lead to a temperature rise, which may increase losses and heat up the core even further. This results in a thermal runaway, which is one of the most severe problems in power supply design [1]. Thus, the power loss minimum (T_m) should be designed to occur near the operating temperature. It is well known that the chemical composition and the oxidation degree of Mn-Zn ferrites are the most important parameters that modify T_m. At low frequency T_m shifts to lower temperatures with increasing ferrous ion contents and coincides with the temperature at which the secondary maximum of initial permeability occurs [2]. With more oxidation there is a decrease in ferrous ion content. At low frequency, hysteresis loss dominates, while at high frequency (>500 kHz), eddy current losses and residual losses dominate. In this paper, we will discuss the shift of T_m at high frequency by analyzing the relative contribution of residual loss.

II. Experimental

Mn-Zn ferrite cores of the composition $Mn_{0.71}Zn_{0.22}Fe_{2.07}O_4$ were prepared using a conventional ceramic powder processing technique. Toroidal cores were sintered at 1150°C for 3 hr using a tube furnace with atmosphere control. At the cooling region, the atmosphere was controlled using the equation for equilibrium oxygen partial pressure given by Morineau [3], with the oxygen parameter set for 7.8. The dimensions of sintered toroidal samples were about 17 mm outside diameter (o.d.), 8 mm inside diameter (i.d.) and 5 mm height. The power loss was measured by means of an Iwatsu 8232 B-H analyzer with the temperature in the range of 20°C–100°C. A Hewlett-Packard impedance analyzer (4194A LCR meter) was used to measure the frequency dependence of the permeability. For resistivity measurements, the specimen was cut into a disc shape and pasted with In-Ga for electrodes.

Manuscript received February 20, 2000. This work was supported in part by the Korea Ministry of Science and Technology under Grant 6-2-3.

J. J. Suh and B. M. Song are with ISU Ceramic, Wonju, Korea.

Y. H. Han is with the Department of Materials Engineering, Sunkyunkwan University, Suwon, Korea.

Publisher Item Identifier S 0018-9464(00)08038-9.

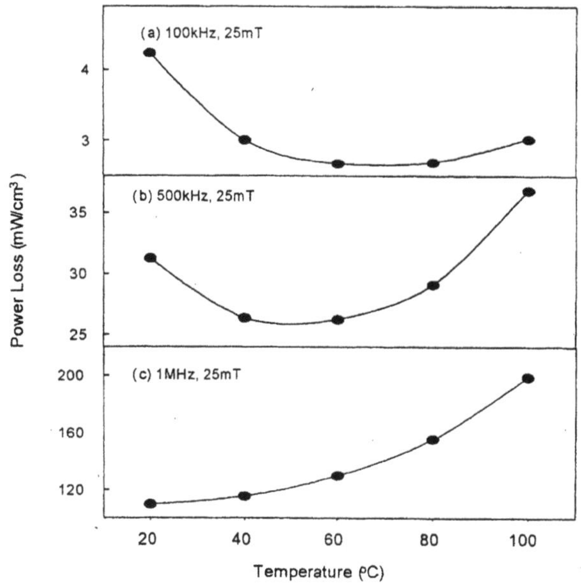

Fig. 1. The temperature dependence of power loss with the measurement frequency (a) 100 kHz, (b) 500 kHz, (c) 1 MHz.

III. Results and Discussion

Fig. 1 shows the temperature dependence of the power loss at various frequencies. Because of the frequency dependence of power losses, the absolute value of the power loss increases with frequency. However, the shift of T_m to lower temperatures with increasing frequency should be noted. The total power loss is divided into three parts, hysteresis loss (P_h), eddy current loss (P_e) and residual loss (P_r). At low frequency, the total power loss is mainly due to the hysteresis loss and T_m occurs at the temperature where the magnetocrystalline anisotropy constant (K_1) is close to zero. As the measurement frequency was increased [Fig. 1(b) and (c)], T_m shifted to lower temperatures.

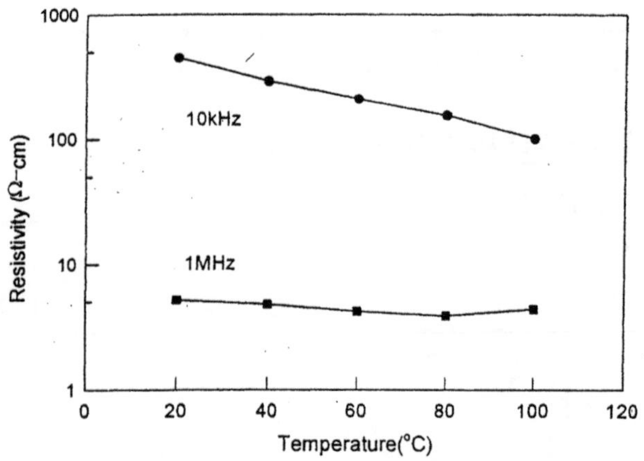

Fig. 2. Temperature dependence of specific resistivity with various frequencies. The resistivity was measured by two point AC technique.

This minimum shift implies that contributions other than hysteresis loss become dominant at high frequency. At frequencies higher than 500 kHz, the contribution of eddy current losses and residual losses become more important [4]. The eddy current loss in ferrites can be expressed as the following equation [5]:

$$P_e = C_e f^2 B_m^2 / \rho \tag{1}$$

where C_e is the dimensional constant and ρ is the resistivity at the measurement frequency. Thus, the eddy current loss is supposed to increase with increasing temperature because the electrical resistivity of Mn-Zn ferrites decreases with increasing temperature. The temperature dependence of the AC resistivity is shown in Fig. 2. There is a significant drop in resistivity as the measurement frequency changes from 10 kHz to 1 MHz. However, the resistivity at 1 MHz is nearly independent of temperature, unlike at 10 kHz. This indicates that the high resistivity grain boundary is short-circuited due to the capacitive property of the insulating layer at the higher frequency. This result suggests that the shift of T_m has nothing to do with the eddy current loss according to (1).

The residual losses are associated with domain wall relaxations, rotational resonances and dimensional resonances. The dimensional resonance is not a concern in this study, because the dimensions of the measured cores are sufficiently small compared with the resonance dimension of the electromagnetic standing wave. Magnetic resonance and relaxation losses give rise to the formation of an imaginary part (μ'') of the magnetic permeability $\mu = \mu' + j\mu''$. The ratio of the imaginary part to the real part is expressed as tan δ. The phase-lag of the magnetic induction (B) with respect to the applied magnetic field (H) can be determined from this value. The magnetic relaxation contribution to these losses is due to domain wall excitation by the driving AC magnetic field, which is not able to follow the applied field instantaneously and thus cause a phase shift between the induction and magnetic field [6]. Rotational resonance occurs when the driving frequency is in resonance with the natural frequency at which the magnetization rotates, leading to a large peak in power absorption. This resonance dominates the ferrite dissipation in the MHz frequency range

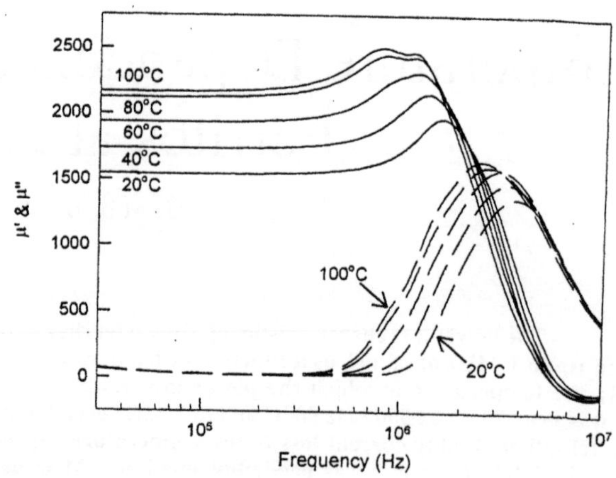

Fig. 3. Frequency spectrum of real part of permeability (μ': solid line) and imaginary part of permeability (μ'': dashed line).

TABLE I
PERMEABILITIES AND RESONANCE FREQUENCIES

Temp. (°C)	Permeability (μ_r)	Resonance Frequency (f_r/MHz)	Product ($\mu_r \times f_r$)
100	2171	2.10	4559.10
80	2127	2.28	4959.56
60	1937	2.45	4745.65
40	1737	2.70	4689.90
20	1545	3.00	4635.00

and the resonance frequency of the ferrite is given by the following equation [7]:

$$f_r = \frac{\gamma B_s}{3\pi\mu_0(\mu_r - 1)} \tag{2}$$

where B_s is the saturation induction, μ_r is the initial magnetic permeability due to the rotation of spins and γ is the gyromagnetic ratio. In the high frequency region (>500 kHz), the contribution of rotational losses is more important than domain wall relaxation [8] and the domain wall resonance is usually of less relevance at frequencies important for power ferrites [6]. Fig. 3 shows the complex magnetic permeability versus frequency at various temperatures. The initial magnetic permeability μ_r increases with temperature. The resonance frequency where tan $\delta = \mu''/\mu' = 1$ and the frequency where the dissipation (μ'') starts to increase are found to decrease with temperature. This is in line with the relationship between f_r and μ_r as given in (2).

Table I summarizes the f_r and μ_r for each temperature. One can see that the product of f_r and μ_r is somewhat constant regardless of the measuring temperature. This is in good agreement with (2). It should also be noted that the resonance frequency is proportional to the anisotropy field that is dependent on temperature.

Fig. 4 shows the total power loss per frequency (P_L/f) at various temperatures as a function of frequency. Hysteresis loss

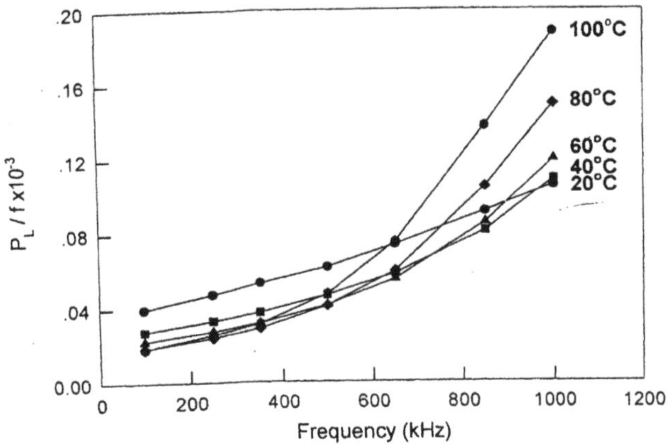

Fig. 4. Frequency dependence of the power loss per cycle at various temperatures.

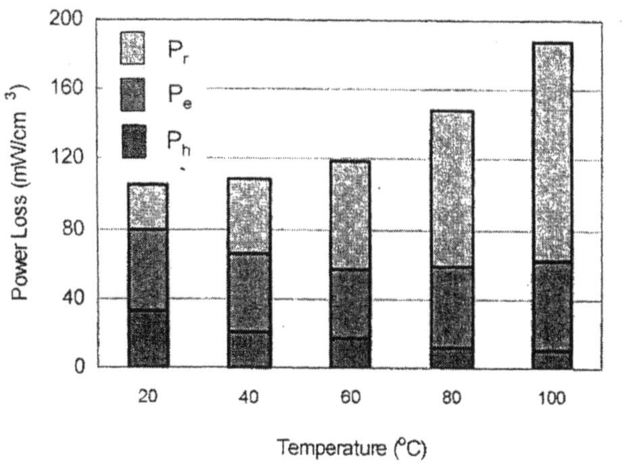

Fig. 5. Power loss separation of residual loss (P_r), eddy current loss (P_e) and hysteresis loss (P_h) at various temperatures (1 MHz).

(P_h) and eddy current loss (P_e) are derived by the zero frequency intercept and the linear part of the initial slope, respectively. The deviation from linearity is due to residual losses. The zero frequency intercepts representing hysteresis losses decrease with temperature, which corresponds to the power loss behavior at 100 kHz as shown in Fig. 1(a), where hysteresis loss is dominant. The frequency at which the nonlinearity starts is dependent on the measurement temperature and decreases with increasing temperature. This onset frequency is in accord with the frequency where the dissipation starts as shown in Fig. 3. The temperature dependence of the complex permeability coincides with the power loss behavior at high frequency, which is attributed to the rotational resonance.

Fig. 5 shows the contribution of each loss mechanism at various temperatures. The contribution of residual loss to the total loss is even greater with temperature at 1 MHz. Since the resonance frequency is proportional to the temperature dependent anisotropy field, some oxide additives should be searched to suppress the residual losses. Furthermore, a sintering technique would be developed to control the grain structure including a single magnetic domain. Zaag reported that a small grained material (<5 μm) with a single domain reduces residual losses [9].

IV. Summary

The power loss characteristics of Mn-Zn ferrite cores with temperature were observed up to 1 MHz at 25 mT. At low frequency, the temperature (T_m) at the power loss minimum coincides with the temperature where the magneto-crystalline anisotropy is close to zero. As the measurement frequency is increased T_m is shifted to lower temperatures. The contribution of eddy current losses to the total loss was negligibly small since the resistivity at 1 MHz is independent of temperature. The contribution of the residual loss due to magnetic rotation was significant and strongly temperature dependent at 1 MHz. The frequency where the residual loss starts to increase was in good agreement with the onset frequency of dissipation of the magnetic permeability.

References

[1] *Handbook of Modern Ferromagnetic Materials*, Kluwer Academic Pub., Netherland, 1999.

[2] H. Inaba, "Nonstoichiometry and its effect on the magnetic properties of a manganese-zinc ferrites," *J. Am. Ceram. Soc.*, vol. 78, no. 11, pp. 2907–2912, 1995.

[3] R. Morineau and M. Paulus, "Chart of Po versus temperature and oxidation degree for Mn-Zn ferrites in the composition range: 50 < Fe_2O_3 < 54: 20 < MnO < 35: 11 < ZnO < 30 (mol{%})," *IEEE Trans. Magn.*, vol. 11, pp. 1312–1314, Sept. 1975.

[4] K. Ishino and Y. Narumiya, "Development of magnetic ferrites: Control and application of losses," *Ceram. Bull.*, vol. 66, no. 10, pp. 1469–1474, 1987.

[5] E. C. Snelling, *Soft Ferrites*, 2nd ed. London: Butterworths, 1988.

[6] D. Stoppels, "Developments in soft magnetic power ferrites," *J. Magn. Magn. Mater.*, vol. 160, pp. 323–328, 1996.

[7] J. L. Snoek, "Dispersion and absorption in magnetic ferrites at frequencies above one megacycle," *Physica*, vol. 14, pp. 207–217, 1948.

[8] J. G. Boerekamp and E. G. Visser, "Grain size dependency of the Steinmetz coefficient soft ferrite power losses," *J. de Phys.*, vol. 7, no. C1, pp. 125–126, 1997.

[9] P. J. van der Zaag, "New views on the dissipation in soft magnetic ferrites," *J. Magn. Magn. Mater.*, vol. 196–197, pp. 315–319, 1999.

Characterization of Dimensional Effects in Ferrite-Core Magnetic Devices

Glenn R. Skutt Fred. C Lee
Bradley Dept. of Electrical Engineering
Virginia Tech, Blacksburg, Va 24061-0111

Abstract - Large ferrite cores used in high-power, high-frequency applications can experience dimensional effects that limit the bandwidth of the device and cause losses higher than those predicted by manufacturer's datasheets. This paper examines such effects in Mn-Zn cores of various sizes and presents corroborating measurement and simulation data that illustrate the impact of both dimensional resonance and eddy-current losses in large cores.

I. Introduction

Power conversion circuits that function at high frequencies and high power levels require magnetic components—both inductors and transformers—with large magnetic core structures. The combination of high frequency excitation and physically large magnetic cores can result in losses and field distributions that are not encountered in lower power and lower frequency devices. This is particularly true for the ferrite core devices that are often used for such applications.

Issues relating to the physical dimension of the core were explored in the early history of ferrites [2] in connection with the use of large MnZn cores for accelerator magnet applications. In recent years the issues encountered in large ferrite cores have been of particular concern in high-power inductive coupling applications [11]. Such high-power applications require devices that can couple large amounts of power in a reasonable size, and this emphasis on size reduction necessitates a relatively high operating frequency. At such frequencies, ferrite materials—and MnZn ferrites in particular—are generally superior to other options such as strip or powder cores due to ferrite's inherently low losses.

While the ferrite materials themselves have a relatively low losses at frequencies in the hundreds of kilohertz, any particular ferrite core may have significantly higher losses than expected for several reasons. Flux may crowd in corners or be unevenly distributed due to uneven core cross-sectional area. This flux crowding causes higher loss hot-spots in some regions of the core, but is only rarely a significant concern. Eddy currents induced in the core can be a significant loss component for very large cores. If the induced currents are large enough they will shield the flux from the inner sections of the core cross-section resulting in a flux skin effect analogous to the skin effect in the conductors of windings. Finally, dimensional resonance due to the propagation of electromagnetic waves within the structure can significantly change the overall flux distribution and lead to additional losses.

In this paper, we focus our attention on primarily the last two of these effects. Through the use of numerical modeling and laboratory measurement we examine the the relative importance of eddy-currents and dimensional resonance on the losses in large MnZn cores.

II. Material Characterization

Ferrites have been examined in great detail since they were first introduced in the middle of the century. Manufacturers and users alike have devoted considerable effort in developing standardized material data sheets[1, 8] that report the losses in ferrites as a function of frequency, flux density, and temperature[4, 3].

While these test data help the ferrite user compare the performance of different materials when used in a common applications, there are still areas where the available material information is lacking. In particular, in order to evaluate the type of dimensional effect encountered in some larger cores it is important to have information not only on the permeability and loss density of the core but also on the electrical conductivity and dielectric constant—or permittivity—of the material.

Figure 1 shows a thin-walled toroidal core and a thin disk of one representative ferrite material[1]. The complex permeability values of this material for various flux density levels and excitation frequencies are extracted from measurements of the inductive impedance of the toroidal core shown in Fig. 1(a) as described in [5]; the electrical characteristics (ϵ and σ) are derived from capacitive measurements on the thin disks, which are plated with platinum contacts on the upper and lower circular surfaces. The capacitive admittance of these plated disks is measured using an HP4194A impedance analyzer with the HP16451B dielectric test fixture.

[1] B2 material from Thompson as pressed and machined by Cermamic Magnetics, Inc.

0-7803-3500-7/96/$ 5.00 © 1996 IEEE

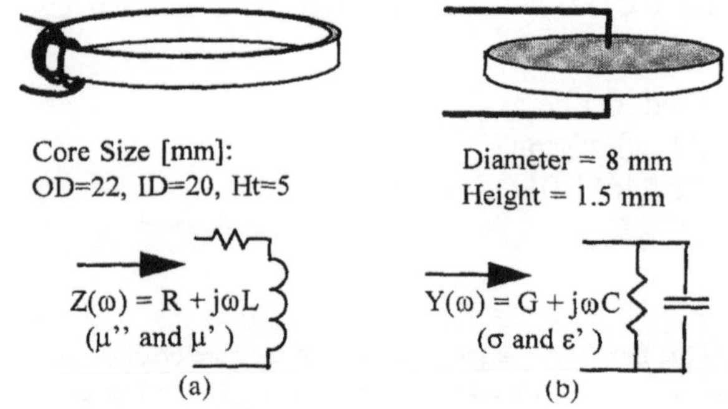

Figure 1: Core samples of B2 ferrite used to measure magnetic and electric characteristics of the material. (a) a thin-walled toroid ensures a uniform flux density, low thermal gradients and low eddy-current effects. (b) a small disk of material with plated contacts ensures an ohmic connection and low fringing effects.

Figure 2(a) shows the values of the real and imaginary components—along with the magnitude—of μ for flux density levels of 10 mT and 100 mT for various excitation frequencies. Figure 2(b) shows the electrical conductivity and relative dielectric constant.

The measured data in Fig. 2 show that this ferrite material has a relatively high magnetic permeability ($\approx$ 2000 times that of free space), relatively low conductivity, and extremely high dielectric constant. The low conductivity is beneficial from a loss perspective and is ferrites' main advantage over other materials. However,the same molecular structure that produces such low conductivity also results in the high value of effective permittivity at lower frequencies.

The combination of high dielectric constant, high permeability and finite resistivity means that ferrite cannot be treated as either a good conductor or a good insulator. It is this combination of material characteristics that leads to concern over geometrical—or dimensional—resonances in combination with significant eddy-current losses. Since the dielectric constant and relative permeability are both rather high, the electrical wavelength in ferrite is significantly shorter than it is in free space. This is detailed further in Section IV.

III. Experimental Measurements

Before proceeding with the field analysis for a large ferrite core, it is worthwhile to examine how dimensional effects manifest themselves in measurements of actual cores. This section presents measurements that illustrate the impact of dimensional resonance on a set of 'block-core' structures. These tests replicate to some degree the measurements presented in [2], and also indicate the limited applicability of simplified analytical expressions to more complex core geometries.

We examine the dimensional effects in these cores by measur-

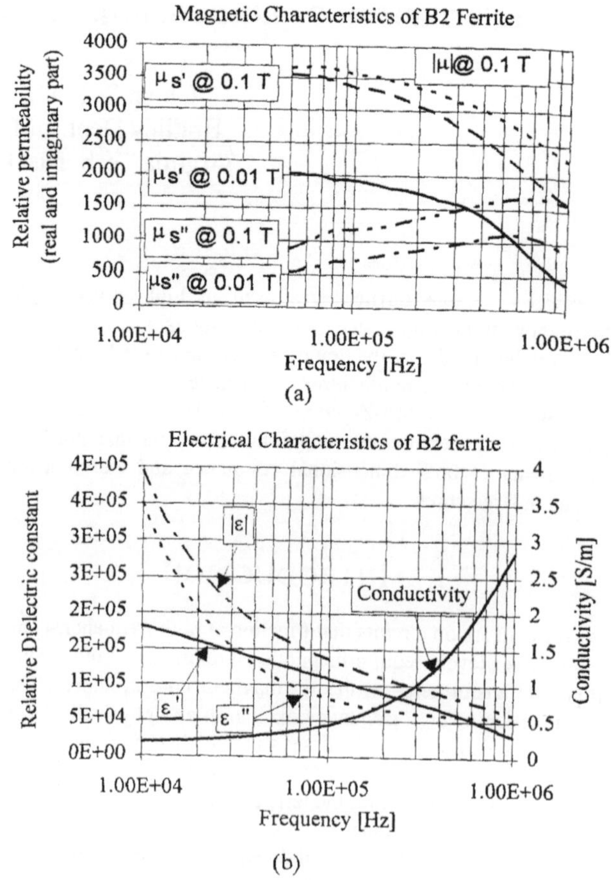

Figure 2: Magnetic and electrical characteristics of B2 MnZn ferrite material. (a)μ' and μ'' for 10 and 100 mT. (b)ϵ and σ vs. frequency.

ing the permeability and loss density of the cores shown in Fig. 3(a) as a function of frequency and flux density at room temperature. The test cores range in size from the thin-walled toroid used for a baseline measurement to a large core constructed from six 2"x3"x4" blocks of material. All of the cores are Ceramic Magnetics pressed B2 material.

Figure 3(b) shows the measured series permeability (real part) as a function of frequency for the different core structures. The flux density in these tests is 5 mT. For all of the large cores, the bandwidth of the apparent permeability is much less than that of the thin toroid; the largest cores have the lowest roll-off frequency since they have the lowest resonant frequency.

Figure 3(c) shows the core loss density of each of the cores for a flux density of 10 mT. Here again, the largest cores have the most significant change from the baseline material data, and in all cases the loss density in the core is significantly higher than predicted by the published material data. The dimensions listed in Fig. 3 are the smallest dimension that is normal to the flux in the core.

IV. Analytical Models

With the measurement data if Fig. 2 and the apparent shift in device performance based on core size illustrated in Fig. 3 as background, we now consider ways of modeling such effects analytically and numerically. In [2], Brockman, et al. present a thorough mathematical model of the dimensional resonance solution for a semi-infinite slab of a given thickness. We use that analysis—in a somewhat modified form—in the discussion below to illustrate how the physical geometry and the material characteristics of the core interact in a large structure.

For time harmonic excitations, the electric and magnetic fields in any given medium can be expressed in terms of the complex wave number, $k = \omega\sqrt{\mu\epsilon}$ where μ and ϵ are the permeability and permittivity of the medium respectively. In general, both μ and ϵ are complex and therefore k can be written in its complex components as,

$$k = \omega\sqrt{\bar{\mu}\bar{\epsilon}} = \omega\sqrt{(\mu' - j\mu'')(\epsilon' - j\epsilon'')} = k' - jk'' \quad (1)$$

The real and imaginary parts of k represent the phase constant and attenuation constant respectively. The wavelength λ is related to the phase constant k' as $\lambda = 2\pi/k'$, and therefore an expression for k' also provides an equation for the wavelength. In terms of the real and imaginary components of the complex permeability and permittivity, k' and k'' can be written as,

$$k' = \frac{\omega}{\sqrt{2}}\sqrt{|\mu||\epsilon| + (\mu'\epsilon' - \mu''\epsilon'')} \quad (2)$$

$$k'' = \frac{\omega}{\sqrt{2}}\sqrt{|\mu||\epsilon| - (\mu'\epsilon' - \mu''\epsilon'')} \quad (3)$$

Equations (2) and (3) are general equations that are valid for all simple media[2] under sinusoidal excitation conditions. These relations are often presented in terms of a real dielectric constant, ϵ, and an explicit conductivity term, σ[3].

In [12], Snelling states that a given material will not support standing waves if certain conditions hold true for the dielectric and magnetic loss tangents. These loss tangents relate the real and complex parts of the complex permittivity and permeability respectively:

$$\tan\delta_m = \frac{\mu''}{\mu'} \quad (4)$$

$$\tan\delta_e = \frac{\epsilon''}{\epsilon'} = \frac{\sigma}{\omega\epsilon'} \quad (5)$$

[2] A simple medium is linear, isotropic, and homogeneous. While ferrite is clearly not a simple medium we can consider it to be so for limited excitation levels.

[3] Equations (2) and (3) can be converted to other forms by substituting σ/ω for ϵ'' and simplifying[11, 6].

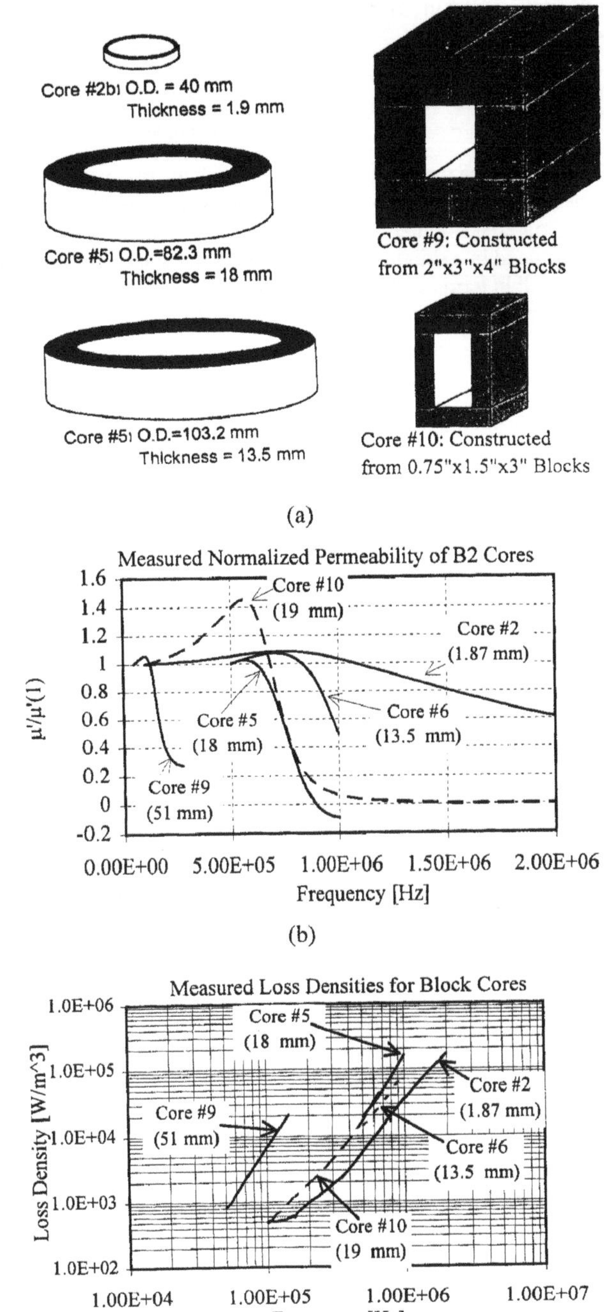

Figure 3: Tests of various BE2 material cores. (a) Core geometry and designation. (b) Series permeability (real part) vs. frequency at 5 mT. (c) Core loss density at 10 mT.

Substituting (4) and (5) in (2) and (3), the half-wavelength, $\lambda/2$, is [12, Eqn. (2.49)],

$$\frac{\lambda}{2} = \frac{\sqrt{2}\pi}{\omega} \frac{1}{\sqrt{|\mu||\epsilon| + \mu'\epsilon'(1 - \tan\delta_m \tan\delta_d)}} \quad (6)$$

From (6) we see that for $\tan\delta_m \tan\delta_d \gg 1$ the wavelength becomes large; this corresponds to the case where the attenuation in the material dominates the field penetration and the field is concentrated in a skin region near the surface. We can write the skin depth in terms of the wave attenuation constant, k'', as

$$\delta = \frac{1}{k''} = \frac{\sqrt{2}}{\omega\sqrt{|\mu||\epsilon| - \mu'\epsilon'(1 - \tan\delta_m \tan\delta_d)}} \quad (7)$$

Figure 4 shows the wavelength and skin depth in the the B2 ferrite based on (6) and (7). For these plots, the wavelength and skin depth are plotted for four different cases: (a) a lossless case where μ and ϵ are purely real[4], (b) a magnetic loss case where the permeability is complex but ϵ is real, (c) an electrical loss case where the permittivity is complex and μ is real, (d) a combined loss case with both μ andϵ complex. In all cases the material constants are are frequency dependent and are taken from the measured data presented in Fig. 2.

The wavelength and skin depth shown in Fig. 4 are on the order of centimeters for frequencies in the range of a few hundred kilohertz. The similarity between the electromagnetic wavelength and skin depth means that there will be a combination of effects within the core at the frequencies of interest. Fig. 4 shows that the combination of different loss components results in different values of skin depth; the combination of magnetic losses and electrical losses results in more significant attenuation than is present for either mechanism acting alone. The wavelength, however, is not as significantly different for the various cases.

The analytical model in [2] gives the solution for the magnetic field within the infinite slab in terms of the wave propagation constant, the thickness of the slab, d, and the value of the magnetic field on the two infinite surfaces of the slab, H_o, as

$$H(x) = H_o \times \frac{\cos(kx)}{\cos(kd/2)} \quad (8)$$

Figure 5 shows this field solution for a line that cuts horizontally through a 19.05 mm (0.75″) thick infinite slab at a variety of excitation frequencies. The magnetic field within the ferrite material is uniform and in phase with the exciting field of 22 [A/m] at low frequencies. At approximately 200 kHz there is a peak in the magnitude of the H field in the center of the core, but as the frequency increases the attenuation dominates and the interior of the core is shielded. At 1 MHz the slab is approximately 3.6 skindepths thick and the surface field is essentially excluded from the core interior.

The flux density, B, in the core is related to the H field by the permeability of the core. As the hysteresis loss in the core is modeled through the use of a complex permeability, the flux density, is shifted in phase by an angle $\delta_\mu = \tan^{-1}(\mu''/\mu')$. The magnitude of the flux density at any point in the core is therefore the result of an in-phase real part, B_{real}, and an out-of-phase imaginary part, B_{imag}.

[4] The skin depth is not defined in this lossless case.

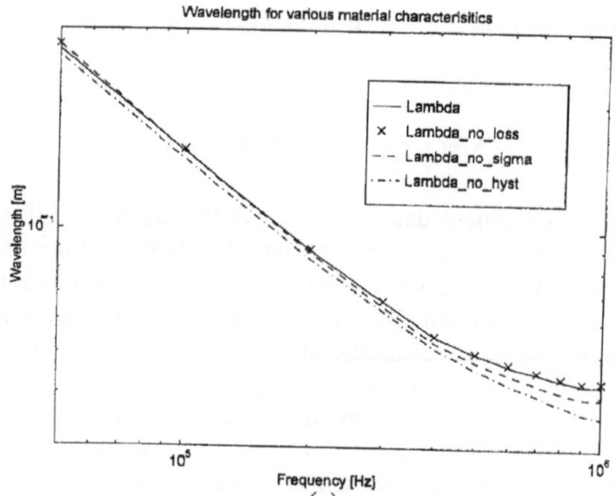

(a)

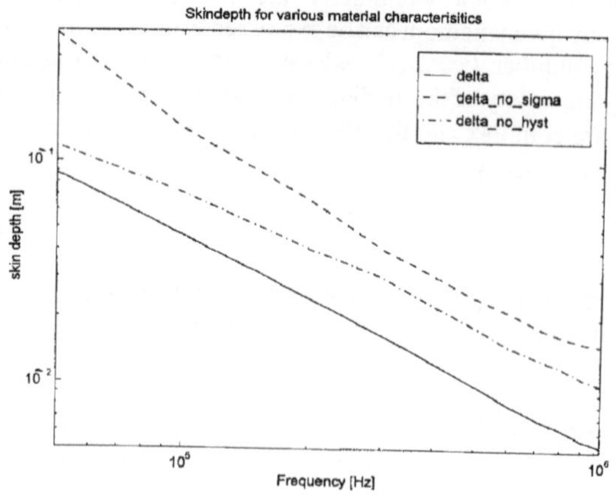

(b)

Figure 4: (a)Electromagnetic wavelength in ferrite vs. frequency for various combinations of loss mechanisms.(b) skin depth for the various combinations of loss mechanisms.

V. SIMULATIONS

The theoretical solution presented above is fundamentally limited to the semi-infinite geometries assumed in the derivation. In practical situations we need to be able to solve for the field distribution in actual physical structures where the assumptions made in the 1D solution do not apply. Numerical solvers such as finite elements can be used to explore the dimensional effects in such cases.

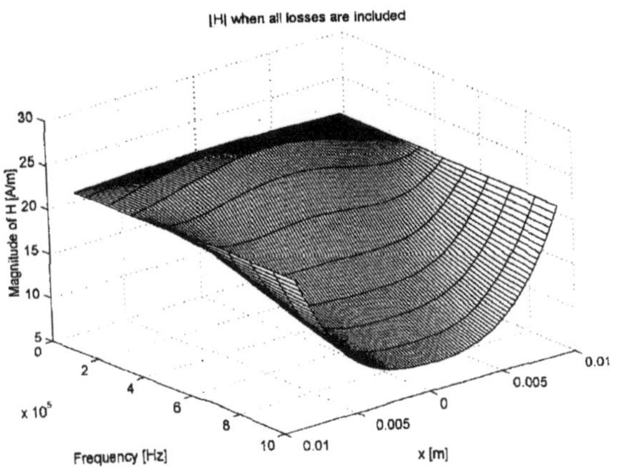

Figure 5: The one dimensional model of the field distribution within a infinite slab of thickness $d = 19.05$ mm (0.75"). The H-field distribution within the ferrite is uniform and in phase with the exciting current at low frequencies. There is a peak in the center of the slab at approximately 200 kHz, and the slab is shielded from the exciting field as the frequency increases

In order to be effective in analyzing large ferrite cores, the numerical solver must be capable of modeling materials with complex characteristics. Specifically, the solver must allow for magnetic loss (complex permeability), eddy-current loss (non-zero conductivity) and displacement current (non-unity dielectric constant). Table 1 shows the FEA input values used in the

Table 1: Values of material constants used in the FEA simulation for a simulation at 200 kHz. For each frequency the core is simulated with eight possible combinations of permeability, permittivity and conductivity

	μ'	μ''	ε'	σ [S/m]
setup2	3040	-1359	1	0.0001
setup1	3330	0	1	0.0001
setup4	3040	-1359	1	0.85
setup3	3330	0	1	0.85
setup6	3040	-1359	90000	0.0001
setup5	3330	0	90000	0.0001
setup8	3040	-1359	90000	0.85
setup7	3330	0	90000	0.85

FEA simulations presented below for one given frequency. At each simulation point the problem is solved eight times with eight different combinations of input material values that include the "ideal" magnetic case with no losses and a dielectric constant of free space as a baseline case. Through such a set of simulations it is possible to examine separately the influence of each material characteristic on the overall losses in the core and just as importantly to see how these variations affect the field distribution within complicated structures.

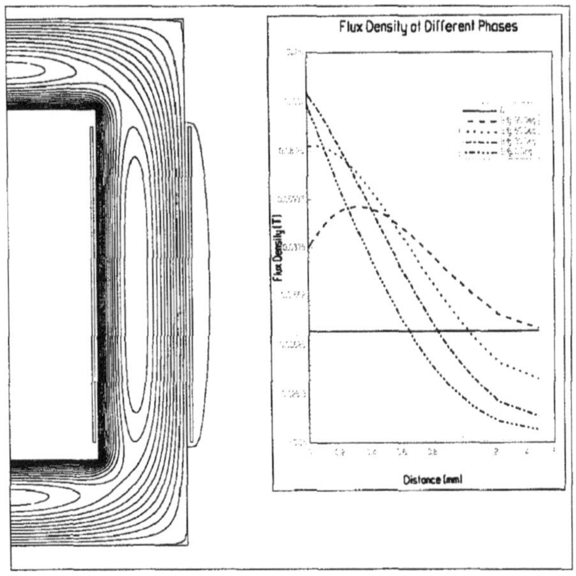

Figure 6: An example 2D simulation of a block core. (a) the flux lines within the core cross section at 300 kHz show localized eddys of magnetic flux not predicted by the one-dimensional model. (b) line plots of the flux at different phase angles along a line cutting through the middle of the core.

Fig. 6 shows one example two-dimensional simulation for a large block core such as presented in Fig. 3. This figure shows the flux lines in the right-hand half of the core at a frequency of 300 kHz where losses are highest. In Fig. 6(b) the value of the flux density is plotted along a line that cuts through the core cross-section. The different lines represent the flux density at different values of phase angle. These results illustrate that the flux distribution in this relatively simple core structure is significantly more complicated than predicted by the simple infinite-slab solution presented above.

In Fig. 7, the various curves of the simulated frequency response for the block core of Fig. 6 show trends very similar to those of the measured cores shown in Fig. 3. The plots of normalized energy storage in Fig. 7 illustrate that for tha same material input data, the performance of a given core is dependent largely on the core cross-sectional size. In addition, the core loss density—which is often assumed to be a property of the material alone—is significantly higher for the larger cores.

VI. Conclusions

Both eddy-current and dimensional-resonance phenomena have been known from a fields perspective for many years. Until this time, however, it has been rare to find cases where these effects become significant for practical size cores. As the power levels of high-frequency switching circuits continue to increase these issues become important and it is critical to understand that the core material characteristics, size and shape all play an important role in determining the performance of the structure.

We have illustrated in this paper that the combination of dimensional resonance and eddy-currents in ferrite cores increases the core loss and thereby decreases the efficiency of the device. The interaction of the various loss mechanisms in the core result in a skin depth that is significantly lower than if the core is modeled without concern for such losses. This dependence of the penetration depth on the dielectric, conduction, and magnetic losses is perhaps the most important point to remember when dealing with large cores. It is not adequate to assume that one loss mechanism will dominate the others.

Eddy-currents tend are generally most significant at higher frequencies where the dielectric constant falls off. In this case, the interior of the core is shielded from the exciting field. Such changes in the flux distribution in the core affects the frequency response characteristics of the inductor or transformer of which it is a part. In particular, as dimensional effects become significant, the effective core permeability decreases rapidly.

In order to adequately design magnetic components for new high-power, high-frequency applications we must start from much more detailed and thorough information about how large ferrites structures actually perform. It is not likely that a reliance on the information available from ferrite manufacturers alone will prove adequate.

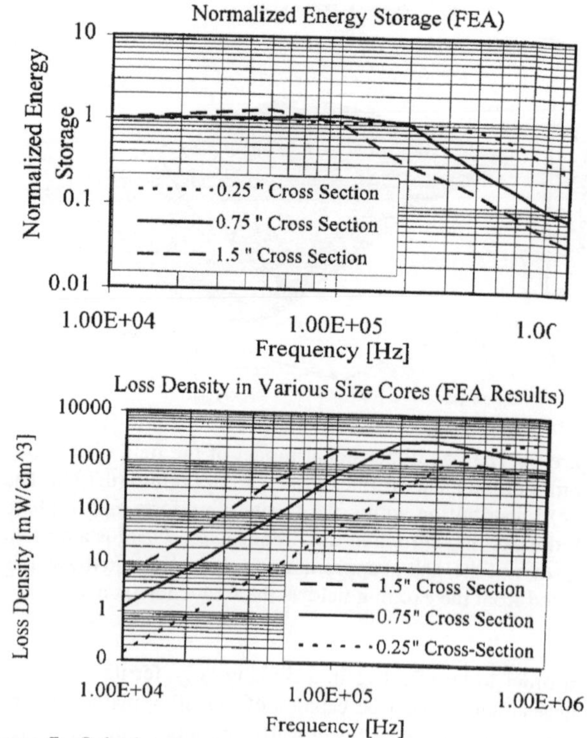

Figure 7: Calculated energy storage and loss density for three block cores with different cross sectional dimensions. Note the similarity to the curves in Fig. 3.

REFERENCES

[1] L.M. Bosley, "Data sheets provide clues for optimizing selection of SMPS ferrite core material," *Powerconversion and Intelligent Motion*, 1993, p. 13-15.

[2] F.G. Brockman, P.H. Dowling and W.G. Steneck, "Dimensional Effects Resulting from a High Dielectric Constant Found in a Ferromagnetic Ferrite", *Physical Review, Vol. 77, NO. 1*, January, 1950.

[3] P.M. Gradzski, *Core Loss Characterization and Design Optimization of High- Frequency Power Ferrite Devices in Power Electronics Applications*, Ph.D. Dissertation, Virginia Polytechnic Institute and State University, 1992.

[4] C. Guillaud, "The properties of Manganese-Zinc Ferrites and the Physical Processes Governing Them", *Proceedings of the IEE, Suppl. 5*, 1957, p. 165-178.

[5] P. Han, G. Skutt and F. Lee, "Finite Element Method for Ferrite Coreloss Calculation", *IEEE Applied Power Electronics Conference Record*, 1995, p. 348–353.

[6] W. H. Hayt, *Engineering Electromagnetics*, 5th Ed., McGraw-Hill, New York, 1989.

[7] C.G. Koops, "On the Dispersion of Resistivity and Dielectric Constant of Some Semiconductors at Audio Frequencies", *Physical Review,Vol. 83 NO. 1*, 1951, p. 121-124.

[8] MMPA, *Magnetic Materials Producers Association Ferrite Users Guide*, 1992.

[9] E. Otsuki, S.Y. Otsuka, K. Shoji and T. Sato, "Microstructure and Physical Properties of Mn-Zn Ferrites for High-Frequency Power Supplies", *Journal of Applied Physics,Vol. 69 NO. 8*, 1991, p. 5942-5944.

[10] D. Polder, "Ferrite Materials", *Proceedings of the IEE,Vol. 97(Part II)*, 1950, p. 246-256.

[11] R. Severns, E. Yeow, G. Woody, J. Hall and J. Hayes, "An Ultra-Compact Transformer for a 100 W To 120 kW Inductive Coupler for Electric Vehicle Battery Charging", *IEEE Applied Power Electronics Conference Conference Record*, 1996, p. 32-38.

[12] E.C. Snelling, *Soft Ferrites*, 2nd Ed., Butterworths, London, 1988.

Physical Meaning of Equivalent Loss Resistance of Magnetic Cores

TOSHIRO SATO AND YO SAKAKI, MEMBER, IEEE

Abstract—The equivalent loss resistance of magnetic cores is analyzed on the basis of the magnetic domain theory. The hysteresis resistance due to hysteresis loss is dependent on the variation of the Steinmetz constant. On the other hand, the eddy current resistance due to eddy current loss is strongly affected by the change of domain configurations and is described by an experimental parameter that depends on the number of domain walls, the resistivity of the material, and the dimensions of the core, all of which control the mechanism generating eddy current losses.

I. Introduction

The authors have discussed the mechanism generating power losses in many kinds of magnetic cores for switching power supplies up to 10 MHz [1], [2]. On the other hand, it is convenient to represent power loss characteristics of magnetic cores as an equivalent loss resistance for circuit design and analysis. However, the equivalent loss resistance varies in a complicated manner with the magnetizing conditions, and power supply engineers confront difficulties as to whether it represents the dynamic behaviors exactly or not.

In this paper, the relationship between power losses and equivalent loss resistance, which is composed of equivalent eddy current resistance and equivalent hysteresis resistance, is discussed. Through this relationship, it is clear that the former is strongly dependent on the domain configurations and the latter depends upon the Steinmetz constant.

II. Magnetic Core Loss and Equivalent Loss Resistance

A. Introduction of Equivalent Loss Resistance

Magnetic core loss W_c is the sum of hysteresis loss W_h and eddy current loss W_e:

$$W_c = W_h + W_e \qquad \mathrm{W/m^3}. \tag{1}$$

From this expression, we obtain a parallel equivalent circuit as shown in Fig. 1 where R_h is the equivalent hysteresis resistance, R_e is the equivalent eddy current resistance, and L_p is the parallel equivalent inductance. From Fig. 1, the power losses generated in each resistance are given by (2) and (3):

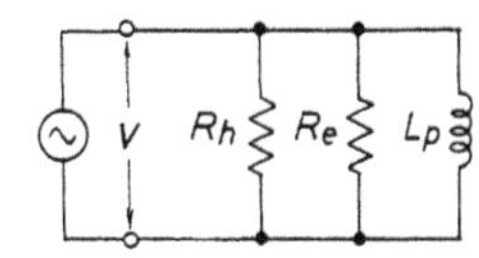

Fig. 1. Equivalent circuit of magnetic core.

$$W_h \times (S \cdot l) = V^2/R_h \qquad \mathrm{W} \tag{2}$$

$$W_e \times (S \cdot l) = V^2/R_e \qquad \mathrm{W} \tag{3}$$

where S and l are the core cross section and the average magnetic path length, respectively, and V is the effective value of the applied sinusoidal voltage. The relation between the applied voltage V, the frequency f, and the maximum flux density B_m is

$$V = \sqrt{2}\pi f B_m S \qquad \mathrm{V} \tag{4}$$

where we assume that the number of coil turns is unity. From (2)–(4) we obtain the following expressions:

$$W_h \times \lambda_h = 2\pi^2 f^2 B_m^2 \tag{5}$$

$$W_e \times \lambda_e = 2\pi^2 f^2 B_m^2 \tag{6}$$

where

$$R_h = \lambda_h S/l \qquad \Omega \tag{7}$$

$$R_e = \lambda_e S/l \qquad \Omega. \tag{8}$$

Both λ_h and λ_e are in ohms per meter, and we define λ_h as the hysteresis parameter and λ_e as the eddy current parameter. Equations (5) and (6) show that the product of magnetic core loss per unit volume and these parameters depend only on the frequency f and the maximum flux density B_m and are independent of the physical constants of magnetic core materials. Equations (7) and (8) show that R_h and R_e are uniquely determined when these parameters are given. These expressions have the same form as the formula for inductance, i.e.,

$$L = \mu S/l \qquad \mathrm{H} \tag{9}$$

where μ is the permeability in henrys per meter. Thus λ plays the same role in the formula for resistance that the permeability plays in the expression for inductance. In the following, we discuss the physical meanings of λ_h and λ_e.

Manuscript received November 7, 1989; revised April 2, 1990. This work was supported in part by the Ministry of Education, Science and Culture of Japan under a Grant-in-Aid for Scientific Research.

T. Sato was with the Department of Electrical and Electronics Engineering, Chiba University, Chiba, Japan. He is now with Toshiba R & D Center, Komukai-Toshiba-Cho, Saiwai-Ku, Kawasaki, 210 Japan.

Y. Sakaki is with the Department of Electrical and Electronics Engineering, Chiba University, 1-33, Yayoi-Cho, Chiba, 260 Japan.

IEEE Log Number 9036776.

0018-9464/90/0900-2894$01.00 © 1990 IEEE

B. Representation of Hysteresis Parameter

Assuming that the hysteresis loss per cycle W_{h0} (in joules per cubic meter) is independent of the exciting frequency, the hysteresis loss W_h (in watts per cubic meter) is represented according to the Steinmetz empirical equation

$$W_h = f \cdot W_{h0} = k_h f B_m^{\eta} \quad \mathrm{W/m^3} \tag{10}$$

where k_h is a constant and η is the Steinmetz constant. By using (5) and (10), λ_h is written as follows:

$$\lambda_h = 2\pi^2 f B_m^{2-\eta}/k_h \quad \Omega/\mathrm{m}. \tag{11}$$

Generally, the Steinmetz constant η is nearly equal to 1.6 for metallic cores except for square loop ones and is larger than 2.0 for polycrystalline Mn–Zn ferrite cores [3].

The induction dependence of the hysteresis parameter λ_h is, therefore, strongly affected by the Steinmetz constant. That is, λ_h of metallic cores increases with increasing the maximum induction B_m; on the other hand, λ_h of polycrystalline Mn–Zn ferrite decreases with increasing B_m.

C. Representation of Eddy Current Parameter

A formula representing eddy current loss in metallic tape cores with bar-like domain walls as shown in Fig. 2 is [4]

$$W_e = \frac{8.4 b d f^2 B_m^2}{\pi \rho n} \quad \mathrm{W/m^3} \tag{12}$$

where b and d are the width and thickness of metallic tape (or the height and width of the ferrite toroidal core), respectively, n is the equivalent number of bar-like domain walls, and ρ is the electrical resistivity of the material. As already reported by one of the authors [4], an empirical parameter $(V_m/H_e)/bd$ is effective for estimating the equivalent number of bar-like domain walls n participating in flux reversal:

$$\frac{V_m/H_e}{bd} = \frac{\pi^3 \rho n}{4.2 bd} \quad \Omega/\mathrm{m} \tag{13}$$

where V_m is the peak value of the induced sinusoidal voltage per coil turn per tape layer and H_e is the effective eddy current field defined as the difference between the coercivities of ac and dc B–H loops. These values are easily measured experimentally as shown in Fig. 3.

Combining (12) and (13), the theoretical product of W_e and $(V_m/H_e)/bd$ is obtained:

$$W_e \times \frac{V_m/H_e}{bd} = 2\pi^2 f^2 B_m^2. \tag{14}$$

Experimental results of $W_e \times (V_m/H_e)/bd$ in most magnetic materials agree well with (14) [5], [6].

Comparing (14) with (6), we find that the eddy current parameter λ_e is the same as the experimental parameter $(V_m/H_e)/bd$, that is,

$$\lambda_e = (V_m/H_e)/bd \tag{15}$$

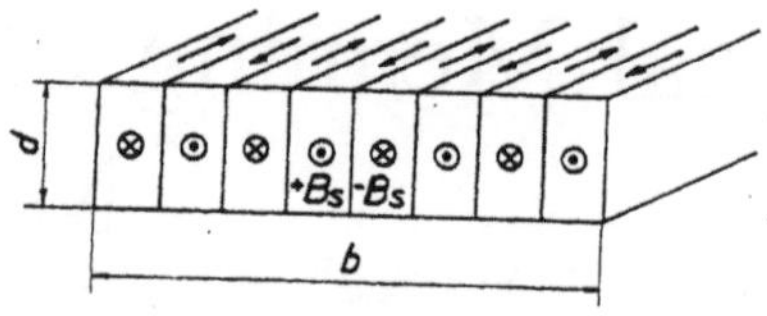

Fig. 2. Bar-like domain model.

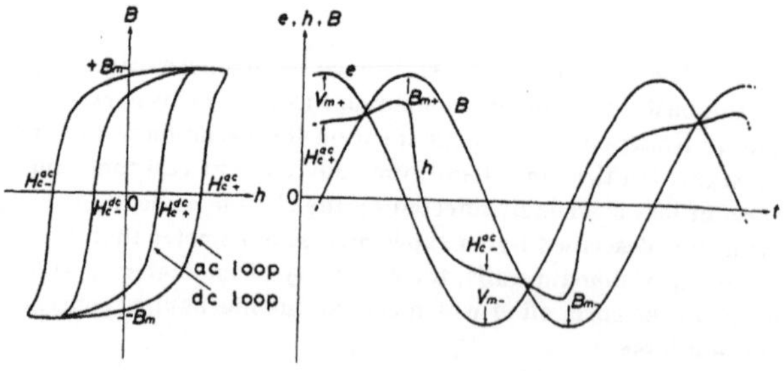

Fig. 3. Definition and measurement of $(V_m/H_e)/bd$.

and that the eddy current parameter λ_e is in proportion to the number of domain walls n and the resistivity ρ of core material and in inverse proportion to the tape cross section. Thus λ_e includes all the factors that control the mechanism generating eddy current loss. Since the parameter $(V_m/H_e)/bd$ is easily measured by observing the waveforms of exciting current and induced voltage, R_e is quantitatively estimated using (8) and (15) without knowing eddy current loss values.

III. Experimental Results and Discussion

A. Equivalent Hysteresis Resistance

Fig. 4 shows relationships between hysteresis losses and the maximum flux densities in Supermalloy tape wound cores (25 μm thick and 5 mm wide) and high permeability Mn–Zn ferrite toroidal cores (nominal specific permeability is 10^4). Also shown are the calculated hysteresis parameter normalized by frequency λ_h/f by using (11). In this experiment, $\eta = 1.55$ for Supermalloy and $\eta = 2.66$ for Mn–Zn ferrite. As a result, the curve of λ_h/f of the former increases with B_m; on the other hand, the one of the latter decreases with B_m. The equivalent hysteresis resistance is, therefore, strongly dependent on the microscopic structures and static behaviors of the domain walls, which results in the change of the Steinmetz constant [3].

B. Equivalent Eddy Current Resistance

The solid curves in Fig. 5 show the frequency dependence of the calculated eddy current parameter λ_e on the basis of measured eddy current losses W_e and (6). Also shown are the plots of measured experimental parameter $(V_m/H_e)/bd$. Both λ_e and $(V_m/H_e)/bd$ increase with increasing frequency. This is attributed to the increase in the number of domain walls.

The maximum flux density dependence of λ_e and $(V_m/H_e)/bd$ are shown in Fig. 6. Note that the λ_e of the

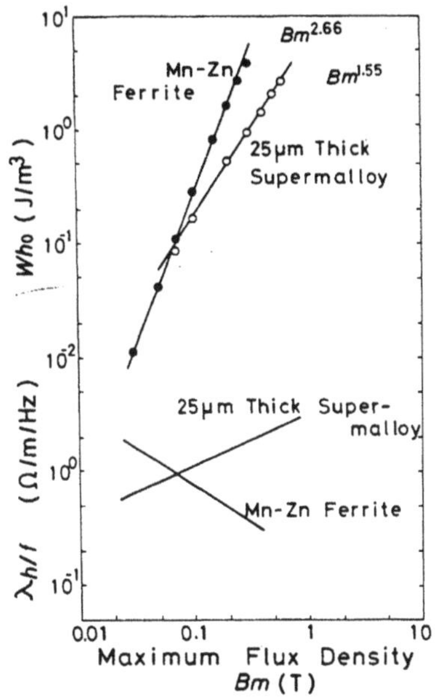

Fig. 4. Relationship between hysteresis loss per cycle W_{h0}, hysteresis parameter per cycle λ_h/f, and maximum flux density B_m.

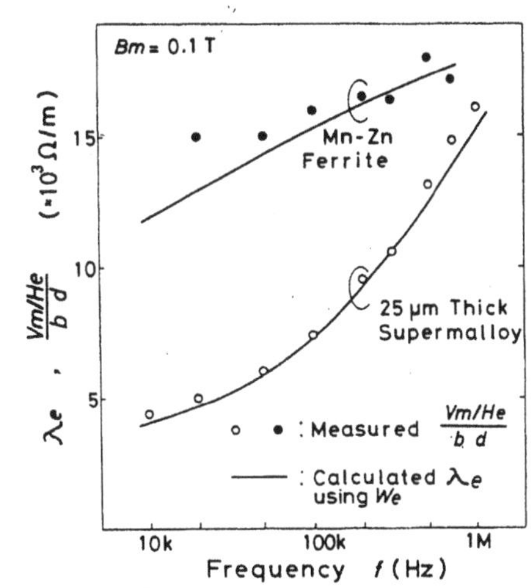

Fig. 5. Relationship between $(V_m/H_e)/bd$, eddy current parameter λ_e and frequency f.

ferrite decreases with B_m, while λ_e for Supermalloy increases with B_m. The decrease of $(V_m/H_e)/bd$ with induction in polycrystalline Mn–Zn ferrite may be due to the domain wall extinction with increasing B_m [7].

In Figs. 5 and 6 the eddy current parameter λ_e agrees well with the experimental parameter $(V_m/H_e)/bd$. This fact supports the theoretical result of (15). The equivalent eddy current resistance R_e can, therefore, be quantitatively evaluated by estimating the parameter $(V_m/H_e)/bd$.

C. Equivalent Core Loss Resistance

The parallel combination of equivalent hysteresis and eddy current resistances yields an equivalent core loss resistance R_c:

$$R_c^{-1} = R_h^{-1} + R_e^{-1} \tag{16}$$

or

$$\lambda_c^{-1} = \lambda_h^{-1} + \lambda_e^{-1} \tag{17}$$

where λ_c is defined as the core parameter.

Table I shows values of λ_h, λ_e, and λ_c in various cores when $f = 500$ kHz and $B_m = 0.1$ T. It is desired that the magnetic cores have large λ_c for high-frequency power applications such as switching power supplies. For this purpose, both λ_h and λ_e should be large. The ultrathin Co-based amorphous ribbon core developed for megahertz power supply [8] has the largest λ_c. The 5-μm-thick Supermalloy core has relatively low λ_c from its high hysteresis loss due to its fine grains which increase the coercivity [9]. It is surprising that the high-resistivity Mn–Zn ferrite core ($\rho = 6.5\ \Omega \cdot$ m) has fairly low λ_c. This may be attributed to high hysteresis loss due to the nucleation and extinction of domain walls at each grain boundary [3] and the increase in eddy current loss due to the extinction of domain walls during magnetization reversal [7].

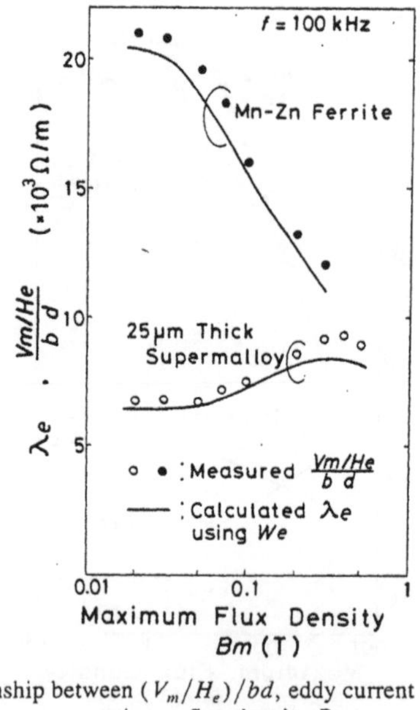

Fig. 6. Relationship between $(V_m/H_e)/bd$, eddy current parameter λ_e, and maximum flux density B_m.

TABLE I
COMPARISON OF THE PARAMETERS λ_h, λ_e, AND λ_c^*

Number	Core Materials	λ_h (Ω/m)	λ_e (Ω/m)	λ_c (Ω/m)
1	25-μm-thick Supermalloy	5.68×10^5	1.23×10^4	1.20×10^4
2	high-μ Mn–Zn ferrite	3.70×10^5	1.72×10^4	1.64×10^4
3	5-μm-thick Supermalloy	2.60×10^5	6.11×10^4	4.95×10^4
4	high-ρ Mn–Zn ferrite	0.82×10^5	5.79×10^4	3.39×10^4
5	7.1-μm-thick Co-based amorphous	15.8×10^5	9.40×10^4	8.87×10^4

*Data are at $f = 500$ kHz and $B_m = 0.1$ T. Numbers 1, 2 are the same cores as discussed in Figs. 4–6. Numbers 3–5 are recommended for high-frequency application ($f > 100$ kHz).

IV. CONCLUSION

The physical meaning of the equivalent loss resistance in soft magnetic cores is discussed. The combination of the equivalent hysteresis resistance and equivalent eddy current resistance results in the equivalent loss resistance. The hysteresis resistance is described by a new parameter λ_h which is dependent on the Steinmetz constant. The eddy current resistance is described by the parameter $\lambda_e = (V_m/H_e)/bd$ already introduced by one of the authors. This parameter is a function of the number of domain walls, resistivity, and the sheet dimensions, all of which control the mechanism generating eddy current loss.

REFERENCES

[1] Y. Sakaki and T. Sato, "Large signal eddy current losses beyond 100 kHz," *IEEE Trans. Magn.*, vol. MAG-20, pp. 1487–1489, Sept. 1984.
[2] T. Sato and Y. Sakaki, "100 kHz-10 MHz iron loss measuring system," *IEEE Trans. Magn.*, vol. MAG-23, pp. 1593–1595, Sept. 1987.
[3] Y. Sakaki and T. Matsuoka, "Hysteresis losses in Mn-Zn ferrite cores," *IEEE Trans. Magn.*, vol. MAG-22, pp. 623–625, 1986.
[4] Y. Sakaki, "An approach estimating the number of domain walls and eddy current losses in grain-oriented 3% Si-Fe tape wound cores," *IEEE Trans. Magn.*, vol. MAG-16, pp. 569–572, 1980.
[5] Y. Sakaki and S. I. Imagi, "Relationship between eddy current losses and equivalent number of domain walls in polycrystalline and amorphous soft magnetic materials and its application to minor loop loss estimation," *IEEE Trans. Magn.*, vol. MAG-18, pp. 1840–1842, Nov. 1982.
[6] T. Katane, T. Yamaguchi, T. Ito, T. Sato, and Y. Sakaki, "A high frequency iron loss measuring system," *IEEE Trans. Magn.*, vol. MAG-20, pp. 1729–1731, 1984.
[7] T. Sato and Y. Sakaki, "Mechanism generating power losses in MHz band in polycrystalline Mn–Zn ferrite cores used for switching power supplies," *Trans. IEICE* (Japan), vol. 71-C, pp. 1121–1127, 1988.
[8] M. Yagi, T. Sato, Y. Sakaki, T. Sawa, and K. Inomata, "Very low loss ultra thin Co based amorphous ribbon cores," *J. Appl. Phys.*, vol. 64, pp. 6050–6052, Nov. 1988.
[9] Y. Sakaki and T. Sato, "Conditions for minimum losses in Supermalloy cores used in the high frequency range," *Trans. IECE* (Japan), vol. 68-C, pp. 202–207, 1985.

Toshiro Sato was born in Niigata Prefecture, Japan, in 1959. He received the B.S., M.S., and Ph.D. degrees from Chiba University in 1982, 1984, and 1989, respectively.

In 1989 he joined Metal and Ceramic Laboratory of Toshiba Research and Development Center, Kawasaki, Japan.

Dr. Sato is a member of the IEICE Japan and the Magnetics Society of Japan.

Yo Sakaki (M'83) was born in Chiba, Japan, on March 9, 1932. He received the B.S. degree in electrical engineering from Ibaraki University and the M.Eng. and D.Eng. degrees from Tohoku University in 1954, 1965, and 1971, respectively.

From 1954 to 1963 he engaged in research and design of magnetic amplifiers and power semiconductor devices in several companies. In 1969 he joined the Department of Electrical Engineering, Chiba University, Chiba, Japan, where he is now a Professor. His current research interest includes the mechanism generating eddy current loss in soft magnetic materials, biomagnetic sensor in migratory fish and application of nonlinear magnetic devices.

Dr. Sakaki is a member of IEE Japan, IEICE Japan, and the Magnetics Society of Japan.

MODELING IRON–POWDER INDUCTORS AT HIGH FREQUENCIES

M. Bartoli, A. Reatti, Member, IEEE
University of Florence,
Department of Electronic Engineering
Via di S. Marta, 3, 50139 Florence, Italy
Phone: (39)(55)4796–389 and Fax: (39)(55)494569
E – Mail: CIRCUITI@VM.IDG.CNR.FI

M. K. Kazimierczuk, Senior Member, IEEE
Wright State University,
Department of Electrical Engineering
Dayton, Ohio 45435
Phone: (513)873–5059 and Fax: (513)873–5009
E – Mail: MKAZIM@VALHALLA.CS.WRIGHT.EDU

Abstract – **A high–frequency model of iron–powder core inductors is studied. The skin and proximity effects that cause the winding parasitic resistance to increase with the operating frequency are considered. The inductor self–resonance due to the parasitic capacitances is taken into account as well. The frequency response of the inductor model is compared to that of an experimentally tested iron–powder core inductor. The first self–resonant frequency is determined from the plot of the measured reactance and allows for the calculation of the parasitic capacitance. Equations for the inductor parasitic resistance are derived in a closed form. Expressions giving the ac resistance as a function of the operating frequency are given. These expressions allow for an accurate prediction of the inductor power loss over a wide frequency range. The measured and calculated values of the inductor impedance magnitude and phase, the real and imaginary parts of the inductor impedance, the inductance, and the inductor quality factor are plotted versus frequency and compared. Theoretical results were in good agreement with those experimentally measured. Therefore, it is demonstrated that the discussed equivalent circuit has a frequency response matching that of the real inductor. Moreover, the circuit model is simple, it allows for an immediate understanding of iron–powder core inductor behavior and can be easily used in computer simulations of electronic circuits.**

I. INTRODUCTION

The increasing demand for power supplies with small volumes and weights has necessitated the development of switching power converters operating at frequencies above several hundred kilohertz to achieve a size reduction of magnetic components. At high operating frequencies parasitic resistances and capacitances of inductors can adversely affect the operation of power converters. Therefore, an accurate prediction of the frequency response of inductors is crucial to the design of high–frequency power circuits. Parasitic resistance due to core losses is drastically reduced if distributed air gap cores with a low permeability such as iron–powder cores are used. However, the winding ac resistance at high frequencies is much higher than the dc resistance because of the skin and proximity effects. Moreover, many turns must be wound on a low permeability core, resulting in a relatively large parasitic capacitance that affects the inductor frequency response. As a result, the first self–resonant frequency of the inductor is low and parasitic resistance is high at frequencies close to the self–resonant frequency, and the inductor behaves like a capacitor if it is operated above the self–resonant frequency. Unfortunately, the parasitic capacitances and resistances are distributed parameters and their values depend on the operating frequency. Therefore, it is very difficult to theoretically predict the frequency response of an inductor.

The problem of high–frequency magnetic components is discussed in [1]–[13]. In [1], only the skin and proximity effect contributions to the ac resistance are considered. References [2]–[3] deal with the problem of high–frequency transformer and inductor design but do not give any model of these components. Equivalent circuits of magnetic components are shown in references [6] and [7], but the effect of parasitic capacitances on the frequency response is studied in detail for wide–band transformers used in transmission of information. Therefore, the results of these studies cannot be transferred to power magnetic components. High–frequency equivalent circuits of power inductors are given in [8]–[13], but resistances of the models are assumed to be frequency independent and have too low values at high operating frequencies.

The purpose of this paper is to study a circuit model of iron powder–core inductors including both parasitic capacitances and frequency–dependent winding and core resistances. The impedance magnitude and phase, equivalent series resistance R_s (ESR) and reactance X_s, inductance, and quality factor of the equivalent inductor model are plotted as functions of the operating frequency and compared with those obtained experimentally for an iron–powder core inductor. Theoretical and experimental results were in good agreement.

The new contribution of this work is the simple lumped–parameter equivalent circuit that accurately describes the frequency response of iron–powder inductors. A procedure for determining the values the model components is also presented in the paper. The significance of the paper is that the introduced equivalent circuit is simple, leads to a straightforward understanding of the behavior of iron–powder core inductors, and describes their frequency response very accurately. For this reason, it can be usefully used in designing electronic circuits such as EMI filters, resonant circuits, filters for power converters, etc. The inductor operation over a wide range of frequencies can be predicted by using the model. From this examination,

[1] This work was supported by *MagneTek S.p.A.*

0-7803-1993-1/94 $4.00 © 1994 IEEE

operation at frequencies too close to the self-resonant frequency and with too low efficiencies can be prevented. Moreover, the model is particularly suitable for applications in computer simulations of electronic circuits.

II. Inductor Equivalent Circuit And Computational Procedure Of Equivalent Circuit Parameters

A. Equivalent Circuit of Iron–Powder Inductors

Fig. 1 shows an equivalent circuit of a two-layer inductor. Parasitic capacitances affecting the high-frequency inductor operation are the turn-to-turn capacitances of each j-th turn $c_{t\text{-}tj}$, the turn-to-core capacitances $c_{t\text{-}cj}$, and the layer-to-layer capacitances $c_{l\text{-}lj}$. Parasitic resistances are the winding resistance r_j, the core resistance r_{cj} associated with the finite resistivity of the inductor core, and the resistance of the dielectric material coated on the wire r_{dj}. Since the distributed-parameter equivalent circuit of Fig. 1 is cumbersome, a lumped-parameter equivalent circuit of Fig. 2(a) is considered to model the inductor. Here, L is the coil inductance, R_w is the winding resistance, and R_c is the core resistance. Both resistances are frequency dependent. In an inductor wound on an iron-powder core, the turn-to-turn and layer-to-layer capacitances represent the major components of the total parallel capacitance C, shown in Fig. 2(a). The total parasitic ac resistance $R_{ac} = R_w + R_c$ is the series combination of R_w and R_c, as shown in Fig. 2(b).

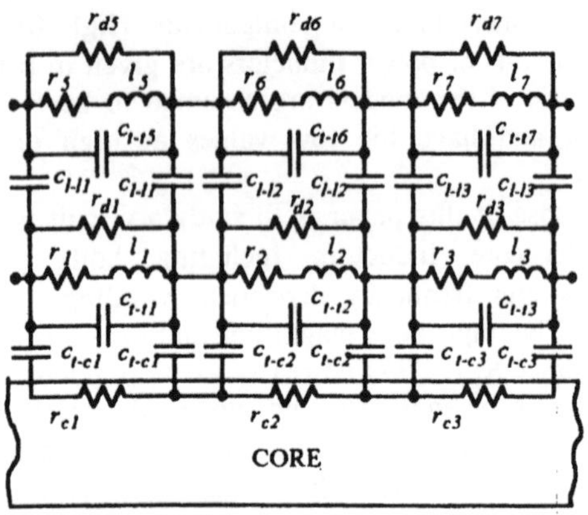

Fig. 1. Model of a two-layer inductor.

The impedance of the equivalent circuits depicted in Figs. 2(a) and (b) is

$$Z_s = \frac{R_{ac} + j\omega L\left(1 - \omega^2 LC - CR_{ac}^2 / L\right)}{\left(1 - \omega^2 LC\right)^2 + \omega^2 C^2 R_{ac}^2}$$
$$= R_s + jX_s = |Z_s| e^{j\phi} \tag{1}$$

where R_s and X_s are the real and the imaginary parts of Z_s, and $|Z_s|$ and f are the magnitude and phase of Z_s, respectively.

By definition, the self-resonant frequency $f_r = \omega_r / 2\pi$ occurs when $X_s = 0$. Since $CR_{ac}^2 / L << 1$, (1) gives the inductor parasitic capacitance

$$C = \frac{1}{\omega_r^2 L} = \frac{1}{(2\pi f_r)^2 L}. \tag{2}$$

Fig. 2(c) shows the resistance R_s and reactance X_s measured by LCR meters when the equivalent series circuit model is selected for the device test. Other impedance parameters are calculated by the LCR meters [12] from $R_s + jX_s$. In particular, the equivalent series inductance (ESL) at a given frequency is

$$L_s = \frac{X_s}{\omega}. \tag{3}$$

Since the series equivalent reactance is positive for $f < f_r$ and negative for $f > f_r$, the series inductance L_s becomes negative for frequencies above the self-resonant frequency.

The inductor quality factor at a given frequency is

$$Q_s = \frac{|X_s|}{R_s}. \tag{4}$$

The quality factor can also be defined by using the energy stored in the magnetic field

$$Q_o = 2\pi f \frac{E_L}{P_R} = \frac{\omega L}{R_{ac}} \tag{5}$$

where E_L is the maximum energy stored in the inductor and P_R is the power loss in R_{ac}.

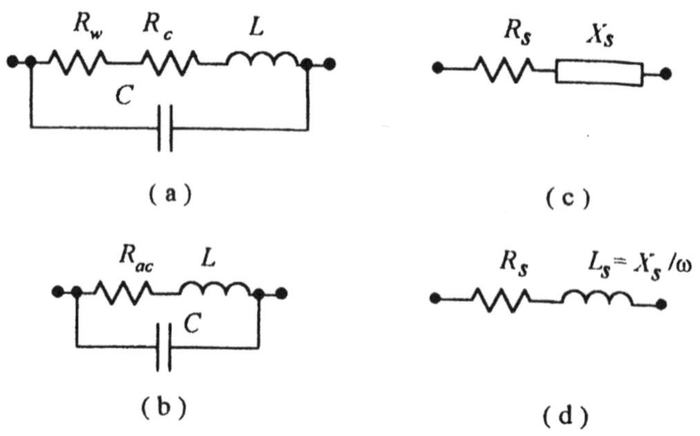

Fig. 2. Equivalent circuits of an inductor.
(a) Lumped parameter equivalent circuit.
(b) Simplified lumped parameter equivalent circuit.
(c) Equivalent series circuit.
(d) Equivalent circuit assumed by many LCR meters.

B. Winding Equivalent Resistance

The winding dc resistance is expressed by

$$R_{dc} = r_L N l_T = \frac{4\rho}{\pi d^2} N l_T \tag{6}$$

where N is the number of turns of the winding, l_T (mm) is the average length of one turn of the winding, r_L (Ω/mm) is the resistance per unit length of the winding wire at a given operating temperature T, d is the conductor diameter in mm, and ρ is the conductor resistivity (ρ = 17.24x10^{-6} Ωmm for a copper conductor at an operating temperature of 20 °C). Wire data sheets usually give two values of r_L at 20 °C and 100 °C; the latter is typically 1.33 times higher than the former.

The ac resistance of a coil winding with an integer number of layers N_ℓ is given by [1]

$$R_w = R_{dc} Re\left\{\frac{A+jA}{tanh(A+jA)} + \frac{2\left(N_\ell^2-1\right)}{3}(A+jA)tanh\left(\frac{A+jA}{2}\right)\right\}$$

$$= R_{dc}A\left\{\frac{e^{2A}-e^{-2A}+2sin2A}{e^{2A}+e^{-2A}-2cos2A} + \frac{2\left(N_\ell^2-1\right)}{3}\,\frac{e^{A}-e^{-A}-2sinA}{e^{A}+e^{-A}+2cosA}\right\} \qquad (7)$$

where [13]

$$A = \left(\frac{\pi}{4}\right)^{3/4} d^{3/2}\left(\frac{\mu_{rw}\mu_o \pi f}{\rho t}\right)^{1/2} = \left(\frac{\pi}{4}\right)^{3/4}\frac{d^{3/2}}{\delta t^{1/2}} \qquad (8)$$

f is the operating frequency in hertz, t is the distance between the centres of two adjacent conductors in mm, μ_{rw} is the relative magnetic permeability of the conductor material (μ_{rw} = 1 for a copper conductor), μ_o = 4π10^{-7} H/m = 4π10^{-10} H/mm, and the skin depth in mm is

$$\delta = \left(\frac{\rho}{\mu_{rw}\mu_o \pi f}\right)^{1/2}. \qquad (9)$$

Fig. 3 shows a plot of the winding ac resistance R_w calculated for a one layer inductor that is made of 95 turns of a solid copper round wire with a diameter d = 0.45 mm and is wound on an Amidon T94 toroidal core made of the iron–powder material #2.

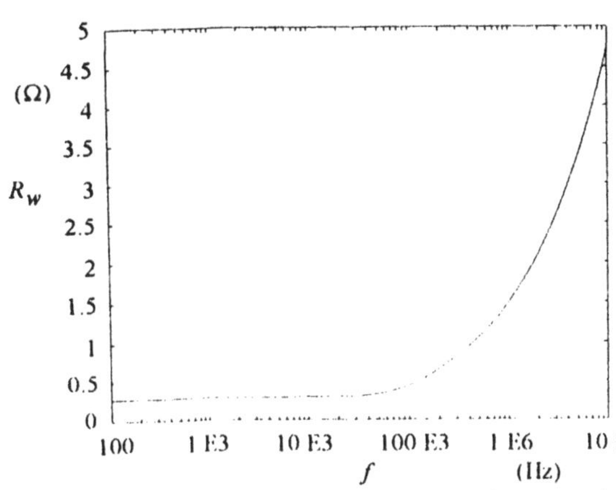

Fig. 3. Winding ac resistance R_w versus frequency f.

The distance between two adjacent turn centers t = 0.51 mm is equal to the external diameter of the round wire including the insulation layer. Since the average length of one turn is l_T = 28 mm and the resistance of a 0.45 mm diameter wire is r_L = 106.2x10^{-6} Ω/mm at 20 °C, a dc resistance R_{dc} = 0.28 Ω is evaluated by using (6). It can be seen from Fig. 3 that the winding resistance is approximately equal to the dc resistance R_{dc} = 0.28 Ω up to 40 kHz and then abruptly increases with frequency. It becomes 5 times higher than R_{dc} at f = 1 MHz and 15 times higher than R_{dc} at f = 10 MHz.

From (7), it follows that a multilayer winding results in a higher ac resistance because the second term of (7) is not zero for $N_l > 1$.

It is shown in the Appendix that the winding ac resistance can be approximated by

$$R_w \approx R_{dc}A\left[1+\frac{2\left(N_\ell^2-1\right)}{3}\right]$$

$$= \left(\frac{4}{\pi}\right)^{1/4} N l_T \left(\frac{\rho\mu_{rw}\mu_o\pi f}{dt}\right)^{1/2}\left[1+\frac{2\left(N_\ell^2-1\right)}{3}\right] \qquad (10)$$

for

$$f \ge f_{bs} = \frac{\rho t}{\mu_{rw}\mu_o \pi d^3}\left(\frac{4}{\pi}\right)^{3/2} \quad \text{if } N_\ell = 1 \qquad (11)$$

and for

$$f \ge f'_{bm} = \frac{16\rho t}{\mu_{rw}\mu_o \pi d^3}\left(\frac{4}{\pi}\right)^{3/2} \quad \text{if } N_\ell > 1. \qquad (12)$$

Moreover, $R_{ac} \approx R_{dc}$ for $f < f_{bs}$ and for

$$f \le f''_{bm} = \frac{\rho t N_\ell}{N_\ell^3 \mu_{rw}\mu_o \pi d^3}\left(\frac{4}{\pi}\right)^{3/2} = \frac{f_{bs}}{N_\ell^2} \quad \text{if } N_\ell = 1. \qquad (13)$$

if $N_\ell > 1$. A single layer and a multilayer windings have the same dc resistance if the same diameter wire is wound and the distance between two adjacent turns is increased proportionally to the number of layers. For a solid wire with d = 0.45 mm, t = 0.51 mm, and N_ℓ = 1, (11) gives f_{bs} = 35 kHz. From (7), the winding ac resistance is $R_w = 1.07R_{dc}$ at f = 35 kHz.

Fig. 4 depicts plots of the exact and approximated ac resistances calculated from (7) and (13), respectively, for the winding with N_ℓ = 2, d = 0.45 mm, and t = 1.02 mm. Observe that both plots become identical for f > 0.8 MHz which is consistent with inequality (12). From (13), we have f''_{bm} = 8.75 kHz. At this frequency, the ac resistance for a two layer winding with d = 0.45 mm and t_{Nl} = 1.02 mm resulting from (7) is $R_w = 1.07\,R_{dc}$.

Comparison of Figs. 3 and 4 shows that the same wire wound on two layers has an ac resistance higher than when it is wound on one layer only. This yields the following design rule: once the number of turns on the winding is known, the wire should have the largest diameter which still allows the winding to fit in one layer.

C. Core Equivalent Resistance

Power losses occurring in magnetic materials are: hysteresis losses, residual losses, and eddy current losses. Power loss due to the hysteresis and residual losses can be specified per unit volume, while eddy current loss not only depends on the core volume but also on its shape. For this reason, eddy current losses should be separately calculated. Since the eddy currents for relatively small toroidal cores have very small amplitude, the contribution of eddy current loss to the core resistance R_C is neglected (this is in agreement with recommendations of Standard IEC 401).

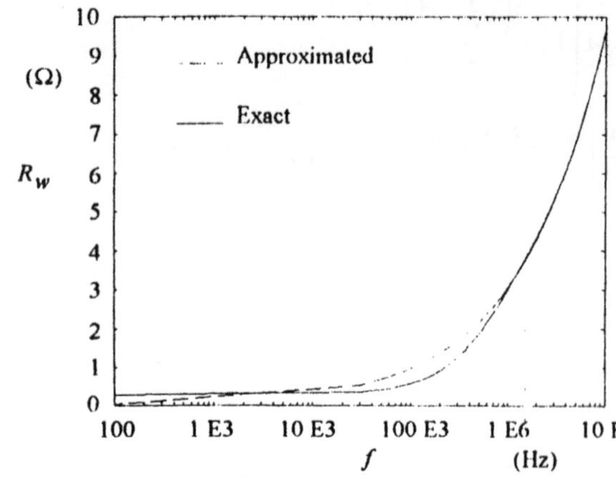

Fig. 4. Winding ac resistance R_w of a two layer winding versus frequency f.

The core loss per unit volume is given by

$$P_{CV} = Kf^x B_m^y \tag{14}$$

were P_{CV} is given in W/m^3 and x, y, and K are coefficients, dependent on the core material. Coefficient K also depends on the inductor operating temperature. The maximum ac flux density the core is operated at is expressed as

$$B_m = \mu_{rc}\frac{\mu_o N I_m}{l_e} = \frac{L I_m}{N A_e}\ (T) \tag{15}$$

where I_m (A) is the amplitude of the current through R_c, μ_{rc} is the core relative magnetic permeability, l_e (m) is the length of the core magnetic path, and A_e is the core cross sectional area (m^2). The power loss in core resistance R_c of Fig. 2(a) is

$$P_{RC} = V_e P_{CV} = \frac{1}{2} R_c I_m^2\ (W) \tag{16}$$

where V_e is the core volume (m^3). Combination of (14) through (16) yields the core resistance

$$R_c = 2Kf^x\left(\mu_{rc}\frac{\mu_o N}{l_e}\right)^y I_m^{y-2} V_e = 2Kf^x\left(\frac{L}{NA_e}\right)^y I_m^{y-2} V_e \tag{17}$$

For a Amidon material #2 T–94 toroidal core, μ_{rc} = 10, K = 8.87 x 10^{-7} at 20 °C, x = 1.14, y = 2.19, l_e = 6 x 10^{-2} m, A_e = 38.5 x 10^{-6} m^2, and V_e = 2.31 x 10^{-6} m^3. Resistance R_c calculated at flux densities of 45 and 250 mT is plotted versus frequency in Fig. 5. Because of the low values of μ_{rc} and K for the Amidon material #2 material, the core resistance R_c is much lower than the winding resistance R_w. As a result, the overall inductor resistance is $R_{ac} \approx R_w$.

III. INDUCTOR DESIGN CRITERIA

According to the Amidon core data sheet, the number of turns to be wound on the core is given by

$$N = 100\sqrt{\frac{L}{A_L}} \tag{18}$$

where the inductance L is in μH and A_L (μH/100 turns) is the inductance factor (A_L = 84 μH/100 turns for a Amidon T94 material #2 core). As a result, 95 turns should be wound on this core to achieve an inductance of 75 μH.

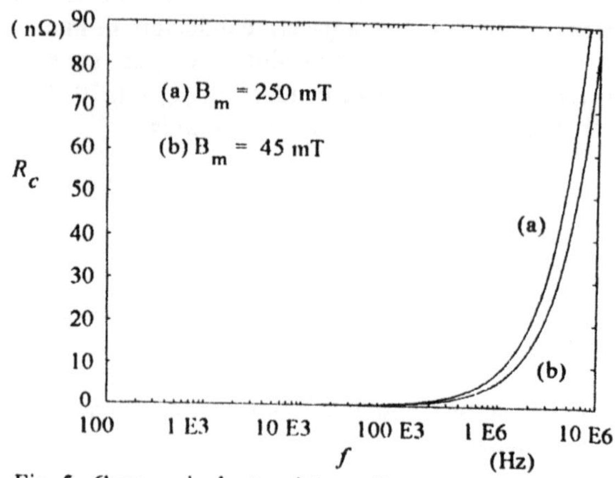

Fig. 5. Core equivalent resistance R_c versus frequency f.

The number of turns that can be wound on one layer is calculated as follows:

$$N_{TL} = \frac{\pi\, d_i}{t} \tag{19}$$

where d_i (mm) is the internal diameter of the ring core. To reduce the parasitic capacitance, increase the first self–resonant frequency, and achieve the lowest ac resistance, only one layer should be wound on the core. Since the internal diameter of a T94 toroidal core is d_i = 14.3 mm, (19) yields a maximum external diameter of the solid round wire t = 0.47 mm. The available wire with an external diameter close to t = 0.47 mm is the AWG#25 with t = 0.51 mm. An AWG#26 solid round wire (with t = 0.46) could be used but it results in a dc resistance 26% higher than the

AVG#25 round solid wire.

Typical design current densities J range from 1 to 5 A/mm^2. Once the rms value of the current I_{rms} and the diameter d of a solid round wire are known, the operating current density is calculated by using $J = 4\ I_{rms}\ /\ \pi\ d^2$.

If a solid round wire with an external diameter $t = 0.51$ mm is used, the number of turns wound on one layer is $N_{TL} = 88$. Since $N = 95$, only 7 turns must be wound on the second layer and the number of layers can be assumed to be $N_\ell = 1$. The dc and winding ac resistance can now be calculated by using (6) and (7). From (17), the core resistance is also calculated.

IV. EXPERIMENTAL RESULTS

The measurements were performed with an HP 4192/A LCR meter for an inductor with a nominal inductance $L = 75$ μH, designed according to the procedure given in the previous section and assembled by winding one layer of 95 turns of a solid round wire with $d = 0.45$ mm on an Amidon toroidal core of iron–powder core material #2. Fig. 7 shows a plot of measured X_s versus frequency. From Fig. 6, it was found that $f_r = 6.2$ MHz and $L_s = L = X_s/\omega = 75$ μH. Hence, from (2), $C = 8.78$ pF. The dc resistance calculated from (6) is $R_{dc} = 0.28\ \Omega$. This was also the measured inductor resistance at $f = 5$ Hz, that is, the lowest operating frequency of the HP 4192/A LCR meter.

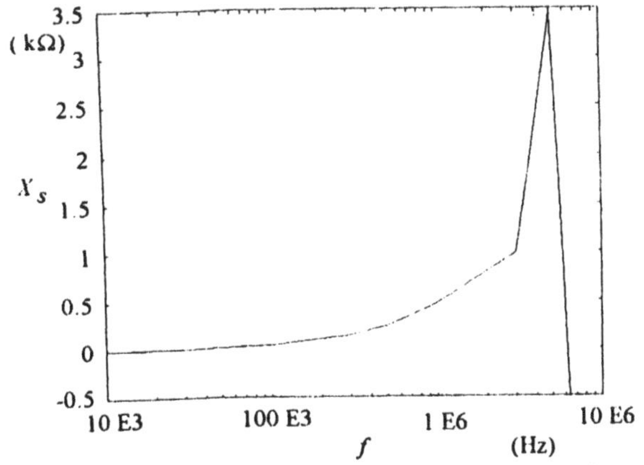

Fig. 6. Measured reactance X_s versus frequency f.

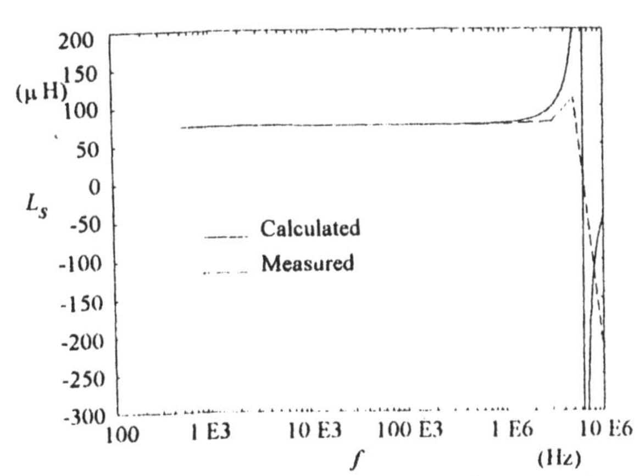

Fig. 7. Apparent inductance L_s versus frequency f.

The measured and theoretical values of the equivalent series inductance L_s are compared in Fig. 8. Both calculated and measured L_s were zero at $f = 6.2$ MHz.

The experimental and theoretical plots of R_s are compared in Fig. 8(a). Note that R_s is approximately ten times higher than R_{dc} at $f = 2$ MHz and R_s increases at higher frequencies. Reactance X_s is depicted in Fig. 8(b). The calculated and measured values of both R_s and X_s are practically the same for $f < f_r$. Note that around the self–resonant frequency accurate measurements cannot be achieved with an LCR meter because the meter can only be operated at factory specified frequencies.

Fig. 9 shows that the impedance magnitude and phase of the equivalent circuits of Fig. 2 closely agree with those of the real inductor. Phase ϕ increased with frequency starting from 0 °C because both real inductor and equivalent circuit behave as a resistor when driven by a dc current and/or voltage.

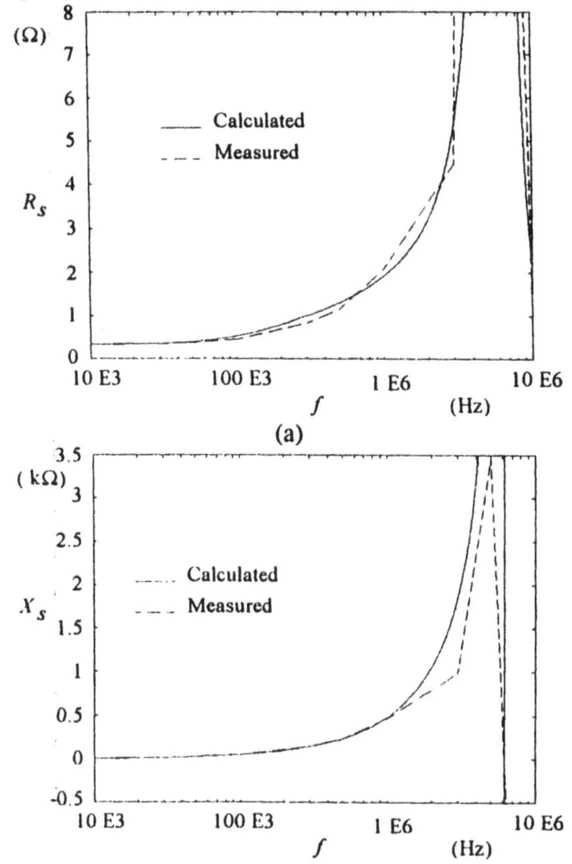

Fig. 8. Inductor impedance Z_s versus frequency f.
(a) Series resistance R_s versus frequency f.
(b) Series reactance X_s versus frequency f.

The plots of the calculated and measured quality factor of the inductor are shown in Fig. 10. The maximum measured value of Q_s was 228 and occurred at $f = 1$ MHz. The theoretical values were in good agreement with those experimentally measured up to about 60 % of the self–resonant frequency f_r.

The plot of Q_o given by (5) is compared with the plot of measured Q_s in Fig. 11. The values of Q_o are nearly equal to the measured ones for frequencies up to f = 500 kHz.

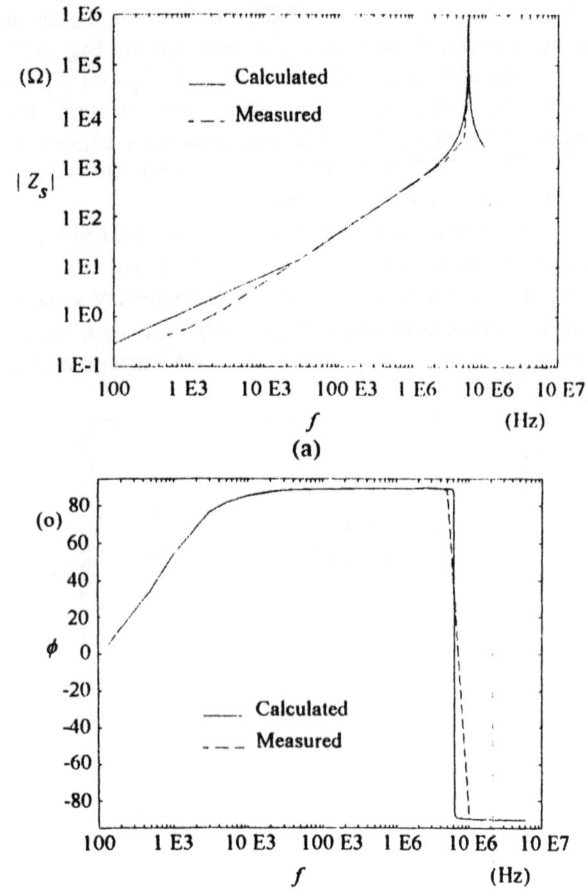

Fig. 9. Inductor impedance Z_s versus frequency f.
(a) Magnitude | Z_s | versus frequency f.
(b) Phase f versus frequency f.

The comparison of the plots of R_s and R_{ac} depicted in Fig. 12 demonstrates that the effect of parasitic capacitance C is negligible up to 1 MHz, that is, up to 20 % of the self resonant frequency. Therefore, for frequencies lower than 1 MHz, the equivalent series resistance of the inductor can be predicted by using the presented equivalent circuit and using the core and wire data sheets. Since the expressions for R_{ac} and Q_o do not require the self–resonant frequency and parasitic capacitance to be known, they can be predicted without the need of any measurement. Therefore, they can be used when designing the inductor. Inductor temperature rise is one important parameter which should be controlled during the inductor design. From Amidon core data sheet, the temperature rise is given by

$$\Delta T_{20^o} = \left(\frac{P_l}{S_a}\right)^{.833} = \left\{\frac{0.5R_{ac}I_m x10^3}{\pi\left[hd_o + hd_i + 0.5\left(d_o^2 - d_i^2\right)\right]}\right\}^{.833} \quad (20)$$

where P_l is the overall power loss in the inductor in mW and S_a is the surface area of the toroidal core in cm^2, h is the core height in cm, d_i is the inner diameter of the core in cm and d_o is outer diameter of the core in cm. From the wire data sheets, it is found that the copper resistance per unit length at T = 100 °C is 33% higher than at T = 20 °C.

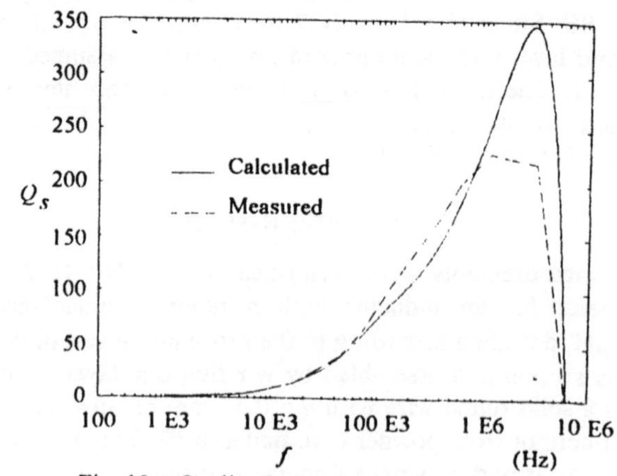

Fig. 10. Quality factor Q_s versus frequency f.

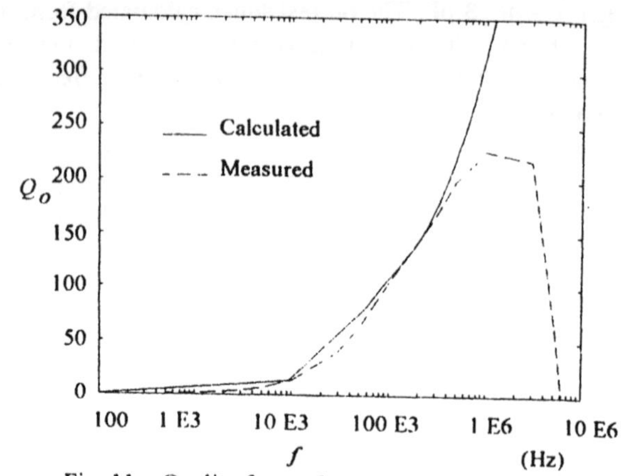

Fig. 11. Quality factor Q_o versus frequency f.

Assuming that the resistance per unit length of the copper is proportional to the operating temperature, the ac resistance R_{ac} as a function of the operating temperature is given by

$$\begin{aligned} R_{ac}(T) &= (1 + 4.125x10^{-3}\Delta T_{20^o})R_{ac}(20^o) \\ &= (912.5x10^{-3} + 4.125x10^{-3}T)R_{ac}(20^o) \end{aligned} \quad (21)$$

where ΔT_{20^o} is the temperature rise with respect to a 20 °C ambient temperature. Substitution of (21) into (20) allows the operating temperature of the inductor to be calculated. For a T94 toroidal core, the surface area is S_a = 15.31 cm^2. If the core is operated at f = 330 kHz, then the ac resistance at 20 °C is R_{ac} = 0.86 Ω. Assuming a sinusoidal current through the winding with an amplitude of I_m = 1.5 A a power loss P_l = 0.95 W is calculated at T = 20 °C, the temperature rise is ΔT_{20^o} = 35 °C, the operating temperature

°C, the operating temperature of the inductor is $T = 55$ °C, the ac resistance at this temperature is $R_{ac} = 0.97\ \Omega$, and the effective power loss is $P_I = 1$ W at $T = 55$ °C.

A solid round wire with a larger diameter can be used to achieve a lower dc resistance. If 95 turns of a solid round wire with $d = 0.9$ mm, $t = 0.99$ mm, and $r_L = 262.6x10^{-6}$ Ω/mm are wound on a T94 toroidal core, a dc resistance $R_{dc} = 0.069\ \Omega$ is achieved. As expected, this resistance is four times lower than the dc resistance of the winding assembled with a 0.45 mm diameter round wire. As shown in Fig. 13, if the coil has one layer only, also the ac resistance R_{ac} is lower over the entire frequency range. However, from (19), the number of turns that can be wound on one layer is $N_{TL} = 45$ and, therefore, two layers are needed to achieve an inductance $L = 75$ μH. Fig. 13 shows that the ac resistance for the winding assembled with two layers of $d = 0.9$ mm is higher than the ac resistance of the winding assembled with a single layer of $d = 0.45$ mm round wire for $f > 30$ kHz.

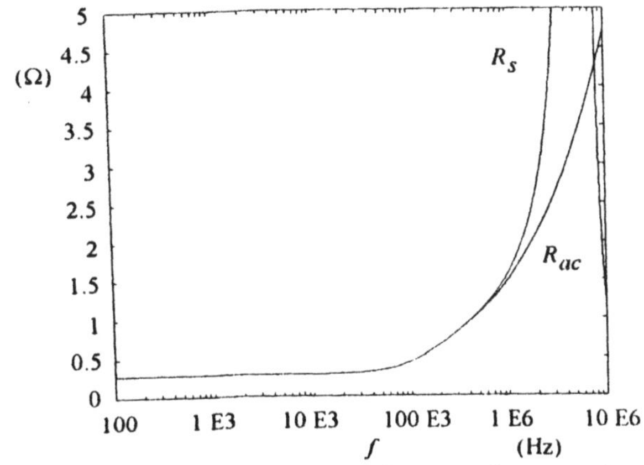

Fig. 12. Resistances R_{ac} and R_s versus frequency f.

V. Conclusions

An equivalent circuit of an iron–powder core inductor including parasitic capacitances and resistances has been examined. The winding ac resistance due to the skin and proximity effects have been considered. The expression for the impedance of the inductor model has been derived along with the procedure for identifying the equivalent circuit components. Experimental and theoretical plots of parameters such as the magnitude and phase of the impedance, the equivalent series resistance, the equivalent series reactance and inductance, and the quality factor have been plotted versus inductor operating frequency and compared. The theoretical values match all of the measured parameters for an operating frequency below the first self–resonant frequency of the inductor. Above the self–resonant frequency, both the equivalent circuit and tested inductor behave as a capacitor. It has also been demonstrated that the quality factor of an inductor highly depends on the operating frequency. The inductor can be modeled by a series combination of an inductance and a resistance for operating frequencies up to 20 % of f_r, while beyond this frequency capacitor C should be taken into account. Core loss is much lower than copper loss also at high flux densities. Moreover, the core resistance is not significantly affected by flux density variations. For these reasons, the overall inductor resistance is primarily determined by the winding resistance.

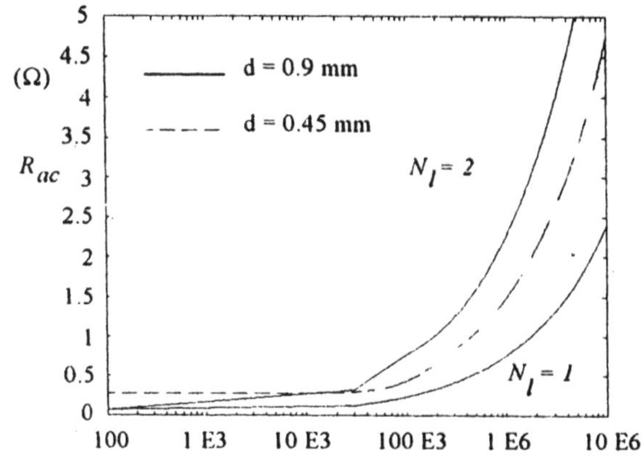

Fig. 13. Resistances R_{ac} of a one layer winding of solid round wire with $d = 0.45$ mm and one layer and two layer winding of solid round wire with $d = 0.9$ mm versus frequency f.

The inductor equivalent series resistance and associated power losses can be calculated for an operating temperature higher than 20 °C by using appropriate values of ρ, r_L, and K. Coefficient K should be given as a function of the inductor operating temperature by core manufacturers to allow the designer for an accurate prediction of inductor power loss.

The iron–powder inductor ac resistance highly increases with the operating frequency and also highly depends on the wound number of layers because of the proximity and skin effects on the winding. As a consequence, the dc resistance is reduced if the solid wire diameter is increased. However, this results in a higher ac resistance when the number of the winding layers is increased.

It has been shown that the lowest ac resistance is achieved when only one layer is wound on the core, if a solid round wire is used. Further studies are recommended for strand and Litz wires.

Approximated expressions of single layer and multilayer equivalent series winding resistances have been introduced. This allows for an immediate evaluation of the high frequency resistance of the winding during the inductor design. The frequency range where the winding resistance does not significatively differs from the dc resistance has been also determined.

The examined equivalent circuit can also be used for inductor operated at non sinusoidal current waveforms. In this case, the equations given in this paper have to be evaluated for a sufficiently large number of harmonics of the inductor current.

The inductor model along with the procedure for extracting the values of the parasitic components can be used by designers to determine the frequency range an iron–

powder core inductor can properly work and prevent its operation close to the self-resonant frequency and/or with too high values of the parasitic resistance. Moreover, the inductor model can be used in computer simulations of electronic circuits to predict their high-frequency operation.

APPENDIX

The winding ac resistance is

$$R_w = R_{dc}A\left[F_{r1} + \frac{2\left(N_\ell^2-1\right)}{3}F_{r2}\right] = R_{dc}F_r \qquad (22)$$

where

$$F_{r1} = \frac{e^{2A} - e^{-2A} + 2sin2A}{e^{2A} + e^{-2A} - 2cos2A} \qquad (23)$$

$$F_{r2} = \frac{e^{A} - e^{-A} - 2sin\,A}{e^{A} + e^{-A} + 2cos\,A} \qquad (24)$$

and

$$F_r = A\left[F_{r1} + \frac{2\left(N_\ell^2-1\right)}{3}F_{r2}\right]. \qquad (25)$$

Fig. 14 clearly shows that F_{r1} and F_{r2} are nearly constant and approach unity for $A \geq 1$ and $A \geq 4$, respectively. If both $F_{r1} = 1$ and $F_{r2} = 1$, (25) gives (10).

Since in one layer winding only F_{r1} contributes to the winding ac resistance rise with frequency, approximation (10) is applicable for $A \geq 1$, that is, for frequencies higher than f_{bs} given by (11). For a multilayer winding, the F_{r2} represents the major contribution to F_r. Therefore, approximation (10) is accurate for $A \geq 4$, which yields (12).

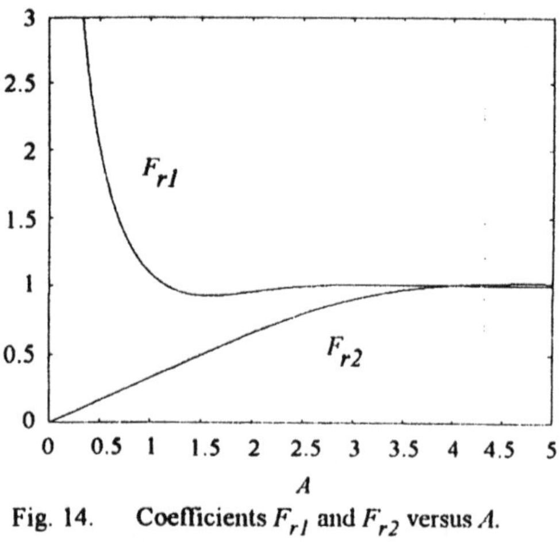

Fig. 14. Coefficients F_{r1} and F_{r2} versus A.

Fig 15 shows that the ratio of the ac and the dc resistance F_r of a one layer winding slightly increases starting from 1 for $A < 1$. Therefore, $R_{ac} \approx R_{dc}$ for $f < f_{bs}$. For $A \geq 1$, F_r linearly increases with A. Using (10), a one layer winding ac resistance ($N_\ell = 1$) is approximated by

$$R_{ac} \approx \left(\frac{4}{\pi}\right)^{1/4} Nl_T\left(\frac{\rho\mu_{rw}\mu\pi f}{dt}\right)^{1/2} \qquad \text{for } f < f_{bs}. \qquad (26)$$

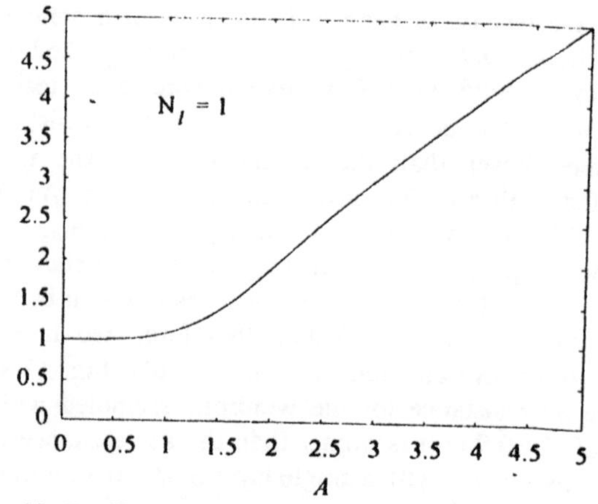

Fig. 15. Coefficient F_r of a one layer winding versus A.

REFERENCES

[1] P. L. Dowell, *"Effects of eddy current in transformer windings"*, Proc. IEE, Vol. 113, No. 8, August 1966, pp. 1287–1394.

[2] C. Wm. T. Mc. Lymann, *"Transformer and Inductor Design Handbook"*, New York: Marcel Decker Inc., 1978.

[3] Colonel Wm. T. Mc. Lymann, *"Magnetic Core Selection for Transformers and Inductors"*, New York: Marcel Decker d Inc., 1982.

[4] W. J. Moore and P. N. Miljanic, *"The Current Comparator"*, United Kingdom: Peter Peregrinus Ltd., 1988.

[5] O. Kiltie, *"Design Shortcuts and Procedures for Electronic Power Transformers and Inductors"*, Cleveland,: Harris Publishing Company, 1975.

[6] W. M. Flanagan, *"Handbook of Transformer Applications"*, New York: Mc. Graw Hill, 1986.

[7] N. R. Grossner, *"Transformers for Electronic Circuits"*, Second Edition, New York: Mc. Graw Hill, 1983.

[8] R. Lee, L. Wilson, and C. E. Carter, *"Electronic Transformers and Circuits"*, Third Edition, New York: John Wiley and Sons, 1988.

[9] M. K. Jutty, V. Swaminathan, and M. K. Kazimierczuk, *"Frequency Characteristics of Ferrite Core Inductors"*, Proc. of *the Electrical Manufacturing & Coil Winding Symposium '93*, Chicago, IL, 1993, pp 369–372.

[10] J. R. Fluke, *"Controlling Conducted Emission by Design"*, New York: Van Nostrand Rehinold, 1991.

[11] K. H. Billings , *"Switchmode Power Supply Handbook"*, New York: Mc. Graw Hill, 1989.

[12] Hewlett-Packard Co., *LCR Meter Model 4192/A Operating Manual.*

[13] J. Jongsma, *"High Frequency Ferrite Power Transformer and Choke Design" – Part 3 "Transformer Winding Design"*, Philips Technical Publication 207, Ordering code: 939805330011.

High-Frequency Small-Signal Model of Ferrite Core Inductors

Marian K. Kazimierczuk, *Senior Member, IEEE*, Giuseppe Sancineto, Gabriele Grandi, Ugo Reggiani, *Member, IEEE*, and Antonio Massarini

***Abstract*— A circuit model of ferrite core inductors is presented. The behavior of the model parameters versus frequency is considered. The total power loss in inductors consisting of the winding resistance loss and the core loss, is modeled by a frequency-dependent equivalent series resistance. The total inductance given by the sum of the main inductance and the leakage inductance is obtained as a function of frequency. In order to study the core equivalent resistance and main inductance versus frequency, the magnetic field distribution in the core is derived from Maxwell's equations for a long solenoid. The complex permeability and permittivity of the ferrite core are introduced in the electromagnetic field equations. Experimental results are also given.**

***Index Terms*— Circuit modeling, coils, eddy currents, ferrite inductors, frequency response, magnetic cores, magnetic losses, skin effect.**

I. Introduction

One method to reduce the overall size and weight of power converters is to increase the switching frequency above several hundred kilohertz. This allows the volume of inductors, capacitors, and transformers to be much smaller. To achieve this goal, it is important to obtain high efficiency at high operating frequencies. Moreover, parasitic components in the converter circuits must be accurately predicted to prevent unwanted behavior. The purpose of this paper is to explore the behavior of ferrite core inductors and present a circuit model valid at high frequencies in which parasitic capacitances, winding and core losses, and the frequency-dependent inductance are taken into account. Expressions are derived for the core equivalent resistance and main inductance as functions of frequency. The Dowell's equations [1] are used for the winding resistance and leakage inductance. Experimental results are also given.

II. Equivalent Circuit of Ferrite Core Inductors

Fig. 1(a) shows a lumped parameter model of a ferrite core inductor [2], [3], where R_w is the winding resistance, R_c is the core equivalent series resistance, L_{ac} is the total inductance, and C is the total parasitic capacitance. In general, R_w, R_c, and L_{ac} are dependent on frequency f. Fig. 1(b) shows the total ac resistance $R_{ac} = R_w + R_c$. The impedance of the inductor model is $\hat{Z}_s = R_s + jX_s$. Fig. 1(c) depicts a series equivalent circuit of the inductor, where the equivalent series resistance (ESR) is [2], [3]

$$R_s = \frac{R_{ac}}{(1-\omega^2 L_{ac}C)^2 + (\omega C R_{ac})^2} \tag{1}$$

and the equivalent series reactance is

$$X_s = \frac{\omega L_{ac}\left(1 - \omega^2 L_{ac}C - \dfrac{CR_{ac}^2}{L_{ac}}\right)}{(1-\omega^2 L_{ac}C)^2 + (\omega C R_{ac})^2} \tag{2}$$

where $\omega = 2\pi f$. Hence, the equivalent series inductance (ESL) is given by

$$L_s = \frac{X_s}{2\pi f} \tag{3}$$

which is also significant because most network analyzers and impedance meters measure R_s and L_s. The self-resonant frequency f_r of an inductor is defined as the frequency at which its reactance X_s becomes zero. Hence, from (2), the inductor total parasitic capacitance can be calculated as

$$C = \frac{1}{(2\pi f_r)^2 L_{ac}(f_r) + \dfrac{R_{ac}^2(f_r)}{L_{ac}(f_r)}} \tag{4}$$

where $L_{ac}(f_r)$ and $R_{ac}(f_r)$ are the total inductance and total resistance at the resonant frequency f_r, respectively. The capacitance C is assumed to be independent of frequency.

III. Winding Resistance

The winding resistance of an inductor increases with frequency because of the skin and proximity effects due to the eddy current in the conductor generated by the time-varying electromagnetic field. The final result of these two combined effects is a reduction in the effective cross-sectional area of the wire available for the current flow. Therefore, the ac resistance at high frequencies becomes greater than the dc resistance. The winding ac resistance of an inductor with N_l layers is [1]

$$R_w = R_{wdc}A\left[\frac{e^{2A} - e^{-2A} + 2\sin(2A)}{e^{2A} + e^{-2A} - 2\cos(2A)} + 2\frac{N_l^2 - 1}{3} \cdot \frac{e^{A} - e^{-A} - 2\sin A}{e^{A} + e^{-A} + 2\cos A}\right] \tag{5}$$

Manuscript received December 7, 1998; revised May 19, 1999.
M. K. Kazimierczuk and G. Sancineto are with the Department of Electrical Engineering, Wright State University, Dayton, OH 45435 USA (e-mail: mkazim@cs.wright.edu).
G. Grandi and U. Reggiani are with the Department of Electrical Engineering, University of Bologna, 40136 Bologna, Italy.
A. Massarini is with the Department of Engineering, Sciences University of Modena and Reggio Emilia, 41100 Modena, Italy.
Publisher Item Identifier S 0018-9464(99)08096-6.

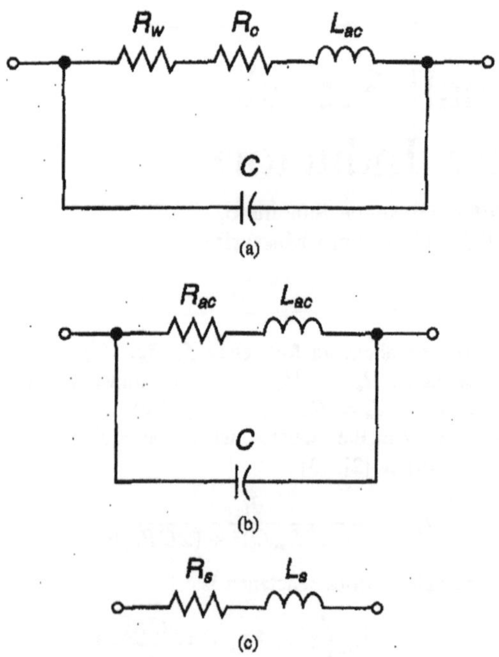

Fig. 1. Equivalent circuits of the inductor: (a) lumped parameter equivalent circuit, (b) equivalent circuit with $R_{ac} = R_w + R_c$, and (c) equivalent series circuit.

where R_{wdc} is the dc winding resistance at a given temperature T, and A for a round wire is

$$A = \left(\frac{\pi}{4}\right)^{3/4} \frac{d^{3/2}}{\delta t^{1/2}} \tag{6}$$

in which d is the conductor diameter, t is the distance between the centers of two adjacent conductors, and δ is the skin depth of the wire given by

$$\delta = \sqrt{\frac{\rho_w}{\pi \mu_0 \mu_{rw} f}}. \tag{7}$$

In (7), ρ_w is the wire resistivity ($\rho_w = 17.24 \times 10^{-9}$ Ωm for a copper wire at temperature of 20 °C), μ_{rw} is the relative magnetic permeability of the conductor material ($\mu_{rw} = 1$ for a copper conductor), and $\mu_0 = 4\pi \times 10^{-7}$ H/m.

IV. Main Inductance and Core Equivalent Resistance

Let us consider a coil uniformly wound around a ferrite core of length l and circular cross section of radius r_o. Assuming that the length of the cylindrical core is very long compared with its radius ($l \gg r_o$), edge effects can be neglected. Thus, the magnetic field can be considered uniform along the z axis of a cylindrical coordinate reference system, as shown in Fig. 2. Because of the cylindrical symmetry, the electric and magnetic fields are independent of φ and z coordinates. In particular, the magnetic field vector is given by

$$\boldsymbol{H}(r, \varphi, z, t) = H_z(r, t)\boldsymbol{a}_z. \tag{8}$$

For the sinusoidal steady state, we can introduce the magnetic field complex vector

$$\hat{\boldsymbol{H}}(r) = \hat{H}(r)\boldsymbol{a}_z \tag{9}$$

where the magnetic field phasor $\hat{H}(r)$ is given by

$$\hat{H}(r) = \hat{H}_z(r) = H_m(r)e^{j\alpha(r)}. \tag{10}$$

The present model takes into account the core magnetic hysteresis loss by a frequency-dependent complex magnetic permeability [4]

$$\hat{\mu}_c = \mu_o \hat{\mu}_{rc} = \mu_o(\mu'_{rc} - j\mu''_{rc}). \tag{11}$$

Therefore, the magnetic induction is described by a complex vector

$$\hat{\boldsymbol{B}} = \hat{\mu}_c \hat{\boldsymbol{H}} = \mu_o(\mu'_{rc} - j\mu''_{rc})\hat{\boldsymbol{H}}. \tag{12}$$

To take into account the core dielectric loss that occurs in grain boundary regions due to the electric hysteresis [4], [5], the electric permittivity is also expressed as a complex quantity dependent on frequency. Hence

$$\hat{\varepsilon}_c = \varepsilon_o \hat{\varepsilon}_{rc} = \varepsilon_o(\varepsilon'_{rc} - j\varepsilon''_{rc}). \tag{13}$$

The relationship between the electric displacement field complex vector $\hat{\boldsymbol{D}}$ and the electric field complex vector $\hat{\boldsymbol{E}}$ is given by

$$\hat{\boldsymbol{D}} = \hat{\varepsilon}_c \hat{\boldsymbol{E}} = \varepsilon_o(\varepsilon'_{rc} - j\varepsilon''_{rc})\hat{\boldsymbol{E}}. \tag{14}$$

In order to derive the main inductance and core equivalent resistance from the voltage induced across the winding by the time-varying magnetic flux, the magnetic field distribution in the core must be derived from Maxwell's equations. The Maxwell's equations in terms of complex vectors can be expressed as

$$\begin{aligned} \nabla \times \hat{\boldsymbol{H}} &= \hat{\boldsymbol{J}} + j\omega\hat{\boldsymbol{D}} \\ &= \gamma_c \hat{\boldsymbol{E}} + j\omega\hat{\varepsilon}_c\hat{\boldsymbol{E}} \\ &= (\gamma_c + j\omega\hat{\varepsilon}_c)\hat{\boldsymbol{E}} \\ &= \hat{\gamma}_c \hat{\boldsymbol{E}} \end{aligned} \tag{15}$$

$$\nabla \times \hat{\boldsymbol{E}} = -j\omega\hat{\boldsymbol{B}} = -j\omega\hat{\mu}_c\hat{\boldsymbol{H}} \tag{16}$$

where $\hat{\boldsymbol{J}}$ is the complex vector of the conduction current density, γ_c is the electric conductivity of the core, and $\hat{\gamma}_c = \gamma_c + j\omega\hat{\varepsilon}_c$ is its complex conductivity. In (15), $\gamma_c \hat{\boldsymbol{E}}$ is responsible of the eddy-current loss due to the nonzero conductivity γ_c of the core, $\omega\varepsilon_o\varepsilon''_{rc}\hat{\boldsymbol{E}}$ accounts for the dielectric loss, and the term $\omega\mu_o\mu''_{rc}\hat{\boldsymbol{H}}$ in (16) is related to the magnetic hysteresis loss. From (15) and (16), we can write

$$\nabla \times \left(\nabla \times \hat{\boldsymbol{H}}\right) = \hat{\gamma}_c \nabla \times \hat{\boldsymbol{E}} = -j\omega\hat{\mu}_c\hat{\gamma}_c\hat{\boldsymbol{H}} = -j\hat{k}^2\hat{\boldsymbol{H}} \tag{17}$$

where

$$\hat{k}^2 = \omega\hat{\mu}_c\hat{\gamma}_c. \tag{18}$$

From (9) and (17), it follows that

$$\frac{d^2\hat{H}}{dr^2} + \frac{1}{r}\frac{d\hat{H}}{dr} - j\hat{k}^2\hat{H} = 0. \tag{19}$$

This is the well-known Bessel's differential equation with the unique solution for $r \to 0$

$$\hat{H}(r) = GJ_0(j^{3/2}\hat{k}r) \tag{20}$$

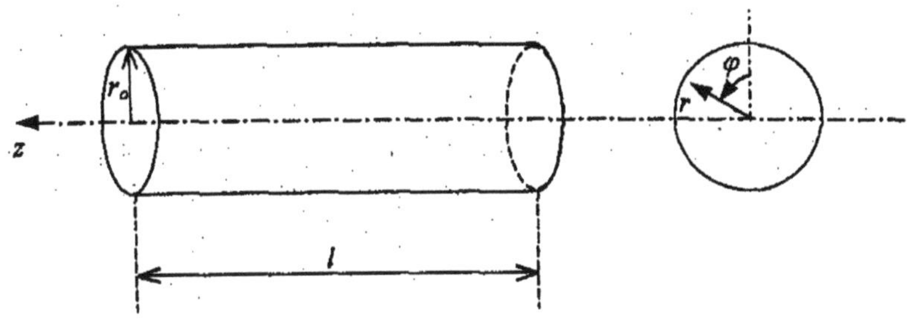

Fig. 2. Core geometrical view.

where J_0 is the Bessel's function of the first kind of zero order and G is an integration constant. Applying the boundary condition to the magnetic field phasor at the surface of the cylinder

$$\hat{H}_o = \hat{H}(r_o) = GJ_0\left(j^{3/2}\hat{k}r_o\right) \tag{21}$$

we can obtain

$$\hat{H}(r) = \hat{H}_o \frac{J_0\left(j^{3/2}\hat{k}r\right)}{J_0(j^{3/2}\hat{k}r_o)} \tag{22}$$

with

$$\hat{H}_o = \frac{N\hat{I}}{l} \tag{23}$$

where N is the total number of turns wound around the cylinder, and $\hat{I}$ is the phasor of the current in the coil. The magnetic flux phasor through a cross section is calculated as

$$\begin{aligned}\hat{\Phi} &= \int_S \hat{\mu}_c \hat{H}\, dS \\ &= 2\pi \int_0^{r_o} \hat{\mu}_c \hat{H}(r) r\, dr \\ &= \frac{2\pi \hat{\mu}_c \hat{H}_o}{J_0\left(j^{3/2}\hat{k}r_o\right)} \int_0^{r_o} rJ_0\left(j^{3/2}\hat{k}r\right) dr. \end{aligned} \tag{24}$$

Through the manipulations shown in the Appendix, (24) becomes

$$\hat{\Phi} = L_{\rm dc}^m \hat{I} \frac{2}{N} \frac{1}{\mu_o \mu'_{rc}} \frac{\hat{\mu}_c}{j^{3/2}\sqrt{2}\frac{r_o}{\hat{\delta}}} \frac{J_1\left(j^{3/2}\sqrt{2}\frac{r_o}{\hat{\delta}}\right)}{J_0\left(j^{3/2}\sqrt{2}\frac{r_o}{\hat{\delta}}\right)} \tag{25}$$

where $L_{\rm dc}^m$ is the low-frequency main inductance, J_1 is the Bessel's function of the first kind of order one, and $\hat{\delta}$ is given by (40). Neglecting the leakage flux, the magnetic flux linking the coil is

$$\hat{\Phi}_c = N\hat{\Phi}. \tag{26}$$

Hence, the voltage induced across the winding by the magnetic flux is

$$\hat{V} = j\omega\hat{\Phi}_c \tag{27}$$

and the impedance of the inductor is given by

$$\hat{Z} = \frac{\hat{V}}{\hat{I}} = j2\omega L_{\rm dc}^m \frac{1}{\mu_o \mu'_{rc}} \frac{\hat{\mu}_c}{j^{3/2}\sqrt{2}\frac{r_o}{\hat{\delta}}} \frac{J_1\left(j^{3/2}\sqrt{2}\frac{r_o}{\hat{\delta}}\right)}{J_0\left(j^{3/2}\sqrt{2}\frac{r_o}{\hat{\delta}}\right)}. \tag{28}$$

The real and imaginary parts of $\hat{Z}$ represent the core equivalent series resistance R_c (which accounts for the eddy current loss and both magnetic and electric hysteresis losses in the core) and the coil main reactance $\omega L_{\rm ac}^m$, respectively. These parameters are

$$\begin{aligned} R_c &= {\rm Re}\left(\hat{Z}\right) \\ &= -\frac{4\pi}{\mu'_{rc}} f L_{\rm dc}^m\, {\rm Im}\left[\frac{\hat{\mu}_{rc}}{j^{3/2}\sqrt{2}\frac{r_o}{\hat{\delta}}} \frac{J_1\left(j^{3/2}\sqrt{2}\frac{r_o}{\hat{\delta}}\right)}{J_0\left(j^{3/2}\sqrt{2}\frac{r_o}{\hat{\delta}}\right)}\right] \end{aligned} \tag{29}$$

and

$$\begin{aligned} L_{\rm ac}^m &= \frac{{\rm Im}\left(\hat{Z}\right)}{2\pi f} \\ &= \frac{2}{\mu'_{rc}} L_{\rm dc}^m\, {\rm Re}\left[\frac{\hat{\mu}_{rc}}{j^{3/2}\sqrt{2}\frac{r_o}{\hat{\delta}}} \frac{J_1\left(j^{3/2}\sqrt{2}\frac{r_o}{\hat{\delta}}\right)}{J_0\left(j^{3/2}\sqrt{2}\frac{r_o}{\hat{\delta}}\right)}\right]. \end{aligned} \tag{30}$$

V. Leakage and Total Inductances

An expression for the leakage inductance derived by Dowell [1] and simplified here is described by

$$L_{\rm ac}^l = R_{\rm wdc}\frac{A}{2\pi f}\left[\frac{e^{2A} - e^{-2A} - 2\sin(2A)}{e^{2A} + e^{-2A} - 2\cos(2A)} + 2\frac{N_l^2 - 1}{3} \cdot \frac{e^A - e^{-A} + 2\sin A}{e^A + e^{-A} + 2\cos A}\right]. \tag{31}$$

Thus, the total inductance of the coil is given by

$$L_{\rm ac} = L_{\rm ac}^m + L_{\rm ac}^l. \tag{32}$$

VI. Experimental Results

To verify the validity of the circuit model, an inductor was wound on a Siemens EC52 ferrite core (part no. B66341) with a circular cross section of the central leg and no air gap. This core uses N27 magnetic material. The diameter of the central

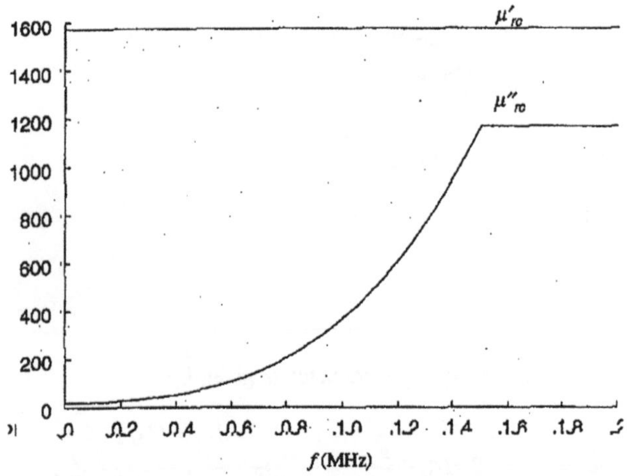

Fig. 3. Real part μ'_{rc} and imaginary part μ''_{rc} of the relative permeability of the Siemens N27 core material versus frequency.

leg was $2r_o = 13.75 \times 10^{-3}$ m. A solid copper wire with $d = 0.66 \times 10^{-3}$ m and $t \approx d$ (AWG#22) was wound on the central leg to obtain a one layer winding inductor ($N_l = 1$). The required number of turns of the winding inductor was $N_t = 39$. From data sheets, the inductance factor was $A_L = 3.4$ μH, resulting in the calculated low-frequency main inductance $L^m_{dc} = (N_l N_t)^2 A_L = 5.15$ mH. The dc resistance calculated for the winding was $R_{wdc} = 8\rho_w r_o N_t / d^2 = 84.89$ mΩ. The core resistivity was $\rho_c = 3$ Ωm. The measurements were performed using an HP 4194A impedance analyzer equipped with an HP 16047D test fixture [6] to achieve a higher accuracy by minimizing residual parameters such as lead inductances and contact resistances.

Using the data sheets in the Siemens catalogue, the real and imaginary parts of the complex permeability $\mu'_{rc}(f)$ and $\mu''_{rc}(f)$ for low values of magnetic induction B were approximated in the frequency range from 100 Hz to 2 MHz as shown in Fig. 3. The core ferromagnetic resonant frequency was about 2 MHz. Most inductors (except chokes) are operated well below the ferromagnetic resonance. For the tested core, the electrical hysteresis was neglected. Thus, only the real part of the relative permittivity was taken into account, so that $\hat{\varepsilon}_c = \varepsilon_o \varepsilon'_{rc}$. The core relative electric permittivity ε'_{rc} was calculated from measurements of a capacitance whose dielectric was a core material sample, using the same HP 4194A impedance analyzer. Fig. 4 shows the measured real part of the core relative permittivity by a dashed line, which was approximated by a solid line and used in computations. It can be seen that the real part of the permittivity decreases with frequency. In Fig. 5, the winding resistance R_w is plotted using the Dowell's equation (5). The winding resistance increased rapidly with frequency for $f > 20$ kHz. The equivalent series core resistance R_c given by (29) is plotted versus frequency in Fig. 6. As expected, the core resistance increased with frequency and reached very high values of the order of 40 kΩ. The main inductance L^m_{ac} versus frequency given by (30) is plotted in Fig. 7. The main inductance was independent of frequency up to 1 MHz and then rapidly decreased. At

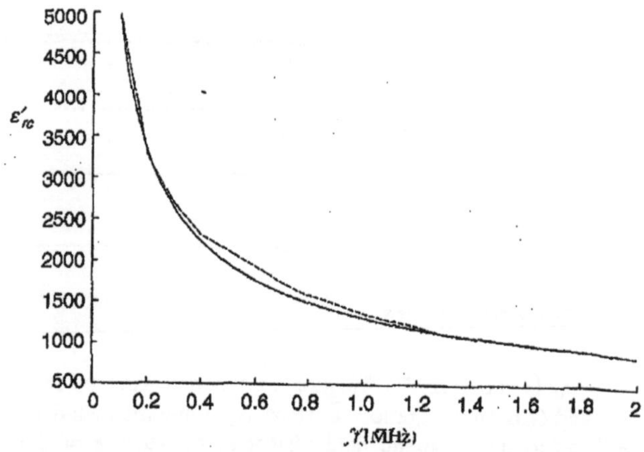

Fig. 4. Measured (dashed line) and approximated (solid line) relative permittivity ε'_{rc} of the Siemens N27 core material versus frequency.

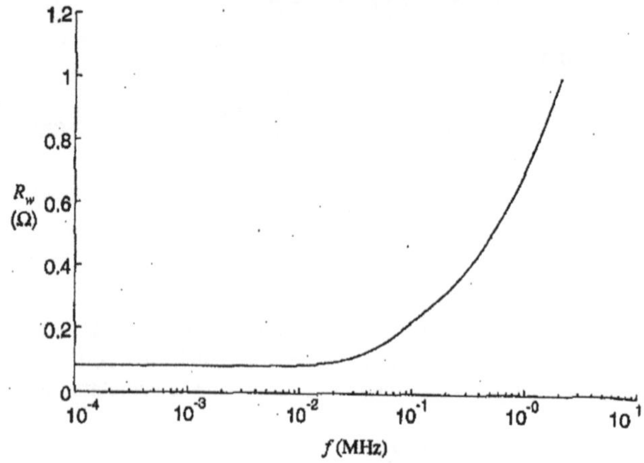

Fig. 5. Winding resistance R_w versus frequency.

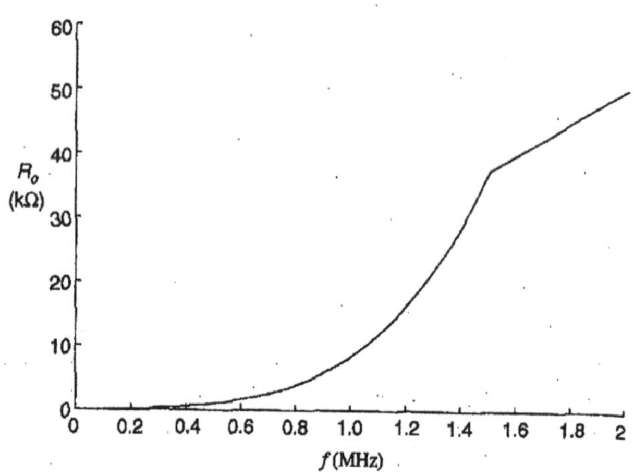

Fig. 6. Core equivalent series resistance R_c versus frequency.

high frequencies, in spite of the ferrite low conductivity, eddy currents strongly weaken the magnetic field in the core and, consequently, the main inductance decreases. Fig. 8 shows the calculated and measured ESR R_s versus frequency.

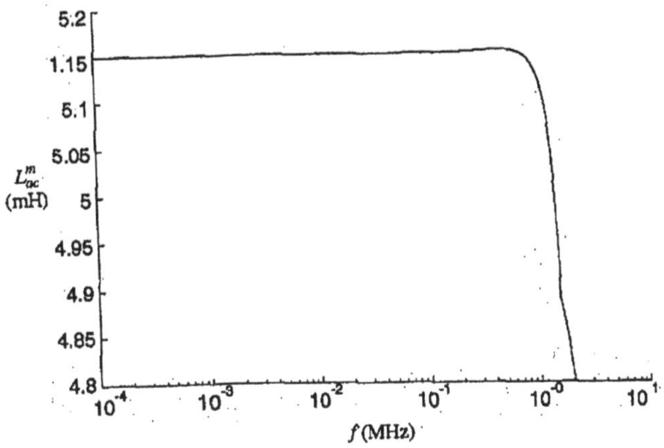

Fig. 7. Main inductance L^m_{ac} versus frequency.

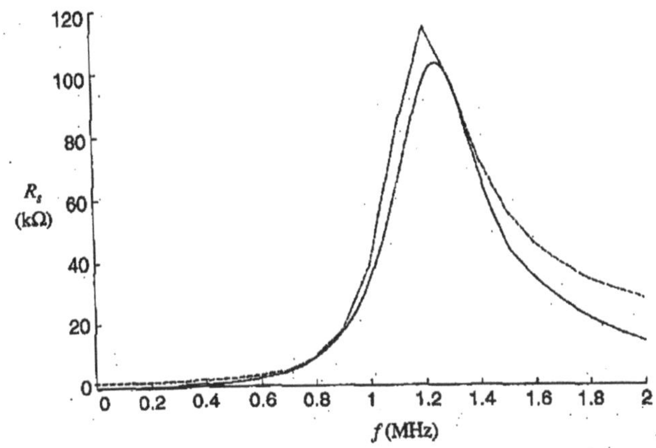

Fig. 8. Measured (dashed line) and calculated (solid line) ESR R_s versus frequency.

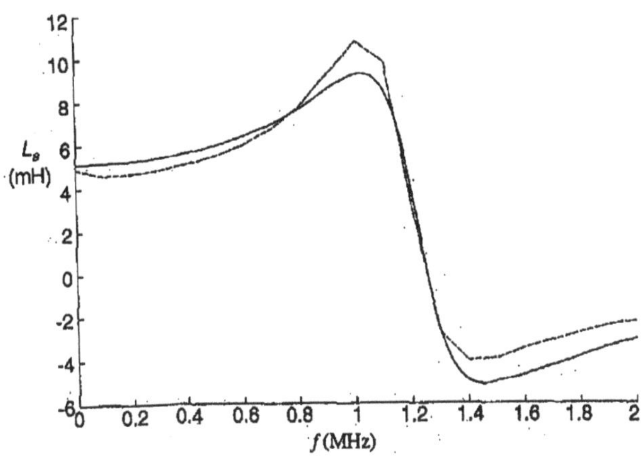

Fig. 9. Measured (dashed line) and calculated (solid line) ESL L_s versus frequency.

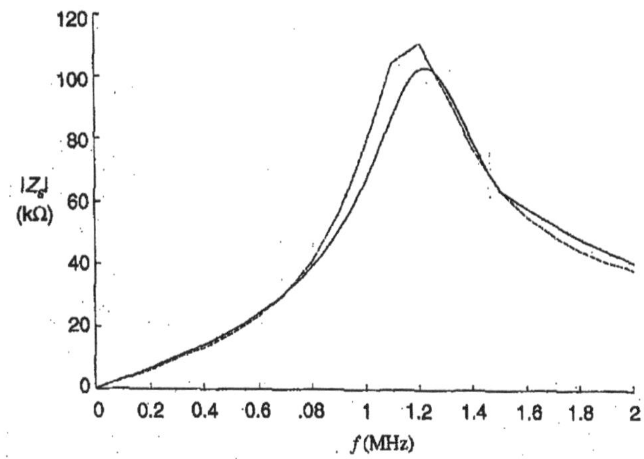

Fig. 10. Measured (dashed line) and calculated (solid line) magnitude of the inductor impedance $|Z_s|$ versus frequency.

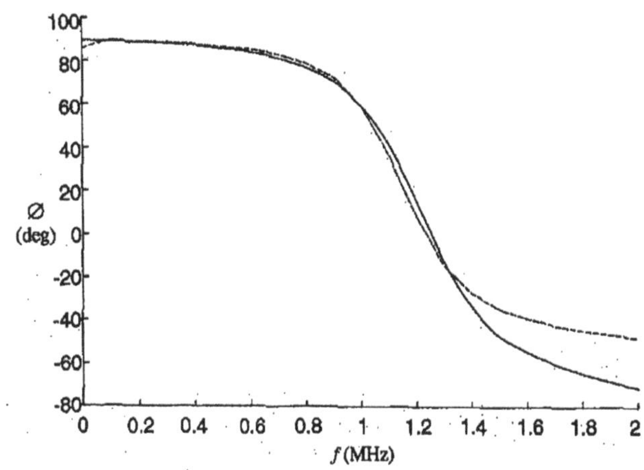

Fig. 11. Measured (dashed line) and calculated (solid line) phase of the inductor impedance ϕ versus frequency.

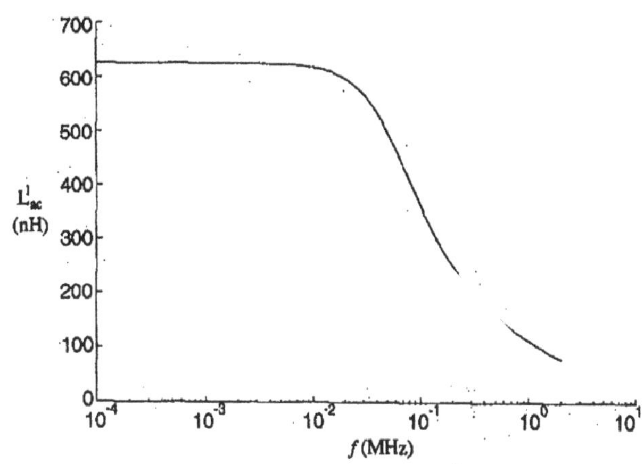

Fig. 12. Leakage inductance L^l_{ac} versus frequency.

Fig. 9 depicts the calculated and measured ESL L_s versus frequency. In Fig. 10, the calculated and measured values of the magnitude of the inductor impedance are compared. Fig. 11 shows the behavior of the calculated and measured phase of the inductor impedance versus frequency. Fig. 12 shows a plot of the leakage inductance L^l_{ac} versus frequency. The leakage inductance is almost independent of frequency at low frequencies and decreases to zero at high frequencies. At high frequencies, the current tends to concentrate close

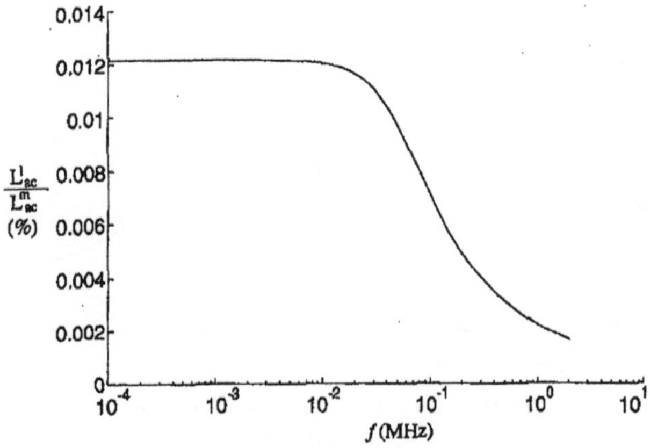

Fig. 13. Ratio of the leakage inductance to the main inductance L^l_{ac}/L^m_{ac} versus frequency.

to the surface of the conductor and the internal magnetic field approaches zero. The ratio L^l_{ac}/L^m_{ac} versus frequency is shown in Fig. 13. We can conclude that the leakage inductance is negligible as compared to the main inductance L^m_{ac} in the examined case of the ferrite core inductor with a single layer. For this reason, L^l_{ac} is neglected in this paper. The measured reactance was zero at $f = f_r = 1.25$ MHz. Hence, from (4) and Figs. 5–7, $C = 2.61$ pF was calculated. The choice of $N_l = 1$ in the experimental tests allows us to verify better the accuracy of the proposed formulas (29) and (30). In fact, (29) and (30) do not depend on the number of layers N_l, whereas Dowell's formula (5) and (31) do. Thus, it is desirable that the contributions of the winding resistance and leakage inductance given by (5) and (31) to the total ac resistance and inductance are reduced to a minimum. This minimum occurs when $N_l = 1$.

VII. Conclusion

A small-signal circuit model of ferrite core inductors has been examined and analytical functions have been derived to describe the behavior of the lumped parameters versus frequency. Experimental and calculated results have also been compared. It seems that the model is sufficiently accurate up to the self-resonance frequency. The model components are sensitive to the physical functions $\varepsilon'_{rc}(f)$, $\mu'_{rc}(f)$, and $\mu''_{rc}(f)$. A large-signal model is recommended for future research.

Appendix

Starting from (24), we can obtain (25) through the following manipulations. Letting

$$\hat{\eta} = j^{3/2}\hat{k} \tag{33}$$

$$\hat{\eta} r = \hat{v} \tag{34}$$

we can write

$$dr = \frac{d\hat{v}}{\hat{\eta}}. \tag{35}$$

Hence, the phasor of the magnetic flux through a cylindrical cross section given by (24) becomes

$$\begin{aligned}\hat{\Phi} &= 2\pi\frac{N\hat{I}}{l}\frac{\hat{\mu}_c}{J_0(\hat{v}_o)}\int_0^{\hat{v}_o}\frac{\hat{v}}{\hat{\eta}}J_0(\hat{v})\frac{d\hat{v}}{\hat{\eta}}\\ &= 2\pi\frac{N\hat{I}}{l}\frac{\hat{\mu}_c}{J_0(\hat{v}_o)}\frac{1}{\hat{\eta}^2}\int_0^{\hat{v}_o}\hat{v}J_0(\hat{v})\,d\hat{v}.\end{aligned} \tag{36}$$

Using the Bessel's function property [7]

$$\int u^n J_{n-1}(u)\,du = u^n J_n(u) \tag{37}$$

(36) becomes

$$\begin{aligned}\hat{\Phi} &= 2\pi r_o^2\frac{N\hat{I}}{l}\frac{\hat{\mu}_c\hat{v}_o}{\hat{\eta}^2 r_o^2}\frac{J_1(\hat{v}_o)}{J_0(\hat{v}_o)}\\ &= 2\pi r_o^2\frac{N\hat{I}}{l}\frac{\hat{\mu}_c}{\hat{v}_o}\frac{J_1(\hat{v}_o)}{J_0(\hat{v}_o)}\\ &= L^m_{dc}\hat{I}\frac{2}{N}\frac{1}{\mu_o\mu'_{rc}}\frac{\hat{\mu}_c}{\hat{v}_o}\frac{J_1(\hat{v}_o)}{J_0(\hat{v}_o)}\\ &= L^m_{dc}\hat{I}\frac{2}{N}\frac{1}{\mu_o\mu'_{rc}}\frac{\hat{\mu}_c}{j^{3/2}(\omega\hat{\mu}_c\hat{\gamma}_c)^{1/2}r_o}\\ &\quad\cdot\frac{J_1\left[j^{3/2}(\omega\hat{\mu}_c\hat{\gamma}_c)^{1/2}r_o\right]}{J_0\left[j^{3/2}(\omega\hat{\mu}_c\hat{\gamma}_c)^{1/2}r_o\right]}\\ &= L^m_{dc}\hat{I}\frac{2}{N}\frac{1}{\mu_o\mu'_{rc}}\frac{\hat{\mu}_c}{j^{3/2}\sqrt{2}\frac{r_o}{\hat{\delta}}}\frac{J_1\left(j^{3/2}\sqrt{2}\frac{r_o}{\hat{\delta}}\right)}{J_0\left(j^{3/2}\sqrt{2}\frac{r_o}{\hat{\delta}}\right)}\end{aligned} \tag{38}$$

where the low-frequency main inductance is

$$L^m_{dc} = N^2\frac{\mu_o\mu'_{rc}\pi r_o^2}{l}. \tag{39}$$

Expression (38) coincides with (25), where $\hat{\delta}$ is given by

$$\begin{aligned}\hat{\delta} &= \sqrt{\frac{1}{\pi\hat{\mu}_c\hat{\gamma}_c f}}\\ &= \sqrt{\frac{1}{\pi f\mu_o(\mu'_{rc} - j\mu''_{rc})(\gamma_c + j2\pi f\hat{\varepsilon}_c)}}\\ &= \sqrt{\frac{1}{\pi f\mu_o(\mu'_{rc} - j\mu''_{rc})[\gamma_c + j2\pi f\varepsilon_o(\varepsilon'_{rc} - j\varepsilon''_{rc})]}}.\end{aligned} \tag{40}$$

References

[1] P. J. Dowell, "Effects of eddy currents in transformer windings," *Proc. Inst. Elect. Eng.*, vol. 113, pp. 1387–1394, Aug. 1966.
[2] M. Bartoli, M. K. Kazimierczuk, and A. Reatti, "Predicting the high-frequency ferrite-core inductor performance," presented at the Elect. Manufacturing Coil Winding Conf., Chicago, IL, Sept. 27–29, 1994, pp. 409–413.
[3] ——, "Modeling iron-powder inductors at high frequency," presented at the 29th IEEE Industry Applicat. Soc. Annu. Meeting Tech. Conf., Denver, CO, Oct. 2–5, 1994, pp. 1225–1232.
[4] J. K. Watson, *Applications of Magnetism*. New York: Wiley, 1980.
[5] H. Saotome and Y. Sakaki, "Iron loss analysis of Mn-Zn Ferrite cores," *IEEE Trans. Magnetics*, vol. 33, pp. 728–734, Jan. 1997.
[6] M. Honda, *The Impedance Measurement Handbook*. Hewlett Packard Co., 1994.
[7] G. N. Watson, *A Treatise on the Theory of Bessel Functions*. Cambridge, U.K.: Cambridge Univ. Press, 1966.

Marian K. Kazimierczuk (M'91–SM'91) received the M.S., Ph.D., and D.Sci. degrees in electronics engineering from the Department of Electronics, Technical University of Warsaw, Warsaw, Poland, in 1971, 1978, and 1984, respectively.

He was a Teaching and Research Assistant from 1972 to 1978 and an Assistant Professor from 1978 to 1984 with the Department of Electronics, Institute of Radio Electronics, Technical University of Warsaw, Poland. In 1984, he was a Project Engineer for Design Automation, Inc., Lexington, MA. In 1984–1985, he was a Visiting Professor with the Department of Electrical Engineering, Virginia Polytechnic Institute and State University, VA. Since 1985, he has been with the Department of Electrical Engineering, Wright State University, Dayton, OH, where he is currently a Professor. His research interests are in high-frequency high-efficiency switching-mode tuned power amplifiers, resonant and PWM dc/dc power converters, dc/ac inverters, high-frequency rectifiers, electronic ballasts, magnetics, power semiconductor devices, high-frequency high-efficiency switching-mode power tuned amplifiers, and applied controls. He is the coauthor of the book *Resonant Power Converters* (New York: Wiley, 1995). He has published more than 200 technical papers, 70 of which have appeared in IEEE TRANSACTIONS and JOURNALS.

Dr. Kazimierczuk received the IEEE Harrel V. Noble Award for his contributions to the fields of aerospace, industrial, and power electronics in 1991. He is also a recipient of the 1991 Presidential Award for Faculty Excellence in Research, the 1993 College Teaching Award, the 1995 Presidential Award for Outstanding Faculty Member, and the Brage Golding Distinguished Professor of Research Award from Wright State University. He has been an Associate Editor of the IEEE TRANSACTIONS ON CIRCUITS AND SYSTEMS I and serves as an Associate Editor for the *Journal of Circuits, Systems, and Computers*. He was a member of the Superconductivity Committee of the IEEE Power Electronics Society. He is a member of Tau Beta Pi.

Giuseppe Sancineto was born in Crotone, Italy, in 1968. He received the M.Eng. degree in electrical engineering from the Department of Electrical Engineering, University of Bologna, Bologna, Italy, in 1998, where he is currently working toward the Ph.D. degree.

In 1998, he was a visiting research scholar at the Department of Electrical Engineering, Wright State University, Dayton, OH. His research interests are in the areas of power electronics and electromagnetic compatibility.

Gabriele Grandi was born in Bologna, Italy, in 1965. He received the M.Sc. degree in 1990 (cum laude) and the Ph.D. degree in 1994, both in electrical engineering, from the Department of Electrical Engineering, University of Bologna, Bologna, Italy.

Since 1995, he has been an Assistant Professor (Research Associate) at the Department of Electrical Engineering, University of Bologna, Italy. His research interests are in the areas of modeling, simulation, and design of electronic power converters with particular reference to power conditioning systems. Currently, he studies EMC problems related to electric power apparatus and systems, mainly switching converters and high-frequency circuit models of passive components.

Ugo Reggiani (M'92) received the Laurea degree (cum laude) in electrical engineering from the University of Bologna, Bologna, Italy, in 1969.

He was appointed Assistant at the Department of Engineering, University of Bologna, in 1972, Assistant Professor in 1974, and Full Professor of electrotechnics in 1980. Presently, he is also Head of the Department of Electrical Engineering, University of Bologna. His main research interests concern electromagnetic field theory, electrical machines and drives, electromagnetic compatibility, circuit analysis, and modeling of switched networks.

Dr. Reggiani is a member of the Italian Electrotechnical and Electronic Association (AEI).

Antonio Massarini received the Laurea degree (cum laude) in nuclear engineering and the Ph.D. degree in electrical engineering from the Department of Electrical Engineering, University of Bologna, Bologna, Italy, in 1987 and 1992, respectively.

Since 1993, he has been a Teaching and Research Assistant with the Faculty of Engineering, Department of Engineering Sciences, University of Modena and Reggio Emilia, Modena, Italy. He also cooperates with the Electrical Engineering Department of the University of Bologna, and was a Visiting Professor with the Department of Electrical Engineering, Wright State University, Dayton, OH. His research interests include MHD flow simulation, switched network simulation, magnetics, numerical methods for circuit analysis and design, and switching power converter simulation and design.

Derivation of a Complex Permeability From the Preisach Model

K. Hollaus and O. Bíró

Abstract—**An attractive way to deal with hysteresis nonlinearity in case of time harmonic excitation of an eddy current problem is to introduce an effective material representing the ferromagnetic material. This fictitious model describes the relationship of the field quantities approximately. Two methods to create an effective magnetization curve on the basis of the Preisach model of different soft magnetic materials are presented. The complex magnetization curves obtained are applied to a simple one-dimensional finite-element model.**

Index Terms—**Ferromagnetic materials, finite-element methods, hysteresis, losses.**

I. Introduction

THE AIM of this paper is to develop an efficient approximate technique to simulate the eddy current distribution and the associated losses, i.e., eddy current losses and hysteresis losses, by the finite-element method (FEM). Due to the great practical importance of large electrical devices with laminated cores made of ferromagnetic materials, a robust method facilitating their analysis is desired. A transient analysis in terms of a time stepping procedure does not represent a suitable solution because of its time-consuming behavior. A very common and widespread way to model the hysteresis is the Preisach model (PM). It is well known in this context that the PM suffers from its extremely high demand on memory as well as from its expensive handling. Numerous models treating nonlinear materials have been developed and applied successfully; see [1]–[4], for instance. However, the generalization of the models to materials with hysteresis are rather scarce; see [5]–[9]. Therefore, two different ways to derive a nonlinear complex permeability from classical Preisach models (CPM) are shown. The ferromagnetic materials with significant hysteresis and pronounced saturation will be modeled by an effective material. The losses in a sheet are calculated by an approximate technique involving a one-dimensional (1-D) finite-element model and applying effective magnetization curves.

II. Material Properties

The CPM [10] describes the hysteresis nonlinearity by a set of elementary rectangular loops with different up and down switching levels. The state of the loops can be only $+1$ or -1, and they are weighted by the Preisach distribution function (PDF), which determines a certain material uniquely. These distribution functions for three different soft magnetic materials are assumed to be known. The Preisach distribution function represents the starting point of the derivation of the complex magnetization curves. The magnetic properties of the materials are represented by their B–H loops and their initial curves in Fig. 1. Dynamic effects are neglected.

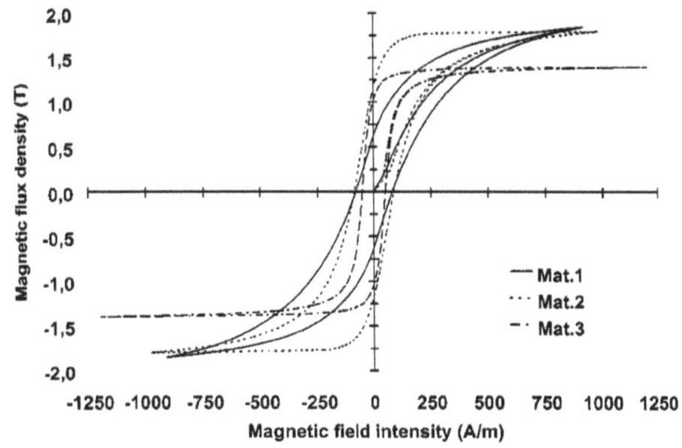

Fig. 1. B–H loops of three soft magnetic materials along with their initial curves.

III. Constitutive Law

The complex formalism can be exploited favorably for problems with linear material properties and time harmonic excitation in the steady state because they have to be solved only once to obtain the exact solution. To also preserve the advantage of the complex formalism for nonlinear problems, equivalent material characteristics have to be introduced. In this equivalent material, the field quantities are assumed to be sinusoidal and obey criterions, which are described below. A constitutive law

$$B = \mu H \tag{1}$$

is introduced. In (1), B represents the magnetic flux density and H the magnetic field intensity, respectively. They are complex numbers. Without hysteresis, a real number suffices for the magnetic permeability μ to represent the original material [2]. Due to the phase shift in materials with hysteresis between the magnetic flux density $B(t)$ and the magnetic field intensity $H(t)$ in the time domain and due to the saturation effect, the magnetic permeability μ has to be complex, and a nonlinear function of the field quantities have to be able to approximate the relationship between $B(t)$ and $H(t)$. Since the nonlinear complex magnetization curve has to be calculated only once, a very fine

Manuscript received July 5, 2001. This work was supported by the "Fonds zur Förderung der wissenschaftlichen Forschung," Vienna, Austria, under the Grant P 13551-TEC.

K. Hollaus and O. Bíró are with the Institute for Fundamentals and Theory in Electrical Engineering, Technical University of Graz, Graz, Austria (e-mail: hollaus@igte.tu-graz.ac.at; biro@igte.tu-graz.ac.at).

Publisher Item Identifier S 0018-9464(02)00908-1.

0018-9464/02$17.00 © 2002 IEEE

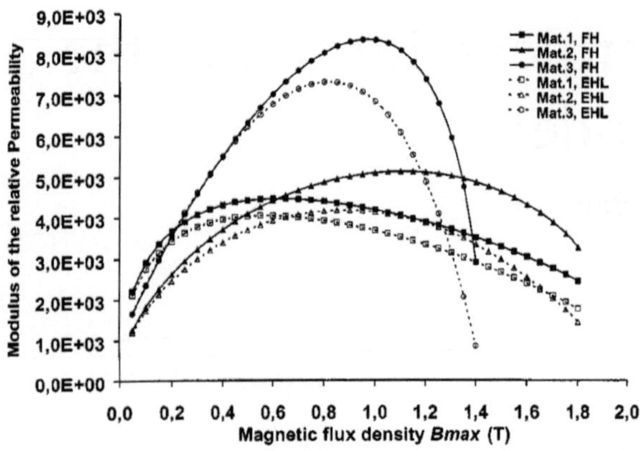

Fig. 2. Modulus of the relative permeability of the three soft magnetic materials.

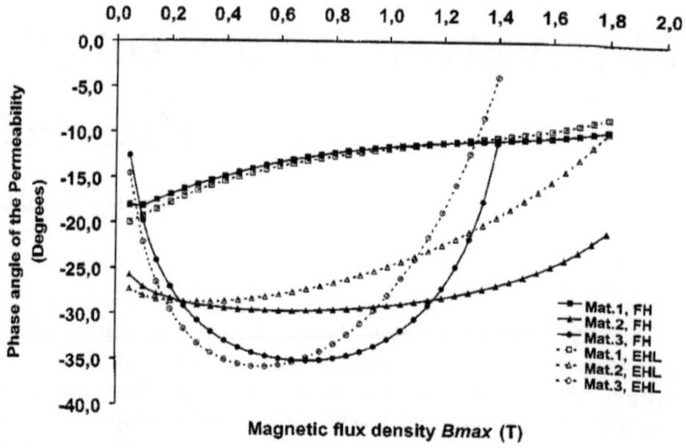

Fig. 3. Phase angle of the relative permeability of the three soft magnetic materials.

discretization of the limiting triangle in the Preisach plane can be carried out.

IV. Nonlinear Complex Permeability

The solution by this technique is, of course, not exact. The task is to devise a criterion to derive an effective material in order to obtain solutions as accurate as possible by the approximate technique.

Subsequently, two possibilities for the creation of an effective material are discussed. To create a complex permeability, only the magnetic field density B is assumed to be sinusoidal and given because this can be expected in most practical problems. Prescribing the time behavior of B implies that the PM has to be treated in the reverse mode.

Fundamental Harmonics (FH): The first method is based on a very simple idea that derives a complex permeability μ by considering only the fundamental harmonics of the magnetic field quantities.

Equivalent Hysteresis Losses (EHLs): In the second approach, a complex permeability μ is derived to yield the same hysteresis losses P_{hyst} as the real material. The hysteresis losses P_{hyst} are proportional to the enclosed area of the B–H loop. The area of the approximating ellipse can be calculated by

$$P_{\mathrm{hyst}} \sim -\pi|\mu||H|^2 \sin\varphi_\mu. \tag{2}$$

Here, $|H|$ and $|\mu|$ mean the moduli of the magnetic field intensity and of the magnetic permeability. φ_μ is the phase angle between B and H. The second condition used is

$$|\mu| = \frac{B_{\max}}{H_{\max}} \tag{3}$$

which ensures that the maximal value of H is retained.

From (2) and (3), it can be easily seen that the losses P_{hyst} determine the phase angle φ_μ, whereas the saturation is responsible for the modulus $|\mu|$. Figs. 2 and 3 show the complex permeability derived for the three materials.

V. Approximated Hysteresis Loops

The constitutive law (1) describes, in the time domain, an ellipse approximating the original hysteresis loop. Figs. 4 and 5 represent the ellipses obtained by the approximating techniques FH and EHL together with the corresponding original hysteresis loop of the two materials Mat.2 and Mat.3.

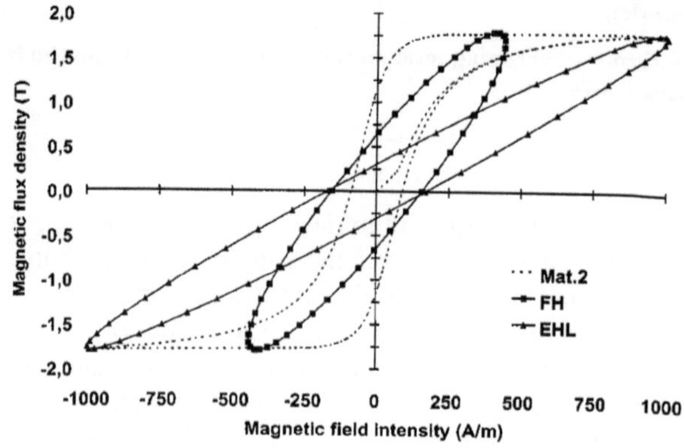

Fig. 4. Approximating ellipses and the original hysteresis loop of the material Mat.2, $B_{\max} = 1.8$ T.

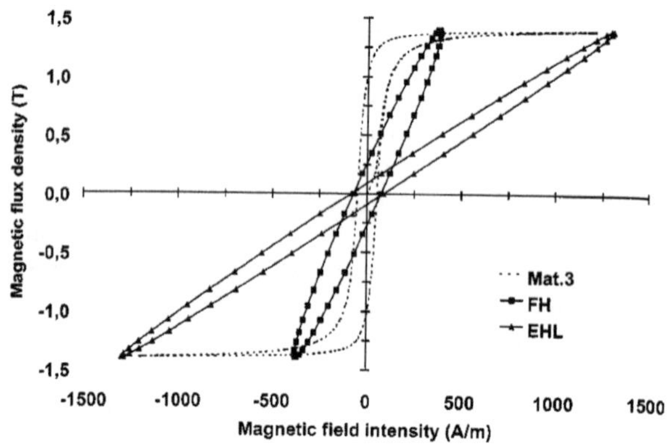

Fig. 5. Approximating ellipses and the original hysteresis loop of the material Mat.3, $B_{\max} = 1.4$ T.

VI. Results

The time functions of the transient and the approximate field quantities at a distance of $x = 0.25$ mm from the surface are shown in Figs. 6–8.

To validate the accuracy and efficiency of the approximating technique, the hysteresis losses and the eddy current losses in

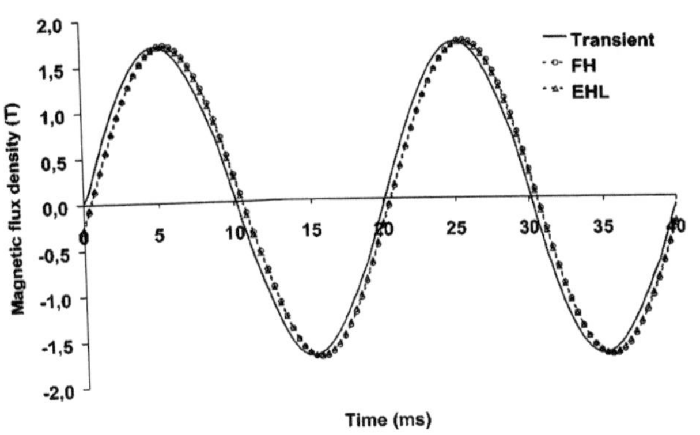

Fig. 6. Magnetic flux density of the material Mat.1 at $x = 0.25$ mm.

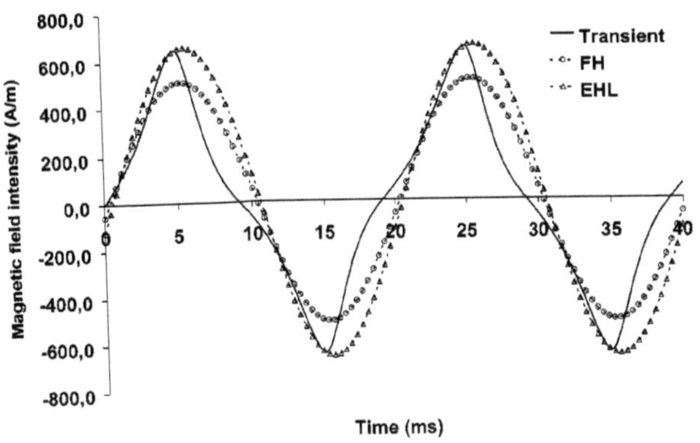

Fig. 7. Magnetic field intensity of the material Mat.1 at $x = 0.25$ mm.

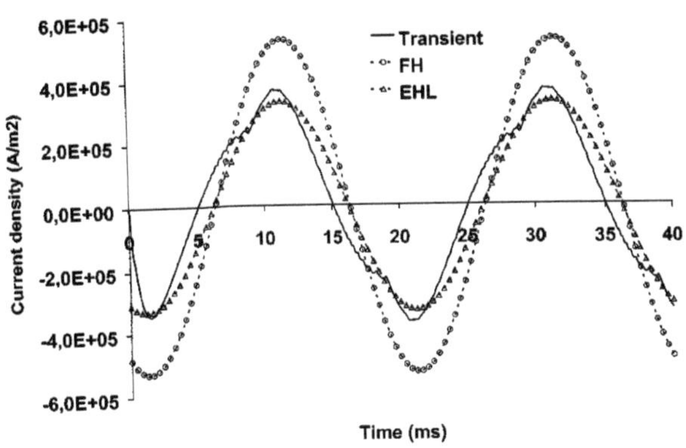

Fig. 8. Current density of the material Mat.1 at $x = 0.25$ mm.

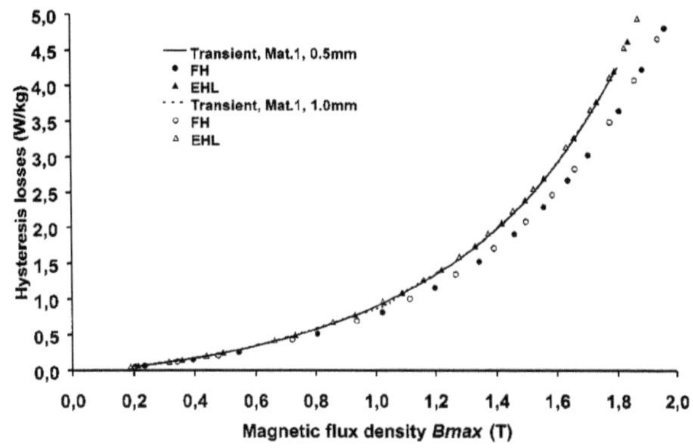

Fig. 9. Hysteresis losses of the material Mat.1.

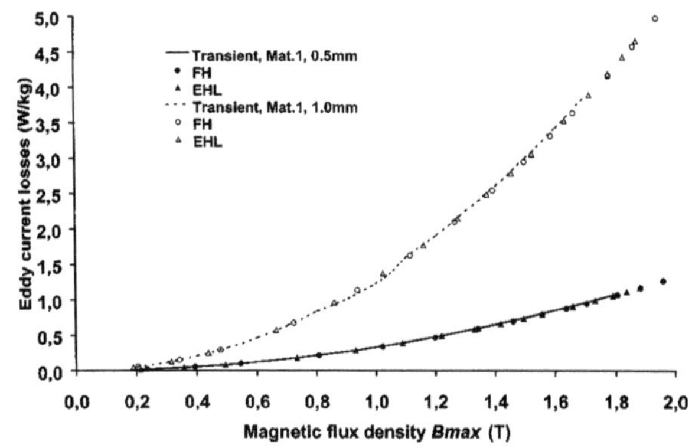

Fig. 10. Eddy current losses of the material Mat.1.

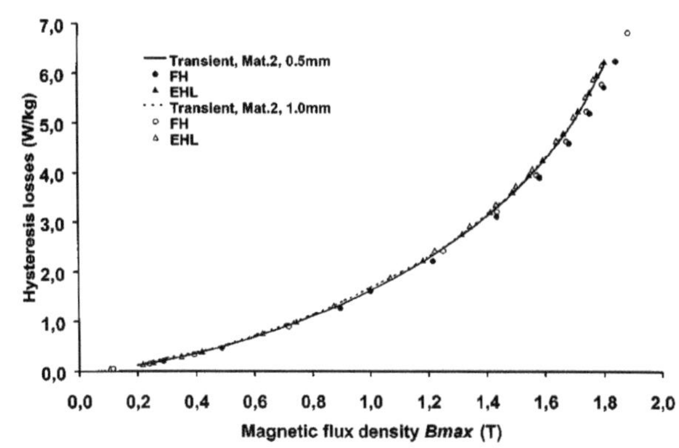

Fig. 11. Hysteresis losses of the material Mat.2.

thin ferromagnetic sheets (0.5 and 1.0 mm thick) have been computed by a 1-D finite-element model of the three materials. In Figs. 9–14, the specific eddy current and hysteresis losses are shown. $B_{\max}$ is the average maximum value of the magnetic flux density across the sheet.

The frequency of the excitation was $f = 50$ Hz. In the transient case, the magnetic vector potential A has been applied. The transient simulations are two periods long, and each period is subdivided in 400 time steps. For the complex analysis, the magnetic field intensity H has simply been used as the solution variable. Appropriate boundary conditions for the two formulations have been prescribed on the surface.

For the 0.5-mm-thick sheet, only two finite elements (FE) have been required in both cases. However, the number of required FEs had to be increased drastically for the 1.0-mm-thick sheet in the transient case, especially for material Mat.3. This might be due to the structure of the PDF of Mat.3, which possesses an additional term valid only along the hypotenuse of the Preisach triangle [10]. To achieve a stable transient simulation, 20 FEs were found to be sufficient. In the complex analysis, four FEs were used.

In general, the losses are reproduced by the approximating technique very satisfactorily. The eddy current losses increase extremely in the 1.0-mm-thick sheet compared with that in the 0.5-thick sheet.

The complex magnetization curves are approximated by 29 sampling points and linear interpolation. The maximal field values of the complex magnetization curves are restricted by

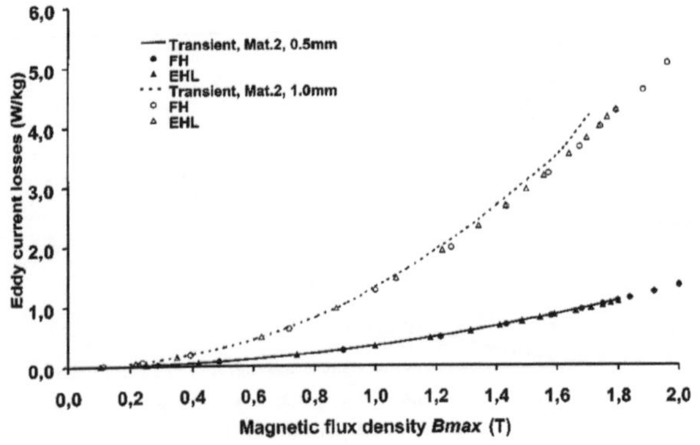

Fig. 12. Eddy current losses of the material Mat.2.

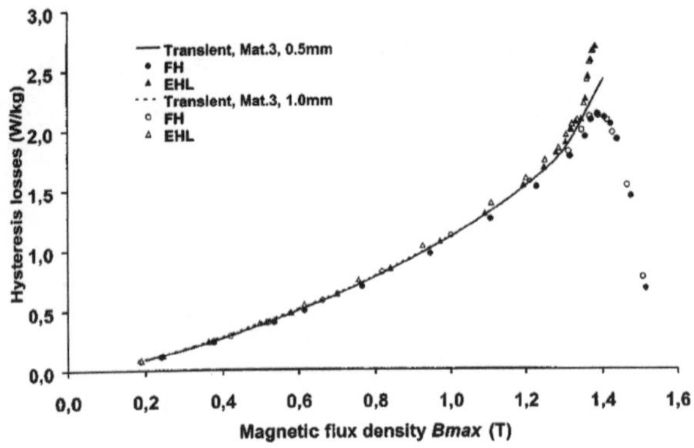

Fig. 13. Hysteresis losses of the material Mat.3.

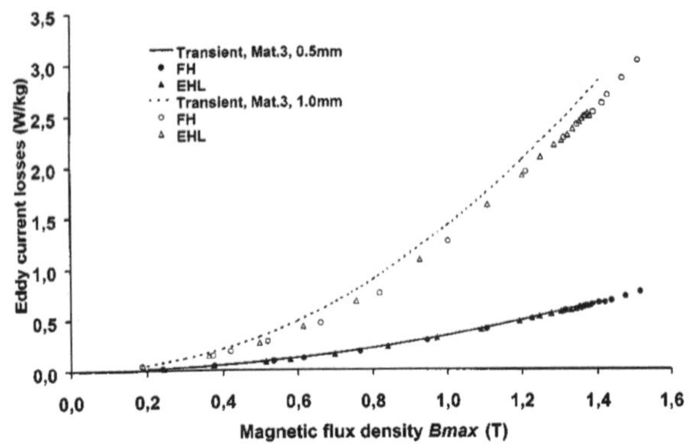

Fig. 14. Eddy current losses of the material Mat.3.

the limiting triangle in the Preisach plane. The corresponding upper bound in the case of the materials Mat.1 and Mat.2 is $B_{\max} = 1.8$ T and of Mat.3 $B_{\max} = 1.4$ T. Therefore, a linear expansion of the curves has been used. This is the reason for the deviation in Fig. 13.

The approximation by EHL is slightly better than that by FH.

VII. Conclusion

The accuracy of the approximating technique is very satisfactory, and the time and memory savings are extremely high. A single time step in the transient analysis needs about the same number of nonlinear iterations as the corresponding complex computation. The complex simulation is about 9500 times faster then the transient one. Only a complex number instead of the PM has to be stored for an integration point of the FEM.

References

[1] G. Paoli, O. Bíró, and G. Buchgraber, "Complex representation in nonlinear time harmonic eddy current problems," *IEEE Trans. Magn.*, vol. 34, pp. 2625–2628, Sept. 1998.

[2] J. D. Lavers, "Finite element solution of nonlinear two dimensional TE-mode eddy current problems," *IEEE Trans Magn.*, vol. 19, pp. 2201–2203, Sept. 1983.

[3] Y. Du Terrail, J.-C. Sabonnadiere, P. Masse, and J. L. Coulomb, "Nonlinear complex finite elements analysis of electromagnetic field in steady-state AC devices," *IEEE Trans. Magn.*, vol. MAG-20, pp. 549–552, July 1984.

[4] D. Labridis and P. Dokopoulos, "Finite element computation of eddy current losses in nonlinear ferromagnetic sheaths of three-phase power cables," *IEEE Trans. Power Delivery*, vol. 7, pp. 1060–1067, July 1992.

[5] D. O'Kelly, "Flux penetration in a ferromagnetic material including hysteresis and eddy-current effects," *J. Phys. D (Appl. Phys.)*, vol. 5, pp. 203–213, 1972.

[6] M. S. El-Markaby, T. H. Fawzi, and M. T. Ahmed, "Approximate treatment of nonlinear eddy currents problems," *IEEE Trans. Magn.*, vol. MAG-18, pp. 1737–1739, Nov. 1982.

[7] B. Leutke-Daldrup, "Comparison of exact and approximate finite-element solution of two-dimensional nonlinear eddy-current problem with measurements," *IEEE Trans. Magn.*, vol. MAG-20, pp. 1737–1739, Sept. 1984.

[8] A. G. Jack and B. C. Mecrow, "Methods for magnetically nonlinear problems involving significant hysteresis and eddy currents," *IEEE Trans. Magn.*, vol. 26, pp. 424–429, Mar. 1990.

[9] H. Hedia, P. D. Henrotte, A. Genon, and W. Legros, "Determination of equivalent material characteristics for parasitic effects in electrical apparatus," *IEEE Trans. Magn.*, vol. 32, pp. 168–1170, Sept. 1996.

[10] I. D. Mayergoyz, *Mathematical Models of Hysteresis*. New York: Springer-Verlag, 1991.

IEEE TRANSACTIONS ON MAGNETICS, VOL. MAG-5, NO. 3, SEPTEMBER 1969

Noise of Ferromagnetic Materials

HEINZ BITTEL

Abstract—The main types of noise in ferromagnetic materials are thermal or Nyquist noise, excess noise, and Barkhausen noise. They differ in magnitude as well as in the frequency dependence of their associated power spectra. The comparatively weak Nyquist noise is generated by the thermal energy of the conduction electrons and spin waves acting on the domain configuration in the ferromagnetic materials. Excess noise occurs during temperature changes of the sample. It is generated by magnetization processes caused by a changing mechanical strain inside the sample. Strongest by far can be the Barkhausen noise. This is due to a pronounced tendency to form clusters of Barkhausen jumps. The temporal and spatial behavior of such clusters can be well described as a diffusion process; a simple model of this process is given.

Introduction

The flux in a ferromagnetic core exhibits temporal fluctuations. They can be observed as an open circuit noise voltage $U(t)$ at the terminals of a coil surrounding the core (Fig. 1). Three different types of noise may be distinguished. ($I(t)$ is the current flowing in the exciting windings.)

1) $I(t)$ constant and temperature T constant: Nyquist noise, which in this case is mainly caused by the thermal energy of the spin system and in metallic cores also by the thermal energy of the conduction electrons. This thermal noise is important if the core is at a constant temperature and the winding carries only a dc current. The demagnetized state in zero field belongs to this case.

2) $I(t)$ constant and T not constant: excess noise during a change in core temperature as is observed with ferromagnetic metals and also with ferrites.

3) $I(t)$ not constant: noise due to discontinuous magnetization processes caused by a changing field excitation. This Barkhausen noise is produced by periodically sweeping along the hysteresis loop or also by pulsed fields.

The induced noise voltage, Fig. 1, according to Parseval's theorem, can be characterized by its power spectrum

$$\overline{U^2}(t) = \int W(f)\, df. \tag{1}$$

The function $W(f)$ is mainly determined by the statistical law which controls the sequence of the individual magnetization processes. $W(f)$ can be measured with the usual setup consisting of an amplifier, a wave analyzer, and a squaring device with integrating performance for averaging. Such an analysis provides essential information about the dynamic behavior of the magnetization processes. In addition, it is possible to represent the open circuit noise

Manuscript received March 24, 1969; revised June 15, 1969. Paper 12.1, presented at the 1969 INTERMAG Conference, Amsterdam, The Netherlands, April 15–18.
The author is with the Institut für Angewandte Physik, Universität Münster, Münster, Germany.

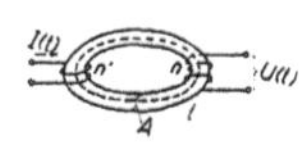

Fig. 1. Ferromagnetic core, cross section A with mean length l. Current $I(t)$ is from a high-impedance generator. Both windings are assumed to be evenly distributed around core.

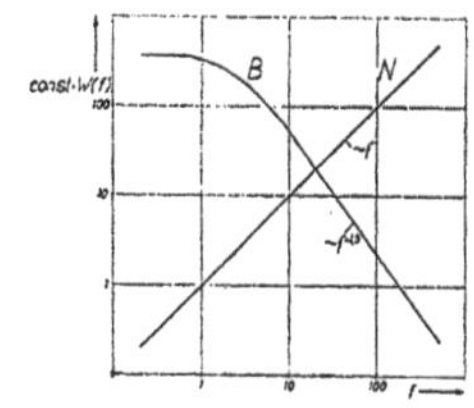

Fig. 2. N: low-frequency Nyquist noise. B: Barkhausen noise. Different scales are used in both cases.

voltage so measured by an equivalent noise generator. It can, as already shown by Warren [1], be included in the equivalent circuit of a transformer in order to describe completely the ferromagnetic noise also in the loaded case.

The mathematical treatment of stochastic processes is simple if the process 1) is stationary 2) has a normal distribution of amplitudes, and 3) belongs to the class of Markoff processes. In the case of ferromagnetic noise, quite often none of the three conditions is satisfied.

While sweeping through the hysteresis loop ($I(t)$ not constant), the Barkhausen noise is generated above all in the steep part of the loop near the coercive field H_c. $W(f)$ thus is systematically dependent on time (nonstationary noise). If one is only interested in the integral noise obtained from one magnetization cycle of duration T, the energy spectrum $E(f)$ is used.

$$\int_0^T U^2(t)\, dt = \int E(f)\, df. \tag{2}$$

In general, the discontinuities during reversal of magnetization are coarse and tend to form clusters. The condition for a normal amplitude distribution therefore is not fulfilled. Such formation of clusters is caused by coupling between the single elementary processes, for instance, by an interaction between neighboring Bloch walls or between parts of one and the same wall. Thus a statistical dependence between the individual processes that finds its expression in the function $W(f)$ is produced. This dependence between the processes is influenced by the type of ferromagnetic noise and by the external condition under which the noise is generated. The power spectrum of Nyquist noise, for instance, is different from that of Barkhausen noise, as indicated schematically in Fig. 2.

Nyquist Noise

The n coil windings of Fig. 1 have an impedance Z that, when measured with small amplitude, is given by

$$Z = j\omega\mu L_0 \tag{3}$$

with $L_0 = \mu_0 n^2 A/l$.

The permeability $\mu = \mu' - j\mu''$ depends on the magnetic state of the core, i.e., on the given value of $I(t)$. According to the Nyquist theorem, if the winding resistance is ignored, one can expect, at temperature T, a thermal noise

$$W(f) = 4kT\omega\mu'' L_0. \tag{4}$$

Equation (4) was checked by experiment and found to be valid [2], and not only for the demagnetized state. This is important because the core is not in thermal equilibrium for any given magnetic state (e.g., the remanent one), an assumption made to derive (4).

The losses in the core, described by μ'', are caused in part 1) by excitation of the spin system, 2) in general by after effects, and 3) in metallic cores by eddy currents. According to (4), each of these three effects generates a contribution to the thermal noise.

It is well known that the Bloch walls can move elastically. Without external excitation such movements are already present due to the thermal energy. They are activated either by thermal spin waves (effect 1)) or by the thermal motion of conduction electrons (effect 3)). One can imagine in the latter case that the motion of the electrons has a stochastic circular component with respect to each arbitrarily given axis, thereby generating local magnetic fields of all directions that cause the movement of a Bloch wall.

Besides elastical movements, a Bloch wall can perform jumps. Such jumps also can be triggered thermally, the causes again being spin waves and conduction electrons. This is the mechanism of the thermal fluctuation aftereffect. It gives for many ferromagnetic materials a contribution to μ'', which is independent of frequency in a wide range [3]. The noise generated by this mechanism according to (4) has a power spectrum $W(f)$, which, because μ'' is constant, is proportional to f (Fig. 1). This spectrum can be explained from the following elementary process. When thermally activated, a Bloch wall can jump almost only over weak potential barriers. With a high probability, the next jump of the Bloch wall will be in the opposite direction since it is not very likely that the adjacent barrier is also small. As a consequence, pulses with alternating signs are generated in the coil winding. A large number of such pulse pairs yields [4] the spectrum $W(f)$ proportional to f.

Excess Noise

A noise corresponding to the Nyquist theorem no longer is observed in most cases if the core temperature is changed during the measurement [5]. Fig. 3 shows the excess noise for a ferrite core in the demagnetized state. Presumably this noise has its origin in changes of the domain configuration. "Invar" (36 percent Ni–Fe) does not show the effect, therefore, thermal strain can be regarded as the basic cause of the effect.

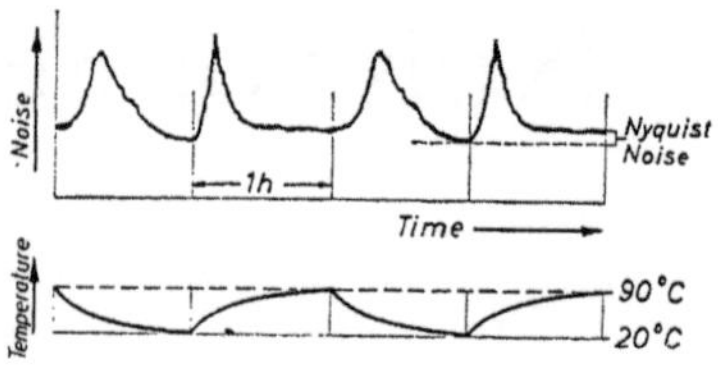

Fig. 3. Excess noise of a toroid of Mn–Zn ferrite during change of temperature. Noise is measured at 1.3 kHz, Curie temperature 150° C. Core is in demagnetized state.

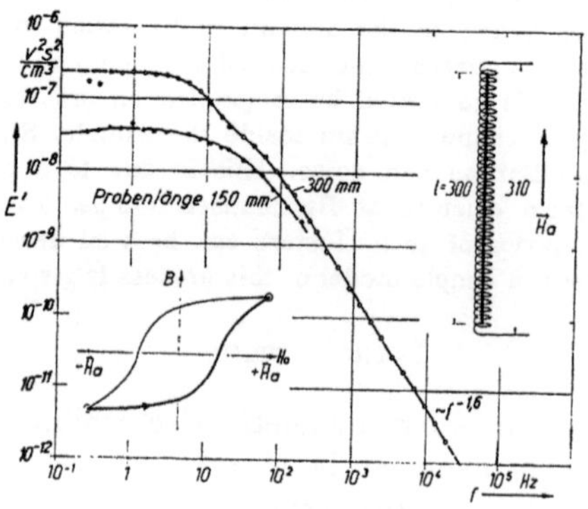

Fig. 4. Energy spectrum of Barkhausen noise for two different lengths l of sample.

Barkhausen Noise

Fig. 4 gives a typical example for the energy spectrum of Barkhausen noise as was investigated by Grosse-Nobis on grain-oriented 3-percent Si–Fe [6]. The specimen in the form of a thin strip with cross section $A = 2 \times 0.1$ mm² was located inside a cylindrical coil which generated the field H_a. A pick-up coil wound tightly around the specimen served for the noise measurement (shown in the insert of Fig. 4). This and the other results, which will be reported first, were obtained while sweeping the field linearly along the hysteresis loop with a very low field rate. The value $dH_a/dt = 2.5$ mOe·s^{-1} corresponds to a time of 8000 seconds for a run from $-\hat{H}_a$ to $+\hat{H}_a$ with $\hat{H}_a = 10$ Oe. E' is the value referred to the core volume, $n = 3860$. Notice the weaker noise at low frequencies from the sample with a length of 150 mm.

Fig. 5 shows the power density $W(f)$ plotted for some selected frequencies as a function of field. The strongest noise occurs in the neighborhood of the coercive field H_c or near that point of the hysteresis loop where the differential permeability dB/dH_a has a maximum.

Fig. 5 clearly indicates a relation between dB/dH_a and $W(f)$. This is shown in Fig. 6. The high-frequency noise W(18 kHz) is approximately proportional to permeability, and can be explained by Campbell's theorem. The power density is proportional to the number of jumps per unit time. In addition, there is a formation of clusters, which is expressed by the stronger increase $W(f)$ proportional

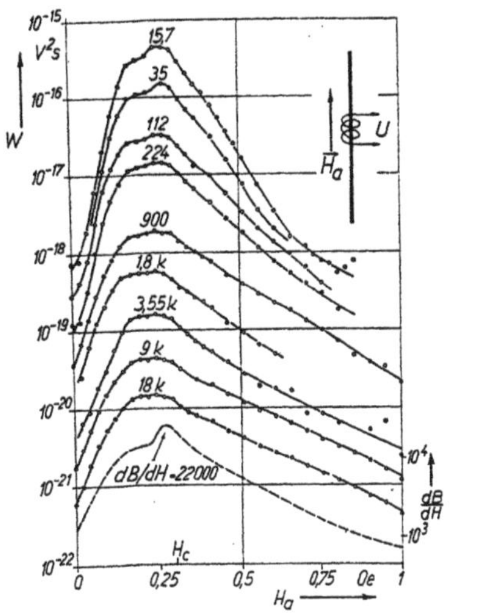

Fig. 5. $W(f)$ for some selected frequencies between 15.7 Hz and 18 kHz as a function of field H_a. Same material as in Fig. 4. l = 300 mm, $\hat{H}_a$ = 3.22 Oe, dH_a/dt = 27 mOe·s⁻¹; bottom curve gives differential permeability dB/dH_a.

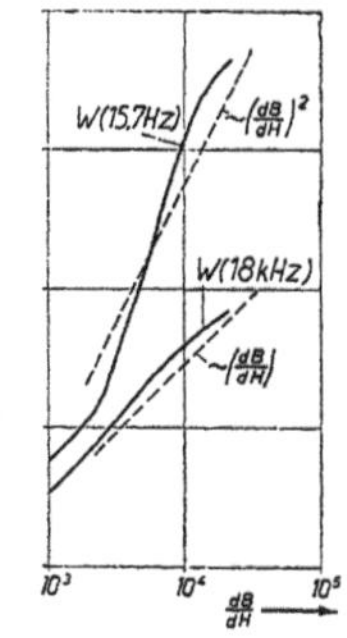

Fig. 6. Relation between $W(f)$ and dB/dH_a, taken from Fig. 5. Different arbitrary scales for both curves.

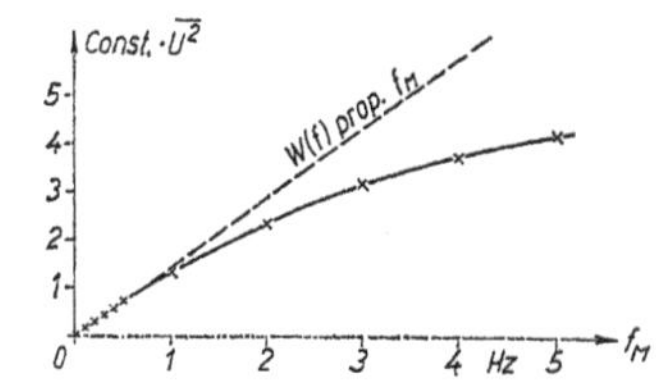

Fig. 7. Mean square $\overline{U^2}$ measured on 0.015-inch sheet metal of Stalloy in frequency range 40 Hz–10 kHz as function of magnetizing frequency f_M, B = 2000 gauss, taken from [1].

$(dH_a/dt)^2$ at low frequencies. At low frequencies W(15.7 Hz) on the other hand, the noise becomes much stronger, it increases approximately with the square of permeability.

There are many publications about Barkhausen noise, [1], [7]–[13], [18], [19]. It cannot be the subject of this paper to consider all the details that have been found. Rather we will try to put forward the essential aspects of the hitherto existing results and relate them to our knowledge of the dynamics of ferromagnetism. It is also not possible to deal with the mathematical relations that exist between the observed power spectrum and the average temporal behavior of the jumps (see [14]–[16]). In addition, it should be mentioned that our knowledge of Barkhausen noise has been increased to a large extent by applying methods of pulse and digital electronics to the bigger voltage jumps occurring in $U(t)$ [17].

Considering these data, we can extract the following results.

1) The form of the noise spectrum is independent of the field rate as long as one sweeps slowly through the hysteresis loop. In this region the power density $W(f)$ increases at any frequency f proportional to the magnetizing frequency f_M whereas the total noise $E(f)$ obtained by integration along the hysteresis loop is independent of f_M.

Remark: The different meanings of the two introduced frequencies should be noticed. f is the selected Fourier component of the noise, but f_M is the frequency with which the core is periodically magnetized. For instance, magnetizing frequency f_M = 0.1 Hz (cf. Fig. 7), or a linear sweep dH_a/dt = 2.5 mOe·s⁻¹ with $\hat{H}_a$ = 10 Oe which corresponds to $f_M = 6.3 \times 10^{-5}$ Hz.

2) $W(f)$, and also $E(f)$, already decrease in the region of a few hertz, which only is possible if elementary events of long duration are present in the whole process. The spectrum is not of the type to be expected from comparison with the shot noise of a diode, if one regards each local jump of a Bloch wall as an event that is independent of what happens elsewhere in the sample. Rather, the magnetization processes occur in big clusters; this can be verified by direct observation of the voltage $U(t)$ on the oscilloscope.

3) The formation of clusters is predominant at the steep parts of the hysteresis loop and finds its expression by enhancing the noise in the region of low frequencies. Here, Fig. 6 may be recalled. (If one observes the noise at 18 kHz, then $W(f)$ is always proportional dB/dH_a along the hysteresis loop.) There are more elementary events per unit time in the steep part of the loop, which enhance the noise. If one looks at the noise at 15.7 Hz, then $W(f)$ is about proportional $(dB/dH_a)^2$. With increasing dB/dH_a, not only is the number of elementary events enlarged but, in addition, they tend to form clusters.

4) Not all ferromagnetic materials exhibit such a formation of clusters. This ability was found to be weak in hard magnetic samples of 50-percent Ni–Fe. These samples had only a weak noise as shown in Table I. $E(f)/n^2A$ is the noise energy obtained from a half-period sweep $-\hat{H}_a \rightarrow +\hat{H}_a$, referred to a bandwidth of 1 Hz, one turn of the pick-up coil, and a sample cross section of 1 cm². Although they are taken at somewhat different frequencies, the results can be compared because of their weak frequency dependence (Fig. 4). f_M is magnetizing frequency. No Barkhausen noise could be detected in Mumetal [8]. Here one has to assume that the already small irreversible magnetization consists of extremely small discontinuities, which are to a considerable extent independent of each other.

TABLE I
EXPERIMENTAL RESULTS OF LOW-FREQUENCY BARKHAUSEN NOISE

Reference	Material	H_c (Oe)	Sweep Rate	H_a (Oe)	f (Hz)	$E(f)/n^2A$ $V^2 \cdot s^2 \cdot cm^{-2}$ turn^{-2}
[7]	Fe wire $l = 35$ mm 0.5-mm diameter	≈ 0.25	$f_M = 0.2$ Hz		100	4×10^{-15}
[7]	6-percent Si–Fe toroid	≈ 0.2	$f_M = 0.1$ Hz		10	0.3×10^{-13}
[13]	50-percent Ni–Fe hard wire $l = 200$ mm 1-mm diameter (weak coupling)	4.0	$\dot{H}_a = 20$ mOe·s^{-1}	34	45	2.5×10^{-15}
[13]	81-percent Ni–Fe hard wire $l = 200$ mm 1-mm diameter	8.3	$\dot{H}_a = 50$ mOe·s^{-1}	34	90	6.4×10^{-13}
[6]	3-percent Si–Fe grain oriented	0.35	$\dot{H}_a = 2.5$ mOe·s^{-1}	10	1	5×10^{-13}
Independent reversal of the single spins ($l = 1$ cm)						2×10^{-30}

5) The tendency to form clusters only comes to its full extent if the magnetic circuit does not give rise to a demagnetizing effect. Each individual process controlled by the demagnetizing factor generates a counterfield, which acts like a negative feedback on the further development of a cluster. Therefore, shortening the sample reduces the low-frequency noise because of the increased demagnetizing factor (see Fig. 4).

6) The power spectrum that can be observed with metallic materials in the range of high frequencies is mainly determined by the electrodynamics of the magnetic circuit. Even if the local magnetizing processes are very fast this nevertheless cannot be detected by the pick-up coil because of the shielding eddy currents, which act as a low-pass filter. Only when the Barkhausen noise of different materials are compared might it be regarded as meaningful if the total noise is not considered, that is the noise integrated over a large frequency range, but only the noise in the region of the lowest possible frequencies (see Table I).

7) Up to now, we only considered the noise generated while sweeping along a complete hysteresis cycle from saturation to saturation. If we focus our interest on the noise from small minor loops, we can state that in the region where Rayleigh's law is valid, i.e., for amplitudes $\hat{H}_a$ smaller than say, $0.5H_c$, the noise is extremely weak. The noise rather abruptly sets in if the field is swept beyond this limit. Obviously, the Rayleigh law is valid as long as the individual processes are independent of each other. Exceeding this region introduces coupling between the processes, which in turn leads to the formation of clusters.

BEHAVIOR AT HIGHER MAGNETIZING FREQUENCIES f_M

$E(f)$ remains independent of dH_a/dt and thus of f_M as long as the local magnetization processes are completed before they can be influenced by a further increase in the field. Warren found in his measurements on Stalloy that $E(f)$ remains independent of f_M, and thus $W(f)$ proportional to f_M is valid only for $f_M \leq 0.5$ Hz. For faster magnetization cycles the noise remains below this value (Fig. 7); in the range 200 Hz $\leq f_M \leq$ 1000 Hz, the measured noise again is assumed to be proportional to f_M, but now with a much smaller proportionality constant.

The noise shown in Fig. 7 is obtained with a wide-band amplifier. A difficulty arises if one wishes to know the function $W(f)$ of this noise in detail. When sweeping the field periodically with higher magnetizing frequencies f_M, harmonic components of the magnetizing field enter the frequency range of the noise that one wishes to analyze. In addition, the noise is modified by the periodically changing density of the Barkhausen pulses. This problem is discussed in detail by Mazzetti and Montalenti [7] and [8], using the fundamental results of Bunkin [20].

ESTIMATE OF THE BARKHAUSEN NOISE

As will be shown in the following, the formation of clusters, which is responsible for the strong noise of a ferromagnetic material, involves a very complicated spatial and temporal interaction of magnetization processes. Strictly speaking, the noise cannot be described simply by a material constant and a factor for the geometric dimensions of the core. This nevertheless is done in the following because, with a simple model, it is possible to understand, at least qualitatively, the essential relationships.

If one assumes that the elementary process is the fast reversal of a magnetic moment from position $-m$ to $+m$ referred to the coil axis, then a great number of equal and independent processes produce a white noise in the low-frequency region

$$W(f) = 2 \cdot (n^2/l^2) \cdot (2m)^2 z \tag{5}$$

(shot-noise formula with the average number z of element-

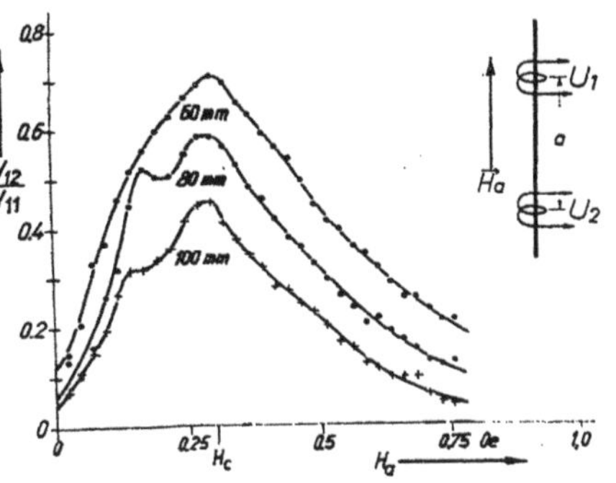

Fig. 8. Ratio of W_{12}/W_{11} versus H_a for three different coil distances a.

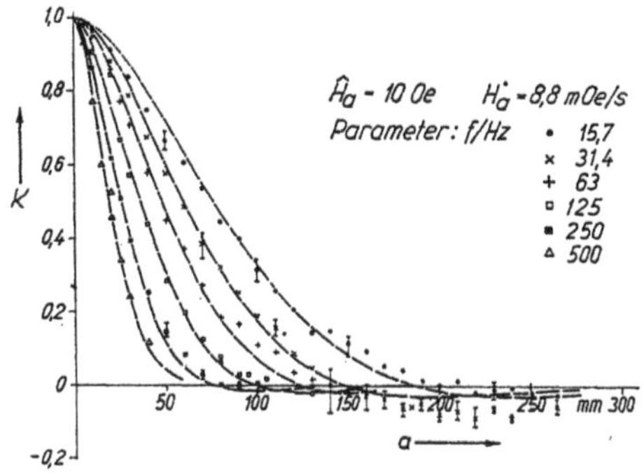

Fig. 9. Normalized cross spectrum at different frequencies f and coil distances a.

ary processes per second; notation after Fig. 1; frequency $f > 0$).

Because

$$\int_0^T 2mz\,dt = 2I_sAl \tag{6}$$

when changing the magnetization from saturation to saturation in the time T, we obtain,

$$E(f) = 8(n^2/l)AmI_s. \tag{7}$$

Referring to the unit of cross section, to one turn and a length $l = 1$ cm, we get, with the saturation magnetization, $I_s = 2 \times 10^{-4}\,\mathrm{V\cdot s\cdot cm^{-2}}$,

$$\frac{n^{-2}A^{-1}E(f)}{\mathrm{V^2\cdot s^2\cdot cm^{-2}\ turn^{-2}}} = 16 \times 10^{-4}\frac{m}{\mathrm{V\cdot s\cdot cm}}. \tag{8}$$

The total integrated noise $E(f)$ is directly proportional to the magnetic moment m of the single individual process. Three typical values may be given.

1) Statistically independent reversal of the single spins:

$$m = m_{\mathrm{Bohr}} = 1.2 \times 10^{-27}\,\mathrm{V\cdot s\cdot cm}$$

$$n^{-2}A^{-1}E(f) = 2 \times 10^{-30}\,\mathrm{V^2\cdot s^2\cdot cm^{-2}\ turn^{-2}}. \tag{9}$$

This value is included in Table I for comparison with the observed Barkhausen noise.

2) The elementary process consists of the jump of a 180° wall over a volume $v = 10^{-12}$ cm³. With a favorable direction of magnetization we obtain

$$m = vI_s = 2 \times 10^{-16}\,\mathrm{V\cdot s\cdot cm} \tag{10}$$

and with (8),

$$n^{-2}A^{-1}E(f) = 3 \times 10^{-19}\,\mathrm{V^2\cdot s^2\cdot cm^{-2}\ turn^{-2}}. \tag{11}$$

This value still is considerably smaller than the observed noise as can be seen by comparing it with Table I.

3) A cluster of 10^6 jumps of the type described in 2) is formed in the elementary process. At low frequencies, where the Fourier analysis does not resolve all the minor processes involved in the cluster formation, we obtain

$$n^{-2}A^{-1}E(f) = 3 \times 10^{-13}\,\mathrm{V^2\cdot s^2\cdot cm^{-2}\ turn^{-2}}. \tag{12}$$

Now the value given by (12) has the same magnitude as the observed noise (see Table I). A number of 10^{11} and, if cluster formation is taken into account, a number of 10^{17} spins are involved in each magnetic discontinuity.

These are macroscopic effects, which, as could be assumed, should not exhibit a random behavior. But each discontinuity consists of a large number of smaller processes that are strongly coupled. This coupling removes the apparent contradiction.

Mechanism of Cluster Formation in Grain-Oriented 3-Percent Si–Fe

The observed noise spectra $W(f)$ and $E(f)$ show that a large number of single magnetization processes participate in a strongly coupled way to build clusters whose formation needs a time of up to 100 ms. One can assume that these clusters have a certain spatial extension. This can be determined, as by Heiden [13], by comparing the magnetization changes at different locations on the sample. Two examples are given in Figs. 8 and 9. Fig. 8 shows that the two voltages $U_1(t)$ and $U_2(t)$ are dependent on each other if induced while sweeping through the hysteresis loop. The ratio of the cross-power spectrum W_{12} to the power spectrum W_{11} is a measure for the correlation. Results are for $f = 35.5$ Hz and for three different coil distances a. Reducing this distance increases the correlation between both voltages. This correlation is very pronounced in the steep part of the hysteresis loop. The material is the same as in Fig. 4, $l = 300$ mm, $\dot{H}_a = 27$ $\mathrm{mOe\cdot s^{-1}}$. Fig. 9 shows the normalized cross-spectrum $K = E_{12}/E_{11}$ at different frequencies f and coil distances a. The correlation of the signals induced in both coils increases with decreasing a and f. Negative K values occur near $a = 200$ mm. The curves are obtained from the diffusion calculations.

Measuring on the strip-like sample with two equal pick-up coils $U_1(t)$ and $U_2(t)$ gives, on the one hand, the power spectrum $W_{11}(f)$ with

$$\overline{U_1^2} = \int W_{11}(f)\,df \tag{13}$$

and on the other hand, the cross-power spectrum $W_{12}(f)$

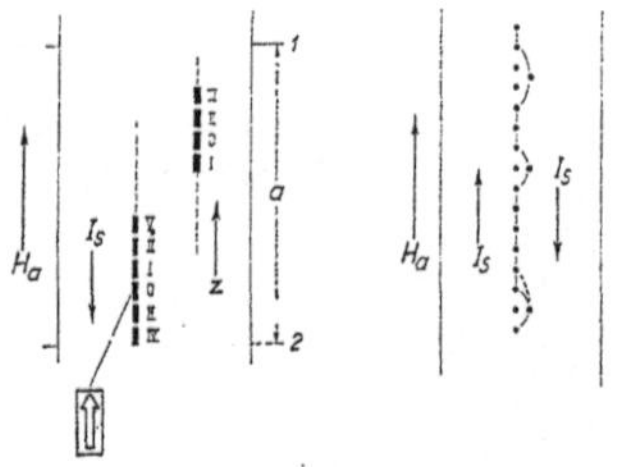

(a) (b)

Fig. 10. (a) Model for magnetization process. (b) Magnetization changes from a single Bloch wall.

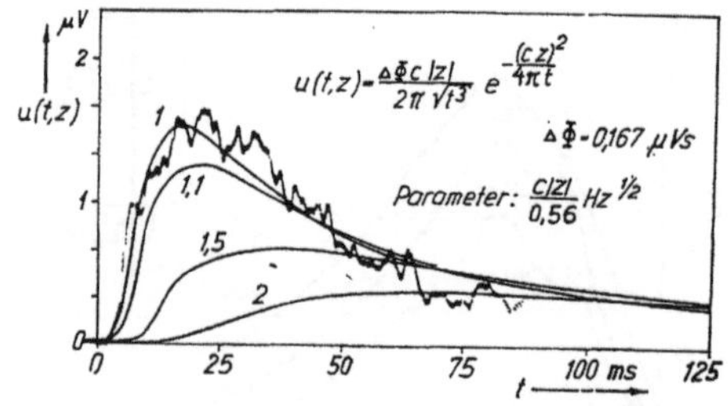

Fig. 11. Pulse transfer by irreversible magnetization processes. Four pulses $u(t)$ are calculated with diffusion theory for different distances. They are compared with an observed pulse.

with

$$\overline{U_1 U_2} = \int W_{12}(f)\, df. \tag{14}$$

An analog situation is valid for the energy spectra E_{11} and E_{12} with

$$K = \frac{E_{12}}{E_{11}} \tag{15}$$

$K(f,a)$ being the normalized cross-energy spectrum for the coil-distance a and the frequency f.

The ratio $W_{12}/W_{11} \leq 1$ is a measure for the correlation that exists between the two voltages. It is a very far-reaching correlation; a strong dependence between the two voltages can be seen in Fig. 8 for a distance $a = 100$ mm although the cross section of the sample amounts to only 0.1×2 mm^2. In addition, the frequency dependence in Fig. 9 shows that time is needed for the correlation to take its maximum effect. This means we are dealing with a spatial and temporal transient phenomenon.

The Propagation of Magnetization Changes as a Diffusion Problem

Fig. 10(a) gives a model for the magnetization process. It is a strip-like 3-percent Si–Fe sample with its main extension in z direction. One assumes that by the external field a first process 0 is initiated, which is followed by further processes in the sequence I, II, III, IV, $\cdots$. An irregular propagation in z direction leads to long threads whose magnetization is reversed. The signals $U_1(t)$ and $U_2(t)$, induced in the two short pick-up coils at the positions 1 and 2 of the sample, will be correlated to such an extent as single thread-like structures penetrate both cross sections at the coil positions.

If the starting points 0 are assumed to be evenly distributed, the normalized cross spectrum $K(f,a)$ can be calculated, if the law that controls the average growth of the threads is known.

The sum of all threads gives rise to a macroscopic magnetic flux $\phi(z,t)$. Since the behavior of the model reminds us of a diffusion process, we shall try to describe the average behavior of the thread-like structures by a diffusion equation,

$$\frac{\partial^2 \phi}{\partial z^2} = \frac{c^2}{\pi} \frac{\partial \phi}{\partial t}. \tag{16}$$

Diffusion coefficient π/c^2. The periodic solution gives:

$$\phi = \phi_0 \exp\left(-c\sqrt{f}\, z\right) \exp\left[j(\omega t - c\sqrt{f}\, z)\right]. \tag{17}$$

This damped wave has the phase velocity

$$\frac{\omega}{c\sqrt{f}} = 2\pi c^{-1}\sqrt{f} \tag{18}$$

and the range $c^{-1} f^{-1/2}$. With $c = 5 \times 10^{-3}$ mm^{-1} Hz$^{-1/2}$ and a frequency $f = 100$ Hz, a phase velocity of 12.7 ms^{-1} is obtained. However, it should be noticed that the propagation velocity for pulses is not constant but decreases with increasing z since a dispersion law is active in (17).

Equation (16) was checked by Grosse-Nobis [6] in two different ways.

1) *Pulse Transfer:* At the position $z = 0$, a voltage pulse

$$u(t, z = 0) = \Delta\phi \cdot \delta(t) \tag{19}$$

is applied and the voltage pulses at position $z \neq 0$ are observed. Fig. 11 shows the good agreement between theory and experiment.

2) *Cross Spectrum $K(f,a)$:* Using (16), the cross spectrum is calculated and compared with the observed values (Fig. 9).

Fig. 10(b) shows a possible mechanism for the propagation of magnetization changes from a single Bloch wall. The wall is torn off locally from defects by the magnetic-field pressure thus favoring the magnetization change of the neighboring wall section. The characteristic relaxation time for the single process is made by local eddying currents, which act against further propagation. The time needed for formation of the whole cluster is calculated by adding in a complicated way the relaxation times of a great number of single processes.

Acknowledgment

The author is grateful to Dr. Grosse-Nobis, Dr. Heiden, and Priv.-Doz. Dr. Storm for their fruitful cooperation, and thanks the Vakuumschmelze GmbH Hanau for the sample materials.

References

[1] K. G. Warren, "Barkhausen noise in transformer cores," *Electron. Technol.*, vol. 38, p. 89, March 1961.
[2] H. Bittel and H. Lütgemeier, "Über spontane zeitliche Schwankungen der Magnetisierung eines Ferromagnetikums," *Z. Angew. Phys.*, vol. 15, p. 476, 1963.
[3] W. Hampe, "Relaxationen und Resonanzen im Spektrum der Permeabilität ferromagnetischer Werkstoffe," *Z. Angew. Phys.*, vol. 18, p. 395, 1965.
[4] H. Bittel and C. Heiden, *Kopplungserscheinungen zwischen ferromagnetischen Elementarprozessen*, Forschungsbericht des Landes Nordrhein-Westfalen, no. 1663. Köln und Opladen: Westdeutscher Verlag, 1966.
[5] H. Lütgemeier, "Magnetisierungsvorgänge im pauschal unmagnetischen Zustand, die als Folge einer Temperaturänderung auftreten," *Z. Angew. Phys.*, vol. 16, p. 167, 1963.
[6] W. Grosse-Nobis, "Untersuchung der zeitlichen und räumlichen Korrelation von Ummagnetisierungsvorgängen in Silizium-Eisen," doctoral dissertation, Universität Münster, Münster, Germany, 1968.
[7] P. Mazzetti and G. Montalenti, "Sul calcolo del rumore di fondo in circuiti con elementi utilizzanti materiali ferromagnetici," *l'Energia Elettrica*, vol. 39, p. 562, 1962.
[8] ——, "Power spectrum of the Barkhausen noise of various magnetic materials," *J. Appl. Phys.*, vol. 34, p. 3223, 1963.
[9] H. Lütgemeier, "Über das Frequenzspektrum des Barkhausen-Rauschens," *Z. Angew. Phys.*, vol. 16, p. 153, 1963.
[10] L. Storm and C. Heiden, "Untersuchungen über das Frequenzspektrum des Barkhausen-Rauschens," *Z. Angew. Phys.*, vol. 17, p. 161, 1964.
[11] L. Storm, C. Heiden, and W. Grosse-Nobis, "The power spectrum of the Barkhausen noise of grain-oriented three-percent silicon iron," *IEEE Trans. Magnetics*, vol. MAG-2, pp. 434–438, 1966.
[12] H. Lütgemeier, "Über das Rauschen von Spulen mit ferromagnetischen Kernen," doctoral dissertation, Universität Münster, Münster, Germany, 1962.
[13] C. Heiden, "Untersuchungen über Kopplungen zwischen irreversiblen Magnetisierungsprozessen," doctoral dissertation, Universität Münster, Münster, Germany, 1964.
[14] P. Mazzetti, "Correlation function and power spectrum of a train of nonindependent overlapping pulses having random shape and amplitude," *Nuovo Cimento*, vol. 31, p. 88, 1964.
[15] P. Mazzetti and G. Montalenti, "The theory of the power spectrum of Barkhausen noise," *Proc. 1964 Internatl. Conf. on Magnetism* (Nottingham, England), p. 701.
[16] P. Y. Arques, "Sur certains problèmes statistiques liés a l'effet de Barkhausen," *J. Phys.*, vol. 29, p. 369, 1968.
[17] R. Kassing, "Messungen der Größen-, Zeitdauer- und Abstandsverteilungen der Barkhausen-Impulse von dreiprozentigem Silizium-Eisen," doctoral dissertation, Universität Münster, Münster, Germany, 1968.
[18] L. Storm, "Irreversible Ummagnetisierungsvorgänge in Ferromagnetika und der Barkhauseneffekt," *Z. Angew. Phys.*, vol. 26, p. 91, 1969.
[19] J. J. Brophy, *Fluctuation Phenomena in Solids*, R. E. Burgess, Ed. New York: Academic Press, 1965.
[20] F. V. Bunkin, "On the statistical nature of remagnetization in ferromagnets," *Soviet Phys.—Tech. Phys.*, vol. 1, p. 1727, 1956.

TECHNICAL FEATURE

INTERMODULATION CHARACTERISTICS OF FERRITE-BASED DIRECTIONAL COUPLERS

The advent of new broadband digital cable communications systems has led to exciting new developments in the areas of interactive television, Internet services, Internet protocol telephony over cable and a host of other features being offered by cable operators. This impressive digital revolution has brought with it some major challenges to cable equipment suppliers. Foremost among these challenges is the issue of intermodulation (or cross-modulation) distortion. Of course, this problem has always been present in the older analog systems, but the higher power levels and increased numbers of digital channels in the new-generation systems have exposed unseen problems. The concept of intermodulation in active devices such as amplifiers and mixers is familiar. Intermodulation products are expected to be generated in such nonlinear components, and systems have always been designed with this in mind. Less common and equally less documented are the effects of intermodulation in passive components such as couplers and splitters.

Evidence of the limitations of existing passive component designs was first identified in 1997. It was determined that, under certain power levels and signal conditions, intermodulation was seen after signals were passed through one of the standard coupler products used in a new cable modem application. Once alerted to the potential problem, a research program aimed at identifying the causes of intermodulation in passive ferrite devices was begun and solutions that would alleviate the problem were sought. More recently, evidence has emerged that the issue of passive component intermodulation is also having an impact in other network equipment, such as line amplifiers and extenders, fiber nodes and network interface units. Two primary causes of intermodulation or cross-modulation in ferrite-based passive components have been identified.

COMPONENT CHARACTERISTICS

The type of passive component determined to be the cause of the intermodulation problem is a ferrite core-based component constructed by winding a number of turns of wire around a toroid or balun core. The windings are configured to create a variety of different devices, the most common of which are transformers, couplers and power splitters. The wound element is most commonly attached to a PCB or ceramic substrate for convenient automatic handling and board level insertion.

SHANE COLLINS AND KIERAN FLYNN
M/A-COM Eurotec Operations

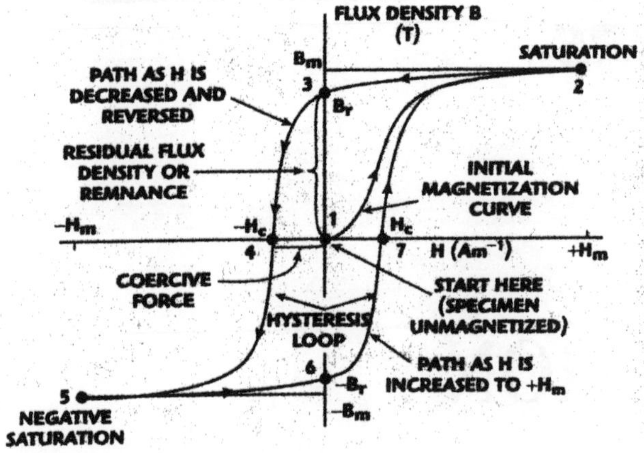

▲ *Fig. 1 The B vs. H hysteresis curve for magnetic devices.*

FERRITE CORE PROPERTIES

One cause of cross-modulation is related to the basic material properties of the ferrite core and the level of power being transmitted through the device. Two important terms in relation to ferrite-based components are magnetic field intensity (H) and magnetic flux density (B). The magnetic field intensity is proportional to the level of electrical signal voltage or power level that is input to the device. The magnetic flux density induced by the applied electrical signal is described by a formula that includes physical properties of the core as well as the applied electrical signal:

$$B = \frac{E \times 10^8}{4.44 \times F \times N \times A}$$

where

E = applied voltage or signal level
F = frequency of operation
N = number of turns of wire
A = cross-sectional area of the core

The relationship of B vs. H is described by the characteristic hysteresis curve for magnetic devices, shown in ***Figure 1***. The precise size and shape of the curve are dependent on core properties; the most significant of these is the B/H ratio, referred to as the permeability. For a given material, the hysteresis varies with excitation levels, that is, the shape of the curve changes with higher voltage levels, thus altering the nonlinear behavior.

Intermodulation in active devices such as mixers and amplifiers occurs due to nonlinear characteristics in the active element such as a diode or FET. The B/H hysteresis curve also exhibits nonlinear characteristics under high input signal levels. It can be seen that as the input signal is increased the magnetic flux density also increases, linearly at first, but eventually reaches saturation. It is this area of the B/H plot, where the curve tends toward saturation at a high level, that gives rise to the cross-modula-

TABLE I

SECOND HARMONIC PERFORMANCE

	Permeability	Cross-sectional Area (Type 1 or 2)	Material	Second Harmonic Level (dBc) 20 dBm Input	24 dBm Input	27 dBm Input
Core 1	4400	1	MnZn	102	99	97
Core 2	2500	1	MnZn	> 115	> 115	112
Core 3	7500	1	MnZn	100	97	95
Core 4	370	1	NiZn	> 115	> 115	> 115
Core 5	5000	2	MnZn	> 115	> 115	111
Core 6	850	2	NiZn	> 115	> 115	> 115

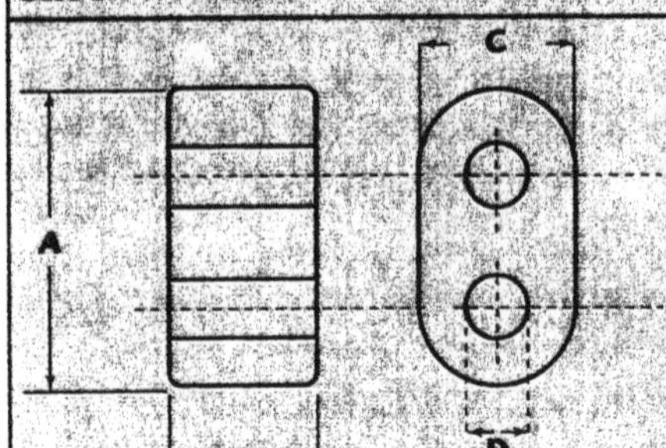

	Type 1 Dimensions (mm)	Type 2 Dimensions (mm)
A	3.50	7.00
B	2.54	3.00
C	2.00	4.00
D	0.86	0.86

tion characteristics that previously went unnoticed in the lower power analog systems. A series of experiments were conducted to investigate the relationship among B/H ratio (permeability), cross-sectional area and cross-modulation performance. Cores with a number of different permeabilities and two different cross-sectional areas were selected for the experiments.

To achieve good performance at 5 MHz in low coupling designs, a high permeability core is normally selected. This high permeability is necessary to compensate for the lower impedance due to the reduced turns-squared ratio. (Minimal turns ratios are necessary to achieve the relevant coupling responses.) However, the undesirable cross-modulation products have been found to be more significant in high permeability cores. To illustrate the problem, all of the cores were wound in the form of a 10 dB coupler and soldered into a fixture. A high power signal below 40 MHz (F1) was applied at the input port and the second harmonic was recovered at the output port (2F1). ***Table 1*** lists the second harmonic performance as a function of permeability and cross-sectional area for three different power levels of the input signal.

Core 1 has a permeability of 4400 and is relatively small. It displays a high degree of distortion even at 20 dBm. Maintaining type 1 dimensions and increasing the permeability raise the distortion levels further (core 3) while lowering the permeability (core 2) reduces the amount of harmonics. Increasing the cross-sectional area lowers the nonlinear effects. Core 5 has a higher cross-sectional area than core 1, but a slightly higher permeability. However, while this core has better harmonic performance it would be an impractical choice due to the higher insertion loss that would result in the high frequency response. The nickel zinc (NiZn) cores exhibit excellent distortion characteristics. Even at power levels up to 2 W, the coupler is linear. However, the disadvantage is that the 5 MHz loss increases considerably due to the lack of permeability.

The distortion problem is also turns-dependent, for example, core 1 wound as a 12 or 14 dB coupler exhibits less distortion while, in contrast, wound as an 8 dB coupler, it produces an increased harmonic level. Thus, it can be seen that careful selection of the core material, permeability and physical size is critical to achieving good cross-modulation performance. Equation 1 demonstrates that choosing the right permeability can help in achieving the optimum linear region for the hysteresis curve. Similarly, choosing a core with a higher cross-sectional area also leads to a lower flux density for a given input, thus increasing the saturation power level. However, proper core selection means keeping the right balance between the cross-modulation performance and the performance of the other RF parameters.

FERRITE CORE RESISTIVITY

A second cause of cross-modulation is related to core resistivity. Traditional passive components have used manganese zinc (MnZn) low resistivity cores. In order to prevent accidental shorting of the wires to the core, the core is often coated with an insulating material known as paraylene. The wire is also insulated to prevent interwinding shorts that would impact the performance of the component. It has been found that standard manual assembly operations can cause the insulation in the wires and the paraylene on the cores to become damaged, leading to a short in a percentage of units. While not detrimental to standard RF parameters such as insertion loss and isolation, this short does have a significant impact on cross-modulation performance. Tests have shown that performance can be degraded by as much as 30 dB if such a short exists. Unlike the previous distortion effect, this occurrence is independent of excitation levels and turns ratio. The applied signal creates nonlinearities depending on resistivity.

The data plots illustrate the effects of removing insulation on a MnZn core. ***Figure 2*** shows the third-order intermodulation product before the insulation is removed. ***Figure 3*** shows the distortion that results when the wire is exposed to the core surface. This effect occurs due to the low resistivity of the core that allows the signal to pass through the core, creating nonlinear effects.

An alternative material, NiZn, which has a high resistivity, is the most appropriate solution to the problem described here. Tests have demonstrated that shorts to the high resistivity material have little or no impact on the cross modulation of the device. Distortion using the NiZn core is shown in ***Figure 4***; distortion using the NiZn core after tuning with the wire shorting to the core is shown in ***Figure 5***.

▼ *Fig. 2 The coupler's distortion.*

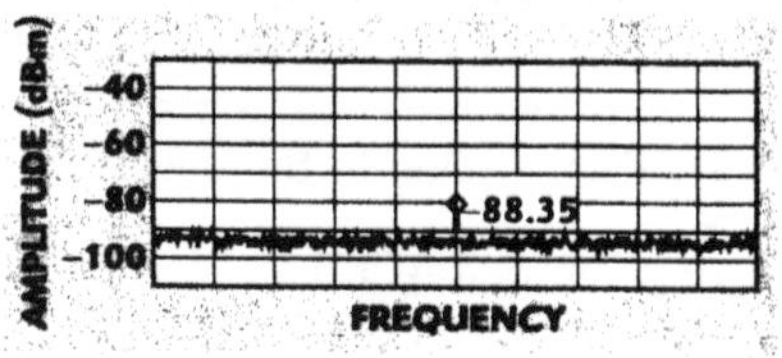

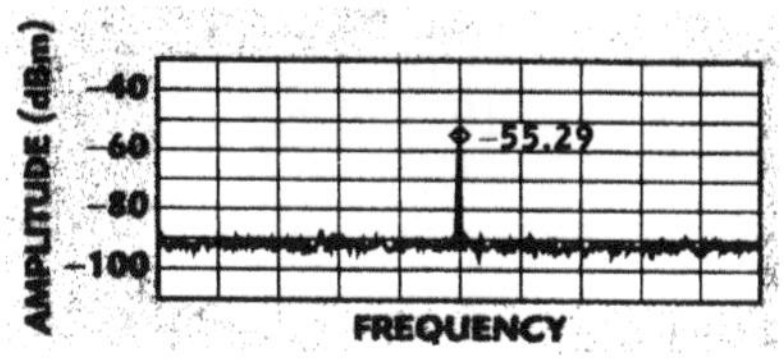

▲ *Fig. 3 Distortion of coupler with wire insulation removed.*

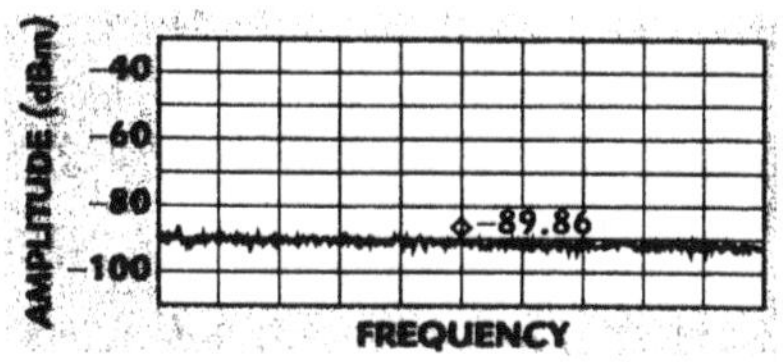

▲ *Fig. 4 Distortion using the NiZn core.*

CHARACTERIZATION OF CROSS-MODULATION PERFORMANCE

A representative two-tone, third-order intermodulation test was devised to show the relative merits of various core materials and parameters. ***Figure 6*** shows the test setup that was used to demonstrate the effect of using the high resistivity NiZn cores. The generator is set to a relatively low signal to reduce the harmonics created by the amplifier. The bandpass filters attenuate distortion from the amplifiers and pass only the required fundamental frequencies.

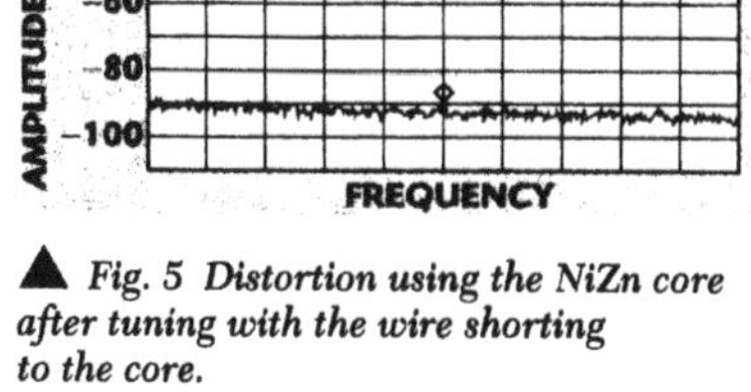

▲ *Fig. 5 Distortion using the NiZn core after tuning with the wire shorting to the core.*

The power level from the two amplifiers is approximately 23 dBm, which, after passing the two filters and splitter, results in two tones of approximately 14 dBm applied to the device under test.

It has been found that the core resistivity test is somewhat independent of power level. The lowpass filter at the output of the coupler is chosen to pass the desired IF, which, for the purposes of this test, is 40 MHz. If available, a bandpass filter also can be used at the IF output and may provide improved results. The 10 dB pad is used to reduce the signal at the spectrum analyzer front end. Signals above +5 dBm may create unwanted harmonics that could add to the measurement error. The amplifier used to create the high power signals must have low distortion and care must be taken to avoid operating the amplifiers near saturation, which can cause unwanted spurious responses in the test system.

THE SOLUTION

Recognizing the significant impact that cross-modulation has to play in the design of the new-generation system and line amplifiers, research aimed at identifying the most appropriate core characteristics to achieve excellent cross-modulation as well as maintaining good RF performance on other critical parameters has been conducted. The result is a range of low intermodulation couplers specifically designed for broadband amplifier and other system applications. These new devices use NiZn cores with a typical resistivity above 10^8. Cross-sectional area and permeability are selected to optimize cross-modulation, isolation and bandwidth performance. ***Table 2*** lists the performance of the standard RF coupler products now available as a result of this effort.

CONCLUSION

Levels of intermodulation and cross-modulation performance have been identified in ferrite-based passive components that create performance problems in today's broadband digital communications systems. Proper selection of ferrite core material with regard to permeability and size has been determined to minimize these effects, and a series of ferrite-based RF couplers have been designed that demonstrate excellent distortion performance as a result of this investigation. ■

Shane Collins *received his BEng degree in electrical engineering from the University of Ulster, Belfast. He received his MSc in electronics from Queens University, Belfast. Currently, Collins is a design engineer with M/A-COM Eurotec Operations in Cork, Ireland, where he has been designing active and passive components for the wireless and broadband communications market for the past three years.*

Kieran Flynn *received his BE in electrical engineering from University College, Cork in 1980. In 1995, he received a certificate in marketing from the Irish Marketing Institute. He worked as a design engineer for eight years in the UK defense industry before transferring to a career in sales and marketing. Flynn is currently global sales and marketing manager for the M/A-COM Eurotec E-Series product line in Cork, Ireland.*

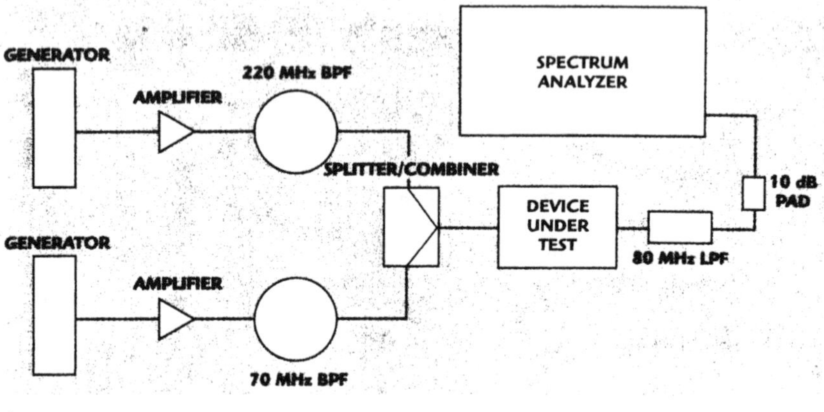

▲ *Fig. 6 The test setup.*

TABLE II
RF COUPLER PERFORMANCE

Frequency Range (MHz)	Insertion Loss (dB)	Coupling (dB)	Return Loss (dB)	Distortion at 27 dBm (dBc)
40 to 870	1.5	8.5 ±0.4	18	> 110
40 to 870	1.3	10.4 ±0.6	20	> 110
5 to 870	0.6	17.5 ±0.5	24	> 110

Chapter 3

BALUNS/HYBRIDS/ANTI-PHASE POWER SPLITTERS

3.1 INTRODUCTION

In an unbalanced port, one of its two terminals is connected to the ground, an example being the output of a conventional signal generator. A balanced port, on the other hand, is one where both terminals are floating with respect to ground [e.g., the input to a dipole antenna as shown in Figure 3.1(a)]. A BALUN is a device for interconnecting a BALanced port to an UNbalanced one. The ideal balun is a lossless, perfectly matched, two-port network whose properties are independent of frequency, power level and the like, and may also provide some impedance transformation as well. The ideal balun can be realized by the ideal transformer shown in Figure 3.1(b), and deviations from the ideal depend solely on how closely one can realize the ideal transformer in practice, as for example, the inevitable inter-winding capacitance limits the high frequency performance.

The dipole antenna shown in Figure 3.1(a) has a virtual ground at the mid-point between the terminals, but an important class of applications has a real physical ground between the terminals, for example, a balanced mixer, [1] or push-pull transistor, shown in Figure 3.2. Although it is common practice to refer to the device interconnecting a push-pull transistor to an unbalanced port as a balun, it is debatable whether this is the most appropriate terminology. Another term in common use is hybrid,[1] but, again, this term is not always an appropriate description. A better choice might be an anti-phase power splitter. Certainly the term "anti-phase power splitter" captures one essential aspect, namely that it is a three-port device, and it is necessary to realize that a three-port device is being used to understand some of the basic circuit properties. However, in the interest of simplicity, this chapter will retain the use of the term balun when there is no ambiguity. Most, if not all, baluns can be converted into anti-phase power splitters by grounding the center point of the balanced load as shown in Figure 3.3.

In the same way that an ideal transformer is an idealized realization of the two-port balun, a center-tapped transformer is an idealized realization of an anti-phase power splitter. It is instructive to analyze the ideal center-tapped transformer in detail

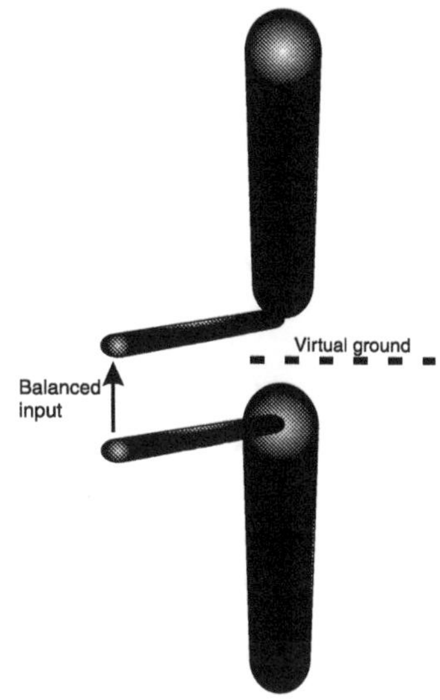

Figure 3.1(a) *A dipole antenna—an example of a balanced port.*

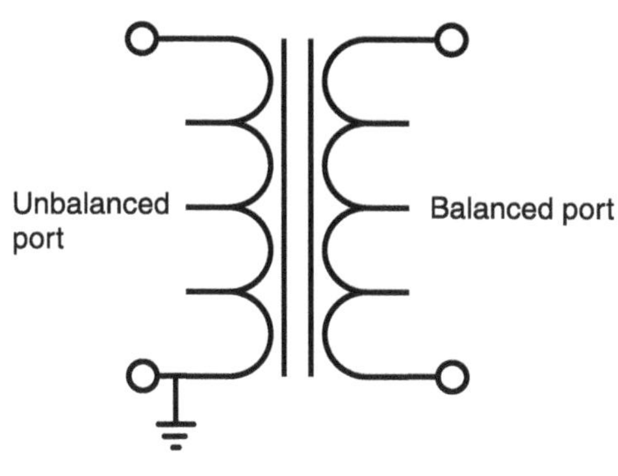

Figure 3.1(b) *Realization of an ideal balun by an ideal transformer.*

Figure 3.2 *A push-pull transistor.*

1. The term "hybrid" refers to a matched four-port which can combine two signals into sum and difference signals. The term comes from the old telephone "hybrid coil" which functions as an in-phase/anti-phase coupler when all four ports (including the center tap) are terminated properly [2].

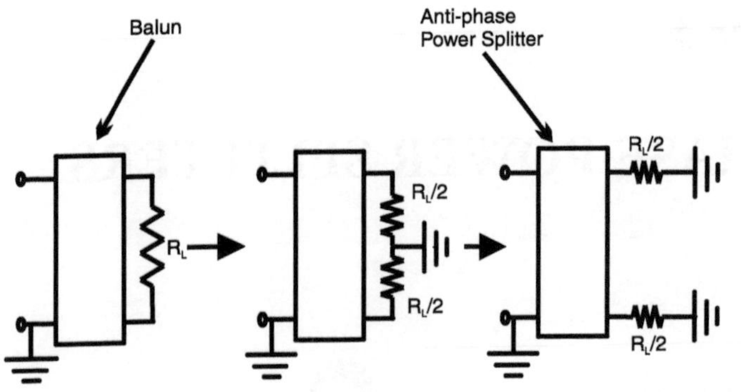

Figure 3.3 *Conversion of a balun into an anti-phase power splitter.*

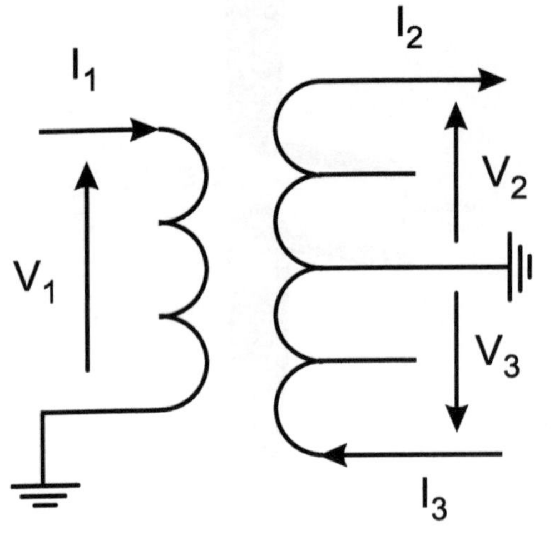

Figure 3.4 *Center-tapped transformer as an anti-phase power splitter.*

to understand some basic properties of this type of network. Figure 3.4 shows the voltage and current relationships. Straightforward analysis shows that the magnitudes of the S parameters are given by

$$\begin{pmatrix} \frac{1-2n^2}{1+2n^2} & \frac{2n}{1+2n^2} & \frac{2n}{1+2n^2} \\ \frac{2n}{1+2n^2} & \frac{1}{1+2n^2} & \frac{2n^2}{1+2n^2} \\ \frac{2n}{1+2n^2} & \frac{2n^2}{1+2n^2} & \frac{1}{1+2n^2} \end{pmatrix} \tag{3.1}$$

It can be seen from (3.1) that a perfect match can be achieved at port 1 (i.e., $S_{11} = 0$ if $n = 1/\sqrt{2}$), but that it is impossible to achieve a perfect match at ports 2 or 3 under any circumstances. Also, from energy conservation,

$$|S_{22}|^2 + |S_{21}|^2 + |S_{23}|^2 = 1 \tag{3.2}$$

but since $|S_{21}|^2 = 0.5$ if $S_{11} = 0$, then $|S_{22}|^2 + |S_{23}|^2 = 0.5$ and so the better the match at port 2 then the worse the isolation between ports 2 and 3 (and vice versa) with a worst case scenario of only 3 dB isolation. If $n = 1/\sqrt{2}$, then the isolation is 6 dB and ports 2 and 3 have 6 dB return loss. Figure 3.5 shows a commercially available center-tapped transformer; this type of component has limited current and power handling capability coupled with relatively high insertion loss that restricts its use to relatively low power applications such as balanced mixers.

Figure 3.5 *An example of a center-tapped transformer. (Photograph courtesy of Tyco Electronics. Reprinted with permission.)*

The limitations on port match and isolation described above are inherent properties of any lossless, reciprocal, three-port network [3]. However, these limitations can be overcome using either a lossless reciprocal four-port network or a lossy three-port network. Obviously, a lossless four-port network can be considered to be a lossy three-port network, the choice of which description is chosen largely depends on whether or not one has easy access to the terminals to which the resistor is connected to make measurements at. For example, a Wilkinson power divider [4] would normally be considered a lossy three-port network since it is difficult to gain access to the resistor's terminals as both are floating with respect to ground, whereas a conventional parallel coupled-line directional coupler [5] would normally be considered a lossless four-port network. It is very important to note that the presence of one or more resistors in a three-port network does not necessarily imply that energy will be dissipated in the resistor. As an example, no energy is dissipated in the internal resistor of a Wilkinson power divider under normal operation with all ports terminated in matched loads.

All the baluns to be considered can be categorized into three groups, namely lossless two or three-port networks, lossy-three port networks, and lossless four-port networks. The first category includes the well-known Guanella [6], Marchand [7], and lattice baluns [8], whereas the second category includes the Wilkinson power splitter followed by a 180° phase shifter network, and the final category includes, among others, the rat-race coupler and its variants.

3.2 LOSSLESS TWO- OR THREE-PORT BALUNS

3.2.1 Bridge (Lattice) Baluns

One of the simplest of all baluns is the bridge type (sometimes called a lattice balun) shown in Figure 3.6(a). For a lossless two-port network, the Z parameters are purely reactive, and hence, $Z_{11}=jX_{11}$ and $Z_{12}=jX_{12}$. Straightforward analysis shows that if port 2 is terminated in a load impedance R_L then the input impedance at port 1 is given by

$$Z_{in} = \frac{X_{12}^2 R_L}{R_L^2 + X_{11}^2} + j\frac{X_{11}}{R_L}(R_L - \mathrm{Re}Z_{in}) \tag{3.3}$$

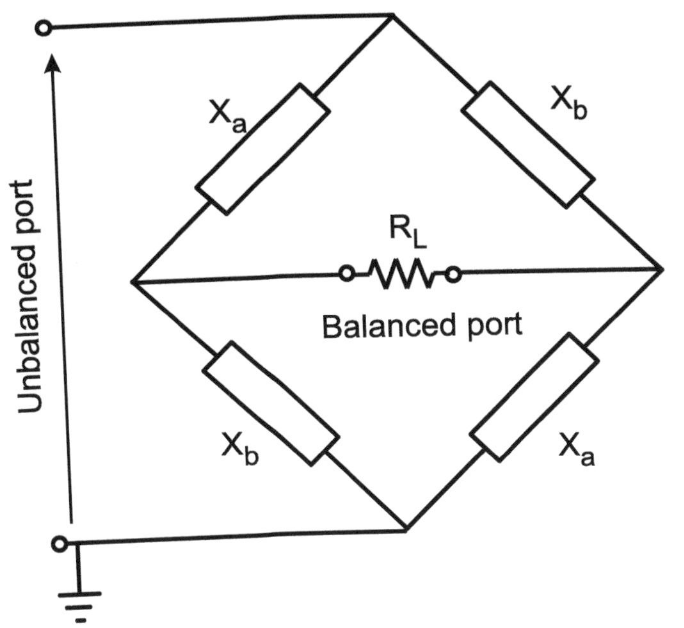

Figure 3.6(a) *A bridge (lattice) balun.*

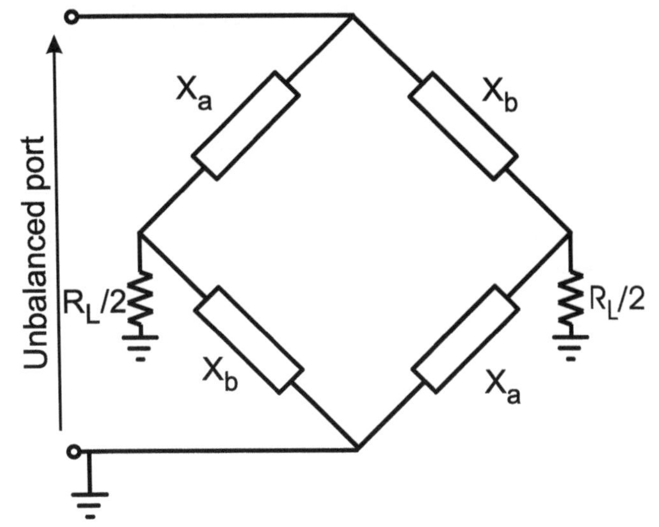

Figure 3.6(b) *Conversion of a bridge balun into an anti-phase power splitter.*

Z_{in} will be purely real if R_L=Re(Z_{in}), in which case $R_L^2=X_{12}^2-X_{11}^2$. Now $X_{11}=(X_a+X_b)/2$ and $X_{12}=(X_a-X_b)/2$, and hence

$$R_L = \sqrt{-X_a X_b} \tag{3.4}$$

In order for R_L to be a real number then either X_a or X_b must be negative which, in the case of a lumped network, means that X_a must be an inductor and X_b must be a capacitor, or vice versa, and $R_L = \sqrt{L/C}$. Alternatively, X_a can be a short-circuited transmission line of characteristic impedance Z_{0a} and X_b an open-circuited transmission line of characteristic impedance Z_{0b} and $R_L = \sqrt{Z_{0a}Z_{0b}}$. For both the lumped and distributed cases the balun is lossless and perfectly matched over an infinite bandwidth with ideal circuit elements (i.e., it is an all-pass network). Trifunović and Jokanović [9] have reported on a distributed bridge type balun covering the frequency range 300 kHz–4 GHz. However, distributed bridge baluns are difficult to implement in practice.

In the case of a lumped element balun, at one particular frequency the inductive and capacitive reactances will be equal in magnitude, but opposite in sign, in which case the voltage with respect to ground of the two nodes of the balanced port will also be of equal magnitude but opposite sign. At this critical frequency, the bridge balun can be redrawn as in Figure 3.6(b) and a lossless three-port balun is created suitable for use in a push-pull amplifier. This simple derivation leads inevitably to the two output ports having a load impedance that is exactly half the input impedance, but a more detailed analysis of the bridge shown in Figure 3.6(b) shows that, in fact, the input and output ports can have any arbitrary values, all that is necessary is that $X_a = -X_b = \sqrt{R_{in}R_L}$. At the frequency where $X_a = -X_b$ the bridge will be perfectly matched at input and output and the two output voltages will be of equal magnitude but opposite phase. However, this situation occurs at only one frequency if X_a and X_b are just single element reactances. Bakalski et al. [8] report results for a push-pull amplifier using a simple lumped-element bridge balun. This paper is included in the reprints. However, if X_a and X_b are multielement networks then it is possible to achieve a finite bandwidth; Kuylenstierna and Linnér [10] describe a balun having an octave bandwidth for the case where X_a and X_b are both three-element networks.

3.2.2 Marchand Balun

Marchand's now-famous paper [7], which is included in the reprints, was published in December 1944. Marchand described three types of balun of increasing complexity and performance, the most sophisticated of the three being shown in Figure 3.7(a). The equivalent circuit for this balun can be deduced from direct inspection and is shown in Figure 3.7(b). In order to permit an analytic design of this balun the electrical lengths of the open and short-circuit stubs as well as the lengths of transmission lines must be the same, in which case the lengths of transmission line are referred to as Unit Elements (u.e.) in filter terminology. In December 1957 Roberts [11] published a paper describing the coaxial balun shown in Figure 3.8. This balun has exactly the same equivalent circuit as Marchand's but as the author made no reference to Marchand in his paper, he may have been unaware of it. Nevertheless, the Roberts balun is a little easier to construct than Marchand's. Edward and Rees [12] describe a Roberts-type balun integrated with a dipole antenna that is used in some phased array radars. Unlike the bridge balun, the Marchand and Roberts baluns are inherently band-pass networks. The simplest design technique for these two baluns is to set Z_1 and Z_4 to the unbalanced and balanced terminal impedances, respectively, and then the characteristic impedances of the two stubs can be determined from standard lumped element filter theory [13]. The center frequency of the balun occurs when the stubs are a quarter wavelength long. If the load and source impedances are the same then the filter response can be maximally flat (Butterworth), whereas if they are different then the response can be equiripple (Tchebychev). The equivalent circuit shows that the

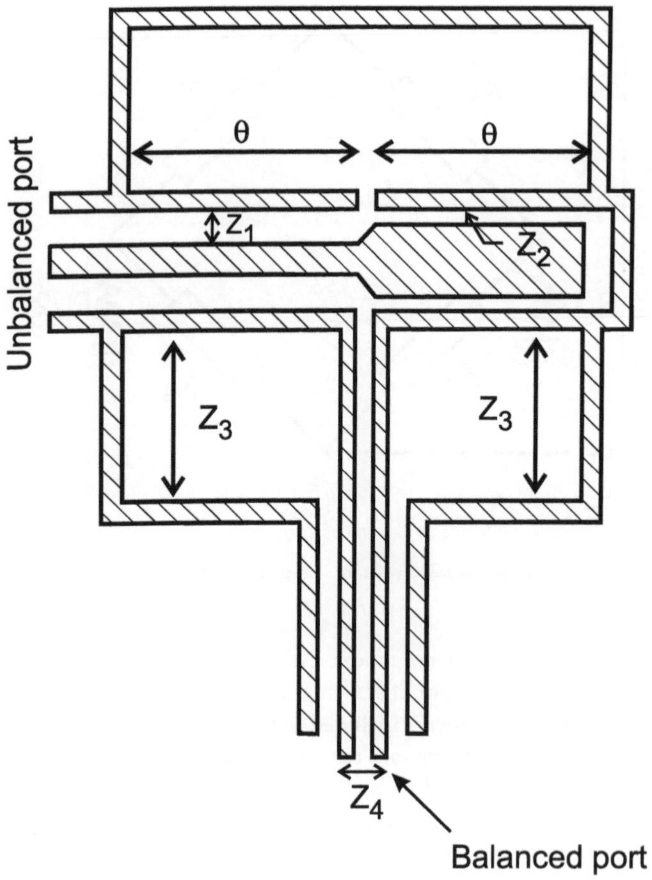

Figure 3.7(a) *Marchand balun.*

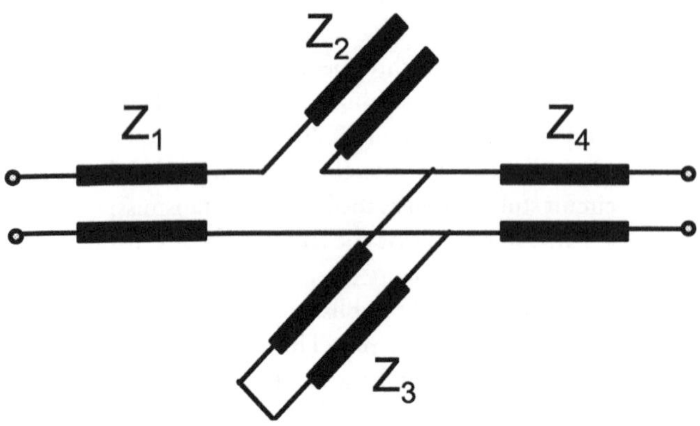

Figure 3.7(b) *Equivalent circuit of Marchand balun.*

balun has perfect transmission at band center, and so the worst-case in-band return loss occurs at band center and is determined only by the difference between the balanced and unbalanced port impedances (e.g., for a 50Ω unbalanced port and a 100Ω balanced port the VSWR is 2:1). As observed by Cloete [14], constraining the two unit elements to have the same impedance as the adjacent external port impedance results in a nonoptimum design. If one or both of the u.e.s are allowed to have impedances that differ from the external port impedances, then they contribute directly to the balun's overall frequency response. In this case the frequency response of the balun is given by [15]

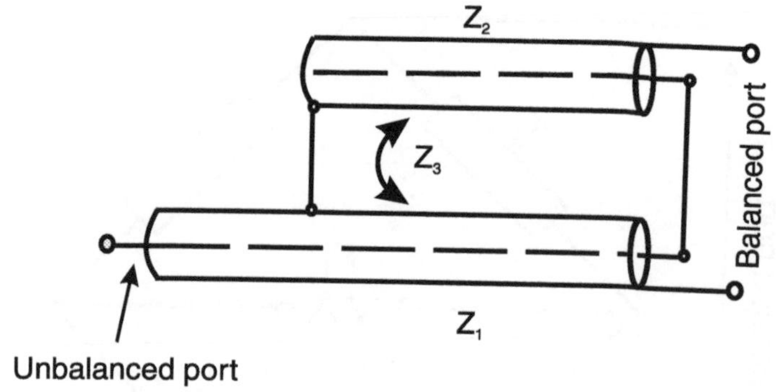

Figure 3.8 *Roberts balun.*

$$|S_{12}|^2 = \frac{1}{1+\varepsilon^2 \cos^2\left(m\cos^{-1}\left(\frac{\cos\theta}{\cot\theta_0}\right)+n\cos^{-1}\left(\frac{\cos\theta}{\cos\theta_0}\right)\right)} \tag{3.5}$$

for an equiripple (Tchebychev) response where m is the number of stubs, in this case, two, and n is the number of u.e.s. If Z_1 and Z_4 are both set to equal the corresponding external port impedance then n=0 while if both are allowed to vary for the optimum bandwidth then n=2 and the fourth order balun is created. Unfortunately, there are no closed-form explicit element value formulae if $n\neq0$. However, Cloete [14, 16], whose papers are included in the reprints, has published graphs of the element values as a function of bandwidth and passband return loss. Cloete reports a fourth order Marchand balun with 15 dB return loss over a decade bandwidth. The only downside to having $n\neq0$ is that the balanced and unbalanced ports can't have the same load resistance, but a 2:1 ratio works well. However, this is very convenient if an anti-phase power splitter is wanted instead of a balun since the 100Ω balanced load can be replaced by a 50Ω load connected between each of the balanced port's terminals and ground.

There has been much interest in recent years in developing a planar version of the Marchand or Roberts balun for use in monolithic and hybrid integrated circuits. One of the first papers to address this issue is that by Pavio and Kikel [17]. Their paper shows a plan view of the structure and its similarity to the Marchand balun. They used a broadside-coupled stripline configuration, as shown in Figure 3.9. The upper dielectric is very thin compared with the lower one and so they argued that coupling between the upper conductor and the ground plane could be ignored and that the structure could be considered as simply two transmission lines with the upper and middle conductor forming one transmission line and the middle and lower conductor the other one. Nevertheless, this structure is inherently a three-conductor coupled-line network and this needs to be taken into account in any detailed analysis. Since then several authors [18–24] have considered an edge-coupled version of

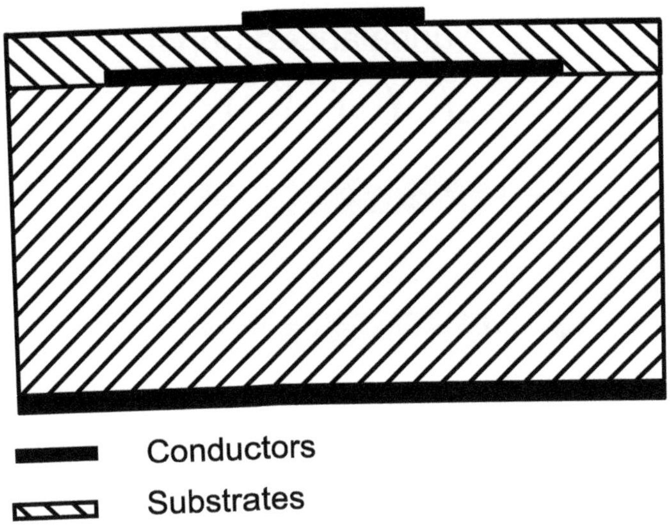

Figure 3.9 *Broadside coupled stripline configuration used by Pavio and Kikel [17].*

Pavio and Kikel's balun and have explicitly taken into account the coupled-line effects. Figure 3.10 shows the generic form of the edge-coupled planar Marchand balun. Both Lange [25] and three coupled-line [24] implementations for the edge coupled-line section have been used. Clearly, the equivalent circuit of Figure 3.10 will differ from that of the Marchand balun, and it cannot be deduced immediately from inspection of Figure 3.10. Goldsmith et al. [26], whose paper is included in the reprints, published the first comprehensive analysis of Figure 3.10. They showed that, provided the two coupled line sections had the same coupling coefficient, then Figure 3.7(b) is also the equivalent circuit of the network in Figure 3.10 except for the inclusion of an ideal transformer at the unbalanced input port. Thus the planar and original versions of the Marchand balun have the same equivalent circuit apart from the presence of an ideal transformer that has no effect on the frequency response apart from altering the peak passband return loss. The planar Marchand balun has four design parameters, namely the coupling coefficient and characteristic impedance (or, equivalently, the even and odd mode characteristic impedances) of the two coupled-line sections which are, in turn, determined by the conductor width to substrate thickness (W/H) and conductor separation to substrate thickness (S/H) ratios. However, the requirement that the coupled-line sections have equal coupling coefficients means that the planar Marchand balun has only three independent design parameters whereas its equivalent circuit has four elements, hence the elements in the equivalent circuit are not all independently variable. Consequently, it is necessary to resort to numerical techniques to design the planar Marchand balun. Goldsmith et al. present graphs of the even and odd mode characteristic impedances required to realize baluns of various bandwidths and these show that extremely high and low impedances are required if an exact equiripple response is desired that are unrealizable in practice for edge-coupled structures and can only be achieved using a broadside multilayer configuration. Also, as for a conventional Marchand balun, the balanced and unbalanced ports cannot in general have the same impedance if an exact equiripple response is required. An additional complication is that irrespective of whether an edge or broadside coupled-line configuration is used, the resulting structure is inhomogeneous and so the even and odd mode phase velocities differ. Engels and Jansen [20], Schwindt and Nguyen [27], and Tsai and Gupta [28] have extended the analysis of Goldsmith et al. [26] to take account of the differing phase velocities, but the results are extremely complex.

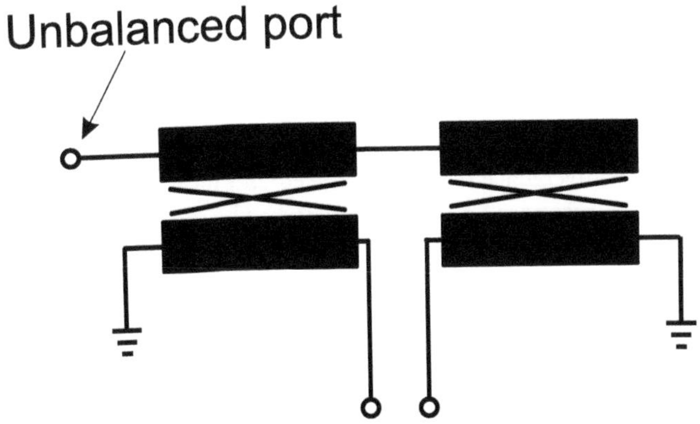

Figure 3.10 *Generic form of planar Marchand balun.*

3.2.3 Guanella Balun

Three months before Marchand's paper was published, Guanella [6] published his now equally famous paper, which is included in the reprints. Guanella's device can also be used as an impedance transformer and this is discussed in more detail in Chapter 4. While the bridge (lattice) balun is an all-pass network and Marchand's is bandpass, Guanella's is a high-pass network. Guanella considered only the two-port balun situation but since this balun is used extensively in practice as an anti-phase power splitter, this introduction will consider this latter configuration. The simpler two-port balun situation is a limiting case of the anti-phase power splitter analysis. The ideal Guanella balun consists of just a length of transmission line that has one conductor grounded at one end, as shown in Figure 3.11, which also shows the voltages and currents. It is important to realize that both the upper and lower conductors of the transmission line have a distributed inductance per unit length; the lower load resistance would be shorted out if only the upper conductor had a distributed inductance. The operation of this circuit can be understood as follows: The voltage source at the unbalanced input port forces a current to flow along the upper conductor of the transmission line and, in the ideal case, none of this current flows to ground at the center point of the two load resistors but instead it all flows into the lower conductor and, ultimately, back through the voltage source. I_1 and I_2 will, in general, have different phases and so not be equal, but it can be immediately seen from Figure 3.11 that if I_2 flows through both loads then a perfect anti-phase power splitter is formed. When used as a balun the balanced and unbalanced ports as well as the transmission

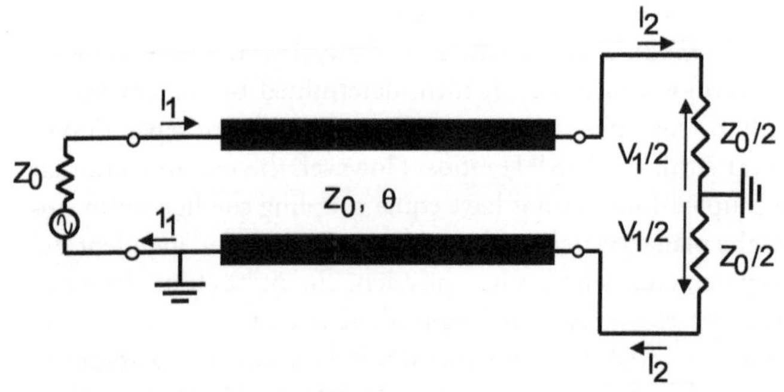

Figure 3.11 *Guanella balun.*

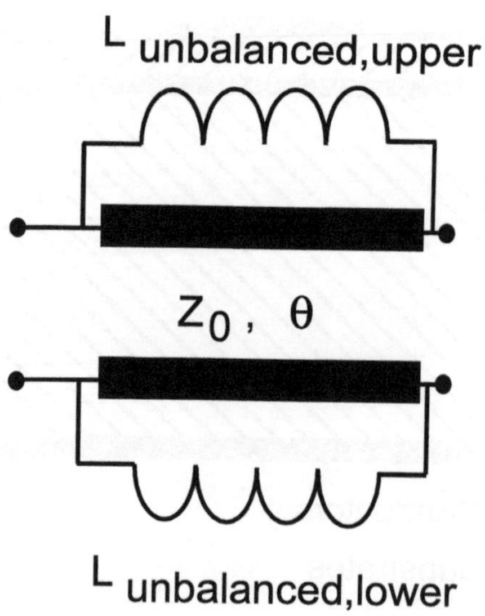

Figure 3.12 *Model of Guanella balun.*

line should all have the same impedance, but when used as an anti-phase power splitter the two unbalanced loads need to be half this value. This results in a perfect input match but $|S_{22}| = |S_{33}| = 0.5$, that is, 6 dB return loss. The isolation between ports 2 and 3 is also 6 dB.

In addition to this wanted mode of operation, which is termed the balanced mode, there is also an unwanted unbalanced mode in which some of the current bypasses the lower load resistor and returns to the voltage source via the ground connection at the junction of the two load resistors. This unwanted current is determined solely by the impedance presented by the upper load resistance in series with the total inductance of the upper conductor, but this inductance is not the distributed inductance per unit length multiplied by the length of line but is the inductance of a length of wire l and radius r (r being the outer radius of the outer conductor of the coaxial transmission line and not the inner radius of the outer conductor which is used to determine the distributed inductance per unit length) and is given by [29]

$$L = 2l[\ln(2l/r) - 1] \tag{3.6}$$

This unbalanced current tends to zero as the frequency increases since the reactance ωL increases indefinitely and hence at high frequencies the Guanella balun behaves as desired, but it causes severe imbalance in V_1 and V_2 at low frequencies since it effectively shorts out the upper conductor of the transmission line. The lower conductor also has a lumped inductance associated with it that shorts it out at low frequencies.

From the preceding paragraph it can be seen that the Guanella balun can be modeled as a length of transmission line with lumped inductors shunting the upper and lower conductors, as shown in Figure 3.12. This model is implemented in several commercially available CAD programs, but they all assume that the two inductors have the same value. This is not, in general, true, especially if ferrite beads are placed around the outside of the coax, which is often the case in practice in order to extend the low frequency performance of the device by increasing the inductance associated with the outer conductor. Figure 3.13 shows a comparison between modeled and measured input impedance based on (3.6) at the unbalanced port for a 2-foot length of RG58 coaxial cable without ferrite beads terminated in a two ordinary 25Ω carbon resistors over the frequency range

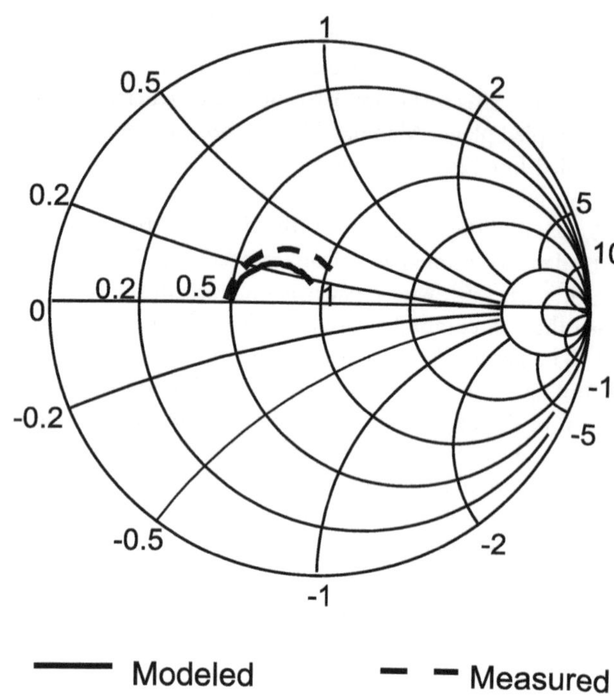

Figure 3.13 *Comparison of modelled and measured unbalanced port input impedence of Guanella balun.*

300 kHz to 25 MHz. At low and high frequencies the input impedance approaches 25Ω and 50Ω, respectively, as expected, the progressive divergence between modeled and measured impedance as the frequency increases being caused by the considerable parasitics associated with the load resistors used in this experiment. Figure 3.14 shows the predicted transmission performance of this balun both with and without ferrite around the outer conductor. From energy conservation

$$|S_{11}|^2 + |S_{12}|^2 + |S_{13}|^2 = 1 \tag{3.7}$$

and so the 25Ω input impedance at low frequencies results in 0.5-dB insertion loss for the sum of the two transmission coeffi-

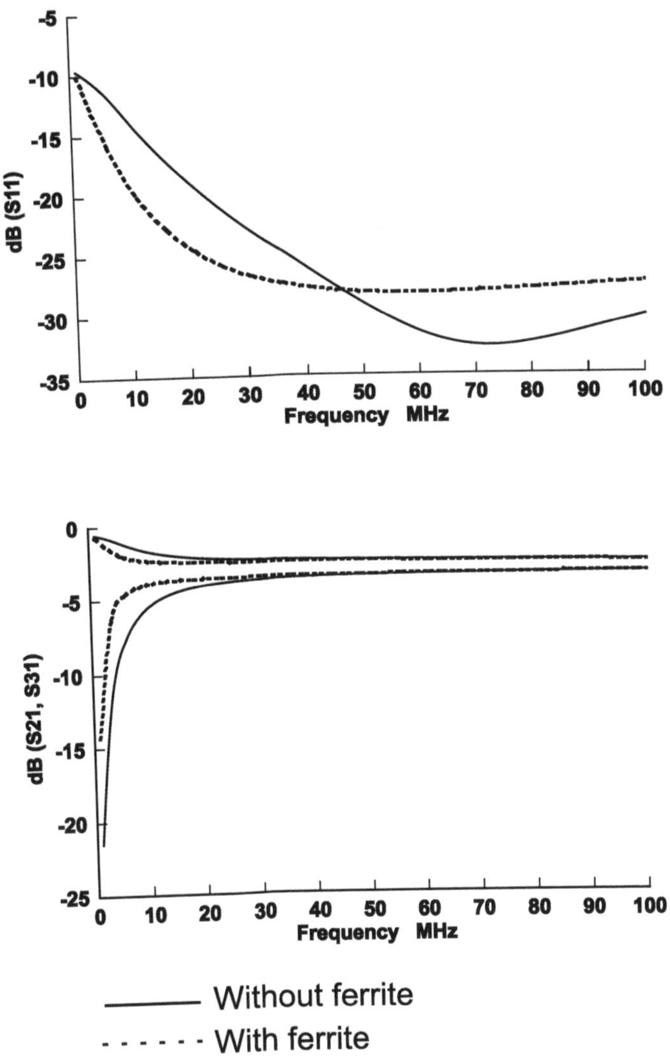

Figure 3.14 *Effect of ferrite loading on calculated performance of Guanella balun (excitation at unbalanced port).*

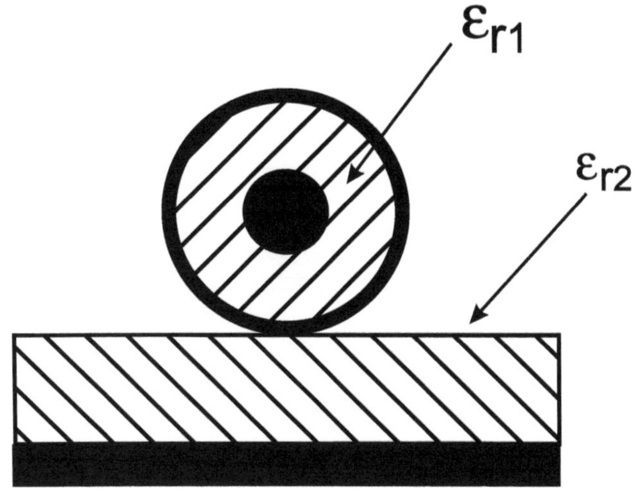

Figure 3.15(a) *Guanella balun above a printed circuit board.*

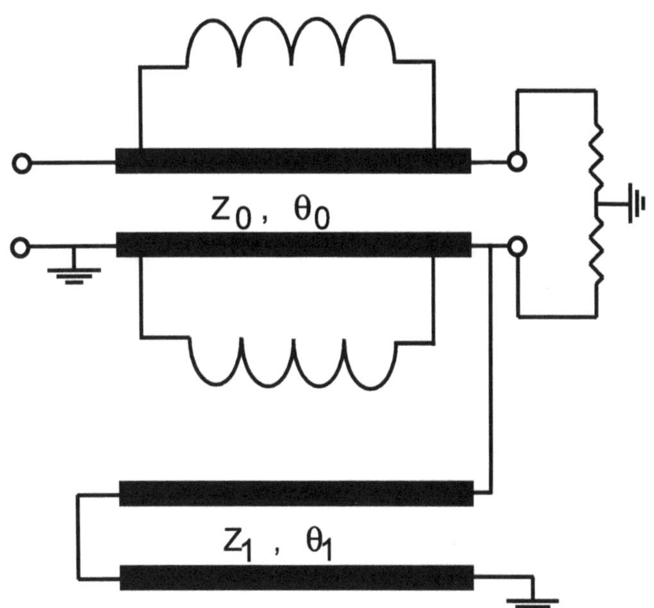

Figure 3.15(b) *Equivalent circuit of Guanella balun above a printed circuit board.*

cients, but this is insignificant in comparison to the severe amplitude imbalance resulting from the unwanted unbalanced current components.

In many practical applications the Guanella anti-phase power splitter is placed above a printed circuit board as shown in Figure 3.15(a). The presence of the lower ground plane creates a shunt short-circuit stub across one of the loads as shown in Figure 3.15(b) and this converts the high-pass Guanella structure into a bandpass one. In fact, for a typical piece of UT85 coax on top of 1.6-mm thick laminate with ε_r=2, this can be modeled by a shunt stub of 75Ω characteristic impedance, in which case the balun performance is as shown in Figure 3.16. It can be seen that this stub has a dramatic effect on the balun performance, the bandwidth being reduced to about an octave based on phase imbalance. There are two solutions to this problem, the first being to attach a compensating stub to the other load which results in perfect amplitude and phase balance above the low frequency cutoff region but this degrades the input VSWR far away from the quarter wave frequency resulting in 1 dB insertion loss at band-edges for a 9:1 bandwidth. The second method is to raise the transmission line above the printed circuit board as high as possible and make both conductors symmetrical with respect to the lower ground plane. Darwish et al. [30] have described this concept in detail, and an example of its use can be found in the MRF374A test fixture [31].

Finally, the analysis up to this point has assumed that all ports are purely resistive, but the Guanella anti-phase power splitter is frequently used in push-pull amplifiers, in which case one transistor is in the off state when the other is conducting. The effect of this is that the Guanella power splitter is not terminated in resistive loads at all ports simultaneously, but, in fact, the push-pull operation creates a commutating load at the output ports with the load looking almost purely reactive in the off-state and a parallel RC network in the on-state. As an example,

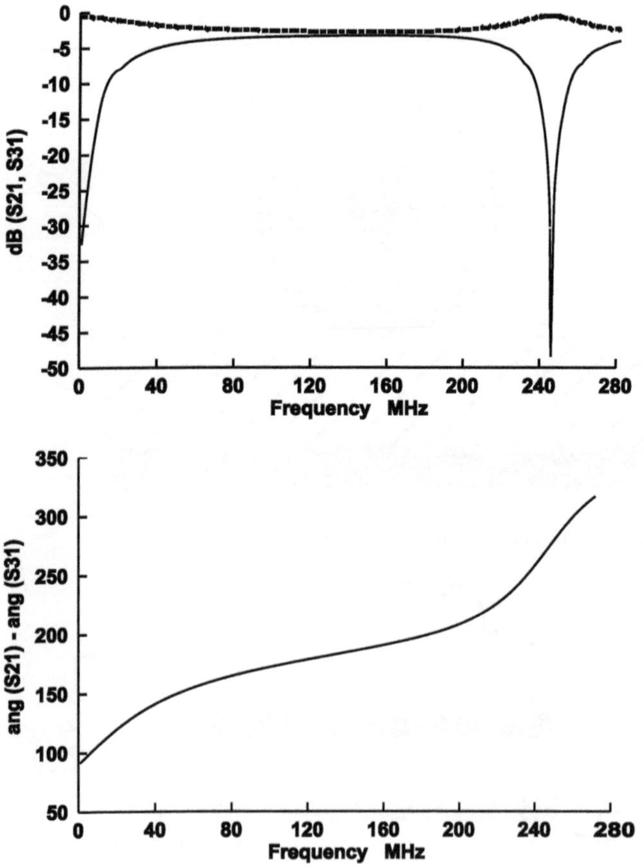

Figure 3.16 *Calculated performance of Guanella balun above a printed circuit board.*

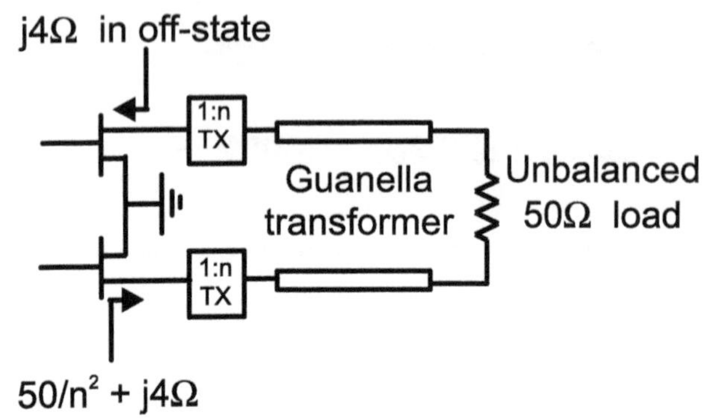

Figure 3.17 *Use of a Guanella balun in a push-pull amplifier.*

consider the block diagram shown in Figure 3.17. In a typical application an n:1 impedance transformer is used to convert the external load impedance down to the low level needed by a power transistor and hence the transistor in the off state presents a load of $j4n^2\Omega$ to the anti-phase power splitter which, in turn, presents an impedance of $50+j4n^2\Omega$ to the other transformer. This is then transformed down to $50/n^2+j4\Omega$ to the other half of the push-pull transistor in the on state.

3.2.4 Anti-Phase Power Splitters Derived from Symmetrical 4-Port Networks

A limitless number of different types of anti-phase power splitter can be designed by taking a symmetrical four-port network and terminating one of the ports in a perfectly reflecting load as shown in Figure 3.18. The networks can be either coupled or connected transmission line structures. Leong et al. [32], whose paper is included in the reprints, have determined the conditions that the four-port network must satisfy in order that the resulting structure will behave as a perfect anti-phase power splitter. The challenge is to devise a structure that has excellent wide-bandwidth performance but which is easy to implement with realisable impedance values. The planar Marchand balun is an example of this general class of network. Other examples include the multisection branch-arm coupler described by Mayer and Knoechel [33], the log-periodic balun [34], and the multisection coupled-line balun [35]. While all the authors classified their devices as baluns, they were all constructed and tested as anti-phase power splitters. The log-periodic balun is basically a wider bandwidth version of Mayer and Knoechel's balun. However, the problem with all these structures is that they use multiple quarter-wave transmission lines and hence are physically large at HF, VHF, and UHF frequencies. Also, the bandwidth is typically about an octave whereas many HF, VHF, and UHF applications need at least decade bandwidths.

3.3 LOSSY THREE-PORT AND LOSSLESS FOUR-PORT BALUNS

Once again, although these devices are usually referred to as baluns, they are invariably used as anti-phase power splitters. In fact, several of these devices are designed on the basis of separating the functions of power splitting and phase shifting. The power splitter may either have in-phase outputs (e.g., the Wilkinson [4] variety) and then be followed by a 180° phase shifter in one arm, or have quadrature-phase outputs and then be followed by a 90° phase-shifter in the quadrature arm of the coupler. For narrow-band applications a simple 50Ω quarter- or half-wavelength transmission line can be used to realize the 90° or 180° phase shift. Good amplitude balance will occur over a

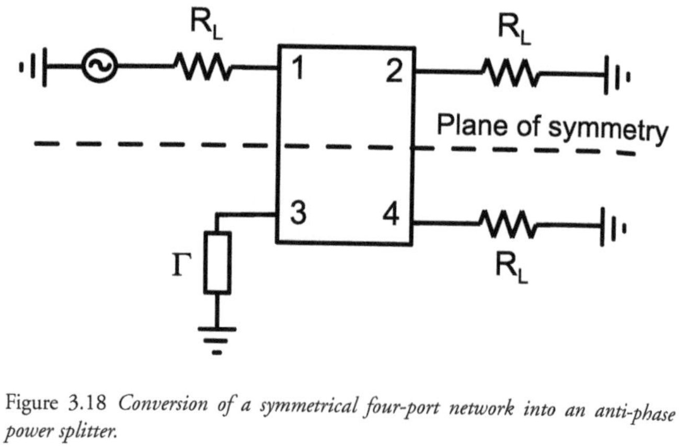

Figure 3.18 *Conversion of a symmetrical four-port network into an anti-phase power splitter.*

broad bandwidth but the desired 180° phase shift will only occur at band center and will deviate rapidly away from band-center. For wide bandwidths it is essential to use a broadband phase shift network. Bharj and Thompson [36] used a Wilkinson power splitter followed by networks in both arms that have an inherent exact 180° phase shift at all frequencies. Their anti-phase power splitter theoretically has perfect amplitude balance and 180° phase difference at all frequencies, but the input VSWR at the center port rises away from band center resulting in 1dB excess insertion loss at a 4:1 bandwidth. Inserting a matching network at the center port can extend the bandwidth. The presence of the resistor in the Wilkinson splitter ensures that there is a minimum of 12-dB isolation between the two output ports, which is a significant improvement over all the other baluns considered so far. This anti-phase power splitter has the desirable attribute of being planar as well as offering excellent wide-band performance, but unfortunately it is physically large at HF/VHF/UHF frequencies due to the use of multiple lengths of transmission line.

Reference [37] used a 3-dB quadrature coupler as the power splitter and placed a half-wavelength 50Ω transmission line at the through port and another 3-dB quadrature coupler with its through and coupled ports terminated in open circuits at the coupled port. The two output networks provide an additional 90° phase shift so that the overall unit functions as an anti-phase power splitter. Walker's device offers inferior performance to Bharj and Thompson's in terms of amplitude and phase balance and it is no smaller, so it won't be considered further.

The final device to be considered is the lossless four-port rat-race coupler shown in Figure 3.19, which can function as either an in-phase power splitter if the input is applied to port 1 or anti-phase power splitter if the signal is applied to port 4. In either case, one port is isolated from the input port. The presence of the load at the fourth port ensures that the two coupled ports are isolated from each other at all frequencies. The simplest approach to realizing the ideal 1:-1 transformer is to use a half wavelength transmission line, but this results in narrow-band performance. March [38] extended the bandwidth by using a quarter-wave coupled line section, but by far the best approach is to construct it using coplanar waveguide (CPW) as the ideal transformer can be realized by a crossover between the ground and signal lines [39]. The only imperfection in the 1:-1 transformer realization is the inductance associated with the bond wires used for the crossover, but this effect is insignificant until millimeter wave frequencies. This form of anti-phase power splitter has perfect amplitude balance, perfect isolation between the two output ports, and perfect 0° to 180° phase split over an infinite bandwidth. The only factor that limits its performance is that the input VSWR gradually increases away from band center causing an excess insertion loss in the two outputs. A 3:1 bandwidth can be achieved with a 10-dB return loss resulting in 0.45-dB excess insertion loss at the band edges. Chang and Yang [40] subsequently extended the bandwidth to 4:1 by adding a matching section at the four ports, and this paper is included in the reprints.

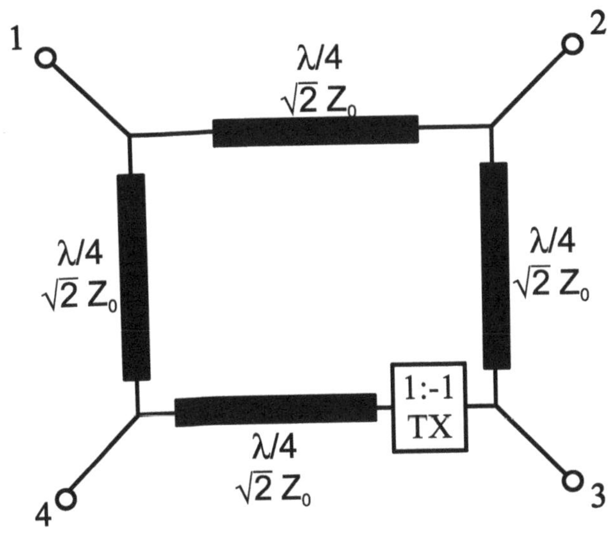

Figure 3.19 *Rat-race coupler as an anti-phase power splitter.*

3.4 CONCLUSIONS

This short introduction has described the properties of baluns and shown the intimate relationship between baluns and anti-phase power splitters. Decade bandwidths are often required at HF/VHF/UHF frequencies and baluns that use multiple lengths of transmission line are generally physically too large, thus lumped element baluns and anti-phase power splitters constructed in bridge (lattice) form offer a number of benefits at HF. Transmission lines are more common at VHF/UHF and the Guanella balun is an excellent choice for this frequency range provided adequate attention is given to suppressing the unwanted mode. However, the Guanella balun is a three-dimensional structure and is incompatible with microwave integrated circuit construction technology. If a truly planar balun at VHF/UHF is required then the planar version of the Marchand balun is probably the best choice. Above UHF or if size is not an issue then the rat-race is the preferred choice for an anti-phase power splitter, especially if realized in CPW, because of its inherent perfect amplitude balance, exact 0/180° phase split and complete isolation between the two outputs over an infinite bandwidth. Finally, the paper by McLean [41] that is included in the reprints provides an alternative introduction to baluns and anti-phase power splitters from an alternative but complementary perspective.

REFERENCES

[1] Maas, S.A, *Microwave Mixers,* Second Edition, Norwood, MA: Artech House, 1993.

[2] Smith, J., *Modern Communication Circuits,* New York: McGraw-Hill, 1986, pp. 203–208.

[3] Collin, R.E., *Foundations for Microwave Engineering,* New York: McGraw-Hill, 1966, p. 308.

[4] Wilkinson, E.J., "An N-Way Hybrid Power Divider," *IEEE Trans. Microwave Theory and Techniques,* Vol. MTT-16, February 1968, pp. 110–116.

[5] Levy, R., *Directional Couplers* in *Advances in Microwaves,* Vol. 1, L.Young (ed.), New York: Academic, 1966.

[6] Guanella, G., "New Method of Impedance Matching in Radio-Frequency Circuits," *Brown Boveri Review*, September 1944, pp. 327–329.

[7] Marchand, N., "Transmission-Line Conversion Transformers," *Electronics*, Vol. 17, December 1944, pp. 142–145.

[8] Bakalski, W., W. Simbürger, H. Knapp, H.-D. Wohlmuth, and L. Scholtz, "Lumped and Distributed Lattice-Type LC-Baluns," *IEEE Int. Microwave Symposium*, Seattle, WA, June 3–7, 2002, pp. 209–212.

[9] Trifunović, V., and B. Jokanović, "Four Decade Bandwidth Uniplanar Balun," *Electronics Letters*, Vol. 28, No. 6, March 1992, pp. 534–535.

[10] Kuylenstierna, D., and P. Linnér, "Design of Broadband Lumped Element Baluns," *IEEE Int. Microwave Symposium*, Fort Worth, TX, June 7–11, 2004, pp. 899–902.

[11] Roberts, W.K., "A New Wide-Band Balun," *Proc. IRE*, Vol. 45, December 1957, pp. 1628–1631.

[12] Edward, B., and D.Rees, "A Broadband Printed Dipole with Integrated Balun," *Microwave Journal*, Vol. 30, No. 5, May 1987, pp. 339–344.

[13] Rhodes, J.D., *Theory of Electrical Filters*, London, U.K.: Wiley, 1976, p. 48.

[14] Cloete, J.H., "Exact Design of the Marchand Balun," *Microwave Journal*, Vol. 23, No. 5, May 1980, pp. 99–102.

[15] Horton, M.C., and R.J. Wenzel, "General Theory and Design of Optimum Quarter-Wave TEM Filters," *IEEE Trans. Microwave Theory and Techniques*, Vol. MTT-13, No. 5, May 1965, pp. 316–327.

[16] Cloete, J.H., "Graphs of Circuit Elements for the Marchand Balun," *Microwave Journal*, Vol. 24, No. 5, May 1981, pp. 125–128.

[17] Pavio, A.M., and A. Kikel, "A Monolithic or Hybrid Broadband Compensated Balun," *IEEE Int. Microwave Symposium*, Dallas, TX, May 8–10, 1990, pp. 483–486.

[18] Nguyen, C., and D. Smith, "Novel Miniaturised Wideband Baluns for MIC and MMIC Applications," *Electronics Letters*, Vol. 29, No. 12, June 1993, pp. 1060–1061.

[19] Tsai, M.C., "A New Compact Wideband Balun," *IEEE Int. Microwave Symposium*, Atlanta, GA, June 14–18, 1993, pp. 141–143.

[20] Engels, M., and R.H. Jansen, "Design of Integrated Compensated Baluns," *Microwave and Optical Technology Letters*, Vol. 14, No. 2, February 1997, pp. 75–81.

[21] Mahmoud, S.F., P. Pieters, and E. Beyne, "Analysis and Design of Two Types of Microwave Baluns," *The 12th Int. Conf. on Microelectronics*, Tehran, Iran, October 31–November 2, 2000, pp. 303–306.

[22] Ang, K.S., and I.D. Robertson, "Analysis and Design of Impedance-Transforming Planar Marchand Baluns," *IEEE Trans. Microwave Theory and Techniques*, Vol. MTT-49, No. 2, February 2001, pp. 402–406.

[23] Sun, J-W., and T-L. Lee, "Design of a Planar Microstrip Balun at S-Band," *Microwave Journal*, Vol. 44, No. 8, August 2001, pp. 132–136.

[24] Lee, J-W., and K.J. Webb, "Analysis and Design of Low-Loss Planar Microwave Baluns Having Three Symmetric Coupled Lines," *IEEE Int. Microwave Symposium*, Seattle, WA, June 3–7, 2002, pp. 117–120.

[25] Lange, J., "Interdigitated Stripline Quadrature Hybrid," *IEEE Trans. Microwave Theory and Techniques*, Vol. MTT-17, No. 12, December 1969, pp. 1150–1151.

[26] Goldsmith, C.L., A. Kikel, and N.L. Wilkens, "Synthesis of Marchand Baluns Using Multilayer Microstrip Structures," *Int. J. of Microwave and Millimeter-Wave Computer-Aided Engineering*, Vol. 2, No. 3, 1992, pp. 179–188.

[27] Schwindt, R., and C. Nguyen, "Computer-Aided Analysis and Design of a Planar Multilayer Marchand Balun," *IEEE Trans. Microwave Theory and Techniques*, Vol. MTT-42, No. 7, July 1994, pp. 1429–1434.

[28] Tsai, C.M., and K.C. Gupta, "A Generalized Model for Coupled Lines and Its Applications to Two-Layer Planar Circuits," *IEEE Trans. Microwave Theory and Techniques*, Vol. MTT-40, No. 12, December 1992, pp. 2190–2199.

[29] Rosa, E.B., "The Self and Mutual Inductances of Linear Conductors," *Bulletin of the Bureau of Standards*, Vol. 4, No. 2, 1908, p. 301.

[30] Darwish, A., A. Ezzeddine, H.C. Huang, K. Bumman, L. Joonyoul, Y. Sungwhan, M. Mah, and J. Cook, "Vertical Balun and Wilkinson Power Divider," *IEEE Int. Microwave Symposium*, Seattle, WA, June 3–7, 2002, pp. 109–112.

[31] Freescale Semiconductor Inc., *MRF374A* datasheet, Rev. 3, 2003, p. 3.

[32] Leong, Y.C., K.S. Ang, and C.H. Lee, "A Derivation of a Class of 3-Port Baluns from Symmetrical 4-Port Networks," *IEEE Int. Microwave Symposium*, Seattle, WA, June 3–7, 2002, pp. 1165–1168.

[33] Mayer, B., and R. Knoechel, "Biasable Balanced Mixers and Frequency Doublers Using a New Planar Balun," *European Microwave Conference*, Budapest, Hungary, September 10–13, 1990, pp. 1027–1032.

[34] Basraoui. M., and S.N. Prasad, "Wideband, Planar, Log-Periodic Balun," *IEEE Int. Microwave Symposium*, Long Beach, CA, June 13–15, 1989, pp. 959–962.

[35] Ang, K.S., Y.C. Leong, and C.H. Lee, "Multisection Impedance-Transforming Coupled-Line Baluns," *IEEE Trans. Microwave Theory and Techniques*, Vol. 51, No. 2, February 2003, pp. 536–541.

[36] Bharj, S.S., S.P. Tan, and B. Thompson, "A 2-18GHz 180 Degree Phase Splitter Network," *IEEE Int. Microwave Symposium*, Baltimore, MD, June 7–12, 1998, pp. 785–788.

[37] Walker, S., "Broadband Stripline Balun Using Quadrature Couplers," *IEEE Trans. Microwave Theory and Techniques*, Vol. MTT-16, No. 2, February 1968, pp. 132–133.

[38] March, S., "A Wideband Stripline Hybrid Ring," *IEEE Trans. Microwave Theory and Techniques*, Vol. 17, No. 6, June 1968, p. 361.

[39] Chang, C-Y., C.-C. Yang and D.-C Niu, "A Multioctave Bandwidth Rat-Race Singly Balanced Mixer," *IEEE Microwave and Guided Wave Letters*, Vol. 9, No.1, January 1999, pp. 37–39.

[40] Chang, C.-Y., and C.-C. Yang, "A Novel Broad-Band Chebyshev-Response Rat-Race Ring Coupler," *IEEE Trans. Microwave Theory and Techniques*, Vol. MTT-47, No. 4, April 1999, pp. 455–462.

[41] McLean, J.S., "Balancing Networks for Symmetric Antennas—I: Classification and Fundamental Operation," *IEEE Trans. Electromagnetic Compatibility*, Vol. 44, No. 4, November 2002, pp. 503–514.

Lumped and Distributed Lattice-type LC-Baluns

Student paper

Winfried Bakalski[1,2], Werner Simbürger[1], Herbert Knapp[1],
Hans-Dieter Wohlmuth[1] and Arpad L. Scholtz[2]

[1] Infineon Technologies AG,
Otto-Hahn-Ring 6, D-81730 Munich, Germany,
E-Mail: winfried.bakalski@infineon.com

[2] Institute of Communications and Radio-Frequency Engineering,
Vienna University of Technology, Gusshausstrasse 25, 1040 Vienna, Austria

Abstract **This paper presents two balun circuits derived from the lumped Lattice-type LC-balun. First the lumped LC-balun bridge elements are substituted by microstrip lines. This results in an improved performance at the 2nd and 3rd harmonic frequency for RF power amplifier output baluns. Secondly, the lumped Lattice-type LC-balun is extended to a dual band balun. Independent impedance transformation and balun conversion can be done at two different frequencies. The design equations are derived.**

I. Introduction

The lumped Lattice-type LC-balun (Fig. 1) was originally known from the early 1930's as antenna balun [1], [2], [3], [4].

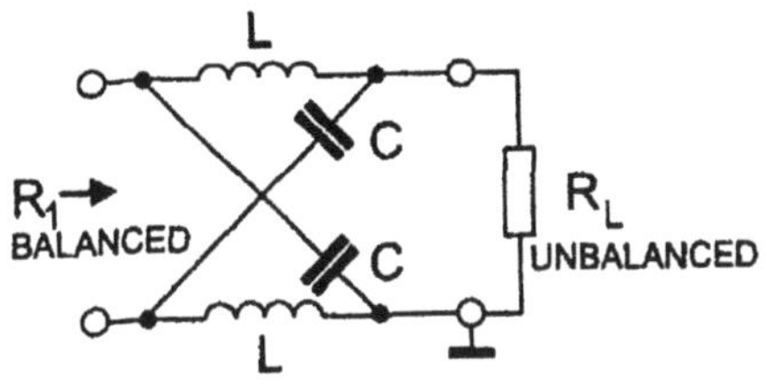

Fig. 1. Lattice-type LC-balun network.

This circuit transforms the balanced input into an unbalanced output by shifting the phase of one input port by +90° and the second input port by −90° and vice versa. Other balun concepts are based on a $\lambda/2$-transmission line ($\hat{=}\,180°$) added to a power divider or a filter structure. [5] shows a microstrip balun with a T-junction with two arms differing in their length by $\lambda/2$ and an even-mode suppression structure.

Baluns like these are required for push-pull amplifiers. As the structure proposed by [5] is rather large for a low dielectric constant ε_r, lumped LC-balun are widely used. Fig. 2 shows the schematics for a push-pull power amplifier usage. The lattice-type balun consists of two

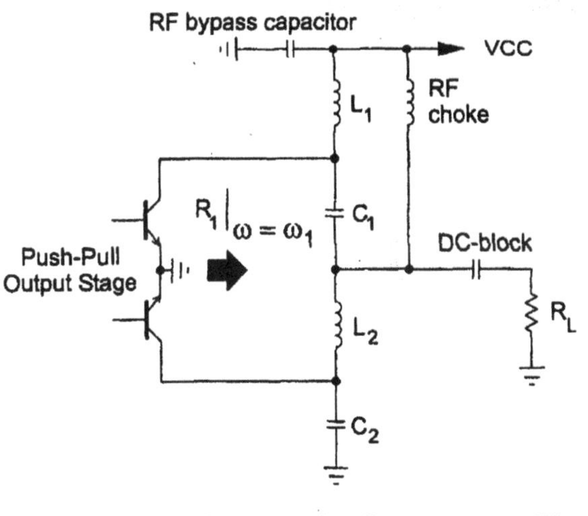

Fig. 2. Lumped LC-balun as push-pull type power amplifier output matching network.

inductances $L_1 = L_2$ and two capacitors $C_1 = C_2$. A RF-choke and a DC-block capacitor are used to feed the supply voltage. R_1 is the balanced input impedance of the bridge. Each collector is loaded by $R_1/2$. R_L is the load resistor, usually $50\,\Omega$. L and C can be calculated by

$$L_1 = L_2 = \frac{Z_1}{\omega_1} \tag{1}$$

0-7803-7239-5/02/$10.00 © 2002 IEEE

$$C_1 = C_2 = \frac{1}{\omega_1 Z_1} \quad (2)$$

where $Z_1 = \sqrt{R_1 \cdot R_L}$ is the characteristic impedance of the bridge. $\omega_1 = 2\pi f_1$ is the frequency of operation.

II. Microstrip line balun

The four lumped elements of the LC-balun in Fig. 2 can be substituted by transmission lines. We designed a balun where L1, L2 and C2 were substituted by transmission lines while C1 remains as a lumped element. Fig. 3 shows a balun layout for f = 2.4 GHz for a substrate with a thickness of 0.51 mm and an ε_r of 3.38. The radial stub is not necessary if it is replaced by a RF bypass capacitor. However, the radial stub is a good RF short-circuit [6], [7].

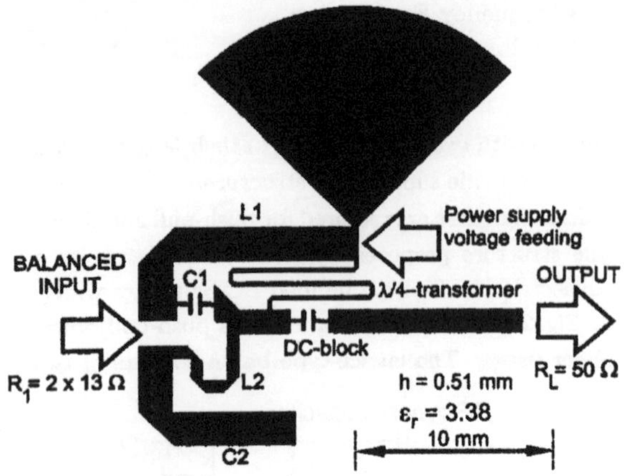

Fig. 3. Microstrip line balun for 2.45 GHz and a substrate with a thickness of 0.51 mm and $\varepsilon_r = 3.38$.

The necessary equations in addition to the Hammerstad [8] formulas are given by:

$$\omega L_1 = Z_0 \tan\theta \quad , \quad \theta \leq \frac{\pi}{2} \quad (3)$$

$$\omega L_2 = Z_0 \sin\theta \quad , \quad \theta \leq \frac{\pi}{2} \quad (4)$$

$$\omega C_2 = \frac{\tan\theta}{Z_0} \quad , \quad \theta \leq \frac{\pi}{2} \quad (5)$$

with $\frac{\pi}{2} \triangleq \frac{\lambda}{4}$ and λ as wavelength [9]. Designing such a balun starts by calculating L and C using Eqn. 1 and Eqn. 2. Next step is the substitution of the lumped elements by transmission lines (Eqns. 3 - 5). C1 is not substituted for the reason, that a long transmission line larger than $\lambda/4$ would be necessary. This would lead to a narrow bandwidth and large outer dimension for low frequencies.

This balun structure shows several advantages to the standard LC-balun solution:

- Only one lumped element is required.
- The use of transmission lines gives two variables per element to adjust the outer dimensions. However this complicates the design procedure and requires simulation tools.
- The second harmonic load impedance shows very low input impedances while the third harmonic load impedance shows very high impedances. Especially for nonlinear operation this leads to high efficient amplifiers [10].
- The outer dimensions are significantly smaller than other microstrip based baluns [5].

Fig. 3 shows the layout of such a balun. It was designed for a center frequency of 2.45 GHz and a load impedance of 26 Ω. Fig. 4 shows the simulated results for fundamental, second and third harmonic frequency.

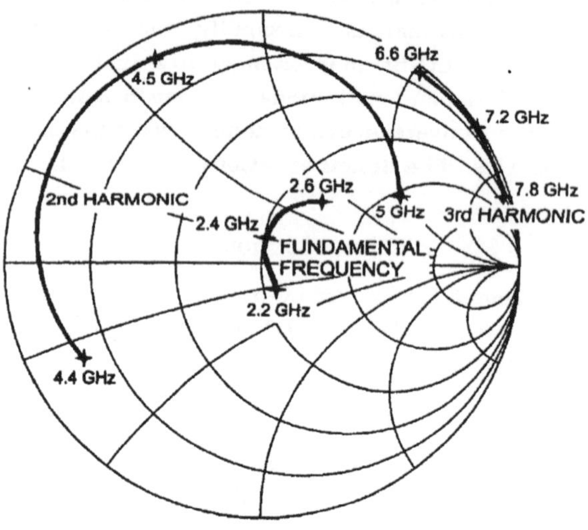

Fig. 4. Simulted results for the load impedance at the fundemental frequency of 2.45 GHz and the second and third harmonic frequency for $Z_0 = 26\,\Omega$.

This balun was tested with a 2.45 GHz power amplifier chip. The power amplifier is presented in [11]. Fig. 6

shows the frequency response of this module. The substrate parameters are $\varepsilon_r = 3.38$, $\tan\delta = 0.0027$ and the dielectric thickness is 0.51 mm. The metallization layers consist of 18 μm copper with a nickel diffusion barrier and 5 μm gold on top for bonding. The lumped capacitor C1 is a 1.2 pF 0805 AVX ACCU-P type capacitor (effective capacity $\approx$ 1.4 pF).

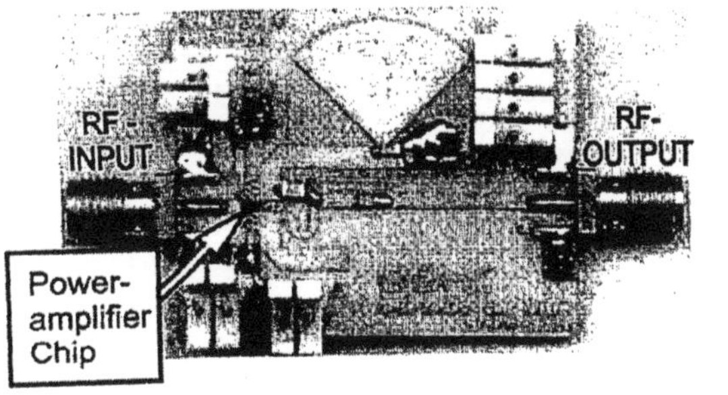

Fig. 5. Photograph of the power amplifier test-board including microstrip line balun (size: 36 x 29 mm^2).

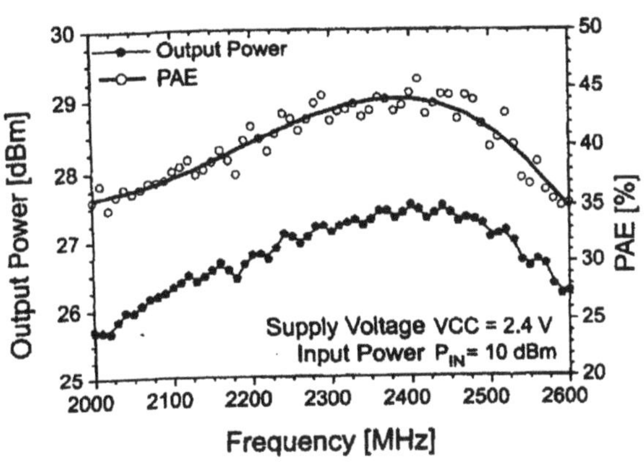

Fig. 6. Measured power amplifier frequency characteristic.

III. Dual band LC-balun

For dual band applications an extension to the standard LC-balun can be made: If the inductors are replaced by a parallel resonant circuit and the capacitors are replaced by a series resonant circuit in Fig. 1, then a lumped dual-band LC-balun, shown in Fig. 7, is available.

Using this as an output matching network for a push-pull type power amplifier leads to a schematic shown in Fig. 8. The circuit provides a balanced input impedance

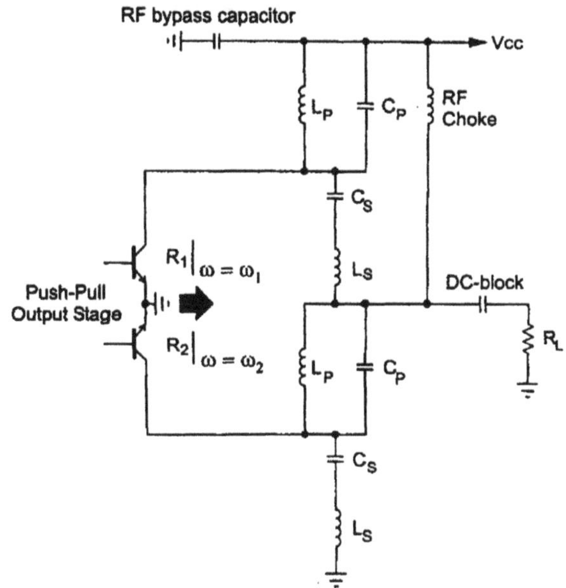

Fig. 8. Dual-band lumped LC-balun as push-pull type power amplifier output matching network.

R_1 at $\omega_1 = 2\pi f_1$ and R_2 at $\omega_2 = 2\pi f_2$. Independent matching and balun conversion at two different frequencies can be done. L_S, C_S, L_P and C_P can be calculated by

$$L_S = \frac{\omega_1 \cdot Z_1 + \omega_2 \cdot Z_2}{\omega_2^2 - \omega_1^2} \tag{6}$$

$$C_S = \frac{\frac{\omega_2}{\omega_1} - \frac{\omega_1}{\omega_2}}{\omega_1 \cdot Z_2 + \omega_2 \cdot Z_1} \tag{7}$$

$$L_P = \frac{\left(\frac{\omega_2}{\omega_1} - \frac{\omega_1}{\omega_2}\right) \cdot Z_1 \cdot Z_2}{\omega_1 \cdot Z_1 + \omega_2 \cdot Z_2} \tag{8}$$

$$C_P = \frac{\omega_1 \cdot Z_2 + \omega_2 \cdot Z_1}{(\omega_2^2 - \omega_1^2) \cdot Z_1 \cdot Z_2} \tag{9}$$

where $Z_1 = \sqrt{R_1 \cdot R_L}$ and $Z_2 = \sqrt{R_2 \cdot R_L}$ are the characteristic impedances of the bridge at ω_1 and ω_2. R_1, R_2, Z_1 and Z_2 are assumed to be real valued. Note, that

$$\omega_2 = k\omega_1 \tag{10}$$

with k > 1. For example $\omega_1 = 2\pi\,900$ MHz and $k = 2$, so that $\omega_2 = 2\pi\,1800$ MHz, would be well suited for a GSM dual-band push-pull power amplifier.

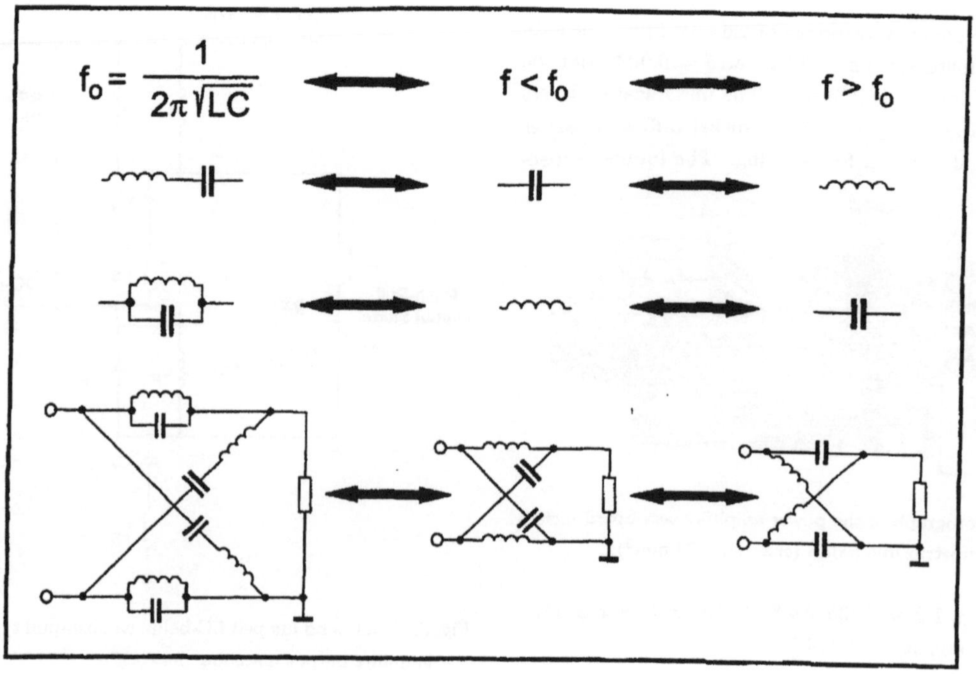

Fig. 7. Equivalent circuit of the dual-band LC balun vs. the frequency.

IV. Conclusions

Two balun circuits derived from the conventional LC-balun bridge are presented. The lumped LC-balun is transformed in a microstrip line balun for the use as a power amplifier output matching network. This results in a low impedance at the 2nd harmonic frequency and a high impedance at the 3rd harmonic frequency to improve the power-added efficiency (PAE). Finally, the lumped LC-balun is extended to a dual band balun. The design equations are derived. Independent impedance transformation and balun conversion can be done at two different frequencies.

References

[1] Peter Vizmuller, *RF Design Guide - Systems, Circuits, and Equations*, Artech House, Norwood, MA 02062, first edition, 1995.

[2] C.LORENZ AG Berlin-Tempelhof, "Schaltungsanordnung zum Übergang von einer symmetrischen, elektrischen Anordnung zu einer unsymmetrischen, insbesondere bei Hochfrequenzanwendungen," *German Patent*, , no. 603816, April 1932.

[3] Steve C. Cripps, *RF Power Amplifiers for Wireless Communications*, Artech House, Norwood, MA 02062, first edition, 1999.

[4] Alois Krischke, *Rothammels Antennenbuch*, Franck-Kosmos, Stuttgart,Germany, 11th edition, 1995.

[5] Dan Raicu, "Design of planar, single-layer microwave baluns," in *IEEE MTT-S International Microwave Symposium Digest*. IEEE, 1998, Cat.No.98CH36192.

[6] J.R. Vinding, "Radial line stubs as elements in strip line circuits," *Nerem Record*, pp. 108–109, 1967.

[7] Agilent, "Broadband microstrip mixer design - the butterfly mixer," *Agilent Technologies Application Note No.976*, pp. 4–6, 1988.

[8] E.O. Hammerstad, "Equations for microstrip circuit design," *Proc. of the European Microwave Conference*, pp. 261–272, 1975.

[9] Mongia, R., Bahl, I., and Bhartia, P., *RF and Microwave Coupled-Line Circuits*, Artech House, Norwood,MA 02062, first edition, 1999.

[10] S. A. El-Hamamsy, "Design of High-Efficiency RF Class-D Power Amplifier," *IEEE Transactions on Power Electronics*, vol. 9, no. 3, pp. 297–308, May 1994.

[11] Bakalski,W., Simbürger,W., Kehrer,D., Wohlmuth,H.-D., Rest,M., and Scholtz,A.L., "A monolithic 2.45 GHz, 0.56 W power amplifier with 45 % PAE at 2.4 V in standard 25 GHz f_T Si-bipolar," in *Proceedings of IEEE International Symposium on Circuits and Systems*, Scottsdale, Arizona, May 2002, IEEE.

Transmission-Line CONVERSION

Methods for joining a balanced two-wire line to a coaxial line are discussed. Conversion transformers which operate at a single frequency or over a band of frequencies are described. Constant impedance is possible over a one-to-three frequency range

By NATHAN MARCHAND
Federal Telephone and Radio Laboratories
New York, N. Y.

In high-frequency power transmission it is frequently necessary to use a balanced two-wire line over part of the system, and a coaxial transmission line over the remainder of the distribution system. The natures of the generator output and the load input will dictate the particular combination of balanced two-wire and coaxial transmission lines which must be used.

At the juncture of the two types of transmission line a conversion transformer is necessary to maintain the currents on the two types of line in their proper relations. The conversion from one line to the other can be made ideal only at one frequency, but methods will be described which provide satisfactory conversion over a range of frequencies.

In converting from a balanced two-wire transmission line to a coaxial line, it is necessary to understand what is meant by equilibrium conditions in a coaxial line and in a balanced two-wire shielded transmission line. The term unbalanced is often used for coaxial feed, but in this analysis it will be used for departure from the normal equilibrium conditions.

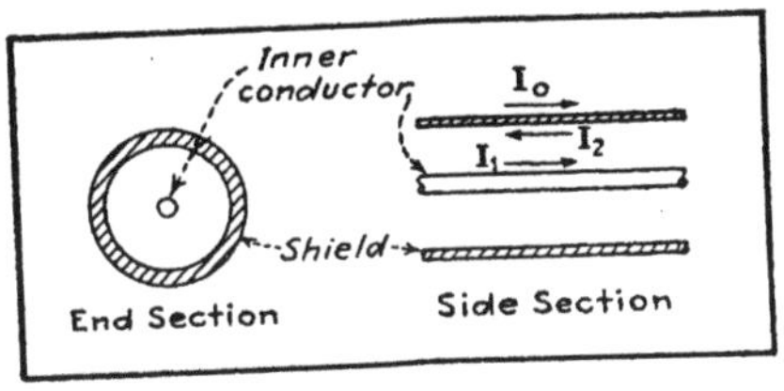

FIG. 1—Section of coaxial line, showing the currents which can flow

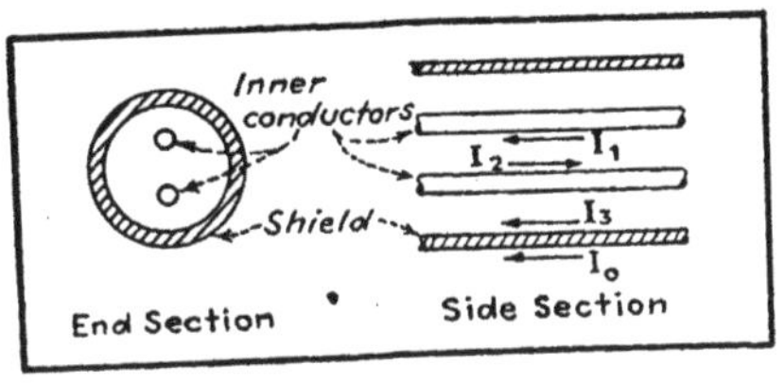

FIG. 2—Section of a balanced two-wire transmission line, showing the line and shield currents which can flow

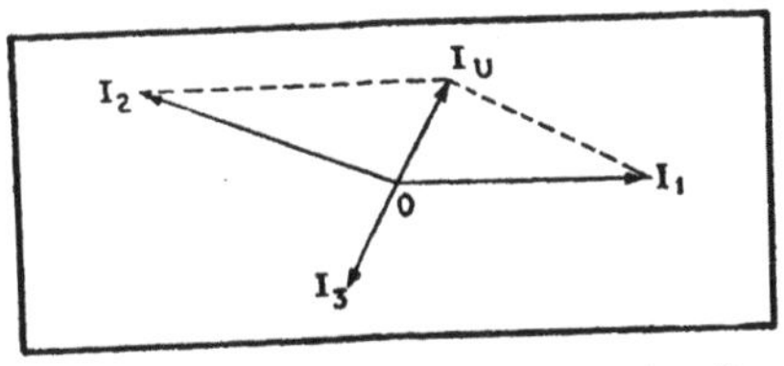

FIG. 3—If the currents of a two-wire line are not equal, an unbalance current I_3 flows. In this case the inequality of the line currents is in the nature of a small phase shift. The two line currents should be 180 degrees apart for balance

Currents on a Coaxial Line

A coaxial transmission line consists of a conducting wire concentrically disposed in a hollow conducting tube, the space between being filled with a dielectric which may be air. Figure 1 shows two sections of a coaxial transmission line. In the side section there are shown three currents, I_1, I_2, and I_0. When a transmission line of this type has a shield which is well constructed, and is used above 50 megacycles, it can be assumed without loss of generality that the shield is perfect. This does not mean that it is a perfect conductor, but rather that there is no coupling between the current I_2 on the inside of the shield and the current I_0 on the outside of the shield. This also means that the current I_2 on the inside of the shield must be exactly equal and opposite to the current I_1 on the inner conductor.

The problem in the coaxial line, as far as conversion is concerned, is to prevent any coupling between the true currents I_1 and I_2 and the interfering or unbalancing current I_0 that might be induced on the outside of the shield. It is also important from power considerations to maintain the surge impedance along the line constant.

Currents on a Two-Wire Line

Balanced two-wire transmission line consists of two parallel conductors in a dielectric which may or may not be surrounded symmetrically by a shield. Since for shielding purposes all balanced transmission lines should be enclosed in a shield, a shield will be included in the discussion.

Figure 2 shows an end and side section of a balanced transmission line. In the side section there are shown four currents, I_1, I_2, I_3, and I_0. I_3 is the total resultant current on the inside of the shield and I_0 is the total resultant current on the outside of the shield. In a perfectly balanced line I_1 would be equal and opposite to I_2, and I_3 would be zero.

If I_1 is not equal in amplitude and opposite in phase to I_2, then I_3 will be equal and opposite to the vector sum of I_1 and I_2. This is shown in Fig. 3, where I_1 and I_2 are not exactly 180 deg out of phase. Their sum is equal to the unbalanced current I_U. I_U behaves as though it were a current flowing in the same direction along both of the transmission lines. That is, the two wires are acting in parallel as the inner conductor of a coaxial line, the return current flowing on the inside of the shield. This return current is designated as I_3. Again I_0 is the

TRANSFORMERS

extraneous current induced on the outside of the shield.

The first problem encountered in maintaining the normal equilibrium conditions in a balanced transmission line is to keep I_1 equal and opposite to I_2 so that I_3 will be zero; the second problem is to prevent I_0 from coupling into the transmission line at any point.

The problem is to convert from a balanced transmission line to a coaxial transmission line by introducing some type of transmission line transformer. First, in order to prevent I_0 from coupling into the lines the outside shields of both lines should be kept continuous so that there can be no way in which I_0 can get into the line. If this method is not possible, some other means must be used to prevent I_0 from coupling into the line. Secondly, in order to maintain I_1 equal to I_2 in the balanced transmission line, the impedances from each conductor to the shield must be maintained equal. Any difference in impedance will cause I_1 to differ from I_2. Third, the continuity of surge impedance, or impedance matching, should be maintained as close as possible, to obtain maximum power transfer.

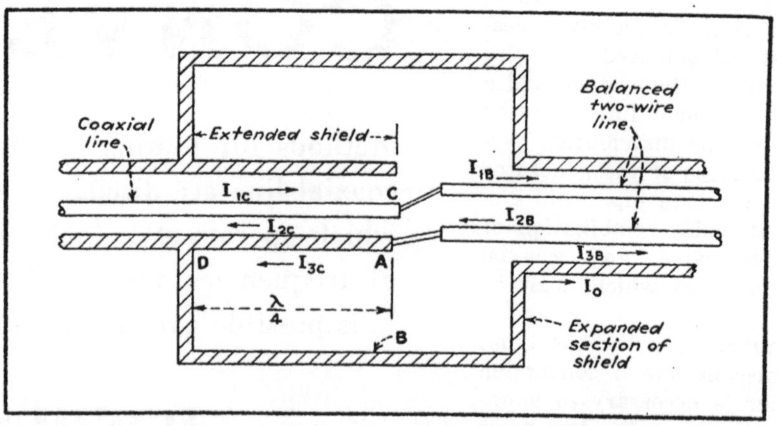

FIG. 4—Cross-section of a single-frequency conversion section, showing how the coaxial shield can be extended to reduce I_3 to zero at resonance

Single-Frequency Transformation

In Fig. 4 is shown one type of transformer that can be used for single-frequency transformation. A large box, an expanded section of the shield, is used. This box must totally enclose the interior circuits to avoid coupling between I_0 and the inside currents. The coaxial line shield is extended a distance of one quarter wavelength inside the box; the dual line is brought in on the other side. One conductor of the dual line is connected to the inner conductor of the coaxial line and the other to the coaxial shield extension.

In order to determine the equivalent circuit of this conversion transformer, the current paths have to be traced. Current I_{1B} of Fig. 4 will flow directly from one side of the balanced two-wire line to the inner conductor of the coaxial line shown as I_{1C}. The current I_{2B} will flow from the other side of the balanced two-wire line and divide into two currents. One will flow along the inside surface of the coaxial shield (shown as I_{2C}), and the other will flow around the outside of the coaxial shield (shown as I_{3C}). In order to maintain the normal equilibrium conditions, I_{3C} should be zero. This is true since I_{1C} is always equal and opposite to I_{2C}; thus for I_{1B} to be equal and opposite to I_{2B}, I_{3C} must be zero. In this case it will be zero since the outer side of the coaxial shield *AD* and the inner side of the box *B* form a coaxial transmission line which is shorted at the end *D* and is one quarter wavelength long, resulting in an extremely high impedance between point *A* and the surface at *B*.

The equivalent circuit is shown in Fig. 5. The wire representing

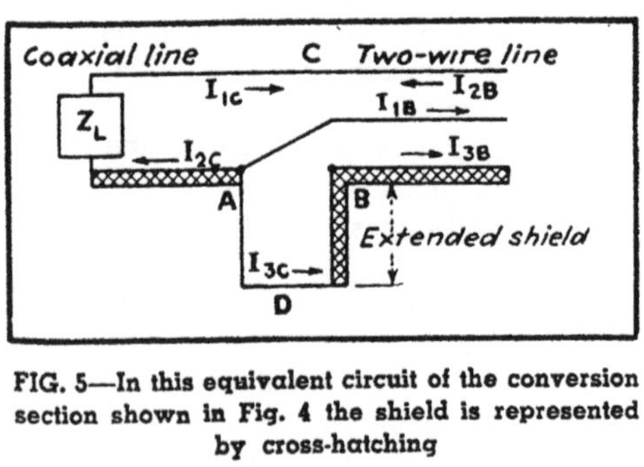

FIG. 5—In this equivalent circuit of the conversion section shown in Fig. 4 the shield is represented by cross-hatching

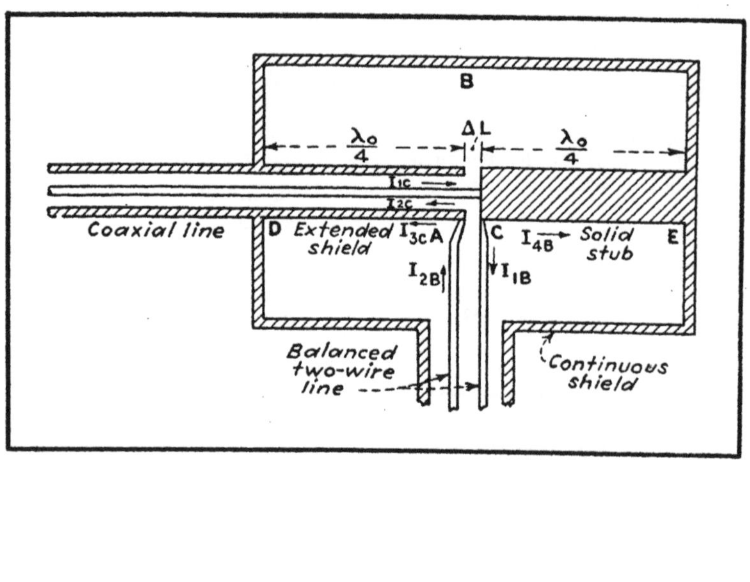

FIG. 6—The addition of a solid stub inside the conversion section improves the off-resonant action of the transformer

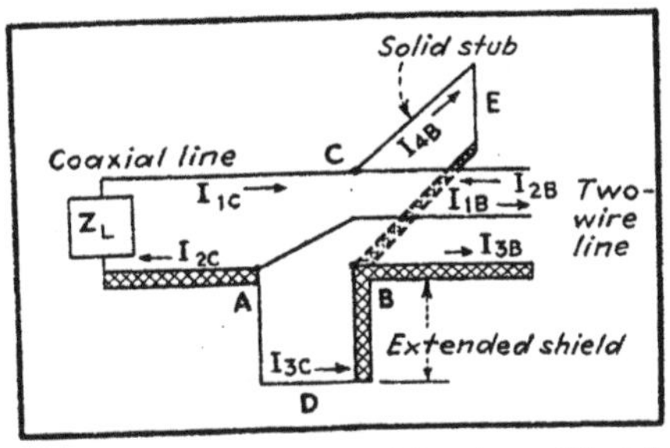

FIG. 7—The loop formed by the solid stub of Fig. 6 is shown on a plane extending into the page in this equivalent circuit

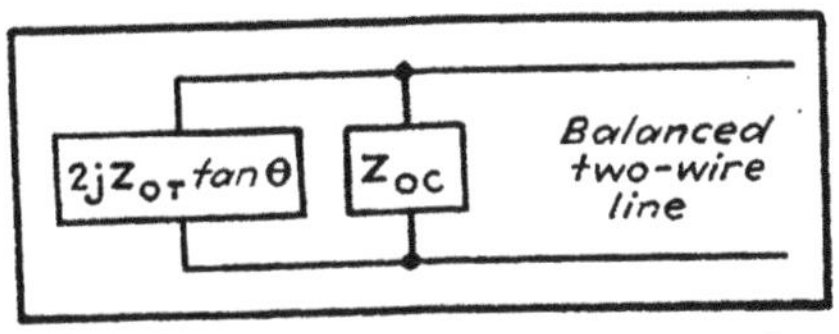

FIG. 8—The lumped constant equivalent of Fig. 6 shows a reactive impedance in parallel with the coaxial impedance. It is this reactive impedance which limits the frequency band over which the conversion section will operate

the inside of the shield is shown with shading. It can be seen here that I_{3C} is being fed into a shorted transmission line ADB which, if it is one quarter wavelength long, will result in an infinite impedance at AB so that I_{3C} will be zero. Whatever impedance is presented across AC looking into the coaxial line will be the load across the end of the balanced line. Thus, if both the coaxial line and the balanced line have the same surge impedances, and if Z_L is matched to the coaxial line, the balanced two-wire line will also be matched.

The one bad feature of this type of conversion transformer is that it will give good balance at only one frequency. As soon as the frequency is shifted so that ADB is no longer a quarter of a wavelength long, I_{3C} will no longer be zero, and hence I_{1B} will no longer equal I_{2B} and the line currents will no longer be balanced. I_{3B} will then no longer be zero, but rather will equal I_{3C}.

Wide-Band Conversion-Transformer

The next logical step is to put an impedance from C to B that will equal the impedance at AB so that even if the frequency is shifted, it will only introduce a phase mismatch, but will not disturb the current balance. This will be true because the voltage at C is equal and opposite to the voltage at A. In Fig. 6 is shown a wide-band conversion transformer. This is the same as the single-frequency transformer (Fig. 4) except that at point C where the inner conductor is directly connected to one side of the balanced line, a solid stub, equal in length and outside diameter to the extended shield, is connected as shown. Their electrical lengths are equal to $\lambda_0/4$ which would be one-quarter wavelength at the mean frequency if the stray capacitance between A and C were negligible. Because of this stray capacitance, the inner conductor of the coaxial line should be removed and the lengths of shield and stub so chosen that the impedance across AC is a maximum at the mean frequency for which the unit is to be used. ΔL should be kept as small as possible since it is an extra distance that I_{1C} must travel to reach C over that which I_{2C} must travel to reach A.

In Fig. 7 is shown the equivalent circuit for this type of conversion transformer or balance box. It can be seen that this is similar to the equivalent circuit of Fig. 5, except that a reactance CEB, equal to the reactance ADB, is inserted. If this transformer is detuned from its mean frequency, currents I_{4B} and I_{3C} will flow. However, because of the quarter-wave extended shield and solid stub construction, I_{4B} will equal I_{3C} in magnitude and they will cancel at B. This will result in I_{3B} being zero and I_{1B} being equal to I_{2B}. However, the transformer introduces a reactive component across the balanced transmission line that will result in a mismatch.

If the coaxial line is matched, its input impedance would be equal to

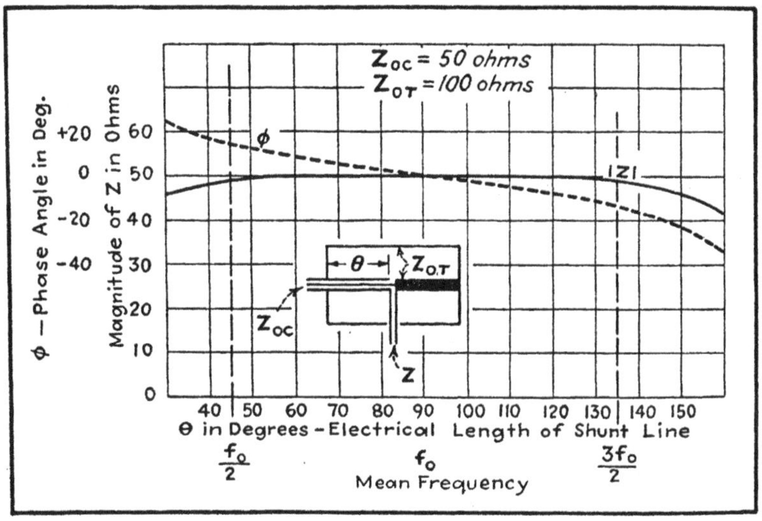

FIG. 9—Theoretical plot showing the variation in magnitude and angle of the impedance presented to the balanced two-wire line by the coaxial line and conversion transformer of Fig. 6 for conditions given in the text

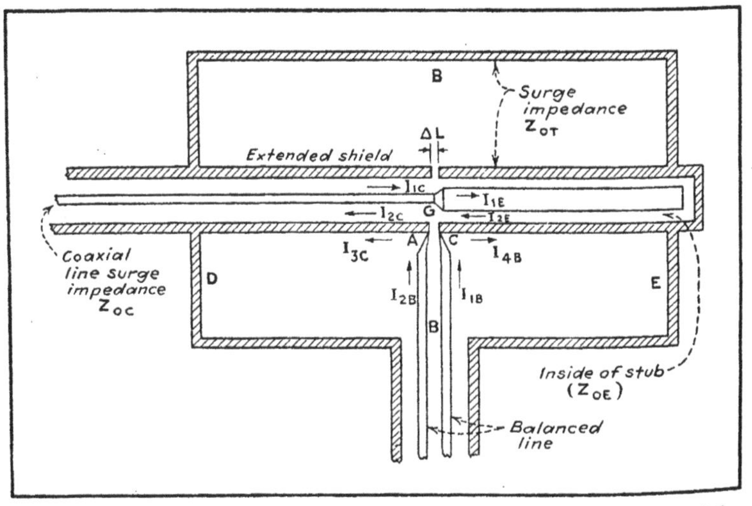

FIG. 10—To decrease the angular variation in impedance as seen by one of the lines being joined, the coaxial line is extended into the stub of Fig. 6.

Z_{0C} where Z_{0C} is the surge impedance of the coaxial line. If the surge impedance of the transmission line, composed of the inside of the conversion box and the outside of the extended inner conductor, is Z_{0T} then the impedance across the balanced line is

$$Z = \frac{(Z_{0C})\,(2jZ_{0T}\tan\theta)}{Z_{0C} + 2jZ_{0T}\tan\theta} \qquad (1)$$

where θ is the electrical length AD which is equal to CE. The lumped constant equivalent circuit is shown in Fig. 8. θ of course will vary with frequency. In Fig. 9 is shown the impedance characteristic presented at the end of the balanced two-wire line in this type of transformer with an impedance Z_{0C} of 50 ohms and an impedance Z_{0T} of 100 ohms. Here θ is picked to be 90 deg at the mean frequency f_0. The balance remains very good over the whole range. For a three-to-one range in frequency the impedance will vary between the points noted as $f_0/2$ to $3f_0/2$. The value of the phase angle ϕ at these points will be about 15 deg. This calculated curve neglects the capacitance between the edges of the extended shield and stub. This capacitance will increase the impedance variation slightly.

Phase-Shift Correction

It is now desirable to make the conversion transformer wideband without the shift in phase. This can be done by constructing the transformer as shown in Fig. 10. The only difference between this transformer and the transformer shown in Fig. 6 is that the coaxial line is extended into the solid stub. This extended line has a surge impedance of Z_{0E} which is equal to

$$Z_{0E} = \frac{(Z_{0C})^2}{2Z_{0T}} \qquad (2)$$

where Z_{0C} is the coaxial line surge impedance and Z_{0T} is the surge impedance of the coaxial shield and box inside the transformer box. Thus the inner conductor of the extended line must be made large since Z_{0T} is always larger than Z_{0C}. The length of the extended shield is resonant as a shorted quarter wavelength at the mean frequency f_0. If the capacitance acros ΔL were neglected, the length from the box to the break would be 90 deg. The extended line is left open at the end and is made resonant as a quarter-wavelength open stub.

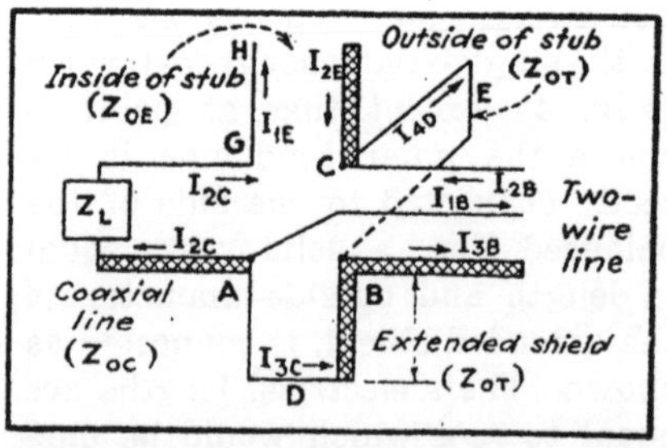

FIG. 11—As this three-dimensional equivalent circuit shows, the extension of the coaxial line into the stub is equivalent to adding an open quarter-wave line in series with the inner conductor of the coaxial line

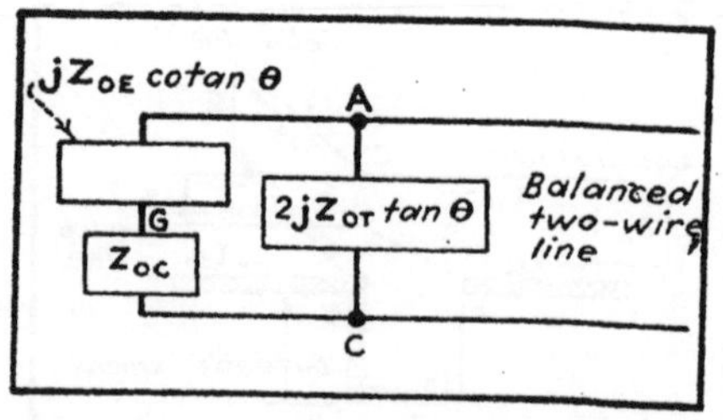

FIG. 12—The lumped-constant equivalent circuit of the conversion section shown in Fig. 10 shows how the addition of an open line in series with the coaxial line will correct the angular variation of impedance of the conversion section

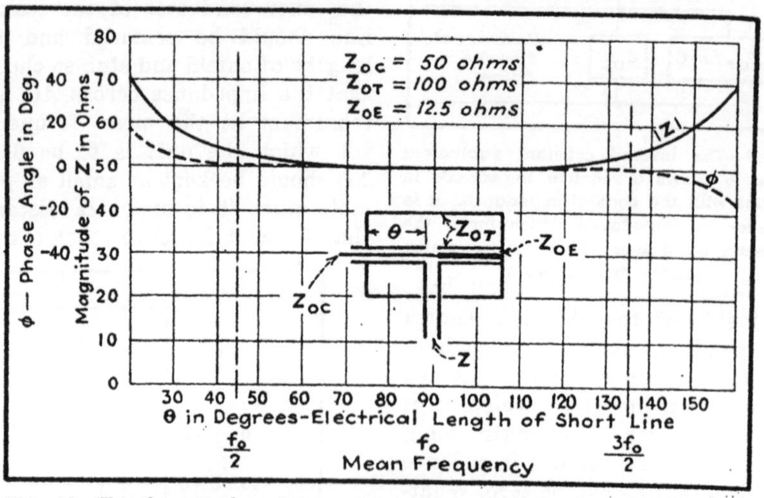

FIG. 13—The theoretical variation of impedance seen by the balanced two-wire line, when converted to a coaxial line by the transformer of Fig. 10, shows but slight change in magnitude and negligible change in angle over a one-to-three change in frequency

Calling the currents in the extended line I_{1E} and I_{2E}, I_{2E} is always equal and opposite to I_{1E}. From Fig. 10 it can be seen that I_{1E} is equal to I_{1C}, which will make I_{2E} equal and opposite to I_{2C}. Since the extended shield and stub have the same diameter, the circuit will be symmetrical, making I_{3C} equal and opposite to I_{4B}. This means that the balanced line currents I_{1B} and I_{2B} will also be equal and opposite, yielding balance. In Fig. 11 is shown the equivalent circuit of this type of transformer. This can be further modified to the lumped constant circuit of Fig. 12.

The input impedance from the balanced line across AC is

$$Z = \frac{(Z_{0C} - jZ_{0E}\cot\theta)\,(2jZ_{0T}\tan\theta)}{Z_{0C} - jZ_{0E}\cot\theta + 2jZ_{0T}\tan\theta} \qquad (3)$$

where θ is the electrical length of the extended line. This will also be equal to the length of the stub line inside the box when the capacitance across ΔL is neglected. The coaxial line is assumed to be matched so that the impedance presented at GA is the surge impedance. Simplifying Eq. (3),

$$Z = Z_{0C}\frac{1 - j(Z_{0E}/Z_{0C})\cot\theta}{1 - (Z_{0E}/2Z_{0T})\cot^2\theta - j(Z_{0C}/2Z_{0T})\cot\theta} \qquad (4)$$

Substituting the value of Z_{0E} given in Eq. (2), Z becomes

$$Z = Z_{0C}\frac{1 - j(Z_{0C}/2Z_{0T})\cot\theta}{1 - [(Z_{0C}/2Z_{0T})\cot\theta]^2 - j(Z_{0C}/2Z_{0T})\cot\theta} \qquad (5)$$

As long as Z_{0T} is kept large this circuit is an excellent transformer.

The calculated impedance of a transformer where $Z_{0C} = 50$ ohms, $Z_{0T} = 100$ ohms and $Z_{0E} = 12.5$ ohms is shown in Fig. 13, where the capacitance across ΔL is neglected.

This type of transformer is excellent for a transfer from a stationary to a rotating member. Since the coaxial line does not make contact with anything, it can be kept stationary while the whole transformer and balanced two-wire line rotates around it.

Technical Note

Exact Design of the **Marchand Balun**

J. H. CLOETE
National Institute for Aeronautics and Systems Technology
Council for Scientific and Industrial Research
Pretoria, South Africa

Published in Proceedings of the Ninth European Microwave Conference, Brighton, England, 17-20 September, 1979.

The Marchand, or compensated, balun[1–9] in **Figure 1** is widely used with multi-octave antennas such as the Archimedes spiral. This paper deals with three implementations of the Marchand balun. The first has the conventional second-order equivalent circuit consisting of a series open-circuit stub and a parallel short-circuit stub, both of quarter-wavelength at the center frequency. Oltman[6] analyzed the second-order balun and presented equations of element values intended to maximize the bandwidth. A 10:1 bandwidth design based on Oltman's equations was found to exhibit significant bandwidth contraction. In order to achieve a specified response the exact method described by Wenzel[10] was used. The method is straightforward and requires only simple calculations. The element values of a balun designed according to this procedure are given in **Table 1** and the frequency response in **Figure 2**.

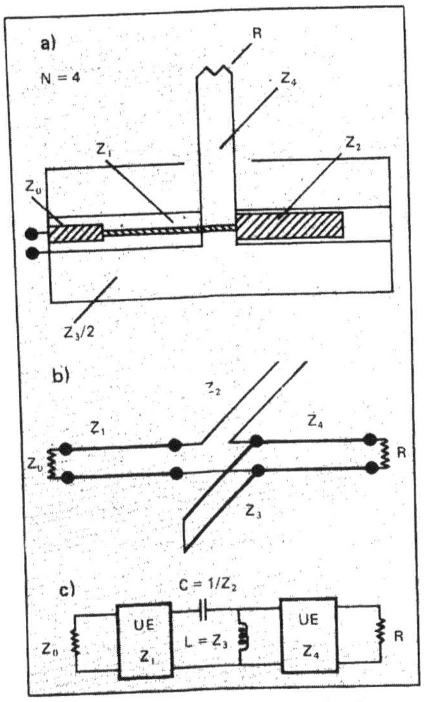

Fig. 1 The fourth-order Marchand balun.
a) Cross section through physical circuit.
b) Transmission line equivalent circuit.
c) Equivalent circuit using distributed element notation.

RESULTS

For applications when the load resistance is greater than approximately 75 Ω the second-order balun may provide inadequate impedance match. Although most authors[2,4,6–9] recognize the need for transformer action when dealing with such loads, and describe the use of various transformers with the second-order balun, only Laughlin[8] takes the interaction between the balun and transformer into account from the start. Failure to do this can lead to unacceptably degraded frequency response for broadband designs.

In the second implementation a quarter-wave line is inserted before the balanced load. The resultant third-order balun has been dealt with by Laughlin[8] who derived approximate equations for the element values. For this paper the element values for Tchebysheff passband response were computed using the exact, nonredundant method for the syntheses of commensurate length distributed networks described by Horton and Wenzel.[1] To simplify the design procedure, graphs of the element values as a function of passband reflection coefficient were produced for selected bandwidths. For example, **Figure 3** gives the graphs for 10:1 bandwidth ratio. A comparison with the second-order design is made in **Table 1** and **Figure 2**. A 0.2 to 2 GHz balun was constructed according to this design and **Figure 4** shows the agreement between the theoretical and measured responses.

For bandwidths up to approximately two octaves, the third-order balun offers marked improve-

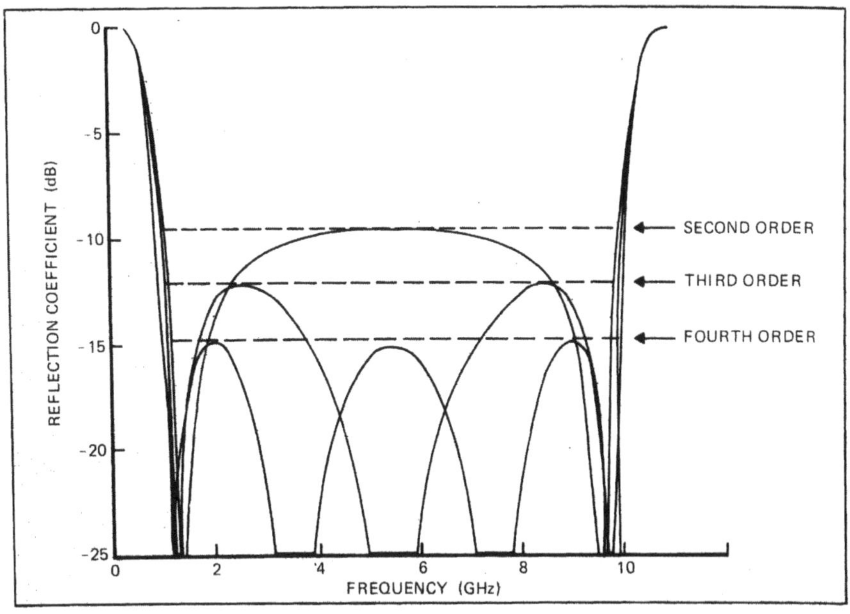

Fig. 2 Theoretical frequency response of the second, third and fourth order Marchand baluns with 50 Ω source resistance and 100 Ω load resistance.

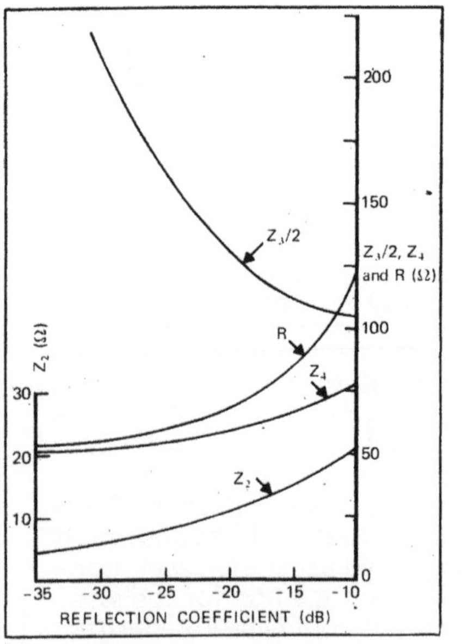

Fig. 3 Graphs of element values for the third-order balun with bandwidth ratio 10:1 and source resistance 50Ω.

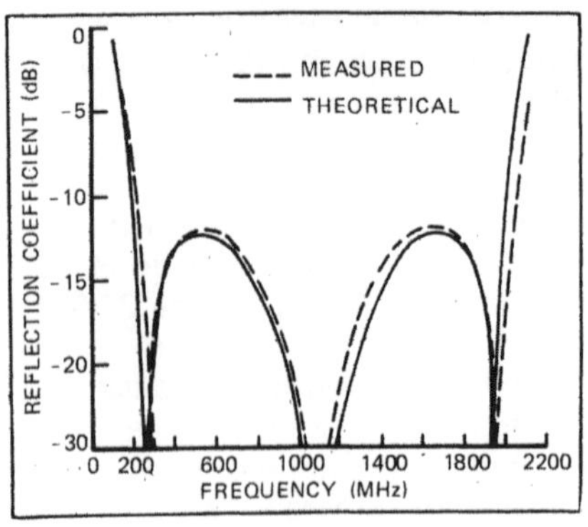

Fig. 4 Measured and theoretical response of a 10:1 bandwidth third-order Marchand balun with 50 Ω source resistance and 100 Ω load resistance.

ment over the second-order balun. However, effectiveness of quarter-wave lines decreases as bandwidth increases.[1] This trend is illustrated in **Figure 2** where the third-order balun offers only 2.9

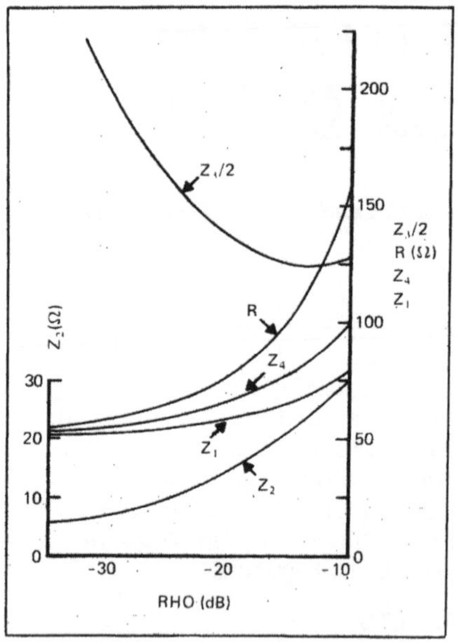

Fig. 5 Graphs of element values for the fourth-order balun with bandwidth ratio 10:1 and source resistance 50Ω.

dB improvement in reflection coefficient.

To counter this effect a new fourth-order implementation, shown in **Figure 1**, with quarter-wave lines included at both the balanced and unbalanced ports, was investigated. The element values for Tchebysheff response were computed using nonredundant synthesis[11] and, as for the third-order case, graphs of element values were prepared. For an example, see **Figure 5**. **Table 1** and **Figure 2** compare the fourth-order design to the previous two. **Figure 6** shows the agreement between the theoretical and measured responses of a fourth-order design. It is evident that there is no significant difference in realizability between the three circuits,

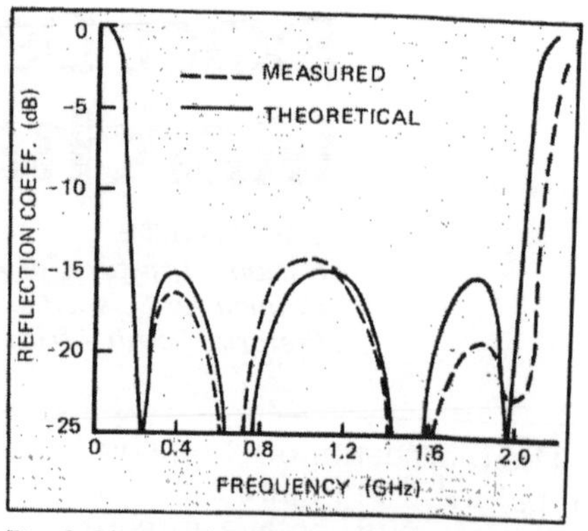

Fig. 6 Measured and theoretical response of a 10:1 bandwidth fourth-order balun with 50 Ω source resistance and 100 Ω load resistance.

but that an improvement of 5.4 dB in reflection coefficient over the second-order balun has been achieved.

The fourth-order design described here extends the operating range of the Marchand balun without compromising on its compact physical realization. For many practical applications the additional expense of manufacturing the input quarter-wave line will be justified by the improved performance.

REFERENCES

1. Marchand, N., "Transmission-Line Conversion," *Electronics*, 1944, Vol. 17, pp. 142-145.
2. Fubini, E. G. and P. J. Sutro, "A Wide-Band Transformer from an Unbalanced to a Balanced Line," *Proc. IRE*, 1947, Vol. 35, pp. 1153-1155.
3. Roberts, W. K., "A New Wide-Band Balun," *Proc. IRE*, 1957, Vol. 45, pp. 1628-1631.
4. McLaughlin, J. W., D. A. Dunn, and R. W. Grow, "A Wide-Band Balun," *IRE Trans.*, 1958, Vol. MTT-6, pp. 314-316.
5. Bawer, R. and J. J. Wolfe, "A Printed Circuit Balun for Use with Spiral Antennas," *IRE Trans.*, 1960, Vol. MTT-8, pp. 319-325.
6. Oltman, G., "The Compensated Balun," *IEEE Trans.*, 1966, Vol. MTT-14(3), pp. 112-119.
7. Phelan, H. R., "A Wideband Parallel Connected Balun," *IEEE Trans.*, 1970, Vol. MTT-18(5), pp. 259-263.
8. Laughlin, G. J., "A New Impedance-Matched Wide-Band Balun and Magic Tee," *IEEE Trans.*, 1976, Vol. MTT-24(3), pp. 135-141.
9. Bartholomew, D.M.L., "Optimum Design for a Broadband Microstrip Balun," *Electronics Letters*, 1977, Vol. 17(13), pp. 510-511.
10. Wenzel, R. J., "Exact Design of TEM Microwave Networks using Quarter-Wave Lines," *IEEE Trans.*, 1964, Vol. MTT-12, pp. 94-111.
11. Horton, M. C. and R. J. Wenzel, "General Theory and Design of Optimum Quarter-Wave TEM Filters," *IEEE Trans.*, 1965, Vol. MTT-13(3), pp. 316-327.

Bandwidth ratio, B = 10 : 1
Source resistance, Z_0 = 50 Ω
Load resistance, R = 100 Ω

Order	Z_1 (Ω)	Z_2 (Ω)	Z_3 (Ω)	Z_4 (Ω)	Maximum passband reflection coefficient (dB)
2	–	21.0	241.0	–	–9.5
3	–	17.5	215.0	70.7	–12.4
4	65.0	20.0	250.0	76.0	–14.9

TABLE I Comparison of Tchebysheff Response Marchand Baluns

Graphs of Circuit Elements for the Marchand Balun

J. H. CLOETE
National Institute for Aeronautics and Systems Technology, CSIR, Pretoria, South Africa

INTRODUCTION

The Marchand balun[1] exhibits excellent amplitude balance for bandwidth ratios in excess of a decade. In addition, it can be designed to incorporate transformer action for applications where the load and source resistances differ.[2] The Marchand balun has been used in a variety of physical implementations, including coaxial and planar, for applications like spiral antennas[3] and balanced mixers.[4]

In typical antenna applications the balanced load resistance differs significantly from the source resistance and it is necessary to implement transformer action in the balun in order to achieve an acceptable match. For example, in the case of spiral antennas the balanced load impedance generally has a nominal real part in excess of 100 Ω which must be matched to 50 Ω. A popular design approach to this problem is to design a balun for a 50 Ω load, and then to incorporate a tapered line section to transform the balanced load resistance to approximately 50 Ω. However, failure to take the interaction between the balun and transformer into account properly can lead to unnecessarily degraded frequency response for broadband designs.

An alternative approach to achieve transformer action is the use of quarter-wave sections instead of a tapered line. This method has the advantage that exact synthesis techniques can be used to include the transformer sections in the balun, thus allowing designs with accurately predictable performance to be implemented. In a recent paper the existence of graphs for the design of a third order balun with a single section quarter-wave output transformer, and a fourth order balun with single section quarter-wave input and output transformers was announced.[2]

The purpose of this paper is to publish a set of these graphs for the design of the fourth order balun shown in **Figure 1**. Graphs for the third order balun will not be published since approximate closed form equations for the element values are available.[5]

Both the third and fourth order baluns have maximum bandwidth ratios in the order of 12:1 due to the high characteristic impedance Z_3, required for the shunt stub, becoming practically unrealizable. The advantage of the fourth order balun over the third order balun is that for a given load resistance, a significant reduction in passband reflection coefficient can be obtained, depending on the required bandwidth. The improved match is achieved without compromising on the compact physical structure of the Marchand balun, but does involve increased manufacturing expense due to the input quarter-wave line, Z_1.

THE GRAPHS OF ELEMENT VALUES

The graphs of fourth order Marchand balun circuit element values are given in **Figures 2** to **6**

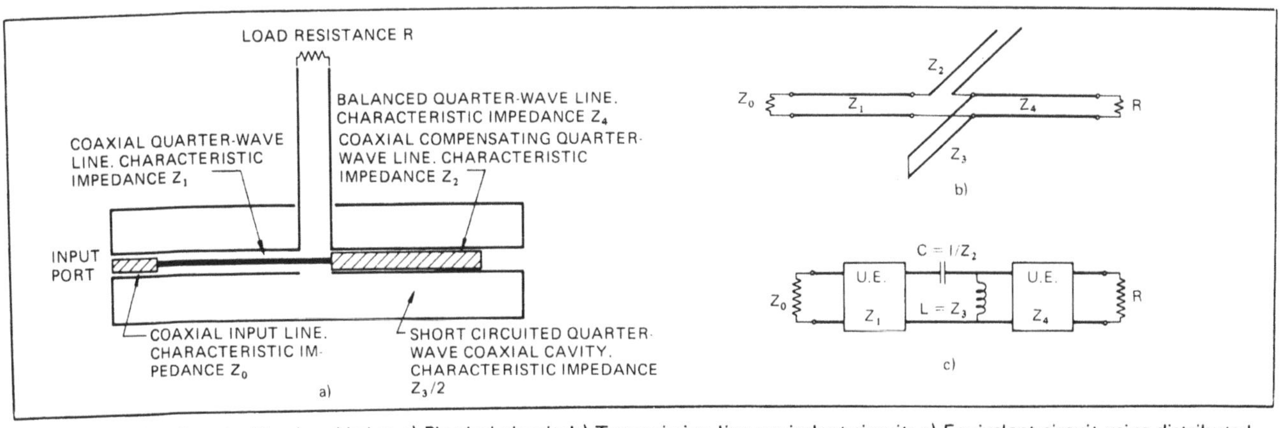

Fig. 1 The fourth order Marchand balun a) Physical circuit; b) Transmission line equivalent circuit; c) Equivalent circuit using distributed element notation.

for five bandwidth ratios ranging from 4:1 to 12:1, and realize the Tchebysheff bandpass response.

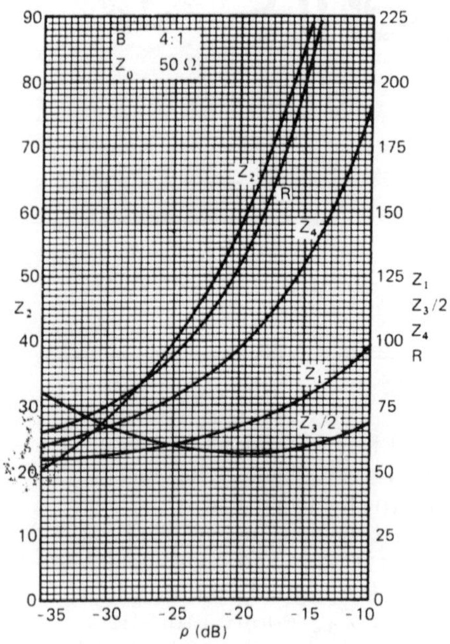

Fig. 2 Graphs of element values for the fourth order balun with source resistance $Z_0 = 50\ \Omega$ and bandwidth ratio B = 4:1.

For each bandwidth ratio, the element values are plotted as a function of maximum passband reflection coefficient. The notation used in the graphs is consistent with that of **Figure 1**.

The graphs were produced using the exact, non-redundant method for the synthesis of commensurate length distributed networks described by Horton and Wenzel.[6] The synthesis program is highly accurate, however, the price to pay for the convenience of the graphs is a loss of accuracy due to read-off errors. In the case of Z_2, the characteristic impedance of the series open circuit stub, the error should typically be less than ± 0.5 Ω, and for the other elements the error should typically be less than ± 1 Ω.

A simple check on the accuracy of the values obtained from the graphs can be made by using the following symmetry equations[7] for the element values of the balun,

$$Z_1 Z_4/R = Z_2 Z_3/R = Z_0$$

where $Z_0 = 50\ \Omega$ for the graphs. (see **Table II**)

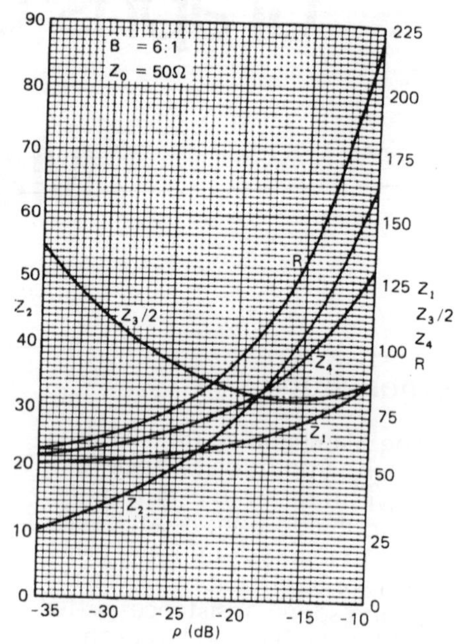

Fig. 3 Graphs of element values for the fourth order balun with source resistance $Z_0 = 50\ \Omega$ and bandwidth ratio B = 6:1.

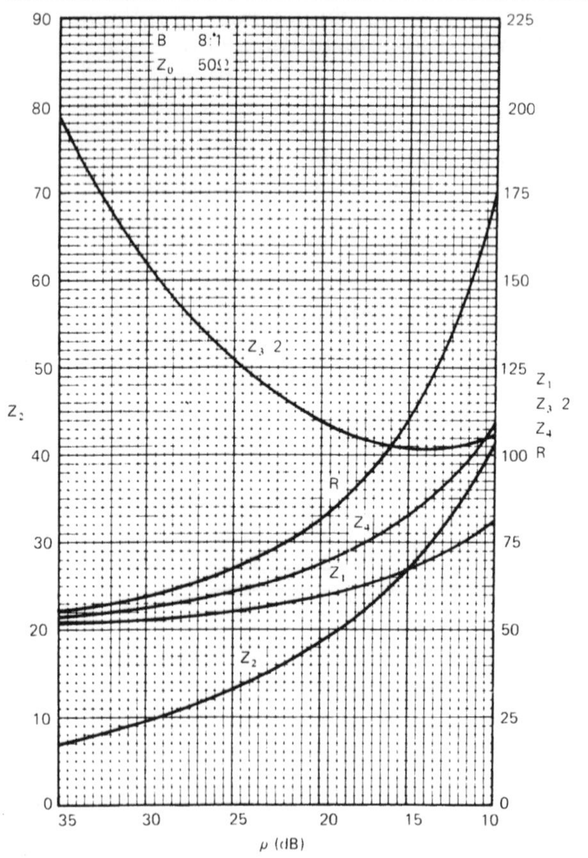

Fig. 4 Graphs of element values for the fourth order balun with source resistance $Z_0 = 50\ \Omega$ and bandwidth ratio B = 8:1.

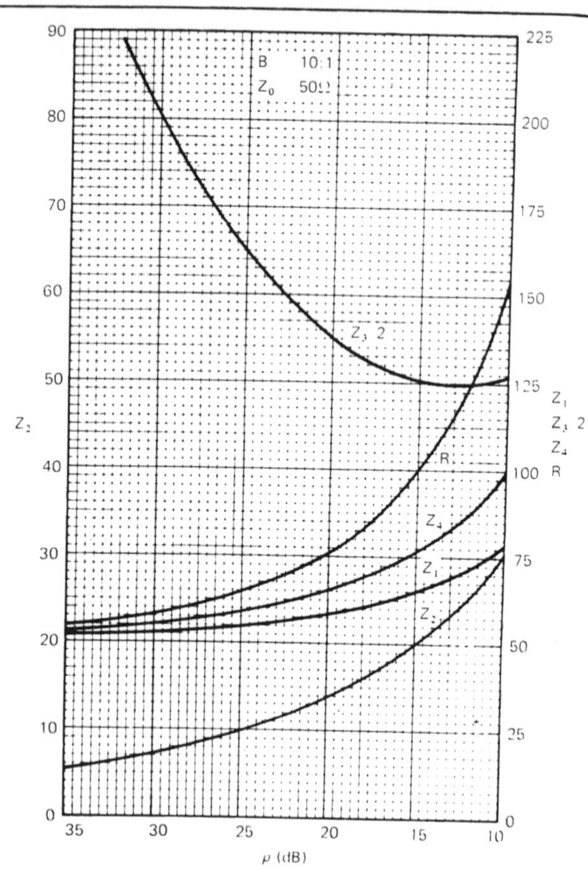

Fig. 5 Graphs of element values for the fourth order balun with source resistance $Z_0 = 50\ \Omega$ and bandwidth ratio B = 10:1.

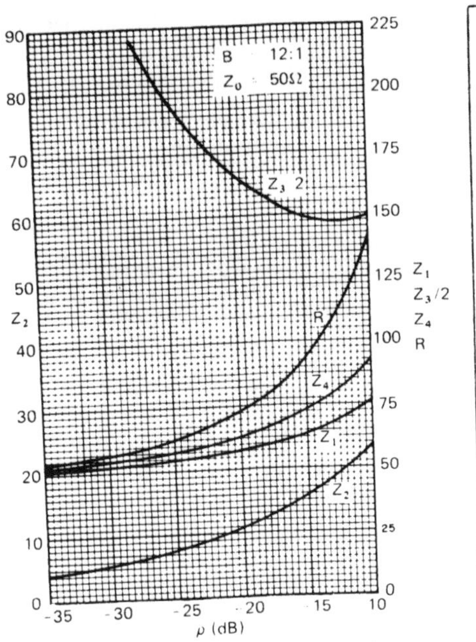

Fig. 6 Graphs of element values for the fourth order balun with source resistance Z_0 = 50 Ω and bandwidth ratio B = 12:1.

DESIGN EXAMPLE

To illustrate the use of the graphs consider the design of a balun with bandwidth ratio 10:1, source resistance 50 Ω, and load resistance 100 Ω. The element values and maximum passband reflection coefficient are found from **Figure 5** and given in Table 1. For comparison, the exact values as computed by the synthesis program are also tabulated. It is seen that the errors are insignificant from a practical viewpoint when manufacturing tolerances are considered. In addition, the

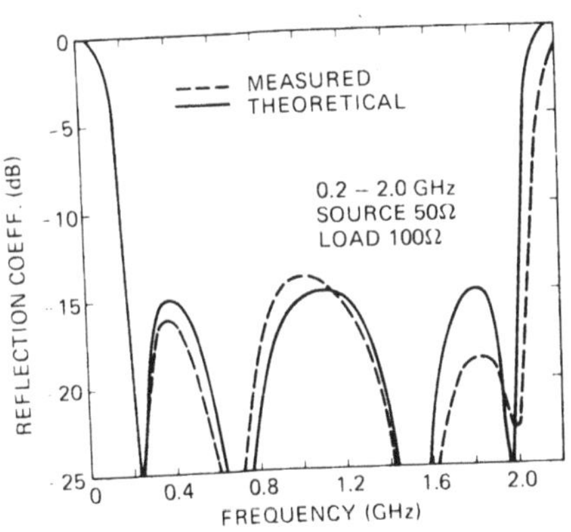

Fig. 7 Measured and theoretical frequency response of an experimental 10:1 bandwidth fourth order Marchand balun with 50 Ω source resistance and 100 Ω load resistance.

TABLE 1

ELEMENT VALUES FOR A FOURTH ORDER BALUN

	Z_1 (Ω)	Z_2 (Ω)	Z_3 (Ω)	Z_4 (Ω)	ρ(dB)
Exact values	65.14	19.99	250.18	76.75	-14.87
Values from Fig. 5	65.0	20.0	250.0	76.0	-14.9
Error magnitude	0.14	0.01	0.18	0.75	0.03

Bandwidth ratio B = 10:1, Source resistance Z_0 = 50 Ω, Load resistance R = 100 Ω

TABLE II

ELEMENT VALUES FOR FOURTH ORDER MARCHAND BALUN
AS COMPUTED BY THE SYNTHESIS PROGRAM

FOURTH ORDER MARCHAND BALUN WITH IN AND OUTPUT TRANSFORMERS

BANDWIDTH RATIO = 10

RHO(DB)	Z_1	Z_2	$Z_3/2$	Z_4	R
-35.00	51.58	5.35	255.44	52.97	54.65
50.00	50.00				
-34.00	51.77	5.68	242.96	53.32	55.20
50.00	50.00				
-33.00	51.97	6.03	231.29	53.70	55.82
50.00	50.00				
-32.00	52.21	6.41	220.37	54.12	56.51
50.00	50.00				
-31.00	52.46	6.81	210.17	54.59	57.28
50.00	50.00				
-30.00	52.75	7.25	200.66	55.12	58.15
50.00	50.00				
-29.00	53.00	7.71	191.81	55.71	59.13
50.00	50.00				
-28.00	53.44	8.20	183.58	56.36	60.23
50.00	50.00				
-27.00	53.84	8.73	175.95	57.08	61.47
50.00	50.00				
-26.00	54.29	9.30	168.90	57.89	62.86
50.00	50.00				
-25.00	54.79	9.92	162.40	58.79	64.43
50.00	50.00				
-24.00	55.36	10.58	156.44	59.80	66.20
50.00	50.00				
-23.00	55.99	11.29	151.00	60.92	68.21
50.00	50.00				
-22.00	56.69	12.06	146.06	62.17	70.49
50.00	50.00				
-21.00	57.49	12.90	141.62	63.56	73.08
50.00	50.00				
-20.00	58.38	13.81	137.67	65.13	76.04
50.00	50.00				
-19.00	59.39	14.80	134.20	66.88	79.43
50.00	50.00				
-18.00	60.52	15.88	131.21	68.84	83.33
50.00	50.00				
-17.00	61.80	17.06	128.72	71.05	87.82
50.00	50.00				
-16.00	63.26	18.35	126.72	73.54	93.04
50.00	50.00				
-15.00	64.91	19.79	125.25	76.36	99.12
50.00	50.00				
-14.00	66.80	21.37	124.32	79.55	106.27
50.00	50.00				
-13.00	68.96	23.14	123.98	83.20	114.75
50.00	50.00				
-12.00	71.46	25.12	124.29	87.37	124.88
50.00	50.00				
-11.00	74.37	27.36	125.32	92.19	137.14
50.00	50.00				
-10.00	77.79	29.91	127.18	97.81	152.16
50.00	50.00				

values satisfy the symmetry equations.

The element values of **Table 1** were used to design a 0.2 GHz to 2 GHz bandwidth balun and **Figure 7** shows good agreement between the theoretical and measured frequency responses of the experimental balun.

CONCLUSIONS

A set of graphs has been presented which enables fourth order Marchand baluns with Tchebysheff passband responses to be designed rapidly for a large range of load resistances and bandwidth ratios, without the designer having to resort to the numerically complicated synthesis procedure.

ACKNOWLEDGEMENT

This paper is published with the permission of the CSIR.

REFERENCES

1. Marchand, N., "Transmission Line Conversion," *Electronics*, Vol. 17, 1944, pp. 142-145.
2. Cloete, J. H., "Exact Design of the Marchand Balun," Proceedings Ninth European Microwave Conference, September 1979, pp. 480-484. (Reprinted in *Microwave Journal*, Vol. 23, No. 5, May 1980, pp. 99-102.
3. Morgan, T. E., "Reduced Size Spiral Antenna," *Proceedings Ninth European Microwave Conference*, September, 1979, pp. 181-185.
4. Hallford, B. R., "A Designer's Guide to Planar Mixer Baluns," *Microwaves*, Vol. 18, No. 12, December, 1979, pp. 52-57.
5. Laughlin, G. J., "A New Impedance-Matched Wide-Band Balun and Magic Tee," *IEEE Trans. Microwave Theory Tech.*, Vol. MTT-24, No. 3, March, 1976, pp. 135-141.
6. Horton, M. C., and R. J. Wenzel, "General Theory and Design of Optimum Quarter-Wave TEM Filters," *IEEE Trans. MTT,* Vol. MTT-13, No. 3, March, 1965, pp. 316-327.
7. Riblet, H. J., "General Synthesis of Quarter-Wave Impedance Transformers," *IRE Trans. Microwave Theory Tech.*, Vol. MTT-5, No. 1, pp. 36-43.

J. H. Cloete (M'77) was born in Clocolan, South Africa, on June 29, 1945. He received the B.Sc., B. Eng. degrees in electrical engineering from the University of Stellenbosch, Stellenbosch, South Africa, in 1968. Mr. Cloete received the M. Sc. degree in electrical engineering from the University of California, Berkeley, in 1972. Since 1970, he has been with the South African Council for Scientific and Industrial Research, Pretoria, S.A. working on microwave components and systems. From August 1972 to March 1973, he worked at Scientific-Atlanta, Inc., on the design and manufacture of a multi-octave shaped beam reflector antenna. From March 1977 to February 1978, he was on leave at the University of Stellenbosch working on microwave linear-phase filters.

Synthesis of Marchand Baluns Using Multilayer Microstrip Structures

Charles L. Goldsmith,[1] Anthony Kikel,[2] and Nancy L. Wilkens[1]

[1]Texas Instruments Incorporated, P.O. Box 655474, MS 245, Dallas, Texas 75265

[2]Andrew Corporation, 1850 N. Greenville Avenue, Suite 100, Richardson, Texas 75081

Received October 10, 1991; revised February 24, 1992.

ABSTRACT

New equations and nomographs are derived for the synthesis of Marchand baluns using coupled-line structures. These equations explicitly determine the coupled-line parameters necessary to design Marchand baluns for specific load impedances and bandwidths. Practical MIC and MMIC circuit topologies are presented for realizing the difficult coupling parameters required by these baluns. © 1992 John Wiley & Sons, Inc.

INTRODUCTION

The Marchand balun [1] continues to be one of the most popular forms of broadband microwave balun used today. The Marchand balun is a device which, by virtue of its construction and symmetry, is able to transfer energy between balanced and unbalanced transmission media. In many circuit applications, the balanced output is used as a means to generate two equal and out-of-phase signals with respect to ground. With proper design, the balun will operate over large bandwidths and provide impedance transformation between the input and outputs. This structure has many important applications in the design of broadband microwave mixer circuits and antenna feeds. Numerous articles [2,3,4,5] have been written defining and extending the science of Marchand balun design. Most recently, Cloete [6,7] simplified the design of Marchand baluns by applying modern filter synthesis techniques to the balun equivalent circuit. His technique clearly determines the design values for Marchand baluns built with coaxial transmission lines. Today, Marchand baluns are finding new applications in planar structures for microwave integrated circuits (MICs) and monolithic microwave integrated circuits (MMICs) [8]. These applications require planar balun realizations not possible with coaxial transmission lines. This article presents, for the first time, a procedure for synthesizing Marchand baluns using coupled transmission lines. With coupled transmission lines, many new possibilities exist for realizing Marchand baluns using planar architectures. Unfortunately, analysis shows that typical microstrip or stripline topologies are mostly unsuitable for realizing the coupling parameters necessary to building broadband Marchand baluns. Therefore, a methodology is presented for using available CAD tools to design multilayer, coupled transmission lines appropriate for building Marchand baluns. These procedures pave the way for designing broadband microwave baluns suitable for thin-film integration.

NETWORK SYNTHESIS

The standard coaxial implementation of a Marchand balun is shown in Figure 1a. Alternatively, equivalent performance can be obtained by replacing the coaxial quarter-wave transmission

International Journal of Microwave and Millimeter-Wave Computer-Aided Engineering, Vol. 2, No. 3, 179–188 (1992)

CCC 1050-1827/92/030179-10$04.00

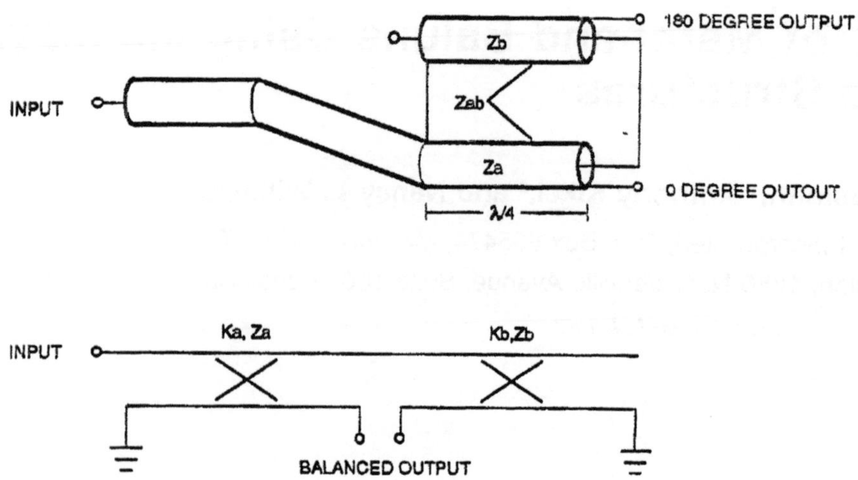

Figure 1. (a) Coaxial and (b) coupled-line Marchand baluns.

lines with quarter-wave coupled transmission lines as shown in Figure 1b. Properly designing the coupling and impedance of each set of parallel-coupled lines yields a balun structure which provides a proper transmission response. Designing a coupled-line balun is accomplished by finding the distributed network equivalent of the balun and equating its element values to those of a bandpass filter which exhibits the desired network response.

In order to obtain the network equivalent of the coupled-line Marchand balun, both sets of coupled transmission lines must be replaced by their four-port equivalent circuits. The equivalent circuit of a coupled transmission line can be derived using capacitance matrix transformations [9]. This derivation results in the equivalent network shown in Figure 2, where Z_1 and Z_2 are the characteristic impedances of distributed unit elements, and N is the turns ratio for the ideal impedance transformers. Coupled transmission lines are often described electrically by their characteristic impedance, z, and voltage coupling constant, k, which is related to the even-mode and odd-mode excitation impedances by [10]:

$$z = \sqrt{Zoe * Zoo} \tag{1}$$

$$k = \frac{Zoe - Zoo}{Zoe + Zoo} \tag{2}$$

The characteristic impedance and coupling coefficient are related to the network equivalent circuit elements by the following equations:

$$N = \frac{1}{k} \tag{3}$$

$$Z_1 = \frac{z}{\sqrt{1 - k^2}} \tag{4}$$

$$Z_2 = z\,\frac{\sqrt{1 - k^2}}{k^2} \tag{5}$$

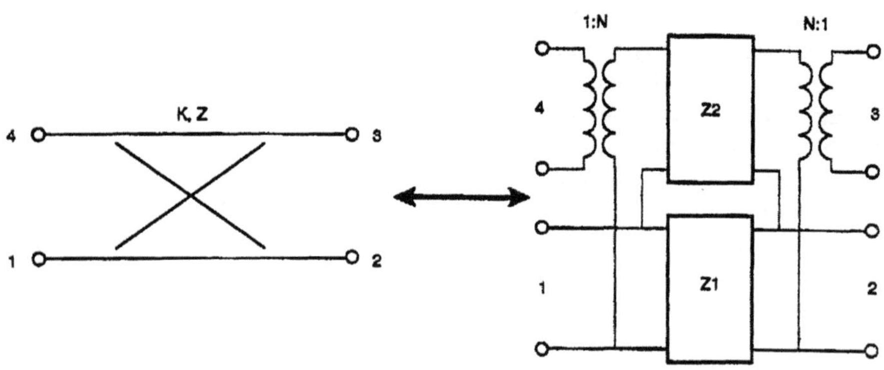

Figure 2. Coupled line four-port equivalent circuit.

The coupled-line balun, like the coaxial balun, inherently has three reactive elements and exhibits a third-order filter response. It has been shown [6] that incorporating an additional transmission line section into the Marchand balun structure can convert the balun to a fourth-order response and significantly improve its broadband performance. Adding an extra transmission line between the coupled lines and the load adds an extra degree of performance while providing a practical means of physically connecting the balun to the load. The schematic of this fourth-order balun is shown in Figure 3a. Incorporating the coupled-line network equivalent into the Marchand circuit schematic yields the overall balun equivalent circuit shown in Figure 3b. A few simplifications reduce the circuit to that shown in Figure 3c. Unfortunately, this network is still relatively complex and difficult to analyze. Eq. (3) shows that forcing the two coupling factors of the Marchand balun to be identical results in equal transformer ratios (i.e., $N_a = N_b = N$). Therefore, the two transformers in the middle of the balun equivalent circuit are equal and cancel each other out. Rearranging and combining elements, and absorbing the remaining transformer into the circuit elements yields the final network equivalent circuit of Figure 3d. This equivalent circuit is considerably simpler and easier to work with than that of Figure 3c. This circuit is now simple enough to be equated with standard filter topologies to obtain specific filter responses. A disadvantage to this technique is that one degree of freedom in the network had to be sacrificed in order to achieve this simplicity.

Equating the impedances of the final equivalent element values Z_1', Z_2', Z_3', Z_4', and R_5' of Figure 3d to the coupled transmission line parameters z_a, z_b, and k yields:

$$Z_1' = \frac{Z_{2a}}{N^2} = z_a\sqrt{1 - k^2} \quad (6)$$

$$Z_2' = \frac{Z_{2b}}{N^2} = z_b\sqrt{1 - k^2} \quad (7)$$

$$Z_3' = \frac{Z_{1a} + Z_{1b}}{N^2} = (z_a + z_b)\frac{k^2}{\sqrt{1 - k^2}} \quad (8)$$

$$Z_4' = \frac{Z}{N^2} = Zk^2 \quad (9)$$

$$R_5' = \frac{R}{N^2} = Rk^2 \quad (10)$$

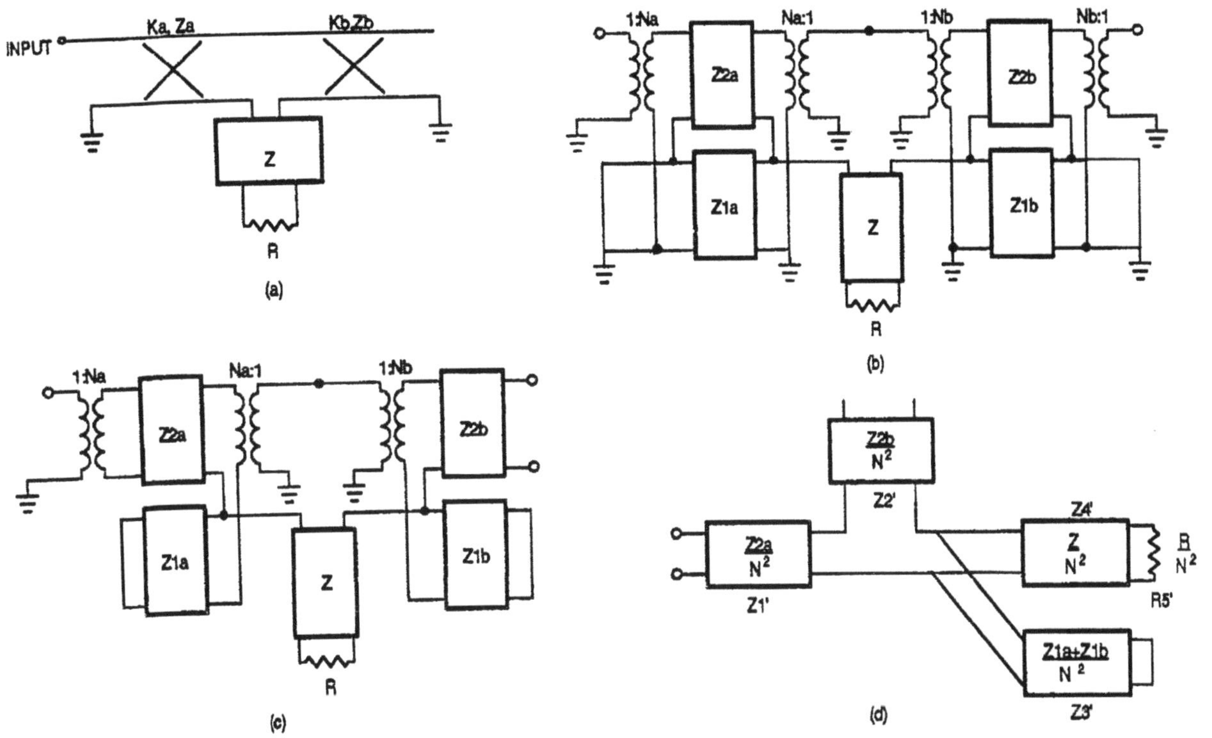

Figure 3. Coupled-line Marchand balun equivalent circuit.

These equations show the specific relationship between the filter element values and the coupled-line characteristics. It should be noted that all the circuit elements, including the transmission line, Z, and load resistance, R, are scaled by N^2 when absorbing the input transformer into the circuit elements. By rearranging the above equations, the coupled-line parameters can be solved in terms of the equivalent element values. In this case:

$$k = \sqrt{\frac{Z_3'}{Z_1' + Z_2' + Z_3'}} \tag{11}$$

$$z_a = \frac{Z_1'}{\sqrt{1 - k^2}} \tag{12}$$

$$z_b = \frac{Z_2'}{\sqrt{1 - k^2}} \tag{13}$$

Choosing element values for Z_1', Z_2', Z_3', Z_4', and R_5' which together exhibit a specific bandpass network response will then permit the design of a coupled transmission line balun which will have an identical network response.

The solution for fourth-order Chebyshev bandpass filters, with the topology shown in Figure 3d, can be accomplished using the direct synthesis techniques of Horton and Wenzel [11], by using design nomographs furnished by Cloete [7], or by using filter design software such as EEsof's E-Syn. Graphs of element values for fourth-order Chebyshev filters having 2:1, 3:1, and 5:1 bandwidths are given in Figure 4. Applying this data to the above equations makes it possible to generate the design nomographs presented in Figure 5. The coupled-line parameters given in this figure use the more traditional even- and odd-mode parameters, which are related to characteristic impedance and coupling factors by eqs. (1) and (2). Given a bandwidth requirement, and a desired amount of in-band return loss, the coupled transmission line parameters Zoe_a, Zoo_a, Zoe_b, Zoo_b, as well as line impedance Z and load resistance R can be extracted. Similarly, picking the bandwidth and load impedance determines the element values and in-band return loss.

As a design example, a distributed filter with a 50-Ω source impedance and element values of $Z_1' = 53.4\ \Omega$, $Z_2' = 23.7\ \Omega$, $Z_3' = 134.9\ \Omega$, $Z_4' =$

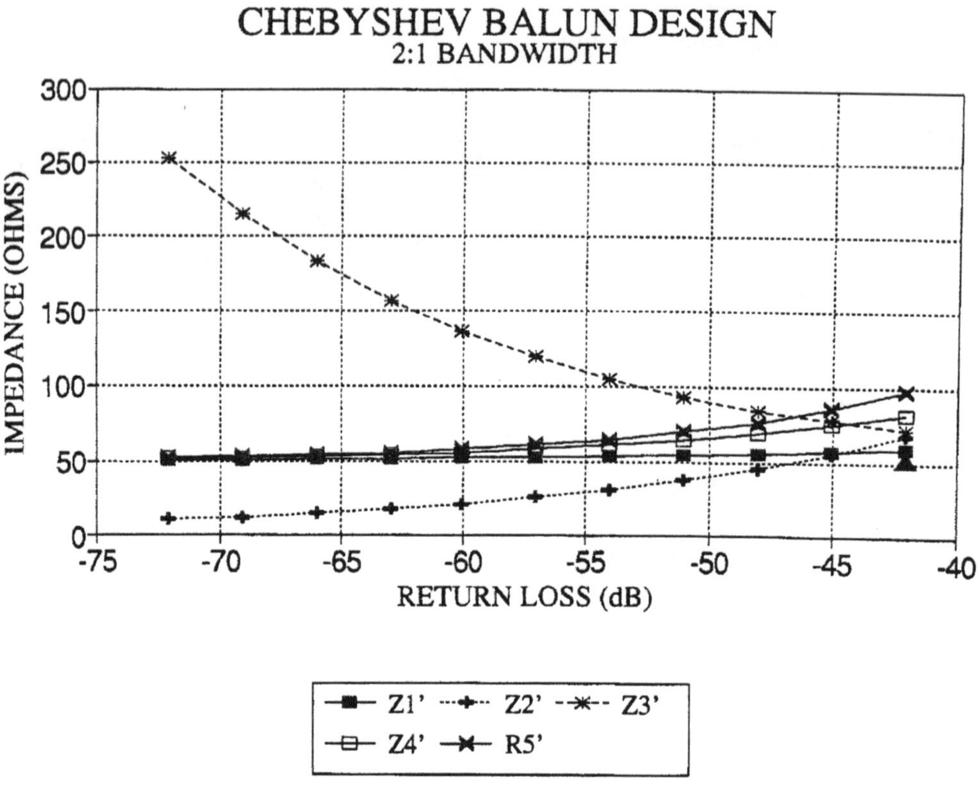

(a)

Figure 4. Balun equivalent circuit parameters as a function of load impedance (Rs = 50 Ω) for (a) 2:1 bandwidth, (b) 3:1 bandwidth, and (c) 5:1 bandwidth.

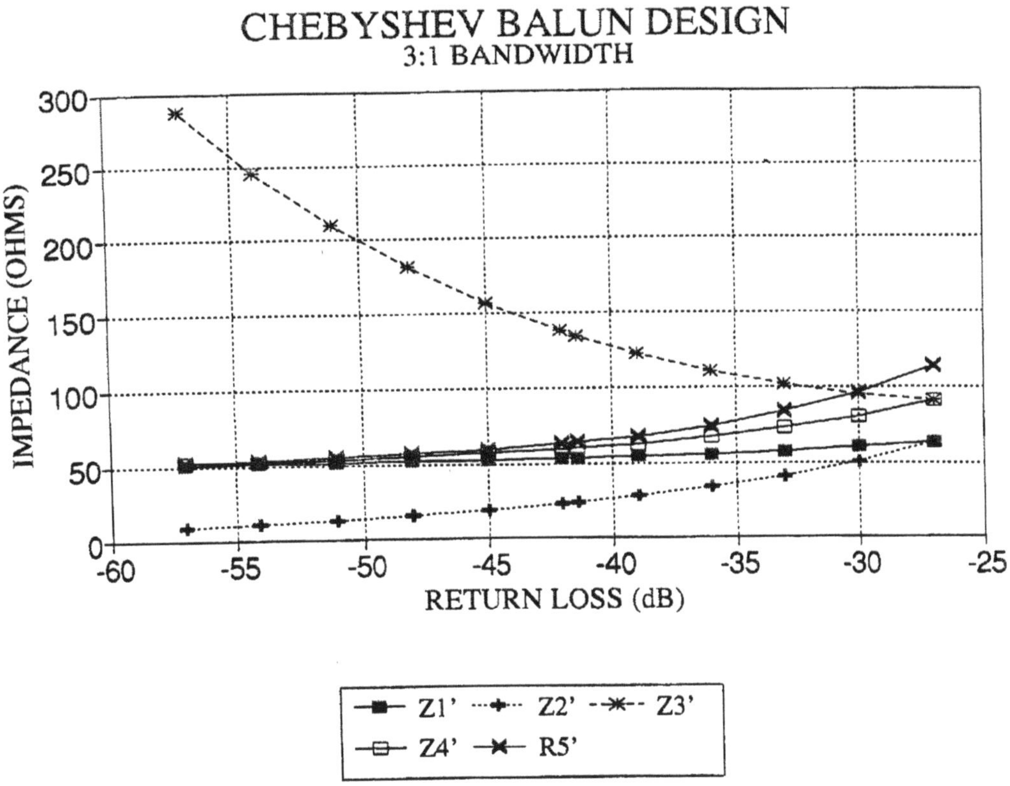
CHEBYSHEV BALUN DESIGN
3:1 BANDWIDTH
IMPEDANCE (OHMS)
RETURN LOSS (dB)
300
250
200
150
100
50
0
-60
-55
-50
-45
-40
-35
-30
-25
Z1'
Z2'
Z3'
Z4'
R5'

(b)

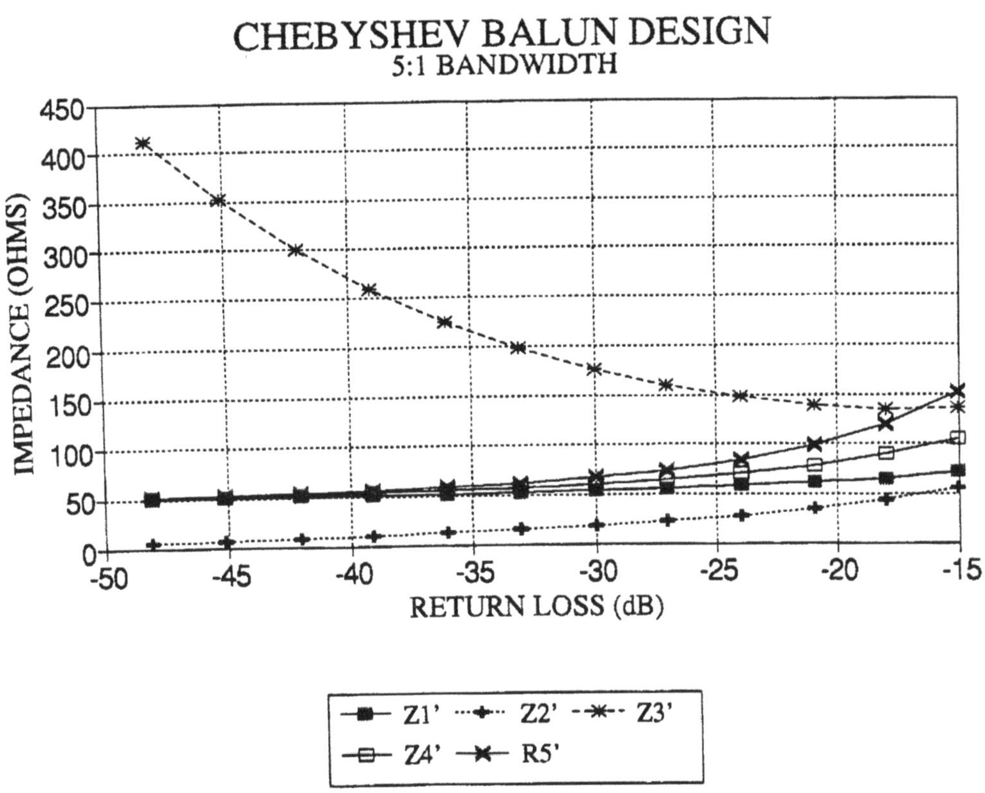
CHEBYSHEV BALUN DESIGN
5:1 BANDWIDTH
IMPEDANCE (OHMS)
RETURN LOSS (dB)
450
400
350
300
250
200
150
100
50
0
-50
-45
-40
-35
-30
-25
-20
-15
Z1'
Z2'
Z3'
Z4'
R5'

(c)

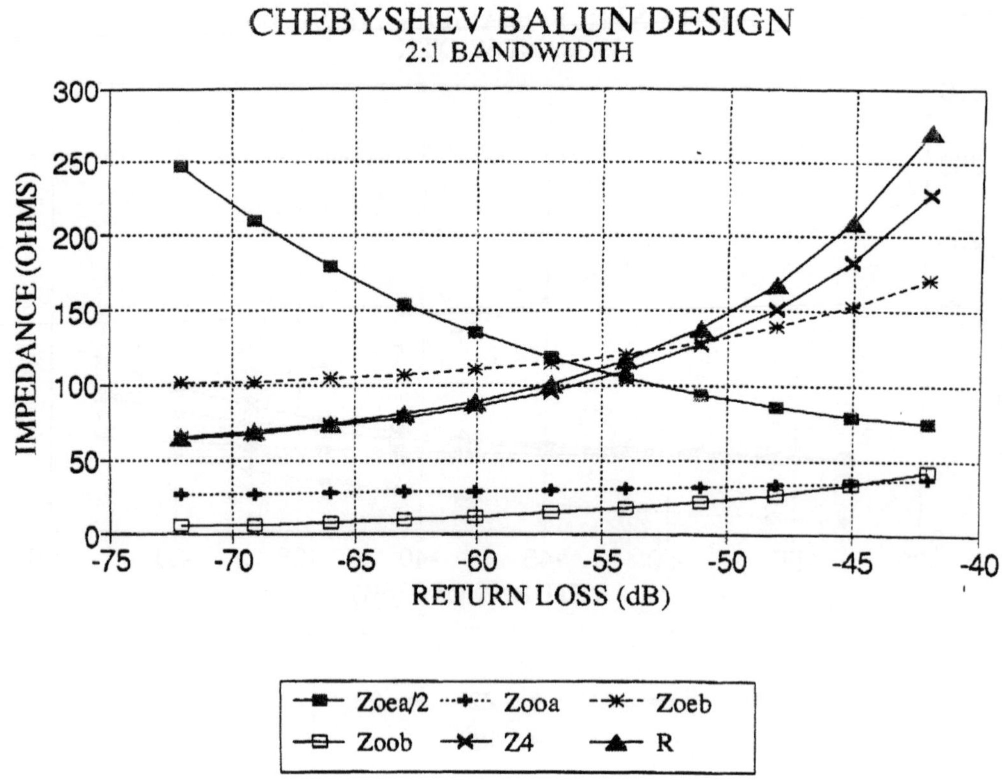

(a)

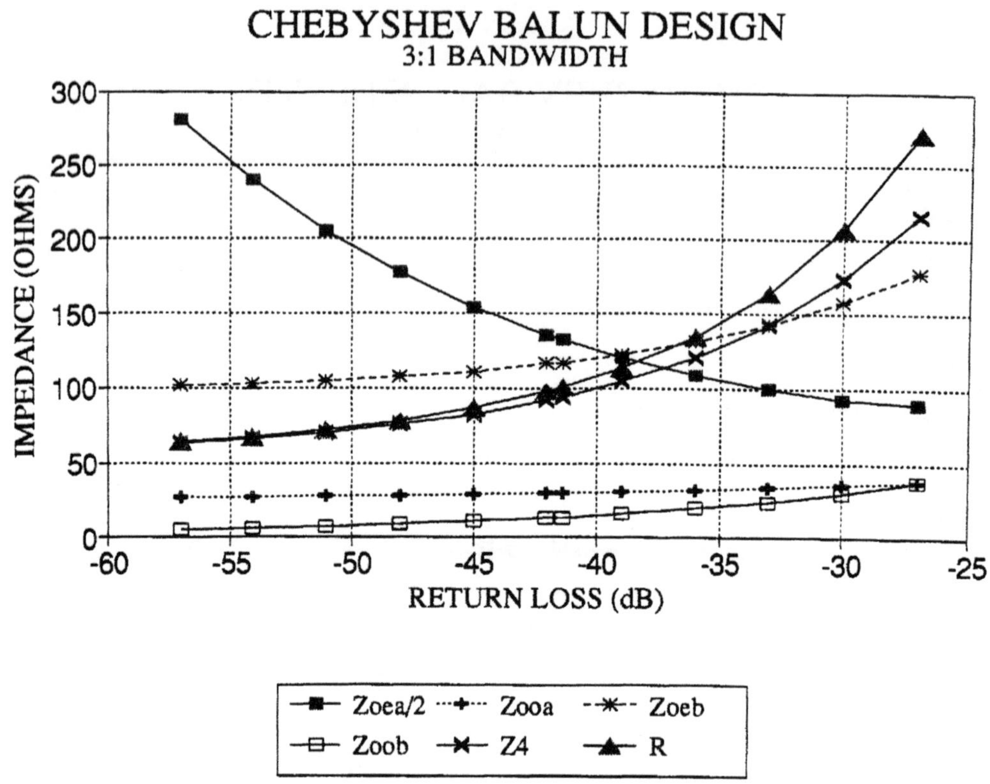

(b)

Figure 5. Balun coupled-line parameters as a function of load impedance (R_s = 50 Ω) for (a) 2:1 bandwidth, (b) 3:1 bandwidth, and (c) 5:1 bandwidth.

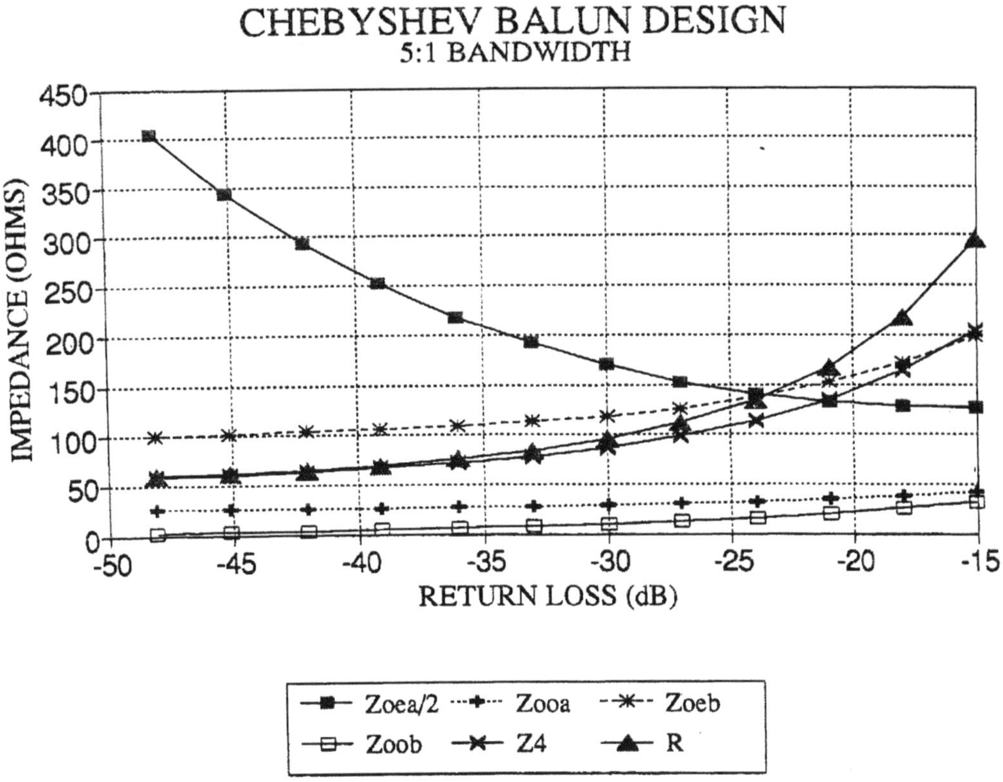

(c)

59.8 Ω, and $R_5' = 63.9$ Ω covers a 3:1 bandwidth with a 41.4-dB in-band return loss. Applying eqs. (11), (12), and (13) yields coupled line parameters $k = .7977$ (−1.96 dB), $z_a = 88.5$ Ω, and $z_b = 39.3$ Ω. Using eqs. (9) and (10) allows calculation of the impedance of the extra (fourth-order) transmission element as $Z = 94.0$ Ω and the balun load impedance of $R = 100.4$ Ω. Converting the coupled-line impedances and coupling factor to even- and odd-mode parameters gives $Zoe_a = 263.9$ Ω, $Zoo_a = 29.7$ Ω, $Zoe_b = 117.1$ Ω, and $Zoo_b = 13.2$ Ω and completes the design. Given a desired bandwidth and load impedance, the nomographs of Figure 5 allow direct evaluation of the parameter values for Zoe_a, Zoo_a, Zoe_b, Zoo_b, and Z. These nomographs allow a speedy evaluation of the coupled transmission line requirements necessary to build a Marchand balun of given specifications. A linear circuit simulation of the above balun design values over a 3:1 bandwidth of 6–18 GHz is given in Figure 6. This simulation verifies the results of the above derivation.

TOPOLOGY SYNTHESIS

Inspection of the coupled-line Marchand balun nomographs reveals that the main difficulties with designing coupled-line baluns are: (1) achieving the high even-mode impedances; and (2) realizing the tight couplings. In general, realization of coupled lines with both (1) and (2) is difficult, if not impossible, with conventional microstrip due to limitations in thin-film integrated circuit processing. For example, manufacture of microstrip on 15-mil thick alumina (Al_2O_3) typically requires linewidths and line spacings greater than 0.5-mil

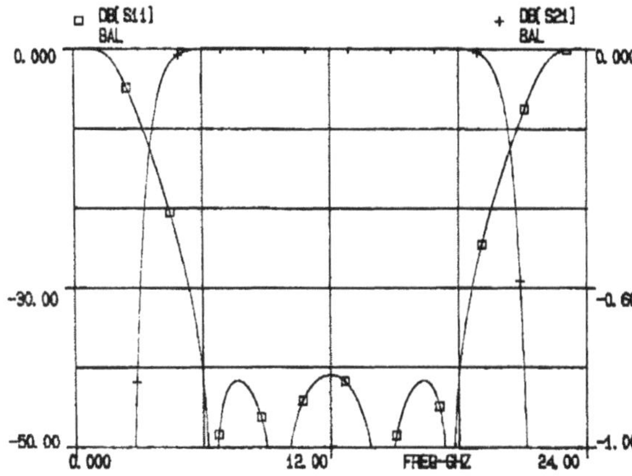

Figure 6. Linear circuit simulation of 3:1 bandwidth Marchand balun with ideal element values: $Z_{oea} = 263.9$ Ω, $Z_{ooa} = 29.7$ Ω, $Z_{oeb} = 117.1$ Ω, $Z_{oob} = 13.2$ Ω, $Z_4' = 94$ Ω, and $R' = 100.4$ Ω.

wide for reproducibility. A set of coupled lines, 0.5-mil wide and 0.5-mil apart, result in impedance values of Z_{oe} = 158 Ω and Z_{oo} = 42 Ω (k = −4.7 dB) at 10 GHz. Both the even-mode impedance and the coupling are far from those values requisite for realizing the 3:1 balun designed above. When working with gallium arsenide (GaAs) dielectrics, somewhat tighter processing tolerances make possible 0.4 mil linewidths and 0.15-mil spaces. Microstrip coupled lines with these physical dimensions on 6 mil thick GaAs yields Z_{oe} = 138 Ω and Z_{oo} = 45 Ω (k = −4.5 dB). Once again, these values are insufficient to meet the requirements needed in building a broadband Marchand balun. However, by using multilayer microstrip topologies, it is possible to achieve practical realizations for the coupled-line Marchand balun on substrate materials such as alumina, quartz, or gallium arsenide [8, 12, 14].

Using multilayer dielectric structures allows much greater freedom in designing the sizes and spacings of coupled lines to satisfy process constraints. Employing a low-dielectric film underneath transmission lines permits conductors to achieve a much higher even-mode impedance than is possible on a single, high-dielectric substrate. A dielectric film between the coupled lines permits greater coupling than is normally possible with an air/dielectric interface in microstrip. These techniques can be implemented by using multilayer microstrip structures such as the re-entrant mode coupler [13,14] shown in Figure 7. The re-entrant mode coupler employs a dielectric material and a third conductor beneath the coupled lines to realize both tight coupling and a higher even-mode impedance. These techniques make possible numerous multidielectric, multiconductor configurations for raising even-mode impedances and increasing coupling between conductors. The ultimate limitations are determined by the minimum plating geometries and maximum dielectric deposition thicknesses used to fabricate the baluns.

The main difficulty in designing with multilayer microstrip topologies is two-fold. First, microstrip is inherently inhomogeneous and therefore propagates in a quasi-TEM fashion. Second, there exists no simple analytic means for determining coupled-line parameters from the physical dimensions of a multilayer topology.

Propagation in an inhomogeneous medium means that the even and odd modes have different phase velocities. A circuit model has been developed [15] to account for differing phase velocities in inhomogeneous coupled lines. This model incorporates unit elements with differing line lengths to account for differing propagation velocities. While this model is more accurate, it is a bit more complicated than the homogeneous model used above. The increased complexity of the inhomogeneous model makes it exceedingly difficult to use when synthesizing filter networks to achieve a desired response. Therefore, this study uses the homogeneous equivalent model to obtain ideal element values which will deliver an appropriate network response.

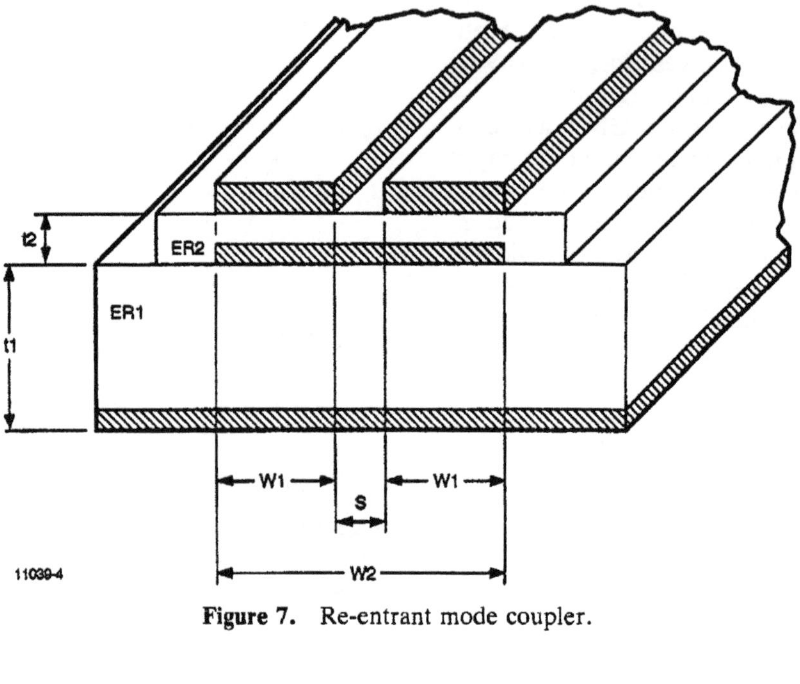

Figure 7. Re-entrant mode coupler.

The second problem associated with designing multilayer microstrip circuits has been lessened with the advent of static electromagnetic solvers. Solvers such as Ansoft's Maxwell allows a direct means of determining approximate physical dimensions necessary to achieve a specific set of coupled-line parameters. Static solvers evaluate Maxwell's equations throughout a topology cross-section and output capacitance matrices necessary for calculating even- and odd-mode impedances. For a given microstrip multilayer structure, variations in physical dimensions and dielectric constants can be simulated until the proper coupled-line parameters have been achieved. The results of these simulations can also be reinserted into eqs. (6)–(10) to evaluate the resulting response when the simulation results are not precisely on target.

As a design example, Maxwell simulations were run for the re-entrant mode coupler configuration of Figure 7 using a 15-mil alumina substrate (ϵ_{r1} = 9.8) under a 0.2 mil thick polyimide dielectric layer (ϵ_{r2} = 3.2). The result of these simulations for the first coupled lines yielded approximate values of $W_1 \sim 0.7$ mil, $S \sim 0.5$ mil, and $W_2 \sim 3.0$ mil. These physical dimensions produced Z_{oeu} = 200 Ω and Z_{ooa} = 28 Ω compared to the design values of Z_{oea} = 264 Ω and Z_{ooa} = 30 Ω. Similarly, for the second coupler, electromagnetic solver simulations determined dimensions of $W_1 \sim 2.5$ mil, $S \sim 1.0$ mil, and $W_2 \sim 14.0$ mil. These dimensions yielded Z_{oeb} = 111 Ω and Z_{oob} = 11 Ω compared to Z_{oeb} = 117 Ω and Z_{oob} = 13 Ω. Inserting the new coupled-line parameters into the balun equivalent circuit, and analyzing the response, showed better than 20-dB return loss and less than 0.1-dB insertion loss variation over the 3:1 bandwidth. Experience from running these simulations shows that it is necessary to come within a few ohms of the target odd-mode impedance, while there is considerably more room for deviation of the even-mode impedance. Simulations were also run for a re-entrant coupler on a 25 mil GaAs substrate (ϵ_{r1} = 12.9) with a 0.2 mil thick polyimide dielectric. For Z_{oea} = 264 Ω and Z_{ooa} = 30 Ω, the approximate dimensions are $W_1 \sim 0.7$ mil, $W_2 \sim 2.0$ mil, and $S \sim 0.5$ mil. Similarly, for Z_{oeb} = 117 Ω and Z_{oob} = 13 Ω, appropriate widths and spaces are $W_1 \sim 2.0$ mil, $W_2 \sim 17.0$ mil, and $S \sim 1.0$ mil. Both of these examples demonstrate practical multilayer balun structures which would not have been realizable on single-layer microstrip substrates.

As a word of caution, static electromagnetic solvers treat these multilayer problems as TEM-static solutions, and do not account for any dispersion due to step dielectric discontinuities. This approximation appears to hold in practice but, at times, these results may need to be taken with additional engineering judgment. A full-scale electromagnetic simulation does provides a more accurate analysis, but usually requires a considerable amount of computation time and memory to run. Fabrication and measurement still remain the most accurate and effective means of verification.

CONCLUSION

We have derived the network equivalent circuit of a coupled-line balun using a homogeneous equivalent circuit model and applying restrictions to the coupling values. From this, we determined the exact parameters necessary for the design of a planar Marchand balun. While it is relatively easy to design baluns with coupled-line structures to yield good amplitude and phase balance, maintaining a good impedance match at the balun ports over broad bandwidths requires high even-mode impedances and tight coupling. Unfortunately, achieving these coupling characteristics is quite difficult with normal microstrip geometries. Multilayer topologies, such as the re-entrant mode coupler, can overcomes these obstacles. The use of electromagnetic solvers provide the means for innovating new multidielectric, multiconductor microstrip configurations while taking into account practical process limitations. The techniques mentioned make possible the design and fabrication of high-performance coupled-line Marchand baluns for MIC and MMIC applications.

REFERENCES

1. N. Marchand, "Transmission line conversion transformers," *Electronics,* Vol. 17, Dec. 1944, pp. 142–145.
2. W. K. Roberts, "A new wide-band balun," *Proc. IRE,* Vol. 45, Dec. 1957, pp. 1628–1631.
3. J. W. McLaughlin, D. A. Dunn, and R. W. Grow, "A wide-band balun," *IRE Trans. Microwave Theory Tech.,* Vol. MTT-6, July 1958, pp. 314–316.
4. G. Oltman, "The compensated balun," *IEEE Trans. Microwave Theory Tech.,* Vol. MTT-14, March 1966, pp. 112–119.

5. G. Laughlin, "A new impedance-matched wideband balun and magic tee," *IEEE Trans. Microwave Theory Tech.*, Vol. MTT-24, March 1976, pp. 135–141.
6. J. Cloete, "Exact design of the marchand balun," *Proc. 9th Eur. Microwave Conf.* Sept. 1979, pp. 480–484.
7. J. Cloete, "Graphs of circuit elements for the Marchand balun," *Microwave J.*, May 1981, pp. 125–128.
8. A. Pavio and A. Kikel, "A monolithic or hybrid broadband compensated balun," *IEEE MTT Symp. Dig.*, June 1990, pp. 483–486.
9. J. A. G. Malherbe, *Microwave Transmission Line Filters,* Artech House, Dedham, MA, 1979.
10. D. M. Pozar, *Microwave Engineering*. Addison-Wesley, Reading, MA, 1990.
11. M. Horton and R. Wenzel, "General theory and design of optimum quarter-wave TEM filters, *IEEE Trans. Microwave Theory Tech.*, Vol. MTT-13, May 1965, pp. 316–327.
12. T. H. Chen, et al., "Broadband monolithic passive baluns and monolithic double-balanced mixer," *IEEE MTT Symp. Dig.*, June 1991, pp. 861–864.
13. S. Cohn, "The re-entrant cross section and wideband 3-dB hybrid couplers," *IEEE Trans. Microwave Theory Tech.*, Vol. MTT-11, July 1963, pp. 254–258.
14. A. Pavio and S. Sutton, "A microstrip re-entrant mode quadrature coupler for hybrid and monolithic circuit applications," *IEEE MTT Symp. Dig.*, June 1990, pp. 573–576.
15. G. I. Zysman and A. K. Johnson, "Coupled transmission line networks in an inhomogeneous dielectric medium," *IEEE Trans. Microwave Theory Tech.*, Vol. MTT-17, Oct. 1969, pp. 753–759.

BIOGRAPHY

Chuck Goldsmith was born in Chicago, Illinois. He received BSEE and MSEE degrees from the University of Arizona, Tucson, Arizona in 1980 and 1982. He is currently pursuing his PhD degree at the University of Texas at Arlington.

From 1982 to 1986 Mr. Goldsmith was a project engineer in the Amplifier Business Center at M/A COM Active Assemblies Division in Tempe, Arizona. At M/A COM, he was involved in the design of broadband amplifiers and receivers for EW applications. Since 1986, Mr. Goldsmith has been employed as a microwave design engineer in the GaAs Electronics Department of Texas Instruments in Dallas, Texas. There he has been involved in the design of both hybrid and monolithic receivers and components in the microwave and millimeter-wave frequency ranges. His main research interests include microwave receiver and receiver component design, as well as integrated photonic/microwave circuits for microwave applications.

Anthony Kikel received a BSEE degree from Case Western Reserve University in 1987 and a MSEE degree from Southern Methodist University in 1990.

From 1987 to 1991 he was with the Microwave Laboratory at Texas Instruments where he was responsible for the design of various monolithic and hybrid microwave circuits. These designs included monolithic low-noise amplifiers for RADAR applications, broadband monolithic downconverters for transponder decoys, and W-band monolithic downconverters for millimeter-wave receivers. In 1991, he joined the Distributed Data Collection Systems Group at Andrew Corporation as a System Engineer assigned to the development of a full duplex microwave data communications system using spread spectrum and pulse code modulation technology. Mr. Kikel is coauthor of several papers in monolithic circuit design.

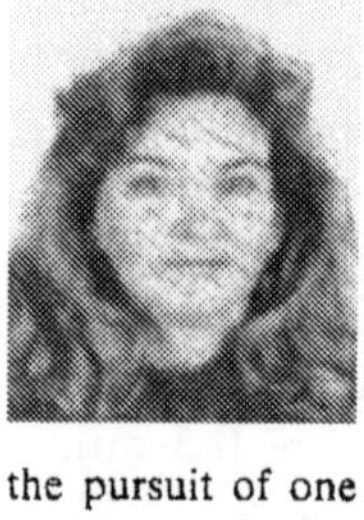

Nancy Wilkens was born in Alexandria, Virginia. She received a BS degree in Mathematics from the University of North Carolina, Chapel Hill in 1983 and an MSEE from North Carolina State University in Raleigh in 1986.

She works in the Defense Systems and Electronics Group of Texas Instruments in Dallas, Texas where she is "involved in the pursuit of one microwave simulator CAD program that can accurately simulate an entire MMIC." Her other areas of interest include the analysis, design, and testing of circuits and systems for airborne applications.

NEW METHOD OF IMPEDANCE MATCHING IN RADIO-FREQUENCY CIRCUITS.

Decimal Index 621.396.611.39

A new transformer method is described which is suitable both for matching circuits of unequal impedance and coupling symmetrical and unsymmetrical radio-frequency circuits. In contradistinction to conventional methods of impedance matching the frequency of the oscillations being transmitted can be varied over a wide range without the necessity of re-tuning.

THE impedances of the individual circuits of radio-frequency equipment are frequently unequal. In order to obviate the reflections and losses involved by mismatching, special matching devices have to be inserted between such dissimilar circuits for the transmission of energy. For instance, matching is necessary between the tubes of a transmitter output stage with high load resistance and the low-impedance antenna transmission line or feeder system. In the case of low frequencies transformers with a corresponding turns ratio can be employed. By reason of the unavoidable leakage inductance of the coupled transformer coils, high frequencies generally involve tuning by means of additional condensers, and should the working frequency be varied, corresponding re-tuning is therefore entailed.

For impedance matching purposes a quarter-wave Lecher wire system having a surge impedance which is the geometric mean between the two impedances to be matched can likewise be employed. Such matching sections must naturally also be re-tuned in the event of the frequency being altered, to correspond to the changed wave-length. Small frequency deviations are, however, permissible when the impedance transformation takes place in several steps adjusted to the mean frequency. — Another method of matching, the line with exponential taper, permits large frequency variations without re-tuning, but has amongst other things the drawback of taking up a large amount of space.

Special couplers are also necessary for transition from symmetrical to unsymmetrical circuits, e. g. between the symmetrical output of a push-pull transmitter stage and a coaxial antenna cable with earthed sheathing. Here, too, variation of the frequency generally involves re-tuning.

A new coupler which obviates re-tuning is shown in Fig. 1a. It comprises two superposed windings W_1 and W_2 separated by an insulating tube R. Given symmetrical currents i_1 (full-lined arrows) the magnetic fields produced by two closely-spaced superposed sections of conductor practically neutralize each other, i.e. the mutual inductance of two successive turns of a coil can be neglected, while it is possible to replace the two windings by two straight conductors having the same cross-section, length, and spacing as the two developed windings. This Lecher wire system is represented in the equivalent diagram (Fig. 1b) by the equivalent line A.

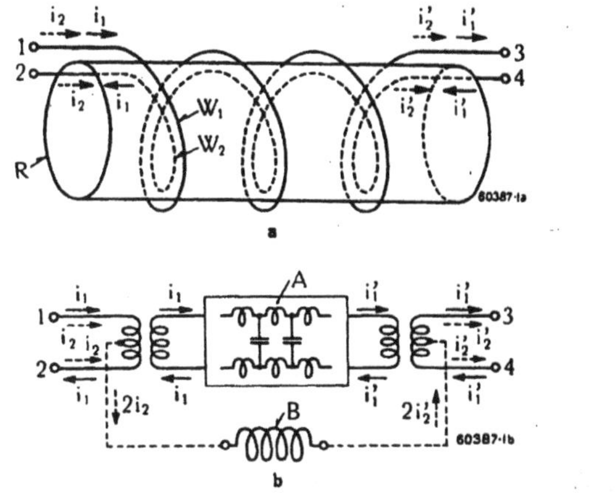

Figs. 1a and 1b. — Double-wire coil system with equivalent diagram.

(a) The coil system comprises two superposed windings W_1 and W_2 separated by an insulating tube R.

(b) According to this equivalent diagram, where symmetrical currents i_1 are concerned, the coil has the effect of a Lecher wire system A, but with unsymmetrical currents i_2 the nature of a choke coil B. The symmetrical and unsymmetrical currents are segregated by ideal centre-tapped transformers.

On the other hand, with unsymmetrical currents i_2 (dotted arrows), the field vectors produced by two superposed sections of the conductors are added together, with the result that the mutual inductance between the individual turns of the coil becomes an important factor. The double-wire coil system behaves here like a conventional choke coil, represented in the equivalent diagram by B. In this diagram the symmetrical and unsymmetrical currents i_1 and i_2, respectively, are segregated by centre-tapped ideal transformers T. Given an adequate number of turns on the windings W_1 and W_2 the impedance of the equivalent choke coil B becomes so high that, even assuming unequal potentials between the centre tappings of the input and output coils, the unsymmetrical current i_2 can be neglected. In this case the described coil system forms an ideal transformer combined with an ideal line.

In view of the effect of this ideal transformer such a system S can now be employed, as shown for example in Fig. 2a, to couple a physically symmetrical circuit (connected to terminals 1 and 2) to a load resistance R_a having one pole earthed. By making the coil of suitable dimensions the surge impedance Z_o of the matching line (A in the equivalent diagram Fig. 1b) represented by the coil system can be adapted to the pure load resistance R_a. In this case the input impedance R_e occurring between terminals 1 and 2 is equal to the surge impedance Z_0 and in consequence also to the load resistance R_a, immaterial of the actual working frequency.

The curves in Fig. 3 give the input impedance computed from the coil dimensions for conditions of short circuit and no-load. The measured impedance values are also given and agree with the curves to a high degree. These measurements, which demand great care, were made by a method specially developed for the purpose (cf. Fig. 1, page 293). The characteristic surge impedance can be determined from

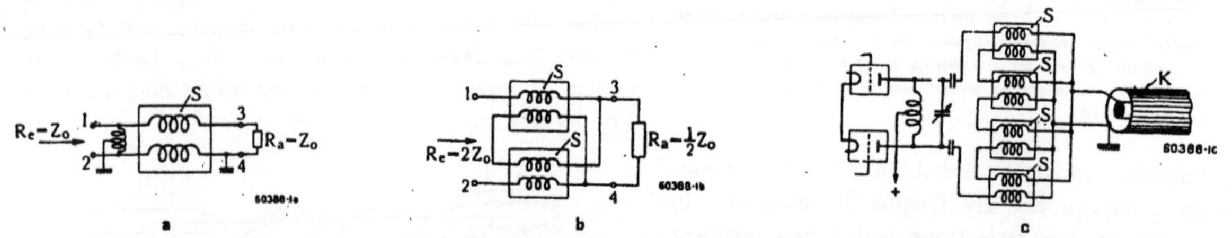

Fig. 2a, b, and c. — Employment of double-wire coil systems for coupling and impedance matching purposes.

(a) Due to the suppression of the unsymmetrical currents by the series inductance of the coils such units can be used for coupling physically symmetrical circuits (connected to terminals 1 and 2) to circuits having one pole earthed (connected to terminals 3 and 4).

(b) By series-parallel connection of two coil systems S the load resistance $R_a = {}^1/_2\, Z_o$ is transformed to the input impedance $R_e = 2\, Z_o$ (Z_o = surge impedance of a coil system).

(c) The antenna cable K and output stage are "matched" by the four coil systems S.

Surge impedance of coil systems = 240 Ω.
Surge impedance of cable = 240 Ω : 4 = 60 Ω.
Load impedance of output stage = 240 Ω×4 = 960 Ω.

By series-parallel connection of two or more coil systems impedance matching is now also possible in a simple manner, independent of the frequency. Fig. 2b shows by way of example the input terminals of two systems of coils S connected in series and the output terminals in parallel. No objections can be raised to this practice provided the series inductance (B in the equivalent diagram in Fig. 1b) is large enough. The load resistance $R_a = {}^1/_2\, Z_o$ is thus transformed to the input impedance $2\, Z_o$. Analogously, with n coil systems impedance transformation in the ratio $1 : n^2$ can be achieved.

In Fig. 2c, for instance, four coil systems are shown connected between a transmitter output stage and the high-frequency antenna cable K, the resulting impedance transformation being in the ratio $4^2 : 1 = 16 : 1$. With a coil system having a surge impedance $Z_0 = 240\ \Omega$, for example, a transmitter output stage with a load impedance of $4 \times Z_0 = 960\ \Omega$ can be coupled to an antenna cable of $Z_0 : 4 = 60\ \Omega$. The coupled coil systems have the same effect as a transformer with separate windings, i. e. the symmetry of the anode circuit at the input end is not affected by single-pole earthing of the cable connected to the other end. Furthermore, the coupled coil systems behave like a Lecher wire system, i. e. the input impedance must follow a tangential function of the frequency when the terminals at the other end are open or short-circuited.

the geometric mean of the measured or computed short-circuit and no-load input impedances. In the present case it is about 240 Ω. Fig. 4 gives the curve of the input impedance for a load impedance of about 53 Ω. From the test points it is clear that the desired impedance transformation in the ratio of 1:16 is actually possible over a very wide frequency range. The deviation of the plotted mean-value curve R_e from the theoretical curve 1 is due to the load

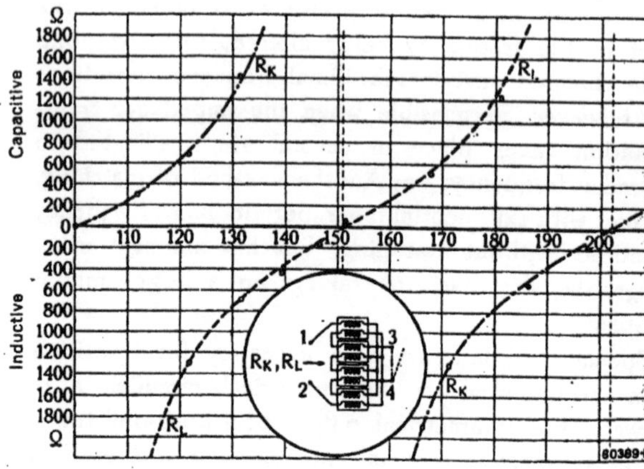

Fig. 3. — Input impedance of matching unit when output terminals short-circuited or open.

The matching unit comprises four double wire coils in series-parallel connection. The computed and measured primary impedances are plotted as a function of the frequency with the secondary terminals open and short-circuited.

—·—·— / — — — } Computed curves. o. Test points.

R_K. Input impedance with secondary terminals short-circuited.
R_L. Input impedance with secondary terminals open.

impedance being slightly lower than the theoretical value, as well as to the inherent capacitance of the circuit.

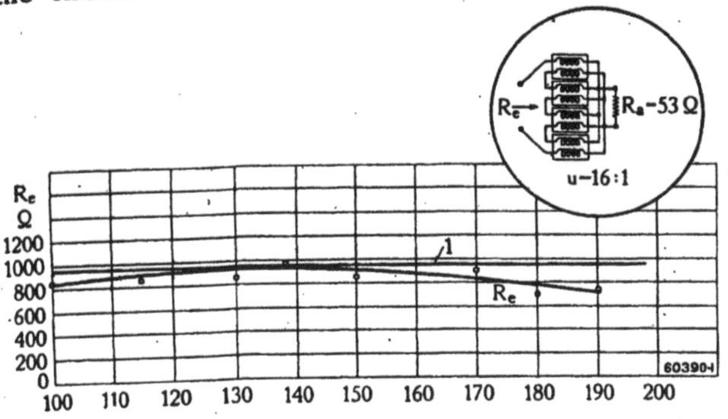

Fig. 4. — Theoretical and measured input impedance of a matching unit with a pure resistive load.

The matching unit comprises four double-wire coils in series-parallel connection. A surge impedance of 240 Ω was computed from the coil data and the measurements in Fig. 3, whence, assuming a pure resistive load of 60 Ω, the theoretical value of the input impedance is 960 Ω. The measured values of the input impedance are somewhat lower owing to the load impedance having been somewhat lower than theoretically assumed.

Theoretical value of $R_a = Z_0$. o. Test points for R_a = 53 Ω.
Impedance transformation ratio 16:1.

The described method of matching is particularly suitable for application in the ultra-short-wave field, where it represents a big simplification compared to conventional tuned matching devices. Fig. 5 shows the external appearance of an impedance transformer with four coils, employed as antenna coupler in a

Fig. 5. — Matching unit with double-wire coils.

The system contains four double-wire coils for impedance transformation from 60 Ω to about 1000 Ω in the case of metre waves. With a power of over 100 W the losses are negligible.

medium-power transmitter. It requires little space and its losses are very low. This new component greatly simplifies the construction and operation of the equipment marketed by the Company.

(MS 564) *G. Guanella. (E. G. W.)*

A Derivation of a Class of 3-port Baluns from Symmetrical 4-port Networks

Yoke Choy Leong, Kian Sen Ang, and Chee How Lee

DSO National Laboratories, 20 Science Park Drive, Singapore (118230)

Abstract **— This paper describes an approach in deriving the general conditions that need to be satisfied for a class of 3-port baluns. This is done by analyzing the behavior of a general symmetrical 4-port network when one of the ports is terminated with an arbitrary impedance. These conditions will be useful in devising new balun structures and in deriving exact design equations for such baluns. Examples will be presented for some known and new baluns in this class. These insights have also made the cascading of multi-section baluns possible by specifying the overall requirements for the cascaded structures. Based on these results, a 3-section, Marchand-type, coupled-line balun has been designed and fabricated. Good agreements between simulation and measured results have been obtained, thereby verifying the validity of the design equations.**

I. INTRODUCTION

Three-port baluns are important circuit components to realize critical circuit functions such as double-balanced mixers, frequency multipliers, push-pull amplifiers, 180^0 hybrid [1], etc. Many of these 3-port baluns belong to the same class, where they are composed of symmetrical 4-port networks with one of the ports terminated with an arbitrary impedance. Some examples include the coupled-line Marchand balun [2] and the N-section halfwave balun [3]. Although there are previous works [4]-[5] analyzing specific configurations of this class of baluns, a generalized analysis is not available. A technique for analyzing 3-port baluns as generalized symmetrical 4-ports is presented in this paper. This technique gives valuable insight to the synthesis of the baluns by specifying the requirements of the odd and even mode parameters of the 4-port symmetrical networks. This results in the development of new balun structures and exact, closed-form design equations at the center frequency of operation for known baluns in this class. Some examples of these baluns will be illustrated.

II. THEORY

For the class of baluns that we are interested in, the 3-port baluns are formed by terminating one of the ports of symmetrical 4-port networks with an arbitrary impedance, Γ, as shown in Figure 1.

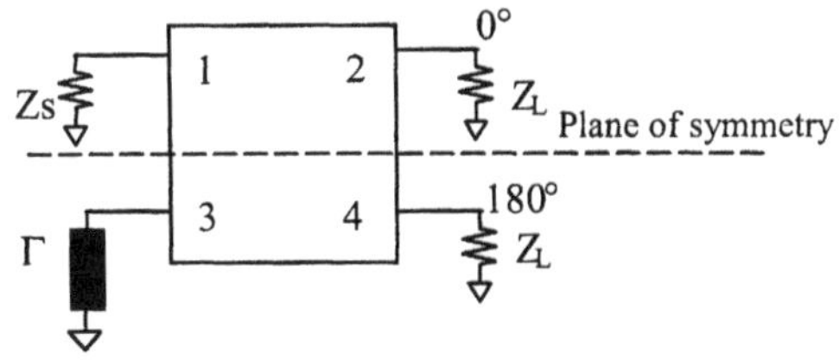

Fig. 1. Configuration of the 3-port balun that is analyzed in this work. It is constructed from a 4-port network that is symmetrical about the dashed line.

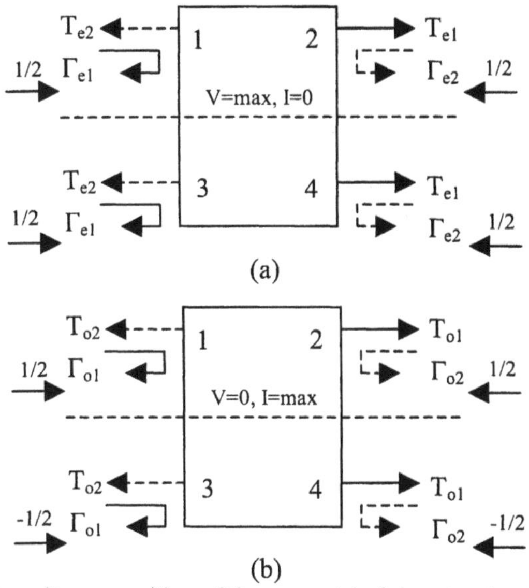

Fig. 2. Decomposition of the symmetrical 4-port network into even- and odd-mode excitations. (a) Even mode; (b) Odd mode.

The 4-port symmetrical network is first analyzed by using even- and odd-mode excitations, as shown in Figure 2, to derive its S-matrix in terms of the even/odd-mode reflection and transmission coefficients. The 4-port S-matrix is reduced to a 3-port S-matrix, by imposing the boundary condition at the reflection port of the 4-port network as shown in Figure 1. The following conditions need to be satisfied in order for the final 3-port network to behave as a balun:

$$S_{21} = -S_{31} \tag{1}$$

and

$$S_{11} = 0 \tag{2}$$

where port 1 is the input port and port 2, 3 are the output ports of the final 3-port.

In addition, if | Γ |=1, and if the 4-port network is lossless, the power at each output port will be –3 dB with respect to the input power. By substituting the derived 3-port S-Matrix elements into (1) and (2), requirements for the final 3-port network to behave as a balun can be specified in terms of the even- and odd-mode parameters of the symmetrical 4-port network.

III. CONDITIONS FOR SYNTHESIZING 3-PORT BALUNS

The above approach has been used to derive the conditions for synthesizing 3-port baluns as illustrated in Figure 1 and the conditions are listed as follows:

$$\frac{T_{el}\cdot(1-\Gamma_{ol}\cdot\Gamma)}{2-\Gamma\cdot(\Gamma_{el}+\Gamma_{ol})}=0 \quad (3)$$

and

$$\frac{\Gamma_{el}+\Gamma_{ol}-2\cdot\Gamma_{el}\cdot\Gamma_{ol}\cdot\Gamma}{2-\Gamma\cdot(\Gamma_{el}+\Gamma_{ol})}=0 \quad (4)$$

IV. APPLICATIONS

Different values of Γ have been evaluated with the above conditions. The following 2 sets of conditions have been found to be most practical and useful:

$\Gamma=+1$: $\Gamma_{el}=-1$, and $\Gamma_{ol}=+1/3$ (5)

$\Gamma=-1$: $\Gamma_{el}=+1$, and $\Gamma_{ol}=-1/3$ (6)

This implies that a balun can be formed if the 4-port symmetrical network behaves as transmission-stop network in the even-mode excitation and as an impedance transformer in the odd-mode excitation. Based on these conditions, 6 baluns have been synthesized and tabulated in Figure 3.

The topology of the balun listed in Figure 3f is new and the exact design equations and design curves for most of the baluns listed have not been published previously. The baluns listed in Figure 3a and 3b are easy to implement in planar form but they are limited in bandwidth as compared to the coupled-line baluns. The balun described in Figure 3b can be viewed as a dual of the balun in Figure 3a. It provides a convenient point at the RF-short-circuit terminal to introduce DC biasing circuitry. The balun in Figure 3e has the characteristic of having lower coupling requirement on the coupled-lines as the load resistance decreases. This is useful when the balun is used to construct push-pull amplifiers, as the devices' impedances are usually low.

All the baluns listed in Figure 3 have band-pass type of response, as the odd-mode topologies of their symmetrical 4-port networks are band-pass in nature. The bandwidth of the baluns will be limited by the choice of topology of the balun. All the baluns listed in Figure 3 can be cascaded with more sections of the same kind provided that the overall structure obeys condition (3) and (4). This will enhance the rejection of the balun at out-of-band frequencies. As it will be shown below, the multi-section balun integrates the functions of a balun and a filter into a single circuit.

V. EXPERIMENTAL VERIFICATION

A 3-section Marchand-type, coupled-line microstrip balun operating from 2.3 GHz to 2.5 GHz, has been designed and fabricated with a substrate of Er=2.2 and 31 mil in height. Each section of the balun has the same topology as listed in Figure 3c and the even- and odd-mode characteristic impedances of each section are listed in Table 1. Section 1 denotes the section that is closer to the input of the balun.

TABLE 1
EVEN- AND ODD-MODE CHARACTERISTIC IMPEDANCES OF THE 3-SECTION BALUN

	Section1	Section2	Section3
Z_{oe}	135	94	95
Z_{oo}	61	66	43

In the odd-mode configuration, the 3-section coupled-line pairs transform the output 50-Ohm load towards a 100-Ohm input resistance across the frequencies of operation. In the even-mode configuration, a short-circuit termination is presented at the input of the balun. Simulation was done in ADS™ by using microstrip coupled-line model, taking into account of the fabrication errors. Care has to be taken in cascading to prevent forward-wave coupling from propagating in-band in the even-mode configuration, which is caused by the different odd- and even-mode phase velocities. Figure 4 shows the photograph and schematic of the balun and Figure 5(a) shows the measured and simulated transmission and reflection response of the balun. Figure 5(b) illustrates the measured and simulated phase balance of the balun. Close agreements have been achieved between the simulated and the measured response.

VI. CONCLUSION

This paper describes an approach to derive the conditions for synthesizing a class of 3-port baluns from symmetrical 4-port networks. These conditions will result in devising new balun structures and in deriving the exact design equations for these baluns in the center frequency of operation. Useful design equations and design curves for some known and new baluns have been derived by using this approach, and the results are presented in the

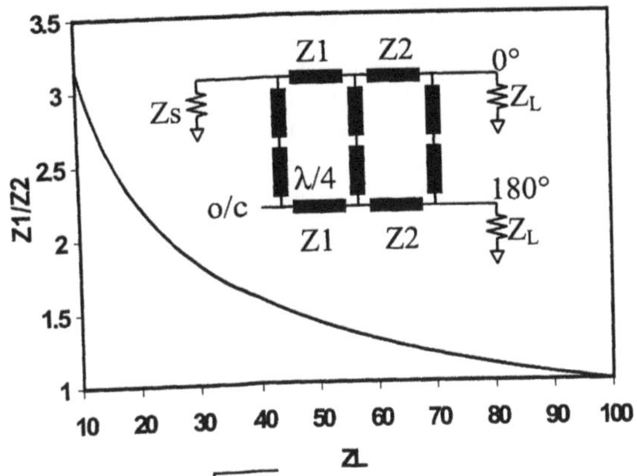

(a) $\frac{Z1}{Z2} = \sqrt{\frac{2 \cdot Zs}{Z_L}}$

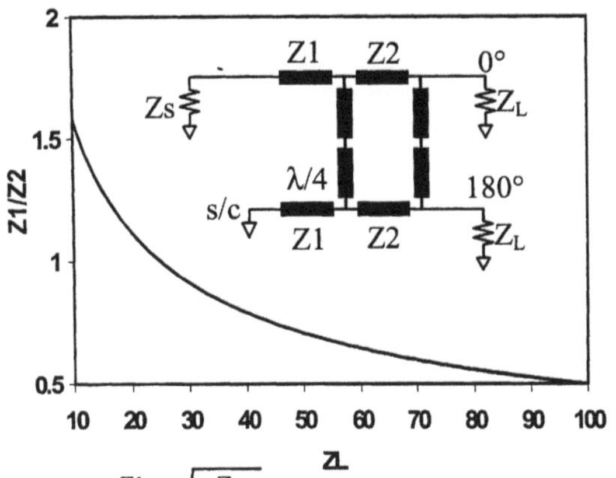

(b) $\frac{Z1}{Z2} = \sqrt{\frac{Zs}{2 \cdot Z_L}}$

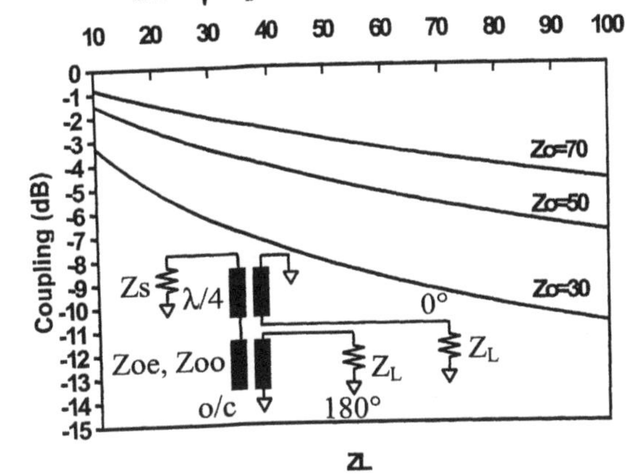

(c) $Zoo = \frac{-Zo^2 + Zo \cdot \sqrt{Zo^2 + 2 \cdot Zs \cdot Z_L}}{\sqrt{2 \cdot Zs \cdot Z_L}}$ $\quad Zoe = \frac{Zo^2}{Zoo}$

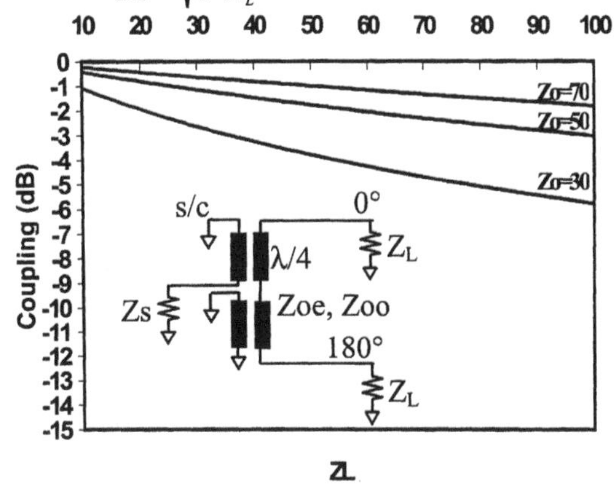

(d) $Zoo = \frac{-\sqrt{2} \cdot Zo^2 + Zo \cdot \sqrt{2 \cdot Zo^2 + Zs \cdot Z_L}}{\sqrt{Zs \cdot Z_L}}$ $\quad Zoe = \frac{Zo^2}{Zoo}$

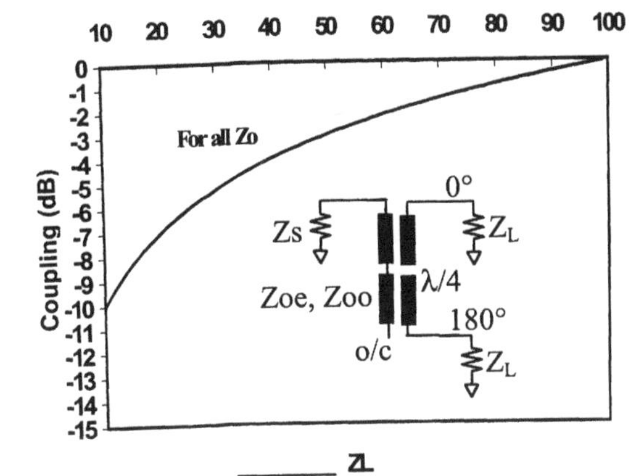

(e) $Zoo = Zo \cdot \sqrt{\frac{\sqrt{\frac{2 \cdot Zs}{Z_L}} - 1}{\sqrt{\frac{2 \cdot Zs}{Z_L}} + 1}}$ $\quad Z_L < 2 \cdot Zs \quad Zoe = \frac{Zo^2}{Zoo}$

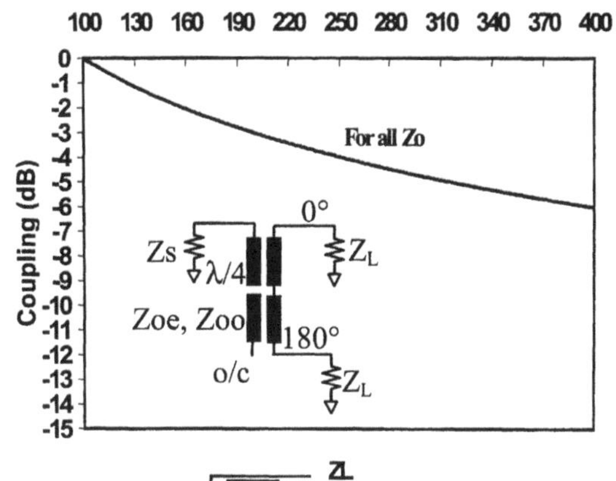

(f) $Zoo = Zo \cdot \sqrt{\frac{\sqrt{\frac{Z_L}{2 \cdot Zs}} - 1}{\sqrt{\frac{Z_L}{2 \cdot Zs}} + 1}}$ $\quad Z_L > 2 \cdot Zs \quad Zoe = \frac{Zo^2}{Zoo}$

Fig. 3. Topologies, design curves, and design equations for 6 balun synthesized from eqn (5) and (6). The design curves are for cases with Zs=50 Ohm and various Zo.

paper. A 3-section, Marchand-type, coupled-line balun, centered at 2.4 GHz have been designed and fabricated. Good agreements between simulation and measured results have been obtained, thereby verifying the validity of the design equations. The valuable insights given by this analysis approach will lead to the development of more novel balun structures and their design equations.

ACKNOWLEDGEMENT

The authors wish to acknowledge the support of DSTA, Singapore.

REFERENCES

[1] Kian Sen Ang, and Yoke Choy Leong, "Converting Baluns into Broadband Impedance-Transforming 180^0 Hybrids," To be published in MTT-T, Aug 2002.

[2] Stephen A. Maas, and Kwo Wei Chang, "A Broadband, Planar, Doubly Balanced Monolithic Ka-Band Diode Mixer," IEEE Trans. Microwave Theory and Techniques, vol. 41, no.12, pp. 2330-2335, Dec. 1993.

[3] Rick Sturdivant, "Balun Designs for Wireless, ...Mixers, Amplifiers and Antennas," Applied Microwave, pp. 34-44, Summer 1993.

[4] Kian Sen Ang, and Ian D. Robertson, "Analysis and Design of Impedance-Transforming Planar Marchand Baluns," IEEE Trans. Microwave Theory and Techniques, vol. 49, no.2, pp. 402-406, Feb. 2001.

[5] Randal Schwindt, and Cam Nguyen, "A CAD Procedure For The Double-Layer Broadside-Coupled Marchand Balun," MTT-S International Microwave Symposium Digest, pp. 389-391, 1994.

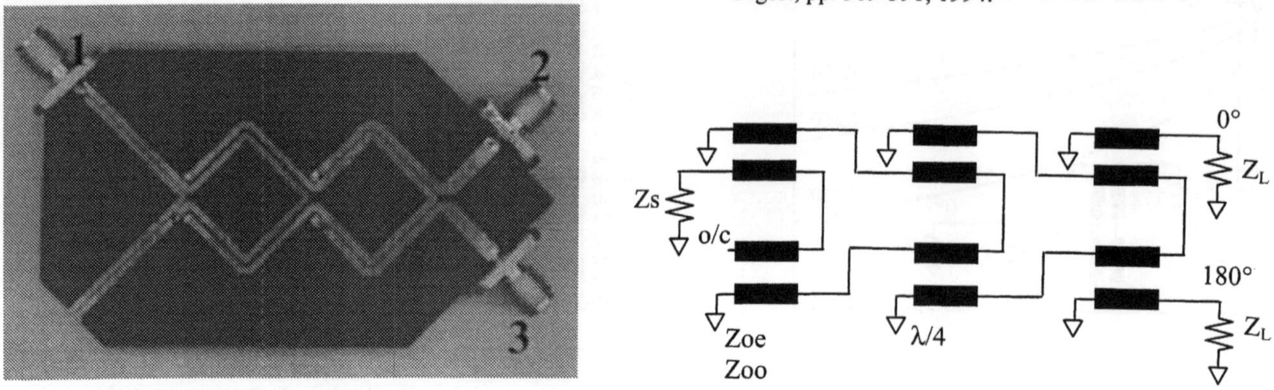

(a) (b)

Fig. 4. (a) Photograph of a 3-section Marchand balun centered at 2.4 GHz; (b) Schematic of the 3-section Marchand balun.

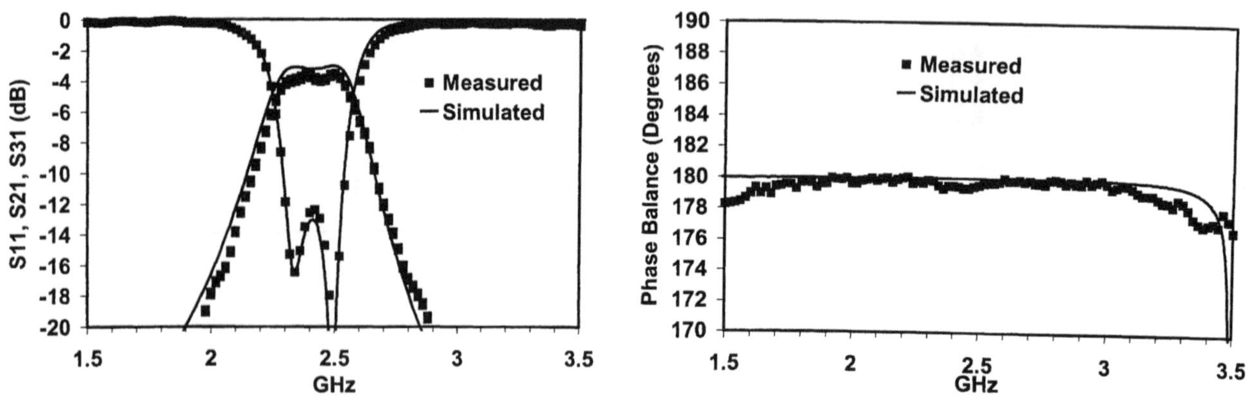

(a) (b)

Fig. 5. (a) S11, S21 and S31; (b) Phase balance of the 3-section Marchand balun. The response of S21 and S31 are overlapping on each other.

A Novel Broad-Band Chebyshev-Response Rat-Race Ring Coupler

Chi-Yang Chang, *Member, IEEE,* and Chu-Chen Yang

***Abstract*— A novel broad-band rat-race ring coupler with a unit element at each port and an ideal phase inverter at one of the ring arms is proposed and analyzed. Design equations based on Chebyshev equiripple functions are derived. The design curves of equal or unequal power division are also presented. Theoretical data show that the bandwidth of unequal power division with a specific return-loss value increases as the power division ratio increases or decreases from unity. For a 180° hybrid with equal power division (3 dB) and 15-dB return loss, the proposed rat-race ring may have a bandwidth of 4 : 1. Three circuits of this novel rat-race ring are realized using finite-ground-plane coplanar waveguide. The measured results match very well with the theory.**

***Index Terms*—Finite-ground plane coplanar waveguide, phase inverter, uniplanar ring coupler.**

I. Introduction

THE 180° hybrid ring coupler is a very useful circuit element in microwave circuits, such as mixer, multiplier, amplifier, beamformer, etc. However, the bandwidth of the conventional rat-race ring coupler is narrow. The conventional rat-race coupler comprises four ring arms. One of the ring arms contains a 180° phase shifter. Traditionally, this 180° phase shifter is formed by a $\lambda/2$ line section, and its phase shifting is 180° only at the center frequency. Many efforts have been made to make the bandwidth of rat-race ring coupler larger [1]–[8]. Most of them pay their attention by making a broad-band phase inverter (180° phase shifter). March [1] replaced the $3\lambda/4$ line section with a short-circuited $\lambda/4$ coupled-line section to make the bandwidth of the rat-race ring larger. The uniplanar type of circuit has many advantages for a broad-band rat-race ring. In [2]–[5], the combinations of coplanar waveguide (CPW) and coplanar strips (CPS's) or slot line were adopted to form a phase inverter, and it increases the bandwidth further. In those types of designs, the CPS direct cross-type phase inverter [2], [5] seems to be an ideal phase inverter (with infinity bandwidth theoretically). Even if an ideal phase inverter is used, the conventional rat-race ring coupler is still limited to a bandwidth of about 74% (2.2 : 1 bandwidth) for return loss better than 15 dB. When a rat-race ring coupler is realized with an ideal phase inverter, it physically looks like a 90° branch-line coupler. It is well known that the bandwidth of a 90° branch-line coupler can be broadened by applying a multiple-section structure [9]. Increasing the number of sections, the bandwidth of a branch-line coupler increases accordingly. It seems like that same concept could be adopted here. Unfortunately, in this study, we find that the rat-race ring coupler could not have a similar multiple-section structure as a branch-line coupler has. The only possible way is to add a unit element with normalized admittance Y_t to each input/output (I/O) port, as shown in Fig. 1. With these unit elements, the response of a rat-race ring coupler could be increased from order two to order four. In [6], although the coupler has a unit element at each I/O port, the circuit after optimization does not show the fourth-order Chebyshev response and the bandwidth is only 50.7% with about 21-dB return loss. This is due to the fact that the phase inverter in [6] is a narrow-band phase inverter. By theoretical analysis, two criterions should be met to create the coupler showing a broader bandwidth. The first criterion is to realize an ideal phase inverter (or a phase inverter with enough bandwidth) with relatively low impedance. The second criterion is to increase the order of circuit response. When the order of freedom increases, the coupling function may fit to a specific frequency response with higher order. The bandwidth of the circuit may, therefore, be broadened. In short, we can say that narrow-band phase inverter limits the bandwidth of a conventional rat-race ring coupler. After solving the bandwidth problem of the phase inverter, increasing the order of the circuit response may still increase the bandwidth of a rat-race ring coupler. In this paper, we develop the solutions for the above-mentioned problems and provide a rigorous procedure to synthesize the circuit to a specified Chebyshev response. Theoretical analysis shows that a 15-dB return-loss bandwidth of 120% (4 : 1 bandwidth) can be achieved with this novel 3-dB rat-race ring coupler. Unequal power division may even have much broader bandwidth. Two practical problems

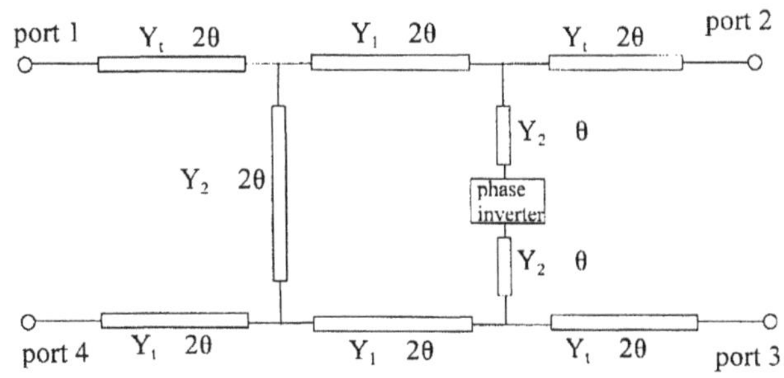

Fig. 1. Schematic circuit of the broad-band rat-race ring coupler.

Manuscript received March 18, 1998; revised January 26, 1999. This work was supported in part by the National Science Council under Grant NSC 87-2213-E-009-107.

The authors are with the Department of Communication Engineering, National Chiao-Tung University, Hsinchu 300, Taiwan, R.O.C.

Publisher Item Identifier S 0018-9480(99)03001-X.

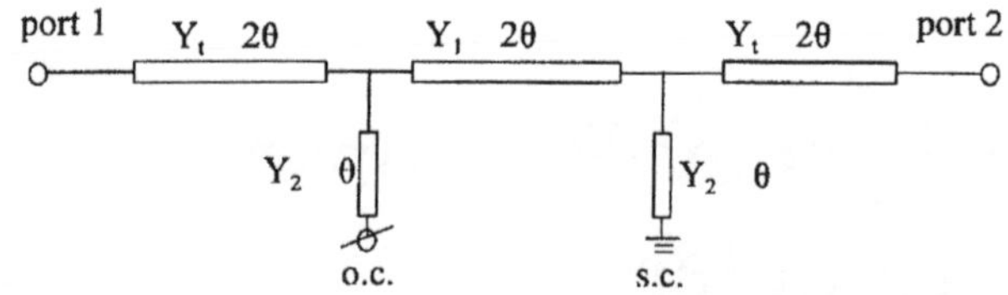

(a)

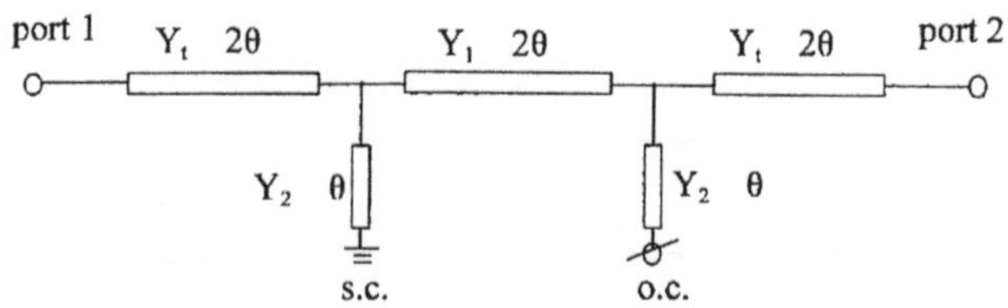

(b)

Fig. 2. (a) Even- and (b) odd-mode equivalent circuits of the broad-band rat-race ring coupler.

should be solved in the realization of our rat-race ring. Realization of low impedance line (3-dB coupler with 15-dB return loss having the line impedance as low as 32 Ω) is the first problem. The design of ideal phase inverter with low impedance line is the second problem. Among the planar transmission lines, the finite-ground-plane CPW (FCPW) and broadside line can solve these problems simultaneously. Three prototype 3-dB rat-race ring hybrids using FCPW are realized with center frequency of 5 GHz. Measured results match very well with the theoretical predictions.

II. Theoretical Analysis of the Coupler

Based on the normal-mode analysis and using the even–odd-mode analysis technique, the even- and odd-mode equivalent circuits of the proposed rat-race ring coupler are shown in Fig. 2(a) and (b). In Figs. 1 and 2, the ring admittance Y_1, Y_2, and the unit element admittance Y_t are normalized values, which is normalized to system admittance Y_0. The $ABCD$ matrices for the even- and odd-mode circuits shown in Fig. 2(a) and (b) are

$$\begin{bmatrix} A_e & B_e \\ C_e & D_e \end{bmatrix} = \frac{1}{(1-t^2)^3}\begin{bmatrix} A_3(t^2) & tB_2(t^2) \\ tC_3(t^2) & D_3(t^2) \end{bmatrix} = \begin{bmatrix} A & B \\ C & D \end{bmatrix} \quad (1)$$

$$\begin{bmatrix} A_o & B_o \\ C_o & D_o \end{bmatrix} = \begin{bmatrix} D_e & B_e \\ C_e & A_e \end{bmatrix} = \begin{bmatrix} D & B \\ C & A \end{bmatrix} \quad (2)$$

where $t = j\tan\theta$ and A_3, C_3, D_3, and B_2 represent the third- and second-order polynomials of t^2, respectively. The S-parameters of the four-port network are

$$S_{11} = S_{22} = \frac{B-C}{A+B+C+D} = \Gamma \quad (3a)$$

$$S_{21} = S_{12} = \frac{2}{A+B+C+D} = T \quad (3b)$$

$$S_{31} = S_{42} = 0 \quad (3c)$$

$$S_{41} = \frac{A-D}{A+B+C+D} = \frac{2(Y_1/Y_2)}{A+B+C+D} \quad (3d)$$

$$S_{32} = \frac{D-A}{A+B+C+D} = \frac{-2(Y_1/Y_2)}{A+B+C+D}. \quad (3e)$$

It is obvious that the output-power division ratio is

$$R = \frac{|S_{41}|^2}{|S_{21}|^2} = \frac{|S_{32}|^2}{|S_{12}|^2} = \left(\frac{Y_1}{Y_2}\right)^2. \quad (4)$$

Assuming that the input is matched well in the passband, for conservation of energy, we have

$$|T|^2\left[(Y_1/Y_2)^2 + 1\right] \approx 1. \quad (5)$$

Now, consider the function

$$F = \frac{|\Gamma|}{|T|\sqrt{R+1}} \quad (6)$$

which approximately equals to the reflection coefficient in the passband because the denominator $|T|\sqrt{R+1}$ is very close to unity [see (5)]. The deviation is small if the ripple in the passband is small. The insertion-loss function can be shown as follows:

$$P_L = 1 + |F|^2 = 1 + \left\{\frac{|B-C|}{2\sqrt{R+1}}\right\}^2. \quad (7)$$

The insertion-loss function can be specified by a different mathematical function. The responses of the coupler will follow the characteristics of the mathematical function.

The coupler in Fig. 1 could be synthesized to have a equiripple response. According to Levy and Lind's [9] analysis, if S_{31} and S_{42} (the isolation of the coupler) are assumed to be zero in the passband, the synthesis procedure can be largely simplified. In our circuit, however, this is already satisfied for all frequencies [see (3c)]. Now, let us define the center frequency of our hybrid ring by setting Y_t, Y_1, and Y_2 length to be $\lambda/4$ (i.e., $\theta = 45°$ in Fig. 1). Using the method similar to Riblet [10], the return loss and coupling could be synthesized simultaneously. The equiripple response, which was described by Riblet [10], Carlin, and Kohler [11], has the form as shown in

$$P_L = 1+h^2 \cdot \left\{\frac{\left(1+\sqrt{1-x_c^2}\right)T_n(x/x_c) - \left(1-\sqrt{1-x_c^2}\right)T_{n-2}(x/x_c)}{2\sqrt{1-x^2}}\right\}^2 \quad (8)$$

where $x = \cos 2\theta$, $x_c = \cos 2\theta_c$, and h is the parameter to control the ripple level.

In order to fit (7) to this equiripple response, we must set $n = 4$. Therefore, we have four return-loss dips in the passband of $-x_c \le x \le x_c$. After some manipulations, we have

$$\frac{Y_t^4 - Y_1^2 - Y_2^2}{Y_2 Y_t^2\sqrt{R+1}} = 2h \quad (9a)$$

$$\frac{2Y_t(Y_t^2-1) + (Y_1^2+Y_2^2)(Y_t^2-2) - Y_t^2 - 2Y_1Y_2 + 2Y_t^4}{Y_2Y_t^2\sqrt{R+1}} = \frac{6\sqrt{1-x_c^2}+10}{x_c^2}h \quad (9b)$$

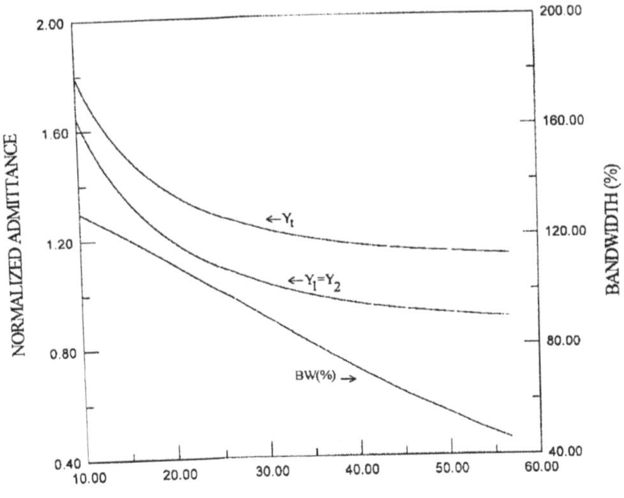

Fig. 3. Normalized admittance Y_1, Y_2, Y_t, and bandwidth (%) versus return loss (dB) for equiripple response (four return-loss dips) 3-dB ring coupler.

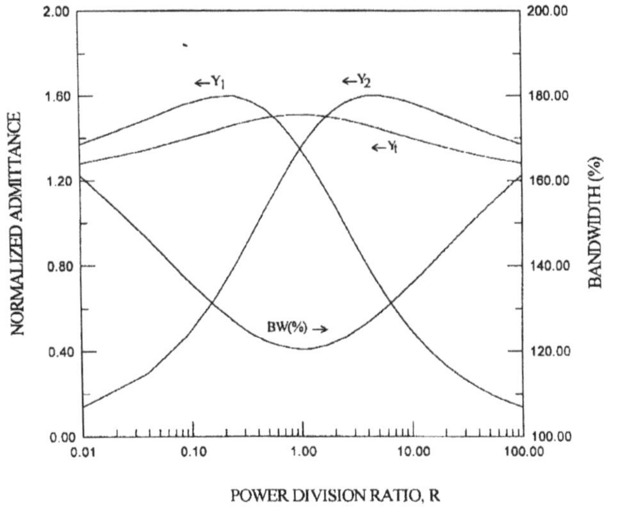

Fig. 4. Normalized admittance Y_1, Y_2, Y_t, and bandwidth (%) versus power division ratio R for 15-dB return loss and equiripple response (four return-loss dips).

$$\frac{(Y_t^2-1)(Y_1+Y_2)(2Y_t+Y_1+Y_2)-Y_t^2+Y_t^4}{Y_2Y_t^2\sqrt{R+1}} = \frac{8\sqrt{1-x_c^2}+8}{x_c^4}h. \tag{9c}$$

If we fix the ring admittance Y_1, the admittance Y_2 is determined by the power-division ratio corresponding to (4). There are three unknown variables related to three equations in (9), which can be solved numerically. Fig. 3 shows the plot of Y_1, Y_2, Y_t, and bandwidth (%) versus return loss for 3-dB power-division (i.e., $Y_1 = Y_2$) rat-race ring. The 120% bandwidth for return loss better than 15 dB or 130% bandwidth for return loss better than 10 dB can be achieved. The relationship between h and return loss is

$$RL = 10\log(1+1/h^2)\ \text{dB}. \tag{10}$$

Fig. 4 shows the computed results for different power-division ratio and for 15-dB return loss. It is clear that the bandwidth increases as the power-division ratio increases or decreases from unity (equal power division) and, of course, the realization of circuit is more difficult. Figs. 3 and 4 provide the useful information for designing the 180° ring hybrid.

Riblet [10] also described an optimum, even, equiripple, function having a double zero at zero. If this type of function is chosen to specify the coupler, there will be three return-loss dips instead of four, as in previous derivation. We summarize the procedures as follows.

The values of the function

$$\frac{\left(a+\sqrt{a^2-1}\right)T_n(x)-\left(a-\sqrt{a^2-1}\right)T_{n-2}(x)}{2\sqrt{a^2-x^2}} \tag{11}$$

vary between ± 1 over the range $-1 \le x \le 1$. Let x_0 be the zero of this function nearest to the origin, then when $n = 4$,

$$T_2(x_0) = \cos 2\cos^{-1} x_0 = \frac{1-\sqrt{1+8Z^4}}{4Z^2} \tag{12}$$

where $Z = a + \sqrt{a^2-1}$.

The function

$$R_n\left[(a^2-x_0^2)x^2+x_0^2\right] \Big/ 2\sqrt{a^2-x_0^2}\,\sqrt{1-x^2} \tag{13}$$

varies between ± 1 over the range $\pm\sqrt{1-x_0^2}/\sqrt{a^2-x_0^2}$ with a double zero at $x = 0$, where $R_n(y^2) = ZT_n(y) - Z^{-1}T_{n-2}(y)$.

Therefore, the insertion-loss function for the optimum, even, equiripple, response has the form

$$P_L = 1 + h^2 \cdot \left\{ R_n\left[(a^2-x_0^2)x^2+x_0^2\right] \Big/ 2\sqrt{a^2-x_0^2}\,\sqrt{1-x^2} \right\}^2. \tag{14}$$

Applying this function to the insertion-loss function of a rat-race ring coupler, we find that

$$\frac{Y_t^4-Y_1^2-Y_2^2}{Y_2Y_t^2\sqrt{R+1}} = h\left[8Zx_0^4-(8Z+2Z^{-1})x_0^2+Z+Z^{-1}\right] \Big/ \sqrt{a^2-x_0^2} \tag{15a}$$

$$\frac{2Y_t(Y_t^2-1)+(Y_1^2+Y_2^2)(Y_t^2-2)-Y_t^2-2Y_1Y_2+2Y_t^4}{Y_2Y_t^2\sqrt{R+1}} = h[16Z(a^2-x_0^2)x_0^2-(8Z+2Z^{-1})(a^2-x_0^2)] \Big/ \sqrt{a^2-x_0^2} \tag{15b}$$

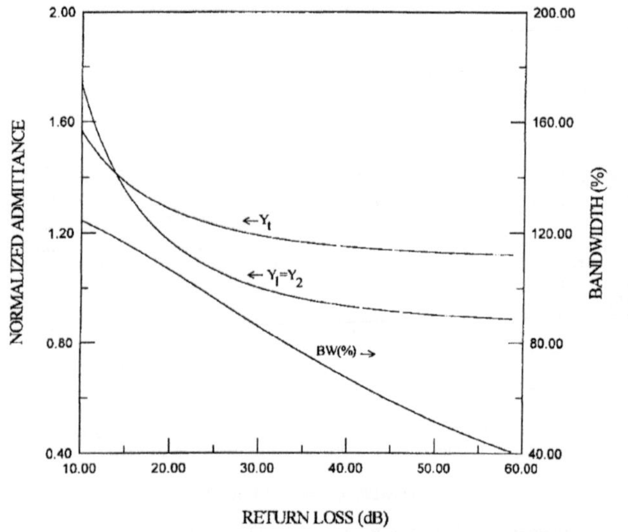

Fig. 5. Normalized admittance Y_1, Y_2, Y_t, and bandwidth (%) versus return loss (dB) for equiripple response (three return-loss dips) 3-dB ring coupler.

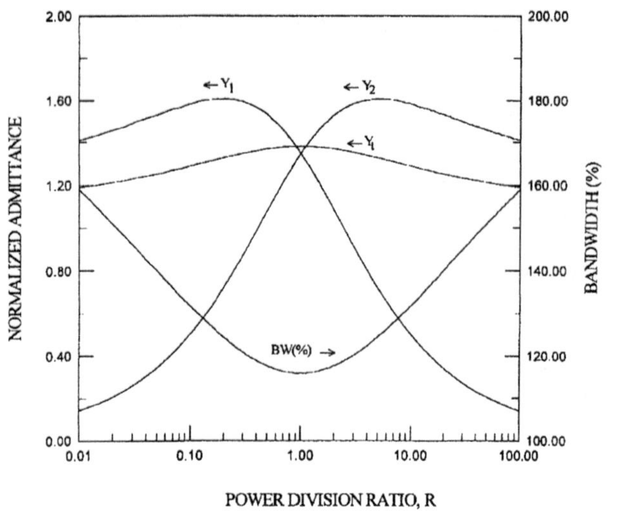

Fig. 6. Normalized admittance Y_1, Y_2, Y_t, and bandwidth (%) versus power-division ratio R for 15-dB return loss and equiripple response (three return-loss dips).

$$\frac{(Y_t^2-1)(Y_1+Y_2)(2Y_t+Y_1+Y_2)-Y_t^2+Y_t^4}{Y_2Y_t^2\sqrt{R+1}} = 8hZ\left(a^2-x_0^2\right)^2\Big/\sqrt{a^2-x_0^2}. \tag{15c}$$

These equations can also be solved numerically. The numerical result of the equal (3-dB) power division ($Y_1 = Y_2$) is shown in Fig. 5, which indicates that the 116% (124%) bandwidth for return loss better than 15 dB (10 dB) can be achieved. Fig. 6 shows the plot of admittance Y_1, Y_2, Y_t, and bandwidth (%) versus power-division ratio R for 15-dB return loss. It is observed that, in Figs. 5 and 6, the admittance Y_1, Y_2, and Y_t satisfy the relation $Y_t^2 = \sqrt{Y_1^2+Y_2^2}$. This relation can be easily derived by letting $|S_{11}| = 0$ or $S|_{22}| = 0$ (i.e., $|B-C| = 0$) at the center frequency. The bandwidth of the design by optimum equiripple function is a little bit narrower than that of the four dips equiripple design.

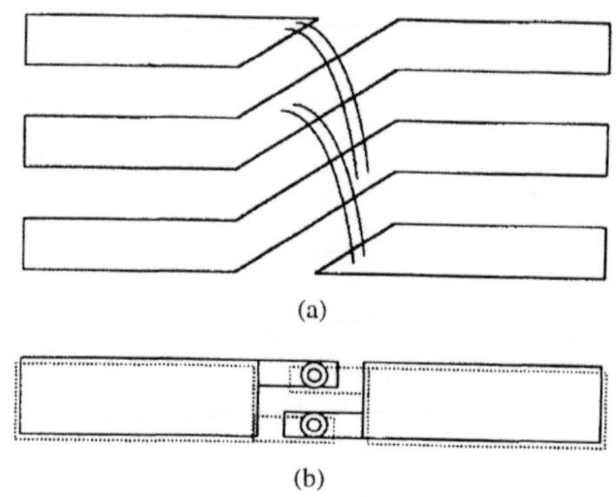

Fig. 7. Circuit configuration of ideal phase inverter. (a) Realized with FCPW. (b) Realized with broadside line.

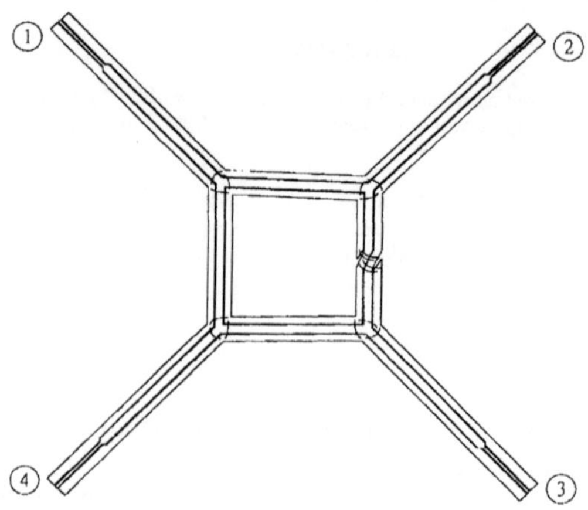

Fig. 8. Circuit layout of the FCPW ring coupler.

III. Experimental Results

The theory of the proposed rat-race ring coupler is based on an ideal phase inverter, as shown in Fig. 1. As we mentioned previously, the FCPW and broadside line are good candidates to realize this ideal phase inverter. Fig. 7 shows the circuit layout of the ideal phase inverter realized in these two types of lines.

Here, we choose an FCPW to realize our circuits because of the avoidance of the via hole. The line impedance calculation is based on a sequence of conformal mappings [12]. The substrate to realize the circuits is 25-mil Al_2O_3 alumina ($\varepsilon_r = 9.8$). Fig. 8 shows the circuit layout of the coupler. Three novel rat-race ring couplers are fabricated using an FCPW with system impedance of 50 Ω. The design parameters and circuit dimensions are listed in Table I. The length of the bonding wires used to form the phase inverter is about 20–30 mil and the diameter of the bonding wires is 1 mil. The effects of the bonding wires at crossover can be compensated by shortening the length of the ring arm by 9 mil for circuits I and

TABLE 1

DESIGN PARAMETERS AND CIRCUIT DIMENSIONS OF THREE NOVEL RAT-RACE RING COUPLERS (THE $\lambda/4$ LINE LENGTH IS 255 mil FOR 5 GHz)

Circuit	Normalized admittance	Strip width (mil)	Slot width (mil)	Finite ground width (mil)
I	$Y_1 = Y_2 = 1.4$	15	1.5	10
	$Y_r = 1.56$	16	1.5	11
II	$Y_1 = Y_2 = Y_r = \sqrt{2}$	15	1.5	12
III	$Y_1 = Y_2 = 1/\sqrt{2}$	3	3.3	3
	$Y_r = 1$	4.2	1.9	15

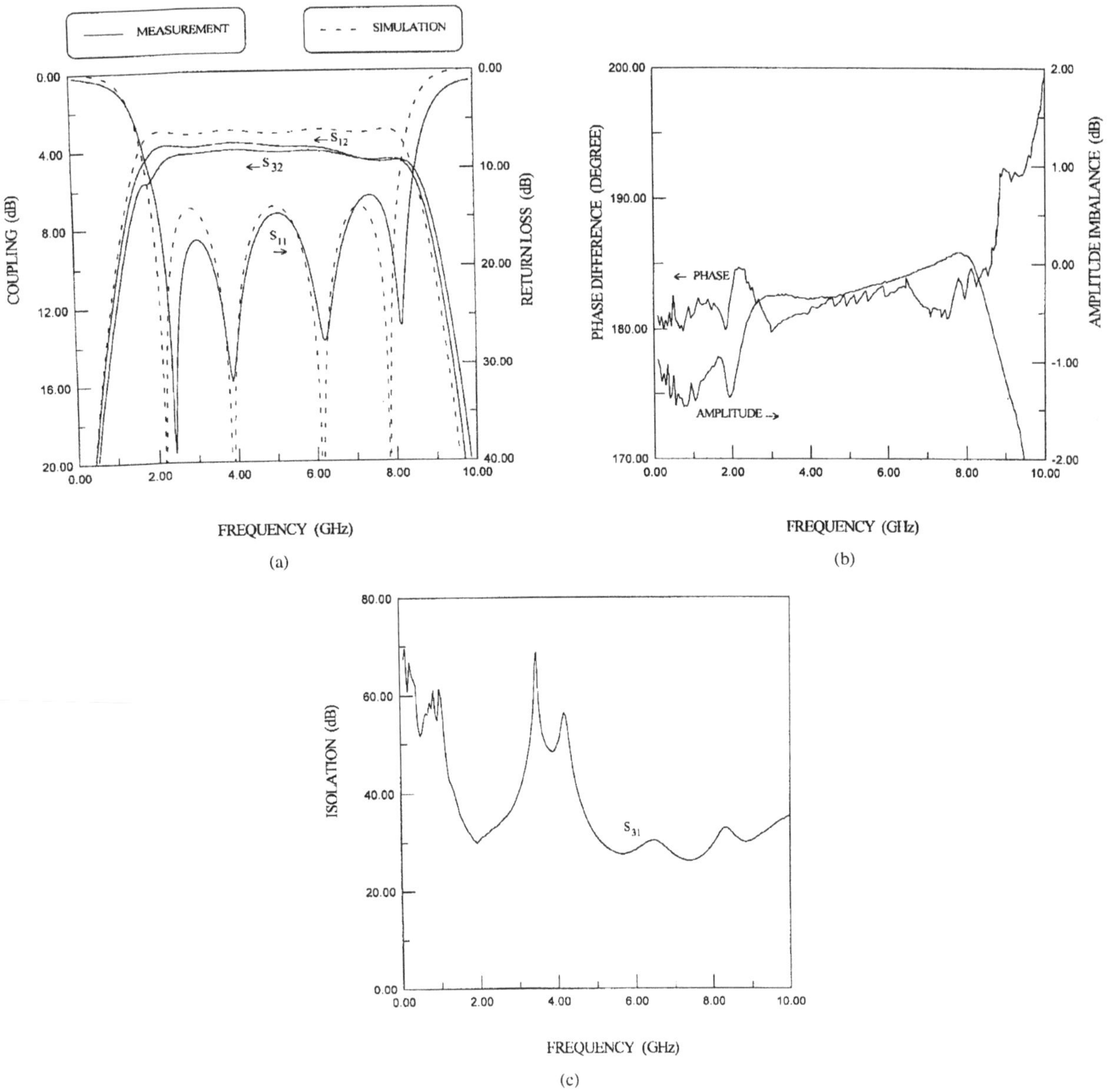

Fig. 9. Measured and simulated frequency response of equiripple (four return-loss dips) rat-race ring coupler. (a) Return loss and coupling. (b) Phase difference and amplitude imbalance. (c) Isolation.

II and 4 mil for circuit III, respectively. Circuit I is designed to satisfy a 13.8-dB return loss corresponding to four dips equiripple function. Circuit II is designed to satisfy a 13.8-dB return loss corresponding to three dips optimum equiripple function. Circuit III is a conventional rat-race ring with an ideal 70.7-Ω FCPW phase inverter.

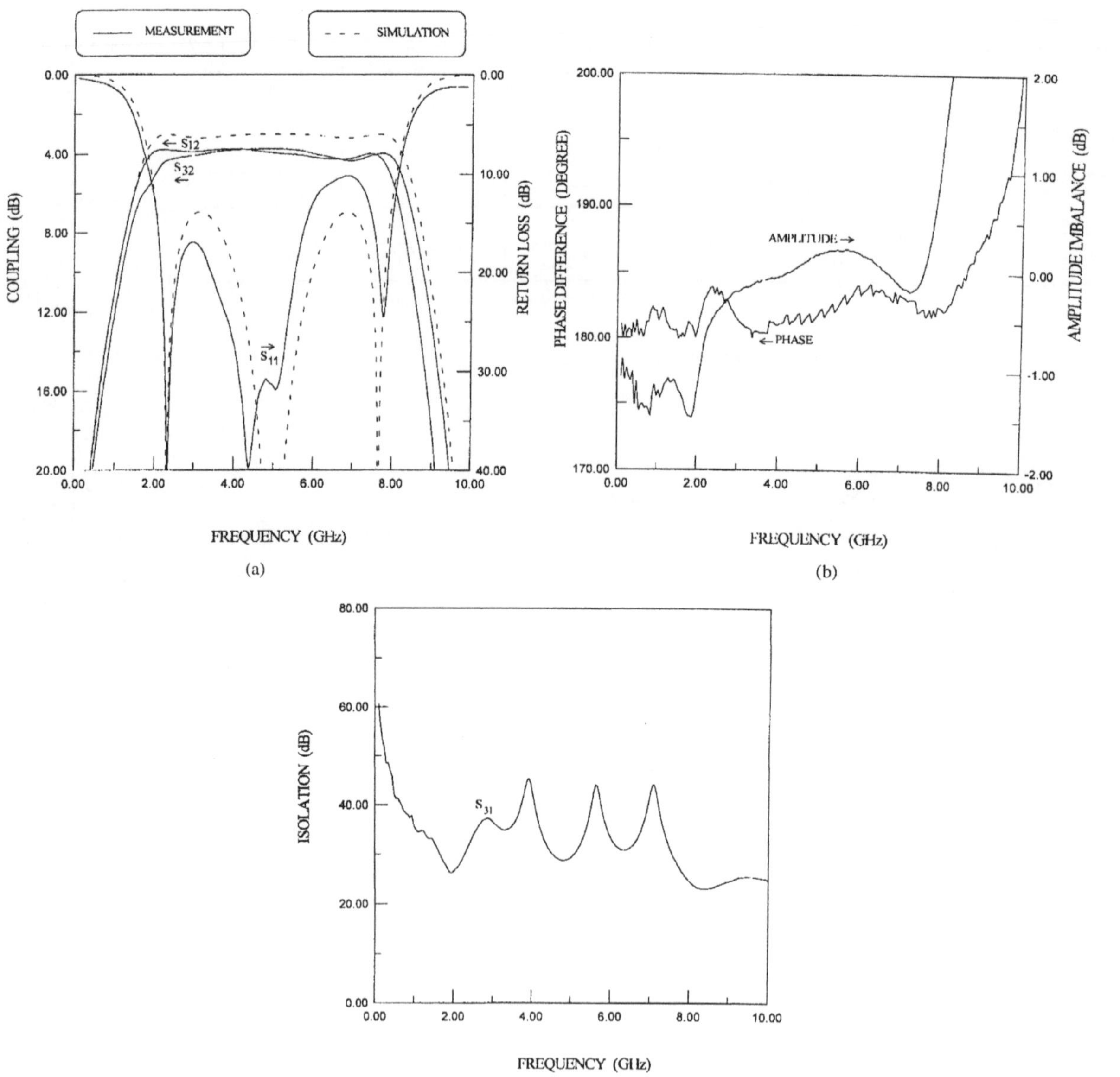

Fig. 10. Measured and simulated frequency response of equiripple (three return-loss dips) rat-race ring coupler. (a) Return loss and coupling. (b) Phase difference and amplitude imbalance. (c) Isolation.

Circuit I has line impedances of $Z_1 = Z_2 = 35.7\ \Omega$ and $Z_t = 32.05\ \Omega$. The CPS, although suitable for the phase inverter, is impossible to realize a line with such a low impedance. The simulated and measured results are shown in Fig. 9. The return-loss data, shown in Fig. 9(a), match very well to the theoretical data. The measured return-loss bump is higher than 13.8 dB at the high frequency end, and the passband is shifted a little bit to higher frequency (the desired passband is 1.93–8.07 GHz and has 123% bandwidth). This is due to the parasitics of four Y junctions, and can be compensated by tuning the line length of Y_t sections. The measured coupling, also shown in Fig. 9(a), is about 4 dB (theoretically, it should be 3 dB), which is due to the line losses of the FCPW. The output amplitude imbalance, shown in Fig. 9(b), is less than 1 dB from 2.16–8.8 GHz. The perfect 180° phase difference between S_{32} and S_{12} in Fig. 9(b) and good isolation in Fig. 9(c) imply good performance of the FCPW phase inverter. The phase error of the difference port is less than 5° from 45 MHz to 8.65 GHz. The sum port phase error and amplitude imbalance are not shown, but the performance is better than that of difference port. The measured isolation is better than 25 dB from 45 MHz to 10 GHz.

Circuit II has line impedances of $Z_1 = Z_2 = Z_t = 35.36\ \Omega$. The simulated and measured results are shown in Fig. 10. The designed passband is 2.05–7.95 GHz (118% bandwidth). The measured return loss and coupling agree well with the simulation. The output amplitude imbalance is less than 1 dB

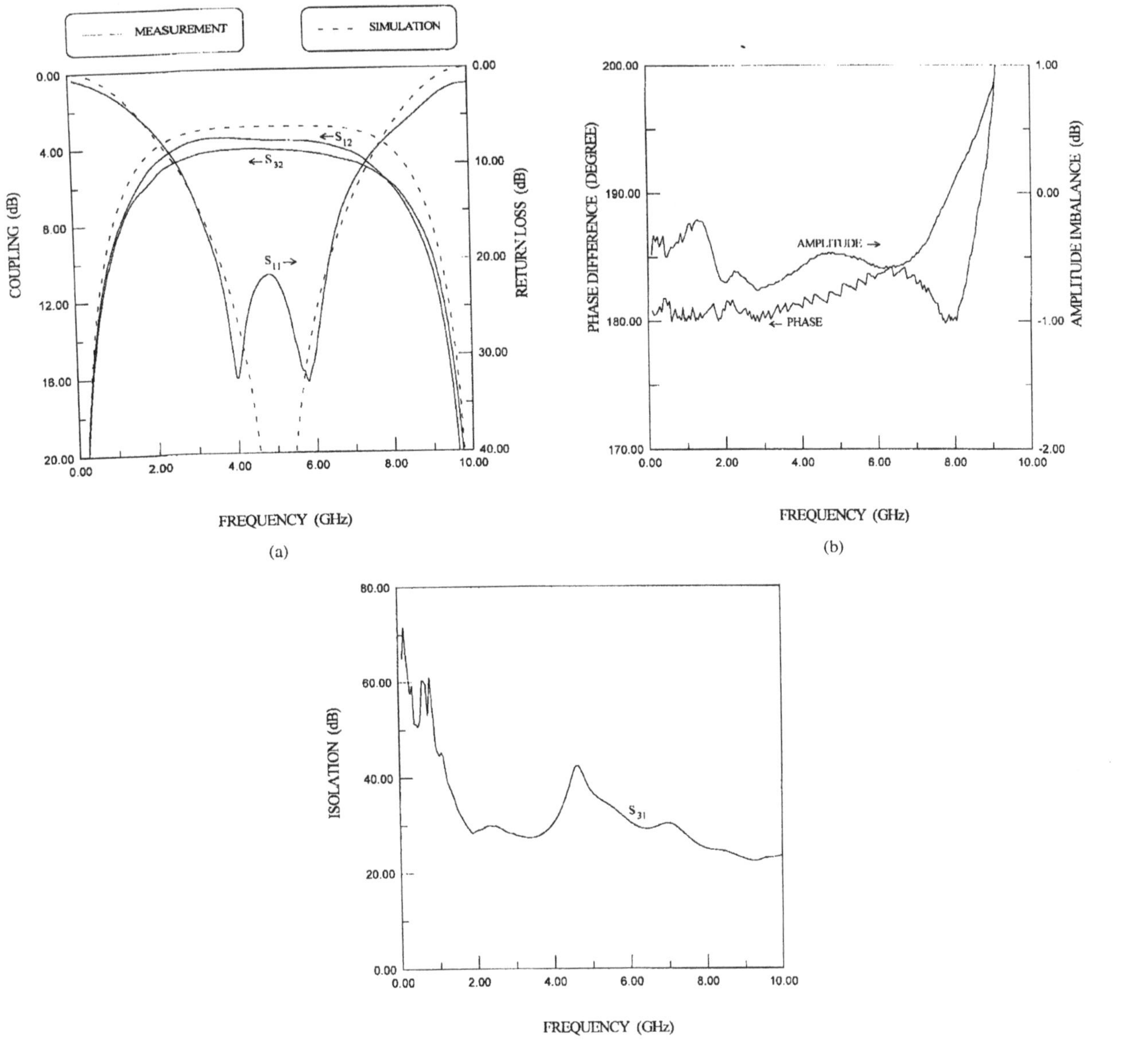

Fig. 11. Measured and simulated frequency response of conventional rat-race ring coupler (Butterworth response). (a) Return loss and coupling. (b) Phase difference and amplitude imbalance. (c) Isolation.

from 2.09 to 7.95 GHz and the phase error of difference port is less than 5° from 45 MHz to 8.6 GHz. The measured isolation is also better than 23 dB from 45 MHz to 10 GHz.

Circuit III has line impedances of $Z_1 = Z_2 = 70.7\ \Omega$ and $Z_t = 50\ \Omega$. The simulated and measured results are shown in Fig. 11. This conventional rat-race ring coupler is fabricated to show that without the unit element at each I/O ports, the bandwidth can only reach 79% (13.8-dB return loss) even with an ideal phase inverter. In Fig. 11, it is actually a Butterworth response of order two. The bandwidth of a Butterworth response is obviously less than that of a Chebyshev response for the same return loss. In Fig. 11(a), the measured return loss has two dips as compared to one dip of the simulation. This is due to the fabricated ring impedance is slightly lower than 70.7 Ω. It can be identified that if the ring impedance is lower than 70.7 Ω (normalized admittance higher than $1/\sqrt{2}$), the return loss will change from Butterworth response to Chebyshev response of order two.

The bonding wires in the phase inverter seem to have little influence on the performance of the couplers. The authors had fabricated a 30 GHz FCPW rat-race ring 3-dB coupler with a Chebyshev response of order two. The measured performance is still very close to theoretical value.

IV. Conclusion

In this paper, the broad-band rat-race ring for Chebyshev characteristics with a bandwidth much better than any previously published magic-T hybrids has been developed.

To achieve the broad bandwidth, both the unit elements of admittance Y_t and the ideal phase inverter are required. The design curves for two types of equiripple response are given. The design curves for different coupling values are also shown. Based on these curves, a broad-band rat-race ring coupler can be easily designed. The FCPW is used to realize the prototype broad-band rat-race ring. The experimental data show very good agreement with the theoretical prediction.

ACKNOWLEDGMENT

The authors would like to thank C.-S. Wu and D.-C. Niu, Chung-Shan, Institute of Science and Technology (CSIST), Lung-Tang, Taiwan, R.O.C., for their technical support and valuable suggestions.

REFERENCES

[1] S. March, "A wide-band stripline hybrid ring," *IEEE Trans. Microwave Theory Tech.*, vol. MTT-16, p. 361, June 1968.

[2] M. H. Murgulescu, E. Penard, and I. Zaquine, "Design formulas for generalized 180° hybrid ring couplers," *Electron. Lett.*, vol. 30, pp. 573–574, Mar. 1994.

[3] C. H. Ho, L. Fan, and K. Chang, "New uniplanar waveguide hybrid-ring couplers and magic-T's," *IEEE Trans. Microwave Theory and Tech.*, vol. 42, pp. 2440–2448, Dec. 1994.

[4] L. Fan, S. Kanamaluru, and K. Chang, "A new wide-band and reduced-size uniplanar magic-T," in *IEEE MTT-S Int. Microwave Symp. Dig.*, 1995, pp. 667–670.

[5] B. R. Heimer, L. Fan, and K. Chang, "Uniplanar hybrid couplers using asymmetrical coplanar striplines," *IEEE Trans. Microwave Theory Tech.*, vol. 45, pp. 2234–2240, Dec. 1997.

[6] D. I. Kim and Y. Naito, "Broad-band design of improved hybrid-ring 3-dB directional couplers," *IEEE Trans. Microwave Theory Tech.*, vol. MTT-30, pp. 2040–2046, Nov. 1982.

[7] D. I. Kim and G. S. Yang, "Design of new hybrid-ring directional coupler using $\lambda/8$ or $\lambda/6$ section," *IEEE Trans. Microwave Theory Tech.*, vol. 39, pp. 1779–1783, Oct. 1991.

[8] J. L. B. Walker, "Improvements to the design of the 0°–180° rat-race coupler and its application to the design of balanced mixers with high LO to RF isolation," in *IEEE MTT-S Int. Microwave Symp. Dig.*, 1997, pp. 747–750.

[9] R. Levy and L. F. Lind, "Synthesis of symmetrical branch-guide directional couplers," *IEEE Trans. Microwave Theory Tech.*, vol. MTT-16, pp. 80–89, Feb. 1968.

[10] H. J. Riblet, "The application of a new class of equal-ripple functions to some familiar transmission-line problems," *IEEE Trans. Microwave Theory Tech.*, vol. MTT-12, pp. 415–421, July 1964.

[11] H. J. Carlin and W. Kohler, "Direct synthesis of band-pass transmission line structures," *IEEE Trans. Microwave Theory Tech.*, vol. MTT-13, pp. 283–297, May 1965.

[12] G. Ghione and C. U. Naldi, "Coplanar waveguides for MMIC applications: Effect of upper shielding, conductor backing, finite-extent ground planes, and line-to-line coupling," *IEEE Trans. Microwave Theory Tech.*, vol. MTT-35, pp. 260–267, Mar. 1987.

Chi-Yang Chang (S'88–M'95) was born in Taipei, Taiwan, R.O.C., on December 20, 1954. He received the B.S. degree in physics and the M.S. degree in electrical engineering from National Taiwan University, Taiwan, R.O.C., in 1977 and 1982, respectively, and the Ph.D. degree in electrical engineering from the University of Texas at Austin, in 1990.

From 1979 to 1980, he was with the Department of Physics, National Taiwan University, as a Teaching Assistant. From 1982 to 1988, he was with the Chung-Shan Institute of Science and Technology (CSIST) as an Assistant Researcher, where he was in charge of development of monolithic integrated circuits (MIC's), microwave subsystems, and millimeter-wave waveguide E-plane circuits. From 1990 to 1995, he returned to CSIST as an Associate Researcher, where he was in charge of the development of uniplanar circuits, ultra broad-band circuits, and millimeter-wave planar circuits. In 1995, he joined the Department of Communication, National Chiao-Tung University, Hsinchu, Taiwan, R.O.C., as an Associate Professor. His research interests include microwave and millimeter-wave passive and active circuit design, planar miniaturized filter design, and monolithic-microwave integrated-circuit (MMIC) design.

Chu-Chen Yang was born in Hualien, Taiwan, R.O.C., on May 20, 1967. He received the B.S. degree in chemical engineering from Tamkang University, Tansui, Taiwan, R.O.C., in 1989, and the M.S. degree in nuclear engineering from National Tsing-Hua University, Hsinchu, Taiwan, R.O.C., in 1991, and is currently working toward the Ph.D. degree at National Chiao-Tung University, Hsinchu, Taiwan, R.O.C.

From 1993 to 1995, he was with the Hexawave Company, Hsinchu, Taiwan, R.O.C., as a Microwave Engineer for MMIC design and measurement. His research interests include planar filter design and MMIC design.

IEEE TRANSACTIONS ON ELECTROMAGNETIC COMPATIBILITY, VOL. 44, NO. 4, NOVEMBER 2002

Balancing Networks for Symmetric Antennas—I: Classification and Fundamental Operation

James S. McLean, *Member, IEEE*

Abstract—**Emphasis has been placed on the quality of a metrology antenna system that allows it to be represented accurately using analytical or numerical models. Central to this capability is the efficacy of the balancing networks, sometimes referred to as balanced to unbalanced transformers (baluns). In this paper, we classify three fundamental types of baluns and show that all three can be derived from the 180°, four-port hybrid network. Balanced antennas driven from coaxial feed lines and operated in the presence of an asymmetric scatterer present an intrinsically unbalanced load to the balun. We show that in such situations the current balun is the only appropriate balun to employ. For numerous antenna metrology applications, in particular, site attenuation measurements, 180°, four-port hybrid networks have been employed as baluns. Network relations are derived relating the so-called voltage and current baluns to the 180°, four-port hybrid network. It is shown that, in addition to acting as a 0° or 180° power divider, the hybrid network can serve as either a voltage or a current balun depending on the termination at the isolated port. In contrast to the traditional approach, it is shown that when using a 180° hybrid as a balun, the isolated port should be terminated in such a way that the hybrid behaves as a current balun as opposed to a 180° power divider. This will yield experimental results that can be more readily reconciled with models.**

Index Terms—**Balancing network (balun), biconical antenna, calculable dipole, dipole, hybrid, site attenuation (SA).**

I. Introduction

SOME electromagnetic measurements such as normalized site attenuation (NSA) [1]–[3] are by nature difficult because they require the complete removal of the influence of the measurement antennas from the measured data. That is, NSA is defined as the NSA divided by the antenna factors of the transmit and receive antennas. Thus, the antenna factors of the measurement antennas must be very accurately known. Moreover, NSA measurements often involve operation of measurement antennas near a conducting ground plane, so near that the ground plane influences the current distribution on the antennas and, hence, their antenna factors. Given the inherent difficulty in accurately carrying out such measurements as SA, it is highly desirable to employ antennas for which accurate, reliable analytical or numerical models are available. Such antennas that are specifically designed to be able to be modeled accurately using limited analytical or numerical models have been termed "calculable" antennas [4]–[9]. Linear dipoles are the traditional reference standard in such measurements because of the degree of

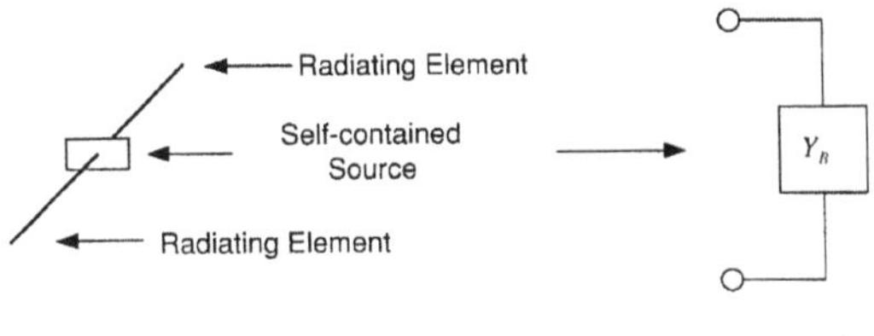

Fig. 1. Equivalent circuit of dipole in free-space with self-contained source.

certainty to which a linear dipole and its interaction with its environment (such as a ground plane) can be characterized analytically and numerically [4]. More recently, because good correlation between numerical predictions and measured results has been obtained [5], the wire-cage biconical broadband dipole antenna has also become accepted for such measurements. Here, we show that the ability of a symmetric or balanced antenna to provide repeatable measurements and to be accurately represented using a limited model is strongly dependent on the type of balancing network, sometimes referred to as balanced to unbalanced transformer (balun) used. When symmetric or balanced antennas such as linear or biconical dipoles are used for these exacting measurements, precision baluns of the appropriate type are necessary if the antennas are to be connected to external sources and/or receivers via coaxial transmission lines. The efficacy of the balun has an overriding influence on the accuracy of such measurements. Because the radiating elements themselves can be made to exceedingly tight tolerances, the balun is usually the principal contributor to any idiosyncratic or anomalous behavior of a symmetric or balanced antenna.

In Section II, we use a numerical simulation to illustrate and quantify the externally induced imbalance in two typical symmetric antennas caused by the presence of a ground plane. To the author's knowledge, this is the first time such results have been presented.

In Section III, we, for the first time, classify the three fundamental types of baluns, provide examples of each type and describe the relationship between them. In Section IV, we discuss the interaction of each balun type with symmetric antennas under generalized environmental conditions and determine the appropriateness of each fundamental balun type for use in various environmental conditions. We conclude that only the current balun should be used in situations in which externally induced imbalance exists. In Section V, we discuss the extensive, traditional use of 180°, four-port hybrids, especially so-called connectorized hybrids, as baluns. In Section VI, we show for the first time that all three fundamental types of baluns may be obtained from a 180°, four-port hybrid network by placing the proper termination at the sum port. We also present, for the first time, the scattering matrix of each fundamental balun

Manuscript received February 1, 2002; revised August 14, 2002. This work was supported in part by TDK Corporation, Cedar Park, TX.
The author is with TDK RF Solutions, Cedar Park, TX 78613 USA (e-mail: jmclean@tdkrf.com).
Digital Object Identifier 10.1109/TEMC.2002.804767

0018-9375/02$17.00 © 2002 IEEE

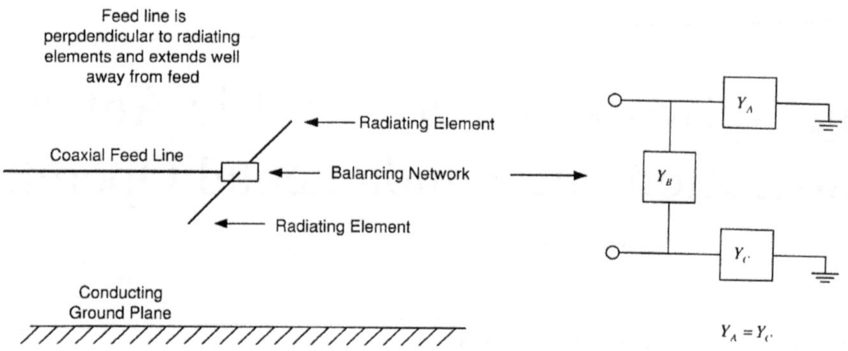

Fig. 2. Equivalent circuit of horizontally polarized dipole with coaxial feed line situated over a conducting ground plane ($Y_A = Y_C$).

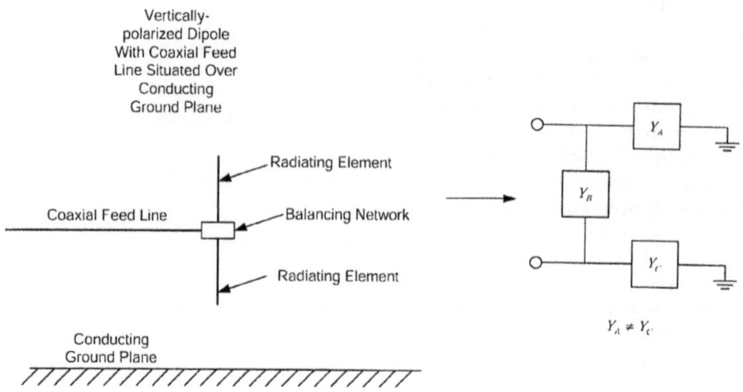

Fig. 3. Equivalent circuit vertically-polarized dipole with coaxial feed line situated over a conducting ground plane ($Y_A \neq Y_C$).

type. This facilitates comparison as well as the development of balun calibration procedures based on vector network analyzer techniques.

It is significant that of the three fundamental types of baluns, only the current balun is appropriate for use in situations in which externally induced imbalance exists such as when vertically-polarized symmetric antennas are operated over a conducting ground plane. This finding stands in stark contrast to the manner in which 180° hybrid networks have previously been configured for use as baluns for antennas.

II. Equivalent Circuits for Symmetric Antennas

In the absence of any scattering objects or mechanisms, symmetric antennas such as linear and biconical dipoles act as isolated two-terminal loads, as shown in Fig. 1. In the presence of a symmetric scatterer such as a perfectly flat ground plane (conducting or otherwise but homogeneous and isotropic in the lateral dimensions) parallel to the antenna, the dipole presents a perfectly symmetric load as shown in Fig. 2. However, in the presence of asymmetric scattering, the dipole presents a three-terminal, unbalanced load as shown in Fig. 3. Any antenna connected to a distant source or receiver via a transmission line must be described by Fig. 2 or 3, depending on the orientation of the feed and the antenna. Some additional discussion of this most general situation shown in Fig. 3 is presented in [10]. Only a self-contained source such as a so-called reference radiator [11] can be represented by the isolated load shown in Fig. 1. The principal concern here is a dipole antenna fed by a coaxial feed line and oriented for vertical polarization over a ground plane. This is a typical arrangement for NSA measurements for open area test sites (OATS) [1]–[3]. Such sites utilize a metallic, conducting ground plane because it provides a well-behaved, consistent reflection in almost all weather conditions.

In Figs. 2 and 3, the two terminals can be taken as the feed ends of the two dipole elements. However, the ground terminal is more difficult to conceptualize. Instead of ground terminal, it is perhaps better to simply refer to this terminal as the reference terminal. Because of the retardation of currents, it is necessary that all three terminals including the reference terminal be located electrically close to each other. That is, the distance between any two terminals must be a small fraction of a wavelength. The approach taken here is to take the reference terminal of the equivalent circuit as the outer surface of the shielded balun at a point electrically near the feed. This is made clear in the following discussions.

A. Simulations of Externally Induced Imbalance

Because the externally induced imbalance is related to feed transmission line arrangement as well as the antenna, there are infinitely many possible examples of this externally induced imbalance. However, we present two situations which are similar to those that might be encountered in performing NSA measurements with vertically-polarized fields on an OATS. The two situations were modeled using the numerical electromagnetics code (NEC-2), a thin-wire method-of-moments simulation program which is very well adapted for the simulation of wire antennas and has been verified many times.

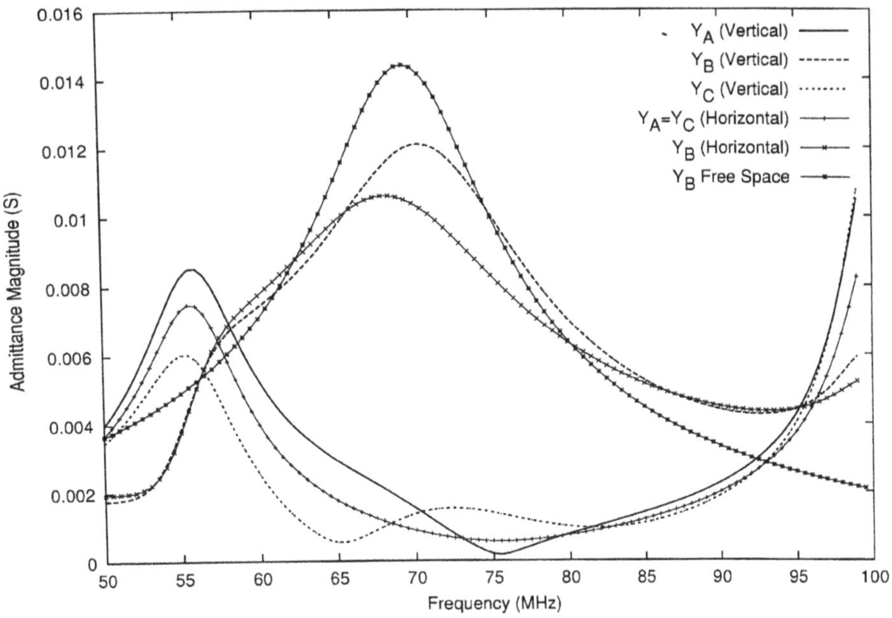

Fig. 4. Magnitude of Π equivalent network elements for linear dipole in free space, horizontal polarization over conducting ground, and vertical polarization over conducting ground.

1) 2-m Linear Dipole Antenna: The first simulation involved a center-fed linear dipole of length 2.0 m and diameter 1 cm. The simulation predicted that the dipole is resonant just below 71 MHz in free space and the input admittance is about 1.4×10^{-2} S at resonance. These values for resonance frequency and resonant admittance are in good agreement with other moment method based analyses of linear dipole antennas such as [13]. Thus, at resonance, the elements of the equivalent Π network for the antenna in free space are $Y_A = 0$ S, $Y_B = 1.4 \times 10^{-2} + j0$ S, and $Y_C = 0$ S. If the same dipole is oriented horizontally, located 1.5 m above a conducting ground plane and fed from a feed line of the same diameter as the dipole which extends normally and horizontally to a point 1.5 m away from the center of the antenna and then vertically down to the ground plane, the elements of the Π network at 71 MHz become $Y_A = 7.55 \times 10^{-4} - j4.69 \times 10^{-4}$ S, $Y_B = 1.02 \times 10^{-2} + j1.76 \times 10^{-3}$ S, and $Y_C = 7.55 \times 10^{-4} - j4.69 \times 10^{-4}$ S. If the dipole is then oriented vertically with the center at the same height with the same feed arrangement, the elements of the Π network at 71 MHz become $Y_A = 8.60 \times 10^{-5} - j1.22 \times 10^{-3}$ S, $Y_B = 1.19 \times 10^{-2} + j1.79 \times 10^{-3}$ S, and $Y_C = 1.32 \times 10^{-3} + j6.76 \times 10^{-4}$ S, Thus, the disparity between elements Y_A and Y_C is significant even for a linear dipole at resonance with the magnitude of Y_A being about 80% of that of Y_C. The frequency dependence of these equivalent circuit elements is shown in Figs. 4 and 5 for three situations, free space, horizontal polarization over a ground plane, and vertical polarization over a ground plane. Note that, for horizontal polarization, the graphs of the network elements Y_A and Y_C overlay each other showing the perfect balance that exists in this situation. As can be seen, the imbalance in vertical polarization is large between 50 and 75 MHz. The frequency dependence of the equivalent network elements is quite complicated and depends on the height of the antenna over the ground plane as well as the length of the horizontal section of feed line. However, at the series resonance, the magnitude of element Y_B is significantly larger than that of either Y_A or Y_C, making the overall imbalance small at this frequency. Thus, we can conclude that when the linear dipole is operated very near its series resonance as is usually done, imbalance induced by the ground plane is small.

2) 1.3-m Wire-Cage Biconical Antenna: The second simulation is of a typical wire-cage biconical antenna, such as is almost universally used for EMC applications in the frequency range of 20–200 MHz. The geometry of the wire-cage biconical antenna was chosen to match as closely as possible that of [12] in which a wire-cage biconical antenna operating in a free-space environment was thoroughly analyzed. Where possible, we have used the same segmentation as [12]. The simulation performed here agrees with that of [12] for the free-space situation. The free-space fundamental series resonance of the biconical antenna occurs very near 72.6 MHz and, thus, is similar to that of the 2-m linear dipole examined here. Also, for simulations including the feed structure and a ground plane, we have used a feed-line arrangement identical to the one used with the linear dipole. The center of the antenna is 1.5 m above the ground plane, and the feed line extends horizontally 1.5 m from the antenna at which point, it extends vertically down to the ground plane.

The simulated values for the Π network elements are shown in Figs. 6–8. Again, for horizontal polarization, the curves for Y_A and Y_C overlay each other. The behavior of the biconical antenna is similar to that of the linear dipole. However, it must be kept in mind that the biconical antenna is used over a much broader frequency range, generally 20–200 MHz. Below the series resonance of the antenna, near 55 MHz, the magnitudes of Y_A and Y_C are of the same order as that of Y_B. In this frequency

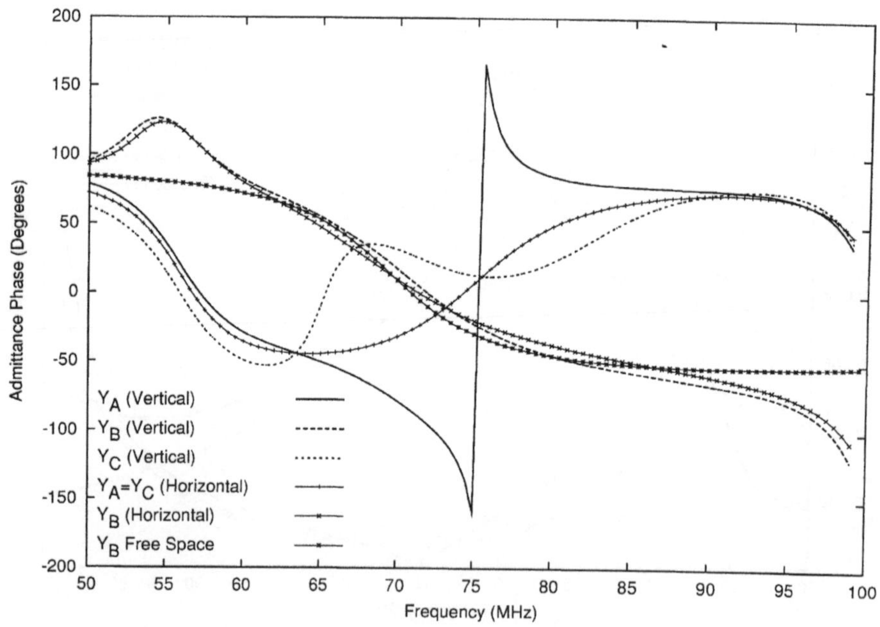

Fig. 5. Phase of Π equivalent network elements for linear dipole in free space, horizontal polarization over conducting ground, and vertical polarization over conducting ground.

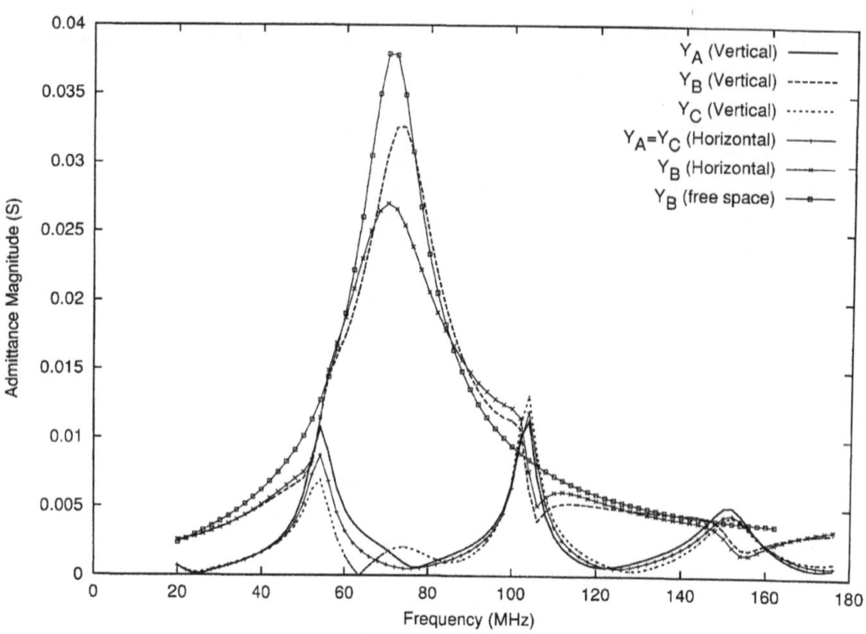

Fig. 6. Magnitude of Π equivalent network elements for biconical dipole in free space, horizontal polarization over conducting ground, and vertical polarization over conducting ground.

range, the disparity between Y_A and Y_C for vertical polarization might cause imbalance. Again, the frequency dependence of the network elements is quite complicated and depends both on the height of the antenna over the ground plane as well as the length of the horizontal section of the feed line.

From Fig. 8, it can be seen that Y_A and Y_C have negative phase angles below 30 MHz. Thus, these equivalent circuit elements are inductive. The reason for this is that the common mode circuit, consisting of driving both dipole elements in phase but against the feed line, exhibits a fundamental series resonance near 15 MHz. Because of the length of the feed line, the common-mode equivalent structure is very large.

Finally, the phase behavior is quite complicated. In Fig. 8, it was necessary to "unwrap" the phase of Y_C. The phase slope appears to be discontinuous near 50 MHz. It is not, however; it is simply very steep. This corresponds to a very sharp nulling in the magnitude of Y_C. In Fig. 5, the phase of Y_C was not unwrapped. Of course, the action of unwrapping (subtracting 360° phase) does nothing to change the quantity. However, it makes it easier to see that for either antenna the parameters Y_A

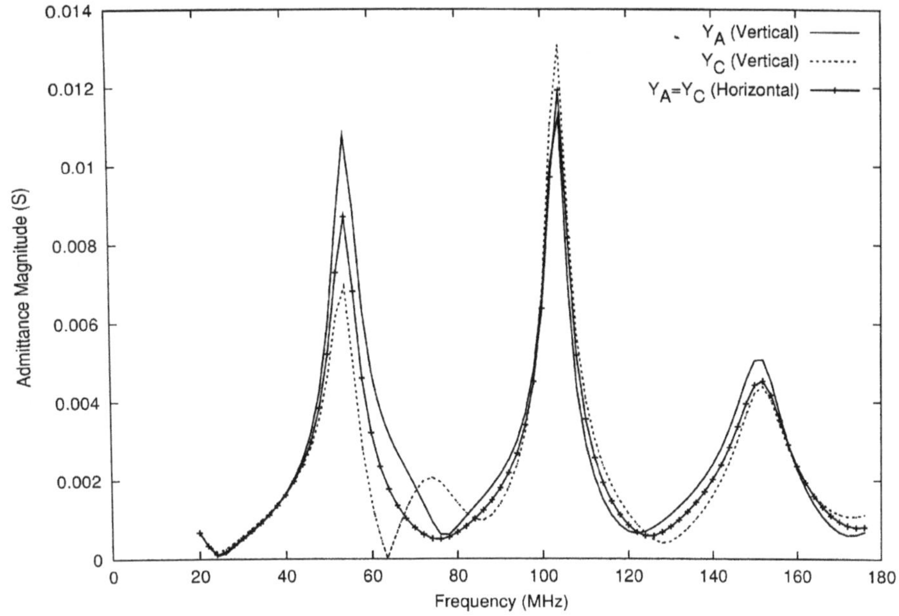

Fig. 7. Magnitude of Π equivalent network elements Y_A and Y_C only for biconical dipole in free space, horizontal polarization over conducting ground, and vertical polarization over conducting ground. Expanded vertical scale is used to elucidate difference in Y_A and Y_C for vertically polarized antenna.

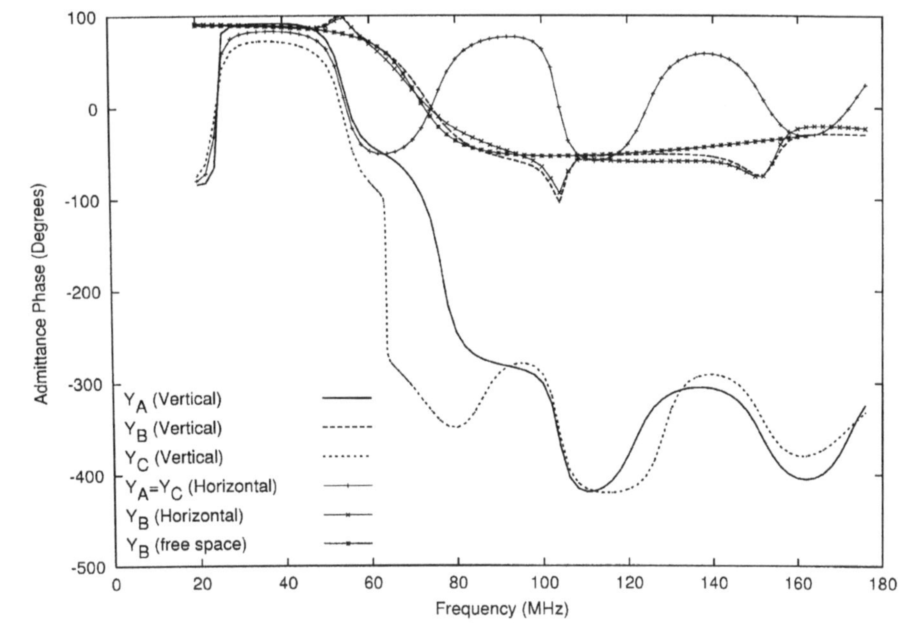

Fig. 8. Phase of Π equivalent network elements for biconical dipole in free space, horizontal polarization over conducting ground, and vertical polarization over conducting ground.

and Y_C tend to become equal both in magnitude and phase at higher frequencies.

The numerical simulations provided here are merely intended to provide some insight about the imbalance induced in a vertically-polarized antenna operating over a ground plane. Obviously, more investigation is needed in order to provide comprehensive prediction of the effects of feed line configurations on such measurements.

III. Classification of Baluns

Baluns are universally employed to drive intrinsically balanced antennas such as linear or biconical dipole antennas from coaxial or single-ended sources. The efficacy of the balun determines the symmetry of the antenna's radiation pattern. Moreover, the presence of common-mode current on the feed transmission line can result in cross-polarized radiation [14]. While

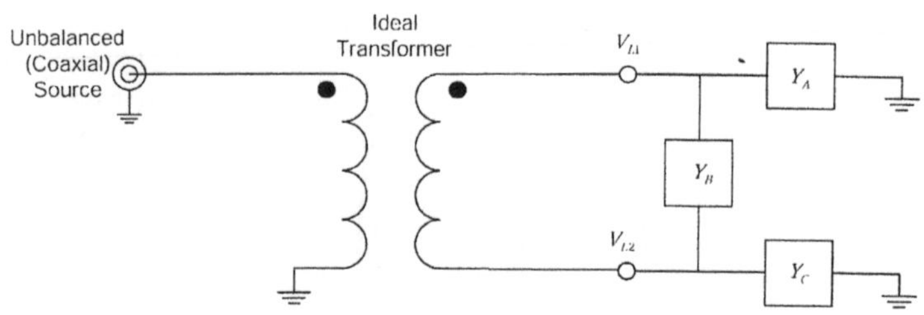

Fig. 9. Ideal transformer employed as a current balun.

the effect of unbalance can be quantified for any specific case [15], the presence of common-mode current on the feed transmission line destroys the repeatability of a measurement since the radiated fields then depend not only on the antenna but also the arrangement of the feed transmission line [16]. In the literature, there exists mention of so-called voltage baluns as well as current baluns [17], [18]. Although strict definitions (the definition of the term balun is noticeably absent from the 1983 IEEE Standard Definitions for Antennas) have not been given, it can be deduced that a voltage balun attempts to enforce a balancing of voltages with respect to a particular reference while a current balun attempts to enforce a balancing of currents (see references [17] and [18]). Practical baluns may or may not fall into either of these classes.

While the first baluns were designed for feeding balanced antennas such as dipoles [19], modern uses for baluns include balanced mixers and push–pull amplifiers. These latter applications are far more prevalent than the first. Moreover, baluns acceptable for use in push–pull amplifiers or balanced mixers may not be appropriate for driving balanced antennas. In most applications, a balun in a push–pull amplifier or balanced mixer "sees" a symmetric load similar to the antenna described in Fig. 2.

Finally, it is inappropriate to discuss balun performance without quantifying the balancing behavior. Common performance tests often involve connecting baluns "back-to-back" and measuring the insertion loss of the cascaded networks. Unfortunately, this test is not adequate for the quantification of performance. The author of [20] first suggested a Π network is necessary to elucidate a balun's behavior. However, use of a more generalized asymmetic Π load to classify baluns was not considered in this reference.

A. Current Baluns

The action of the ideal or canonical current balun can be described using an ideal transformer as shown in Fig. 9. The currents on the balanced side are equal in magnitude and opposite in phase, regardless of the potentials at the output terminals with respect to the ground connection on the unbalanced side.

The most common type of practical RF current balun is the common-mode choke [17]. Such a balun is usually implemented at HF and VHF frequencies by winding a bifilar winding or transmission line around a magnetic core. The action of the current balun combined with a nonunity impedance transformation ratio may be realized in a Guanella, or equal-delay transformer [17]. Other types of current baluns, often employed at UHF and above, are the Harvard Radio Research Laboratories (RRL) Type I, II, and III baluns [19] as well as the Marchand balun [21] and [22]. The Marchand balun is similar to the Harvard RRL Type II balun, except for an additional series-compensating stub. All of these baluns behave as current baluns to the extent that parasitic effects are insignificant. While the balun employed in the Roberts dipole [23] and [24] appears very similar to the Marchand balun, its operation is significantly different; instead of choking off the common mode, the operation hinges on conversion of common-mode energy to a transmission-line mode which is, in turn, choked off. It is interesting to notice that, if this conversion process is effective, the balun will operate well away from the center frequency at which the parallel stub provides an open circuit at the antenna terminals. Nevertheless, the Roberts balun is also a current balun. Another common design, the lattice balun [25] or double-Y balun [26], [27], can be shown to behave as a current balun when the reactances in the lattice arms are equal in magnitude. In fact, in this case, the lattice network can be shown to be the equivalent of an ideal transformer. However, despite the fact that the lattice network upon which such a balun is based is an all pass network, the current balancing behavior can really only be perfect at a single frequency for lumped element realization or an infinite set of harmonic frequencies for a commensurate distributed structure (lumped elements replaced with open-circuited and short-circuited transmission lines).

B. Voltage Baluns

The voltage balun maintains the voltages at the output terminals (with respect to ground) equal in magnitude and opposite in phase, regardless of the load. The idealized or canonical voltage balun is an ideal transformer with a center-tapped secondary winding with the center tap being connected to ground, as shown in Fig. 10. A common type of voltage balun often used at HF through VHF is the bootstrap or Ruthroff balun [28]. This balun is also usually realized using a bifilar winding wrapped around a magnetic core. Another common voltage balun can be implemented using two bifilar windings, one as a phase inverter and one as a delay line [17].

C. 180° Power Dividers

The third fundamental type of balun is sometimes referred to (especially in product literature) as a 180° power divider. Such a device can be thought of as fixing the voltage and current associated with the outward-traveling waves at the two output ports to be equal in magnitude and opposite in phase. Hence, it is called a 180° power divider. Strictly speaking, this type of balun fixes a linear combination of the voltage and current at the two output ports to be equal in magnitude and opposite in phase. Such a

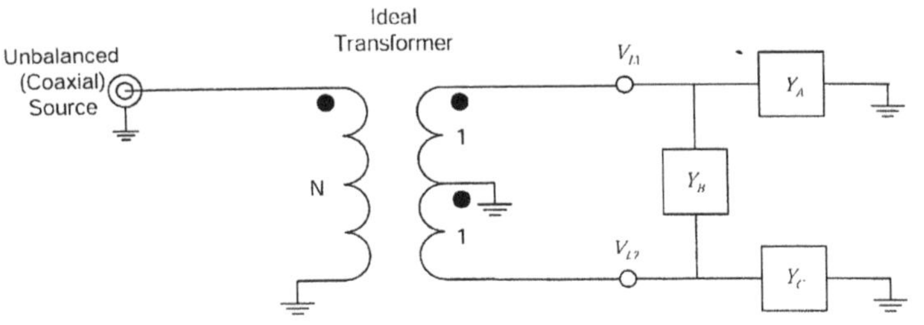

Fig. 10. Ideal transformer employed as a voltage balun.

device can be derived from a 180°, four-port hybrid when the sum port is terminated in a matched resistive load. In general, it is possible to make a device which fixes a nearly arbitrary linear combination of voltage and current at the output ports to be equal in magnitude and opposite in phase, provided some restriction is made on the real part of the constant of proportionality. Thus, all three types of baluns can be derived from this most general form. This will be discussed further in conjunction with hybrid networks.

IV. Baluns for Symmetric Antennas

Having cataloged balun types and developed equivalent circuits for antennas, it is now possible to discuss which balun type is most appropriate for a symmetric antenna. As long as the load in question is isolated from ground, or is perfectly symmetric with respect to ground, there is no difference between the behaviors of these three devices. That is, for the symmetric grounded load, all three devices will provide output voltages equal in magnitude and opposite in phase with respect to the ground, and all three devices will drive currents equal in magnitude and opposite in phase. However, when a more generalized asymmetric load is considered, the behavior of the current balun, voltage balun and the terminated 180° hybrid differ significantly.

When faced with an unbalanced load such as a vertically-polarized dipole situated over a conducting ground plane, the voltage balun will maintain balanced voltages, the current balun will maintain balanced currents, and the 180° power divider will provide equal forward power into each port.

A great deal of emphasis is put on the reconciliation of measurements and numerical simulations. Aside from increasing confidence in measurements, numerical, or analytical models are extremely useful for isolating physical phenomena. For example, numerical simulation can be used to determine if dissipation loss in a ground plane is significant simply by performing the simulation twice, once including the effects of dissipation and once without. Because of limitations in most simulations, it is necessary to take steps to make the antennas is such a way that they can be accurately represented in numerical models. Some authors have coined the adjective "calculable" in reference to an antenna that can be accurately represented in a straightforward manner by an analytical or numerical model [4]. The NEC (NEC-2, NEC-3, NEC-4, and the numerous commercial versions of the code) represents possibly the most verified numerical model in existence. It has been used to model linear dipoles as well as wire cage biconical dipoles. In particular, Chen [5] has developed a wire cage biconical antenna that is intentionally designed to be able to be accurately modeled using the NEC code or other thin wire method of moments model. That is, the design avoids the inclusion of physical features that are known not to be able to be accurately represented due to limitations in the NEC model.

In any "calculable" antenna, careful attention must be paid to the balun. Fitzgerrall proposed using a 180°, four-port hybrid as a balun for a linear dipole antenna. These dipoles were used as standards from 30–1000 MHz. Later, in the specification CISPR 16-1 [34], a similar arrangement was called out for a reference dipole. In the CISPR 16-1 specification, the term "calculable" dipole was used to describe the linear reference dipole. The performance of this dipole was intended to be predictable via the quasi-analytical model given in the specification or a numerical model such as NEC. This dipole design was intended to be used to calibrate a test site that, in turn, was to be used to calibrate other antennas. The test site was therefore referred to as a calibrated test site (CALTS). Chen [5] used a 180° hybrid as a balun for the broadband wire biconical dipole mentioned above. He also terminated the sum port of the hybrid in a matched load. While Chen achieved excellent agreement between experimentally obtained data and a numerical model for both horizontal and vertical polarizations, he noted that the horizontal polarization experiments consistently provided better agreement with the numerical model.

There are many possible source formulations available for numerical simulation. However, when the simulation of the antenna does not include the feed transmission line (that is, a representation of the common mode of the feed transmission line) the simulation inherently implies a current balance at the feed. The most widely verified source formulation in NEC consists of a single delta gap source that is placed across the feed of a linear dipole antenna or wire-cage biconical dipole antenna. Here again, without the inclusion of some representation of the feed line, it is implicit that a current balance exists at the feed. Thus, the most appropriate balun to employ in measurements would be a current balun. This is especially true for vertical polarization measurements. As noted earlier, in a vertical polarization NSA test, the ground plane influences the currents on the antennas in an asymmetric fashion. Again, it is noted that the CISPR-16 CALTS document does not call for vertical polarization measurements. Nevertheless, vertically polarized NSA measurements are necessary as emissions tests (the principal

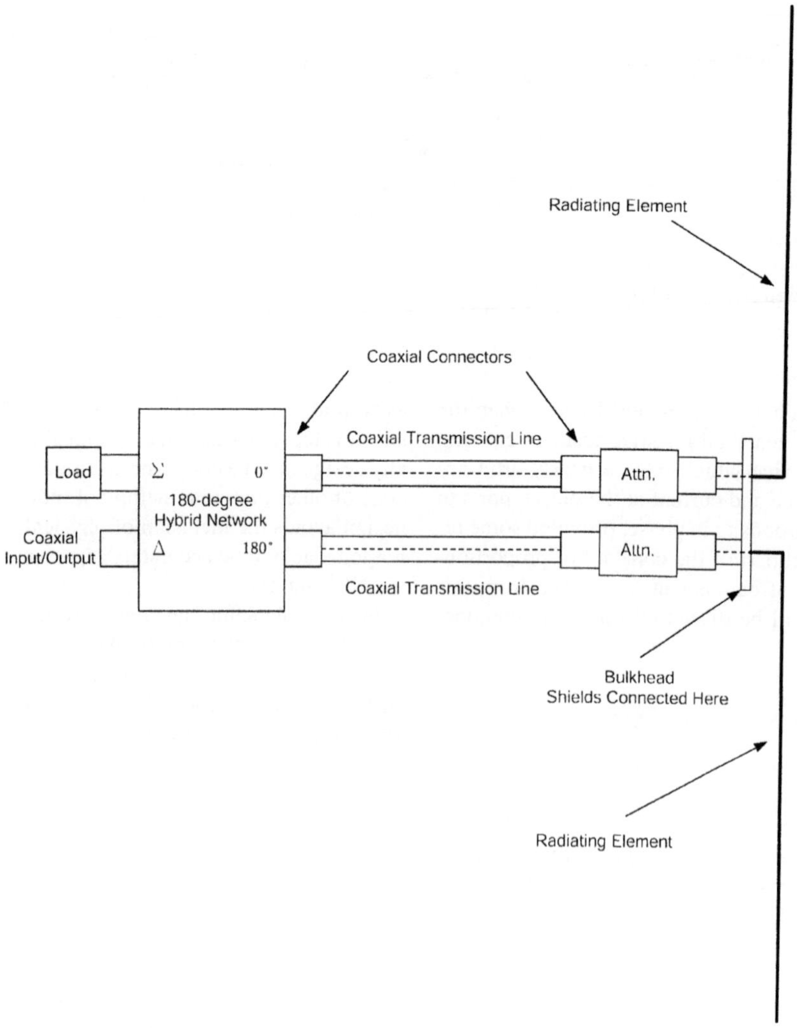

Fig. 11. Balancing approach taken in CISPR 16-1 (CALTS).

function of an OAT) must be performed for both horizontal and vertical polarizations.

Of course, it is possible to develop a comprehensive numerical model of an imperfect balun and a vertically polarized antenna operating over ground (conducting or otherwise) as was done in [15]. However, because such a model would necessarily have to include the effects of current on the feed line, the model would be extremely specific to a particular experimental arrangement. That is, the exact placement of feed lines, mounting hardware, etc., would have to be duplicated in the model. Moreover, the experimental and numerical results could be quite sensitive to the feed line arrangement. Thus, such an approach is not very desirable for measurements such as NSA. Nevertheless, an approach for the rigorous numerical modeling of antennas driven with all three classes of baluns mentioned earlier is presented using the analysis developed in the next section.

By enforcing a current balance at the feed of a linear or biconical dipole antenna, the current balun allows a sharp demarcation between the radiating and nonradiating portions of the system. Thus, the remainder of the system can be modeled using circuit theory and calibrated as a nonradiating network. This is not true if a current balance does not exist at the antenna feed.

V. Traditional Use of 180°, Four-Port Hybrid Networks as Baluns

For antennas [4]–[8], [29]–[33], especially for metrology applications with exacting tolerance requirements, 180°, four-port hybrid networks have been employed extensively as baluns. Apparently, Dyson [29], [30] was the first to formalize the use of coaxial, 180°, four-port hybrid networks for making measurements with balanced transmission line systems. One compelling reason for doing this is that precise, highly refined devices are commercially available. These devices represent years of engineering research and optimization and, thus, it is attractive to the metrologist to leverage off of this effort. Moreover, if it is possible to employ a device with standard 50-Ω coaxial ports such as SMA or APC 3.5 connectors, modern vector network analyzers with error correction techniques may be used to calibrate the system [34]. Thus, the entire nonradiating portion of

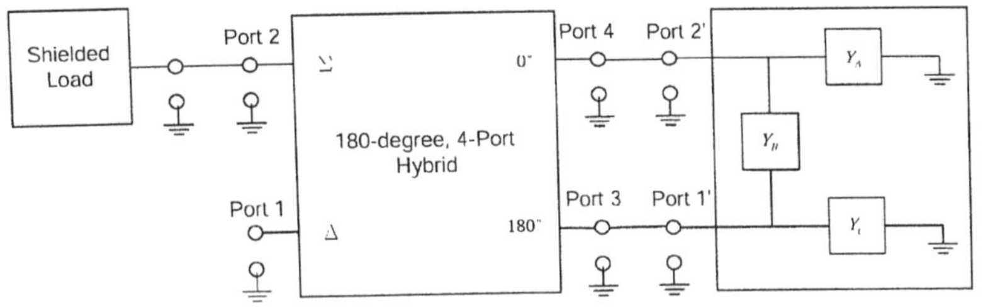

Fig. 12. Interconnection of 180°, four-port hybrid network and network representation of antenna.

the system may be characterized using the network analyzer, leaving only the radiating elements to be considered. The capability to employ a vector network analyzer in this manner is a great advantage because the requirement that feed lines be routed away from the antennas such that they remain orthogonal to the radiated field for a distance that is significant in terms of wavelengths necessarily requires that long lines and possibly multiple connections are involved. The effects of these long lines and connectors can be accurately removed from measured data using a vector network analyzer.

To employ a 180°, four-port hybrid as a balun, the delta port of the hybrid is connected to the single-ended portion of the system (source or receiver), and the 0° and 180° ports are connected as shown in Fig. 11 to the balanced antenna. Note that it is essential to have the connection between the shields of the 0° and 180° port to allow the shield current to circulate. The sum port is the isolated port in this case and has been traditionally terminated in a matched load [4]–[8], [29]–[33]. This causes the hybrid to behave as a 180° power divider as discussed above.

Here, we show that by proper selection of the load at the isolated port of the 180°, four-port hybrid, the network may act as: 1) a 180° power splitter; 2) a voltage balun, or 3) a current balun. It is shown in the next section that the current balun is the most appropriate balun topology to employ in order to reconcile measured data with numerically simulated data. Therefore, the isolated port should be terminated so that current balun behavior is obtained. This is in contrast with all previously published approaches for utilizing 180°, four-port hybrids as baluns [4]–[8], [29]–[33].

VI. Ideal 180° Hybrid Network as a Balun

The scattering matrix for an ideal 180° hybrid network exhibiting perfect isolation between ports 1 and 2, and ports 3 and 4, perfect match at all four ports, perfect amplitude tracking, and perfect phase tracking, can be written as follows [35]:

$$\begin{pmatrix} b_1 \\ b_2 \\ b_3 \\ b_4 \end{pmatrix} = \frac{1}{\sqrt{2}} \begin{pmatrix} 0 & 0 & -1 & 1 \\ 0 & 0 & 1 & 1 \\ -1 & 1 & 0 & 0 \\ 1 & 1 & 0 & 0 \end{pmatrix} \begin{pmatrix} a_1 \\ a_2 \\ a_3 \\ a_4 \end{pmatrix} \quad (1)$$

where $a_1 \ldots a_4$ are the incoming complex wave amplitudes and $b_1 \ldots b_4$ are the outgoing complex wave amplitudes [35]. Here, ports 1 and 3 are the Δ- or E-ports, and ports 2 and 4 are the $\sum$- or H-ports. For operation as a balun, port 1 is used as the input (connection to single-ended coaxial system) port, port 2 is then the isolated port, and ports 3 and 4 are used as the 180° and 0° output ports, respectively. The two center conductors of ports 3 and 4 are taken as the balanced port, and the two shields of these two ports are connected together.

The most general representation of a symmetric antenna operating over a ground plane is the three-terminal network shown in Fig. 3. This three-terminal network may be taken as a two-port network and represented with the following scattering matrix:

$$\begin{pmatrix} b'_1 \\ b'_2 \end{pmatrix} = \begin{pmatrix} S'_{11} & S'_{12} \\ S'_{21} & S'_{22} \end{pmatrix} \begin{pmatrix} a'_1 \\ a'_2 \end{pmatrix}. \quad (2)$$

When the antenna elements are connected to the center conductors of the coaxial ports 3 and 4 and the shields of ports 3 and 4 are connected together (with an electrically short connection length), the interaction of the antenna and hybrid may be represented by the interconnection of networks shown in Fig. 12. Note that the two-port representation of the antenna provides coupling between ports 3 and 4. For the very general situation of an antenna coupled asymmetrically to a ground plane, the two-port network is reciprocal but not symmetric. Thus, while $S'_{21} = S'_{12}$, $S'_{11} \neq S'_{22}$.

A generalized (but isolated) complex load at port 2 may be represented by a single complex reflection coefficient Γ_2. This representation does not include coupling from ports 3 and 4 back to the sum port, but such coupling cannot occur if the sum port is completely shielded.

Referring to Fig. 12, it can be seen that $a_3 = b'_1$, $a_4 = b'_2$, $a'_1 = b_3$, and $a'_2 = b_4$. Thus, the incoming wave amplitudes at ports 2, 3, and 4 can be written in terms of the outgoing wave amplitudes as follows:

$$a_2 = \Gamma_2 b_2 \quad (3)$$

$$a_3 = S'_{11} b_3 + S'_{12} b_4 \quad (4)$$

$$a_4 = S'_{21} b_3 + S'_{22} b_4. \quad (5)$$

By combining these equations with the four-port scattering matrix for the ideal hybrid network, the outgoing wave amplitudes at all four ports may be written in terms of the incoming wave amplitude at port 1

$$\frac{b_1}{a_1} = \frac{1}{2}\left[\frac{(-S'_{12} + S'_{22} + S'_{11} - S'_{21} - 2\Gamma_2 \Delta_S)}{1 - \frac{\Gamma_2}{2}(S'_{11} + S'_{12} + S'_{21} + S'_{22})}\right] \quad (6)$$

$$\frac{b_2}{a_1} = \frac{1}{2}\left[\frac{(S'_{12} + S'_{22} - S'_{11} - S'_{21})}{1 - \frac{\Gamma_2}{2}(S'_{11} + S'_{12} + S'_{21} + S'_{22})}\right] \quad (7)$$

$$\frac{b_3}{a_1} = \frac{-1}{\sqrt{2}}\left[\frac{[1 - \Gamma_2 (S'_{12} + S'_{22})]}{1 - \frac{\Gamma_2}{2}(S'_{11} + S'_{12} + S'_{21} + S'_{22})}\right] \quad (8)$$

$$\frac{b_4}{a_1} = \frac{1}{\sqrt{2}}\left[\frac{[1-\Gamma_2(S'_{11}+S'_{21})]}{1-\frac{\Gamma_2}{2}(S'_{11}+S'_{12}+S'_{21}+S'_{22})}\right]. \quad (9)$$

Now, the terminal or total voltages at ports 3 and 4 are given by

$$V_3 = a_3 + b_3 \quad (10)$$

$$\frac{V_3}{a_1} = \frac{1}{\sqrt{2}}\left[\frac{S'_{12}-S'_{11}-1+\Gamma_2(\Delta_S+S'_{12}+S'_{22})}{1-\frac{\Gamma_2}{2}(S'_{11}+S'_{12}+S'_{21}+S'_{22})}\right] \quad (11)$$

$$V_4 = a_4 + b_4, \text{ and} \quad (12)$$

$$\frac{V_4}{a_1} = \frac{-1}{\sqrt{2}}\left[\frac{S'_{21}-S'_{22}-1+\Gamma_2(\Delta_S+S'_{21}+S'_{11})}{1-\frac{\Gamma_2}{2}(S'_{11}+S'_{12}+S'_{21}+S'_{22})}\right] \quad (13)$$

where $\Delta_S = S'_{11}S'_{22} - S'_{12}S'_{21}$. The terminal or total currents into ports 3 and 4 are given by

$$I_3 = \frac{1}{Z_0}(a_3 - b_3) \quad (14)$$

$$\frac{I_3}{a_1} = \frac{1}{\sqrt{2}Z_0}\left[\frac{S'_{12}-S'_{11}+1+\Gamma_2(\Delta_S-S'_{12}-S'_{22})}{1-\frac{\Gamma_2}{2}(S'_{11}+S'_{12}+S'_{21}+S'_{22})}\right] \quad (15)$$

$$I_4 = \frac{1}{Z_0}(a_4 - b_4) \quad (16)$$

$$\frac{I_4}{a_1} = \frac{-1}{\sqrt{2}Z_0}\left[\frac{S'_{21}-S'_{22}+1+\Gamma_2(\Delta_S-S'_{21}-S'_{11})}{1-\frac{\Gamma_2}{2}(S'_{11}+S'_{12}+S'_{21}+S'_{22})}\right]. \quad (17)$$

From the above equations, we can see that making $\Gamma_2 = 0$ (matched load) results in the outgoing wave amplitudes, b_3 and b_4, to be equal in amplitude and opposite in phase. That is, $b_3/a_1 = -1/\sqrt{2}$ and $b_4/a_1 = 1/\sqrt{2}$. Thus the hybrid behaves as a so-called 180° power divider when the isolated port is terminated in a matched load. In this case, neither the terminal voltages nor the currents are balanced in the case of an asymmetric load. Also, from the above equations, we can see that making $\Gamma_2 = -1$ (short circuit) causes the terminal voltages, V_3 and V_4, to be equal in magnitude and opposite in phase regardless of the values of the admittances and Y_A and Y_C. Thus, in the case of an unbalanced load, the 180° hybrid with the isolated port terminated in a short circuit will cause the two output voltages to be equal in magnitude and opposite in phase

$$\frac{V_3}{a_1} = -\frac{1}{\sqrt{2}}\left[\frac{S'_{11}+S'_{22}+1+\Delta_S}{1+\frac{1}{2}(S'_{11}+S'_{12}+S'_{21}+S'_{22})}\right] \quad (18)$$

$$\frac{V_4}{a_1} = \frac{1}{\sqrt{2}}\left[\frac{S'_{11}+S'_{22}+1+\Delta_S}{1+\frac{1}{2}(S'_{11}+S'_{12}+S'_{21}+S'_{22})}\right]. \quad (19)$$

The currents, however, will not be balanced, and common mode current will exist on the outer surface of the shielded hybrid and the feed cable. On the other hand, making $\Gamma_2 = 1$ (open circuit) causes the terminal currents, I_3 and I_4 to be equal in magnitude and opposite in phase, regardless of the values of the admittances Y_A and Y_C. In this case, the currents will be balanced, and no common mode current will flow even if the load is asymmetric

$$\frac{I_3}{a_1} = -\frac{1}{\sqrt{2}Z_0}\left[\frac{S'_{11}+S'_{22}-1-\Delta_S}{1-\frac{1}{2}(S'_{11}+S'_{12}+S'_{21}+S'_{22})}\right] \quad (20)$$

$$\frac{I_4}{a_1} = \frac{1}{\sqrt{2}Z_0}\left[\frac{S'_{11}+S'_{22}-1-\Delta_S}{1-\frac{1}{2}(S'_{11}+S'_{12}+S'_{21}+S'_{22})}\right]. \quad (21)$$

If we consider the coupled coil implementation of a 180° hybrid, the so-called hybrid coil, the foregoing results can be readily visualized since they correspond to opening or shorting the center tap of the secondary winding of the coil [36]. Note that the hybrid coil configuration shown in [36, Fig. 7.12] always requires different port impedances to provide hybrid action. That is, in order to obtain the exact equivalent of a four-port 180° hybrid with all four ports having the same terminating impedances, another transformer is required at the center-tap, sum port. Nevertheless, it can immediately be seen that a voltage or current balun can be obtained by short circuiting or open circuiting the sum port.

Note that if we simply take the scattering equations describing the 180°, four-port hybrid, assume an open circuit termination at port 2, and reduce the network to a three-port we obtain the following scattering matrix:

$$\begin{pmatrix} S_{11} & S_{12} & S_{13} \\ S_{21} & S_{22} & S_{23} \\ S_{31} & S_{32} & S_{33} \end{pmatrix} = \begin{pmatrix} 0 & -\frac{1}{\sqrt{2}} & \frac{1}{\sqrt{2}} \\ -\frac{1}{\sqrt{2}} & \frac{1}{2} & \frac{1}{2} \\ \frac{1}{\sqrt{2}} & \frac{1}{2} & \frac{1}{2} \end{pmatrix}. \quad (22)$$

Thus, the 180°, four-port hybrid behaves as an ideal transformer as shown in Fig. 9 with a $\sqrt{2}:1$ turns ratio or a 2:1 impedance ratio. The three-port scattering matrix for an ideal transformer with an $n:1$ turns ratio arranged as shown in Fig. 9 is given by

$$\begin{pmatrix} S_{11} & S_{12} & S_{13} \\ S_{21} & S_{22} & S_{23} \\ S_{31} & S_{32} & S_{33} \end{pmatrix} = \begin{pmatrix} \frac{2n^2-1}{2n^2+1} & -\frac{2n}{2n^2+1} & \frac{2n}{2n^2+1} \\ -\frac{2n}{2n^2+1} & \frac{1}{2n^2+1} & \frac{2n^2}{2n^2+1} \\ \frac{2n}{2n^2+1} & \frac{2n^2}{2n^2+1} & \frac{1}{2n^2+1} \end{pmatrix}. \quad (23)$$

This is the most general form of the scattering matrix for a current balun. The explicit scattering matrix is useful in the synthesis of current baluns. It is also useful to note that the lattice [25] and double-Y baluns [26], [27] have exactly the same form of scattering matrix as does the current balun. If the magnitudes of the arm reactances are equal to the port impedances, the turns ratio is $1:\sqrt{2}$. This confirms that the lattice and double-Y baluns behave as current baluns at least over a narrow bandwidth. In fact, lumped element, 180° hybrids directly synthesized from transmission parameters as presented in [37] can be shown to be simply an interconnection of lattice baluns with the interconnection being exactly the same as that required for the implementation of a coupled-coil hybrid.

The three-port scattering matrix of an ideal transformer with turns ratio $n:1$ when arranged as a voltage balun as in Fig. 10 is given by

$$\begin{pmatrix} S_{11} & S_{12} & S_{13} \\ S_{21} & S_{22} & S_{23} \\ S_{31} & S_{32} & S_{33} \end{pmatrix} = \begin{pmatrix} \frac{2n^2-1}{2n^2+1} & -\frac{2n}{2n^2+1} & \frac{2n}{2n^2+1} \\ -\frac{2n}{2n^2+1} & \frac{-1}{2n^2+1} & \frac{-2n^2}{2n^2+1} \\ \frac{2n}{2n^2+1} & \frac{-2n^2}{2n^2+1} & \frac{-1}{2n^2+1} \end{pmatrix}. \quad (24)$$

This is the most general expression for the scattering matrix for a voltage balun. Note that, in any of the above cases-current balun, 180° power splitter, or voltage balun,the impedance or admittance parameter matrices do not exist because of action of the ideal transformer. Also, note that the elements on the principal diagonal S_{ii} are not zero with the exception of S_{11} which can be zero depending on the turns or impedance ratio. This is in keeping with the well-known fact that a lossless, three-port network cannot be matched simultaneously at all three ports [35].

The scattering matrices of the voltage and current baluns can be contrasted to the reduced three-port scattering matrix obtained from the 180° hybrid when the isolated sum port is terminated in a matched resistive load [34]

$$\begin{pmatrix} S_{11} & S_{12} & S_{13} \\ S_{21} & S_{22} & S_{23} \\ S_{31} & S_{32} & S_{33} \end{pmatrix} = \begin{pmatrix} 0 & -\frac{1}{\sqrt{2}} & \frac{1}{\sqrt{2}} \\ -\frac{1}{\sqrt{2}} & 0 & 0 \\ \frac{1}{\sqrt{2}} & 0 & 0 \end{pmatrix}. \qquad (25)$$

This three-port network is the 180° power splitter. As noted above, it enforces an equality of forward power at the output ports. Here, the diagonal elements are all zero. However, the network is no longer lossless now that the matched load has been absorbed into it.

VII. Conclusions

A comprehensive examination of balun topologies for the interfacing of symmetric antennas with coaxial transmission lines has been presented. The 180° hybrid or magic T provides a precise balun for dipoles and other balanced antennas. However, unlike previous experimental arrangements, it is suggested that by properly terminating the isolated sum port, the 180° hybrid will behave as a true current balun. The enforcement of current balance at the feed of an antenna allows the most repeatable experiments and the most straightforward numerical modeling. The use of a shielded, "connectorized" 180° hybrid as a balun, will not, by virtue of being shielded, prevent radiation from the feed line or the external structure of the hybrid when the antenna is unbalanced by the presence of an asymmetric environment. Particularly if the sum port is terminated in a matched load, an externally induced imbalance can cause common mode current to flow on the exterior of the feed line. Only by proper termination of the isolated port can true current balance be obtained.

Acknowledgment

The author would like to gratefully acknowledge the extremely helpful comments of the anonymous reviewers.

References

[1] A. A. Smith, Jr., R. F. German, and J. B. Pate, "Calculation of site attenuation from antenna factors," *IEEE Trans. Electromagn. Compat.*, vol. EMC-24, pp. 301–316, Aug. 1982.

[2] A. A. Smith Jr., "Standard-site method for determining antenna factors," *IEEE Trans. Electromagn. Compat.*, vol. 24, pp. 316–322, Aug. 1982.

[3] *National Standard for Electromagnetic Compatibility-Radiated Emission Measurements in Electromagnetic Interference (EMI) Control-Calibration of Antennas 9 kHz–40 GHz*, ANSI C-63.5-1998, 1998.

[4] M. J. Salter and M. J. Alexander, "EMC antenna calibration and the design of an open-field site," *J. Phys. E. Meas. Sci. Tech.*, vol. 2, pp. 510–519, 1991.

[5] Z. Chen and A. Cook, "Low uncertainty broadband EMC measurement using calculable precision biconical antennas," in *Proc. IEEE Int. Symp. Electromagnetic Compatibility*, vol. 2, 2000, pp. 505–510.

[6] M. J. Alexander, M. J. Salter, D. G. Gentle, and K. P. Holland, "Advances in measurement methods and reduction of measurement uncertainties associated with antenna calibration," *Proc. IEE-Sci. Meas. Technol.*, vol. 141, no. 4, pp. 283–286, 1994.

[7] M. J. Alexander and M. J. Salter, "Low measurement uncertainties in the frequency range 30 MHz to 1 GHz using a calculable standard dipole antenna and national reference ground plane," *Proc. IEE-Sci. Meas. Technol.*, vol. 143, no. 4, pp. 221–228, 1996.

[8] ——, "The design of dipole and monopole antennas with low uncertainties," *IEEE Trans. Instrum. Meas.*, vol. 46, pp. 539–543, Apr. 1997.

[9] B. J. Loader, M. J. Alexander, and M. J. Salter, "Reduced measurement uncertainty in the frequency range 500 MHz to 1 GHz using a calculable standard dipole antenna," presented at the IEE EMC Conf., Warwick, Sept. 1997.

[10] W. T. Blackband, "Coaxial transmission lines and components," in *The Handbook of Antenna Design*, A. W. Rudge, K. Milne, A. D. Olver, and P. Knight, Eds. London, U.K.: Peregrinus, 1986, pp. 1611–1612.

[11] H. F. Garn, W. Müllner, and H. Kremser, "A new accurate and fully automatic calibration facility for radiated-emission test antennas," in *Proc. 8th Int. Conf. Electromagnetic Compatibility*, 1992, pp. 177–180.

[12] S. M. Mann and A. C. Marvin, "Characteristics of the skeletal biconical antenna as used for EMC applications," *IEEE Trans. Electromagn. Compat.*, vol. 36, pp. 322–330, Nov. 1994.

[13] A. Sugiura, T. Morikawa, T. Tejima, and H. Masuzawa, "EMI dipole antenna factors," *IEICE Trans. Commun.*, vol. E78-B, no. 2, pp. 134–139, Feb. 1995.

[14] W. L. Weeks, "Antenna engineering," in *Balance Problems in Antennas*. New York: McGraw-Hill, 1968, sec. 4.5, pp. 167–173.

[15] K. Iizuka and R. W. P. King, "The effect of an unbalance on the current along a dipole antenna," *IRE Trans. Antennas Propagat.*, vol. AP-10, no. 6, pp. 702–707, Nov. 1962.

[16] M. J. Alexander, M. H. Lopez, and M. J. Salter, "Getting the best out of your biconical antennas for emissions and test site evaluation," in *Proc. IEEE Int. Symp. Electromagnetic Compatibility*, 1997, pp. 84–89.

[17] J. Sevick, "Transmission line transformers," American Radio Relay League, Newington, CT, 1990.

[18] ——, "Building and using baluns and ununs," CQ, Hicksville, NY, 1996.

[19] J. A. Nelson and G. Stavis, "Impedance matching devices and baluns," in *Very High Frequency Techniques*, H. J. Reich, Ed. New York: McGraw-Hill, 1947, pp. 86–88.

[20] R. E. Matick, "Transmission line pulse transformers—Theory and applications," *Proc. IEEE*, vol. 56, pp. 47–62, Jan. 1968.

[21] N. Marchand, *Ultrahigh Frequency Transmission and Radiation*. New York: Wiley, 1947, pp. 278–285.

[22] ——, "Transmission line conversion transformers," *Electronics*, no. 12, 1944.

[23] W. K. Roberts, "A new wide-band balun," *Proc. IRE*, vol. 45, pp. 1628–1631, Dec. 1957.

[24] G. H. Stevens Sr, E. T. Lang, R. G. Hendrickson, and J. Hartman, "Construction of general purpose receiving antenna dipole sets and balun assemblies," Fed. Commun. Comm. Offic of Science and Technology Authorization and Stds. Div., Washington, DC, Proj. No-3235-33, 1980.

[25] D. F. Bowman, "Impedance matching and broadbanding," in *Antenna Engineering Handbook*, R. C. Johnson, Ed. New York: McGraw-Hill, 1993, pp. 23–27.

[26] B. Jokanovic and V. Trifunovic, "Double-Y baluns for MMIC's and wireless applications," *Microwave J.*, no. 1, pp. 70–92, 1998.

[27] V. Trifnovic and B. Jokanovic, "Review of printed marchand and double-Y baluns: Characteristics and applications," *IEEE Trans. Microwave Theory Tech.*, vol. 42, pp. 1454–1462, Aug. 1994.

[28] C. L. Ruthroff, "Some broadband transformers," *Proc. IRE*, vol. 47, pp. 1337–1342, Aug. 1959.

[29] J. D. Dyson, "Measurements of near fields of antennas and scatterers," *IEEE Trans. Antennas Propagat.*, vol. AP-21, pp. 446–460, July 1973.

[30] J. D. Dyson and R. Ginyovsky, "Balanced transmission line measurements using coaxial equipment," *IEEE Trans. Microwave Theory Tech.*, vol. MTT-19, pp. 94–96, Jan. 1971.

[31] R. G. Fitzgerrell, "Site attenuation," in *Proc. IEEE Symp. Electromagnetic Compatibility*, 1985, pp. 612–617.

[32] ——, "Standard linear antennas, 30 to 1000 MHz," *IEEE Trans. Antennas Propagat.*, vol. AP-34, pp. 1425–1429, Dec. 1986.

[33] H. Garn, M. Buchmayr, and W. Müllner, "Tracing antenna factors of precision dipoles to basic quantities," *IEEE Trans. Electromagn. Compat.*, vol. 40, pp. 297–310, Nov. 1998.

[34] "Specification for radio disturbance and immunity measuring apparatus and methods," Int. Electrotechnical Commission, Geneva, Switzerland, CISPR 16-1, 2nd ed., pt. 1, vol. 1999–10, 1999.
[35] R. E. Collin, *Foundations for Microwave Engineering*, 2nd ed. New York: McGraw-Hill, 1992, pp. 435–441.
[36] H. J. Carlin and A. B. Giordano, *Network Theory: An Introduction to Reciprocal and Nonreciprocal Circuits*. Englewood Cliffs, NJ: Prentice-Hall, 1964.
[37] S. J. Parisi, "180° lumped element hybrids," in *Dig. IEEE Int. Microwave Symp.*, vol. 3, 1989, pp. 1243–1246.

James S. McLean (M'90) received the B.S., M.S., and Ph.D. degrees in electrical engineering from the University of Texas, Austin, in 1984, 1986, and 1990, respectively.

He was an Engineer at TRW Antenna Systems Laboratory, Redondo Beach, CA. Since 1998, he has been with the TDK Corporation, Cedar Park, TX, performing research in electromagnetic compatibility and electromagnetic metrology. His technical interests include electrically small and low-profile antennas, and he has carried out research in these areas in the Electrical and Computer Engineering Departments at the University of Wisconsin-Madison, and The University of Texas-Pan American.

Chapter 4

TRANSMISSION LINE TRANSFORMERS

4.1 INTRODUCTION

Efficient transfer of radio frequency (RF) power from one network to another is a design task that has faced RF engineers for the larger portion of the last century. The maximum power transfer theorem governs this process, requiring a load impedance to match or be the complex conjugate of the source impedance. The fundamental issue is that rarely do device source and load impedances naturally coexist that are conjugate matches. For example, antenna input impedances are on the order of a complex 50 +/–jX ohms, yet the output impedance of RF transistors used to drive these networks have complex values around 1 to 10 +/–jY ohms. If maximum transfer of power is to occur between these, or any two networks of mismatched characteristics, it is essential for a low loss, two-port, impedance-transforming network to be inserted between the source and load.

A variety of networks are available to the circuit designer to facilitate maximum power transfer. Depending on the desired frequency range of operation, a choice can be made from classic wire-wound transformers, LC (inductor-capacitor) matching networks, and quarter-wave transformers. For the upper audio range frequencies (>10 kHz) to the lower high frequency (HF) range of 1 to 30 MHz, wire-wound transformers offer some degree of suitable performance. LC matching networks can be utilized anywhere in the HF/VHF/UHF and microwave frequency ranges of 1 MHz to well over 2 GHz. Above VHF (300 MHz) the LC elements can be physically realized with printed equivalent circuits for ease of fabrication. Quarter-wave transformers find extensive use in this range as well. However, in spite of the usefulness of these networks, all have critical limitations. The classic wire wound transformer has a high level of interwinding capacitance which resonates with, and ultimately reactively shorts out both the primary and secondary windings, rendering the efficacy of these devices problematical above 10 MHz. LC and quarter-wave networks are inherently narrowband and require unrealistic component values when used at lower frequencies. In addition, they are incapable of converting unbalanced networks to balanced ones.

What would occur if an attempt were made to apply any of these matching techniques to a broadband requirement?

For example, consider the need to match an impedance from the upper audio frequency of 10 kHz to the VHF frequency of 220 MHz. This bandwidth requirement is not merely an exercise for illustrative purposes only. The electromagnetic compatibility (EMC) industry regularly requires this range of operation. Attempting to apply any of the three networks yields poor results.

The wire-wound transformer would invariably self-resonate and present unacceptable levels of insertion loss as well as VSWR. The LC and quarter-wave networks will not have the bandwidth capability. Even if wide bandwidth were not a requirement, low frequency operation would require unrealistic networks. For example, a quarter-wave transformer designed for operation at 10 kHz would require over 3 miles of cable!

Fortunately, in September 1944, G. Guanella [1] published a paper describing a then new class of transformer with unique performance characteristics that would prove extremely useful. The use of this device as a balun has been described in Chapter 3. Here, it is being used in a different configuration as an impedance transformer. This new transformer would use interconnected transmission lines instead of a conventional wire-wound transformer with a primary and secondary. These networks became known as transmission line transformers, and gained immediate acceptance, their popularity bolstered further by Ruthroff's [2] paper in 1959. The concept of this transformer type is nothing short of engineering genius. Introduced over 60 years ago, it remains the most powerful impedance matching network available in the 10 kHz to 2 GHz frequency range. When properly designed and constructed, the versatility of this network is unparalleled:

- Bandwidth ratios in excess of 22,000:1;
- Capable of low loss operation with multioctave bandwidths at frequencies from audio to microwave (10 kHz to over 2.0 GHz);
- High power handling ability in the multikilowatt level;
- Configurable as a balun, converting balanced networks to unbalanced and vice versa;
- Highly efficient (>95%).

Briefly stated, no other impedance matching network possesses all these capabilities. Additionally, the physical construction of this network is surprisingly simple. One can take a piece of enamel wire, cut it into two equal length pieces, twist the wires [3] to an appropriate characteristic impedance and interconnect appropriately (Figure 4.1). The 4:1 impedance transformer in its most basic architecture is ready for use. This network can be augmented with ferrite and compensating capacitors/delay lines to improve its operation at low and high frequencies, respectively. While the network synthesis required for LC networks can be extremely complex, both the realizability conditions and synthesis techniques for transmission line transformers are relatively easy. The complexity of this transformer class lies in its network analysis. Due to the extraordinary operating bandwidth capability, no

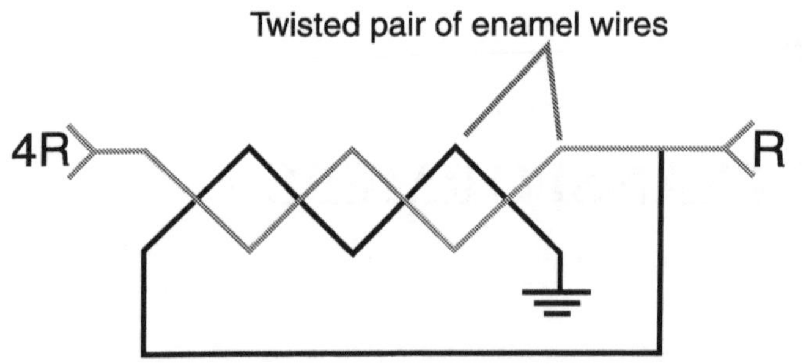

Figure 4.1 *Basic 4:1 transmission line transformer using a twisted pair of enamel insulated wire.*

one single method of network analysis is adequate. Both lumped and distributed network analysis can apply and manual calculation is extremely cumbersome. Analysis, when required, should be executed with the aid of a computer, bearing in mind the analysis complexity is exacerbated with the addition of ferrite materials.

4.2 PHILOSOPHY/ORGANIZATION

While the transmission line transformer is entering its seventh decade of use, only one author has endeavored to write books expressly on this topic [4, 5]. All other relevant publications have been scattered in assorted trade magazines over 60 years, and in many languages. Online searches will yield material on the topic, although this information will not appear in the form of reprints of the older, classic and informative technical papers. It is evident that a collection of reprints on the topic of transmission line transformers is in order. A selection of the more notable reprints have been organized into four sections:

- Historical/original transmission line transformer publications;
- Synthesis of transmission line transformer networks;
- Analysis of transmission line transformer networks;
- Design/construction/applications of transmission line transformer networks.

These four sections cover all the necessary aspects of this device allowing the reader to understand its origin, synthesis and network limitations, how to analyze the network once synthesized and finally how to construct and apply the device to real world requirements.

4.3 HISTORICAL ASPECTS

The first known publication on transmission line transformers was by G. Guanella in September 1944. Originally published in German, the translated version is included in Chapter 3. The fundamental transformer architecture is readily apparent: a specifically arranged series-parallel interconnection of transmission lines. Unlike its conventional wire-wound transformer counterpart, the transmission line transformer has no primary or secondary, only high and low impedance ports. The lines are interconnected such that their boundary (terminal) conditions are satisfied by both the terminating loads and the other coaxial lines. For example, a 4:1 transformer has two 25-ohm lines (Figure 4.2). On the high impedance end of the transformer, the lines are in a series and result in a 50-ohm port, the low impedance port having these two lines in parallel, equaling 12.5 ohms. Connecting a 50-ohm source and 12.5-ohm load resistor on the high and low impedance ports, respectively, produces a system with satisfied boundary conditions. This is paramount since we have produced a flat (frequency independent) system analogous to a single transmission line with satisfied terminal conditions (i.e., Zs = Zl = Zo). We have the same scenario with the 4:1 configuration. The only shortfall occurs at low frequencies when even mode currents flow due to inadequate lumped inductance of the upper transmission line. These currents can be effectively suppressed (to about a 10 kHz low end frequency) by the addition of ferrite cores to the upper line [6]. Also seen in Guanella's paper is the first published application of these transformers, that of matching the output of a push-pull tube amplifier.

Ruthroff's paper [2] takes Guanella's further both in bandwidth capability and introducing the concept of the hybrid transformer. The term "hybrid" comes not from the fact that the transformer might be constructed of two different entities; rather that it is being driven by two signal sources as opposed to only one. The hybrid transformer is a four-port device, two input ports, one sum port and one difference port. Depending on the relative phase and amplitude of the two driving signal sources, either all the power will go to the sum port and zero to the difference port or vice versa. There are also cases where power will exit both ports. The unique and most desirable characteristic of the hybrid transformer is its ability to isolate the two input signal sources. That is, if one signal source either fails or is removed, the remaining source operates unaffected. This feature is particularly useful in combiner/divider design where amplifier outputs can be combined to achieve higher power levels than is possible with a single amplifier. In the event of an amplifier failure, the other amplifiers operate with no degradation in performance. Note that this salient feature comes with a price. For example, if we have two amplifiers combined, if one were to fail, not all of the remaining amplifiers power exits the hybrids sum

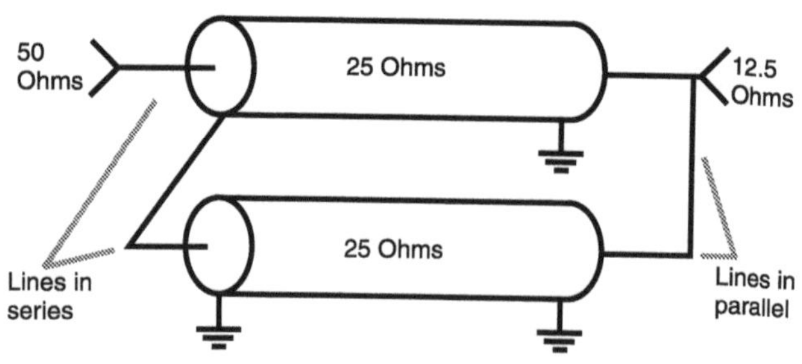

Figure 4.2 *Series-parallel connection of coaxial cables to form a 4:1 transformer.*

port. Half of the remaining functional amplifiers power will go to the difference port, which is usually terminated with an isolation resistor. The hybrid transformer is covered in greater detail in Chapter 5.

4.4 TRANSFORMER NETWORK SYNTHESIS

Synthesis of transmission line transformer networks differs dramatically from classical network synthesis. In classical network synthesis, a certain transfer function is desired, the network architecture is usually known, the circuit component values (inductors and capacitors) are then solved for. With transmission line transformer network synthesis, the component values (i.e., the transmission lines characteristic impedance Zo) are known, and the network topology, or architecture is solved for. The network topology is a graph, or diagram of how the individual transmission lines are interconnected. Both types of network synthesis require necessary and sufficient realizability conditions, as it is essential to be aware, a priori, that the desired network to be synthesized can physically be built. There are a host of realizability conditions in classical network synthesis. For transmission line transformers, there exists only one.

> *Necessary and Sufficient Realizability Condition for Transmission Line Transformers [7]:*
>
> ***The square root of the desired transformation ratio (N) must equate to a rational value.***

By "necessary" it is meant that this condition is fundamentally required to exist in order to determine if a particular transformer network is physically realizable. Here, "sufficient" means that no other condition needs to be met for realization of a particular network. For example, if a 2:1 transformation ratio is desired (N=2), taking the square root of 2 equates to an irrational quantity. This network would require an infinite number of transmission lines to realize *an exact 2:1 transformation ratio.* There is no need to abandon the endeavor to synthesize a transformer in this range of transformation value. A close enough approximation (N=2.25) will yield a transformer comprised of three transmission lines that will provide outstanding performance in terms of impedance match and insertion loss (Figure 4.3).

Once the realizability condition is met, transmission line transformer network synthesis [by way of partial fraction expansions (3,4), etc.] can commence to determine the architecture, or interconnection, of the transmission lines. Note that the synthesis procedure only reveals topology (or architecture); it does not concern itself with line lengths, ferrite loading or capacitive compensation. These are determined in the design process. The synthesis procedure is intrinsically focused on topology, nothing more, and nothing less. Several papers have been included which cover the realizability condition, its proof, as well as a step-by-step guide to topological synthesis procedures that can be used to create functional transformation ratio approximations.

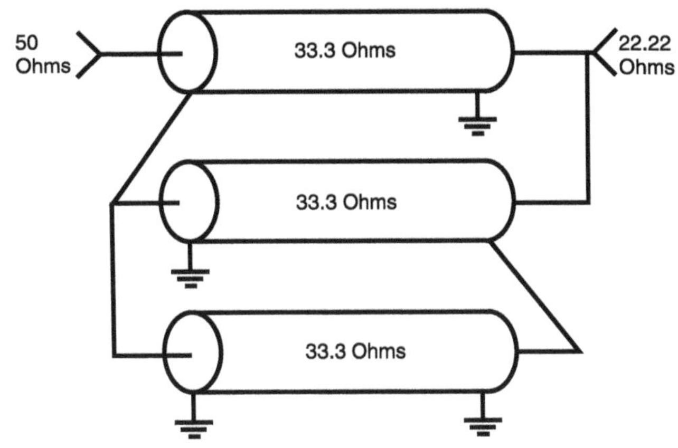

Figure 4.3 *A 2.25:1 transmission line transformer.*

4.5 TRANSFORMER NETWORK ANALYSIS

Having a network synthesis procedure that is so simple could be misleading. LC network synthesis is complex, yet the LC circuits synthesized are readily analyzed with common techniques such as mesh and nodal analysis. The converse is true for transmission line transformers, as network synthesis is simple, yet network analysis is complex. The difficulty arises primarily due to the fact that this network is inherently broadband. Analysis in the low frequency range of the transformer can be approximated with lumped network analysis. Extending into the high end of the operating range, distributed network analysis is invoked as the physical size of the transformer comes within wavelength size. Load ferrites on the transformer lines and this will generate image currents rotating in the ferrite material, which can be calculated using Maxwell's equations. The analysis becomes a mix of lumped, distributed and field equations [8]. It's within reason to purchase an RF software program, create a few models, and click a mouse to perform the analysis. However, most software programs do not have the capability to model ferrites and the ones that do are very costly. Even in these cases, not all programs model a ferrite's complex, frequency-dependent loss characteristics, which can depend on heat transfer/removal in addition to imposed fields at high flux densities. This is not to say analysis is impossible, only that one analysis modality should not be taken for absolute and be expected to take into account all existing phenomena.

The analysis techniques presented in the reprints will serve as a rough guide to assist in analyzing a given transformer network for input/output impedances and insertion loss. Essentially, the four *S* parameters are all that needs to be solved for. While the analysis will provide good results, they assume a low power signal propagating through the transformer. This will result in low insertion loss values. Actual loss values can deviate substantially from calculated ones when high power is applied to a transformer. This is due to the nonlinear properties of ferrites at high flux densities (usually seen at the transformer's lower operating frequency). In addition, at higher frequencies, core loss can elevate substantially and if the core is not properly heat sunk, the ferrite can go into thermal runaway. The thermal runaway mechanism is cyclic: power is applied to the transformer, the ferrite's core temperature increases causing an increase in core loss, now more power is dissipated in the core so the core temperature rises further as does the loss again and so on.

Another useful analysis technique presented in the reprints are methods to determine the input/output impedances of the transformer given nonoptimum transmission line characteristic impedance [9]. This is useful as many times it can be difficult to find coaxial cable with the appropriate impedance. In addition, there are also methods provided to determine input/output impedances with nonoptimum terminating impedances.

4.6 TRANSFORMER DESIGN, CONSTRUCTION, AND APPLICATION

With all the synthesis and analysis information presented, the next logical point of focus is the design, construction and application of the transmission line transformer. The device in and of itself is useless. Only when the transformer is applied to other circuits does it realize its full potential. The applications for these transformers are numerous, and include matching networks for RF transistors, combiners, mixers, antennas, and so on.. The most prevalent is matching networks in RF power amplifiers. Several papers [10, 11] are presented to illustrate how to apply the transformers in actual circuits. Helge Granberg's paper introduces a unique type of transformer that utilizes transmission lines in a conventional transformer configuration. It not a classic transmission line transformer in the respect that the transmission line's boundary conditions are not satisfied. In spite of this, they do yield respectable broadband performance and provide DC isolation. This is a feature not available with classic transmission line transformers.

Finally, a reprint on the design and construction of twisted pair wire transmission lines [3] is included. This will be of use when encountering transformer designs that require unique values of characteristic impedance to achieve matched conditions.

4.7 CONCLUSIONS AND COMMENTARY ON NEXT GENERATION TRANSMISSION LINE TRANSFORMERS

The transmission line transformer is entering its seventh decade of application, yet it remains remarkably unchanged since Guanella first conceived it in 1944. Its longevity can be attributed to its elegantly simple, powerful design concept. There has been substantial and important work over the years that improve the way the device is synthesized and analyzed. However, other than the advent of ferrite and powdered iron technology (please see Chapter 2), little has been done to improve its performance over the years. The problem is that with >95% efficiency and inherent extreme bandwidth capability, there remains, from its inception, little room for improvement.

What, if anything, can be done to better this device? The only electrical parameters that could be improved are the ones that are not ideal. These are insertion loss and bandwidth limitations. The insertion loss of a transmission line transformer is an aggregate of several loss mechanisms: material loss (ferrite, powdered iron), conductor (copper loss), dielectric, radiation and mismatch losses. With this in mind, it can be deduced that improvements in ferrite, conductor, and dielectric materials can yield lower loss transformers. At this point in time, there are advances in high temperature superconducting compounds, (Thallium for example) that can superconduct up to 127 degrees Kelvin (–146°C). This may be helpful in reducing conductor losses, but may reveal the bulk of the loss could be due to ferrites or powdered iron. In addition, the new high temperature superconducting materials may be cost prohibitive. So, improvements in ferrites and powdered iron or some new low-loss, magnetic field augmenting compound may undoubtedly be beneficial. The same applies for dielectric materials. Radiation losses can be minimized by small physical size, however care must be taken to adequately remove excess heat.

Bandwidth limitations of the transmission line transformer stem mostly from parasitics that occur at the interconnection of the transmission line terminals. Breaking out the center conductor from a coaxial line instantly creates a parasitic inductor and positioning a coaxial cable's outer shield above a ground plane generates parasitic capacitance. These are detrimental to extended bandwidth capability as these parasitics will resonate and cause insertion loss peaks. Parasitic series inductance inhibits the proper transformation of very low impedances. Conversely, parasitic shunt capacitance degrades the transformation of high impedances. Improvements in this area are restricted to human creativity, which for the most part, is unlimited. For the past few decades, there have been improvements in fabricating the transmission lines with microstripline and coplanar devices. This has helped in reducing interconnection parasitics.

On a final note, a word of advice to the individual new to transmission line transformers, the best way to develop proficiency with these devices is simply to build a few up, throw them on a vector network analyzer and see what happens. The reprints contained in this section contain a timeless, quintessential collection of information that will facilitate successful designs.

REFERENCES

[1] Guanella, G., "New Method of Impedance Matching in Radio-Frequency Circuits," *Brown Boveri Review*, September 1944, pp. 327–329.

[2] Ruthroff, C.L. "Some Broadband Transformers," *Proc. IRE*, Vol. 47, August 1959, pp. 1337–1342.

[3] Lefferson, P., "Twisted Wire Transmission Line," *IEEE Trans. Parts, Hybrids, and Packaging*, Vol. PHP-7, No. 4, December 1971, pp. 148–154.

[4] Sevick, J., *Transmission Line Transformers*, Fourth Edition, Atlanta, GA: Noble Publishing, 2001.

[5] Sevick, J., *Understanding, Building, and Using Baluns and Ununs*, Hicksville, NY: CQ Communications, Inc., 2003.

[6] Gottlieb, I. M., *Solid State High Frequency Power*, Reston, VA: Reston Publishing Co., Inc., 1982, pp. 173–179.

[7] Myer, D., "Synthesis of Equal Delay Transmission Line Transformer Networks," *Microwave Journal*, Vol. 35, No. 3, March 1992, pp. 106–114.

[8] Myer, D., "Equal Delay Networks Match Impedances over Wide Bandwidths," *Microwaves and RF*, April 1990, pp. 179–188.

[9] Blocker, W., "The Behaviour of the Wideband Transmission Line Transformer for Non-Optimum Line Impedance," *Proc. IEEE*, Vol. 66, No. 4, April 1978, pp. 518–519.

[10] Pitzalis, O., Jr., "Practical Design Information for Broad-Band Transmission Line Transformers," *Proc. IEEE*, Vol. 56, No. 4, April 1968, pp. 738–739.

[11] Granberg, H.O., "New MOSFETS Simplify High Power RF Amplifier Design," *RF Design*, October 1987.

Some Broad-Band Transformers*

C. L. RUTHROFF†, MEMBER, IRE

Summary—**Several transmission line transformers are described which have bandwidth ratios as high as 20,000:1 in the frequency range of a few tens of kilocycles to over a thousand megacycles. Experimental data are presented on both transformers and hybrid circuits.**

Typical applications are: interstage transformers for broad-band amplifiers; baluns for driving balanced antennas and broad-band oscilloscopes; and hybrids for use in pulse reflectometers, balanced modulators, etc.

These transformers can be made quite small. Excellent transformers have been made using ferrite toroids having an outside diameter of 0.080 inch.

SEVERAL transmission line transformers having bandwidths of several hundred megacycles are described here. The transformers are shown in Figs. 1–9. When drawn in the transmission line form, the transforming properties are sometimes difficult to see. For this reason, a more conventional form is shown with the transmission line form. Some winding arrangements are also shown. Certain of these configurations have been discussed elsewhere and are included here for the sake of completeness [1–4].

In conventional transformers the interwinding capacity resonates with the leakage inductance producing a loss peak. This mechanism limits the high frequency response. In transmission line transformers, the coils are so arranged that the interwinding capacity is a component of the characteristic impedance of the line, and as such forms no resonances which seriously limit the bandwidth. Also, for this reason, the windings can be spaced closely together maintaining good coupling. The net result is that transformers can be built this way which have good high frequency response. In all of the transformers for which experimental data are presented, the transmission lines take the form of twisted pairs. In some configurations the high frequency response is determined by the length of the windings and while any type of transmission line can be used in principle, it is

* Original manuscript received by the IRE, February 5, 1959; revised manuscript received, April 1, 1959.
† Bell Telephone Labs., Inc., Holmdel, N. J.

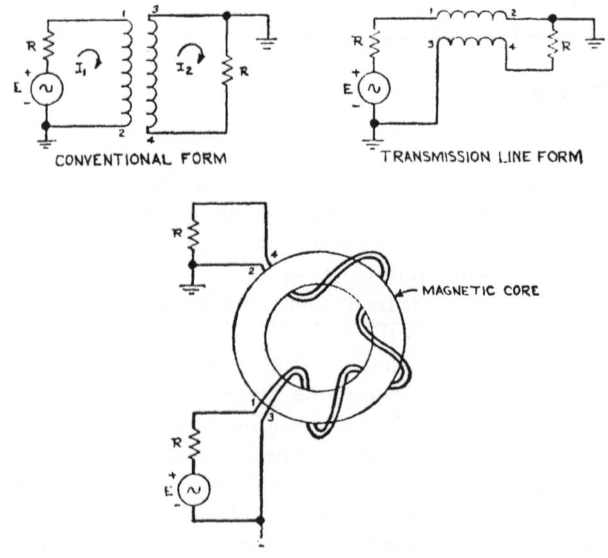

Fig. 1—Reversing transformer.

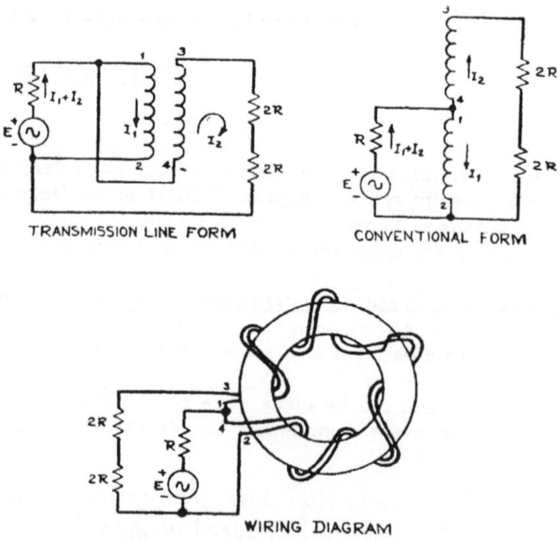

Fig. 3—4:1 Impedance transformer.

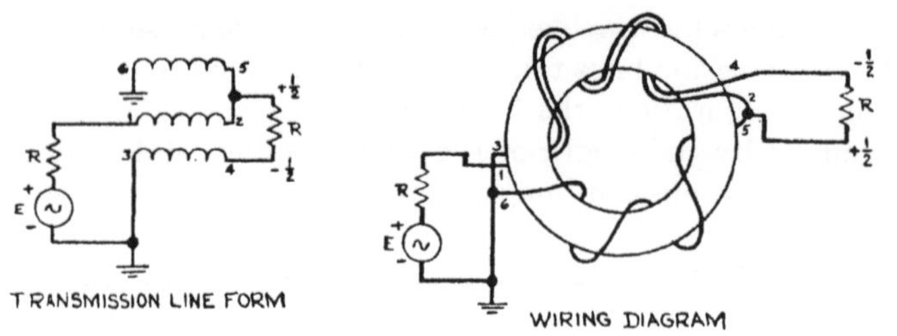

Fig. 2—Unbalanced to balanced transformer.

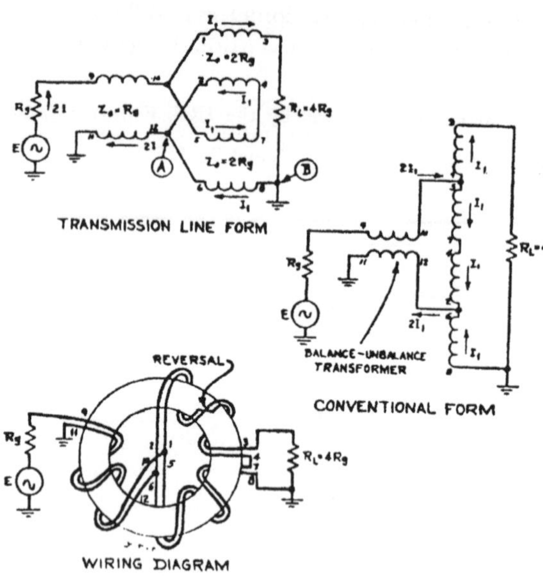

Fig. 4—4:1 Impedance transformer. Unbalanced—symmetrical.

quite convenient to make very small windings with twisted pairs.

The sketches showing the conventional form of transformer demonstrate clearly that the low frequency response is determined in the usual way, *i.e.*, by the primary inductance. The larger the core permeability, the fewer the turns required for a given low frequency response and the larger the over-all bandwidth. Thus a good core material is desirable. Ferrite toroids have been found very satisfactory. The permeability of some ferrites is very high at low frequencies and falls off at higher frequencies. Thus, at low frequencies, large reactance can be obtained with few turns. When the permeability falls off the reactance is maintained by the increase in frequency and good response is obtained over a large frequency range. It is important that the coupling be high at all frequencies or the transformer action fails. Fortunately, the bifilar winding tends to give good coupling. All of the cores used in the experimental transformers described here were supplied by F. J. Schnettler of the Bell Telephone Laboratories, Inc.

Polarity Reversing Transformer—Fig. 1

This transformer consists of a single bifilar winding and is the basic building block for all of the transformers. That a reversal is obtained is seen from the conventional form which indicates current polarities. Both ends of the load resistor are isolated from ground by coil reactance. Either end of the load resistor can then be grounded, depending upon the output polarity desired. If the center of the resistor is grounded, the output is balanced. A suitable winding consists of a twisted pair of Formex insulated wire. In such a winding, the primary and secondary are very close together, insuring good coupling. The interwinding capacity is absorbed in the characteristic impedance of the line.

At high frequencies this transformer can be regarded as an ideal reversing transformer plus a length of transmission line. If the characteristic impedance of the line is equal to the terminating impedances, the transmission is inherently broadband. If not, there will be a dip in the response at the frequency at which the transmission line is a quarter-wavelength long. The depth of the dip

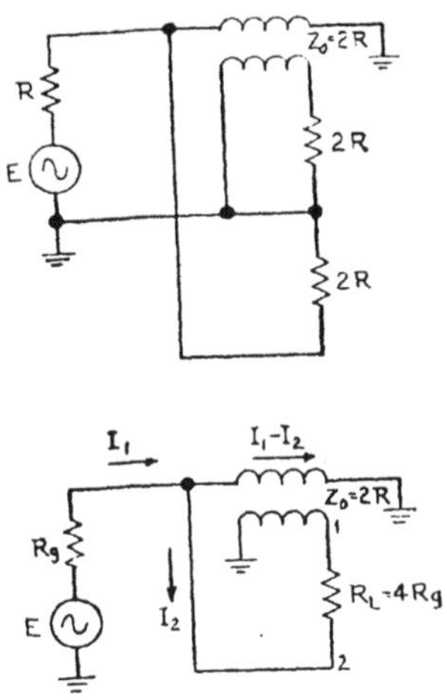

Fig. 5—Balanced—unbalanced 4:1 impedance transformer.

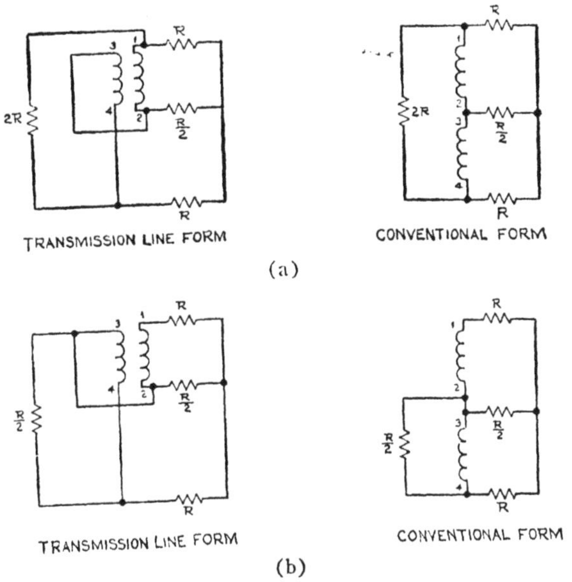

Fig. 6—(a) Basic hybrid. (b) Unsymmetrical hybrid with equal conjugate impedances.

is a function of the ratio of terminating impedance to line impedance and is easily calculated.

Experimental data on a reversing transformer are shown in Figs. 10 and 11. Fig. 10 is the response of a transformer with no extra impedance matching. The return loss of this transformer to a 3 mμ sec pulse is 20 db. The transformer of Fig. 11 has been adjusted to provide more than 40 db return loss to a 3 mμsec pulse. The transformer loss (about 0.5 db before matching) is matched to 75 ohms with the two 3.8-ohm resistors. The inductance is tuned out with the capacity of the resistors to the ground plane. The match was adjusted while watching the reflection of a 3 mμsec pulse.

Balanced-to-Unbalanced 1:1 Impedance Transformer—Fig. 2

This is similar to Fig. 1 except that an extra length of winding is added. This is necessary to complete the path for the magnetizing current.

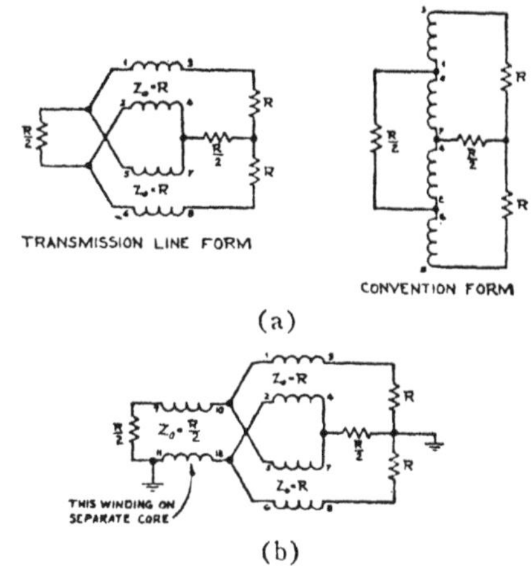

Fig. 7—(a) Symmetrical hybrid with equal conjugate impedances. (b) Unbalanced symmetrical hybrid with equal conjugate impedances.

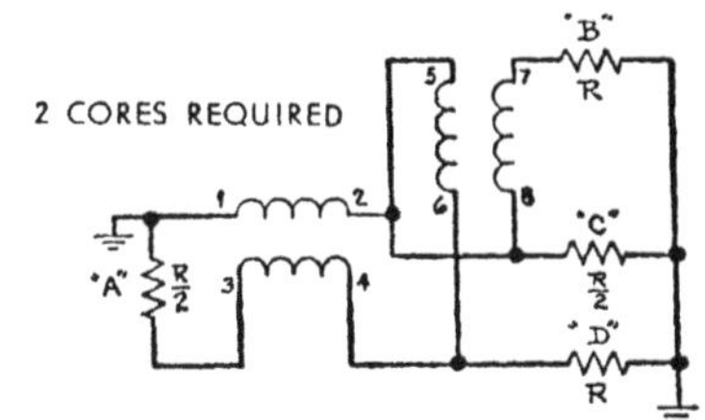

Fig. 8—Hybrid with equal conjugate impedances. Each arm single ended.

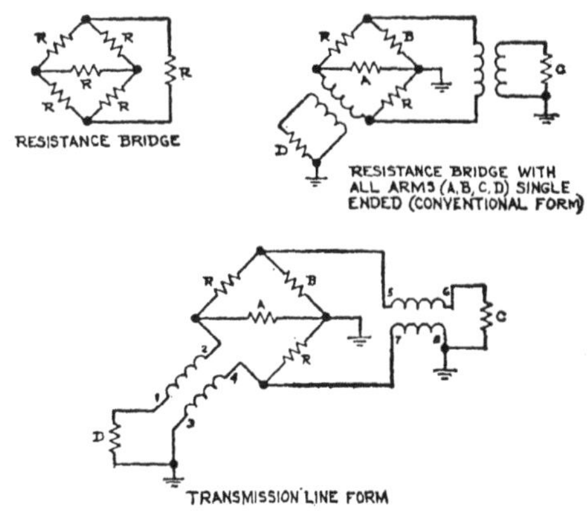

Fig. 9—Resistance hybrid with equal impedance loads. (This hybrid has 3 db loss in addition to transformer loss.)

Unbalanced-Unsymmetrical 4:1 Impedance Transformer—Fig. 3

This transformer is interesting because with it a 4:1 impedance transformation is obtained with a single bifilar winding such as used in the reversing transformer. The transforming properties are evident from Fig. 3. Not so easily seen is the high frequency cutoff characteristic.

The response of this device at high frequencies is derived in the Appendix and only the result for matched impedances is given here.

$$\frac{\text{Power Available}}{\text{Power Output}} = \frac{(1 + 3\cos \beta l)^2 + 4\sin^2 \beta l}{4(1 + \cos \beta l)^2}, \qquad (1)$$

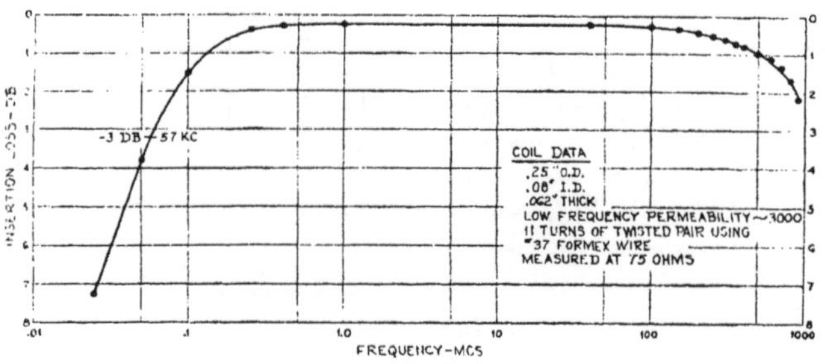

Fig. 10—1:1 Reversing transformer. Insertion loss vs frequency.

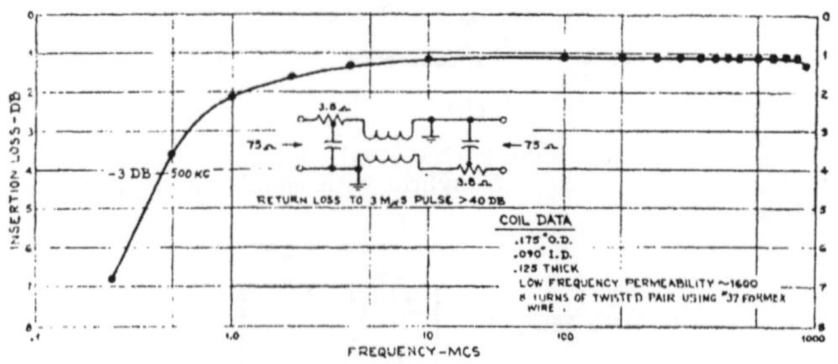

Fig. 11—Matched reversing transformer. Insertion loss vs frequency.

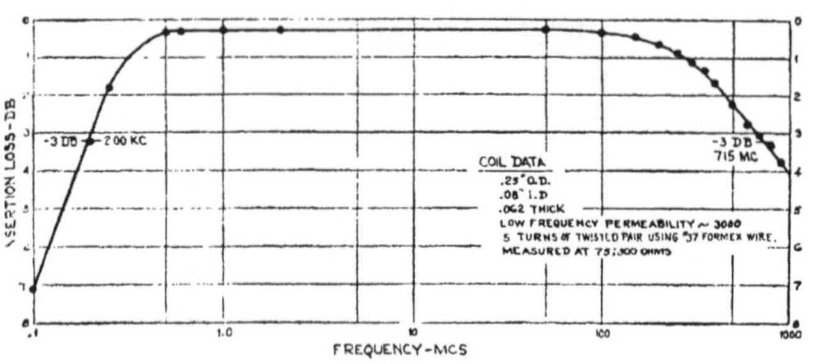

Fig. 12—1 Winding 4:1 impedance transformer unbalanced—unsymmetrical.

where β is the phase constant of the line, and l is the length of the line. Thus, the response is down 1 db when the line length is $\lambda/4$ wavelengths and the response is zero at $\lambda/2$. For wideband response this transformer must be made small. For a plot of (1) see Fig. 16.

Experimental data are given for a transformer of this type in Fig. 12.

Unbalanced-Symmetrical 4:1 Impedance Transformer—Fig. 4

This configuration requires three bifilar windings as shown in Fig. 4. All three windings can be placed on one core, a procedure which improves the low frequency response.[1] When winding multiwinding transformers the following well-known rule should be followed: with the generator connected and the load open, a completed circuit should be formed by the windings so that the core will be magnetized. The fields set up by the currents should be arranged so as to aid each other.

[1] Pointed out to the author by N. J. Pierce of Bell Telephone Labs., Inc., Holmdel, N. J.

Balanced-to-Unbalanced 4:1 Impedance Transformers—Fig. 5

The circuit of Fig. 5 is quite simple. The single bifilar winding is used as a reversing transformer as in Fig. 1. The high frequency cutoff is the same as that for the transformer of Fig. 3.

In some applications it is desirable to omit the physical ground on the balanced end. In such cases, Fig. 5(b) can be used. The high frequency cutoff is the same as for the transformer of Fig. 3. The low frequency analysis is presented in Appendix B.

Hybrid Circuits: Figs. 6–9

Various hybrid circuits are developed from the basic form using the transformers discussed previously. The drawings are very nearly self-explanatory. In all hybrids in which all four arms are single-ended, it has been found necessary to use two cores in order to get proper magnetizing currents.

Two hybrids have been measured and data included here. The response of a hybrid of the type shown in Fig. 8 is given in Fig. 13. For this measurement $R=150$ ohms. In order to measure the hybrid in a 75-ohm circuit, arms B, D were measured with 75-ohm series resistances in series with the 75-ohm measuring gear. This accounts for 3 db of the loss. Under these conditions arms B and D have a 6 db return loss.

The transmission of the resistance hybrid of Fig. 9 is given in Fig. 14. This hybrid has been matched using the technique described previously for the reversing transformer. The results of this matching are included in the figure. This hybrid was designed for use in a pulse reflectometer, the main part of which is a stroboscopic oscilloscope with a resolution of better than 3 mμsec. The oscilloscope was designed by W. M. Goodall.

Applications

Many applications for these transformers will occur to the reader. For purposes of illustration, a few of them are listed here.

1) The reversing transformer of Fig. 1 can be used to reverse the polarity of short pulses, an operation which is frequently necessary. It has also been used in balanced detectors and to drive push-pull amplifiers from single-ended generators.
2) The transformers of Figs. 2 and 5(b) are useful for driving balanced antennas. The circuit of Fig. 5(b) may find application in connecting twin lead transmission line to commercial television receivers.
3) The transformer of Fig. 3 has found wide use in broadband amplifier interstages. It will also be useful in transforming the high output impedances of distributed amplifiers to coaxial cable impedances. They can also be cascaded to get higher turns ratios.

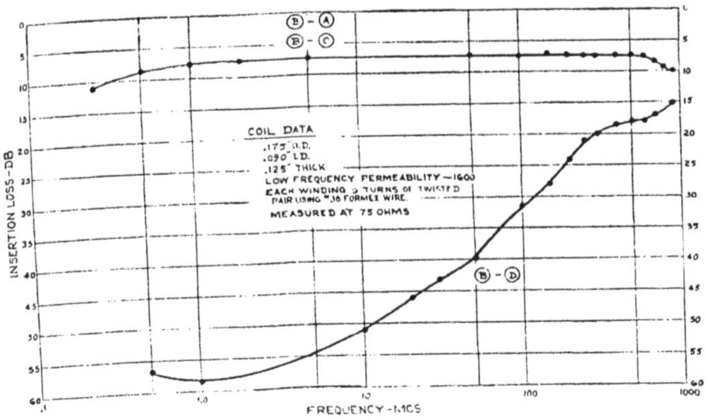

Fig. 13—Hybrid of Fig. 8. Insertion loss vs frequency.

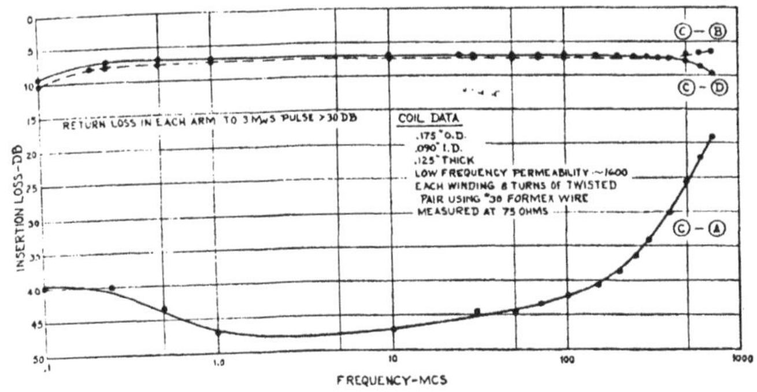

Fig. 14—Matched resistance hybrid. Insertion loss vs frequency.

4) The circuit of Fig. 5(a) has been used to drive broadband oscilloscopes, with balanced inputs, from single-ended generators. It can also find use in balanced detectors.
5) Hybrids have many uses such as in power dividers, balanced amplitude and phase detectors; as directional couplers for pulse reflectometers, IF and broadband sweepers. They might also be used as necessary components in a short pulse repeater for passing pulses in both directions on a single transmission line.

Appendix A

The high frequency response of the circuit of Fig. 3 is derived from Fig. 15. The loop equations are as follows:

$$
\begin{aligned}
e &= (I_1 + I_2)Rg + V_1 \\
e &= (I_1 + I_2)Rg - V_2 + I_2 R_L \\
V_1 &= V_2 \cos \beta l + j I_2 Z_0 \sin \beta l \\
I_1 &= I_2 \cos \beta l + j \frac{V_2}{Z_0} \sin \beta l. \qquad (2)
\end{aligned}
$$

This set of equations is solved for the output power P_0. $P_0 = |I_2|^2 R_L$

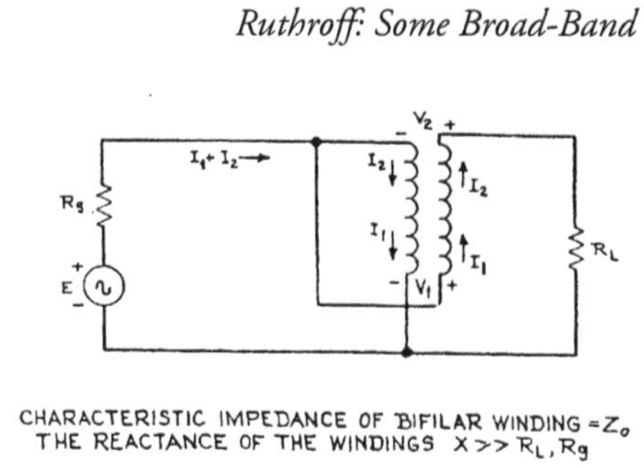

Fig. 15—Transformer schematic.

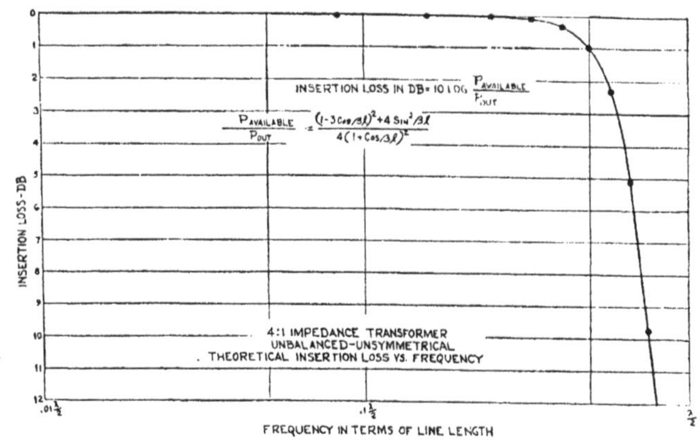

Fig. 16—Theoretical insertion loss vs frequency.

From this expression, the conditions for maximum power transmission are obtained by setting $l = 0$ and setting $dP_0/dR_L|_{l=0} = 0$. The transformer is matched when $R_L = 4Rg$. The optimum value for Z_0 is obtained by minimizing the coefficient of $\sin^2 \beta l$ in (3). In this manner the proper value for Z_0 is found to be $Z_0 = 2Rg$.

Now, setting $R_L = 4Rg$ and $Z_0 = 2Rg$, (3) reduces to

$$
P_0 = \frac{e^2 (1 + \cos \beta l)^2}{Rg[(1 + 3 \cos \beta l)^2 + 4 \sin^2 \beta l]}. \qquad (4)
$$

Also,

$$
P_{\text{available}} = \frac{e^2}{4Rg}, \qquad (5)
$$

and dividing (4) by (3):

$$
\frac{\text{Power Available}}{\text{Power Output}} = \frac{(1 + 3 \cos \beta l)^2 + 4 \sin^2 \beta l}{4(1 + \cos \beta l)^2}. \qquad (6)
$$

This function is plotted in Fig. 16.

The impedances seen at either end of the transformer with the other end terminated in Z_L have been derived. They are:

Z_{in}(low impedance end)

$$
= Z_0 \left(\frac{Z_L \cos \beta l + j Z_0 \sin \beta l}{2Z_0(1 + \cos \beta l) + j Z_L \sin \beta l} \right) \qquad (7)
$$

$$
P_0 = |I_2|^2 R_L = \frac{e^2 (1 + \cos \beta l)^2 R_L}{[+2Rg(1 + \cos \beta l) + R_L \cos \beta l]^2 + \left[\dfrac{Rg R_L + Z_0^2}{Z_0}\right]^2 \sin^2 \beta l}. \qquad (3)
$$

and

Z_{in}(high impedance end)

$$= Z_0\left(\frac{2Z_L(1+\cos\beta l)+jZ_0\sin\beta l}{Z_0\cos\beta l + jZ_L\sin\beta l}\right). \quad (8)$$

Appendix B

In the low frequency analysis of the transformer of Fig. 5 the series impedance of each half of the bifilar winding is denoted by Z. The loop equations are:

$$E = (Rg + Z)I_1 - (Z + kZ)I_2$$

$$E = (Rg - kZ)I_1 + (R_L + Z + kZ)I_2, \quad (9)$$

from which

$$\frac{I_1}{I_2} = \frac{R_L + 2Z(1+k)}{Z(1+k)} \approx 2 \text{ if } Z \gg R_L. \quad (10)$$

We now proceed to calculate the voltages from points 1 and 2 to ground

$$V_{2G} = E = I_1 Rg.$$

When the transformer is matched, $E = 2I_1Rg$ and

$$V_{2G} = I_1 Rg. \quad (11)$$

Similarly,

$$V_{1G} = I_2 Z - kZ(I_1 - I_2).$$

With the aid of (10) this can be rearranged to

$$V_{1G} = ZI_1\left[\frac{Z(1+k)^2 - kR_L - 2kZ(1+k)}{R_L + 2Z(1+k)}\right]. \quad (12)$$

Now let the coupling coefficient $k=1$, then

$$V_{1G} = I_1 Z\left[\frac{-kR_L}{R_L + 2Z(1+k)}\right] \approx -\frac{I_1 R_L}{4}$$

for $Z \gg R_L$.

When the transformer is matched, $R_L = 4Rg$ so that

$$V_{1G} = I_1 Rg = -V_{2G}, \quad (13)$$

and the load is balanced with respect to ground.

From (13) it is clear that the center point of R_L is at ground potential. This point can therefore be grounded physically, resulting in Fig. 5(a).

Acknowledgment

In addition to those mentioned in the text, the author is indebted to D. H. Ring for many stimulating discussions on every aspect of these transformers.

References

[1] Willmor K. Roberts, "A new wide-band balun," Proc. IRE, vol. 45, pp. 1628–1631; December, 1957.
[2] H. Gunther Rudenberg, "The distributed transformer," Raytheon Mfg. Co., Waltham, Mass.
[3] G. Guanella, "New method of impedance matching in radio frequency circuits," *Brown-Boveri Rev.*, vol. 31, p. 327; 1944.
[4] A. I. Talkin and J. V. Cuneo, "Wide-band balun transformer," *Review of Sci. Inst.*, vol. 28, No. 10, pp. 808–815; October, 1957.
[5] C. A. Burrus, unpublished memorandum.

LINE TRANSFORMERS WITH FRACTIONAL TRANSFORMATION FACTOR

S E London and S V Tomashevich

A series-parallel winding connection which enables one to simplify the construction considerably is described.

Line transformers, including matching transformers, with an arbitrary fractional transformation factor n can be constructed by a parallel-series connection of line transformers with integral values of n. This leads to simpler constructions than the cascade connection of line transformers for the same value of n. For certain constructions with a single magnetic core, we give below the characteristics at high frequencies when the phase-compensating lines are eliminated, and the in-phase wave is neglected [1, 2]. The determination of the magnetization inductance and the low-frequency characteristics will be obvious.

The construction of line transformers with n = 1.5 (Fig. 1), formed by parallel-series connection of line transformers with n = 2 and n = 1 is matched for $x_1 = x$. If the phase-compensating line is eliminated ($x_1 = 0$), the g-parameter matrix [3] obtained by summing the $[g]_i$ matrices of the component line transformers has the form

$$[g]=\begin{bmatrix} g_{11} & g_{21} \\ g_{21} & g_{22} \end{bmatrix}=\begin{bmatrix} i\frac{a}{\rho}(\frac{1}{b}+\frac{1}{2b+2}) & \frac{1}{2}+\frac{1}{b} \\ \frac{1}{2}+\frac{1}{b} & -i\rho a(\frac{1}{b}+\frac{1}{2b+2}) \end{bmatrix}$$

where a= sin x; b =cos x.

The construction of a line transformer with n = 1.5 (Fig. 2), having a balanced input and an unbalanced output, is matched if $x_1 = x_2 = x$. Its g-parameter matrices when the phase-compensating lines are eliminated, obtained as for Fig. 1, are

$$\underset{\substack{x_1=0\\x_2=x}}{[g]}=\begin{bmatrix} \frac{i}{\rho}(\frac{a}{2b}+\frac{c}{d}) & \frac{1}{2b}+\frac{1}{d} \\ \frac{1}{2b}+\frac{1}{d} & -i\rho(\frac{a}{2b}+\frac{c}{d}) \end{bmatrix};$$

$$\underset{\substack{x_1=0\\x_2=0}}{[g]}=\begin{bmatrix} \frac{i}{\rho}(\frac{a}{2+2b}+\frac{c}{d}) & \frac{1}{2}+\frac{1}{d} \\ \frac{1}{2}+\frac{1}{d} & -i\rho(\frac{a}{2+2b}+\frac{c}{d}) \end{bmatrix}.$$

where c = sin 2x, d = cos 2x.

The construction of a line transformer with n = 2.5 (Fig. 3), formed parallel-series connection of two line transformers with n = 2, is matched when $x_1/2 = x_3 = x_2/2 = x$. Then the phase-compensating lines are eliminated the [g] matrices are

$$\underset{\substack{x_1=0\\x_2=2x\\x_3=x}}{[g]}=\begin{bmatrix} \frac{i}{\rho}(\frac{a}{2b}+\frac{2f}{k}) & \frac{2}{k}+\frac{1}{2b} \\ \frac{2}{k}+\frac{1}{2b} & -i\rho(\frac{a}{2b}+\frac{2f}{k}) \end{bmatrix};$$

$$\underset{\substack{x_1=0\\x_2=0\\x_3=x}}{[g]}=\begin{bmatrix} \frac{i}{\rho}(\frac{f}{k}+\frac{3a}{2b}) & \frac{1}{k}+\frac{3}{2b} \\ \frac{1}{k}+\frac{3}{2b} & -i\rho(\frac{f}{k}+\frac{3a}{2b}) \end{bmatrix};$$

$$\underset{\substack{x_1=0\\x_2=0\\x_3=0}}{[g]}=\begin{bmatrix} \frac{i}{\rho}(\frac{f}{k}+\frac{a}{b}+\frac{a}{2b+2}) & \frac{1}{2}+\frac{1}{b}+\frac{1}{k} \\ \frac{1}{2}+\frac{1}{b}+\frac{1}{k} & -i\rho(\frac{f}{k}+\frac{a}{b}+\frac{a}{2b+2}) \end{bmatrix}.$$

where f = sin 3x; k = cos 3x.

The modulus of the reflection coefficient, determined from the expression $|\rho|=\left|\frac{R-Z_{in}}{R+Z_{in}}\right|=\left|\frac{g_{21}^2-g_{11}g_{22}-n^2}{g_{21}^2-g_{11}g_{22}+n^2+2n\rho g_{11}}\right|$ for the line transformers considered, is shown in Fig. 4.

REFERENCES

1. Y. Shimade, Rev. of the E.C.L., NTT, Japan, 1970, 18, 97
2. E. Lampert, Arch. Electr. Ubertrag., 1969, 23, No.1
3. G.I. Atabekov, Osnovy teorii tsepei (Principles of Circuit Theory), Energiya Press, 1969.

Submitted July 16, 1973

□♦♦♦□

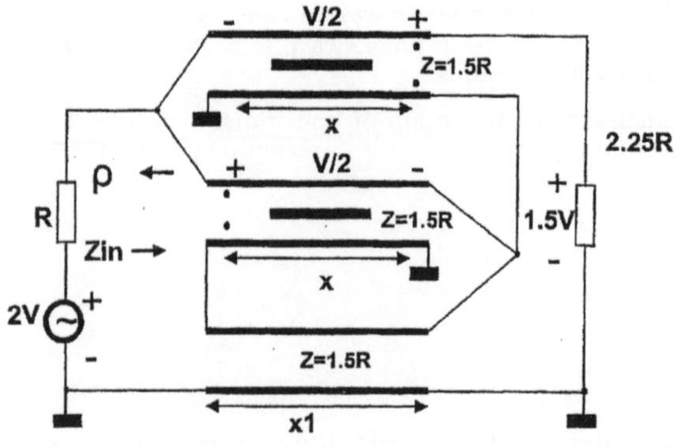

Fig.1

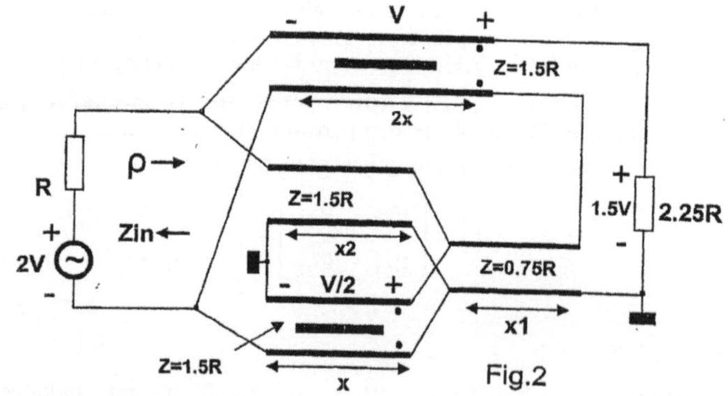

Fig.2

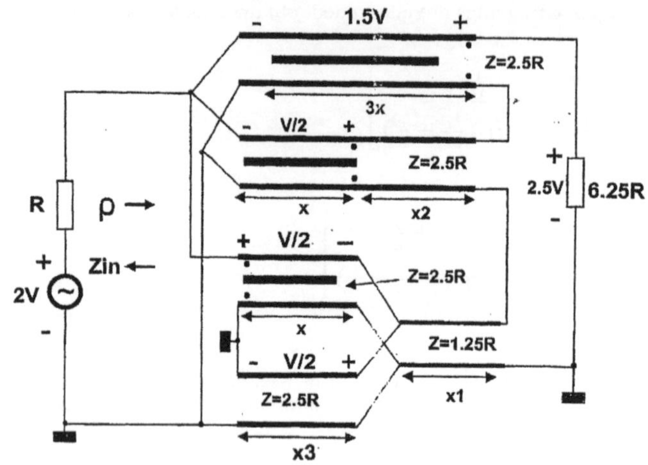

Fig.3

Fig. 4

Editor's Note: This paper had to be re-typed because of photographic quality issues with the original. The content is unchanged but the formatting may differ.

TRANSMISSION LINE TRANSFORMERS WITH INTEGER-RATIO VOLTAGE TRANSFORMATIONS

Thomas J. Gluszczak and John D. Harmer
Raytheon Company
Sudbury, Massachusetts

Abstract

Transmission line transformers with impedance ratio $\left(\frac{1}{m}\right)^2$ are made up of m transmission lines. The general case of an impedance ratio $\left(\frac{n}{m}\right)^2$ could be realized by cascading two transformers of ratios $\left(\frac{1}{n}\right)^2$ and $\left(\frac{m}{1}\right)^2$, and would use n + m transmission lines. It is shown here that the general case can be realized using fewer than n + m lines by a synthesis procedure using the partial fraction expansion of $\frac{n}{m}$.

Transmission line transformers are becoming increasingly popular among circuit designers when broadband (multi-octave) impedance transformation must be achieved at HF and VHF frequencies. In this type of transformer, transmission lines are threaded through or wound around ferrite cores to give the necessary high "line length" inductance so that inputs and outputs are isolated and can be interconnected without excessive losses over a wide frequency range. Each transmission line can be conceived of as a 1:1 turns ratio prototype transformer.[1] Interconnections of the prototype elements result in several useful circuit elements including step up/step down transformers, baluns, and hybrids. The literature has provided the designer with techniques for core selection[2] requirements on transmission line impedance, and line length.[3]

The impedance ratios which can be achieved with transmission line transformers are limited to those of the form $\left(\frac{n}{m}\right)^2$ where n and m are integers. Transformers are usually designed for 1:m turns ratios by paralleling m lines at the input and connecting their outputs in series. The general interconnection of a 1:m turns ratio transformer is shown in Figure 1.

The more general case with both n and m $\neq$ 1 can be realized by cascading two transformers with ratios 1:n and m:1. Such an arrangement requires n + m transmission lines to achieve an n:m turns ratio.

It will be shown here that a general n:m ratio can often be realized in interconnecting fewer than n + m prototype transformers (or lines).

No loss in generality results in requiring n and m to be relatively prime, nor that n < m where n:m is the primary:secondary turns ratio. For n < m, the ratio n:m can be inverted, and the denominator split into an integer plus a new fractional ratio with the smaller integer n as its denominator, and the remainder from m/n in the numerator, (m) modulo (n), which is less than n.

$$\frac{n}{m} = \frac{1}{N_0 + \frac{(m)\bmod(n)}{n}}$$

In the general case that (m)mod(n) $\neq$1, the fraction $\left(\frac{(m)\bmod(n)}{n}\right)$ can be similarly inverted, and a continuous fraction formed. This constitutes the Euclidean algorithm for finding the greatest common divisor (GCD) of n and m. Since the integers n and m are relatively prime, the GCD is 1, and the final fraction in the continuous expansion must have 1 as its numerator.

$$\frac{n}{m} = \cfrac{1}{N_0 + \cfrac{1}{N_1 + \cfrac{1}{N_2 + \dots \frac{1}{N_k}}}}$$

The transformer is built starting with N_k lines connected in series at the input side and in parallel at the output side, and N_{k-1} lines connected in parallel at the input side and in series at the output side. The inputs of the two sets of N_k and N_{k-1} lines are then connected in parallel and their outputs connected in series. This gives a transformer with the fractional ratio:

$$\cfrac{1}{N_{k-1} + \cfrac{1}{N_k}}$$

This transformer (reversed) is then used again with N_{k-2} lines connected in parallel at the input yielding a transformer with the new ratio:

$$\cfrac{1}{N_{k-2} + \cfrac{1}{N_{k-1} + \cfrac{1}{N_k}}}$$

This process is repeated K times using $\sum_{i=0}^{k} N_i$ transmission lines to realize turns ratios of the general form n:m.

For a step up transformer with ratio n:m, the following rules for transmission line interconnections can be deduced from the above:

(a) For each even i, connect N_i transmission lines in parallel at their input (primary) side and in series at their output (secondary) side as shown in Figure 1.

(b) For each odd i, connect N_i transmission lines in series at their input side and in parallel at their output side (the reverse is shown in Figure 1).

(c) For K odd, connect the N_{k-1} element built in step (a) above in parallel with the N_k element built in step (b) above at their inputs and in series at their output. Connect this entire combination in series with the N_{k-2} element on the input side and in parallel at the output side.

(d) For K even, connect the N_{k-1} element built in step (b) above in series with the N_k element built in step (a) above at their inputs and in parallel at their outputs. Connect this entire combination in parallel with the N_{k-2} element at the input side and in series at the output side.

(e) Continue either step (c) or (d) alternating parallel and series connections until the N_0 element is connected (in parallel).

Examples will clarify the technique. Consider a 4:9 impedance transformation between 50 ohms and 112.5 ohms (reasonably close to a 1:2 impedance transformation, which cannot be achieved exactly using transmission line transformers since $\sqrt{2} \neq$ rational number).

Transmission Line Z_0 = 3/2 (Z_{in}) = 3/2 (50) = 75 Ohms.

ΣN_i =3 =number of 75Ω line sections.

The transformer interconnections (primary) require two series 1:1 transformers in parallel with one 1:1 transformer as shown in Figure 2. The secondary interconnections are the complements of those in the primary.

As another example, a 2:5 turns ratio:

$$\frac{n}{m} = \frac{2}{5} = \cfrac{1}{2 + \cfrac{1}{2}}$$

Z_0 = 5/2(Z_{in}) = 5/2(50) = 125 Ohms

$$\Sigma N_i = 4$$

The circuit is shown in Figure 3.

This transformer was built in an unbalanced (50Ω) to balanced (312.5Ω) configuration for broadband testing of television antenna circuits with 50Ω laboratory test equipment. Over the 50-250 MHz band, an insertion loss of less than 0.4 dB and a VSWR of less than 1.3:1 were achieved.

REFERENCES

1. C. L. Ruthroff, "Some Broadband Transformers", Proc. IRE, Vol. 47. pp. 1337-1342; August 1959.
2. Richard E. Matick, "Transmission Line Pulse Transformers- -Theory and Application", Proceedings IEEE, Vol. 56, No. 1, pp. 47-62; January 1968.
3. O. Pitzalis, Jr. , "Practical Design Information for Broadband Transmission Line Transformers", Proceedings IEEE, Vol. 56, No. 1, pp. 738-739; April 1968.

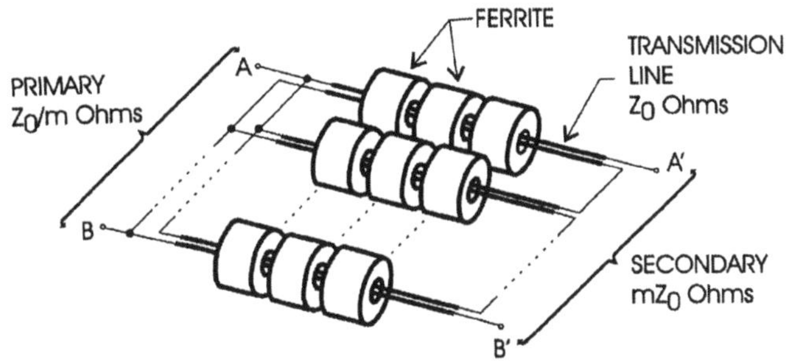

Figure 1. 1:m Turns Ratio Transformer

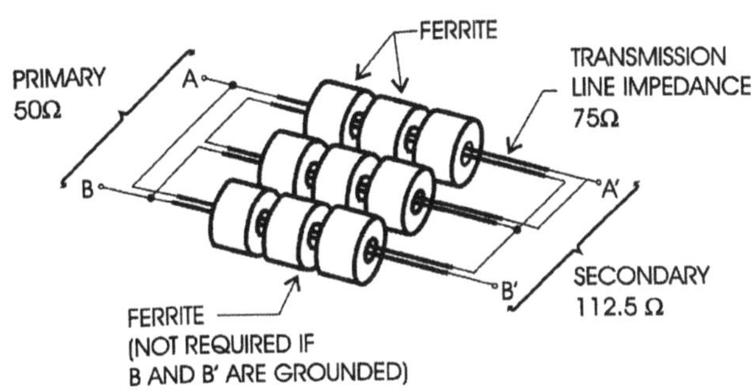

Figure 2. 4:9 Impedance Ratio Transformer

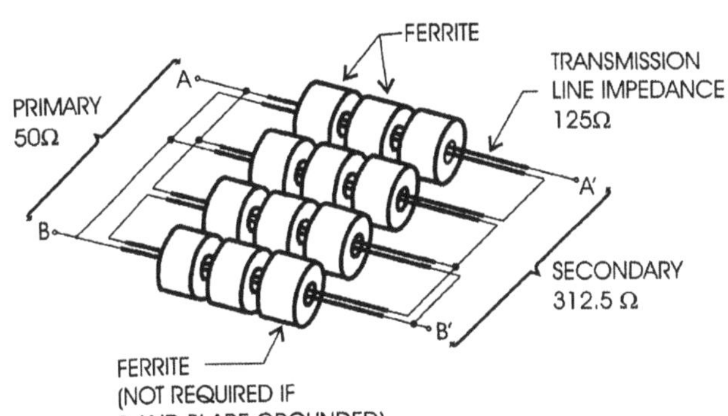

Figure 3. 4:25 Impedance Ratio Transformer

Editor's Note: This article had to be re-typed as it was not possible to obtain a high enough quality copy of the original. The formatting may differ but the content is unchanged.

Synthesis of Equal Delay Transmission Line Transformer Networks

Daniel Myer
Power Systems Technology Inc.
Melville, NY

Introduction

Perhaps the most commonly encountered task in RF engineering is the matching of two impedances. The network most capable of broadband impedance matching is the equal delay transmission line transformer. Wire wound transformers, among other types, can also be used for impedance matching. However, they do not possess comparable loss and frequency response characteristics. The equal delay networks are inherently broadband, since the impedance transformation is accomplished with transmission lines that have satisfied boundary conditions.

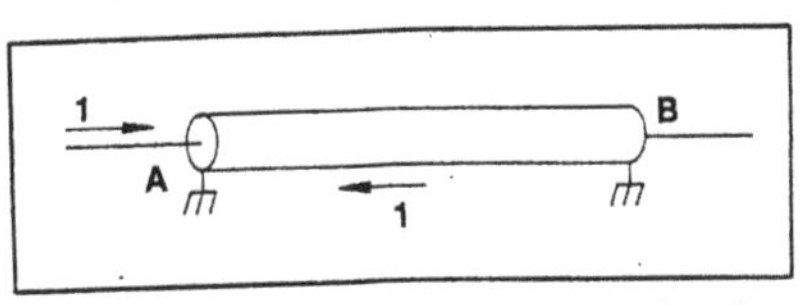

Fig. 1 A length of coaxial transmission line with a current of unit magnitude entering from the left.

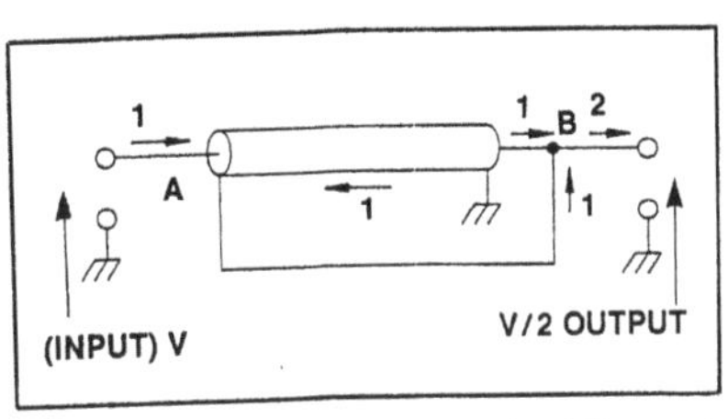

Fig. 2 The normalized network of Figure 1.

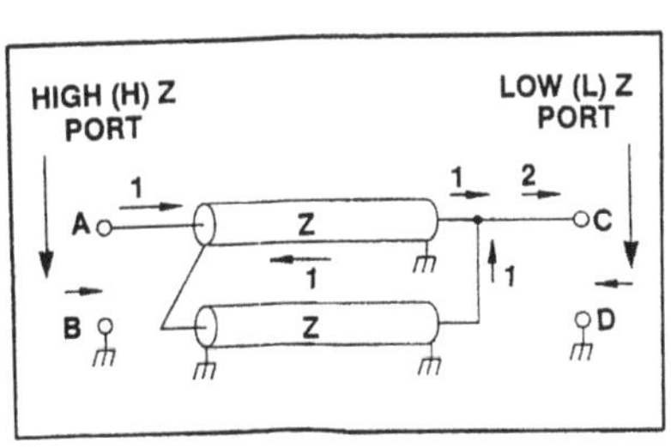

Fig. 3 The normalized network with an additional line added, which is the fundamental building block of all equal delay networks.

Two critical issues of concern when designing equal delay networks are the transformation ratio realizability, that is, whether the desired transformation ratio can be physically built, and network topology synthesis, which is the interconnection of transmission lines to facilitate the proper transformation ratio. Realizability conditions and synthesis technique are the principle concerns in any type of network synthesis. Prior to introducing the network synthesis, it is essential to emphasize certain transmission line characteristics.

Figure 1 is a schematic diagram of the length of coaxial transmission line with a current of unity magnitude entering from the left. If it is assumed that only odd mode currents flow in the line, then an equal and opposite current flows in the outer shield conductor.

The fact that equal and opposite currents flow is a critical point. If an arbitrary value of current enters the network and the current that flows in the shield is normalized to (divided by) this current, it is obvious that integer magnitude currents flow in a normalized transmission line network.

Figure 2 shows the normalized network of Figure 1. The shield at point A is removed from ground and connected to point B. This interconnection forms a 4:1 transformer. This ratio is achieved since one unit current enters the input at the left, yet two units of current exit the output of the network at point B. The network is ideally lossless, so the voltage from point B to ground must be half the voltage at the input. The ratio of voltage to current at the input relative to the ratio at the output is larger by a factor of four.

Figure 3 shows an additional line of equal length added expressly to delay the current from the shield so that both currents enter node B simultaneously, thereby providing a frequency independent transformation. The network shown in Figure 3 is the fundamental building block of all equal delay networks. The characteristics inherent in this network are present in any equal delay configuration.

There are four important network properties. First, integer or integer ratio magnitude currents are the only value of current that flows in a normalized network. By definition, integers or their ratios are defined as rational numbers. Hence, only currents with a rational magnitude flow in a normalized network. Second, the currents at node B will sum according to Kirchhoff's law. Since the currents flowing are only of rational magnitude, then the current exiting the network at node B is the sum of rational numbers. Third, the normalized output current exiting the network is equal to the square root of the transformation ratio N. Last, the transmission lines at the

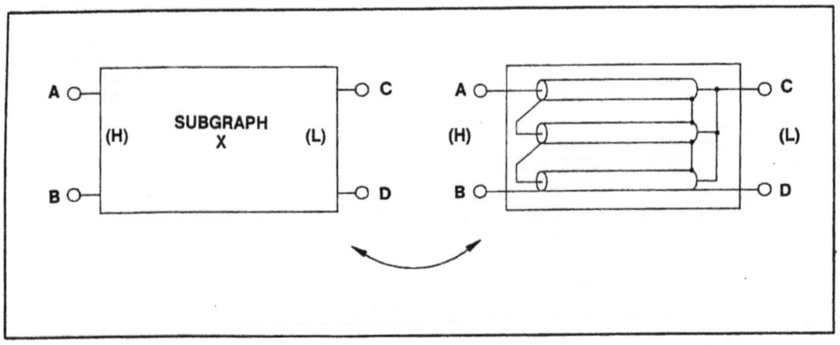

Fig. 4 The basic subgraph consisting of a series-parallel interconnection of transmission lines.

high impedance port of Figure 3 are connected in series, while the lines at the low impedance port are in parallel. Given these properties, the synthesis of equal delay networks may begin. As expected, the realizability conditions are the first consideration.

Necessary and Sufficient Realizability Condition for Finite Element Equal Delay Transmission Line Transformers

The only realizable transformation ratios of finite element equal delay transmission line transformers are ratio values possessing square roots that equate to a rational quantity. Since it was established that the current exiting the output of a normalized network is $\sqrt{N}$ and that this current is the sum of only rational magnitude currents, it is essential to show that the sum of any rational numbers will only yield another rational number. Hence, closure for addition of rational numbers must be proved.

The proof begins with two arbitrary rational numbers M and N, where

$M = A/B$
$N = C/D$

A, B, C and D are all integers.

Summing the two numbers results in

$$M + N = A/B + C/D = (AD + BC)/BD \quad (1)$$

Since integers are closed for addition and multiplication,

AD = an integer
BC = an integer
BD = an integer
AD + BC = an integer + an integer = another integer

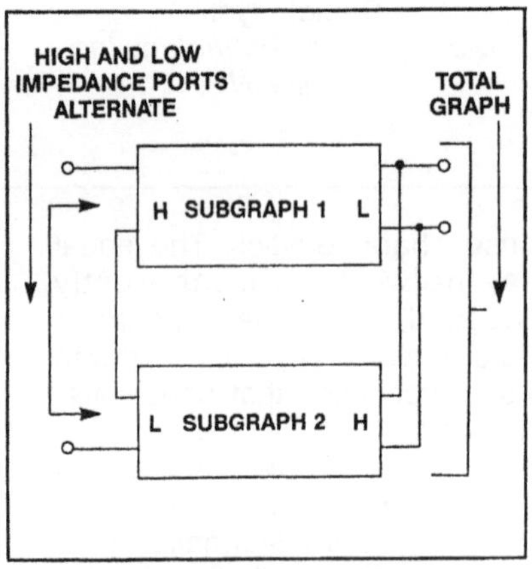

Fig. 5 An expanded version of the basic subgraph.

Therefore, (AD + BC)/BD = integer/integer and by definition, a rational number is the ratio of two integers, so the proof is complete.

For thoroughness, it must be pointed out from a purely theoretical standpoint that ratios with irrational roots require an infinite number of transmission lines to realize, obviously a physical impossibility. Now that the realizability condition is established, the next logical step is to define a procedure to synthesize the network topology.

The complete topology will be defined as a graph. Sections of graph are defined as basic subgraphs. The basic subgraph consists of a series-parallel interconnection of transmission lines, which is the only configuration of a basic subgraph, as shown in Figure 4. However, this network is limited to integer squared transformation ratios (4:1, 9:1, 16:1 etc.). Since it is desirable to have noninteger squared transformation ratios, the basic subgraph can be expanded, as shown in Figure 5. The high and low impedance ports alternate relative to one another while progressing from top to bottom of the graph. The ports must alternate, since nonalternating port subgraphs yield a graph equivalent to the sum of the two, again limited to integer squared transformation ratio capabilities, as shown in Figure 6.

Graphs can be expanded to contain any number of subgraphs provided the expansion is carried out by adding the new subgraph with its low and high impedance ports opposed relative to the subgraph directly above it, as shown in Figure 7; then interconnecting the low impedance port of the new subgraph so that it is in series with the high impedance port of the subgraph, shown in Figure 8; and finally, interconnecting the high impedance port of the new subgraph so that it is in parallel with the low impedance port of the subgraph, as shown in Figure 9.

The graphs can be expanded any number of times. The precise amount of expansions is dependent on the desired transformation ratio. With the capability to expand the graphs, all that remains is to illustrate the iterative network synthesis procedure. This process determines the number of times a graph must be expanded (E) and the number of transmission line elements per subgraph (n_x). The mathematical expressions in Appendix A will be used to solve for these values. The equations given take a square root of the transformation ratio and truncate the result, that is, remove the values to the right of the decimal point. The integer portion then is used to solve for the number of transmission lines in the first subgraph, while the removed decimal portion is utilized to determine the number of lines in the following subgraphs.

Qualitatively speaking, if a realizable ratio is given, then this value is substituted into Equation 1. Equations 1 and 1A are solved. If Equation 1A yields a value of zero, then the procedure terminates. The value given by Equation 1 is the number of transmission lines in subgraph 1.

If Equation 1A does not equal zero, then the graph is expanded once, and Equations 2 and 2A are

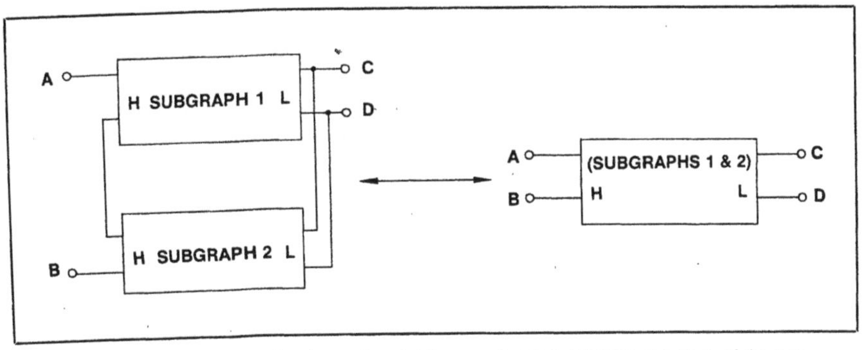

Fig. 6 Non-altering port subgraphs yield a graph equivalent to the sum of the two.

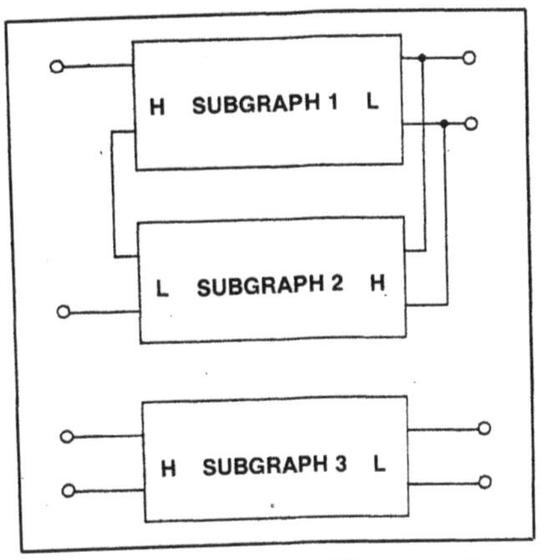

Fig. 7 The first phase of the expanded subgraph.

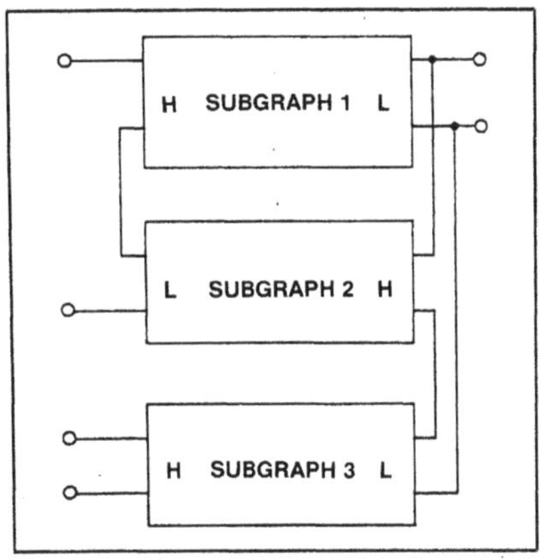

Fig. 8 The second phase of the expanded subgraph.

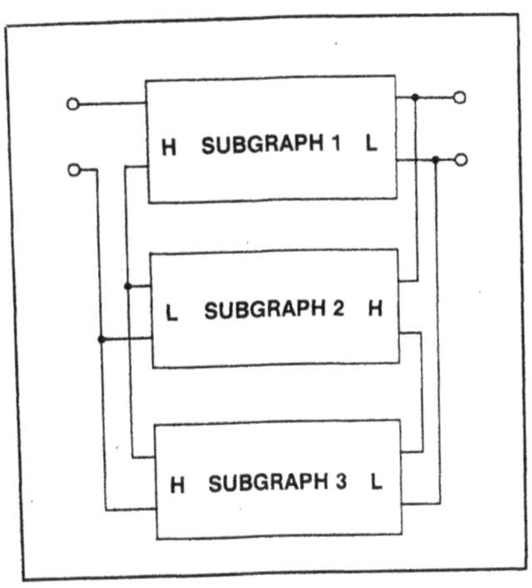

Fig. 9 The completed expanded subgraph, in which the high impedance port of the new subgraph is interconnected such that it is in parallel with the low impedance port.

used in a like manner. As before, the result of Equation 2 is the number of lines in subgraph 2. If Equation 2A equates to a nonzero value, the synthesis continues; otherwise the process terminates.

Sometimes a realizable ratio is not given, so the designer must approximate a realizable value that is close to the desired ratio. For example, an exact 2:1 ratio is not possible since the square root of the number 2 is an irrational quantity. However, the square root of 2.25 is a rational number and is a close approximation of 2. Therefore, the designer can synthesize a 2.25:1 transformer.

The complete synthesis procedure for a given transformation ratio N is summarized in the following steps. The first step is to verify that the desired transformation ratio satisfies the condition for realizability. If it does not, an acceptable close approximation must be chosen. In step two, the numerical value of N is substituted into Equation 1 and progressively the following equations are solved until an N_x value of zero occurs. In step three, the graph is expanded as required by the values determined in step two; then substitute in the proper number and configuration of transmission lines per subgraph. Finally, in step four, the optimum characteristic impedance is determined for the transmission lines from the equation

$$Z_o = \sqrt{(R*R/N)}$$

where

R = 50 Ω (for a 50 Ω system)
N = transformation ratio

All lines in the network will be given this value of impedance.

The procedure is illustrated through examples in Appendices B and C. The resultant networks were constructed and tested over the HF, VHF and UHF bands simultaneously to display the validity of the foregoing theory and the extraordinary bandwidth capabilities.

Transformer Construction

The actual 2.25:1 transformer was fabricated with 35 Ω, 0.085" OD semi-rigid coaxial cable with its center and outer conductors interconnected, as shown in Figure B3. The line lengths must all be equal and, in this example, were cut to four inch lengths. The upper two lines are wound on an 0.5" OD ferrite core with an initial permeability of 850. The lowest line is expressly for phase delay and does not require ferrite loading. The transmission lines are ferrite loaded to suppress even (common) mode currents. These currents cause a degradation of performance in the low frequency region of the operating band. Transmission lines that have both ends of the shield grounded cannot have any potential existing across it. Therefore, these lines have only odd mode currents through them. Since no even mode currents exist in the lowest line, no ferrite loading is necessary. The output port (R/2.25) was connected to a 25 Ω chip resistor. The return loss (S_{11}) was measured at the input port (R) with a vector network analyzer. The parasitic inductance of the exposed coaxial center conductors was empirically compensated for with a 10 pF capacitor in parallel with the 25 Ω resistor.

The 3.0625:1 transformer was constructed in a similar fashion, as shown in Figure C1. The coaxial lines were 25 Ω, 0.085" OD semi-rigid. The output (R/3.0625) port was terminated with a 16 Ω chip resistor. The parasitic inductances also were compensated for with a 10 pF chip capacitor.

Test Data

The two networks were measured for input return loss since this data will provide the best indication of how well the network is transforming the termination resistances. Both networks were tested over the frequency range from 3 to 300 MHz. The data shows an impedance match of better than –16 dB across the entire bandwidth for either network. This corresponds to <2.5 percent of reflected power or 97.5 percent of the incident power reaching the resistive terminations, indicative of a superior broadband impedance match.

Conclusion

A new technique of broadband network synthesis has been presented. A realizability condition and a defined procedure are described. Two networks were synthesized and tested to confirm the theory.

Acknowledgment

The author would like to acknowledge the technical advice of R. Myer from Astex Corp. Manuscript acknowledgments go to S. Lock, L. Falco and D. Johanson; and diagram acknowledgments go to J. Heick. ■

Daniel Myer *is completing his BSEE at Polytechnic University in New York. Currently, he is an RF design engineer at Power Systems Technology Inc. From 1981 to 1984, Myer worked at Telco Products. From 1984 to 1987, he was an RF design engineer at Fonar Corp. From 1987 to 1989, he was at M/A-COM Microwave Power Devices. Myer has worked on the US Air Force's HF-UHF airborne psychological warfare/deception jamming systems, the Royal Dutch Navy's L- and S-band marine search radar amplifier and the FAA's L-band airborne tracking radar test systems. He is a member of Eta Kappa Nu and the AOC.*

Appendix A: Equation Set 1

For a given transformation ratio N,

EQ 1: n_1 = truncation of $\sqrt{N}$
EQ 1A: $N_1 = \sqrt{N} - n_1$
EQ 2: n_2 = truncation of $1/N_1$
EQ 2A: $N_2 = 1/N_1 - n_2$
EQ 3: n_3 = truncation of $1/N_2$
EQ 3A: $N_3 = 1/N_2 - n_3$
EQ x: n_x = truncation of $1/N_{x-1}$
EQ xA: $N_x = 1/N_{x-1} - n_x$

where

n_x = the number of transmission line elements in subgraph x
x = the total number of subgraphs necessary for the given ratio N
E = x – 1 = the number of times a graph will have to be expanded beyond a single basic subgraph

$\sum_{1}^{x} n_x = n_1 + n_2 + n_3 + \ldots\ldots + n_x$ = the total number of transmission lines

Terminate synthesis procedure when $N_x = 0$.

Appendix B: Example 1

The desired transformation ratio is 2:1. $\sqrt{2}$ is not a rational quantity so an approximation of 2.25:1 is chosen since $\sqrt{2.25}$ equals 1.5, which is rational. Substituting into Equation 1,

$$n_1 = \text{truncation of } \sqrt{2.25}$$
$$= \text{truncation of } 1.5 = 1$$
$$N_1 = \sqrt{2.25} - 1 = 1.5 - 1 = 0.5 \qquad \text{(B1)}$$

Since N_1 does not equate to zero, the graph is expanded once and the synthesis procedure continues.

$$n_2 = \text{truncation of } 1/0.5 = 2$$
$$N_2 = 1/0.5 - 2 = 2 - 2 = 0$$

Hence, the procedure terminates. The value of E, from Equation B1, is found to be 1, so the graph must be expanded only once. Configure the graph such that subgraph 1 has its high impedance port on the right hand side, as shown in Figure B1.

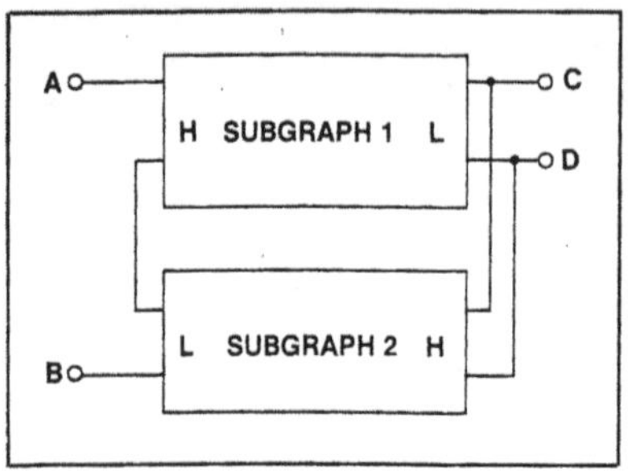

Fig. B1 Subgraph 1 with its high impedance port on the right hand side.

Substituting in the transmission lines the graph is re-configured, as shown in Figure B2. Points B and D are grounded to form an unbalanced-to-unbalanced transformer, as shown in Figure B3. The optimum characteristic impedance is $\sqrt{50*50/2.25}$ = 33 1/3 Ω for a 50 Ω system. The synthesis is now complete.

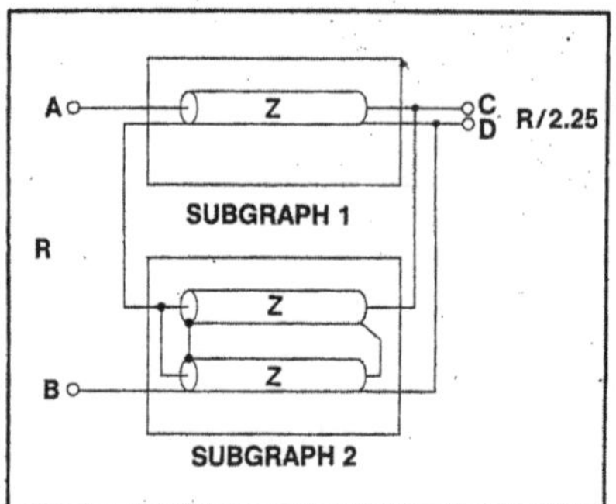

Fig. B2 The expanded subgraph with the transmission lines substituted in.

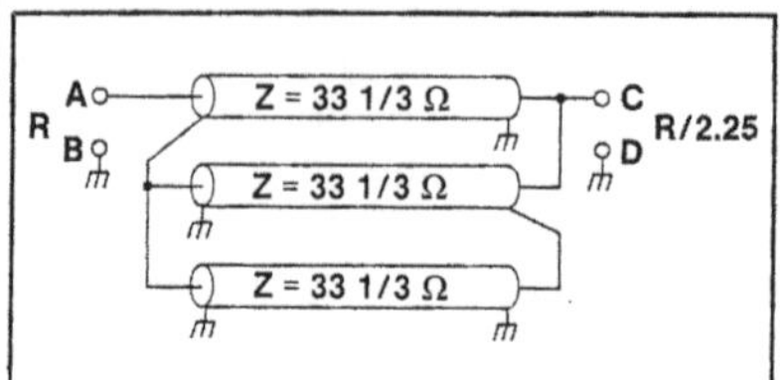

Fig. B3 An unbalanced-to-unbalanced transformer.

Appendix C: Example 2

The desired transformation ratio is 3:1. As in Example 1, the square root of the number 3 is not rational, so an approximation of 3.0625:1 was selected since $\sqrt{3.0625}$ = 1.75, which is rational. The synthesis procedure was carried out and it was found that the graph had to be expanded twice, resulting in the general topology of Figure 9. The number of transmission lines and their characteristic impedance were determined and the resulting 3.0625:1 transformer is shown in Figure C1.

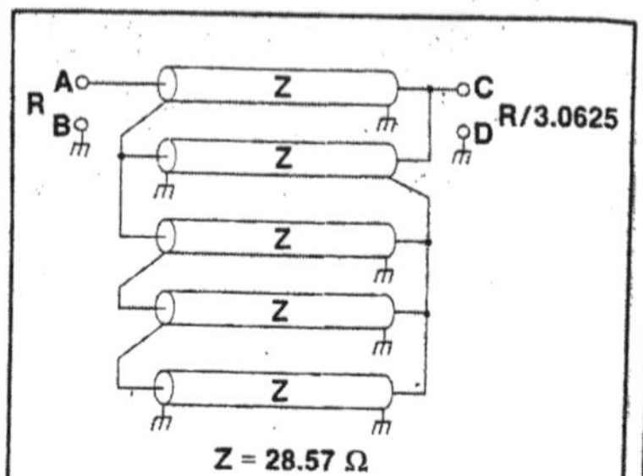

Fig. C1 The resulting 3.0625:1 transformer.

Broadband Transmission Line Transformer Family Matches a Wide Range of Impedances

By Donald A. McClure

The family of broadband transmission line transformers introduced here achieves impedance matching flexibility that is not possible with existing designs, such as the well-known Guanella devices [1]. In 1972, the author prepared a patent disclosure document [2] describing this method, but a patent application was not filed and the information was not made public. After more than twenty years, and with his personal retirement, the author now presents this transmission line transformer family for industry-wide dissemination.

To begin, we should review the example of a typical Guanella transformer shown in Figure 1. A number of transmission lines of equal length and characteristic impedance are all connected in parallel at one end and all in series at the other. These connections provide voltage transformation ratios that are equal to the number of lines used, and impedance transformations that are the squares of those simple integers. The impedance ratios are thus restricted to 4, 9, 16, 25 and so forth, or their inverses. These ratios may be useful, but it is often frustrating that other ratios can not be attained without cascading Guanella structures.

We shall see that Guanella devices constitute a special case of a more general connection scheme that can provide voltage ratios of any rational number without the need to cascade two or more transformers. Rational numbers are defined to be of the form m/n, where m and n are integers. For lack of a better name, it is convenient to term these m/n devices as RAVOR transmission line transformers — the name being an acronym of *RA*tional *VO*ltage *R*atio. In the same vein, devices made in the Guanella connections could be called IVOR transformers for *I*nteger *VO*ltage *R*atio. The IVOR devices are clearly a special case of the RAVOR family in which either the m or n integer is 1. The existence of the RAVOR family has generally been overlooked except for a rather recent article [3] in which some connections were derived in an entirely different manner from the method described here.

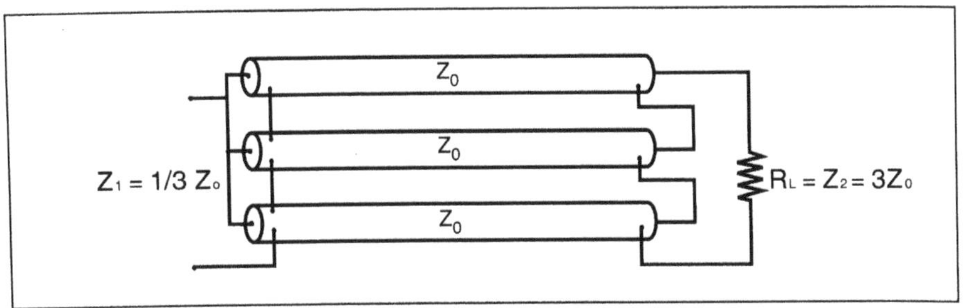

Figure 1. A simple 1/3 voltage ratio Guanella transformer as an example of prior art.

Transmission Line Connections

If a group of equal-length transmission lines of equal characteristic impedance are connected at one end in arbitrary combinations of parallel and series, how should the lines be connected at the other end to obtain frequency-independent behavior of the input and output impedances? Calculations of input impedance based upon equality of input and output powers, or simply physical experimentation, will quickly show that the connections at the output end cannot be arbitrary with respect to the input end without the producing mismatches and frequency-selective behavior. With some possible trivial exceptions, broadband behavior can be achieved in only two ways:

1) Output connections are exactly the same as the input connections. This arrangement results in no voltage or impedance transformation and does not constitute a transformer, but it can be useful to achieve a composite cable of some Z_0 other than that of the constituent cables.

2) Output connections are exactly the opposite of the input connections. That is to say, that any transmission lines or groups of transmission lines which are connected in parallel at one end of the transformer must be connected in series at the other end, and vice versa.

This second method of interconnection describes the family of RAVOR transformers. Figure 2 shows an example of the RAVOR devices, a transformer with a voltage ratio of 5/2, constructed to provide transformation from 50 to 8 ohms. This transformer will operate over decades of frequency with low loss and low standing wave ratio (SWR).

The General Principle

The series-parallel interchange interconnection of transmission lines having equal length and equal characteristic impedance, Z_0, causes the driving port and load port impedances, Z_1 and Z_2, respectively to be related in the following manner (neglecting time delay):

$$Z_1 = KZ_0$$

$$Z_2 = \frac{1}{K} Z_0 = R_{Load}$$; it follows that:

$$Z_1 Z_2 = KZ_0 \frac{1}{K} Z_0 = Z_0^2$$; and:

$$Z_0 = \sqrt{Z_1 Z_2}$$

The characteristic impedance of the lines is the geometric mean of the input and load impedances. The coefficient K is the quotient of two integers, the values of which are determined by the way in which the transmission lines are con-

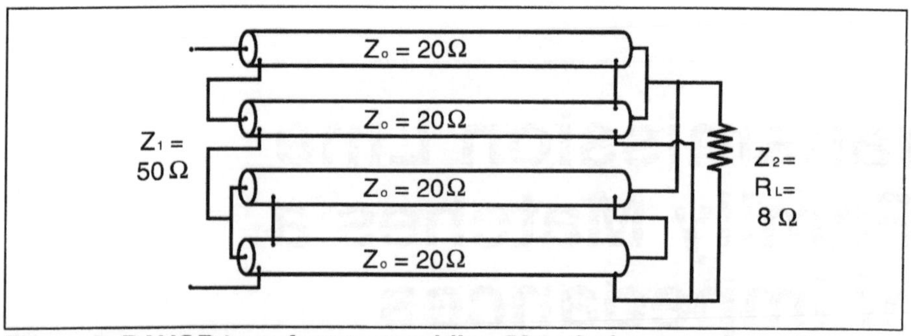

Figure 2. RAVOR transformer providing 50 to 8 ohm transformation.

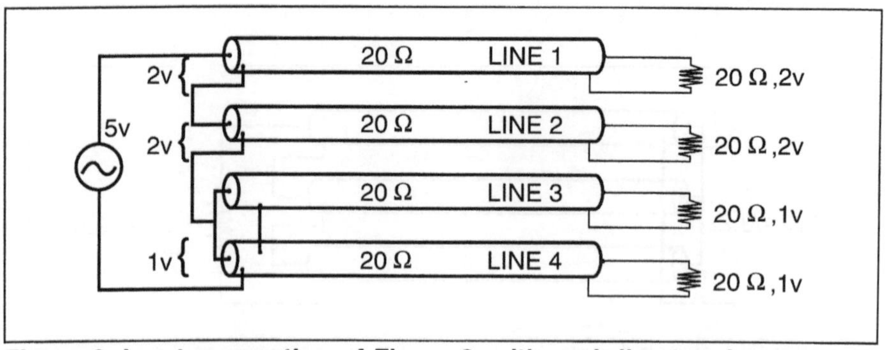

Figure 3. Input connection of Figure 2, with each line terminated in Z_0, showing voltage distribution.

nected. It is also the voltage transformation ratio of the transformer.

The coefficient K can always be expressed in the form m/n, where m and n are integers, and is thus a rational number.

Current-Voltage Duality

A current-voltage duality exists between the two ends of the transformer, resulting from the series-parallel interchange of connections between one end and the other. One way of expressing the duality is that the input current loop equations and the output voltage nodal equations have exactly the same form and coefficients, shown as follows:

Input Loop equation: $V_1 - I_1 KZ_0 = 0$

Output Nodal equation: $I_2 - V_2 KY_0 = 0$

where: I_1 = Input current
I_2 = Load current
$Y_0 = 1/Z_0$

Similar equations apply to a lossless, conventional transformer of turns ratio K, which would also exhibit duality.

The duality relationship causes the voltages and currents to be distributed among the termination ends of the lines so that each line operates as though it were terminated in its characteristic impedance. Thus, the transformers can operate over extreme bandwidths, limited at the high frequencies only by the stray inductances of the end connections. The low frequency limit Is determined by the amount of inductive reactance that can be attained along each conductor so that the input and and output ends are effectively isolated from one another and common-mode currents are nil.

To illustrate how the series-parallel interchange results in properly terminated lines, Figure 3 depicts the same input connection as in Figure 2, but with the output of each line terminated in a resistance equal to the Z_0 of the line. Five volts of RF is applied to the input. Because lines 1 and 2 each present 2 volts to their respective loads, the loads may be paralleled to yield a composite load of 10 ohms with no effect at the driving end. The currents in the loads of lines 3 and 4 are the same, hence the loads and lines may be respectively series connected to obtain 2 volts across 40 ohms. Now, the 3 and 4 series combination provides 2 volts, as does each of lines 1 and 2. Therefore, the series 3-4 combination may be paralleled with lines 1 and 2 to give a total output of 2 volts. The 40 ohm load of the 3-4 combination acts in parallel with the 10 ohm load of the 1-2 combination to yield a composite load impedance of 8 ohms.

It is apparent from the above discussion that the individual 20 ohm resistors of Figure 3 can be replaced by a single 8 ohm load if the lines are connected as in Figure 2. No change will occur in the input voltage distribution or impedance when the substitution of a single load is made, provided that there is sufficient inductive reactance along each conductor to isolate the voltages at the load end from those at the input end.

End-to-End Isolation

If 5 volts were applied to the input of the transformer of Figure 2, 2 volts will appear at the output, leaving a 3 Volt drop along the center-conductor of the upper-most cable, which could cause a large, unwanted current to flow. Such currents are reduced to insignificant values by making each conductor have a high inductive reactance throughout the intended frequency range. The low-frequency reactance can be enhanced by winding the lines into individual coils, or by winding them on, or surrounding them by, high permeability materials such as ferrite. For the 2 MHz to 100 MHz range, winding each coaxial cable on a ferrite toroid having an initial permeability of about 125 works well. For proper operation, there should be no magnetic or electric coupling between the transmission lines.

Impedance Calculations

With confidence that the lines making up the transformers are properly terminated, the driving point impedance may be written by inspection of the input connections. In the case of the connection in Figure 2, writing in general form:

$$Z_1 = 2Z_0 + \frac{Z_0}{2} = \frac{5}{2}Z_0 \text{ ; therefore:}$$

$$K = \frac{5}{2}$$

The above example illustrates the way in which the ratio $K = m/n = V_1/V_2$ appears in the impedance equations. A similar inspection of the output end yields:

$$Z_2 = \frac{\frac{Z_0}{2} \cdot 2Z_0}{\frac{Z_0}{2} + 2Z_0} = \frac{Z_0^2}{\frac{5}{2}Z_0}$$

$$= \frac{2}{5} Z_0 = \frac{1}{K} Z_0$$

Connection Notations

There are so many possible transformer configurations that it is convenient to use a system of shorthand notation to describe the connections without sketching the cables and wiring. The device of Figure 2 is fully described, except for impedance levels, by the symbology:

2S, 2P > S

which means: two lines are connected in series to form a group; two other lines are paralleled to form a group; then the two groups are connected in series. The opposite end of the transformer is connected in the opposite fashion and need not be described. In this case it is denoted by:

2P, 2S > P

The transformer connections are completely defined by either the input or output end notations above.

Table 1 lists voltage ratios and corresponding connection schemes for values of K with a denominator up to 9. Equivalent ratios (e.g. 2/3 and 4/6) are not duplicated. The author has determined connections for ratios up to a denominator of 13, a total of 57 different voltage ratios. Of these, only 12 can be achieved by a single-stage Guanella transformer, which means the RAVOR principle provides 45 additional ratios to choose from!

Also, all of the transformers of the RAVOR family, including the Guanella types, may be used as baluns (BALanced to UNbalanced converters) because of the necessary isolation from input to output.

While all ratios of the form m/n are theoretically realizable with this class of transformer, there are obviously practical limits to the number of transmission lines that can be used. As of this writing, the author has conducted a systematic search for all the connection possibilities up to and including six lines. Some ratios resulting from the use of more lines have also been found for the purpose of knowing the simplest connection for all of the ratios shown in Table 1. Using six lines, there are 53 different

Voltage Ratio K	Connections	Minimum No. Lines
1/2	2P*	2
1/3	3P*	3
2/3	2P, 1 > S	3
1/4	4P*	4
3/4	3S, 1 > P	4
1/5	5P*	5
2/5	2P, 2S > P	4
3/5	2P, 1 > S; 1 > P	4
4/5	4S, 1 > P	5
1/6	6P*	6
5/6	3P, 2P > S	5
1/7	7P*	7
2/7	2S, 3P > P	5
3/7	2P, 3S > P	5
4/7	3P, 1 > S; 1 > P	5
5/7	2S, 2P > S; 1 > P	5
6/7	2P, 1 > S; 2S > P	5
1/8	8P*	8
3/8	2P, 1 > S; 2P > P	5
5/8	2S, 1 > P; 1 > S; 1 > P	5
7/8	2P, 1 > S; 2P > P; 2P > S	7
1/9	9P*	9
2/9	3P, 2S > P; 1 > P	6
4/9	2P, 4S > P	6
5/9	4P, 1 > S; 1 > P	6
7/9	3S, 2P > S; 1 > P	6
8/9	2P, 1 > S; 1 > P; 1 > S; 2S > P	7

*Guanella (integer) ratios

Table 1. Voltage ratio and connection data for RAVOR transformers with fractional ratios through a denominator of 9.

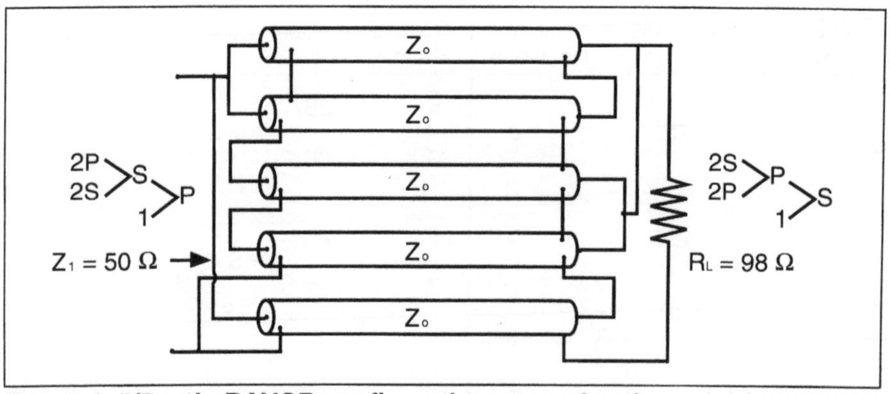

Figure 4. 5/7 ratio RAVOR configuration approximating a 1:2 impedance transformation (Z_0 = 70 ohms).

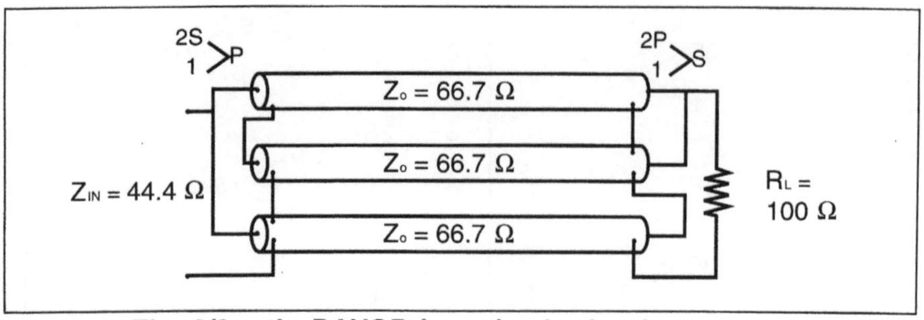

Figure 5. The 2/3 ratio RAVOR is a simpler but less accurate way to approximate a 1:2 transformation.

combinations that can be made, not all of them useful. Using seven and eight lines, there are hundreds of connections and corresponding ratios, but to use such a large number of lines and ferrite cores is to reach beyond the limits of practicality.

Configuration Selection

As an example of the use of Table 1, postulate that it is desired to make a transformer that will provide a 50 ohm to 100 ohm conversion. The voltage ratio, K, is found to be $\sqrt{50/100}$ = 0.707. Look through Table 1 to find a ratio that approximates 0.707. The ratio 5/7, which in decimal form is 0.7143, differs from 0.707 by only about 1 percent and is thus an excellent approximation.

Because 5/7 is a ratio of two prime numbers, it is already in its lowest form. Table 1 indicates the connections for the low impedance end of the transformer and five transmission lines are required. The configuration is sketched in Figure 4. The Z_0 of the lines is chosen to be 70 ohms, the geometric mean between 98 and 50 ohms.

A test transformer using the configuration of Figure 4 was constructed using five equal lengths of 50 Ohm cable (RG-196), each wound on a ferrite toroidal core of Indiana General type Q1 material which has an initial permeability of about 125. The frequency response of the test transformer was essentially flat from 3 to 300 MHz, and the voltage transformation ratio was 5/7, as expected.

For some purposes, a less accurate approximation to a 1/2 impedance ratio than that provided by the five-line configuration of Figure 4 would suffice. Returning to Table 1, it is seen that the simplest structure to approximate the 0.707 voltage ratio is the three-line device which yields a K of 2/3, or 0.667. The connection is sketched in Figure 5.

If the transformer of Figure 5 were loaded with 100 ohms at the output as shown, and the line Z_0 were chosen to be 66.67 ohms, also as shown, impedance at the input (left) end would be 44.44 ohms. In a 50 ohm system, the transformer would thus introduce an SWR of 1.125:1, which may be a tolerable number.

Summary

The article describes the theoretical basis for, and the general principles of, a class of broadband RF transmission-line transformers having voltage transformation ratios of rational form, m/n, where m and n are integers. The described configurations overcome the simple integer transformation ratio limitations of the older Guanella designs which represent a special case of these principles. The theoretical basis of operation is viewed in the light of current-voltage duality due to the series-parallel, parallel-series interchange of connections between the ends.

Design tools are provided in the form of a configuration selection procedure, and a table of ratios and the corresponding connections. These aids enable the reader to determine quickly the required number of transmission lines and the connection scheme to obtain the desired transformation. The table lists 19 rational ratios and their connections, in addition to 8 Guanella integer ratios and connections. Two examples of configuration selection are included.

With regard to synthesis of a transformer configuration, it can be shown that the connections for any ratio can be determined by using a continued fraction expansion, but sometimes connections requiring fewer lines can be found by the use of partial fraction expansions.

Future articles are planned, covering synthesis, construction and measured performance of RAVOR transformers.

Acknowledgement

Many thanks to Dr. Jerry Sevick of Basking Ridge, NJ for his encouragement and useful suggestions. ***RF***

References

1. G. Gaunella, New Method of Impedance Matching in Radio Frequency Circuits, the *Brown Boveri Review*, September 1944, pp. 327-329.
2. Donald A. McClure, "A Family of Broadband Transformers," RCA Corporation patent disclosure document, April 28, 1972, RCA Docket no. 66148. Note: the RCA Patent Operations office decided not to seek patents for the many transformer configurations disclosed.
3. Daniel Myer, "Equal Delay Networks Match Impedances of Wide Bandwidths," *Microwaves & RF*, April 1990, pp. 179-188.

About the Author

Don McClure is an engineer retired from RCA. He can be reached at 12 W. Azalea Lane, Mt. Laurel, NJ 08054; tel. (609) 778-7328.

Broadband Transmission Line Transformer Family Matches a Wide Range of Impedances – Part 2

By Donald A. McClure

This article is the second of a two-part series introducing a family of transmission-line transformers. This family achieves impedance matching flexibility that has not been possible with earlier designs. This second article provides further information on the class of transmission-line transformers dubbed RAVOR devices (for RAtional VOltage Ratio) by the author. The first article was published in the February 1994 issue of RF Design. *To avoid confusion between the two papers, figures and references in this present work are numbered in sequence beyond those in the previous work.*

In the first article [4], the theoretical basis for and the general principles of the RAVOR class of broadband transmission-line transformers were presented. These devices comprise various numbers of equal length transmission lines of equal characteristic impedance connected in various combinations of series and parallel and have voltage transformation ratios of rational form, m/n, where m and n are integers. The basis of operation is viewed in the light of current-voltage duality arising from the series-to-parallel and the reverse interchange of connections between the ends.

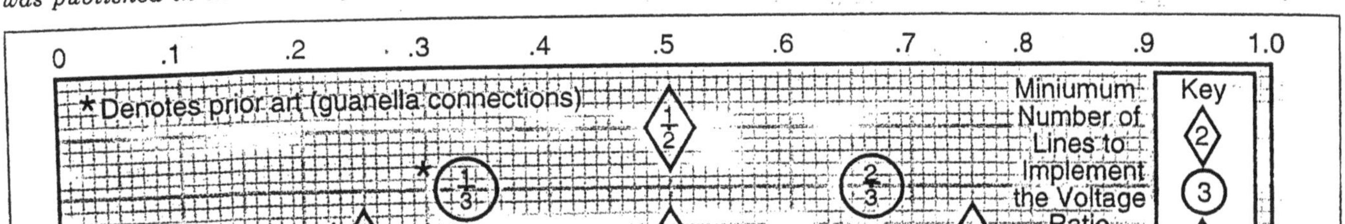

Figure 6. Conversion chart, decimal fractions to rational ratios, denominators 2 through 13.

A system of notation was introduced to describe the end connections without having to sketch the cables and wiring. The notation scheme is re-introduced here for the convenience of the reader:

2S
> S
2P

The above symbology means that two lines are connected in series to form a group; two other lines are connected in parallel to form a group: then the two groups are connected in series. The opposite end of the transformer is connected in the complementary manner and is denoted by:

2P
> P
2S

The transformer end-connections are completely defined by either the input or output end notations above and describe the 2/5 voltage-ratio transformer shown in Figure 9, except that DC isolating capacitors have been added in the figure.

Table 1 of the first article [4] lists 27 ratios and the connection notation for each, along with the number of lines required to implement each ratio. Two design examples were provided to approximate a 1 to 2 impedance transformation.

The transformer input and output impedances are written in terms of the lines' characteristic impedance, Z_0, and the voltage transformation ratio, K, as follows:

At the low impedance end,

$$Z_1 = K \cdot Z_0, \qquad 0 < K < 1 \tag{1}$$

and at the opposite end,

$$Z_2 = \frac{1}{K} \cdot Z_0 \tag{2}$$

The Z_0 of the constituent transmission lines is the geometric mean of the end impedances:

$$Z_0^2 = \frac{Z_1}{K} \cdot KZ_2 = Z_1 Z_2 \tag{3}$$

$$Z_0 = \sqrt{Z_1 Z_2} \tag{4}$$

Voltage Ratio (K)	Connections	Min. # Lines
2/3	2S, 1 > P	3
1/10	10P*	10
3/10	2P, 3S > P, 1 > P	6
7/10	2S, 3P > S, 1 > P	6
9/10	2S, 2P > P, 2P > S	6
1/11	11P*	11
2/11	5P, 2S > P	7
3/11	2P, 1 > S, 3P > P	6
4/11	3P, 1 > S, 2P > P	6
5/11	2P, 3P > S, 1 > P	6
6/11	2S, 3S > P, 1 > P	6
7/11	3S, 1 > P, 1 > S, 1 > P	6
8/11	2S, 1 > P, 2S > S, 1 > P	6
9/11	3S, 2P > S, 1 > S, 1 > P	7
10/11	2S, 1 > P, 1 > S, 2S > P	6
1/12	12P*	12
5/12	2P, 2S > S, 2P > P	6
7/12	2P, 2S > P, 1 > S, 1 > P	6
11/12	2S, 1 > P, 4P > S	7
1/13	13P*	13
2/13	2S, 6P > P	8
3/13	4P, 3S > P	7
4/13	4S, 3P > P	7
5/13	2S, 1 > P, 1 > S, 2P > P	6
6/13	2P, 1 > S, 1 > P, 2S > P	6
7/13	2S, 1 > P, 2P > S, 1 > P	6
8/13	2P, 1 > S, 1 > P, 1 > S, 1 > P	6
9/13	2S, 4P > S, 1 > P	7
10/13	3S, 3P > S, 1 > P	7
11/13	5S, 2P > S, 1 > P	8
12/13	3P, 1 > S, 3S > P	7

Table 1 (Part 2). Continuation of the voltage ratio and connection data for the RAVOR transformers, covering denominators 9 through 13. (Plus corrected entry for 2/3). (*) indicates a Guanella connection.

The Ratio Chart

It is a useful fact of number theory that all rational numbers between one and infinity can be positioned on a linear scale in the space between zero and one simply by taking the reciprocal of each. Thus, it is convenient to have a chart showing the gamut of practical fractional voltage ratios, K, positioned according to their decimal values. The chart is presented in Figure 6. Any ratio larger than one may be entered on the chart merely by taking the reciprocal, which is the same as turning the transformer end-for-end. The dot on the underside of each fraction-bar locates the decimal equivalent value of the fraction on the abscissa of the chart.

The ratio chart is a design aid which allows the reader to visuallize easily the m/n voltage ratios that lie close in value to the desired decimal fraction. This decimal fraction is found by taking the square root of the desired impdance ratio. The symbols enclosing the ratios indicate the minimum number of transmission lines to implement each ratio. The key in Figure 6 matches each of these symbols to the minimum number of lines. Ratios in which the numerator and denominator contain some common factor, such as 2/6, or 6/10, are shown as being attained by the minimum number of lines to implement the equivalent, irreducible fraction.

The ratios that are not symbol-enclosed (at the left of the chart) result from Guanella (IVOR) [4] connections having denominators 9 through 13. The number of lines required is simply the same as the value of the denominator.

The design procedure outlined in the first article [4], page 66, "Configuration Selection", should be modified to include the use of the Ratio Chart of Figure 6 to ease the transition from the decimal value to the rational form. After choosing a rational ratio from the chart that is a suitable approximation to the desired decimal value, refer to Table 1 and/or Table 1 (Part 2) to determine a set of end-connections to achieve the selected ratio.

A continuation of Table 1 from the first article is presented here, and it provides connection data for 30 more voltage ratios. These additional ratios offer finer-grained steps of transformation than were presented previously. The complete Table 1 now contains 57 ratios in irreducible form and covers fractions having denominators 2 through 13 as does the chart of Figure 6. Figure 6 shows 78 ratios, but 21 of them are reducible to the simpler forms presented in the complete Table 1. Also note the corrected connection notation for the voltage ratio 2/3.

Synthesis Techniques

Most of the connections for the ratios enclosed by line-quantity symbols in Figure 6 were first determined by making tables of all possible combinations of a given number of lines and then calculating the resulting voltage ratios. Some mistakes in the calculations were discovered, but were corrected by using a synthesis procedure. This procedure is demonstrated below for the example ratio, 8/11.

Inverting and expanding in the continued fraction form,

$$\frac{11}{8} = 1 + \frac{1}{\frac{8}{3}} = 1 + \frac{1}{2 + \frac{1}{\frac{3}{2}}} \quad (5)$$

and continuing to divide until all the numerators are unity results in a complete expansion,

$$\frac{11}{8} = 1 + \frac{1}{2 + \frac{1}{1 + \frac{1}{2}}} \quad (6)$$

The physical significance of the continued fraction can be more easily understood by multiplying both sides of the equation by Z_0 so that the expressions are for a port impedance of the transformer. In this case that port is the high impedance end because the fractional ratio, 8/11, has been inverted to be greater than one. Bear in mind that $Y_0 = 1/Z_0$:

$$Z_2 = \frac{1}{K} Z_0 = \frac{11}{8} Z_0 \quad (7)$$

$$Z_2 = \frac{11}{8} Z_0 = Z_0 + \frac{1}{2Y_0 + \frac{1}{Z_0 + \frac{1}{2Y_0}}}$$

Thus, the continued fraction expansion (c.f.e.) here is an impedance expression written in terms of Z_0 and its reciprocal, Y_0, and the integer coefficients on each. The author has algebraically manipulated the positions of Z_0 and Y_0 to achieve unity in all numerators so that a complete expansion is obtained.

To visualize the resultant transformer end-connections, it should be borne in mind that $2Y_0$, or $Z_0/2$ both mean two transmission lines in parallel and that

$$Z_0 + \frac{1}{2Y_0} \quad (8)$$

and

$$Z_0 + \frac{Z_0}{2} \quad (9)$$

both mean one line in series with two lines that are in parallel. The plus sign in a Z expression indicates a series connection.

Conversely,

$$Y_0 + \frac{1}{2Z_0} \quad (10)$$

and

$$Y_0 + \frac{Y_0}{2} \quad (11)$$

both mean one line in parallel with two that are in series. The plus sign in a Y expression indicates a parallel connection. From the above c.f.e. and the indicated interpretations of the Z_0 and Y_0 expressions, a transformer end-structure can be inferred by inspection and is shown in Figure 7a. The lower case letters in the following discussion refer to the cable identifiers, a through f, in Figure 7a.

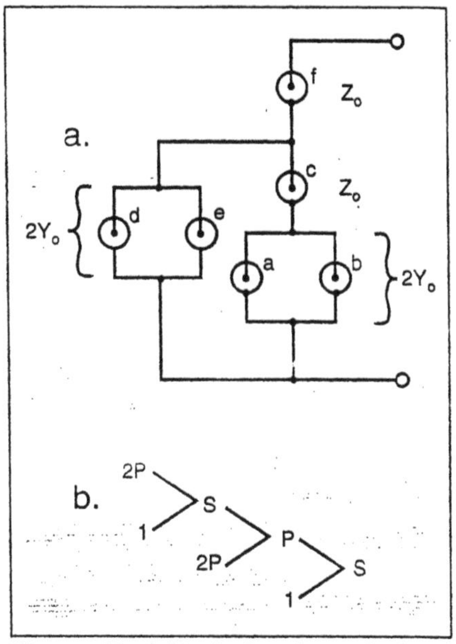

Figure 7. a) End structure derived from continued fraction expansion of $Z_0 \cdot 11/8$. b) Short-hand notation for connections in 7a.

Starting at the lower right of the expansion of equation 7 and looking at the lower two tiers,

$$Y_0 + \frac{1}{2Z_0} \quad (12)$$

means two transmission lines, (a and b), are connected in parallel with each other and the combination is in series with another line, (c). Proceeding up to the next tier of the expansion, $2Y_0$ + means that two lines in parallel, (d and e), are connected in parallel with all that went before. At the uppermost tier, Z_0 + means that one line, (f), is in series with all that went before.

Thus, using the foregoing thought process, the structure of Figure 7a is assembled, starting at the lower part of the figure and ending at the top.

Figure 7b shows the short-hand notation for the end-connections of Figure 7a. These connections are for the high impedance end of the transformer as mentioned above. To describe the structure at the low-impedance end, simply

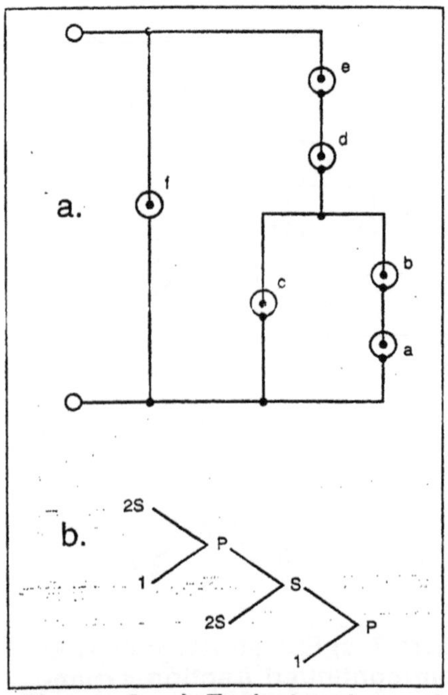

Figure 8. a) End-structure at opposite end of transformer from that shown in Figure 7a. b) Shorthand connection notation.

exchange the S's and P's in the notation of Figure 7b to yield the notation of Figure 8b which is the same as the connection symbol shown for a K of 8/11 in Table 1, (Part 2).

The ends of corresponding transmission lines in Figures 7a and 8a have been labeled with letters, a through f, in each figure so that, in comparing the two figures, the exchange of connections between the two ends of the transformer is readily discernible.

As a check that the synthesis of the end-connections has been done properly, an analysis of the configuration of Figure 7 is easily performed by setting the Z_0 of each transmission line to one ohm and then calculating the resultant output impedance. This impedance is numerically equal to the voltage ratio 1/K as follows:

$$\frac{\left(\frac{1}{2}+1\right)\cdot\frac{1}{2}}{\frac{1}{2}+1+\frac{1}{2}}+1=\frac{\frac{3}{4}}{2}+1=\frac{3}{8}+\frac{8}{8}=\frac{11}{8} \qquad (13)$$

Synthesis using Partial Fraction Expansions

A connection should result for any rational voltage ratio by using the c.f.e synthesis method, but the connection may not be the simplest (i.e., may not have the smallest number of lines possible) to implement the ratio of interest. Some ratios possess partial fraction components that can allow simpler implementations. A case in point is the economy of lines afforded by partial fraction expansion (p.f.e.) of the ratio, 6/7:

Expanding the inverted ratio in continued form,

$$\frac{7}{6}=1+\frac{1}{6}$$

the end connections are

6P > S
1

requiring 7 transmission lines. But in partial fraction form,

$$\frac{7}{6}=\frac{3}{6}+\frac{4}{6}=\frac{1}{2}+\frac{2}{3}$$

for which the end connections are,

2S > P > S
1 2P

The inverse connections at the opposite end are,

2P > S > P
1 2S

which are the connections seen in Table 1 for the ratio 6/7, requiring only five transmission lines instead of the 7 lines needed for the c.f.e case shown above.

Although many of the ratios can be expanded into partial fractions, it is only in a certain few cases, mostly involving ratios toward the right-hand side of the chart of Figure 6, that this kind of expansion yields a simpler structure than the continued fraction case.

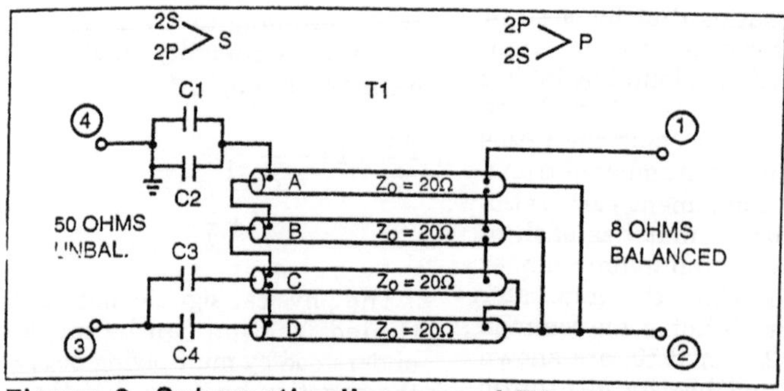

Figure 9. Schematic diagram of T1, 50 to 8 Ohm transformer, 5 to 2 voltage ratio.

Physical Example of a RAVOR Transformer

In the design of a 300 Watt, 2 to 30 MHz, push-pull, solid state power amplifier module, the need arose for a device that would transform the 50 Ohm, unbalanced, external load to an 8 Ohm, balanced, collector-to-collector load. Isolation of the external circuits from the D.C. potential at the collectors was also required.

The configuration shown schematically in Figure 9 was chosen and is labeled T1 here for convenience of reference. This transformer provides a 5/2 voltage ratio and comprises four, 20 Ohm Z_0 coaxial cables, each wound in ten turns on a 1 inch O.D., ferrite toroidal core having an initial permeability of about 125. A separate, push-pull coupling transformer/feed choke (not shown) powered the transistors. The D.C. isolating capacitors, C1 through C4, are placed at the high impedance (50 Ohm) end of T1 where the currents are smaller, thus fewer capacitors are required. Because there is no D.C. return-path through the windings, it doesn't matter that each of the conductors carries about 45 Volts D.C with respect to ground. This transformer is schematically the same as that shown in Figure 2 of the preceding article [4], except that in Figure 2 no D.C. isolation was shown and in T1 all the center-conductor and shield connections have been interchanged relative to those shown in Figure 2. The wiring interchange is of

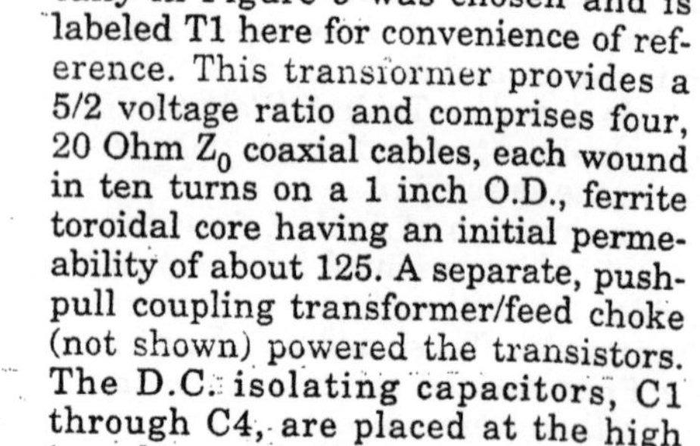

Figure 10. Photograph of physical implementation of T1.

no electrical consequence and was done for convenience of layout. The physical implementation of T1 is shown in the photograph of Figure 10. The substrate of the device is a 2.5 by 3.3 inch glass-epoxy printed circuit board having lands and traces designed to accommodate the cable interconnections and connections to external circuits. In the photograph, the external connection places have been labeled as 1 and 2 at the balanced end, and 3 and 4 at the opposite port. These numbers correspond to terminals of the same numbers on the schematic diagram of Figure 9. The trace labeled 4 is grounded in several places by plated-through holes.

The copper sheet mounted vertically at the rear edge of T1 is a Faraday shield installed to protect a nearby circuit from electric fields. It should be noted that the T1 construction shown here was for an engineering "brassboard" assembly and was not intended to survive rigorous environmental tests, such as shock and vibration.

A test set-up to measure the characteristics of T1 was made as shown in the block diagram of Figure 11. Because the intended use of T1 was over the 2 to 30 MHz range, the transformer and test-load combination was optimized for minimal VSWR in a 50 ohm circuit at just above 30 MHz. The optimization was made by the choice of value of the shunt capacitance trimmer, C, installed across terminals 1 and 2 at the load end. C is a porcelain chip capacitor, 47 pF, type ATC-100B, and compensates for some stray inductances in the test-load and in the cable terminations.

In Figure 12, curves of R and X versus frequency, taken at the high-impedance (50 Ohm) port of T1, are presented. The curves show tolerable variations from the ideal values over the 2 to 30 MHz range. The curve in Figure 13 is for VSWR at the same port and shows that the largest VSWR encountered over the frequency range of interest was about 1.06. That's good enough for most purposes.

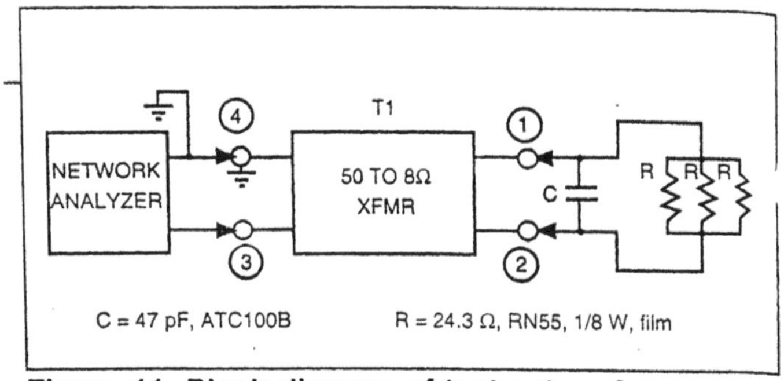

Figure 11. Block diagram of test set-up for measuring performance of T1.

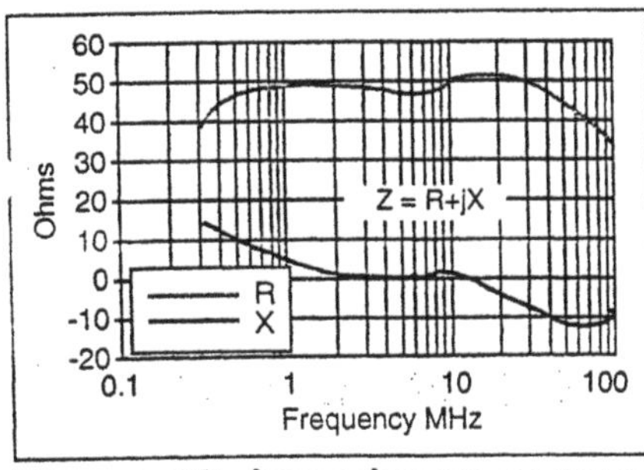

Figure 12. Impedance components, R and X, versus frequency for transformer T1.

Summary and Conclusions

In this second article on the subject of the RAVOR transmission-line transformers, we have presented further information for which there was insufficient space in the first article[4].

The ratio chart (Figure 6) introduced here is an important design tool that overcomes the difficulty in the selection of a rational voltage ratio to approximate the desired decimal fraction. The ratio chart enables the designer to visualize the options which are difficult to see otherwise.

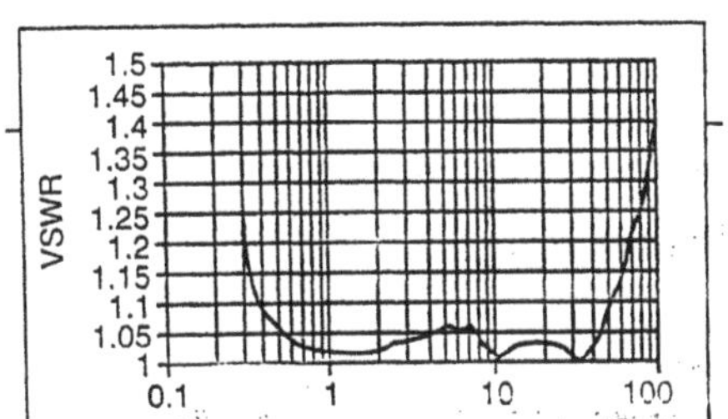

Figure 13. VSWR versus frequency for transformer T1.

The Part 2 augmentation to Table 1 provides 30 more ratios so that finer-grained transformation steps are available. The now complete Table 1 provides connection data for all ratios having denominator 2 through 13, as appear on the chart of Figure 6.

The section on ratio synthesis offers some insight into the kinds of mathematics that apply to the RAVOR devices, thus enabling the reader to check out the connections for himself.

A physical example illustrates the construction of a transformer, T1, which provides balanced-to-unbalanced conversion and a voltage transformation of 2/5. Plots of measured performance data pertaining to T1 show that the RAVOR devices truly work.

Acknowledgments

Many thanks to John Caton of Buckingham, PA, for contributing measured data on T1

References

Note: The first three references were listed at the end of the first article which is here listed as reference 4.

4. McClure, Donald A., "Broadband Transmission Line Transformer Family Matches a Wide Range of Impedances", *RF Design*, February, 1994, pp. 62-66.

About the Author

Don McClure is an E.E. graduate of Kansas State University, Class of '58. He is retired after a long career with the RCA Corporation and resides at 12 W. Azalea Ln., Mt. Laurel, NJ 08054.

IEEE TRANSACTIONS ON MICROWAVE THEORY AND TECHNIQUES, VOL. MTT-29, NO 4, APRIL 1981

Transmission-Line Transformers

ERSCH ROTHOLZ

Abstract—**The radio-frequency transformers described in this paper consist of matched transmission lines of equal length and characteristic impedance. The lines are connected according to rules given in the text. These transformers exhibit a very broad frequency response which can be readily estimated; two methods of analysis are presented. The computations agree well with test results.**

I. Introduction

A. Background

In 1944, Guanella [1] reported a transformer of novel design which consisted of two interconnected transmission lines. The transmission lines were coiled in order to suppress undesirable current modes. Later, when ferrites became available, the transmission lines were wrapped on ferrite cores. Such a transformer has been in use for some time in all television receivers, where it provides an impedance match between the 300-Ω antenna and the 75-Ω tuner input. The device has a wide frequency response, yet its construction is very simple: four wires wrapped on a ferrite core. The wires form two 150-Ω transmission lines that are connected in series at the 300-Ω end, and in parallel at the 75-Ω end. In 1959, Ruthroff [2] reported new transmission line circuits implemented with ferrite toroids.[1]

Innovative companies quickly adopted and perfected the new ideas. During 1964–1965, a generation of ultra-wideband components [3], sometimes referred to as "ferrite-loaded broad-band devices," emerged on the market. In this paper, Guanella's circuit is generalized and the design of arbitrarily complex transmission-line transformers (TLT's), is presented.

B. The Model

The TLT consists of two-wire lines, most frequently implemented as twisted-wire pairs. The wire diameter, the insulation thickness, and, to some extent, the twisting pitch determine the characteristic impedance z_o. Transmission lines relative to ground are unintentional and undesirable. Throughout this paper it is assumed that 1) all transmission lines are of equal length l, 2) the lines are connected at their ends only, and 3) they are not coupled magnetically or capacitively.

Manuscript received September 23, 1980; revised December 2, 1980.
The author is with the MITRE Corporation, P.O. Box 208, Bedford, MA 01730.

[1]A 4:1 impedance transformer discussed in detail by Ruthroff [2, fig. 3 and Appendix A] does not fit into our classification of transmission-line transformers.

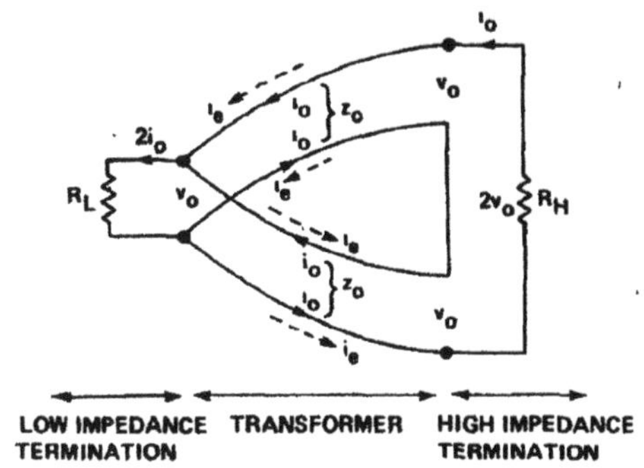

Fig. 1. Current modes in a transmission-line transformer (TLT).

C. Current Modes in a TLT

It is convenient to decompose the currents which circulate in a TLT into odd-mode currents i_o, and even-mode currents i_e, as illustrated in Fig. 1. The odd-mode currents couple with both the generator and the load and provide the desirable transformation ratio ($2i_o/i_o = 2:1$ in Fig. 1). The even-mode currents usually couple with only one termination, shunting the other; in rare cases, they couple with both the generator and the load. In the former instance they restrict the passband of the transformer and must be suppressed. The even-mode currents can be suppressed by increasing the even-mode impedance, which is accomplished by winding the lines on ferrite cores. The odd-mode currents of a tightly twisted transmission line generate a negligible external magnetic field which does not couple with the ferrite. The even-mode currents generate a strong magnetic field which is greatly affected by the ferrite.

II. Design

A. Design Approach

Since the transformation properties of the TLT's are determined by the odd-mode currents, the even-mode currents are ignored during the design procedure. The transformers so designed are then analyzed taking both modes into account.

B. Construction Algorithm

The number of transmission lines which comprise a transformer is termed the *order* of the TLT. An order-m transformer is a two-terminal pair device which consists of m connected lines. A TLT of order-($m+1$) is obtained by connecting a new transmission line to the *terminals* of the order-m transformer, in parallel at one end and in series at the other. Parallel (or serial) connection at both ends is excluded because it does not alter the transformation ratio.

A transformer of order-1 is a single transmission line, whose current and voltage are denoted i_o and v_o, respectively. An order-2 TLT is formed when a second line is connected in series at one end and in parallel at the other end. As a result of the serial connection between the two lines, the current in the second line equals the current in the first line i_o; as a result of the parallel connection the voltage across the second line equals the voltage across the first line v_o. The terminal parameters of the order-2 TLT are v_o, $2i_o$ at one end and $2v_o$, i_o at the other end (Fig. 1). In general, the terminal parameters of a TLT are Lv_o, Hi_o at one end and Hv_o, Li_o at the other end, where L and H are integer numbers.

C. Synthesis Procedure

Transformers of $r:1$ voltage ratio have the simplest configuration; they consist of r transmission lines, all of them connected in series at one end and in parallel at the other.

The synthesis procedure for an arbitrary integer voltage ratio follows. An $H:L$ $(H>L)$ voltage ratio transformer is decomposed into an $(H-L):L$ ratio TLT and a transmission line which is connected in series with the $H-L$ side and in parallel with the L side. The procedure is repeated until a $1:1$ transformer is reached.

For example, suppose that a 2.5:1 *impedance* transformer is desired. This ratio may be approximated by 5:3, 8:5, 11:7, etc., *voltage* ratios. The simplest transformer is selected and is decomposed into $(5-3):3=2:3$, $(3-2):2=1:2$, and $(2-1):1=1:1$ *voltage* ratio transformers. Fig. 2 shows the circuit diagram and illustrates the details of TLT operation. The voltage (current) transformation ratios of the first five orders of TLT's are shown in Fig. 3.

D. Characteristic Impedance

Let z_o denote the characteristic impedance of the bottom line in Fig. 1. Since the voltage and current of the top line are identical with those of the bottom line, the characteristic impedance of the top line must also be z_o. The above observation applies in general; all transmission lines in a TLT must have identical characteristic impedance z_o. The proof is straightforward. Consider a TLT with terminal parameters Lv_o, Hi_o at one end and Hv_o, Li_o at the other end. Suppose that a new line is connected, at one end in parallel with Hv_o, at the other end in series with Hi_o. Its characteristic impedance, again, must be $Hv_o/Hi_o=z_o$.

The value of z_o is determined from the specified terminal impedances, $R_L=z_oL/H$ and $R_H=z_oH/L$. For example, if $L/H=3:5$ and R_L is specified as 50 Ω, $z_o=83$ Ω. Note that such a TLT will exhibit the best frequency response between $R_L=50$ Ω and $R_H=139$ Ω, because only then will the constituent 83-Ω lines be properly terminated.

Before the transformer is constructed, the characteristic impedance of the transmission lines must be adjusted to the design value z_o. This can be done by trial and error

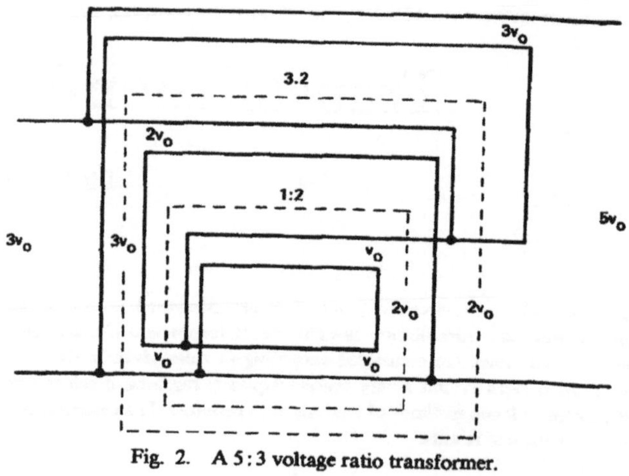

Fig. 2. A 5:3 voltage ratio transformer.

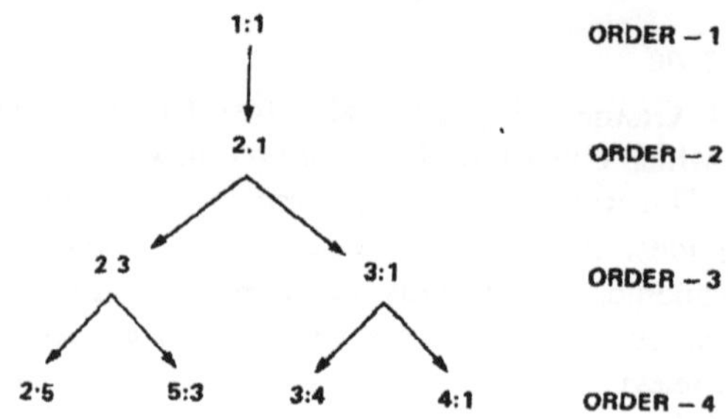

Fig. 3. The voltage transformation ratios of the first four orders of TLT's.

using wires of different diameters and insulation thicknesses. A time-domain reflectometer (TDR) is a suitable instrument for controlling the z_o value attained.

III. Analysis

For the sake of clarity, the analysis methods outlined below are illustrated with simple transformers. However, they can be applied readily to transformers of arbitrary complexity.

A. Method 1

This method employs the transformer's low-frequency equivalent circuit diagram.

First, two basic transmission line structures are analyzed. Fig. 4 shows an *in-phase* 1:1 voltage transformer. The even-mode currents couple with both the generator and the load, and, consequently, the frequency response is unlimited. Fig. 5 shows a *reversing* $1:-1$ voltage transformer. The even-mode currents circulate in a shunting short-circuited stub. For low frequencies, the stub can be replaced with the wire inductance L_w. If ferrite is used, it increases this inductance to L_p and introduces losses. For convenience, these losses are represented with a parallel resistance R_p. In the equivalent circuit shown in Fig. 5(c), the parallel combination L_p–R_p, designated as Z_p, determines the voltage insertion loss

$$\left(1+\frac{R_0}{2R_p}\right)\left(1+\frac{\dfrac{R_0R_p}{R_0+2R_p}}{j\omega L_p}\right). \qquad (1)$$

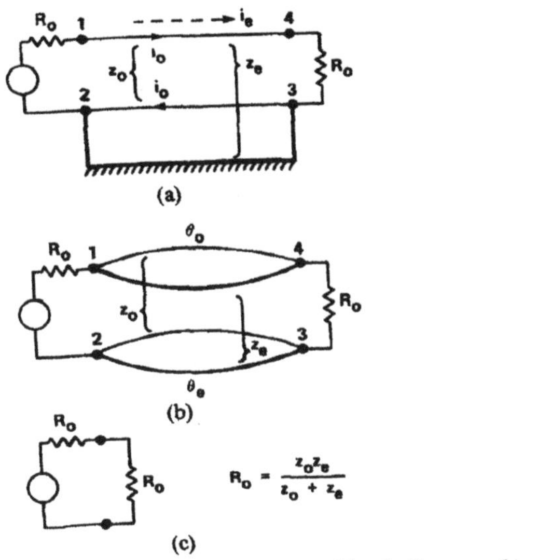

Fig. 4. An *in-phase* 1 : 1 voltage transformer. (a) Circuit diagram. (b) Equivalent circuit. (c) Low-frequency equivalent circuit.

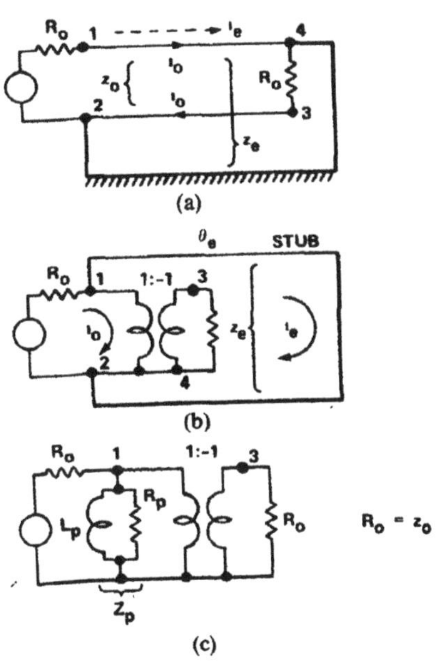

Fig. 5. A *reversing* 1 : −1 voltage transformer. (a) Circuit diagram. (b) Equivalent circuit. (c) Low-frequency equivalent circuit.

L_p and R_p are the equivalent parallel inductance and loss resistance of one wire wound on ferrite core. They can be measured conveniently with an *R–X* Meter (Boonton Radio, Type 250-A) from 0.5 to 250 MHz.

Similar analyses can be applied to higher order TLT's. The TLT shown in Fig. 6 consists of an *in-phase* transformer 1-2-3-4 and a *reversing* transformer 1-2-4-8. Only the latter affects the frequency response. The shunting impedance Z_2 is determined by the F_2 ferrite core.

Another order-2 TLT, shown in Fig. 7, consists of two *reversing* transformers. Both Z_1 (determined by F_1) and Z_2 (determined by F_2) affect the frequency response. The voltage insertion loss is calculated with (1), if appropriate impedance transformations are made in accordance with the equivalent circuit diagrams.

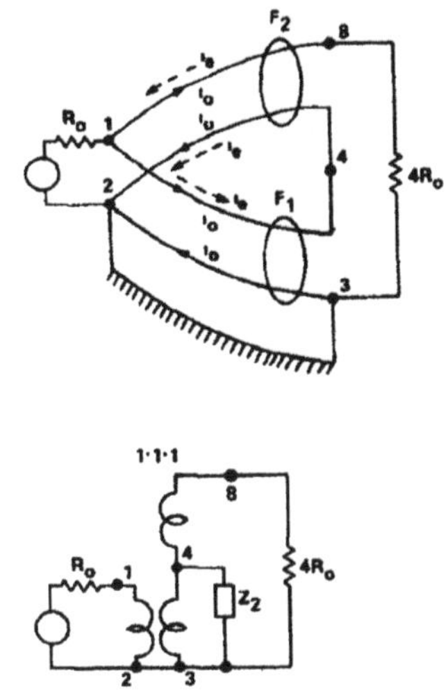

Fig. 6. An order-2 transformer and its low-frequency equivalent circuit.

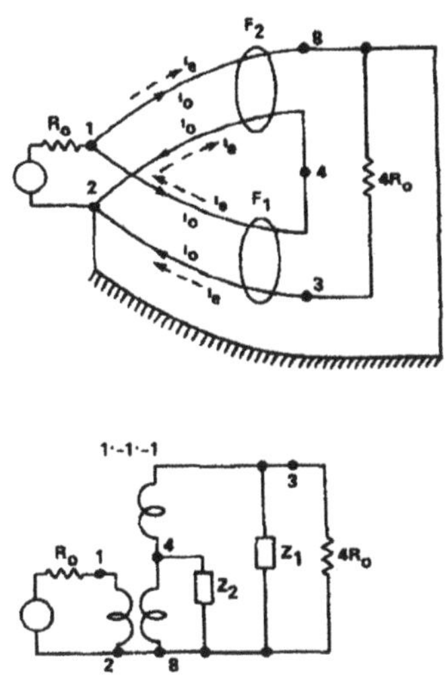

Fig. 7. An order-2 transformer and its low-frequency equivalent circuit.

B. Method 2

This method uses the admittance matrix of the TLT. The insertion loss is calculated from it by familiar matrix manipulations.

Each transmission line (two conductors running parallel to the ground plane) can be represented with a 4×4 admittance matrix [4]. (The ground plane coincides with the reference node.) Let matrix Y_1 represent transmission line 1, Y_2—line 2, etc. In general, $Y_1 \neq Y_2$, etc., because some transmission lines may use ferrite and some may not.

The admittance matrix of the transformer can be derived from $Y_1, Y_2, \cdots$ in two steps. First, the transmission lines are considered disconnected. The assembly of disconnected lines is represented by the following matrix:

$$\begin{bmatrix} Y_1 & 0 & 0 & \cdot \\ 0 & Y_2 & 0 & \cdot \\ \cdot & \cdot & \cdot & \cdot \\ \cdot & \cdot & \cdot & \cdot \end{bmatrix}. \qquad (2)$$

Then, transmission-line terminals are connected in accor-

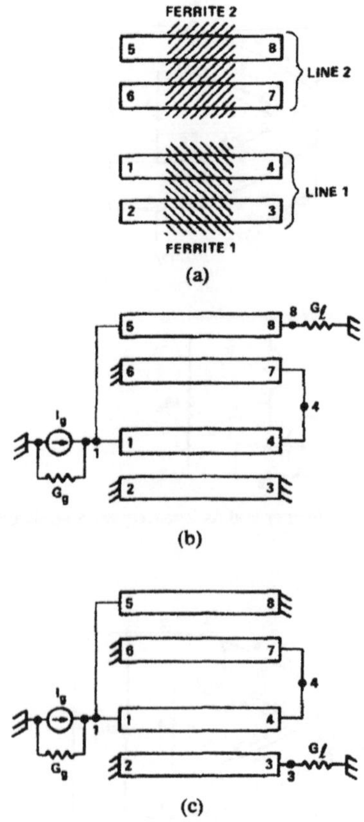

Fig. 8. Two order-2 transformers, (b) and (c), obtained by appropriate connections between disconnected transmission lines, (a). This represents a top view of the transformers. The lines run parallel to the ground plane which coincides with the surface of the drawing.

dance with the desired TLT configuration and the admittance matrix is modified correspondingly.

1) When node j is merged with node i, row j and column j are eliminated and row i and column i are modified [5], $Y_{ik}(\text{new}) = Y_{ik}(\text{old}) + Y_{jk}(\text{old})$, $Y_{ki}(\text{new}) = Y_{ki}(\text{old}) + Y_{kj}(\text{old})$, for all $k \neq i$ and $Y_{ii}(\text{new}) = Y_{ii}(\text{old}) + Y_{jj}(\text{old}) + Y_{ij}(\text{old}) + Y_{ji}(\text{old})$.

2) When a node is grounded, the corresponding row and column are eliminated.

The admittance matrix of the transformer shown in Fig. 8(b) is

$$\begin{array}{c} \\ 1 \\ 4 \\ 8 \end{array}\begin{array}{c} \begin{array}{ccc} 1 & 4 & 8 \end{array} \\ \begin{bmatrix} Y_{11}+Y_{55}+G_g & Y_{14}+Y_{57} & Y_{58} \\ Y_{41}+Y_{75} & Y_{44}+Y_{77} & Y_{78} \\ Y_{85} & Y_{87} & Y_{88}+G_l \end{bmatrix} \end{array}. \tag{3}$$

The voltage insertion loss is $\Delta/(2\Delta_{18}\sqrt{G_g G_l})$, where Δ is the determinant and Δ_{18} is a cofactor of the above matrix.

The admittance matrix of the transformer shown in Fig. 8(c) is

$$\begin{array}{c} \\ 1 \\ 3 \\ 4 \end{array}\begin{array}{c} \begin{array}{ccc} 1 & 3 & 4 \end{array} \\ \begin{bmatrix} Y_{11}+Y_{55}+G_g & Y_{13} & Y_{14}+Y_{57} \\ Y_{31} & Y_{33}+G_l & Y_{34} \\ Y_{41}+Y_{75} & Y_{43} & Y_{44}+Y_{77} \end{bmatrix} \end{array}. \tag{4}$$

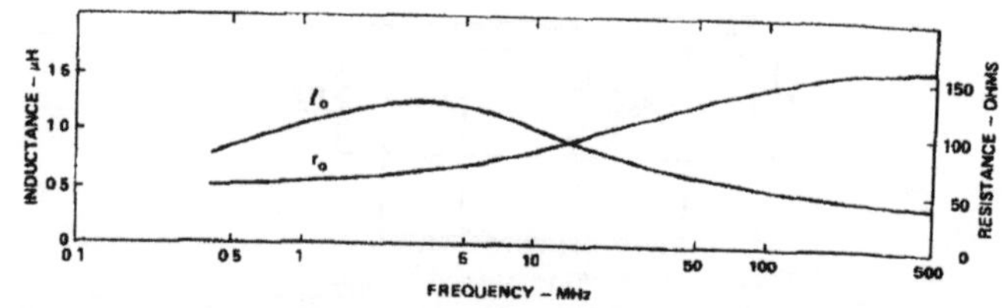

Fig. 9. Measured parallel equivalent inductance and resistance per bead.

The voltage insertion loss is $\Delta/(2\Delta_{13}\sqrt{G_g G_l})$, where Δ and Δ_{13} apply to the above matrix. The admittance matrix of TLT's can also be written by inspection. The admittance matrix elements are[2]

For line 1

$$\begin{aligned} Y_{11} &= Y_{22} = Y_{33} = Y_{44} = Y_{0o1}\coth\gamma_{o1}l + Y_{0e1}\coth\gamma_{e1}l \\ Y_{12} &= Y_{21} = Y_{34} = Y_{43} = -Y_{0o1}\coth\gamma_{o1}l + Y_{0e1}\coth\gamma_{e1}l \\ Y_{13} &= Y_{31} = Y_{24} = Y_{42} = Y_{0o1}\operatorname{csch}\gamma_{o1}l - Y_{0e1}\operatorname{csch}\gamma_{e1}l \\ Y_{14} &= Y_{41} = Y_{23} = Y_{32} = -Y_{0o1}\operatorname{csch}\gamma_{o1}l - Y_{0e1}\operatorname{csch}\gamma_{e1}l. \end{aligned} \tag{5}$$

For line 2

$$Y_{55} = Y_{66} = Y_{77} = Y_{88} = Y_{0o2}\coth\gamma_{o2}l + Y_{0e2}\coth\gamma_{e2}l, \text{ etc.} \tag{6}$$

Hyperbolic functions are used because of losses caused by the ferrite. The odd-mode and even-mode propagation constants are

$$\gamma_o = j\omega\sqrt{L'_w(C'_a+2C'_{ab})} \quad \text{and} \quad \gamma_e = \sqrt{(R'_s+j\omega L'_s)j\omega C'_a}. \tag{7}$$

The odd-mode and even-mode characteristic admittances are

$$Y_{0o} = \sqrt{(C'_a+2C'_{ab})/L'_w} \quad \text{and} \quad Y_{0e} = \sqrt{j\omega C'_a/(R'_s+j\omega L'_s)}. \tag{8}$$

All primed quantities are per unit length. L'_w is wire inductance without ferrite. L'_s and R'_s are series equivalent wire inductance and loss resistance in the presence of ferrite; they depend on the frequency. C'_a is capacitance of one wire to ground and C'_{ab} is capacitance between the transmission line wires.

Method 2 is based on the assumption that the transmission lines to ground are homogeneous. However, this condition is not required for TLT operation. Although the method is rigorous, it requires a ferrite characterization (L'_s and R'_s) usually unavailable above 250 MHz.

C. Ferrite Core

TLT's use ferrite core in two forms, beads and toroids [3]. In the first case, the transmission line is threaded through n_B beads. The equivalent parallel inductance and loss resistance are linearly proportional to n_B: $L_p = n_B l_0$ and $R_p = n_B r_0$, where l_0 and r_0 are inductance and resis-

[2] Sign errors appear in the $Y_{12} = \cdots$ and $Y_{14} = \cdots$ parameters of the Jones and Bolljahn paper, [4].

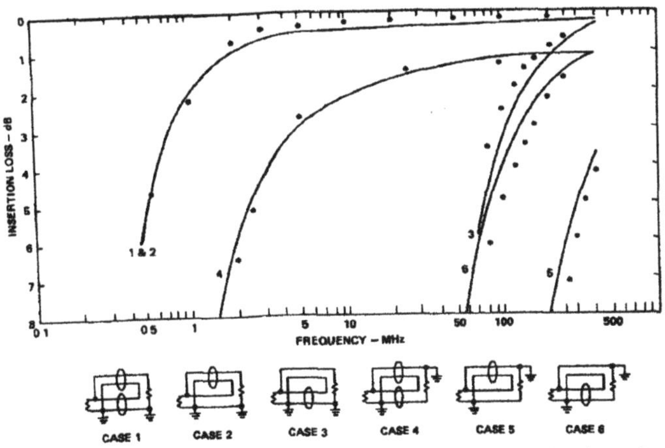

Fig. 10. Calculated (solid line) and measured (dots) insertion loss of an order-2 transformer for various ferrite and grounding combinations, as indicated by the six cases.

tance per bead. Measured values of l_0 and r_0 for a particular ferrite bead are shown in Fig. 9.

If a toroid core is used, the transmission line is wound n_t turns around it. In this case, $L_p = n_t^2 l_0$ and $R_p = n_t^2 r_0$.

Several transmission lines of a transformer may be wound on a single toroid, as demonstrated by Ruthroff [2]. Correct winding sense must be observed, so that the magnetic flux generated by each even-mode current flows in the same direction in the core.

The high-frequency cutoff of TLT's is inversely proportional to the length of the transmission lines. Geometric considerations indicate that for a given L_p, the bead designs require the same wire length as the toroid designs. However, toroid designs are more compact.

D. Insertion-Loss Measurements

In order to assess the feasibility of method 1, an order-2 transformer was tested from 0.5 to 500 MHz.

The transformer operated between 50- and 200-Ω terminal impedances, thus requiring 100-Ω transmission lines. Such a characteristic impedance was obtained with teflon-insulated wire twisted to 30 turns/130 mm. The wire used had a 0.255-mm conductor diameter (size 30) and a 0.483-mm overall diameter. The transmission lines length was 38 mm and the wire inductance L_w was 35 nH. The transmission lines were threaded through six ferrite beads. The inductance and parallel loss resistance per bead are shown in Fig. 9.

The measured and calculated insertion losses for various ferrite and grounding combinations of the transformer are shown in Fig. 10. Good agreement between them indicates that the analytical method is adequate.

Several conclusions can be drawn from this figure: 1) not all transmission lines require ferrite (cf. cases 1 and 2); and 2) grounding of a TLT is not a trivial matter (cases 4, 5, and 6 are clearly inferior to their counterpart cases 1, 2, and 3). For a wide frequency response, grounding must create *in-phase* transformers. Those transmission lines which are *in-phase* transformers do not need ferrite.

IV. Summary

TLT's consist of lines of equal length and characteristic impedance. The characteristic impedance value depends on the specified terminal impedances. The number of transmission lines and their interconnections depend on the required transformation ratio. Any integer voltage ratio may be obtained.

The frequency response is very wide. The low-frequency response is affected by undesirable even-mode currents and can be extended by using ferrite cores on some of the transmission lines. The choice of grounding points is important. In theory, the high-frequency response depends primarily on the length of the transmission lines, because at these frequencies ferrite no longer exhibits magnetic properties. In practice, transmission-line discontinuities (stray reactances) at line junctions are the determining factor.

Several methods can be used for analyzing these transformers. Adequate agreement with measured performance can be obtained with a simple model.

References

[1] G. Guanella, "Nouveau transformateur d'adaptation pour haute fréquence," *Rev. Brown Boveri*, pp. 327–329, Sept. 1944.

[2] C. L. Ruthroff, "Some broad-band transformers," *Proc. IRE*, vol. 47, pp. 1337–1342, Aug. 1959.

[3] Cover Feature, "Two ultra-broadband hybrids," *Microwaves*, pp. 106–107, Mar. 1965.

[4] E. M. T. Jones and J. T. Bolljahn, "Coupled-strip-transmission-line filters and directional couplers," *IRE Trans. Microwave Theory Tech.*, vol. MTT-4, pp. 75–81, Apr. 1956.

[5] J. Shekel, "Voltage reference node—Its transformation in nodal analysis," *Wireless Eng.*, vol. 31, pp. 6–19, Jan. 1954.

Equivalent Representation of Broad-Band Transmission-Line Transformer with Vector Impedances

G. SOUNDRA PANDIAN AND R. AROCKIASAMY

Abstract—A method of analyzing broad-band transmission-line transformers from their open- and short-circuit vector impedances is presented. From a set of open- and short-circuit impedance measurements, the input impedance and insertion loss of the transformer when terminated in any arbitrary load impedance are calculated directly.

The broad-band transmission-line transformers constructed with high permeability ferrite toroids have been extensively used for impedance matching, phase reversal, power combining, and power splitting applications in the frequency range of 100 kHz–100 MHz [1]–[7]. However, the high frequency analysis of these transformers have been dealt with the classical balanced and unbalanced transmission-line theories [2], [3], [5], [7], [8]. Since instruments which measure impedance, voltage, and gain of networks in ordinary circuit theory sense are available in the frequency range up to 110 MHz, it is not necessary to use transmission-line theory in this frequency range. The transmission-line theory poses undesirable limitations by suggesting fixed termination to match the characteristic impedance of the line. Also this theory is not accurate, since it neglects the core loss and magnetic coupling due to the ferrite core in the transformer. Recently, the validity of the transmission-line theory for the 4:1 transmission-line impedance transformer in low impedance applications has been doubted by Sevick [9]. This letter reveals that the transmission-line transformer analysis can be made accurately with simple lumped element transformer equivalent circuit.

Fig. 1(a) shows the usual transformer equivalent circuit, referred to primary. L_M, R_M, and C_T are magnetizing inductance, core loss resistance, and total lumped shunt capacitance, respectively. r_p and r_s' and l_p and l_s' are winding resistances and leakage inductances, respectively. Fig. 1(a) can be used as such for low frequency analysis. However, at high frequencies the circuit parameters of Fig. 1(a) are not unique for different frequencies. The magnetizing inductance L_M and core loss resistance R_M decrease at high frequencies due to the complex frequency dependent behavior of the ferrite core. r_p and r_s' increase with frequency due to skin and proximity effects. Usually, the ferrite core manufacturers do not provide informations about the variation of the ferrite core parameters at high frequencies. Fig. 2 shows typical variations of the relative permeability (μ_r) and the relative dissipation factor (defined as the ratio of the quality factor of a coil wound on a ferrite core and the initial permeability of the core) of a ferrite material of the Manganese-Zinc ferrite type. The curves in Fig. 2 were computed from the impedance measurements of an inductor wound on a toroidal core (1-cm overall diameter) of Mn-Zn ferrite material with relative permeability specification of 2300 ± 25 percent. R_{mp} is the parallel resistance of the inductor, reflecting the core loss. The relative permeability of the core falls from a value of 2000 at 1 MHz–45 at 100 MHz. The relative dissipation factor increases many fold with increase in frequency. Accordingly, the values of L_M and R_M will vary widely at higher frequencies. Fig. 1(a) may not be convenient for the high frequency analysis of the transformer, unless empirical relations of L_M, R_M, r_p, and r_s' as a function of frequency are known beforehand. The equivalent circuit of Fig. 1(d) solves this problem, by incorporating open- and short-circuit vector impedances instead of fixed circuit elements. By performing open- and short-circuit tests at different frequencies, the complete performance characteristics of the transformer can be predicted. Fig. 1(b) and 1(c) show the circuit elements that constitute the open circuit impedance Z_{open} and short-circuit impedance Z_{short}, respectively. In Fig. 1(d), Z_{strue} is the true short-circuit impedance which is equal to $(r_p + r_s') + j2\pi f(l_p + l_s')$. Usually, Z_{strue} is approximately equal to the measured value of Z_{short}.

Table I illustrates the use of the equivalent circuit of Fig. 1(d). The readings of Table I were taken for a bifilar 4:1 transmission-line transformer wound on a toroidal core of Mn-Zn ferrite material having the same characteristics described by Fig. 2. The open circuit vector impedance Z_{open} is inductive at low frequencies and capacitive at high frequencies, as expected from Fig. 1(b). The parallel resonance due to C_T and L_M occurs at some frequency (around 4.5 MHz). From the Z_{short} readings, it may be verified that the leakage inductance $\{(l_p + l_s') = |Z_{\text{short}}| [\sin(\theta_{\text{short}})]/2\}$ remains more or less constant in

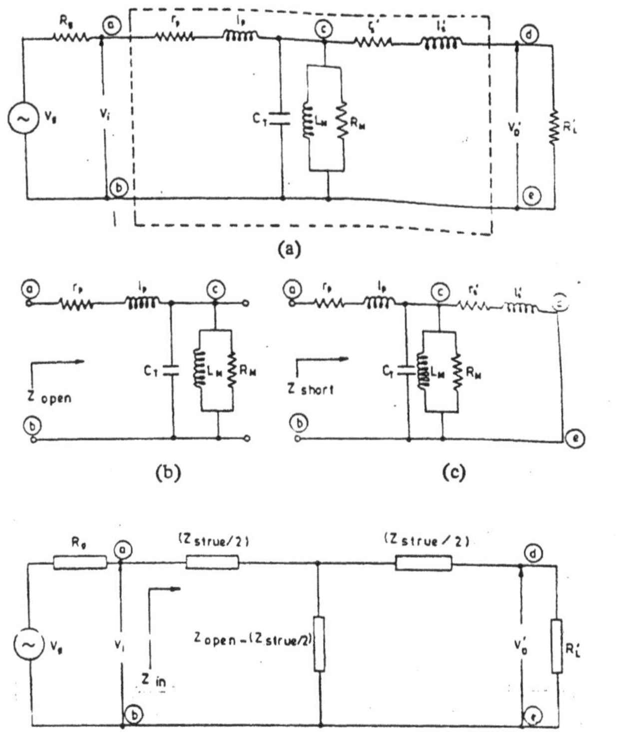

Fig. 1. (a) Basic transformer equivalent circuit. (b) Open-circuit impedance. (c) Short-circuit impedance. (d) Equivalent circuit of the transmission-line transformer with vector impedances.

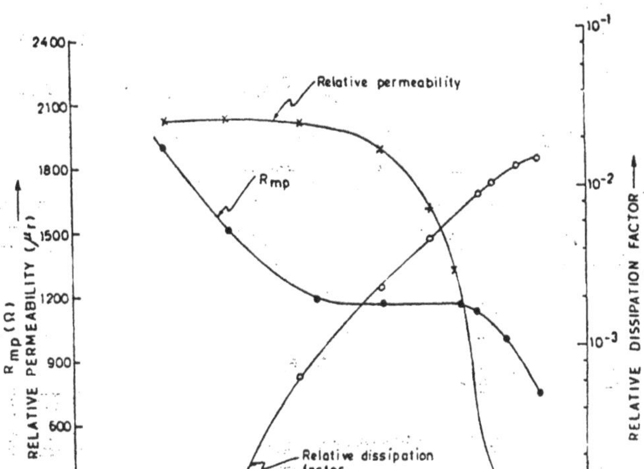

Fig. 2. Ferrite core parameters variation at high frequencies.

Manuscript received April 17, 1981; revised June 12, 1981.
The authors are with Instrument Design Development Centre, Indian Institute of Technology, Hauz Khas, New Delhi-110 016, INDIA.

0018-9219/81/1000-1368$00.75 © 1981 IEEE

Proceedings of the IEEE. Vol. 69, No. 10, October 1981

TABLE I
BROAD-BAND 4:1 IMPEDANCE MATCHING TRANSMISSION-LINE TRANSFORMER CALCULATIONS IN THE FREQUENCY RANGE 500 kHz–108 MHz

Frequency (f) MHz	(Measured) Z_{open}		(Measured) Z_{short}		Calculated Z_{in} for R'_L = 328Ω		Calculated insertion loss (for $R_g = R'_L$=328 (decibles)	Measured Z_{in} for R'_L = 328Ω		Measured insertion loss (for $R_g = R'_L$=328) (decibles)
	$\|Z_{open}\|$ Ω	(θ_{open})	$\|Z_{short}\|$ Ω	(θ_{short})	$\|Z_{in}\|$ Ω	(θ_{in})		$\|Z_{in}\|$ Ω	(θ_{in})	
0.5	1290.0	+77.0°	0.25	73.0°	302.3	+13.2°	0.3053	304	+15.0°	0.33
1.0	2400.0	+59.5°	0.52	79.0°	305.1	+6.33°	0.3100	307	+7.5°	0.33
4.5	5290.0	+11.5°	2.15	80.5°	309.5	+1.04°	0.2651	309	+0.5°	0.26
10.0	4030.0	-40.5°	4.60	82.5°	309.2	-2.06°	0.2730	308	-1.5°	0.26
20.0	2290.0	-59.0°	8.94	85.5°	304.5	-5.00°	0.3292	307	-5.0°	0.33
30.0	1590.0	-66.0°	12.8	86.5°	299.1	-7.63°	0.3858	301	-7.0°	0.35
40.0	1290.0	-68.0°	17.6	88.0°	294.1	-9.11°	0.4321	299	-9.0°	0.39
50.0	1020.0	-72.0°	21.7	88.0°	288.9	-11.7°	0.4763	285	-11.0°	0.42
60.0	851.0	-73.0°	26.1	88.5°	281.5	-13.5°	0.5466	280	-12.5°	0.54
70.0	752.0	-74.0°	30.0	89.5°	275.8	-15.3°	0.5852	278	-14.0°	0.54
80.0	660.0	-75.5°	35.0	89.5°	270.2	-17.0°	0.6245	275	-15.0°	0.63
90.0	570.0	-78.5°	38.7	89.5°	265.7	-20.0°	0.6394	265	-19.5°	0.63
100.0	520.0	-77.0°	43.7	89.5°	256.4	-20.5°	0.7432	261	-19.0°	0.72
108.0	480.0	-77.5°	48.2	89.5°	249.6	-21.6°	0.8155	255	-20.5°	0.72

(Specification of transformer: Ten turns of twisted bifilar 32 SWG magnet wires of length = 16 cm (with 6 twists per centimeter) wound on a toroidal core of 1 cm over all diameter with $\mu_r = 2300 \pm 25$ percent).

the complete frequency range 500 kHz-108 MHz. Also, since the phase angle (θ_{short}) is above 73° for all the frequencies, it is obvious that the effect of $(r_p + r'_s)$ has been completely swamped out by the large leakage reactance at high frequencies. Now referring to Fig. 1(d), the input impedance (Z_{in}) of the transformer is

$$Z_{in} = \frac{[(Z_{strue}/2) + R'_L]\,[Z_{open} - (Z_{strue}/2)]}{Z_{open} + R'_L} + \frac{Z_{strue}}{2} \tag{1}$$

where R'_L is the load resistance referred to primary.

The insertion loss (IL) of the transformer is defined as

$$\begin{aligned} \text{IL} &= 10 \log \left[\frac{\text{Available power output}}{\text{Actual power output}}\right] \\ &= 10 \log \left[\frac{\{v_g/(R_g + R'_L)\}^2 / R'_L}{(v'_o)^2 / R'_L}\right] \qquad (2) \\ &= 20 \log \left[\frac{v_g}{2v'_o}\right] \\ &= 20 \log \left\{\text{Magnitude of} \left[\frac{(R'_L + Z_{in})\,[R'_L + (Z_{strue}/2)]}{2R'_L\,[Z_{in} - (Z_{strue}/2)]}\right]\right\} \qquad (3) \end{aligned}$$

when $R'_L = R_g$.

v_g is the generator voltage and R_g is the internal impedance of the generator. From equations (1) and (3), just by knowing Z_{open} and Z_{strue}, it is possible to predetermine the input impedance and insertion loss of the transformer for any terminating condition. Table I shows the calculated values of Z_{in} and insertion loss (IL) of the transformer when $R_g = R'_L = 328$ ohms, assuming $Z_{strue} = Z_{short}$. The measured values of Z_{in} and IL shown in the last two columns of the table closely agree with the calculated values, for the same terminating condition of $R_g = R'_L = 328\ \Omega$. Table I excellently demonstrates the validity, accuracy, and simplicity of the proposed equivalent circuit of Fig. 1(d).

The impedance measurements were made with a Hewlett-Packard 4815 A vector impedance meter. Corrections were made in the measured readings of the vector impedance meter to account for its probe circuit elements. By measuring the voltages v_g and v_o with a high frequency oscilloscope having high input impedance active probe, and then substituting these values in (2), the insertion loss of the transformer was found out. It may be noted that, due to the presence of stray reactances in the test resistors R_g and R_L, and due to the inaccuracies involved in measuring high frequency voltages, the calculated values of IL will be more accurate than the measured values. Table I includes frequencies in the range of 500 kHz–108 MHz only, but it can be easily extended to lower and higher frequencies.

The advantage of the proposed vector impedance equivalent circuit is that, with only one set of open- and short-circuit impedance measurement, it is possible to calculate directly the Z_{in} and IL of the transformer for any arbitrary load termination. The applications of the equivalent circuit will be twofold; (i) it helps in analyzing, thoroughly, a transformer designed, and (ii) it gives guidelines for the designing of a transformer itself, to obtain desired bandwidth and efficiency.

REFERENCES

[1] G. Guanella, "New method of impedance matching in radio-frequency circuits," *Brown Boveri Rev.*, pp. 327–329, Sept. 1944.

[2] C. L. Ruthroff, "Some broadband transformers," *Proc. IRE*, vol. 47, pp. 1337–1342, Aug. 1959.

[3] H. L. Krauss and C. W. Allen, "Designing toroidal transformers to optimize wideband performance," *Electronics*, pp. 113–116, Aug. 16, 1973.

[4] H. Granberg, "Broadband transformers and power combining techniques for RF," Motorola Semiconductor Products Inc., Application Note, AN-749, pp. 1-10, 1975.

[5] C. W. Allen and H. L. Krauss, "Wideband rotary transformer—Unbalanced current analysis," *Proc. IEEE*, pp. 200–206, Feb. 1977.

[6] O. Pitzalis, R. E. Horn, and R. J. Barnnelio, "Broadband 60W HF linear amplifiers," *IEEE J. Solid-State Circuits*, vol. SC-6, pp. 95–102, June 1971.

[7] Jerry Sevick, "Broadband matching transformers can handle many kilowatts," *Electronics*, pp. 123–128, Nov. 25, 1976.

[8] O. Pitzalis and T. P. Couse, "Practical design information for broadband transmission line transformers," *Proc. IEEE*, pp. 738–739, Apr. 1968.

[9] Jerry Sevick, "Transmission line transformers in low impedance applications," in *MIDCON 78, Midwest Electron. Show and Conv.*, (Dallas, TX), pp. 34.1–34.4, Dec. 12–14, 1978.

TRANSMISSION LINES ACCURATELY MODEL AUTOTRANSFORMERS

A good match is obtained between modeled and measured S-parameter data from 1 to 2500 MHz.

TRANSMISSION-line autotransformers are the most widely-used components for the impedance-matching of two single-ended circuits across wide bandwidths spanning VHF and UHF. Traditionally, these components are modeled using lumped-element equivalent circuits. However, by employing transmission-line theory to autotransformer modeling, accurate results can be obtained over an even greater frequency range.

The transmission-line transformer concept[1] utilizes the principle that the resonances caused by the wire inductances and the capacitances between adjacent coil windings can be reduced by forming a transmission line, thus extending the bandwidth to higher frequencies. In practice, this is done by winding the coil with bifilar wire on a ferrite core (Fig. 1). One strand of the bifilar winding is returned to the beginning of the two-wire system, where it is connected to its twin strand.

K.B. NICLAS, R.R. PEREIRA, and A.P. CHANG, Watkins-Johnson Co., 3333 Hillview Ave., Palo Alto, CA 94304; (415) 493-4141; FAX: (415) 493-1027.

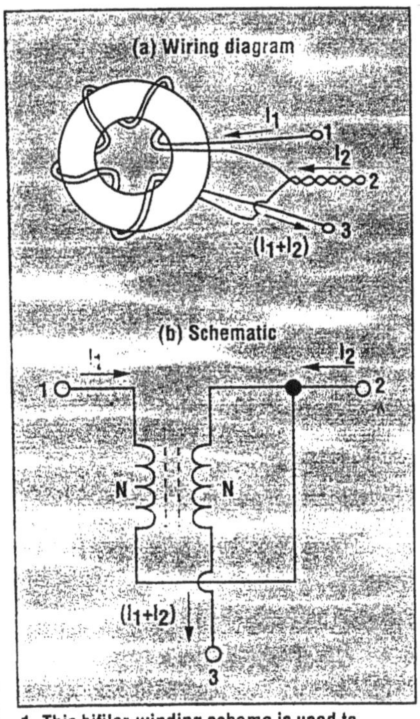

1. This bifilar-winding scheme is used to produce a transmission-line autotransformer.

The concept of reducing the parasitic-capacitance effects in the transmission-line transformer is similar to that of the distributed amplifier.[2] The latter theory proposes to extend an amplifier's frequency band by neutralizing the interelectrode capacitances of the active devices (electron tubes). This is accomplished by placing inductances between the grids as well as between the anodes of the electron tubes, thus forming an active artificial transmission line. The resulting device (known as the distributed amplifier) has, thanks to the invention of the GaAs field-effect transistor (FET), found widespread use up to 100 GHz.

A paper by Ruthroff[3] popularized the concept of the transmission-line transformer. In this analysis, transmission-line theory was used to explain the device's high-frequency performance, but the transformer model was limited to a single lossless line. O. Pitzalis *et al.* later described various forms of broadband transmission-line transformers for use in VHF bands, as well as power hybrids for push-pull amplifiers.[4]

The S-parameters of a five-turn bifilar autotransformer were measured from 1 to 2500 MHz. The measured data (Fig. 2) represents the experimental basis of a Smith-chart analysis. As the first step in the analysis, two equivalent-circuit types are developed for the transmission-line transformer. The first model consists of lumped elements, while the second is primarily composed of lossy transmission lines.

LUMPED-ELEMENT MODEL

The lumped-element model (Fig. 3a) is based on the traditional treatment of transformers using lumped elements such as inductors, resis-

Table 1: Lumped-element model parameters

Frequency (MHz)	N	L_1 (μH)	L_2 (μH)	L_{12} (μH)	R_1 (Ω)	R_2 (Ω)	G_{01} (mS)	C_{01} (pF)	C_{02} (pF)	C_{12} (pF)
5 to 600	5	2.937	2.939	2.929	1.035	0.542	0.962	0.597	1.676	0.557
5 to 800	4	1.650	1.645	1.640	1.727	1.000	1.500	0.702	1.750	0.271
5 to 1000	3	1.028	1.021	1.018	2.320	2.060	2.307	0.659	1.046	0.130

tors, and capacitors. Redistributing the model's elements results in the equivalent circuit of Fig. 3b, which consists of the inner circuit surrounded by the parasitic capacitances (C_1, C_{02}, and C_{12}) and conductance (G_{01}). Assuming linear operation, the inner circuit can be characterized by:

$$V_1 = [R_1 + j\omega(L_1 + L_{12})\times I_1' + [R_2 + j\omega(L_2 + L_{12})]\times (I_1' + I_2') \quad (1a)$$

$$V_2 = -j\omega L_{12} I_2' + [R_2 + j\omega(L_2 + L_{12})](I_1' + I_2') \quad (1b)$$

The open-circuit impedance parameters are:

$$Z_{11}' = R_1 + R_2 + j\omega(L_1 + L_2 + 2L_{12}) \quad (2a)$$

$$Z_{12}' = Z_{21}' = R_2 + j\omega(L_2 + L_{12}) \quad (2b)$$

$$Z_{22}' = R_2 + j\omega L_2 \quad (2c)$$

Conversion to the equivalent short-circuit admittance parameters yields:

$$Y_{11}' = (R_2 + j\omega L_2)/[(R_1 + j\omega L_1)(R_2 + j\omega L_2) + \omega^2 L_{12}^2] \quad (3a)$$

$$Y_{12}' = Y_{21}' = -[R_2 + j\omega(L_2 + L_{12})]/[(R_1 + j\omega L_1)(R_2 + j\omega L_2) + \omega^2 L_{12}^2)] \quad (3b)$$

$$Y_{22}' = [R_1 + R_2 + j\omega(L_1 + L_2 + 2L_{12})]/[(R_1 + j\omega L_1)(R_2 + j\omega L_2) + \omega^2 L_{12}^2)] \quad (3c)$$

Taking into account the parasitic capacitances and conductance G_{01}, the transformer equivalent circuit can be reduced to the equivalent π circuit of Fig. 3c and described in terms of its admittance parameters:

$$Y_{T1} = (G_{01} + j\omega C_{01}) + \ldots$$

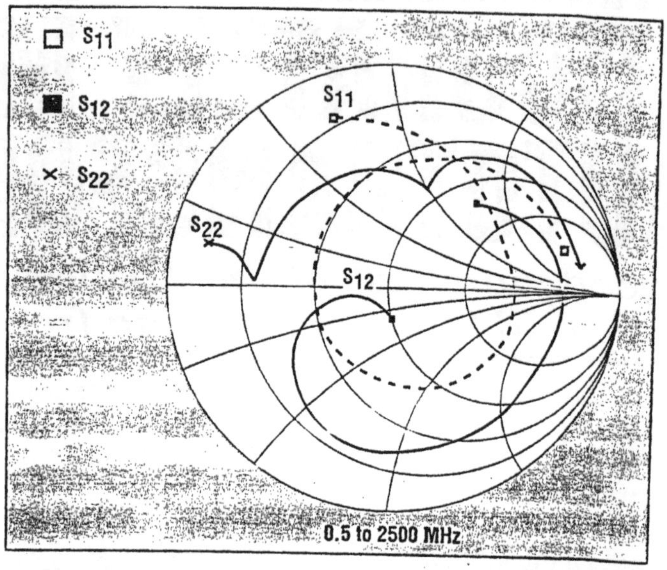

2. A five-turn bifilar transformer's S-parameters were measured for use in lumped-element and transmission-line model development.

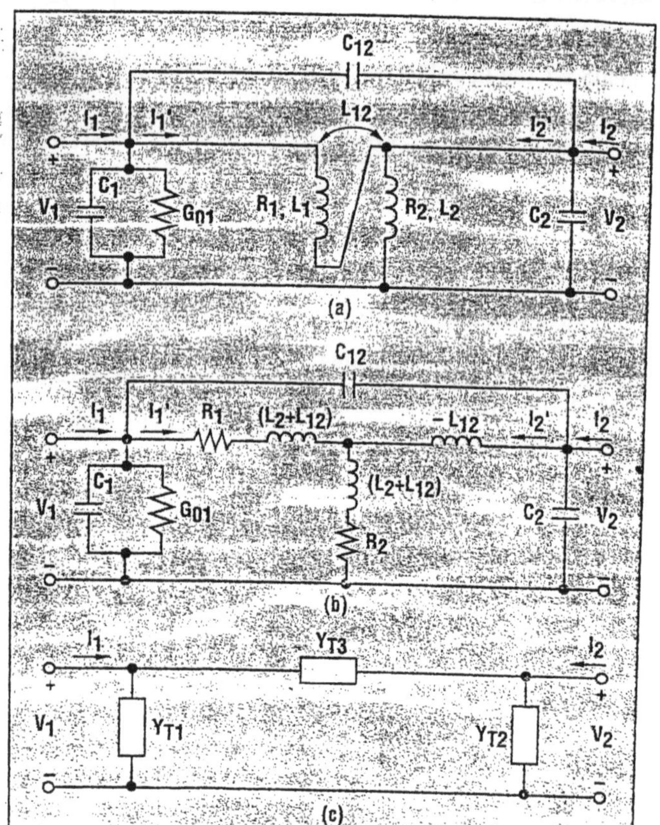

3. The transformer lumped-element model can be simplified by converting the circuit elements to their equivalent admittances.

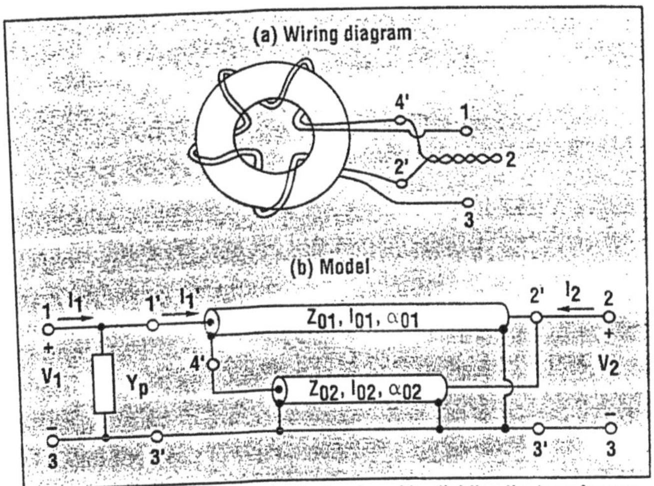

4. The transmission-line model is developed by dividing the transformer into two two-port networks (covering low- and high-frequency operation) which are cascaded.

$$Y_{11}' + Y_{12}' \quad (4a)$$

$$Y_{T2} = j\omega C_{02} + Y_{22}' + Y_{12}' \quad (4b)$$

$$Y_{T3} = j\omega C_{12} - Y_{12}' \quad (4c)$$

The relationship between the currents and voltages is best expressed using the admittance matrix equation:

$$\begin{bmatrix} I_1 \\ I_2 \end{bmatrix} = \begin{bmatrix} Y_{11} & Y_{12} \\ Y_{21} & Y_{22} \end{bmatrix} \begin{bmatrix} V_1 \\ V_2 \end{bmatrix} \quad (5a)$$

Using the admittance parameters defined by Eq. 4, the matrix equation becomes:

$$\begin{bmatrix} I_1 \\ I_2 \end{bmatrix} = \begin{bmatrix} Y_{T1}+Y_{T3} & -Y_{T3} \\ -Y_{T3} & Y_{T2}+Y_{T3} \end{bmatrix} \begin{bmatrix} V_1 \\ V_2 \end{bmatrix} \quad (5b)$$

Once the admittance parameters are determined, they can easily be converted into the corresponding S-parameters using familiar formulas. The set of S-parameters measured at each frequency characterizes the transformer performance and constitutes the data to which the corresponding model S-parameters will be fitted.

Transmission-line theory was first applied to analysis of the bifilar autotransformer by Ruthroff. This work is expanded in order to extend the validity of the transformer model to lower and higher frequencies (1 to 2500 MHz) and significantly improve the model's accuracy. To accomplish this, the transformer is divided into two two-port networks, one covering the very low end and the other covering the rest of the frequency band. Both two-ports are cascaded to form the transformer model (Fig. 4). Between 1 and 50 MHz, the transformer characteristics can be simulated by the admittance:

$$Y_p = G_p + j(\omega C_p - 1/\omega L_p) \quad (6)$$

followed by an ideal transformer with the winding ratio of n = 2. The accuracy of this simple low-frequency circuit is fairly high due to the coupling coefficient of the toroidal transmission line being extremely high. The parallel inductance L_p provides the actual DC short needed to complement the transmission-line model of the second two-port, which incorrectly represents an ideal (n = 2) transformer at zero frequency. The function of the ideal transformer at low frequencies is accurately provided by the transmission-line circuit, which is assumed to have frequency-dependent losses proportional to the square-root of the frequency, and by the admittance Y_p across the input port (1′ to 3′). At frequencies above Y_p's parallel resonance, the second two-port dominates the transformer characteristics.

The high-frequency response of the bifilar autotransformer can be accurately modeled by an equivalent circuit, which consists of two lossy transmission lines in a series-shunt configuration, preceded by the shunt admittance Y_p at the high-impedance port. The transmission-line portion of Fig. 5 adds the voltages of the series port and the currents of the shunt port according to:

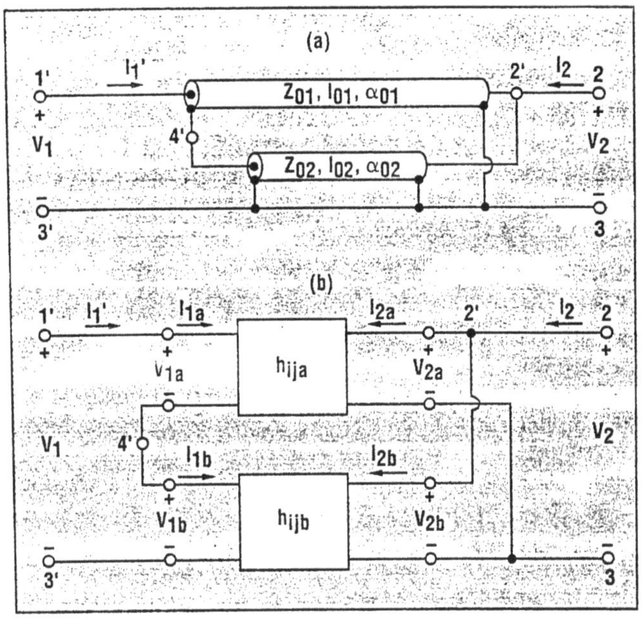

5. The transmission-line model of the bifilar transformer is converted to block-diagram form to characterize it in terms of two-port admittances.

$$V_1 = V_{1a} + V_{1b} \quad (7a)$$

$$I_2 = I_{2a} + I_{2b} \quad (7b)$$

It is therefore convenient to char-

Table 2: Transmission-line model parameters

Frequency (MHz)	N	L_p (μH)	C_p (pF)	G_p (mS)	Z_{01} (Ω)	l_{01} (cm)	α_{01} (dB/cm)	l_{02} (cm)	α_{02} (dB/cm)
1 to 2500	5	9.90	0.447	0.8554	69.6	5.350	0.142	1.160	0.091
1 to 2500	4	6.42	0.441	1.2760	56.0	4.444	0.135	1.170	0.130
1 to 2500	3	3.95	0.424	2.1980	79.6	4.250	0.171	0.856	0.057

Notes: l_{01} and l_{02} are the physical lengths for an air dielectric. α_{01} and α_{02} represent attenuation per unit length at 2000 MHz.

acterize the two-port network by means of the transmission line's hybrid parameters:

$$\begin{bmatrix} V_1 \\ I_2 \end{bmatrix} = \begin{bmatrix} h_{11} & h_{12} \\ h_{21} & h_{22} \end{bmatrix} \begin{bmatrix} I_1 \\ V_2 \end{bmatrix} \quad (8)$$

The resultant hybrid parameters are:

$$h_{11} = Z_{01} \tanh \theta_1 + Z_{02} \tanh \theta_2 \quad (9a)$$

$$h_{12} = (\cosh \theta_1 \cosh \theta_2)/(\cosh \theta_1 + \cosh \theta_2) \quad (9b)$$

$$h_{21} = -(\cosh \theta_1 \cosh \theta_2)/(\cosh \theta_1 \cosh \theta_2) \quad (9c)$$

$$h_{22} = Z_{01}^{-1} \tanh \theta_1 + Z_{01}^{-1} \tanh \theta_2 \quad (9d)$$

where:
$\theta = \gamma l$,
γ = propagation constant = $\alpha + j\beta$,
l = physical length of the transmission line,
α = attenuation constant, and
β = phase constant.

α is given in nepers (or 0.1151 dB per unit length) and assumed to be proportional to the square root of the frequency. β is given in radians per unit length. Converting Eq. 9 into the corresponding Y-parameters yields:

$$Y_{11}' = \cosh \theta_1 \cosh \theta_2/(Z_{01} \sinh \theta_1 \cosh \theta_2 + Z_{02} \sinh \theta_2 \cosh \theta_1) \quad (10a)$$

$$Y_{12}' = Y_{21}' = (\cosh \theta_1 + \cosh \theta_2)/(Z_{01} \sinh \theta_1 \cosh \theta_2 + Z_{02} \sinh \theta_2 \cosh \theta_1) \quad (10b)$$

$$Y_{22}' = [2(1 + \cosh \theta_1 \cosh \theta_2) + (Z_{01}/Z_{02} + Z_{02}/Z_{01}) \times \sinh \theta_1 \sinh \theta_2]/(Z_{01} \sinh \theta_1 \cosh \theta_2 + Z_{02} \sinh \theta_2 \cosh \theta_1) \quad (10c)$$

The admittance Y_p is shunted across Y_{11}, resulting in the following two-port admittance representation of the transformer's equivalent circuit:

$$Y_{T1} = Y_{11}' + Y_{12}' + Y_p \quad (11a)$$

$$Y_{T2} = Y_{22}' + Y_{12}' \quad (11b)$$

$$Y_{T3} = -Y_{12}' \quad (11c)$$

As was the case with the lumped-element model, the current-voltage relationship is best expressed by Eq. 5a. Again, the Y-parameters of the transmission-line model can be transformed into the corresponding S-parameters using familiar conversion formulas.

MODELING EXAMPLE

The circuit-element values are adjusted in order to match the computed model's performance parameters to the actual transformer's measured data. This can be done with any of the commercially-available computer programs that provide an optimization routine, for instance, Touchstone from EEsof, Inc. (Westlake Village, CA) or Supercompact from Compact Software (Paterson, NJ). For convenience, the device performance is characterized at microwave frequencies by measuring the S-parameters. Resistors of 50 Ω are used as the source and load impedances.

The autotransformer to be modeled consists of five turns of bifilar wire wound on a toroidal core of MnZn ferrite material, with outer and inner dimensions of 0.155 in. (0.394 cm) and 0.088 in. (0.224 cm), respectively. The core thickness is 0.050 in. (0.127 cm).

The measured S-parameters from 0.5 to 2500 MHz are shown in the Smith chart of Fig. 2. Since the transformer is a reciprocal device, S_{12} and S_{21} are identical. It is apparent from the S-parameter curves that the transformer has a significant amount of parasitic elements, and its characteristics are by no means those of an ideal transformer widely used in circuit design. While the characteristics exceed the useful operating range at fundamental frequencies, they are needed to accurately determine the transformation of harmonics and intermodulation products, as well as the distortion of nonsinusoidal waveforms. It is important to point out that the transformer analysis assumes linear operation and cannot be applied to power levels that cause the device to approach saturation.

The element values of the lumped-element equivalent circuit of Fig. 3a are first found via computer (Table 1). Smith-chart displays of the modeled and measured S-parameters demonstrate excellent agreement, but make it difficult to trace the equivalence of the corresponding parameters at individual frequencies. A better means of display is to plot the magnitude and angle of the corresponding S-parameters and (due to inordinate bandwidth) compare them using a logarithmic frequency scale. This is done in a system consisting of a 200-Ω source resistor and 50-Ω load resistor appropriate for the intended 4:1 impedance transformation. The comparison (Fig. 6) reveals excellent agreement between measured and modeled data up to at least 600 MHz, and in the case of transducer loss $|S_{21}|$, to 1200 MHz.

Similar agreement was achieved for three- and four-winding transformers, although not at the same low frequencies. A comparison of the primary, secondary, and mutual inductances for each individual case displays the impressive proximity of these parameters to each other. The elements representing ohmic losses (such as G_{01}, R_1, and R_2) move in the direction of increased losses with a decreasing number of windings (N). Of the parasitic capacitances, the feedback capacitance C_{12} shows a definite trend of reduced value with decreasing N.

The transmission-line model of the bifilar autotransformer consists of two lines in a series-shunt configuration, along with the admittance Y_p, across the high-impedance port. In its effort to match the model's performance parameters to measured values, the computer program arrived at a set of admittance and line characteristics (Table 2). Measured and computed parameters were compared for the 200-Ω/50-Ω system. An excellent fit was found between the two sets of S-parameters from 1 to 2500 MHz, demonstrating the quality of the transmission-line model.

The computer program was also able to determine the transmission-line element values for the three- and four-winding autotransformers. The lengths of the individual transmission lines assume an air dielectric (relative permittivity of $\epsilon_r = 1$ and relative permeability of $\mu_r = 1$). Consequently, the measured length of the winding ($l_{01ph} = 2.30$ cm) is shorter than the modeled length ($l_{01} = 5.35$ cm).

The line connecting the points 2′ and 4′ in Fig. 4 is basically suspended in air and has a physical length of $l_{02ph} = 1.20$ cm, remarkably close to the modeled length of $l_{02} = 1.16$ cm. The length of l_{02ph} was dictated by the test fixture and can be reduced for practical applications. While the parallel inductances (L_p) are approximately proportional to N_2, the computer-derived physical lengths generally do not adopt values proportional to N once the optimal fit is obtained. No effort was made to constrain the transmission-line lengths to force compliance in this area, since the tightness of the windings affects the electrical lengths, characteristic impedance, and main-transmission-line loss.●●

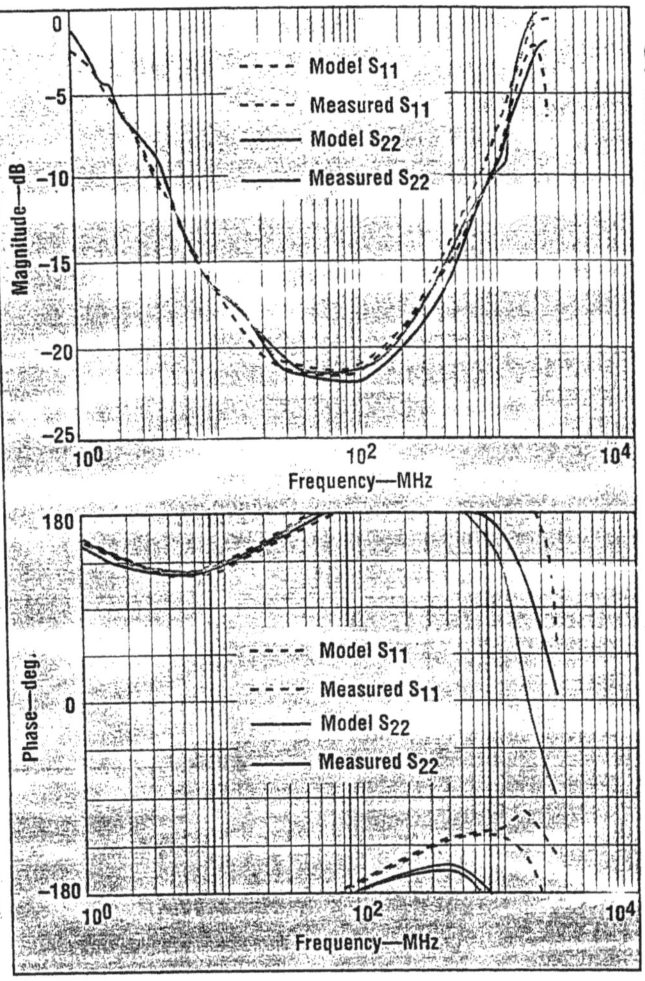

6. Good agreement is found between measured and modeled S_{11} and S_{22} (both magnitude and phase) for the bifilar transformer up to at least 600 MHz.

References

1. G. Guanella, "New Method of Impedance Matching in Radio-Frequency Circuits," *Brown-Boveri Review*, Vol. 31, 1944, p. 327.
2. W.S. Percival, "Thermionic Valve Circuits," British patent 460562, January 25, 1937.
3. C.L. Ruthroff, "Some Broadband Transformers," *Proceedings of the IRE*, Vol. 47, August 1959, pp. 1337-1342.
4. O. Pitzalis and T. Couse, "Broadband Transformer Design For RF Transistor Power Amplifiers," *Electronic Components Conference Proceedings*, 1968, pp. 207-216.

MODEL CHARACTERIZES TRANSMISSION-LINE TRANSFORMERS

A distributed-element analysis is used to derive the transmission-line transformer's input impedance.

TRANSMISSION-LINE transformers are useful circuits for microwave impedance-matching applications due to their broad operating bandwidth. An equivalent model for transmission-line transformers supports accurate calculation of input impedance through a distributed-circuit approach. The complete model accounts for electromagnetic (EM) parameters, including mutual-coupling effects.

The basic expression for the input impedance of a transmission-line transformer (Fig. 1) was first obtained by Ruthroff:[1]

$$Z = R_o[2R_L[1+\cos(\beta l)] + jR_o \sin(\beta l)]/[R_o \cos(\beta l) + jR_L \sin(\beta l)] \quad (1)$$

where:

$\beta = \omega(LC)^{0.5}$,

ω = the radian frequency,

l = the electrical length of the transmission line,

R_o = the characteristic impedance of the transmission line,

R_L = the load impedance, and

L and C = the resonant inductance and capacitance, respectively.

With this expression, an input impedance of approximately four times the load impedance is obtained at the design frequency. This impedance, however, cannot be used at low frequencies. The formula is based on the usual hypothesis of equal but opposite currents on the line.

Recently, Abrie obtained two different equations that provide separate descriptions of the behavior at low and high frequencies.[2] He verified that different currents in the two line conductors must be considered. Shimada derived the input impedance of a 1:1 balun using electric and magnetic coupling coefficients,[3] while Rotholz analyzed transmission-line transformers using even- and odd-mode currents.[4]

Figure 2 shows the electrical model used for the transmission-line transformer analysis. Inductance L/2 and mutual inductance M are related to the system geometry. Since the currents in the two line conductors are not equal, the balanced and

ENZO CARPENTIERI, RF Circuits Designer, Alenia SpA, Via Fusaro 187, Bacoli, 80070 Naples, Italy; (39) 81-527-2242, FAX: (39) 81-868-7552.

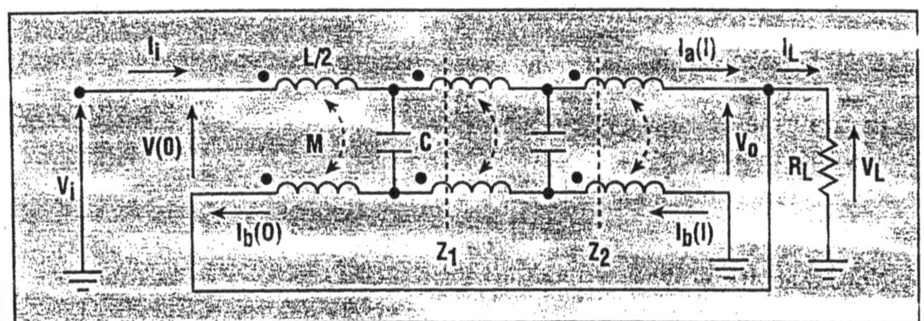

2. This network is used to electrically model the transmission-line transformer.

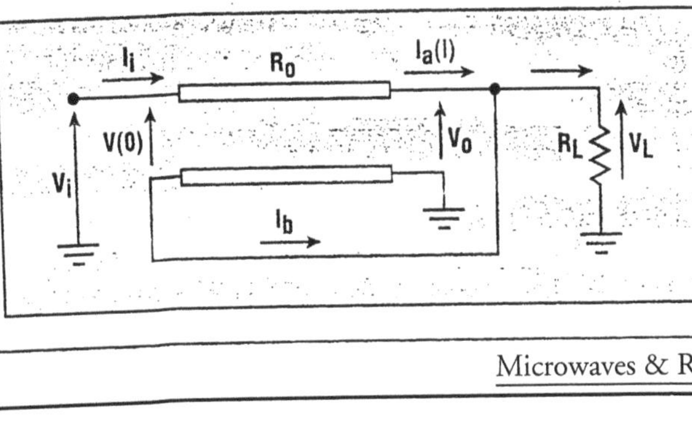

1. This simple transmission-line network delivers a 4:1 impedance transformation.

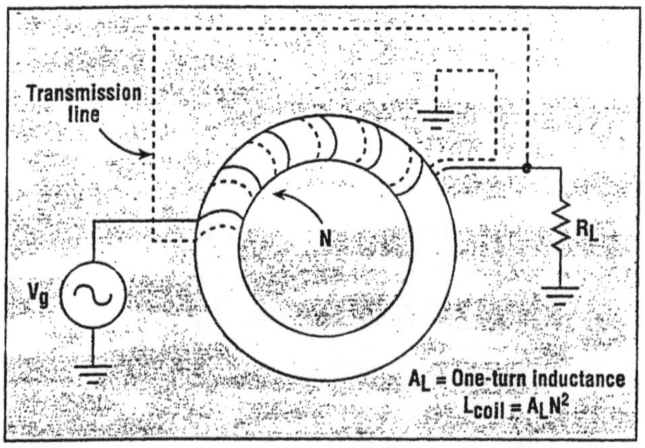

3. A practical transmission-line transformer can be formed by winding a transmission line on a toroid.

unbalanced components should be considered. The inductance seen by the unbalanced currents (which are also called coil-mode currents) will differ from that of the transmission line, as the effect of M is a function of the current verse (or direction).

In the case of equal-verse currents, the equivalent inductance is given by:

$$L_{coil} = L + 2M \quad (2)$$

Note that the network appears to be a coil to the equal-verse currents. In the case of equal but opposite currents, the equivalent inductance is described by:

$$L_{line} = L - 2M \quad (3)$$

L_{line} and L_{coil} are easily evaluated using physical parameters (conductor diameter, dielectric properties, line length, etc.). These two inductances are then used to determine L and M.

In the practical case where a transmission line is wound on a toroid (Fig. 3), the parameters to be considered are the transmission-line inductance (given by L_{line}) and the toroid inductance (described by L_{coil}):

$$L_{line} = 0.921 l \mu_r \log(d/r) \quad (4)$$

$$L_{coil} = \mu N^2 r^2 / 2R \quad (5)$$

where:

l = the line length (m),

μ_r = the relative permeability,

d = the spacing between the centers of the wires (m),

r = the radius of the wires (m),

r_c = the radius of the coil,

N = the number of turns,

μ = the permeability of the medium inside the coil, and

R = the radius of the toroid (m).

Based on Fig. 2, the following equations are used to obtain an expression for the input impedance:

$$V(Z_1,t) = (L/2)\, dZ \frac{dI_a}{dt} - M\, dZ \frac{dI_b}{dt} + V(Z_2,t) + (L/2)\, dZ \frac{dI_b}{dt} - M\, dZ \frac{dI_a}{dt} \quad (6)$$

$$I_a(Z_1,t) = I_a(Z_2,t) + C\, dZ \frac{dV}{dt} \quad (7)$$

$$I_b(Z_1,t) = I_b(Z_2,t) + C\, dZ \frac{dV}{dt} \quad (8)$$

where Z_1 and Z_2 denote spatial abscissas (as shown in Fig. 2). In fact, Eqs. 6 to 8 can be used in conjunction with the basic electrical model in Fig. 2 to characterize conventional transmission lines (see "Conventional lines").

The input and output voltages (V_i and V_o, respectively) are given by:

$$V_i = \int_o^l \left[(L/2)\frac{dI_a}{dt} - M\frac{dI_b}{dt} \right] dZ + V_o \quad (9)$$

$$V_o = \int_o^l \left[-(L/2)\frac{dI_b}{dt} + M\frac{dI_a}{dt} \right] dZ \quad (10)$$

By using these equations in con-

CONVENTIONAL LINES

With some minor modifications, the basic electrical model for the transmission-line transformer can be applied to conventional transmission lines (see figure). The modified circuit can be used to generate the formulas for voltage $V(z_1, t)$ and currents $I_a(z_1, t)$ and $I_b(z_1, t)$ that are given by Eqs. 6-8.

By combining these formulas with the hypotheses $I_a = I_b$ and $M = KL/2$, the equations can be written in the form (substituting I for I_a and I_b):

$$\frac{dV}{dZ} = L(K-1)\frac{dI}{dt}$$

$$\frac{dI}{dZ} = -C\frac{dV}{dt}$$

These show conventional transmission-line equations if the line inductance is given by L(1 – K). ●●

This circuit is used to apply the transmission-line transformer analysis to conventional transmission lines.

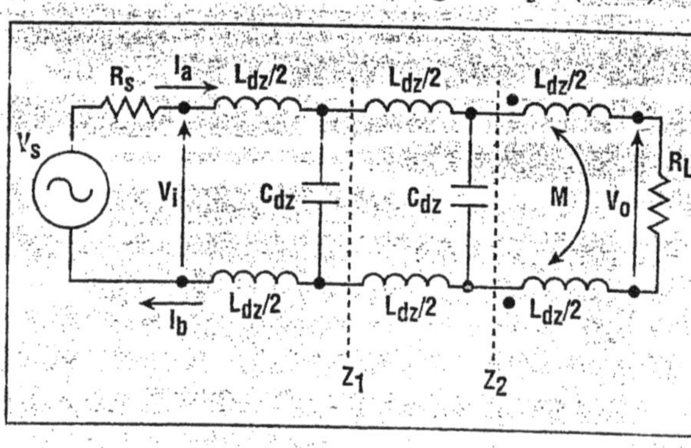

junction with boundary conditions $V(0) = V_i - V_o$ and $I_a(0) = I_i$, the transformer impedance matrix is obtained. This matrix is used to derive the input impedance as:

$$Z_i = Z_{11} = Z_{12}Z_{21}/(Z_{22} + Z_L) \quad (11)$$

where:

Z_L = the load impedance.

MATH DEVELOPMENT

Several equations can be developed to obtain the impedance parameters in Eq. 11 and to facilitate the calculation of Z_i. Using the hypothesis $L_{coil} > L_{line}$ (from Eqs. 2 and 3), it is determined that:

$$0 < M < L/2 \quad (12)$$

Using Eqs. 2 and 3, it is possible to define:

$$K = M/(L/2) \quad (13)$$

Note that K can be considered to be an estimate of the difference between the coil and line inductances. The transformer's working principle is based on this inductance difference. The higher the inductance seen by the coil-mode currents, the more accurate the hypothesis of equal and opposite currents on the line (see "Simplified analysis").

From Eqs. 6, 7, and 8, the following differential equations are obtained:

$$\frac{dV}{dZ} = (L/2)(K-1)\frac{dI_a}{dt} + (L/2)(K-1)\frac{dI_b}{dt} \quad (14)$$

$$\frac{dI_a}{dZ} = -C\frac{dV}{dt} \quad (15)$$

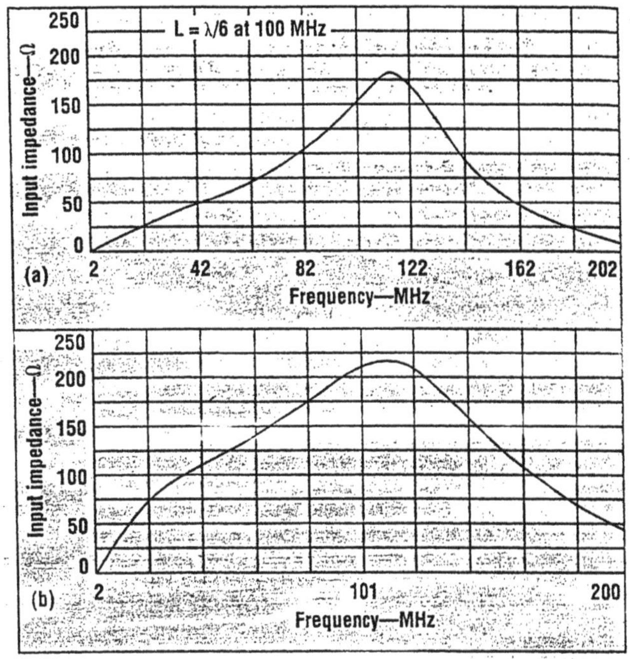

4. Input impedance versus frequency was obtained using the transmission-line transformer model (a) and measurements for an unwound transformer realized with RG-58U coaxial cable (b).

$$\frac{dI_b}{dZ} = -C\frac{dV}{dt} \quad (16)$$

By differentiating Eq. 14 with respect to z and Eqs. 15 and 16 with respect to t, then differentiating Eq. 14 with respect to t and Eqs. 15 and 16 with respect to z, these equations can be combined as:

$$\frac{d^2V}{dZ^2} = LC(1-K)\frac{d^2V}{dt^2} \quad (17)$$

$$\frac{d^2I_a}{dZ^2} = -C\frac{d^2V}{dt\,dZ} \quad (18)$$

$$\frac{d^2I_b}{dZ^2} = -C\frac{d^2V}{dt\,dZ} \quad (19)$$

In the frequency domain, these solutions can be written as:

$$V(Z,\omega) = (V_i - V_o)\cos(\mu Z) - [j(L^{0.5}/C)(1-K)^{0.5}/2] \times [I_i + I_b(0)]\sin(\mu Z) \quad (20)$$

$$I_a(Z,\omega) = (I_i/2)[1+\cos(\mu Z)] + [I_b(0)/2][\cos(\mu Z) - 1] - [j(V_i - V_o)/(L^{0.5}/C)(1-K)^{0.5}] \times \sin(\mu Z) \quad (21)$$

$$I_b(Z,\omega) = [I_b(0)/2][1+\cos(\mu Z)] + (I_i/2)[\cos(\mu Z) - 1] - [j(V_i - V_o)/(L^{0.5}/C)(1-K)^{0.5}] \times \sin(\mu Z) \quad (22)$$

where:

$\mu = \beta(1-K)^{0.5}$,

$\beta = \omega(LC)^{0.5}$, and

$0 < K < 1$.

Note that $L = (L_{coil} + L_{line})/2$ here. In Eq. 1, L was assumed to equal L_{line}.

Equation 11 describes the transmission-line transformer's input impedance. Referring to Fig. 1, the following relationships are derived:

$$V_i = Z_{11}I_i + Z_{12}I_o \quad (23)$$

$$V_o = Z_{21}I_i + Z_{22}I_o \quad (24)$$

with:

SIMPLIFIED ANALYSIS

By applying the hypothesis of equal and opposite current densities, a simplified transformer analysis can be obtained.[1] The equal and opposite currents on the transmission line will be confluent on the load (as shown in Fig. 1). According to the hypothesis, $I_b(0) = I_i$. For a short line, it is possible to assume that $I_a(l) \approx I_i$. From Fig. 1, it follows that $I_L \approx 2I_i$. Thus, the load current is almost twice the input current.

The output voltage can be written as $V_o = V_i - V(0)$. On a short line, this voltage can be approximated as $V_o \approx V(0)$. It follows that $V_o \approx V_i/2$. As a result, input impedance can be approximated as $Z_i = V_i/I_i \approx 4R_L$. ●●

Reference

1. J. Smith, *Modern Communication Circuits*, McGraw-Hill, 1986.

$$I_o = -I_L \quad (25)$$

The circuit's impedance parameters can be written as:

$$Z_{11} = \left.\frac{V_i}{I_i}\right|_{I_o \to 0} \quad (26)$$

$$Z_{12} = \left.\frac{V_i}{I_o}\right|_{I_i \to 0} \quad (27)$$

$$Z_{21} = Z_{12} \quad (28)$$

$$Z_{22} = \left.\frac{V_o}{I_o}\right|_{I_i \to 0} \quad (29)$$

This results in:

$$Z_{11} = \{2jL^{0.5}/C\,\beta l\,(1+K)(1-K)^{0.5}[1+\cos(\mu Z)]\}/\{(1-K)^{0.5}[1+\cos(\mu Z)]^2 - \sin(\mu z)[\beta l\,(1+K) - (1-K)^{0.5}\sin(\mu Z)]\} \quad (30)$$

$$Z_{12} = \eta Z_{11} + \tau \quad (31)$$

$$\eta = \{(1-K)^{0.5}[\cos^2(\mu Z) - 1] + \sin(\mu Z)\,[\beta l\,(1+K) + (1-K)^{0.5} \times \sin(\mu Z)]\}/\{(1-K)^{0.5}[1+\cos\,(\mu Z)^2 + \sin(\mu Z)\,[\beta l\,(1+K) + (1-K)^{0.5}\sin(\mu Z)]\} \quad (32)$$

with:

$$\tau = \{jL^{0.5}/C\,\beta l\,(1+K)(1-K)^{0.5}[1+\cos(\mu Z)]\}/\{(1-K)^{0.5} \times [1+\cos(\mu Z)]^2 + \sin(\mu Z) \times [\beta l(1+K) + (1-K)^{0.5}\sin(\mu Z)]\} \quad (33)$$

$$Z_{22} = (1/2)\{Z_{12} + [j(L^{0.5}/C) \times (1-K)^{0.5}\sin(\mu Z)]/[1+\cos(\mu Z)]\} \quad (34)$$

By solving for Eqs. 28, 30, 31, and 34, Eq. 11 can be synthesized on a personal computer.

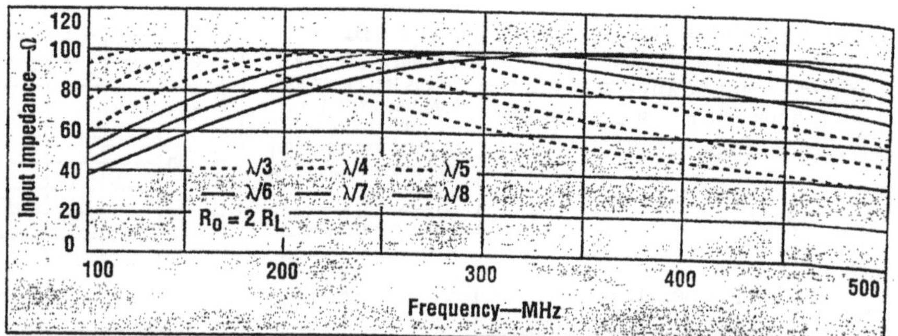

6. Input impedance versus frequency was plotted for various transmission-line lengths in order to determine an optimum length for wideband operation.

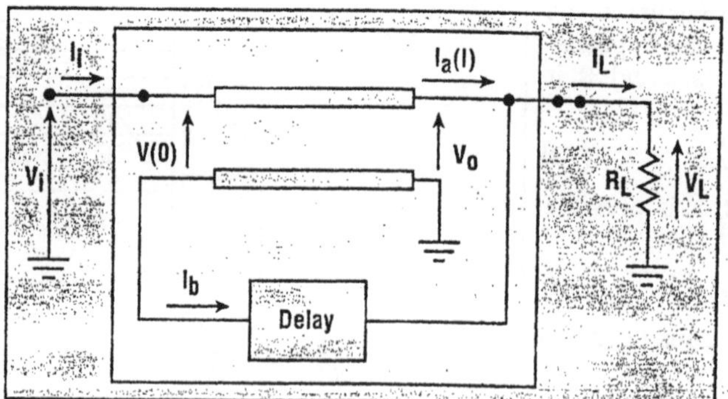

7. Practical transmission-line transformers may use a delay-compensation scheme in order to prevent phase opposition at the load from currents I_b and I_a.

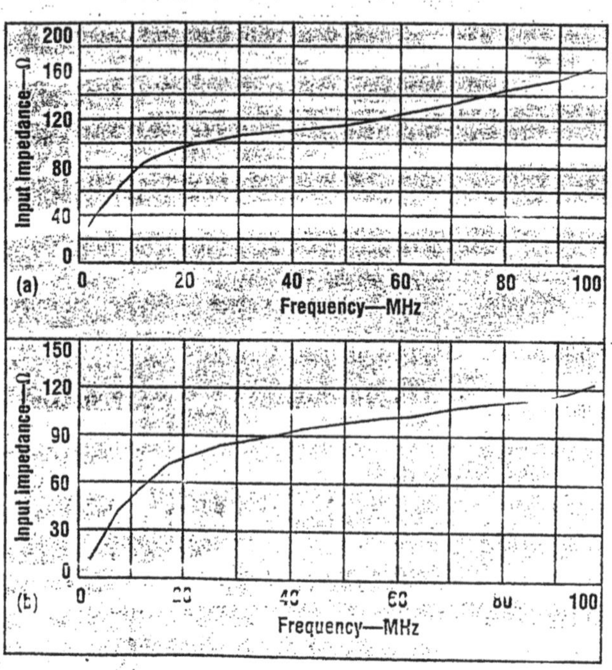

5. Experimental (a) and theoretical input impedance (b) are shown for a transmission-line transformer wound on a toroid with a one-turn inductance of approximately 35 nH.

TEST RESULTS

Equation 11 can be employed to synthesize the transmission-line transformer. Figure 4a shows a plot of the theoretical input impedance versus frequency obtained with Eq. 11. Figure 4b presents the measured response of a real (unwound) transformer realized with RG-58U coaxial cable. Note that if the transmission line is not wound on a toroid, the value of L_{coil} will represent the inductance of two straight parallel wires, which is generally much lower than the inductance obtained with Eq. 5.

Figure 5 presents the experimental and theoretical response of a transformer wound on a toroid whose one-turn inductance is given by $A_L \approx$ 35 nH.

In order to ascertain the optimum choice of line length, Eq. 11 was plotted for various length values (Fig. 6). As the plot illustrates, the line length for a wide operating bandwidth should be kept short (for instance, $l < \lambda/5$).

Regarding the optimum choice of R_o, a good value is almost twice the load impedance.[5] However, the exact determination of this transmission-line parameter involves a design compromise among the bandwidth, the desired reactive response, and the input impedance.

Another important guideline in the construction of wideband transformers[6] is that induced current $I_b(0)$ should be delayed by the same amount as current $I_a(l)$ (as shown in Fig. 1) in order to avoid a phase opposition on the load. A practical transmission-line transformer can meet this requirement by using a delay line to provide delay compensation (Fig. 7). ●●

Calculation of the line impedance involves a trade-off among the bandwidth, the reactive response, and the input impedance.

References

1. C.L. Ruthroff, "Some Broadband Transformers," *Proceedings of the IRE*, August 1959, pp. 1337-1342.
2. L.D. Abrie, *The Design of Impedance-Matching Networks for Radio-Frequency and Microwave Amplifiers*, Artech House, 1985.
3. Y. Shimada, "Input-Impedance Analysis of a 1:1 Balun," *IEEE Transactions on Microwave Theory and Techniques*, May 1970.
4. E. Rotholz, "Transmission-Line Transformers," *IEEE Transactions on Microwave Theory and Techniques*, April 1981.
5. O. Pitzalis, "Practical Design Information for Broadband Transmission-Line Transformers," *Proceedings of the IRE*, 1968.
6. D. Myer, "Equal-Delay Networks Match Impedances Over Wide Bandwidths," *Microwaves & RF*, April 1990, pp. 179-188.

For further reading

Grossner, *Transformers for Electronic Circuits*, 2nd ed., McGraw-Hill, 1983.

K.B. Niclas, R.R. Pereira, and A.P. Chang, "Transmission Lines Accurately Model Autotransformers," *Microwaves & RF*, November 1992, pp. 67-76.

J. Sevick, *Transmission-Line Transformers*, 2nd ed., American Radio Relay League, 1990.

C.W. Allen and H.L. Krauss, "Wideband Rotary Transformer—Unbalanced Current Analysis," *Proceedings of the IEEE*, February 1977.

American Radio Relay League, *The ARRL Handbook for Radio Amateurs*, 69th ed., 1992.

Equations Model Transmission-Line Transformers

A distributed-circuit analysis method is used to characterize solenoid transmission-line transformers.

Enzo Carpentieri
RF Circuits Designer
Alenia SpA, Via Fusaro 187, Bacoli, 80070 Naples, Italy; +39-81-5272242, FAX: +39-81-8687552.

THE broadband performance of transmission-line transformers renders them useful for impedance matching at microwave frequencies. A previous article (see *Microwaves & RF*, November 1996, p. 73) described a model for transmission-line transformers that allows for input-impedance calculation using distributed-circuit analysis and supports electromagnetic (EM) effects such as mutual coupling. This analysis can be extended to predict the performance of solenoid transmission-line transformers.

The previous article defined the parameter μ (not to be confused with the permeability of the toroidal-transformer medium) as:

$$\mu = \beta (1-K)^{0.5} \quad (1)$$

where:

$\beta = \omega(LC)^{0.5}$,

L, C = the resonant inductance and capacitance, respectively, and

K = a parameter used to represent the difference between the coil and line inductances.

It should be noted that μ is, in effect, coincident with the transmission-line phase constant (since the phase velocities of the transformer and transmission line are equal):

$$\mu = \omega (LC)^{0.5} (1-K)^{0.5} = \omega (L_{line} C)^{0.5} \quad (2)$$

where:

L_{line} = the transmission-line inductance, and

ω = the radian frequency.

This relationship is derived from the fact that the phase velocity (v_p) of the transformer equals that of the transmission line. Also, the term given by $z = (L/C)^{0.5}(1-K)^{0.5}$ is coincident with the transmission-line characteristic impedance (R_o), or:

$$Z = (L/C)^{0.5} \ (1-K)^{0.5} = R_o \quad (3)$$

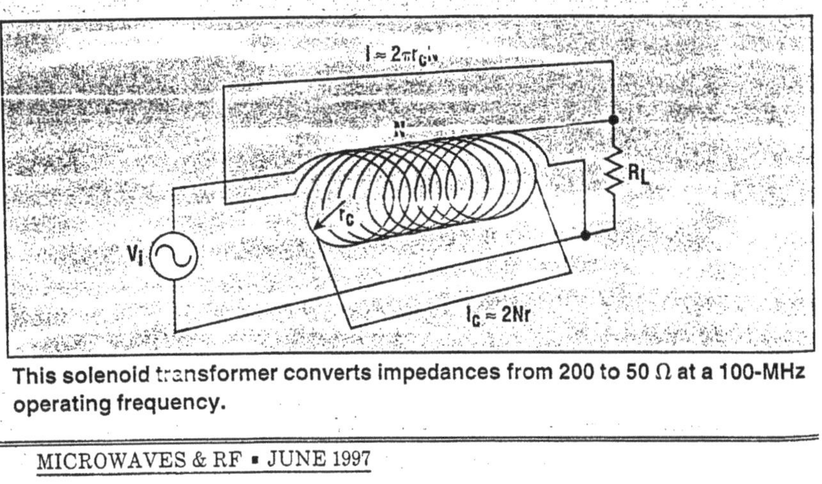

This solenoid transformer converts impedances from 200 to 50 Ω at a 100-MHz operating frequency.

It should be noted that these parameters were expressed (in the previous article) as functions of K in order to evaluate the limit of the input-impedance (Z_i) expression for K → 1. It is therefore appropriate to derive a simpler expression for the transmission-line transformer's impedance parameters (given by Z_{11}, Z_{12}, Z_{21}, and Z_{22}):

$$Z_{11} = \{2j\,(L/C)^{0.5}\,\beta Z\,(1+K)\times [1+cos(\beta Z)]\}/\{2\,(1-K)^{0.5}\times [1+cos(\beta Z)] - \gamma Z\,(1+K)\,sin(\beta Z)\} \quad (4)$$

$$Z_{12} = \eta\, Z_{11} + \tau \quad (5)$$

$$Z_{21} = Z_{12} \quad (6)$$

$$Z_{22} = 0.5\,\{Z_{12} + jR_o\,sin(\beta Z)/[1+cos(\beta Z)]\} \quad (7)$$

with:

$$\eta = [\gamma Z\,(1+K)\,sin(\beta Z)]/\{2\,(1-K)^{0.5}[1+cos(\beta Z)] + \gamma Z\,(1+K)\,sin(\beta Z)\} \quad (8)$$

A REAL TRANSFORMER'S COIL INDUCTANCE IS TYPICALLY MUCH GREATER THAN THE LINE INDUCTANCE AT THE DEVICE'S OPERATING FREQUENCIES.

$$\tau = \{j\,(L/C)^{0.5}\,\beta Z\,(1+K)\times [1+cos(\beta Z)]\}/\{2\,(1-K)^{0.5}\times [1+cos(\beta Z)] + \gamma Z\,(1+K)\,sin(\beta Z)\} \quad (9)$$

$$\beta = \omega\,(L_{line}C)^{0.5} \quad (10)$$

$$\gamma = \omega\,(LC)^{0.5} \quad (11)$$

$$L = (L_{coil} + L_{line})/2 \quad (12)$$

where:

L_{coil} = the inductance that is seen by the equiverse (or equal-direction) currents.

Finally, these expressions can be combined to obtain:

$$Z_i = Z_{11} - Z_{12}Z_{21}/(Z_{22} + Z_L) \quad (13)$$

where:

Z_L = the load impedance.

SOLENOID TRANSFORMERS

It is useful to observe that in a real transformer (at least at working frequencies), the coil inductance is generally much greater than the line inductance (solenoidal transformers can be used as an example). In this analysis, this means that K ≈ 1.

A 200-to-50-Ω transformer operating at 100 MHz can be realized by winding a transmission line to form a long solenoid (see figure). The parameters that must be considered are:[1]

$$L_{line} = (\mu_{coil}\,l/\pi)\,ln(d/r) \quad (14)$$

$$L_{coil} = (\mu_{coil}\,N^2\pi r_c^{\,2}/lc \quad (15)$$

where:

l = the line length (m),

μ_{coil} = the permeability of the medium inside the coil (H/m),

d = the spacing between the centers of the wires (m),

r = the radius of the wires (m),

l_c = the length of the solenoid (m), and

r_c = the radius of the solenoid (m).

Note that Eq. 14 holds true by assuming a constant magnetic field (B) across the entire interior of the solenoid. This condition can be approximated by assuming $l_c \ll r_c$.

The characteristic impedance of the transmission line should be $R_o \approx 2R_L$. As a result, choosing the dimensions d = 1.2 mm and r = 0.5 mm would yield the following result for an air medium:[2]

$$R_o = 276\,log(d/r) \approx 105\,\Omega \quad (16)$$

It can be assumed that l = λ/10 (where λ represents the wavelength). Also:

$$\lambda = v_p/f \quad (17)$$

where:

f = the operating frequency, and:

$$v_p = 1/(LC)^{0.5} \quad (18)$$

L_{line} is given by Eq. 13 with l = 1 m. The corresponding C value can be obtained as:

$$C = L/R_o^{\,2} \quad (19)$$

It follows that at f = 100 MHz, λ = 3.0 m. Choosing an r_c of 0.39 mm, the following results are obtained:

$$N = integer\,(l/2\pi r_c) = 12 \quad (20)$$

$$l_c = 2\,Nr_c = 12\ mm \quad (21)$$

where:

N = the number of turns in the toroid transformer.

Equations 13 and 14 can be used to obtain $L_{coil} \approx 757$ nH and $L_{line} \approx 105$ nH. Finally, the coupling coefficient (k) is determined from the following formula:

$$k = (L_{coil} - L_{line})/(L_{coil} + L_{line}) \quad (22)$$

Substitution of the L_{coil} and L_{line} values yields k ≈ 0.76. ●●

References

1. E.C. Jordan, *Electromagnetic Waves and Radiating Systems*, 2nd ed., Prentice-Hall, 1968.
2. J.D. Kraus, *Electromagnetics*, 4th ed., McGraw-Hill, 1992.

The Behavior of the Wide-Band Transmission-Line Transformer for Nonoptimum Line Impedance

WADE BLOCKER

***Abstract*—The transmission-line transformer can perfectly match at a single frequency any resistance ratio greater than or equal to four. This requires an optimum choice for the impedance and length of the line. For short lines, the transformer behavior is insensitive to small variations from optimum of the line impedance. A simple easily applied theory predicts the transformer behavior as a function of line length, the ratio of resistances to be matched, and the departure of line impedance from the optimum value.**

Ruthroff showed [1] that when a transmission-line transformer is used as in Fig. 1 to connect a load of resistance R_L to a source of resistance R_g, the power P_L delivered to the load is

$$P_L = \frac{e^2(1+\cos\beta l)^2 R_L}{[2R_g(1+\cos\beta l) + R_L\cos\beta l]^2 + \left[\dfrac{R_gR_L + Z_0^2}{Z_0}\right]^2 \sin^2\beta l}. \tag{1}$$

In this equation β is the phase factor of the line, l its length, and Z_0 its characteristic impedance. Ruthroff specialized this equation to a short line for which $\beta l = 0$, approximately, and showed for this case that the maximum power transfer to the load occurs for $R_L = 4R_g$, and that the optimum value for Z_0 is $2R_g$. Theoretically, a perfect four-to-one impedance match occurs for these conditions in the limit $\beta l = 0$ and a good approximate match occurs over a broad frequency band when the line is short. Ruthroff's analysis has been widely followed, and this transformer has seen extensive use as a wideband matching device for a four-to-one resistance ratio.

The greatest load power always occurs for this transmission line transformer when

$$Z_0 = \sqrt{R_gR_L}. \tag{2}$$

It was later shown [2] that for this optimum value of Z_0, the ratio of the available power, $P_{av} = e^2/4R_g$, to the load power is

$$[P_{av}/P_L]_{opt} = \frac{[2(R_g/R_L)(1+\cos\beta l) + 1]^2 - \sin^2\beta l}{4(R_g/R_L)(1+\cos\beta l)^2} \tag{3}$$

and that a perfect match, $[P_{av}/P_L]_{opt} = 1$, results whenever

$$\sec\beta l = (R_L/2R_g) - 1. \tag{4}$$

This last condition can be satisfied for any ratio R_L/R_g equal to or greater than four. So, at least theoretically, by the proper choice of line length, resistance ratios very different from the usual four-to-one ratio can be matched, and this match can be closely maintained over a broad frequency band. The transformer is not restricted to integer values for the ratio R_L/R_g.

The preceding results were for the specialized optimum case, $Z_0 = \sqrt{R_gR_L}$. It may not always be convenient to satisfy this condition, so the question arises of how much the transformer behavior departs from the ideal result of (3) when (2) is not satisfied. By straightforward manipulation Ruthroff's basic equation, our (1), can be put into the form

$$P_{av}/P_L = 1 + \left(\frac{R_g}{R_L}\right)\left[1 - \frac{(R_L/2R_g)}{(1+\sec\beta l)}\right]^2 + \left(\frac{1}{4}\right)\left[\frac{\sqrt{R_gR_L}}{Z_0} - \frac{Z_0}{\sqrt{R_gR_L}}\right]^2 \tan^2\left(\frac{\beta l}{2}\right). \tag{5}$$

The first two terms on the right side are just a simpler way to write (3) for $[P_{av}/P_L]_{opt}$, so that

$$P_{av}/P_L = [P_{av}/P_L]_{opt} + B(\sqrt{R_LR_g}/Z_0, \beta l) \tag{6}$$

where

$$B(\sqrt{R_LR_g}/Z_0, \beta l) = \left(\frac{1}{4}\right)[(\sqrt{R_LR_g}/Z_0) - (Z_0/\sqrt{R_LR_g})]^2 \tan^2\left(\frac{\beta l}{2}\right). \tag{7}$$

The term $B(\sqrt{R_LR_g}/Z_0, \beta l)$ is a correction to $[P_{av}/P_L]_{opt}$ that accounts for any departure of the line impedance from the optimum value of (2). When the line impedance has the optimum value, the correction term is zero. The correction is also identically zero when $\beta l = 0$, and is negligible for a short line as long as Z_0 does not differ too greatly from the optimum value. This insensitivity was investigated earlier for the four-to-one transformer by Pitzalis and Couse [3].

The term $[P_{av}/P_L]_{opt}$ depends only upon βl and the ratio R_L/R_g, and the correction term depends only upon βl and the ratio $\sqrt{R_LR_g}/Z_0$. This is convenient since it means that two tables, one for $[P_{av}/P_L]_{opt}$ and one for the correction term, allow P_{av}/P_L to be

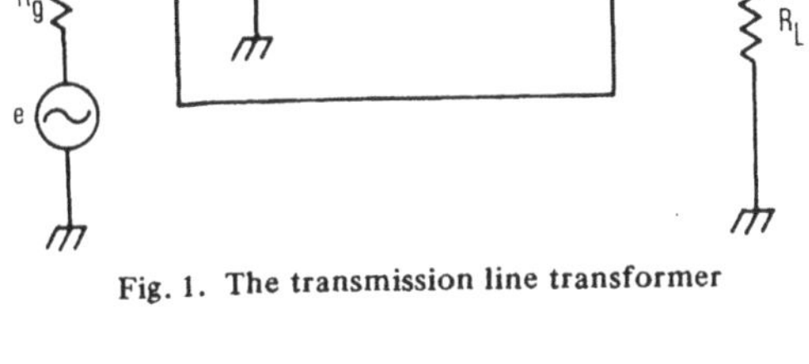

Fig. 1. The transmission line transformer

TABLE I
RATIO OF AVAILABLE POWER TO LOAD POWER FOR THE OPTIMUM CHOICE OF LINE IMPEDANCE

	$[P_{av}/P_L]_{opt}$						
R_L/R_g	4.0	5.0	6.0	9.0	16.0	25.0	∞
$\beta l/2\pi$							
.000	1.000	1.013	1.042	1.174	1.563	2.103	∞
.134	1.010	1.000	1.007	1.071	1.303	1.640	∞
.167	1.028	1.006	1.000	1.028	1.174	1.401	∞
.204	1.077	1.040	1.019	1.000	1.038	1.126	∞
.227	1.141	1.095	1.065	1.021	1.000	1.013	∞
.236	1.176	1.128	1.096	1.046	1.008	1.000	∞
.250	1.250	1.200	1.167	1.111	1.063	1.040	1.000
.333	3.250	3.450	3.667	4.361	6.063	8.290	∞
.500	∞	∞	∞	∞	∞	∞	∞

TABLE II
THE CORRECTION TERM $B(\sqrt{R_LR_g}/Z_0, \beta l)$ FOR NONOPTIMUM CHOICES OF LINE IMPEDANCE

	$B(\sqrt{R_LR_g}/Z_0, \beta l)$						
$\sqrt{R_LR_g}/Z_0$	1.00	1.50	2.00	3.00	4.00	5.00	∞
$\beta l/2\pi$							
.000	.000	.000	.000	.000	.000	.000	--
.134	.000	.035	.112	.356	.703	1.152	∞
.167	.000	.058	.187	.593	1.172	1.920	∞
.204	.000	.096	.312	.988	1.953	3.200	∞
.227	.000	.130	.422	1.333	2.637	4.320	∞
.236	.000	.146	.472	1.493	2.953	4.838	∞
.250	.000	.174	.562	1.778	3.516	5.760	∞
.333	.000	.521	1.687	5.333	10.547	17.280	∞
.500	--	∞	∞	∞	∞	∞	∞

Manuscript received November 8, 1977; revised December 29, 1977.
The author is at 17221 Osborne St., Northridge, CA 91325.

0018-9219/78/0400-0518$00.75 © 1978 IEEE

TABLE III

RATIO OF LOAD POWER TO AVAILABLE POWER FOR NONOPTIMUM LINE IMPEDANCES FOR $R_L = 4R_g$

	$P_L/P_{available}$ for $R_L=4R_g$						
$\sqrt{R_L R_g}/Z_0$	1.00	1.50	2.00	3.00	4.00	5.00	∞
$\beta l/2\pi$							
.000	1.000	1.000	1.000	1.000	1.000	1.000	--
.134	.990	.957	.891	.732	.584	.463	.000
.167	.973	.921	.823	.617	.455	.339	.000
.204	.928	.852	.719	.484	.330	.234	.000
.227	.876	.786	.640	.404	.265	.183	.000
.236	.850	.756	.606	.375	.242	.166	.000
.250	.800	.702	.552	.330	.210	.143	.000
.333	.308	.265	.203	.117	.072	.049	.000
.500	.000	.000	.000	.000	.000	.000	.000

TABLE IV

RATIO OF LOAD POWER TO AVAILABLE POWER FOR NONOPTIMUM LINE IMPEDANCES FOR $R_L = 9R_g$

	$P_L/P_{available}$ for $R_L=9R_g$						
$\sqrt{R_L R_g}/Z_0$	1.00	1.50	2.00	3.00	4.00	5.00	∞
$\beta l/2\pi$							
.000	.852	.852	.852	.852	.852	.852	--
.134	.934	.905	.845	.701	.564	.450	.000
.167	.973	.921	.823	.617	.455	.339	.000
.204	1.000	.912	.762	.503	.339	.238	.000
.227	.979	.868	.693	.425	.273	.187	.000
.236	.956	.839	.659	.394	.250	.170	.000
.250	.900	.778	.598	.346	.216	.146	.000
.333	.229	.204	.165	.103	.067	.046	.000
.500	.000	.000	.000	.000	.000	.000	.000

determined for all values of βl, R_L/R_g, and $\sqrt{R_L R_g}/Z_0$. For the last parameter, only values between one and infinity need be considered since reciprocal values for this parameter give the same result for the correction term. Table I gives the value of $[P_{av}/P_L]_{opt}$, and Table II gives the value of the correction term. Tables III and IV, derived from I and II, give values of P_L/P_{av} for two selected values of R_L/R_g as a function of βl and $\sqrt{R_L R_g}/Z_0$. These tables show that the previous common use of a short line and transformation ratio four-to-one results in the transformer which is least critical. Apparently, however, the transformer should be useful over a much wider set of conditions than previously considered.

REFERENCES

[1] C. L. Ruthroff, "Some broadband transformers," *Proc. IRE*, vol. 47, no. 8, pp. 1337–1342, Aug. 1959.

[2] W. Blocker, "Extended impedance matching capabilities in the wideband transmission line transformer," *Proc. IEEE*, pp. 1405–1406, Sept. 1977.

[3] O. Pitzalis, Jr. and T. P. M. Couse, "Broadband transformer design for RF transistor power amplifiers," in *Proc. 1968 Electronic Components Conf.*, pp. 207–216; "Practical design information for broadband transmission line transformers," *Proc. IEEE*, pp. 738–739, Apr. 1968.

CIRCUIT THEORY AND APPLICATIONS, VOL. 8, 55-64 (1980)

A TRANSMISSION LINE TRANSFORMER HAVING FREQUENCY INDEPENDENT MATCHING PROPERTIES*

SUHASH C. DUTTA ROY†

Department of Electrical Engineering, Iowa State University, Ames, Iowa 50011, U.S.A.

SUMMARY

In this paper, we present a new transmission line transformer which is capable of matching a load resistance R_L to a given source of internal resistance R_g in a frequency independent manner. This matching can be made perfect at all frequencies if $R_L/R_g = 4$; for $R_L/R_g \neq 4$, the matching can still be made frequency independent but the power transferred to the load becomes less than the maximum available power.

INTRODUCTION

The problem of matching a given load resistance R_L to a source of internal resistance R_g, so as to transfer the highest possible amount of power, is a classic one, and arises in numerous practical situations. Various solutions to this problem have been suggested and practised; these include quarter wave lines, stubs, multisection quarter wave transformers and tapered transmission lines. The matching characteristics obtained with the first three devices are of the bandpass type while that realized by the tapered line is of the highpass type. Fairly elaborate accounts of these devices, including the necessary design information, are available in standard textbooks (see, e.g. Reference 1).

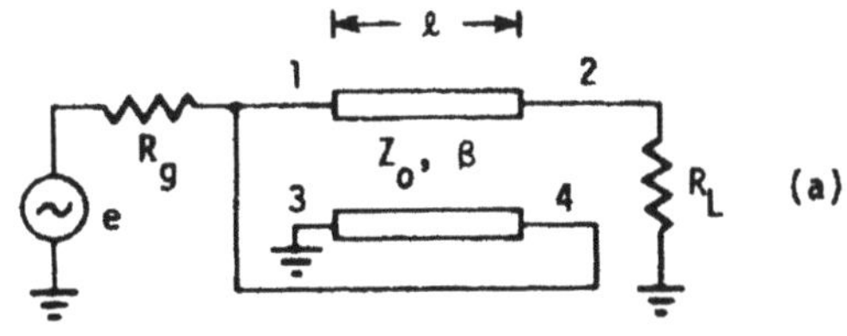

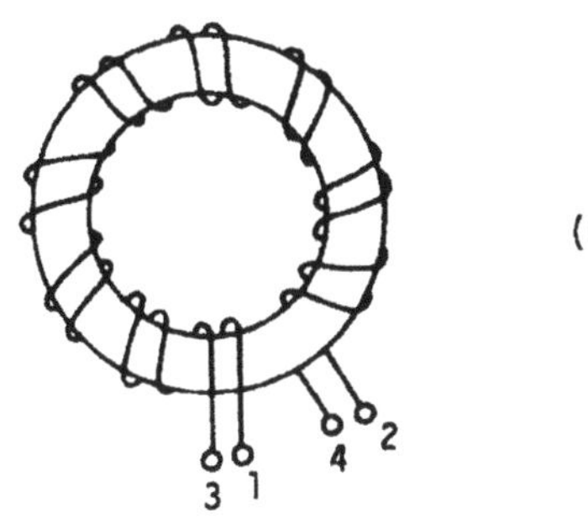

Figure 1. (a) The Ruthroff connection and (b) a possible practical implementation

A matching device with a lowpass type power transfer characteristic was introduced by Ruthroff[2] in 1959, but its versatility has been fully investigated only recently.[3-6] The device is shown in Figure 1(a) while a

* Patent applied for.

† Present address: Indian Institute of Technology, New Delhi 110029, India.

0098-9886/80/0108-0055$01.00

Received 17 July 1979

scheme for its practical implementation is shown in Figure 1(b), where the core consists of a ferrite toroid and the transmission line could be of the twisted wire, coaxial or stripline form. Detailed information on the design and applications of the device is available in References 7–14.

In this paper, we present, for what is believed to be the first time, a transmission line transformer which provides a frequency independent or allpass matching characteristic. For a load to source resistance ratio of 4, perfect matching is obtained, that is, the maximum available power (P_a) is transferred to the load, at all frequencies. For $R_L/R_g \neq 4$, the matching can still be made frequency independent, but the load power (P_L) becomes less than P_a. An important consequence of the result for $R_L/R_g = 4$ is that the device can act as an *ideal* 1:2 transformer.

The sensitivity of the power transfer characteristic to deviations from frequency independence conditions is also investigated.

THE DEVICE AND ITS CHARACTERIZATION

A schematic diagram of the transmission line transformer under consideration is shown in Figure 2(a). It consists of a cascade connection of two transmission lines of equal length, l, but different characteristic

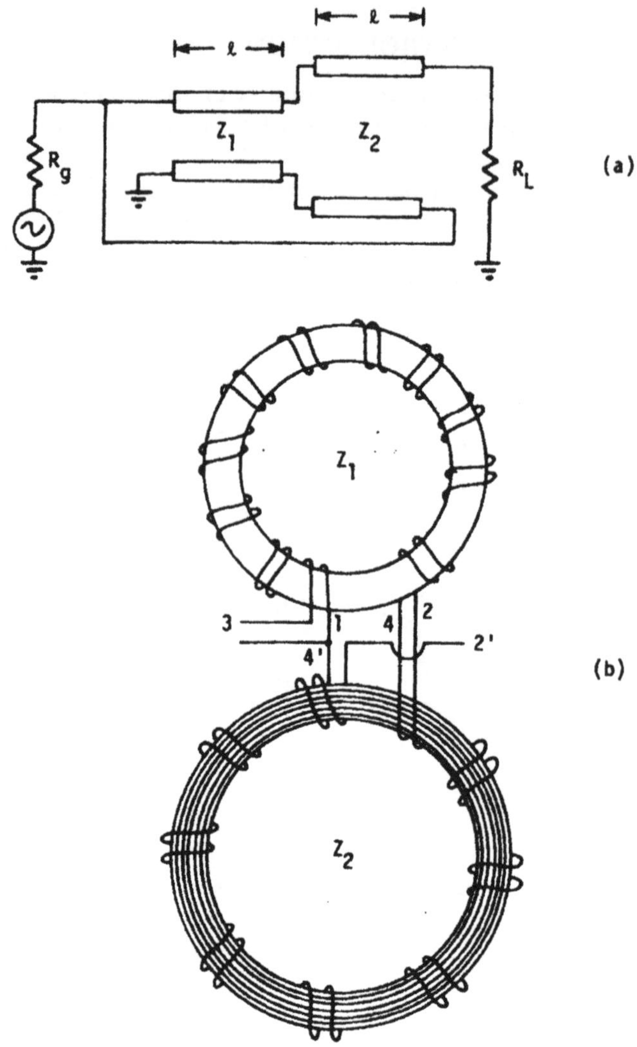

Figure 2. (a) Transmission line transformer under consideration, and (b) a possible practical implementation

impedances Z_1 and Z_2, the composite line being connected in the Ruthroff configuration. A possible practical implementation is shown in Figure 2(b).

In order to make a two-port characterization of the device, note that the $Z_1 - Z_2$ cascade has the following transmission matrix:

$$\begin{bmatrix} A & B \\ C & D \end{bmatrix} = \begin{bmatrix} \cos\frac{\phi}{2} & -jZ_1 \sin\frac{\phi}{2} \\ -jY_1 \sin\frac{\phi}{2} & \cos\frac{\phi}{2} \end{bmatrix} \begin{bmatrix} \cos\frac{\phi}{2} & -jZ_2 \sin\frac{\phi}{2} \\ -jY_2 \sin\frac{\phi}{2} & \cos\frac{\phi}{2} \end{bmatrix}$$

$$= \frac{1}{2} \begin{bmatrix} (1 - Z_1Y_2) + (1 + Z_1Y_2)\cos\phi & -j(Z_1 + Z_2)\sin\phi \\ -j(Y_1 + Y_2)\sin\phi & (1 - Z_2Y_1) + (1 + Z_2Y_1)\cos\phi \end{bmatrix} \tag{1}$$

where

$$\begin{aligned} \phi &= 2\beta l = 4\pi l/\lambda = \text{electrical length of the cascade,} \\ \beta &= \text{phase change coefficient of each line,} \\ \lambda &= \text{wavelength of excitation, and} \\ Y_i &= 1/Z_i,\ i = 1, 2 \end{aligned} \tag{2}$$

Under Ruthroff connection of the $Z_1 - Z_2$ cascade, as in Figure 2(a), it is easily shown that the transmission matrix between R_g and R_L becomes

$$\begin{bmatrix} A' & B' \\ C' & D' \end{bmatrix} = \frac{1}{A+1} \begin{bmatrix} A & B \\ C & A + D + 2 \end{bmatrix} \tag{3}$$

Our device can therefore be characterized by (3) in conjunction with (1).

EXPRESSION FOR LOAD POWER

When terminated in a resistance R_L, as shown in Figure 2(a), the impedance presented to the source becomes

$$Z_{in} = \frac{A'R_L + B'}{C'R_L + D'} = \frac{AR_L + B}{CR_L + (A + D + 2)} \tag{4}$$

The input reflection coefficient is, therefore,

$$\Gamma_{in} = \frac{Z_{in} - R_g}{Z_{in} + R_g} = \frac{(AR_L + B) - R_g[CR_L + (A + D + 2)]}{(AR_L + B) + R_g[CR_L + (A + D + 2)]} \tag{5}$$

Note, from (1), that B and C are purely imaginary while A and D are purely real. Thus, if we denote B/j by b and C/j by c, then

$$|\Gamma_{in}|^2 = \frac{[AR_L - (A + D + 2)R_g]^2 + (b - cR_gR_L)^2}{[AR_L + (A + D + 2)]^2 + (b + cR_gR_L)^2} \tag{6}$$

The insertion power ratio, p, defined by

$$p = P_L/P_a \tag{7}$$

is simply related to $|\Gamma_{in}|^2$ as

$$p = 1 - |\Gamma_{in}|^2 \tag{8}$$

Combining (6) and (8) and simplifying, it is possible to express p in the form

$$p = 1/(1 + \psi) \tag{9}$$

where

$$\psi = \frac{[AR_L-(A+D+2)R_g]^2+(b-cR_gR_L)^2}{4R_gR_L[A(A+D+2)+bc]} \tag{10}$$

As will be seen in the subsequent developments, expressing p in terms of ψ facilitates the derivation of conditions for frequency independent matching. Since we wish to maximize p, i.e. minimize ψ, we should explore the possibility of making one or both of the terms in the numerator of (10) zero. While vanishing of the first term gives a frequency dependent condition, the second term, $(b-cR_gR_L)^2$, vanishes when

$$Z_1Z_2 = R_gR_L \tag{11}$$

which is independent of frequency. In the subsequent analysis, (11) is therefore assumed to be valid. Further, we define

$$m = R_L/R_g \quad \text{and} \quad Z_1/Z_2 = r \tag{12}$$

Then (10) becomes

$$\psi = \frac{[mA-(A+D+2)]^2}{4m[A(A+D+2)+bc]} \tag{13}$$

where, from (1) and (12),

$$\begin{aligned} A &= [(1-r)+(1+r)\cos\phi]/2 \\ D &= [(1-1/r)+(1+1/r)\cos\phi]/2 \\ bc &= (1+r)(1+1/r)\sin^2\phi/4 \end{aligned} \tag{14}$$

Substituting these values in (13) and simplifying, we get

$$\psi = \frac{1}{4m}\left\{\frac{[m(1-r)-6+r+1/r]+[m(1+r)-(2+r+1/r)]\cos\phi}{(3-r)+(1+r)\cos\phi}\right\}^2 \tag{15}$$

CONDITIONS FOR PERFECT MATCHING

We now ask the following question: Does there exist a combination of m and r such that $\psi = 0$, i.e. $p = 1$ (perfect matching) at all frequencies? Obviously, from (15), this will be so if both the constant term and the coefficient of $\cos\phi$ in the numerator of (15) vanish simultaneously. The required conditions are

$$m = (6r-r^2-1)/[r(1-r)] = (1+r)/r \tag{16}$$

From (16), the combination of m and r required for perfect matching is obtained as

$$r = 1/3 \quad \text{and} \quad m = 4 \tag{17}$$

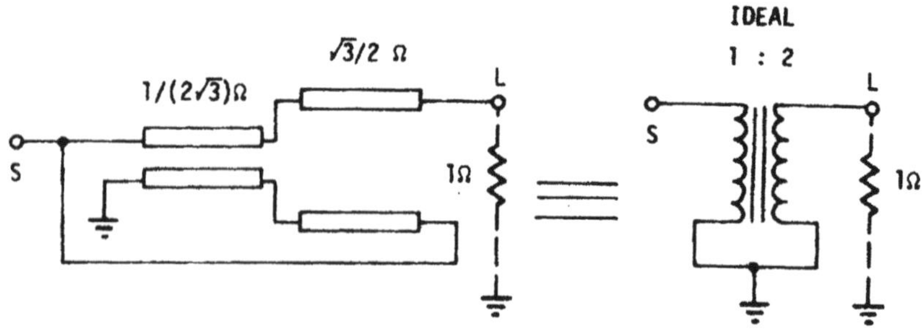

Figure 3. Emphasizing ideal transformer behaviour of the transmission line transformer for the particular combination of load and characteristic impedances

From (12) and (17), we therefore conclude that a load R_L can be perfectly matched to a source of internal resistance $R_L/4$ with a commensurate cascade of two transmission line sections, having the characteristic impedances $R_L/(2\sqrt{3})$ for the line fed by the source and $\sqrt{3}\,R_L/2$ for the other line, connected in the Ruthroff configuration. Another way of expressing the same result is that the preceding transmission line configuration behaves as an ideal transformer with 1 : 2 turns ratio for the particular load R_L. This is shown pictorially in Figure 3.

GENERAL CONDITIONS FOR FREQUENCY INDEPENDENT MATCHING

Referring to (15) again, observe that ψ will be independent of frequency if the condition

$$\frac{m(1-r)-6+r+1/r}{3-r}=\frac{m(1+r)-(2+r+1/r)}{1+r} \tag{18}$$

can be satisfied. On simplification, (18) gives

$$m=2(1/r-1) \quad \text{or} \quad r=2/(m+2) \tag{19}$$

Under (19), either side of (18) becomes $(1/r-3)$ so that from (15), we get

$$\psi_0=\frac{(1/r-3)^2}{8(1/r-1)} \tag{20}$$

where the subscript 'o' is used to indicate the validity of (19). The corresponding value of p can be obtained

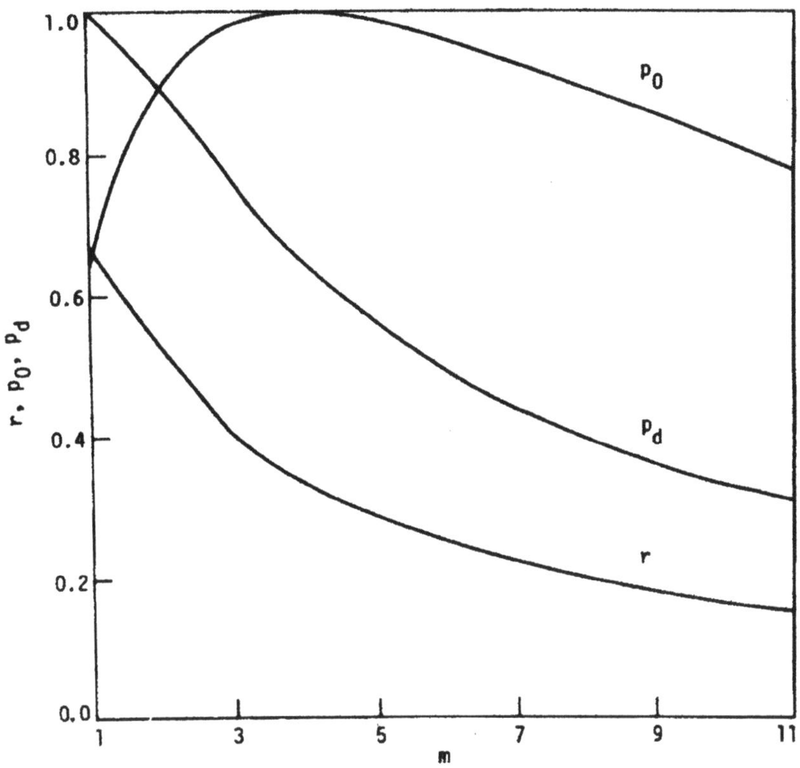

Figure 4. Plots of $r = Z_1/Z_2$ required for frequency independent matching; p_0 = fraction of maximum available power transferred to load under frequency independent matching condition; and p_d = fraction of maximum available power transferred to load under direct connection, with $m = R_L/R_g$

from (9), (19) and (20) as

$$p_o = 16m/(m+4)^2 \tag{21}$$

It should be noted, in passing, that if p_o is to be the power ratio at all frequencies, it should also be the DC value of p. That this is so easily follows from (15) by putting $\cos\phi = 1$.

If the frequency independent transmission line matching section is to be useful, the p_o must be greater than p_d, the power ratio under direct connection of source to load, given by

$$p_d = 4m/(m+1)^2 \tag{22}$$

Thus

$$p_o/p_d = 4[(m+1)/(m+4)]^2 \tag{23}$$

This is greater than unity if $m \geqslant 2$. Figure 4 shows plots of (i) r required for frequency independent matching, (ii) p_o and (iii) p_d versus m. Note that for m ranging from 1·5 to 10·4, the device can transfer more than 80 per cent of the available power to the load in a frequency independent fashion.

It has been observed in the preceding section that the combination $m = 4$, $r = 1/3$ makes our device behave as an ideal transformer, i.e. the input impedance becomes purely resistive. What about the general case $m \neq 4$, but r and m related by (19)? Since, under (19), p becomes a constant, independent of frequency, then so is $|\Gamma_{in}|^2$, by virtue of (8). Does this mean that Γ_{in}, given by (5), is a constant, which, in turn, would mean that Z_{in} is resistive? The answer is no, because the numerator of Γ_{in} is purely real, because of (11) but its denominator has a nonzero imaginary part, equal to $2b$ or $-(Z_1 + Z_2)\sin\phi$. Thus, all that frequency independent power transfer means, in the general case, is that the denominator of (6) becomes a constant times its numerator, so that $|\Gamma_{in}|^2$ becomes independent of ϕ.

EFFECT OF NONOPTIMUM RATIO OF CHARACTERISTIC IMPEDANCES

Our next investigation concerns the power transfer characteristic when (11) is satisfied but r deviates from the optimum value given by (19). Referring to (15), note that perfect matching is possible, i.e. $p = 1$ at a frequency satisfying the equation

$$\cos\phi = \frac{m(1-r) - 6 + r + 1/r}{2 + r + 1/r - m(1+r)} = \frac{1}{1+2q} \tag{24}$$

where

$$q \triangleq (m-4)r/[(m-1)r^2 + (6-m)r - 1] \tag{25}$$

For a real value of ϕ to exist, which satisfies (24), it is required to have $q \leqslant -1$ or $q \geqslant 0$. The range of r for a maximum of unity in the p versus ϕ curve can thereby be established. For example, for $m = 9$, (25) gives

$$q = 5r/(8r^2 - 3r - 1) \tag{26}$$

The variation of q with r is shown in Figure 5(a). Obviously, for $0 \leqslant r < 0{\cdot}25$, q lies between 0 and -1 and hence p will never reach a value of unity. For $r = 0{\cdot}25$, $q = -1$ so that from (24), p becomes unity at $\phi = \pi$; as r increases from 0·25, this maximum is reached at a lower value of ϕ. At $r = 0{\cdot}5877$, q becomes infinite, corresponding to a maximum at $\phi = \pi/2$. As r increases further, q decreases from $+\infty$ so that the maximum point moves towards $\phi = 0$.

What happens when r is in the range in which (24) does not have a real solution? For the example $m = 9$, this range is $0 \leqslant r < 0{\cdot}25$. The value of r satisfying (19) is $2/11 = 0{\cdot}1818$ at which the p versus ϕ curve is flat at the DC value of 0·852. It is therefore reasonable to expect that:

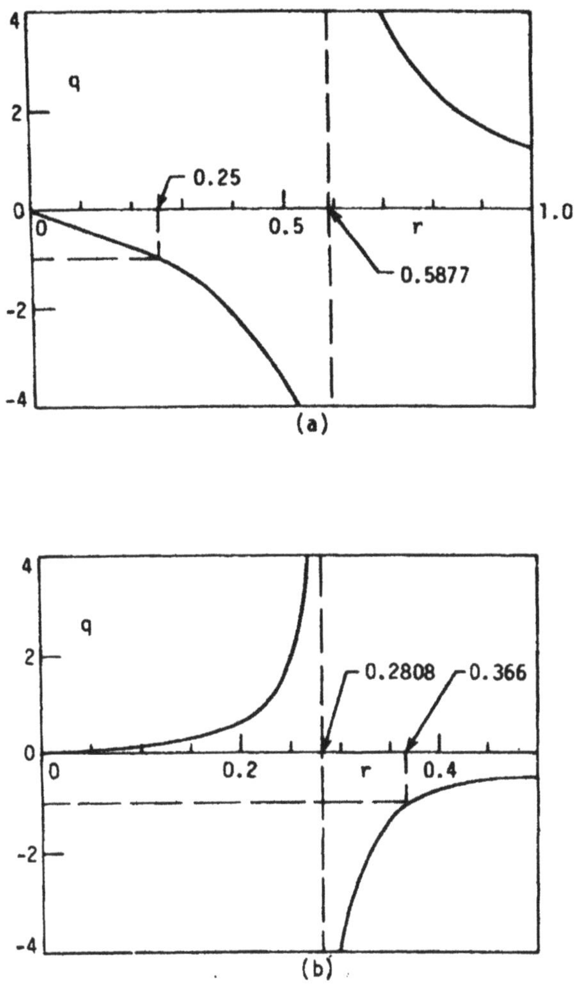

Figure 5. (a) Variation of (26) with r, (b) variation of (28) with r

1. for $0{\cdot}1818 < r < 0{\cdot}25$, p will rise with ϕ from the DC value to a highest value at $\phi = \pi$, given by

$$p_\pi = 1/\{1 + [(1-r)/r - mr/(1-r)]^2/(4m)\} \tag{27}$$

which is less than unity; and

2. for $0 < r < 0{\cdot}1818$, p will monotonically fall with ϕ, reaching the lowest value, given by (27) at $\phi = \pi$.

These conclusions have been verified by computing the variation of p with ϕ for $m = 9$ and various values of r; the results are shown in Figure 6.

As another example, let $m = 3$; then

$$q = -r/(2r^2 + 3r - 1) \tag{28}$$

a plot of which is shown in Figure 5(b). In this case, a maximum of $p = 1$ will be reached if $0 < r < 0{\cdot}366$. At $r = 0{\cdot}366$, the maximum will be reached at $\phi = \pi$. For $r = 0{\cdot}400$, (satisfying (19)), the $p - \phi$ curve is flat at the DC value of 0·98; thus for $0{\cdot}366 < r < 0{\cdot}400$, p will rise with ϕ, reaching the highest value (<1) at $\phi = \pi$, and for $r > 0{\cdot}400$, p will fall monotonically, reaching the lowest value at $\phi = \pi$.

These two examples demonstrate that even if one cannot fabricate a device with the exact optimum value of r corresponding to a given m, as specified by (19), a very wideband matching is still possible.

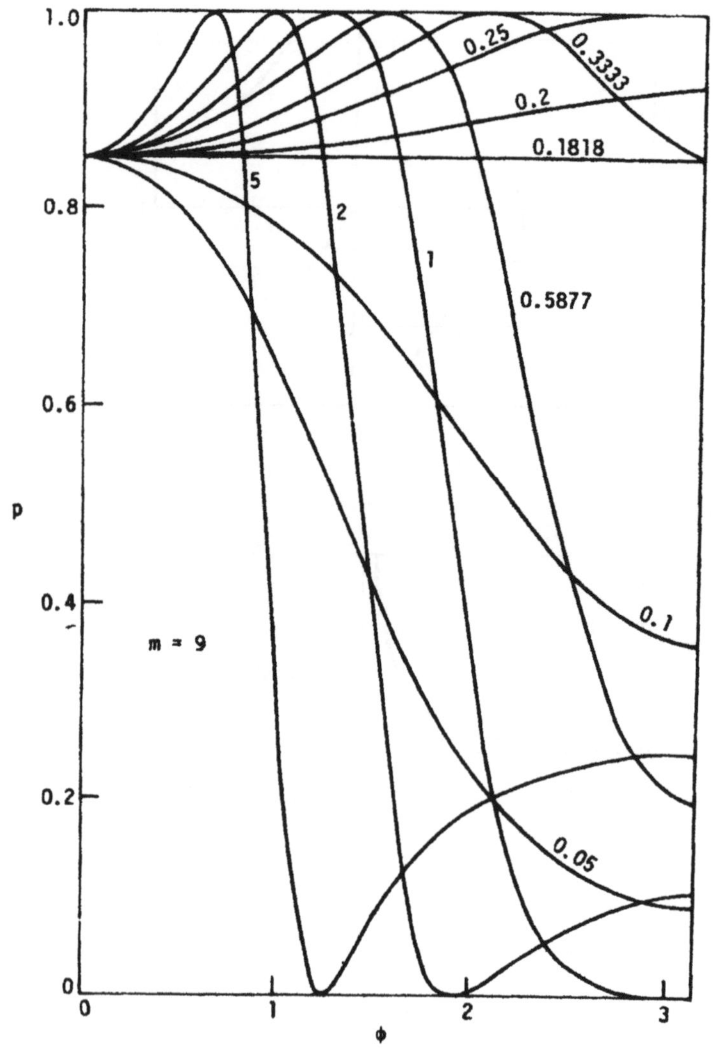

Figure 6. Variation of p with ϕ in the range $0 \leq \phi \leq \pi$. (It suffices to plot in this range only because $p(\pi - \phi) = p(\pi + \phi)$, $0 \leq \phi \leq \pi$; and $p(\phi) = p(2r\pi + \phi)$, where r is an integer) for $m = 9$ with r as a parameter

EFFECT OF DEVIATIONS FROM THE CONDITION $Z_1Z_2 = R_LR_g$

So far in our analysis, we have assumed condition (11) to be valid. In this section, we investigate the effect of deviations from this condition. Let the deviation be characterized by a parameter

$$n = \sqrt{[R_gR_L/(Z_1Z_2)]} \tag{29}$$

so that the ideal value of n is unity. With $n \neq 1$, (1), (12) and (29) give

$$(b - cR_gR_L)^2/R_g^2 = (m/r)(1+r)^2(n - 1/n)^2 \sin^2 \phi \tag{30}$$

Noting the difference between (10) and (13) and using (15) and (30), we get, for a general n,

$$\psi = G/(4mH) \tag{31}$$

where

$$G = \{[m(1-r) - 6 + r + 1/r] + [m(1+r) - (2 + r + 1/r)] \cos \phi\}^2 + m(2 + r + 1/r)(n - 1/n)^2 \sin^2 \phi \tag{32}$$

and

$$H = [(3-r)+(1+r)\cos\phi]^2 \tag{33}$$

If r has been adjusted to the value given by (19), then ψ_0 modifies to

$$\psi_0' = \frac{(1/r-3)^2}{8(1/r-1)} + \frac{(2+r+1/r)(n-1/n)^2 \sin^2\phi}{4[(3-r)+(1+r)\cos\phi]^2} \tag{34}$$

The first term on the right-hand side is ψ_0, as given by (20), while the second term gives the frequency dependent deviation, originating from nonoptimum n. The latter has no effect at DC and at $\phi = \pi$, and is always positive, which means that except at $\phi = 0$ and π, p will always be lower than the value for $n = 1$. By differentiating (34), it can be easily shown that p has a minimum value of

$$p_{\min} = 1/\{1+\psi_0+(1+r)^2(n-1/n)^2/[32r(1-r)]\} \tag{35}$$

at a value of ϕ given by

$$\phi_{\min} = \cos^{-1}\left(-\frac{1+r}{3-r}\right) = \pi - \cos^{-1}\left(\frac{1+r}{3-r}\right) \tag{36}$$

Thus for nonoptimum n, p varies between p_0, given by (21) and $p_{\min}$, given by (35). As an example, let $m = 4$ and $r = 1/3$; then

$$p_0 = 1,\ \phi_{\min} = 2\pi/3 \quad \text{and} \quad p_{\min} = 1/[1+(n-1/n)^2/4] \tag{37}$$

With $n = 0{\cdot}8$ or $1{\cdot}25$, $p_{\min}$ is $0{\cdot}95$, which amounts to a deviation of only 5 per cent from its ideal constant value of unity. As another example, let $m = 9$ and $r = 2/11$; then

$$p_0 = 0{\cdot}852,\ \phi_{\min} = 2\ \text{rad/s},$$

and

$$p_{\min} = 1/[1+25/144+169(n-1/n)^2/576] \tag{38}$$

If $n = 0{\cdot}8$ or $1{\cdot}25$, then $p_{\min} = 0{\cdot}822$, amounting to a deviation of less than 4 per cent from the constant value of $0{\cdot}852$ which exists when $n = 1$.

It is therefore concluded that the performance of the transmission line transformer is not critically dependent on the validity of condition (11).

COMMENTS AND CONCLUSIONS

A new transmission line device has been described in this paper, which can transfer power from a source of internal resistance R_g to a load resistance $R_L = mR_g$ in a frequency independent fashion. For $m = 4$, the maximum available power from the source can be transferred to the load while for other values of m, the load power becomes less than this value. The frequency response of the power transfer characteristic has been shown to be fairly insensitive to small deviations from the optimum conditions.

A systematic procedure for the design of the transformer would be as follows.

1. From the given R_L and R_g, calculate $m = R_L/R_g$. If $1 < m < 2$, a direct connection should be preferred.
2. If $m > 2$, calculate p_0 from (21). This is the fraction of the maximum available power from the source which can be transferred to the load in an allpass manner. If this is not adequate, the device being considered cannot be used.
3. If p_0 is adequate, calculate r from (19).
4. Finally, find Z_1 and Z_2 from

$$Z_1 = \sqrt{(rR_LR_g)}, \quad Z_2 = \sqrt{(R_LR_g/r)} \tag{39}$$

5. The length of each line does not seem to matter so long as conditions are ideal; any deviation of Z_1 or Z_2 from the ideal values given by (39) will, however, make the power transfer frequency dependent. After the best values of Z_1 and Z_2 which can be realized from the available materials have been determined,[14] compute the power transfer characteristic using (9) and (31) through (33) for $0 \leq \phi \leq \pi$. Comparing this with the specified tolerances, one can have an idea of the required length. Consider, for instance, the situation in which $m = 9$ and Z_1 and Z_2 satisfy (11) but r realized is 0·2 instead of the ideal value of 0·1818. If the specification is that power to load is to stay within ±5 per cent of the DC value up to a frequency of 1 GHz, then from Figure 6, we see that this frequency should correspond to $\phi = 2{\cdot}2$, i.e. the required length is $l = (2{\cdot}2)v_p/[(4\pi)(10^9)]$, where v_p is the phase velocity. If $v_p = 1{\cdot}5 \times 10^8$ m/s, then l comes out to be 0·0263 m.

In concluding this paper, it is worth mentioning and emphasizing that the transmission line analysis will not be valid at very low frequencies (VLF). This is easily seen by considering the connection of Figure 2(a) at DC; since the source becomes short-circuited, the DC load power must be zero. In order to obtain an idea of the VLF behaviour, consider the transformer type of construction shown in Figure 2(b), with $m = 4$ and $r = 1/3$. Then at VLF, the device can be adequately represented by[13] a magnetizing inductance L in parallel with a resistance $R_L/4$, across the source. Routine analysis shows that under this condition, the insertion power ratio is given by

$$p = \frac{4\omega^2 L^2}{R_g R_L/4 + 4\omega^2 L^2 (1/\sqrt{m} + \sqrt{m}/4)^2} \tag{40}$$

where ω is the radian frequency. Obviously $p = 0$ at $\omega = 0$; when ω is increased to the value ω_1, at which the second term in the denominator of (40) starts dominating over the first, one obtains $p \cong p_0$ as given by (21). The term 'DC' or the expression '$\phi = 0$' as used in this paper should therefore be taken to correspond to frequencies of the order of ω_1. It must be pointed out that in the construction of Figure 2(b), use of a high permeability core, and the cancellation of core flux in each toroid due to equal and opposite currents contribute to a very large value of L, so that ω_1 will indeed be close to DC. Except at the extreme end of low frequencies, viz. $\omega = 0$, there will then be no substantial loss of power to the load.

ACKNOWLEDGEMENTS

The research reported here was supported by the Engineering Research Institute at the Iowa State University (ISU). The author thanks Dr. J. O. Kopplin, Chairman of the Electrical Engineering Department at ISU for providing the necessary facilities.

REFERENCES

1. R. E. Collin, *Foundations for Microwave Engineering*, McGraw-Hill, New York, Chapter 5, 1966.
2. C. L. Ruthroff, 'Some broadband transformers', *Proc. IRE*, **47**, 1337–1342 (1959).
3. W. Blocker, 'Extended impedance matching capabilities in the wideband transmission line transformer', *Proc. IEEE*, **65**, 1405–1406 (1977).
4. W. Blocker, 'The behavior of the wideband transmission line transformer for nonoptimum line impedance', *Proc. IEEE*, **66**, 518–519 (1978).
5. S. C. Dutta Roy, 'Low frequency wideband impedance matching by exponential transmission lines', *Proc. IEEE*, **67**, to appear.
6. S. C. Dutta Roy, 'Optimum design of exponential transformer for wideband matching at low frequencies', *Proc. IEEE*, to appear.
7. *RCA Designer's Handbook*, Solid State Power Circuits. Technical Series SP-52, 482–488, 504–505 and 520–521 (1971).
8. *RCA Solid State Data Book*, SSD-205B, RF Power Devices, 156, 162, 208, 214, 349–350, 367, 459 and 465–470 (1974).
9. O. Pitzalis and T. Couse, 'Broadband transformer design for RF transistor amplifiers', *Proc. Electronic Components Conf.*, 207–216 (1968).
10. O. Pitzalis and T. Couse, 'Practical design information for broadband transmission line transformers', *Proc. IEEE*, **56**, 738–739 (1968).
11. O. Pitzalis, R. E. Horn and R. J. Baranello, 'Broadband 60-W HF linear amplifier', *IEEE J. Solid State Circuits*, **SC-6**, 95–102 (1971).
12. J. Sevick, 'Simple broadband matching networks', *QST*, **60**, 20–23 (1976).
13. J. Sevick, 'Broadband matching transformers can handle many kilowatts', *Electronics*, **49**, 123–128 (1976).
14. H. L. Krauss and C. W. Allen, 'Designing toroidal transformers to optimize wideband performance', *Electronics*, **46**, 113–116 (1973).

Twisted Magnet Wire Transmission Line

PETER LEFFERSON

Abstract – Transmission line principles are applied to twisted magnet wire lines made of two wires to establish design limits. Expressions are developed to predict the effects of wire film insulation and of twisting. A design procedure is developed to realize a desired characteristic impedance for the design of radio frequency broad-band transformers, signal combiners, and pulse transformers.

INTRODUCTION

BROAD-BAND transformers and signal combiners for HF thru UHF and the digital field using twisted magnet wire transmission line have been discussed in the literature and are widely used. [1]-[3]. The terminal impedances of these passive devices are a function of the transmission-line characteristic impedance, line length, core material, and source impedance at other terminals. If the line impedance can be controlled, these devices can be built with a wide frequency range without requiring additional lumped elements. The terminal reactance can be tailored in the same way to improve matching into active devices.

This paper will consider only the twisted magnet wire transmission line and a design procedure will be given as a function of wire size, wire insulation thickness, insulation relative dielectric constant, and wire twist.

Twisted magnet wire transmission lines using the wire size range from 4 to 44 can be realized with characteristic impedances ranging from 10 to 85 Ω. More practical examples would be a 50-Ω line made with a pair of number 40 wires and a 25-Ω line made with a pair of number 15 wires.

Basic Dimension Standard

The United States military specification for magnet wire, MIL-W-583, will be used for a well-ordered base on which to build a design procedure. Among other things, this specification defines round magnet wire having film insulation. It lists American Wire Gage sizes 4-44 in four insulation thickness groups with their dimensions and tolerances. The minimum insulation thickness for each wire size in groups 2-4 is approximately the minimum thickness for the wire size in group one multiplied by the group number. The maximum thickness is about equal to the minimum of the next higher group. These data from MIL-W-583 is repeated in Table I.

Using the dimensions from this table and the expression for characteristic impedance of two parallel wires [4]

$$Z_1 = \frac{120}{\sqrt{\epsilon_{req}}} \cosh^{-1} \frac{D}{d}. \tag{1}$$

The computer calculated characteristic impedances are plotted in Fig. 1 for a relative dielectric constant of 1.0. D and d are the wire diameters with and without the insulation film, respectively. ϵ_{req} is the equivalent relative dielectric constant. Fig. 1 shows the calculated impedances for the film-thickness extremes.

WIRE TWIST AND PITCH ANGLE

The effect of twisting the line can be normalized for all wire sizes if the twist per inch T is equated to the angle included between each wire and a line drawn down the center of the transmission line (pitch angle θ). For bifilar wire, twist is given by

$$T = \frac{\tan \theta}{\pi D} \tag{2}$$

The relation of twist and pitch angle is plotted in Fig. 2 for wire sizes 4-44 and the insulation thicknesses of classes one and four.

Experimental data suggest that optimum performance is obtained from lines having pitch angles between 20°-45°. When the line twist is loose, it becomes difficult to maintain continuous line geometry as the line is wound on a form. This is seen as a large reflection on a time-domain reflectometer. The transmission line is subjected to excessive stress as the twist angle approaches 50.5° and in the vicinity of 50.5° it will break. Equation 2 and the maximum pitch angle are developed in the Appendix.

A line impedance rise of 1 or 2 Ω can be anticipated when the line is wound of a ferrite form.

FILM DIELECTRIC

The equivalent dielectric constant that must be considered in order to give meaning to the theoretical characteristic impedance of Fig. 1, is a function of the film insulation and pitch angle. It is common to express a system having two dielectrics as

$$\begin{aligned} \epsilon_{req} &= \alpha\epsilon_{r1} + \beta\epsilon_{r2} \quad \alpha + \beta = 1 \\ &= \epsilon_{r1}(1-\beta) + \beta\epsilon_{r2} \\ &= \epsilon_{r1} + \beta(\epsilon_{r2} - \epsilon_{r1}), \end{aligned} \tag{3}$$

where ϵ_{r1} and ϵ_{r2} are the relative dielectric constants of air (or some other surrounding material) and the film insulation,

Manuscript received June 9, 1971; revised September 13, 1971.
The author is with the Milton Roy Company, St. Petersburg, Fla.

[1] C.L. Ruthroff, "Some broad band transformers," *Proc. IEEE*, vol. 47, pp. 1337 - 1342, Aug. 1959.

[2] O. Pitzalis, Jr., "Practical design information for braod-band transmission line transformers," *Proc. IEEE*, vol. 56, pp. 738 - 739, Apr. 1968.

[3] R.E. Matick, "Transmission line pulse transformers – Theory and application," *Proc. IEEE*, vol. 56, pp. 47 - 62, Jan. 1968.

[4] International Telephone and Telegraph Co., *Reference Data for Radio Engineers*, 5th ed. New York: Sames, 1969, ch. 22, p. 22.

TABLE I

MINIMUM INCREASE IN DIAMETER OF BARE ROUND WIRE AND THE MAXIMUM OVERALL DIAMETER DUE TO FILM INSULATION

Wire size (AWG)	Diameter nominal	CLASS 1 Minimum increase in diameter	CLASS 1 Maximum overall diameter	CLASS 2 Minimum increase in diameter	CLASS 2 Maximum overall diameter	CLASS 3 Minimum increase in diameter	CLASS 3 Maximum overall diameter	CLASS 4 Minimum increase in diameter	CLASS 4 Maximum overall diameter
	(In)	(In)	(In)	(In)	(In)	(In)	(In)	(In)	(In)
4---	0.2043	0.0019	0.2093	0.0037	0.2111	0.0049	0.2125	0.0064	0.2148
5---	.1819	.0019	.1867	.0036	.1884	.0048	.1897	.0062	.1920
6---	.1620	.0018	.1665	.0035	.1682	.0047	.1695	.0059	.1717
7---	.1443	.0017	.1485	.0034	.1502	.0046	.1515	.0059	.1537
8---	.1285	.0016	.1324	.0033	.1342	.0045	.1355	.0058	.1377
9---	.1144	.0016	.1181	.0032	.1198	.0044	.1211	.0057	.1233
10---	.1019	.0015	.1054	.0031	.1071	.0043	.1084	.0056	.1106
11---	.0907	.0015	.0941	.0030	.0957	.0042	.0969	.0054	.0991
12---	.0808	.0014	.0840	.0029	.0855	.0040	.0867	.0052	.0888
13---	.0720	.0014	.0750	.0028	.0765	.0039	.0776	.0050	.0796
14---	.0641	.0014	.0670	.0027	.0684	.0038	.0695	.0049	.0715
15---	.0571	.0013	.0599	.0026	.0613	.0037	.0624	.0048	.0644
16---	.0508	.0012	.0534	.0026	.0548	.0036	.0558	.0046	.0577
17---	.0453	.0012	.0478	.0025	.0492	.0035	.0502	.0045	.0520
18---	.0403	.0011	.0426	.0024	.0440	.0034	.0450	.0044	.0468
19---	.0359	.0011	.0382	.0023	.0395	.0033	.0404	.0042	.0422
20---	.0320	.0010	.0341	.0022	.0353	.0031	.0362	.0040	.0379
21---	.0285	.0010	.0306	.0021	.0317	.0030	.0326	.0039	.0342
22---	.0253	.0010	.0273	.0020	.0284	.0029	.0292	.0037	.0308
23---	.0226	.0009	.0244	.0019	.0255	.0028	.0263	.0036	.0279
24---	.0201	.0009	.0218	.0019	.0229	.0027	.0237	.0035	.0252
25---	.0179	.0009	.0195	.0018	.0206	.0026	.0214	.0034	.0228
26---	.0159	.0008	.0174	.0017	.0185	.0025	.0192	.0032	.0206
27---	.0142	.0008	.0156	.0016	.0165	.0023	.0172	.0030	.0185
28---	.0126	.0007	.0139	.0015	.0148	.0022	.0155	.0029	.0166
29---	.0113	.0007	.0126	.0014	.0134	.0021	.0141	.0028	.0152
30---	.0100	.0006	.0112	.0013	.0120	.0020	.0127	.0027	.0137
31---	.0089	.0006	.0100	.0013	.0108	.0019	.0115	.0026	.0124
32---	.0080	.0006	.0091	.0012	.0098	.0018	.0105	.0025	.0113
33---	.0071	.0005	.0081	.0011	.0088	.0017	.0095	.0024	.0102
34---	.0063	.0005	.0072	.0010	.0078	.0015	.0084	.0021	.0091
35---	.0056	.0004	.0064	.0009	.0070	.0014	.0076	.0020	.0082
36---	.0050	.0004	.0058	.0008	.0063	.0013	.0069	.0019	.0074
37---	.0045	.0003	.0052	.0008	.0057	.0012	.0062	.0017	.0067
38---	.0040	.0003	.0047	.0007	.0051	.0011	.0056	.0016	.0060
39---	.0035	.0002	.0041	.0006	.0045	.0010	.0050	.0015	.0053
40---	.0031	.0002	.0037	.0006	.0040	.0009	.0044	.0013	.0047
41---	.0028	.0002	.0033	.0005	.0036	.0008	.0040	.0010	.0043
42---	.0025	.0002	.0030	.0004	.0032	.0008	.0037	.0010	.0038
43---	.0022	.0002	.0026	.0004	.0029	.0007	.0033	.0010	.0035
44---	.0020	.0001	.0024	.0004	.0027	.0006	.0030	.0009	.0032

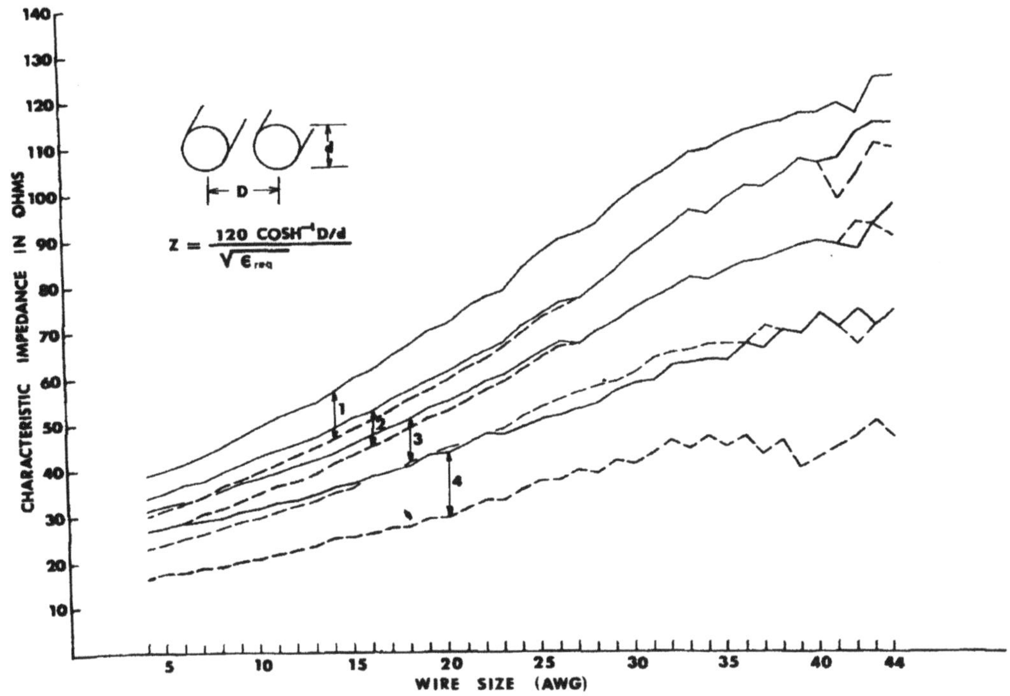

Fig. 1. Characteristic impedance for bifilar magnet wire transmission line based on MIL-W-583 dimensions and with relative dielectric constant of one.

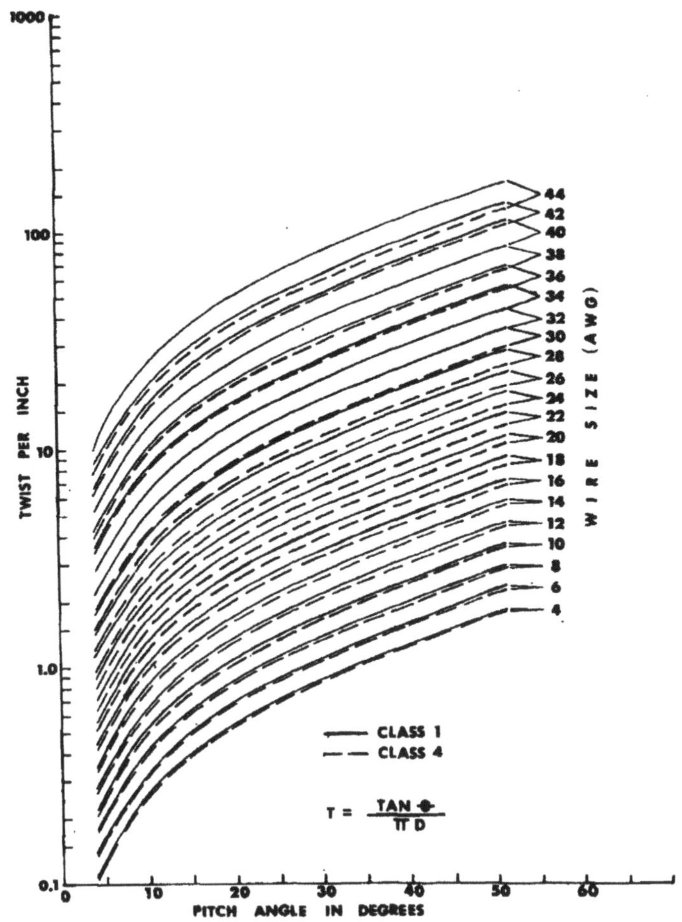

Fig. 2. Bifilar transmission line twist per inch versus pitch angle for wire sizes 4-44 having class-1 and 4 insulation thickness.

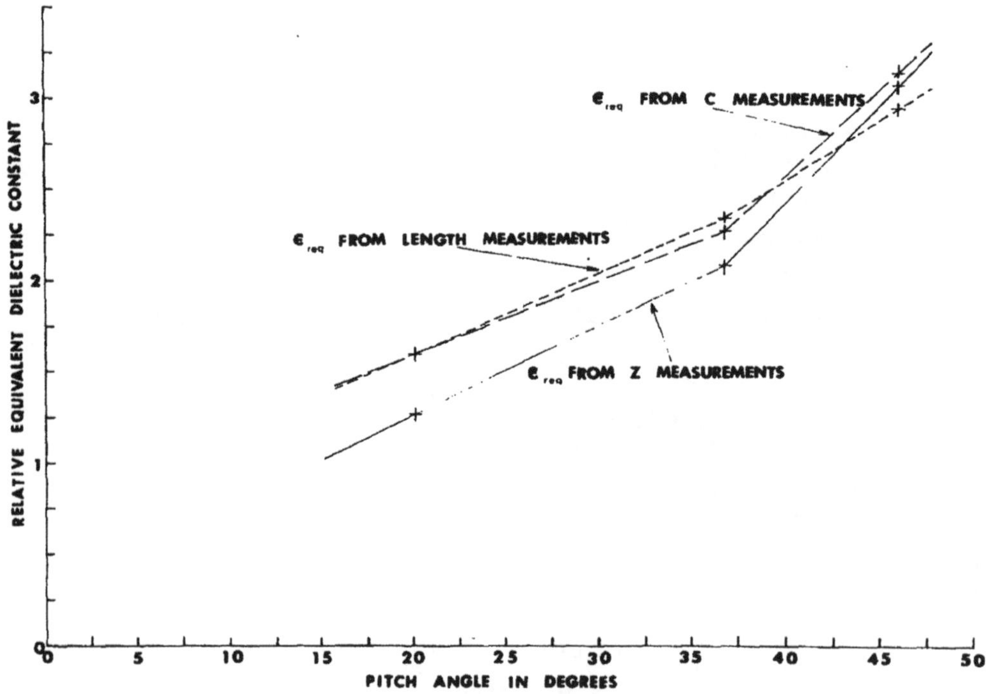

Fig. 3. Measured equivalent dielectric constant versus pitch angle for a sample of no. 29 magnet wire having class-2 insulation thickness.

respectively. ϵ_{r1} and ϵ_{r2} are published constants, but β is a function of pitch angle.

The behavior of β can be seen by observing ϵ_{req} as the pitch angle is increased. If one assumes that the internal, external, and mutual inductance of the line are not changed by twist, ϵ_{req} can be calculated from measurements of the actual characteristics impedance by

$$\sqrt{\epsilon_{req}} = \frac{\text{calculated } Z_o \text{ for flat parallel wires with } \epsilon_{req} = 1}{\text{measured } Z_o} . \quad (4)$$

It can also be calculated from capacitance measurements of the line by

$$\epsilon_{req} = \frac{\text{measured capacitance}}{\text{calculated capacitance for the same length of flat parallel wires with } \epsilon_{req} = 1} . \quad (5)$$

It can be calculated from the electrical line length by

$$\sqrt{\epsilon_{req}} = \frac{\text{measured electrical length}}{\text{measured physical length of the twisted line}} . \quad (6)$$

Laboratory data show agreement among all three of these within the measurement accuracy of wire dimension, characteristic impedance, and relative dielectric constant of the film. Five percent random impedance fluctuation along a twisted line is common. Fig. 3 illustrates an example using number 29 wire having a class-2 thickness of polyester film insulation. The example shows ϵ_{req} calculated from capacitance, electrical length, and impedance measurements of four line samples twisted to different pitch angles.

The characteristic impedance and electrical length data were taken with the Hewlett-Packard model 1415A time domain reflectometer. The film relative dielectric constants were calculated from capacitance measurements of wire samples in a mercury bath.

It was observed that the measured relative dielectric constant can be greater than the published value for the film material. This might be expected because as the twist is made very tight, the wire dimension and the film thickness change so that they no longer fit the model for which the theoretical parameters were calculated.

When the β is calculated from the equivalent dielectric constant measurements, it is found to fit the expression.

$$\beta = 0.25 + 4 \times 10^{-4} \theta^2 . \quad (7)$$

In Fig. 4 this expression is used to predict the characteristic impedances for three magnet wire sizes, as samples. It is seen to consistently predict the correct impedance sufficiently closely to yield a voltage standing-wave ratio (VSWR) within 1.1:1.

The slope of β versus θ in (7) is a function of the softness of the wire insulation. Equation (7) holds for most modern magnet wire insulating films but it will change if the insulation is very soft. Polytetrafluoroethylene (Teflon) covered hook-up wire is a good example that shows how far it can change. The relative dielectric constant versus pitch angle is given in Fig. 5 for a sample of no. 24 19-strand wire. A fit is found by using

$$\beta = 0.25 + 1 \times 10^{-3} \theta^2 . \quad (8)$$

DESIGN PROCEDURE

The following steps form a useful outline for designing a twisted magnet bifilar wire transmission line for a required characteristic impedance.

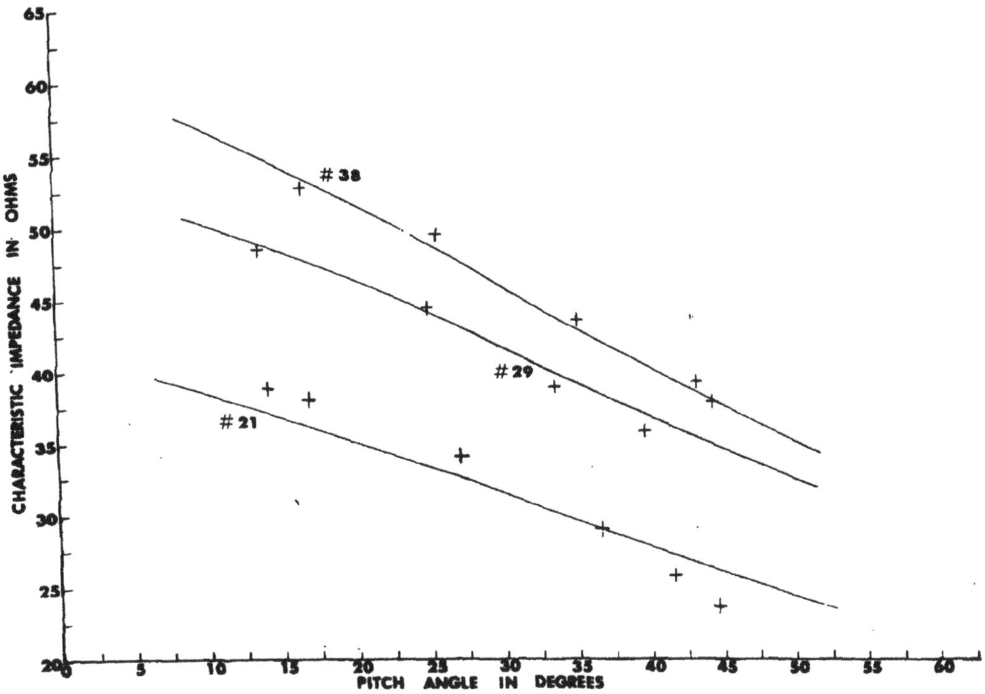

Fig. 4. Comparison of measured to estimated characteristic impedance for 3 wire sizes versus pitch angle.

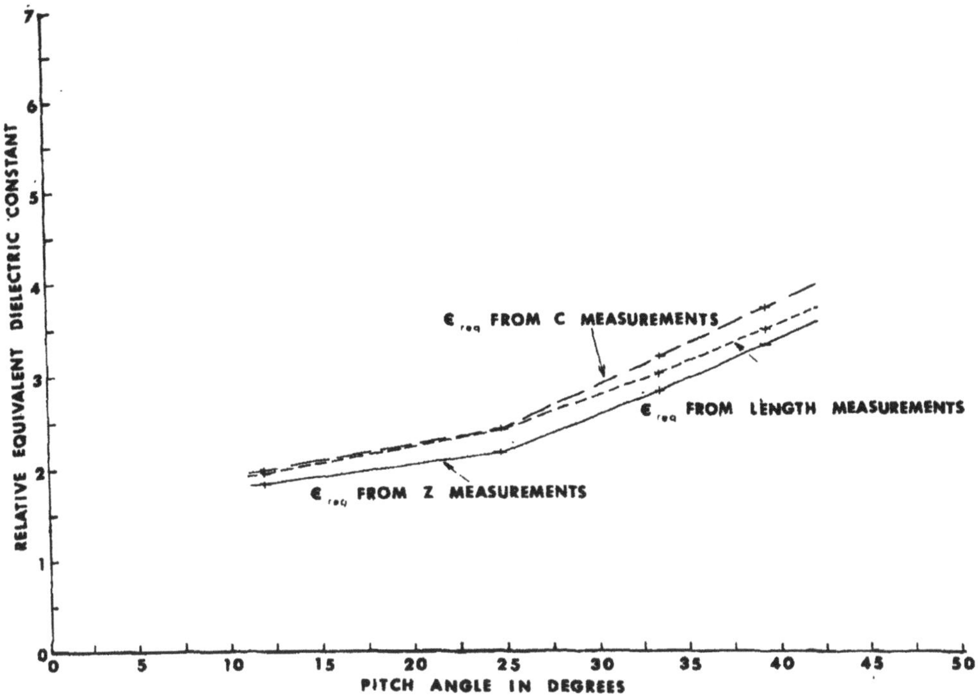

Fig. 5. Measured equivalent dielectric constant versus pitch angle for a sample of Teflon-covered no. 24 19-strand wire.

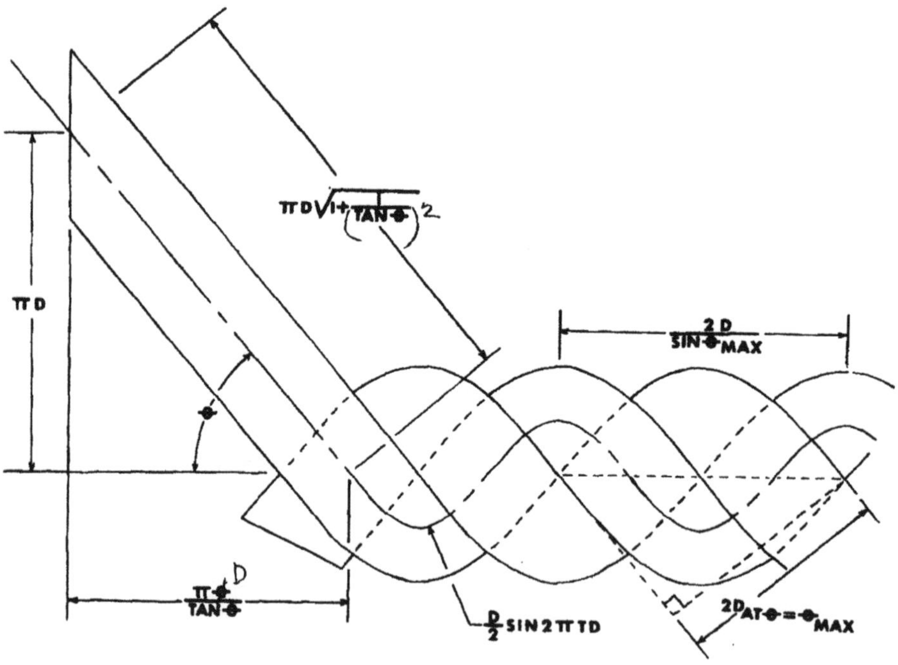

Fig. 6. Bifilar wire transmission line twisted to its maximum pitch angle of 50.5°.

1) Choose a film insulation thickness group and film relative dielectric constant. (These are usually limited by their availability and other physical and electrical considerations.)

2) Calculate the relative dielectric constant assuming pitch angle of 30° from (3) and (7):

$$\epsilon_{req} = \epsilon_{r1} + 0.61\,(\epsilon_{r2} - \epsilon_{r1})$$

where $\epsilon_{r1} = 1$ for air.

3) Calculate the required characteristic impedance for relative dielectric constant of 1 from

$$Z_1 = Z\sqrt{\epsilon_{req}};$$

where

Z_1 impedance when $\epsilon_{req} = 1$;
Z desired impedance.

4) Use Fig. 1 to choose the correct wire size.

5) Use Fig. 2 to determine the required twist per in.

CONCLUSION

The characteristic impedance of any twisted magnet wire transmission line can be found if the dimensions and the dielectric constant are known. Figs. 1 and 6 give the impedance at the expected tolerance limits for standard wire sizes having an insulation dielectric constant of 1. This is modified by the equivalent dielectric constant, which is a function of the relative dielectric constant of the film insulation and increases as the line is twisted. The twist for various pitch angles [determined by (2)] is shown in Fig. 2 for wire sizes 4-44. The equivalent relative dielectric constant is related to the pitch angle by (3) and (7).

The electrical length of a twisted line is related to the square root of the equivalent dielectric constant in (6).

Using these basic expressions the following statement can be made.

1) The impedance increases as the wire dimensions decrease.

2) The impedance decreases as the wire twist is increased. (The pitch angle increases.)

3) The relative dielectric constant for the line is controlled more by the film dielectric constant as the pitch angle increases.

4) For small pitch angles the line impedance can be reduced by immersing the line in a material other than air.

5) For large pitch angles the line impedance can be reduced by using a wire insulating film having a high dielectric constant. The useful pitch angle range is bounded by the limitation on line uniformity at about 20° and the strength of copper at about 45°.

For a film dielectric constant of 3.5 and with air as the surrounding medium, the useful impedance range for twisted magnet wire transmission lines is 10-60 Ω.

Over 60 magnet wire transmission lines of many different forms have verified the conclusions of this paper.

APPENDIX

The center lines of two twisted wires form a helix as shown in Fig. 6. The helix diameter is one wire diameter. It forms a sine wave in the longitudinal plane.

$$A = (D/2)\sin(2\pi Td) \tag{9}$$

where

D wire diameter;
T twist per inch;
d distance along the transmission line.

The pitch angle comes out of this as

$$\tan \theta = \pi TD \tag{10}$$

where θ is the pitch angle defined in Fig. 6.

The maximum pitch angle occurs when the line has been wound to the point where the distance between wire center lines is *D* and twist can be expressed as

$$T_{max} = (\sin \theta)/(2D). \tag{11}$$

The maximum pitch angle of 50.5° is found by equating 9 and 11. It is not a function of wire diameter.

The actual wire length can also be seen in Fig. 6 where one twist is unwrapped. The wire length for one twist is

$$\pi D\sqrt{1 + 1/(\tan \theta)^2} \tag{12}$$

Peter Lefferson received the B.E.E. and M.E. degrees from the University of Florida, Gainesville, in 1962 and 1965, respectively.

He worked for NASA as an Instrumentation Engineer for one year. He spent five and one half years with Electronic Communication Inc., in design of UHF transceivers. Presently, he is with Milton Roy Company, St. Petersburg, Fla.

CORRIGENDUM

In the paper, "Twisted Magnet Wire Transmission Line," by Peter Lefferson, PHP Transactions, Volume PHP-7, Number 4, December 1971, page 148, the following errors should be corrected:

In Figure 6

$\frac{\pi\theta}{\mathrm{TAN}\ \theta}$ should be $\frac{\pi D}{\mathrm{TAN}\ \theta}$

$\pi D\sqrt{1 + \frac{1}{\mathrm{TAN}\ \theta}}$ should be $\pi D\sqrt{1 + \frac{1}{(\mathrm{TAN}\ \theta)^2}}$

In Equation 12

$\pi D\sqrt{1 + 1/\mathrm{TAN}\ \theta}$ should be $\pi D\sqrt{1 + 1/(\mathrm{TAN}\ \theta)^2}$

Practical Design Information for Broadband Transmission Line Transformers

Abstract—**Normalized design curves useful in constructing broadband, 4-to-1 impedance matching, transmission line transformers of predictable performance are presented. Insertion loss plus the magnitude and phase angle of input impedance as functions of the length and characteristic impedance of the transmission line are included. Measurements demonstrate close agreement between measured and calculated performance.**

Recent applications of RF power transistors to untuned broadband operation, from LF through VHF, have primarily employed low-loss transmission line and ferrite structures similar in concept to those described by Ruthroff.[1] These elements are suitable for impedance matching, phase reversal, power splitting, and power adding over several octaves. Impedance matching at the input and output of each transistor stage is frequently accomplished by broadband 4-to-1 impedance transformers used singly or cascaded for greater impedance ratios. Upper frequency bandwidth depends upon two major factors: the physical length of the transmission line sections used and the relationship between the load impedance and the characteristic impedance Z_0 of the transmission line.

Ideally, for use in impedance step-up of the load, the optimum Z_0 should equal twice the value of a purely resistive load being transformed. For step-down transformation, the optimum Z_0 should equal half the resistive load. Ruthroff considered this ideal case in deriving an expression for insertion loss (defined by power output divided by maximum power available from the generator) as a function of line length expressed in wavelength.

Low impedances (upwards from 2 Ω at transistor base inputs) whose values are often vague suggest unusually low Z_0 transmission lines. Exact values for the optimum Z_0 are not often practically realized. It is commonly acknowledged that some deviation from the ideal is tolerable with only minor sacrifices in performance. To date no specific design information relating broadband performance to the choice of Z_0 has been available. Ruthroff developed a simple expression for insertion loss of a 4-to-1 transformer as follows:

$$\frac{P_{out}}{P_{available}} = \frac{4R_L R_g(1+\cos\beta l)^2}{[2R_g(1+\cos\beta l)+R_L\cos\beta l]^2+\left[\dfrac{R_g R_L + Z_0^2}{Z_0}\right]^2\sin^2\beta l}$$

where $P_{available}$ is the maximum available power from a source with internal resistance R_g, and R_L is the load at the output terminals of the transformer. β is the phase constant of the transmission line, and l is the length of the line. Fig. 1 shows curves of insertion loss versus length of line λ for a 4-to-1 transformer generated through computer calculations using the above expression. The upper curve indicates insertion loss with optimum Z_0, as defined above, for any given R_L; the other curves indicate performance compromises with various degrees of deviation of Z_0 from the optimum. Pairs of Z_0 values, related by $KZ_{0\,optimum}$ and $Z_{0\,optimum}\,K$, where K is any given constant, yield identical insertion loss curves.

The curves could be used as an aid to design as follows. Assume a Z_0 such that the practical value available is twice the determined optimum value, and assume an insertion loss of no more than 0.2 dB is desired. The curve of $2Z_0$ optimum crosses the 0.2 dB point at about 0.09 λ. Thus if the length of transmission line employed is 0.09 λ or less at the uppermost frequency of operation, insertion loss will not exceed 0.2 dB. This curve applies equally to the performance of a transmission line with half the optimum value of Z_0.

Measurements to verify calculated results were made over a frequency range of 1 to 108 MHz using a Hewlett-Packard 4815A vector impedance meter. A 9-inch length of RG-58 cable ($Z_0 = 50$ Ω) representing 0.125 λ at 100 MHz was interconnected as a 4-to-1 impedance step-down transformer. Ferrite toroids of Indiana General type Q-1 material covered the transmission line section. The low-frequency falloff is a function of the load impedance, the length of toroid-covered section, and the permeability of the ferrite at these frequencies. The lower 1-dB point corresponds to a frequency of about 1 MHz. Equivalent performance was achieved using a 9-inch length of #22 Formvar twisted-pair with Z_0 of approximately 50 Ω wound on a single toroid. Measured and calculated results are compared in Fig. 2. These results confirm the relevance of the theoretical curves to actual design performance.

Manuscript received December 18, 1968.

[1] C. L. Ruthroff, "Some broadband transformers," *Proc. IRE*, vol. 47, pp. 1337–1342, August 1959.

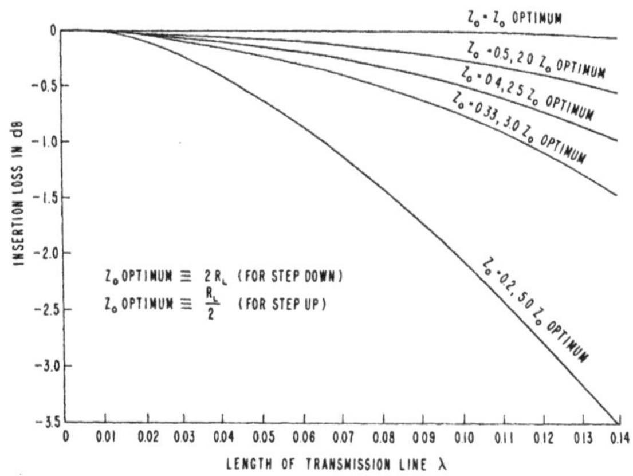

Fig. 1. Insertion loss as a function of transmission line length in a 4-to-1 impedance transformer for various values of Z_0.

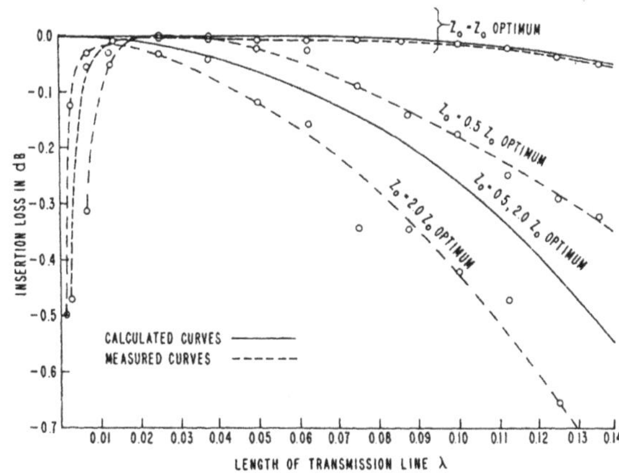

Fig. 2. Calculated and measured curves of insertion loss as a function of transmission line length in a 4-to-1 impedance transformer for several Z_0 values. (Measured curves for the transformer driven from the low-impedance port have: $R_L = 100$ Ω, $Z_0 = 50$ Ω for the Z_0 optimum curve; $R_L = 50$ Ω, $Z_0 = 50$ Ω for the $2Z_0$ optimum curve; and $R_L = 100$ Ω, $Z_0 = 50$ Ω for the 0.5 Z_0 optimum curve.)

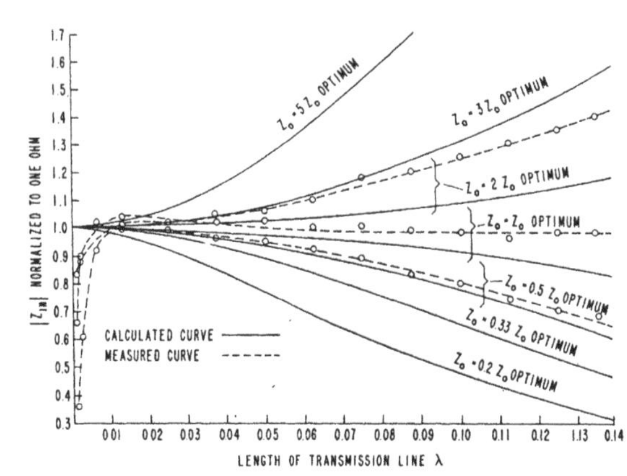

Fig. 3. Calculated and measured curves of the magnitude of input impedance at the low-impedance port of a 4-to-1 transformer as a function of transmission line length for various values of Z_0 (Z_{in} normalized to unity.)

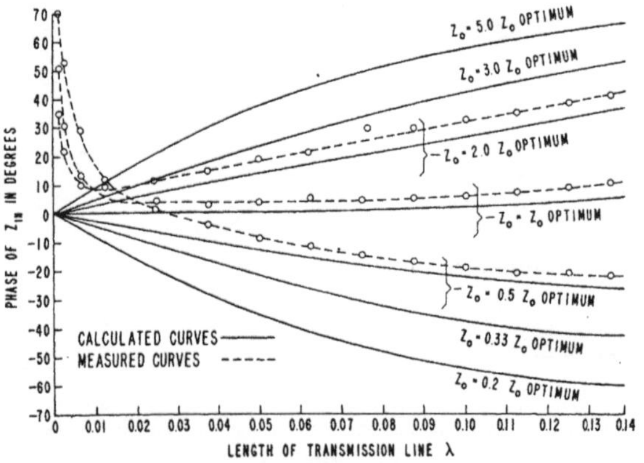

Fig. 4. Calculated and measured curves of the phase of input impedance at the low-impedance port of a 4-to-1 transformer as a function of transmission line length.

Insertion loss, as defined previously, is a function of the input impedance of the transformer. Losses within the transmission line section are insignificant. Fig. 3 shows the variation of the transformer input impedance versus length of line for step-up impedance transformation. The curves of Fig. 3 are normalized impedances and may be converted to actual values by multiplying them by $Z_{0\,\mathrm{optimum}}$ (equal to $2R_L$). Fig. 4 indicates phase variations as a function of line length for cases in Fig. 3. Measured results shown in Figs. 3 and 4 demonstrate excellent agreement with theory.

O. PITZALIS, JR.
U. S. Army Electronics Command
Fort Monmouth, N. J. 07703
T. P. M. COUSE
Electronic Engrg. Dept.
Monmouth College
West Long Branch, N. J.

BROADBAND TRANSFORMER DESIGN FOR RF TRANSISTOR POWER AMPLIFIERS

Octavius Pitzalis Jr.
U. S. Army Electronics Command
Fort Monmouth, New Jersey

Thomas P. M. Couse
Department of Electronic Engineering
Monmouth College
West Long Branch, New Jersey

Summary

Various forms of broadband transmission line transformers for use through VHF, and concepts basic to their operation are summarized and described. Broadband power adding and power dividing hybrids for push-pull amplifiers and parallel additive uses are given similar attention. The design of broadband input and output matching elements for a single ended stage using a 2N5071 transistor for 30 watts of output power from 30 MHz to 80 MHz is illustrated. Useful curves for guiding the design of 4:1 transformers with resistive loads are given. Curves showing typical input impedance versus frequency of a pair of cascaded 4:1 transformers loaded by the large signal R-L-C equivalent input circuit of the 2N5071 are included. The use of hybrids to operation is described.

Introduction

In the past few years RF power transistors have attained performance levels which have made them attractive for implementing many solid state transmitters intended for military and commercial communications. Many of these applications are for selective frequency operation over one or more frequency bands, a band being typically between 0.4 and 0.8 of an octave in bandwidth. With modern trends stressing minimized tuning operations, the use of untuned broadband interstage and output matching networks has become popularly sought. The most suitable broadband impedance matching networks in the realm of lumped components, applicable from LF into UHF, have been broadband transmission line transformers and baluns of the type originally proposed by Guanella[1] and more recently expanded upon by Ruthroff[2]. Short lengths of transmission line (parallel strip, coaxial or twisted pair) provide upper frequency capability free from normal transformer interwinding resonance limitations. High permeability materials, such as ferrites extend the lower bandwidth capabilities. Though the concept of these broadband transformers is not new, to date little analysis of their two port parameters as a function of frequency, or transmission line length, and characteristic line impedance has been available.

Survey of Some Popular Broadband Transmission Line Components

The 4:1 Impedance Transformer

The 4:1 transmission line transformer is a popular element for impedance matching a single ended source to a single ended load. A close analogue among conventional transformers is the autotransformer with a 2:1 turns ratio as depicted in Fig. la, showing a load R_L transformed to $R_L/4$ at the input.

The transmission line counterpart is shown schematically in Fig. lb. The transmission line provided excellent coupling of the magnetic and electric fields. This ensures a minimum of leakage reactance over the useful frequency range of the line. With conventional windings, interwinding capacitance limits the performance at high frequencies. Between transmission line conductors the uniformly distributed capacitance acts harmoniously with series distributed inductance to give reactance free operation over a broad frequency range.

Upper Frequency Cutoff Factors. The transmission line transformer has upper frequency limitations related to the length of the transmission line in wavelengths and to the ratio between the characteristic impedance Z_0 of the line and the load impedance. In most cases, the length of transmission line is chosen to be $\lambda/8$ or less at the uppermost frequency desired. The optimum match is $Z_0 = \frac{1}{2} R_L$ for step down of a purely resistive load and $Z_0 = 2 R_L$ when the transformer is being used to step up the load from R_L to $4 R_L$. As expected with transmission line matching, the shorter the physical length of the line in wavelengths the less important exact matching becomes. The interconnection of the transmission line is important in keeping with the two general matching principles for a flat line: 1) The voltage between conductors must be maintained constant as a wave propagates from the input until it reaches the load. This is illustrated in Fig. 1b by the voltage E, between points 1 and 3 at the driven end and also at the load end between points 2 and 4. Differences in the signal path length to corresponding points along the line cause phase differences increasing with frequency. Such phase differences in the 4:1 connection, (not typical of normal, matched transmission line) determine the final bandwidth capability of the transformer. 2) For a current flowing in one conductor (i.e. I/2 flowing from 1 to 2) an equal current in the opposite direction should exist in the other conductor (i.e. I/2 flowing from 4 to 3).

Low Frequency Cutoff Factors. Transmission line transformers have a low frequency cutoff determined by the falloff of primary reactance as frequency is decreased. This reactance is determined by the series inductance of the transmission line conductors. Therefore, the longer the length of line, the greater the series inductance, extending performance to lower frequency. This is in direct conflict with extending the upper frequency performance which, as previously stated, is enhanced by shortening the physical length of the transmission line. Minimizing the overall length consistent with achieving the suitable reactance for low frequency requirements can be accomplished by using high permeability materials such as ferrites. Inductance of a conductor is directly proportional to the relative permeability of the surrounding medium. A high permeability material placed close to the transmission line conductors acts on the external fringe field present and can magnify the inductance appreciably and thereby provide a lower cutoff frequency. There is no influence upon internal magnetic fields nor upon the characteristic impedance of the line. The power transferred from input to output is not coupled through the ferrite material, but rather through the dielectric medium separating the transmission line conductors. This is an important concept for design. Relatively small cross section ferrite material can perform unsaturated at impressively high power levels. In contrast to this, conventional transformers couple power from primary to secondary entirely through a high permeability core which must be chosen to suitably carry the total power without saturating. Fig. 1c illustrates a popular construction for a 4:1 transformer using twisted pair wound on a ferrite toroid. A basic advantage of this construction is the ability to minimize the length of the interconnection from point 1 to point 4. Fig. 1d is a photograph of several 4:1 transformers made in this manner. The larger ones in the upper row have windings from 3.5 to 4.5 inches long. The transformer at the upper left uses a loosely twisted pair of number 20 formvar coated transformer wire having a Z_o = 50Ω. The middle unit is twisted of two bifilar windings of number 26 formvar wire for a Z_o = 20Ω. The unit at the upper right is constructed of homemade strip transmission line. Flat former transformer wire about 0.10 inches wide paired together with contact bond cement gives a Z_o = 10Ω. These transformers have each carried 100 watts of RF power without saturating the ferrite cores. With properly matched resistive loads they have less than 0.1 dB insertion loss at an upper frequency of 100 MHz and a lower frequency of 2.0 MHz. The lower two transformers are constructed with smaller toroids and have 2.5 inch lengths of transmission line wound about them. The transformer at the lower right uses RG-196 (50Ω coaxial line) with the outer insulation removed. This improves its flexibility and permits winding the shield closer to the ferrite which makes more effective gains from the high permeability material. In the lower left of Fig. 1d is a small unit wound with the strip line of Z_o = 10Ω described earlier. The small toroid demonstrates the relative flexibility of this ribbon type line. The 4:1 transformer can be used singly or in cascade for stepping up or stepping down a load impedance.

Higher Ratio Impedance Transformers

A brief mention of some other possible extensions of the basic 4:1 transformer maybe informative. Fig. 2a shows a 9:1 impedance transformer in schematic form. Two equal lengths of transmission line are used. Construction forms usually follow the style of the 4:1 type described above. Each length of transmission line ordinarily uses a separate ferrite core. The cores may be stacked to keep interconnections as short as possible. The transformer operation should be clear with the voltages and currents as indicated. For optimally matched conditions and negligible phase shift:

$$R_{IN} = \frac{E}{I} \tag{1}$$

$$R_L = \frac{E/3}{3I} = \frac{E}{9I} = \frac{R_{IN}}{9} \tag{2}$$

$$Z_0 = \frac{E/3}{I} = \frac{R_{IN}}{3} = 3R_L \tag{3}$$

Figure 2b shows the next logical extension of the 9:1. One additional length of transmission line is employed to achieve a 16:1 impedance transformer. The general winding approach may be observed from Fig. 2a and 2b and extended to higher ratios by induction.

The available ratios will progress 1:1, 4:1, 9:1, 16:1, 25:1.... n:l (n always an integer). Also, the optimum Z_o choice will always be given by:

$$Z_0 = \sqrt{R_L R_{IN}} \tag{4}$$

Broadband Power Adding and Power Dividing Networks

It is frequently necessary to use two or more transistors to attain some output power beyond that of any available single device. Broadband power adding and dividing can be nicely accomplished with hybrids constructed simply as transmission line structures. Hybrids permit summing the power from two or more sources with insignificant interaction between the individual sources. The hybrid provides excellent isolation between the input ports when separate sources of power are to be added (or alternately isolation between two output ports acting as separate sources when power splitting). The input impedance (or output impedance in the alternate case) at one port remains essentially invariant regardless of conditions of an open circuit or a short circuit at the second port. Obviously, such characteristics will effectively eliminate chain type failures common in multiple transistor output stages with transistors directly paralleled at their outputs.

The Zero Degree Hybrid. A hybrid can be constructed as a zero degree phase type for adding inputs which are approximately in phase. Similarly, it will split a single input into two identical in-phase sources.

Ruthroff described several zero degree phase hybrids. Fig. 3 illustrates schematically in transmission line form one such hybrid together with voltage, current and impedance relationships. The component R_T is a physical resistor which dissipates non-addable input power. This waste power is related to phase and/or amplitude differences between the signals at ports 1 and 2. For the extreme case where one port is open circuited or short circuited, one half of the total input power from the other port is dissipated in R_T and half is delivered to the load R_L.

The 180 Degree Phase Hybrid. The push-pull operation of two power transistors is often desirable in output stages. Push-pull usually reduces the second harmonic output power by about 8 dB over single ended stages, thereby easing the demands for second harmonic rejection in the low pass output filter.

The 180 degree phase hybrid affords all the virtues of isolation between adjacent ports, described previously, plus 180 degree phase splitting (or an adding for two signals 180 degrees out of phase).

Fig. 4a illustrates the interconnections and operating relationships for a 180 degree phase hybrid being used for power adding. Fig. 4b shows the voltage and current relations when one port is either open or shorted. As can be noted, under this condition, one half of the output at the working port is dissipated in R_T. In general for this hybrid:

$$P_{out} = \frac{P_1 + P_2}{2} - \sqrt{P_1 P_2}\cos\theta \qquad (5)$$

where:

P_{out} = power into the load R_L
P_1 = power into port 1
P_2 = power into port 2
θ = phase angle between signals at ports 1 and 2.

When used for power splitting, this hybrid will contribute equal out of phase power to each transistor load. If one load is open circuited, short circuited or in general dissimilar from the alternate load, the input drive to the alternate load will remain unchanged. This is controlled by a change in impedance at the input port.

With $Z_0 = R_L$, the length of the series output line section is unimportant, since it performs as a simple length of transmission line terminated in its characteristic impedance. With very low values of load impedance it is impractical to exactly match Z_0 to R_L. The length of this section must then be maintained quite short relative to $\lambda/8$ if reactive effects due to mismatch are to be avoided.

The Application of Broadband Transmission Line Structures to the Design of Untuned Broadband RF Transistor Amplifiers

Requirements for broadband untuned VHF transistor amplifiers often demand operation over two or more bands representing over an octave of frequency range. For single-ended stages, the sole purpose of the broadband elements is for impedance matching. The 4:1 impedance transformer is a popular broadband component for this. For power additive stages, signal phasing and isolation between transistors must also be considered. The design of broadband components for both a single-ended output stage and a push-pull output stage will be demonstrated.

Extensive characterization of the 4:1 transformer will be described as part of this design.

Single-Ended RF Power Stage Design

Matching for power transistor stages consists of 1) Converting some fixed real antenna load to the appropriate load line for the final power stage such that some specified power output can be realized with good efficiency at all operating frequencies, and 2) Interstage matching, which requires transforming the reactive low input impedance of a transistor into a suitable load line for the driving stage at all operating frequencies

To demonstrate the design of broadband elements to accomplish matching, a typical power output stage will be investigated. The transistor is a 2N5071 operating from a 28 volt d.c. supply. Operation is specified to be 30 watts of RMS output power from 30 MHz to 80 MHz. The antenna load is 50Ω.

Output Matching

The optimum load line for producing 30 watts of output power from this transistor operating from 28 volts d.c. is 13Ω to 15 Ω. One 4:1 transformer will step down the 50Ω antenna load to 12.5Ω. Equations for terminal impedance of the 4:1 transformer can be derived from two Kirchoff loop equations for the circuit of Fig. 1b solved simultaneously with the two general defining equations for voltage and current at one end of a lossless transmission line. At the low impedance port ($R_{IN} = R_L/4$) i.e. the load is stepped down:

$$Z_{IN} = Z_0\left[\frac{Z_L\cos\beta l + jZ_0\sin\beta l}{2Z_0(1+\cos\beta l) + jZ_L\sin\beta l}\right] \qquad (6)$$

At the high impedance port ($R_{IN} = 4R_L$) i.e. the load is stepped up:

$$Z_{IN} = Z_0\left[\frac{2Z_L(1+\cos\beta l) + jZ_0\sin\beta l}{Z_0\cos\beta l + jZ_L\sin\beta l}\right] \qquad (7)$$

Where Z_L is the load impedance, β the phase constant of the transmission line in radians per unit length, and ℓ is the length of the line section. For the broad number of applications where the load impedance is purely resistive, normalized design curves can describe Z_{IN} as a function of frequency and length of line (in terms of $\beta\ell$) and as a function of Z_o.

Figs. 5, 6, and 7 display R_{IN}, X_{IN}, and phase respectively, at the low impedance port (i.e. the transformer used for load step down) as a function of length and Z_o of the transmission line. The R_{IN} and X_{IN} values can be unnormalized by multiplying by $R_L/4$. The various Z_o values are simply related to R_L as shown. The curves apply only to high frequency cutoff effects and assume nonexistence of low frequency effects. With Z_o equal to Z_o optimum, R_{IN} drops off gradually, from a value of $R_L/4$, with increasing length of line (i.e. the equivalent of increasing frequency for a fixed length of line). An inductive X_{IN} becomes apparent beyond 0.10λ. With Z_o values greater than Z_o optimum the inductive character of the input impedance is accented. The behavior closely approaches a fixed R-L series circuit when Z_o is $2Z_o$ optimum or greater. With Z_o values less than Z_o optimum, the input impedance behaves as a parallel R-C circuit.

Fig. 8 shows curves of insertion loss as a function of line length for various values of Z_o. Insertion loss is defined as the ratio of power into the load to power available from a source E_g, with internal resistance R_g, matched to the ideal R_{IN} value of the transformer. It should be clearly understood that the transformer is as internally lossless as the transmission line comprising it. As typical of lossless filters, power transfer is determined by the variation of terminal impedance with frequency. There is insignificant accuracy loss in assuming that real power into the transformer P_{IN} equals real power into the load P_{out}. An expression for insertion loss with any generalized impedance ($Z_{IN} = R_{IN} + jX_{IN}$) represented at a particular value of is:

$$\text{Insertion Loss} = \frac{P_{out}}{P_{available}} = \frac{4R_g R_{IN}}{(R_g + R_{IN})^2 + X_{IN}^2} \qquad (8)$$

Where $R_g = R_L/4$ the value of R_{IN} at $\ell = 0$ for load step down or
$R_g = 4R_L$ the value of R_{IN} at $\ell = 0$ for load step up.

Note that in Fig. 8 pairs of Z_o values, related by KZ_o optimum and Z_o optimum/K, where K can be any given constant, yield identical insertion loss.

For each of the curves described above the abscissa is given in wavelengths λ, which allows relating units of length and frequency for any design care. Knowing the highest operating frequency f_u in MHz and next judging from the chart the largest fractional part of a wavelength n, which affords R_{IN} and X_{IN} within acceptable limits, the physical length of line is given by:

$$l = \frac{7200n}{f_u} \quad \text{inches} \qquad (9)$$

This expression holds for lines constructed with conventional dielectrics of relative dielectric constants ranging from 2.0 to 2.5.

The required minimum length for the lower frequency cutoff requirements should next be examined. A conservative empirical relation which holds quite closely is:

$$l \geq \frac{20R_L}{\left(1 + \frac{\mu}{\mu_0}\right) f_L} \quad \text{inches} \qquad (10)$$

Where R_L is the load resistance in ohms, f_L the lowest operating frequency in MHz, and μ/μ_0 the relative permeability of the ferrite at f_L. This expression relates a length corresponding to the input impedance having a 10 to 15 degree inductive phase angle at f_L when the transformer is terminated in a real load R_L. A typical ferrite for use at 30 MHz (e.g. Q-2 Ferramic) will have $\mu/\mu_0 = 15$. With $R_L = 50\Omega$, and $f_L = 30$ MHz, equation (10) gives $\ell = 2.08$ inches. Reference to the curves for ℓ, a low impedance port (Fig. 7) with $Z_o = Z_o$ optimum $= 25\Omega$ at $.08\lambda$; R_{IN} is about 6% down from the ideal 12.5Ω value and X_{IN} is essentially zero.

Using equation (9) with fu = 80 MHz and n = 0.08 it follows that $\ell = 7.5$ inches. Therefore any length between 2.15 inches and 7.2 inches wound on a ferrite toroid, will offer excellent performance. Arbitrarily let the choice be a length of 4 inches. It is apparent that for this frequency range of 30 MHz to 80 MHz the design of the transformer is quite flexible. If the frequency range were expanded (e.g. 2 MHz to 80 MHz) the length determined from equation (9) would converge upon that of equation (10).

The curves are very useful in other ways. For example, assume it to be more convenient to use transmission line of $Z_o = 50\Omega$ than the optimum value of 25Ω. At n = 0.08 and observing the $Z_o = 2\ Z_o$ optimum curves, R_{IN} is almost ideally equal to $R_L/4$ or $12.5\ \Omega$, X_{IN} is about $5\ \Omega$ inductive. The effect of this value of reactance may be difficult to place in perspective when attempting to estimate performance. The insertion loss curves of Fig. 8 will be useful for this. Note that at 0.08λ on the $Z_o = 2\ Z_o$ optimum curve the insertion loss is less than 0.2 dB. This gives more specific meaning to the effect of the $5\ \Omega$ inductive reactance. Fig. 15 illustrates the output matching circuit in block diagram.

Measurements. Measurements to verify calculated results were made over a frequency range of 1 MHz through 108 MHz using an HP 4815A Vector Impedance Meter. A 9-inch length of RG-58 cable ($Z_o = 50\ \Omega$ and $\ell = 0.25\lambda$ at 100 MHz) with outer insulation removed was interconnected as a 4:1 impedance step down transformer. Ferrite toroids, CF 102 size of type Q-1 material ($\mu/\mu_0 = 100$ up to 10 MHz), covered the entire transmission line section. With the fixed $Z_o = 50\Omega$ line terminated in R_L of 50Ω, 100Ω and 200Ω, the 0.5 Z_o optimum, Z_o optimum and 2.0 Z_o optimum conditions were simulated. Measured and calculated insertion loss with frequency are compared in Fig. 9. Differences between measured and calculated insertion loss with frequency are compared in Fig. 9. Differences between measured and calculated was less than 0.2 dB at 100 MHz for the 0.5 and 2.0 Z_o optimum conditions. The Z_o optimum simulation gave performance hardly distinguishable from the theoretical from 10 MHz to 108 MHz.

The measured curves evidence low frequency effects. As apparent from equation (9) low frequency cutoff is a function of the load impedance, the length of the ferrite covered section, and the permeability of the ferrite at the lower frequencies. The low 1 dB frequencies were 1 MHz or less.

Input Matching

The broadband input matching of the transistor is considerably more difficult than the matching at the output. The transistor base input is a low impedance which makes it impractical to select the value of characteristic impedance considered optimum for the transformer. The input impedance appears to be an R-L-C series circuit from 30 MHz to 80 MHz. Large signal base input impedance measurements of this transistor at 30 MHz, 40 MHz, 60 MHz, and 80 MHz indicate the resistance R_B = 2Ω, the inductance L_B = 3nH and the capacitance C_B = 2500 pF. The L_B is due to parasitic lead inductance in the transistor package plus any contributed by circuit interconnections. Extreme care in circuit construction is required to minimize this series base input inductance. This input circuit is resonant at about 58 MHz and has a Q = 0.55. Minimized input Q is desirable for transistors intended for broadbanding. Fig. 9 depicts the transistor input circuit.

Universal design curves for the 4:1 transformer, with various reactive terminations, are not practical, but design attitudes evident from this example are applicable to other transistors and other frequency ranges. The equations (6) and (7) can always be applied to any particular design situation.

Some general 4:1 transformer characteristics apparent from Fig. 6 and 7 with real loads may be used to guide design attitudes. The transistor input impedance is inductive above 58 MHz. The design curves for reactance at the high impedance port suggest that the use of low Z_0 (i.e. Z_0 optimum or less) would lend an increasingly capacitive phase with frequency. This would offset the increasingly inductive tendency of the transistor input impedance with frequency. Unfortunately, this specifies Z_0 values from 1Ω to 4Ω. Such unusually low values have been achieved with homemade coaxial and strip transmission line, but have not been useful because of the unavoidably bulky construction. The minimum interconnection lengths contribute parasitic inductance that more than offsets the improvements anticipated over higher Z_0 lines with compact interconnections.

The lowest practical values of Z_0 are 10Ω to 12Ω. Assuming the use of 12Ω for design, the previous curves indicate that inductive reactance contributed by the transformer is minimized by minimum length. The smallest practical construction length of transmission line with Z_0 = 12Ω is about 2 inches. Fig. 10 and 11 show the variation of R_{IN1} and and X_{IN1} with Z_0 = 12Ω transmission line of lengths 2.0, 2.5 and 3.0 inches. The differences in performance are seen to be relatively minor. These curves were developed from computer calculations using equations (7) with Z_L being the transistor input impedance.

The transistor input was carefully simulated with a 2Ω carbon resistor, three leadless capacitor chips totaling 2500 pF. With interconnecting inductance of 3 to 4 nH. The dotted curve shows measured performance for a 4:1 transformer constructed with a 2.5 inch length of Z_0 = 12Ω ine. The transformer was identical in structure to the strip line unit at the lower right of Fig. 1d. Measurements for these curves were made with a HP 4815A Vector Impedance Meter. Corrections must be made in low impedance measurements to account for the parasitics in the probe. These values are typically a resistance of 0.6Ω nd an inductance of 12 to 15 nH in the probe. The exceptionally close correlation of measured and calculated curves emphasizes that the calculated values for these transformers are reliably accurate in predicting performance.

Curves of calculated R_{IN1} and X_{IN1} for various values of Z_0 with line length fixed at 2.5 inches are shown in Fig. 12 and 13. These calculations do not include the effects of parasitic series inductances. It is likely that larger parasitics would accompany the lower Z_0 values thereby nullifying much or all of the slight performance edge indicated by the curves.

It is always desirable to maintain an operating load line as purely resistive as possible. Where this is not possible best stability often favors capacitive rather than inductive reactance. Assuming this to be true, the addition of a parallel capacitor added as shown at point 1 in Fig. 15 can maintain effectively the phase of Z_{IN1} capacitive over the entire 30 MHz to 80 MHz range rather than growing increasingly inductive above 50 MHz. For zero phase at 80 MHz:

$$C = \frac{X_{IN1}}{\omega\left(R_{IN1}^2 + X_{IN1}^2\right)} \qquad (11)$$

where ω, X_{IN1}, and R_{IN1} are defined at 80 MHz. This given C close to 100 pF. The use of a variable capacitor would allow trimming to the fixed value for the best overall operation. The cascading of an additional 4:1 transformer to step up Z_{IN1} paralleled by the capacitor gives an input Z_{IN2} at point 2 as shown in Fig. 15.

The large signal dynamic swing of a transistor stage is limited by available voltage. For this reason load lines are conventionally represented by the equivalent parallel resistance for partly reactive loads. Real output power can then be defined easily by $V_{pk}^2/2R_L$ where R_L is the equivalent parallel load resistance.

R_{IN2} represents the equivalent parallel resistance looking in at point 2. Fig. 15 gives a clear representation of this. Fig.14 shows the behavior of R_{IN2} and the phase of Z_{IN2} with frequency for two values of Z_0. The transmission line length is 2.5 inches. A Z_0 of 8Ω gave results hardly distinguishable from the 16Ω value plotted. From Fig. 14 the better choice of Z_0 is the 16Ω value which produces the smaller overall variation in R_{IN2} and phase. These variations must be accepted as a consequence of the low impedance R-L-C input of the transistor. The use of the capacitor has preserved the non-inductive character of this load line.

If the efficiency of the driver stage is critical the value of this load line (between 32Ω and 40Ω) is somewhat low. An additional 4:1 transformer would give a load line resistance too large to develop sufficient drive power for the output stage (P_{IN} = 4 watts). The use of a 9:1 transformer in place of the second 4:1 between points 1 and 2 would be suitable. The resultant load line would have very similar trends to Fig. 14 but with the approximate resistive value between 72Ω and 90Ω. This load line would permit delivering the 4 watts of input power to the 2N5071 transistor at good efficiency.

Push-Pull Design

Assume it is desired to operate two 2N5071 transistors for an output power of 60 watts. Two 180 degree phase hybrids as previously described are necessary; One for splitting single ended input power source into two out of phase sources driving the transistor inputs, the other for adding the outputs from the two transistors. Fig. 16 shows a typical arrangement.

Output Matching

The output matching offers no particular problems. Assuming the antenna load to be 50Ω it must be partially transformer 2:1 by the low pass filter to either a 25Ω or 100Ω value for efficient output matching. Fig. 16 shows the load as 25Ω. The hybrid can easily be constructed with the optimum values of Z_0 as indicated. Under the conditions shown each transistor operates into a 12.5Ω load line suited to 30 watts output power from each. The transmission line lengths are not critical. A length of 4 inches for each winding is adequate.

Input Matching

The input hybrid is constructed of the lowest practical characteristic impedance line Z_0 = 12.5Ω. The shortest practical length of 2.5 inches will minimize mismatch effects. During push-pull operation one transistor is conducting while the second is biased off. The conducting transistor has a low impedance as described previously. The off transistor appears as a relatively high impedance due to the reverse biased base emitter junction. This affects the impedance of the input port of the power splitter. With balanced loads of R_B = 2Ω each the input would be approximately 4Ω. However, with one output port terminated in an open circuit the input impedance approaches three times this value. For transistor loads the input impedance is somewhere between these extremes. Fig. 15 indicates R_{IN} could be between 6Ω and 10Ω typically. The 4:1 transformer would step this up to a 25Ω to 40Ω load line for the driver. This is a suitable load line for developing 10 watts of drive power with good efficiency.

Conclusions

A group of transmission line transformers and hybrids useful for untuned broadband RF transistor amplifiers has been presented together with the fundamental principles underlying their operation. Design curves and expressions for realizing predictable performance from 4:1 transformers terminated in resistive loads were presented. The typical behavior of RF power transistor input impedance over the VHF range of 30 MHz to 80 MHz was described. A means for transforming this input impedance to a non-inductive load line was demonstrated. Finally the push-pull power adding of two transistors through the use of the 180 degree hybrid was illustrated.

Acknowledgements

The authors wish to acknowledge the assistance of Mr. Dave Huisman for the necessary computer programming for this work. Mr. Robert Horn and Mr. Dennis bowman contributed noteworthy ideas and observations during the course of several stimulating discussions.

Key Words

Transmission line transformers, broadband transformers, impedance matching transformers. RF transformers, untuned VHF transistor power amplifiers, broadband hybrids.

References

1. G. Guanella, "New Method of Impedance Matching in Radio Frequency Circuits", Brown-Boveri Review, vol. 31, pp. 327-329, 1944.

2. C. L. Ruthroff, "Some Broad-Band Transformers" Proc. IRE, vol. 47, pp. 1337-1342, August 1959.

3. W. Doherty and W. Hatley, "Broad Transmitter Techniques", ECOM Technical Report Contract Number DA28-043AMC-01829 (E), April 1967.

4. D. R. Lohrman. "High Efficiency CW RF Power Amplifier", USAECOM Technical Report 2836, May 1967.

5. O. Pitzalis and T. Couse, "Practical Design Information for Broad-band Transmission Line Transformers", Proc. IEEE, (to be published).

6. H. M. Schlicke, Essentials of Dielectromagnetic Engineering, N. Y., Wiley 1961.

7. W. C. Johnson, Transmission Lines and Networks, N. Y., McGraw Hill, 1950.

Editor's Note:

This article had to be re-typed as it was not possible to obtain a high enough quality copy of the original. The formatting may differ but the content is unchanged apart from the correction of some obvious typographical errors.

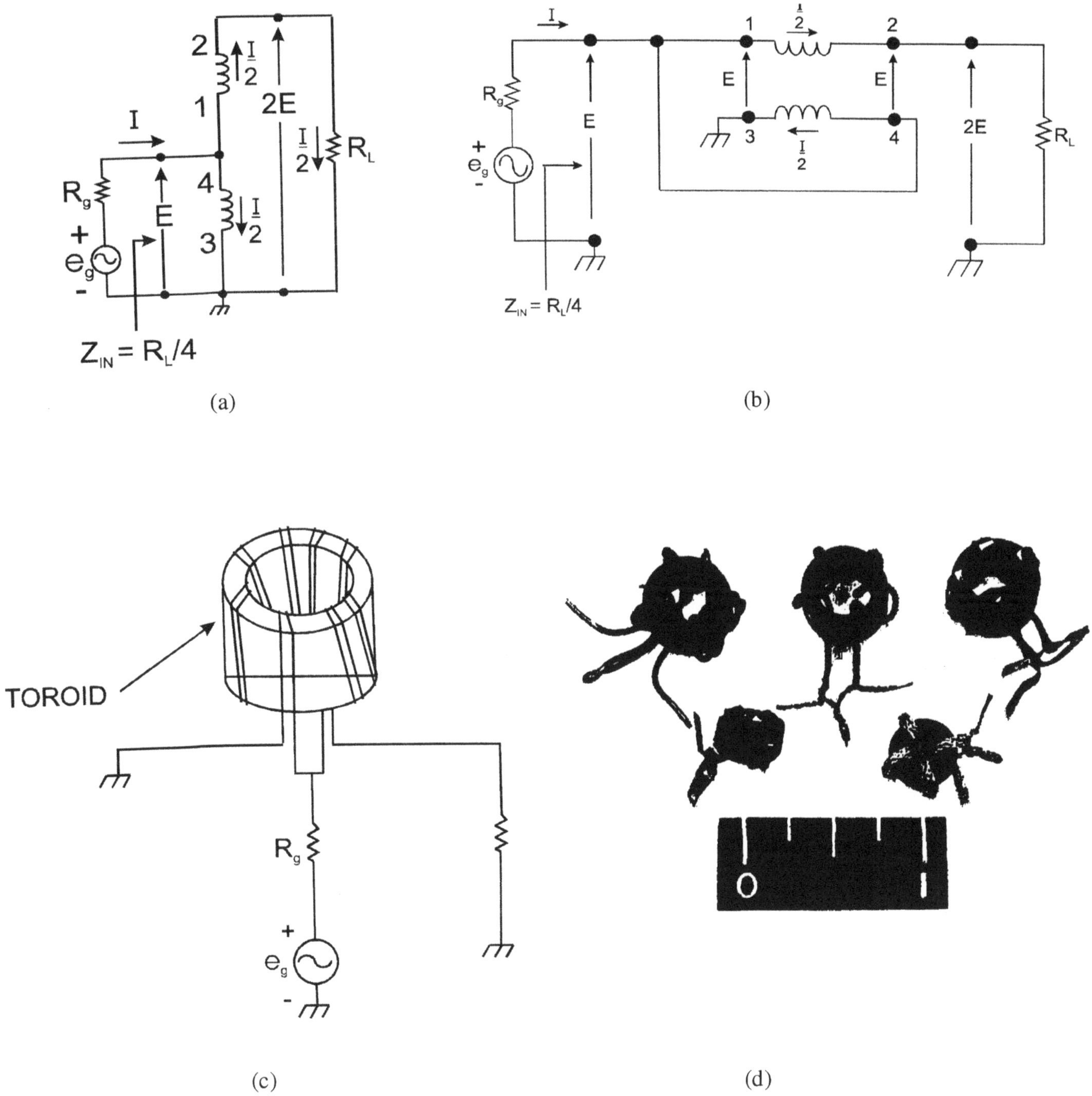

Figure 1. 4:1 Transformer (a) Autotransformer analogue. (b) Equivalent transmission line form. (c) Pictorial representation of construction using twisted pair transmission line. (d) Actual units of twisted pair, strip, and coaxial transmission line.

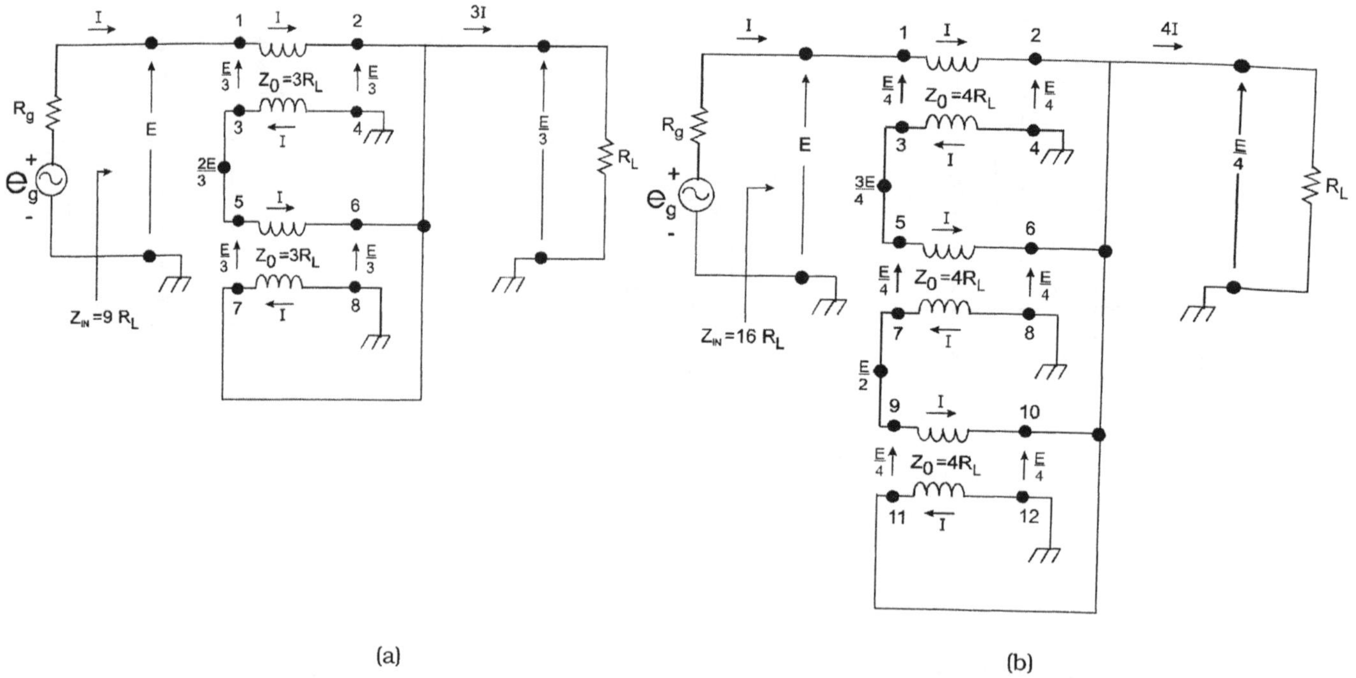

Figure 2. 9:1 Impedance transformer. (b) 16:1 Impedance transformer

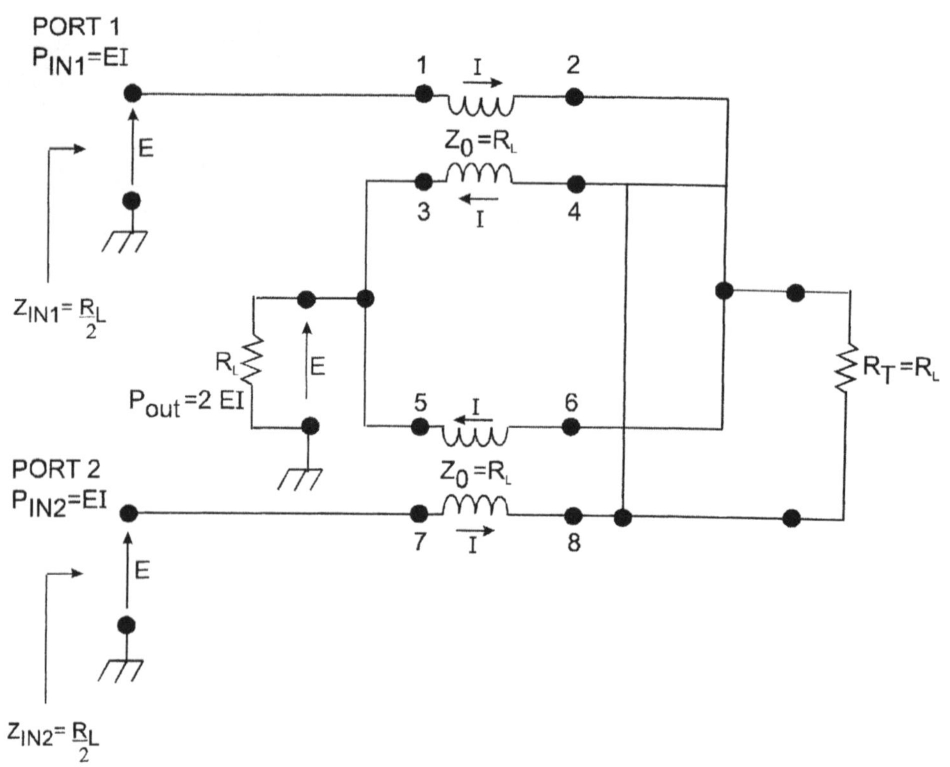

Figure 3. Broadband hybrid for power adding or splitting of in-phase signals.

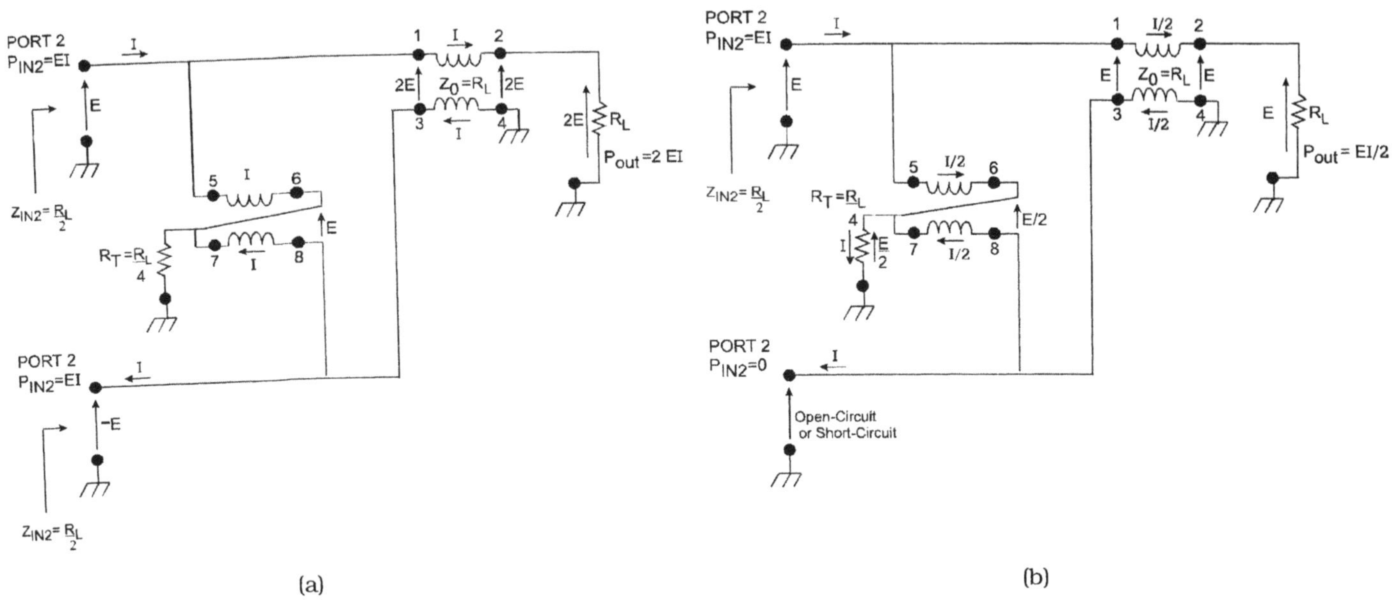

Figure 4. Broadband hybrid for power adding or splitting signals 180° out of phase. (a) Normal balanced conditions for power adding ($P_{IN1} = P_{IN2}$). (b) Severe mismatch operation for power adding (P_{IN1} = EI, P_{IN2} = 0).

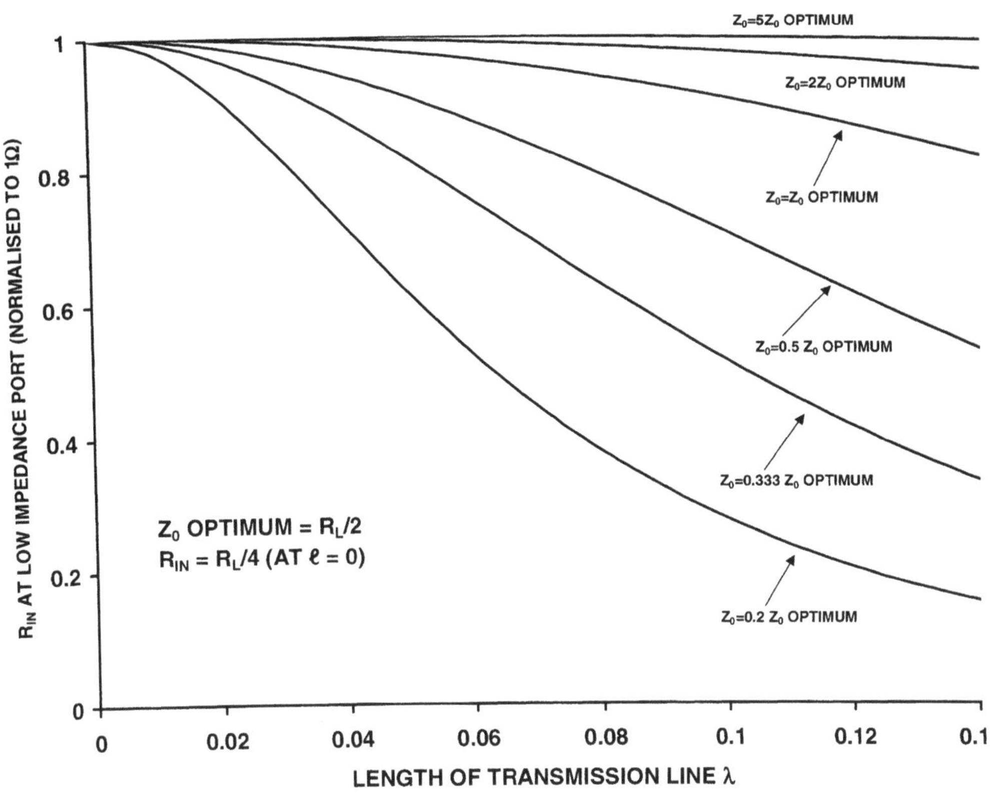

Figure 5. Normalized input resistance R_{IN} at the low impedance port of the 4:1 transformer with real load R_L versus length of transmission line in wavelengths for various values of characteristic impedance Z_o.

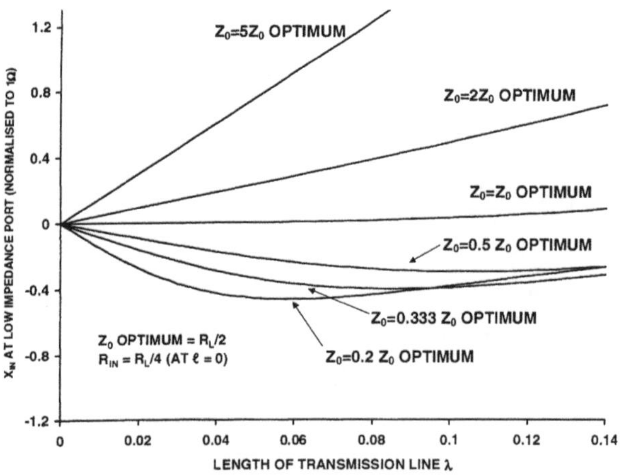

Figure 6. Normalized input reactance X_{IN} at the low impedance port of the 4:1 transformer with real load R_L versus length of transmission line in wavelengths for various values of characteristic impedance Z_0.

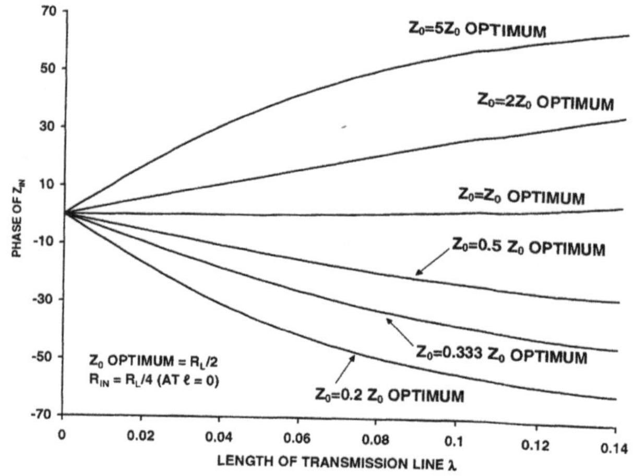

Figure 7. Phase of input impedance at the low impedance port of the 4:1 transformer with real load R_L versus length of transmission line in wavelengths for various values of characteristic impedance Z_0.

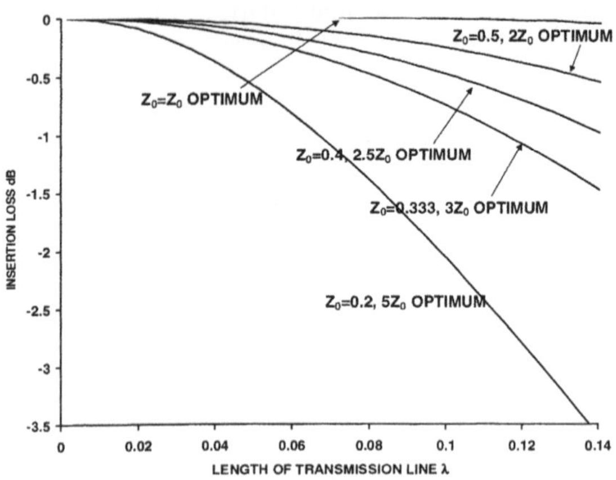

Figure 8. Insertion loss of the 4:1 transformer with real load R_L versus length of transmission line in wavelengths for various values of characteristic impedance Z_0.

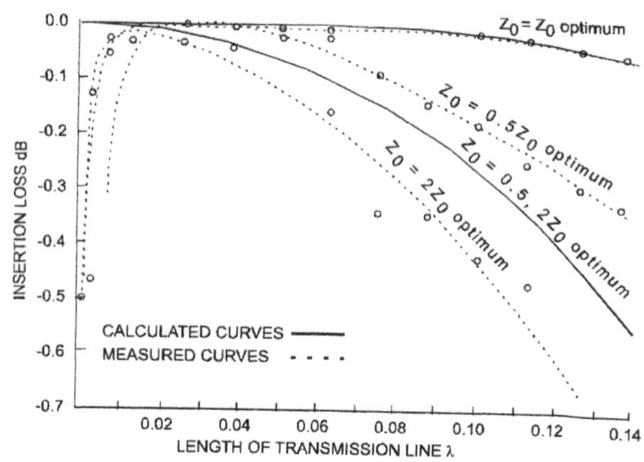

Figure 9. Curves comparing measured and calculated insertion loss for a 4:1 transformer with real load R_L versus length of transmission line in wavelengths for several values of characteristic impedance Z_0.

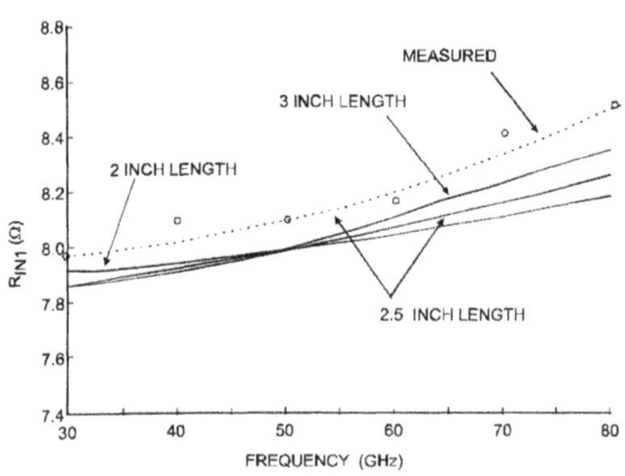

Figure 10. Calculated and measured curves of input resistance R_{IN1} (at the high impedance port of a 4:1 transformer with transistor load) for various lengths of transmission line with Z_0 equals 12Ω.

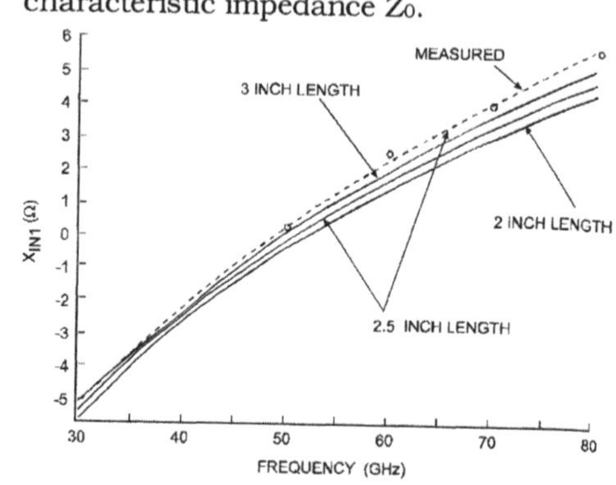

Figure 11. Calculated and measured curves of input reactance X_{IN1} (at the high impedance port of a 4:1 transformer with transistor load) versus frequency for various lengths of transmission line with Z_0 equals 12Ω.

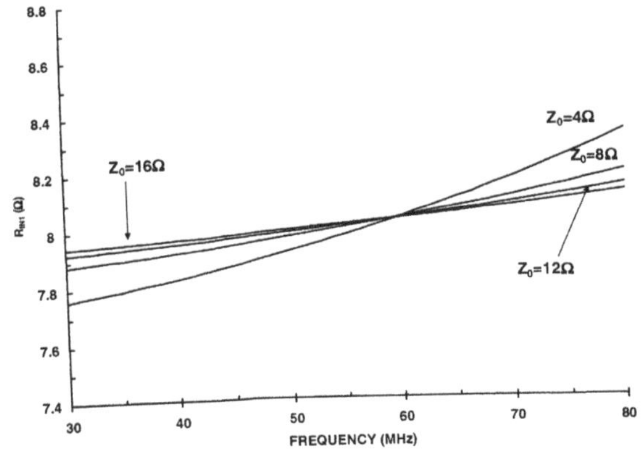

Figure 12. Input resistance R_{IN1} (at the high impedance port of a 4:1 transformer with transistor load) versus frequency for various values of characteristic impedance Z_0. (Line length constant at 2.5 inches).

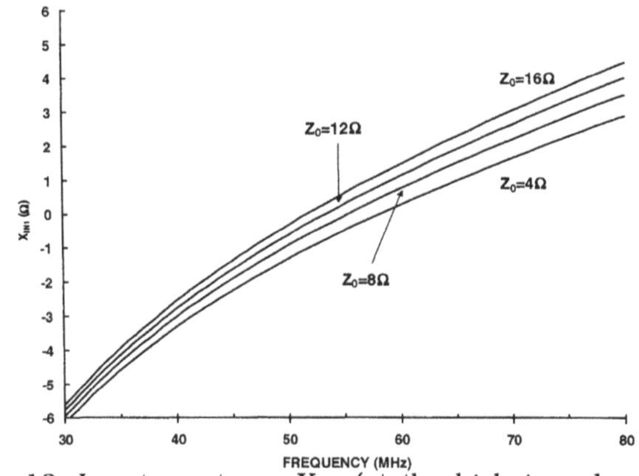

Figure 13. Input reactance X_{IN1} (at the high impedance port of a 4:1 transformer with transistor load) versus frequency for various values of characteristic impedance Z_0. (Line length constant at 2.5 inches).

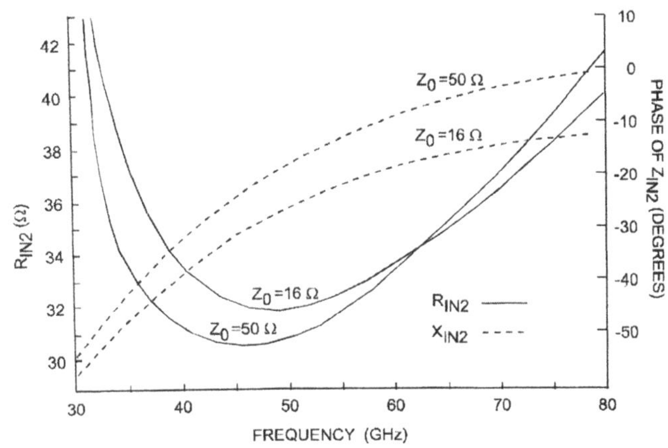

Figure 14. Curves of input resistance R_{IN2} and phase of input impedance Z_{IN2} (at the high impedance port of two cascaded 4:1 transformers with transistor load) versus frequency.

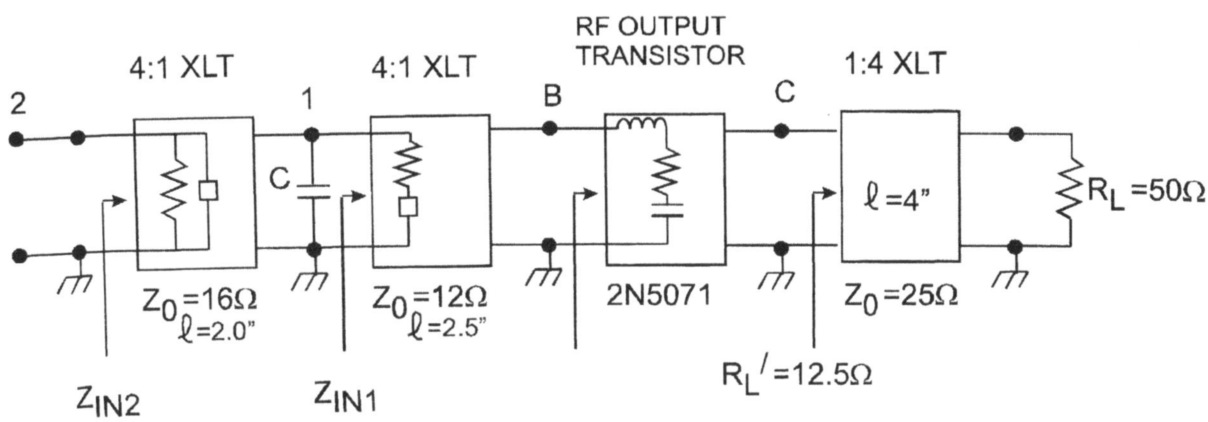

Figure 15. Block diagram of the 30 Watt, 30-80 MHz power amplifier with input and output matching transformers.

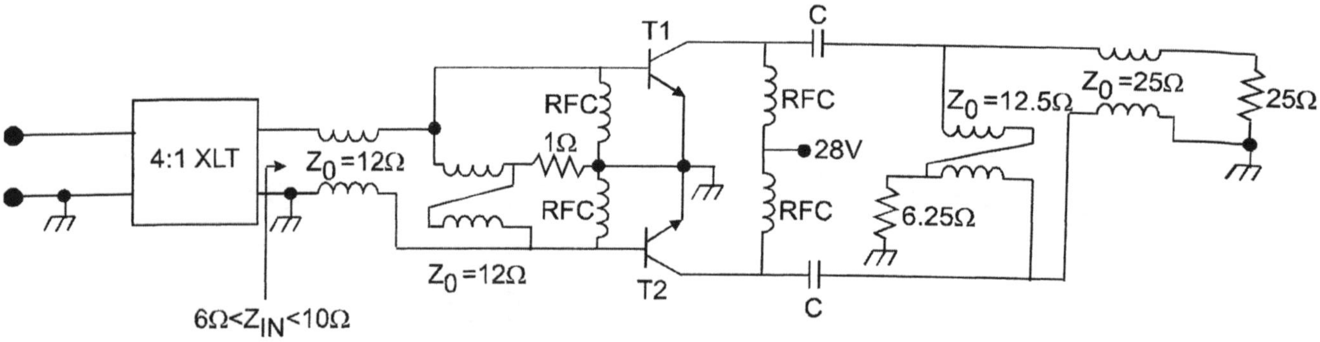

Figure 16. Basic schematic for a pair of 2N5071 transistors operating in a broadband push-pull amplifier with 180 degree hybrids for power splitting and power adding.

New MOSFETs Simplify High Power RF Amplifier Design

By H.O. Granberg
Motorola Semiconductor Products Inc.

There are many applications for high power solid state RF amplifiers in the 1 to 120 MHz range. Past designs have consisted of a number of 200-300 watt modules combined to produce power outputs in the multikilowatt level. Power combiners and splitters are expensive and difficult to design for extremely wide bandwidths, and at low frequencies are bulky as well. It is always desirable to combine as few modules for a given power output as possible, although system cooling becomes more difficult with thermal energy concentrated in a smaller area. There is obviously a practical limit to the point where this philosophy is valid.

Motorola has recently introduced high power MOSFETs MRF153 and MRF154, which are rated for 300 watt and 600 watt power output, respectively. A 600 watt bipolar transistor (MRF430) is also in this family of RF power devices. The MRF154, which is the subject of this article, is usable up to 100 MHz with a power gain of 8-10 dB. Special design techniques result in a junction to ambient thermal resistance as low as 0.13°C/w. The transistor housing is designed for conduction cooling, and has 1.4 square inch flange surface area. A mounting surface of high conductivity material such as copper is recommended, since up to 900 watts of power may be dissipated in each device. The heat dissipator itself can be forced air or liquid cooled.

The 1 kW push-pull amplifier described here is designed to cover a frequency range of 10 to 90 MHz. Its applications include military communications, jammer, low channel TV, etc. Although the point of saturation is well over 1000 watts, the amplifier was tested for linearity at 800 watts. The available output transformer impedance ratios (9:1 or 16:1) are the limiting factor: the 16:1 impedance ratio would be optimum at around 1500 watts power output, but the 9:1 was chosen in order to achieve a better overall CW efficiency at the 1 kW level. In pulsed applications such as Nuclear Magnetic Resonance, linear operation is possible up to 1000 watts per device due to the low average dissipation and lowered thermal limits. In such case the output impedance matching can be modified accordingly.

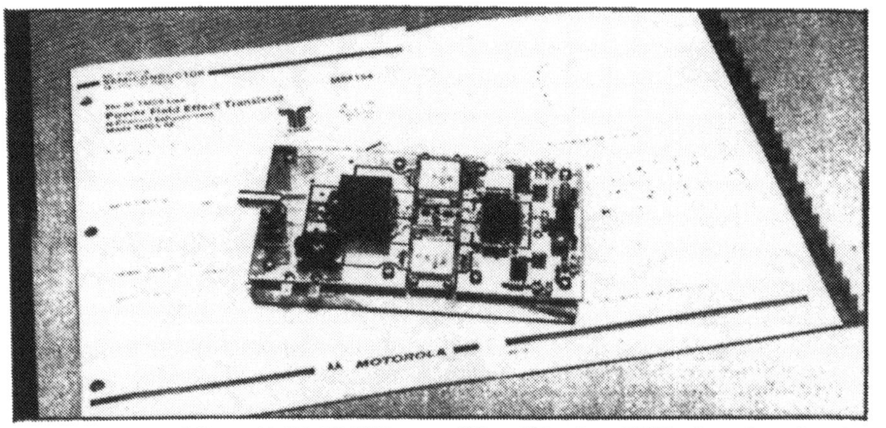

Figure 1. One kilowatt 10-90 MHz amplifier. The two FETs, input and output boards are mounted to a copper plate, which is then attached to the main heat sink.

Circuit Description

In contrast to a single ended amplifier circuit, in a push-pull configuration only the device mutual inductance (source to source in this case) is critical, and must be as low as possible for good high frequency performance. The common mode inductance (from each source to ground) is less important, reducing the requirement for low inductance grounding between the input and the output circuits. Input and output sections of the circuit board can be split and grounded only through metal spacers to the heat sink. The source of the MRF154 is internally connected to the mounting flange, which is also grounded to the heat sink. This provides a good, low inductance path between the two sources. The arrangement (Figure 1) results in a convenient and compact mechanical layout and makes the unit easily serviceable, since each board can be removed separately.

In addition to the matching network, the input circuit board includes the FET bias regulator, making the bias current insensitive to supply voltage variations. With the component values shown in Figure 2, excursions of 30 to 50 volts result in less than 1 percent changes in the bias current. The regulator also provides a convenient point for connecting a thermistor for bias current temperature tracking purposes. The thermistor must be of NTC type, and can be thermally connected to one of the FETs or to any central location at the heat sink, depending on the ther-

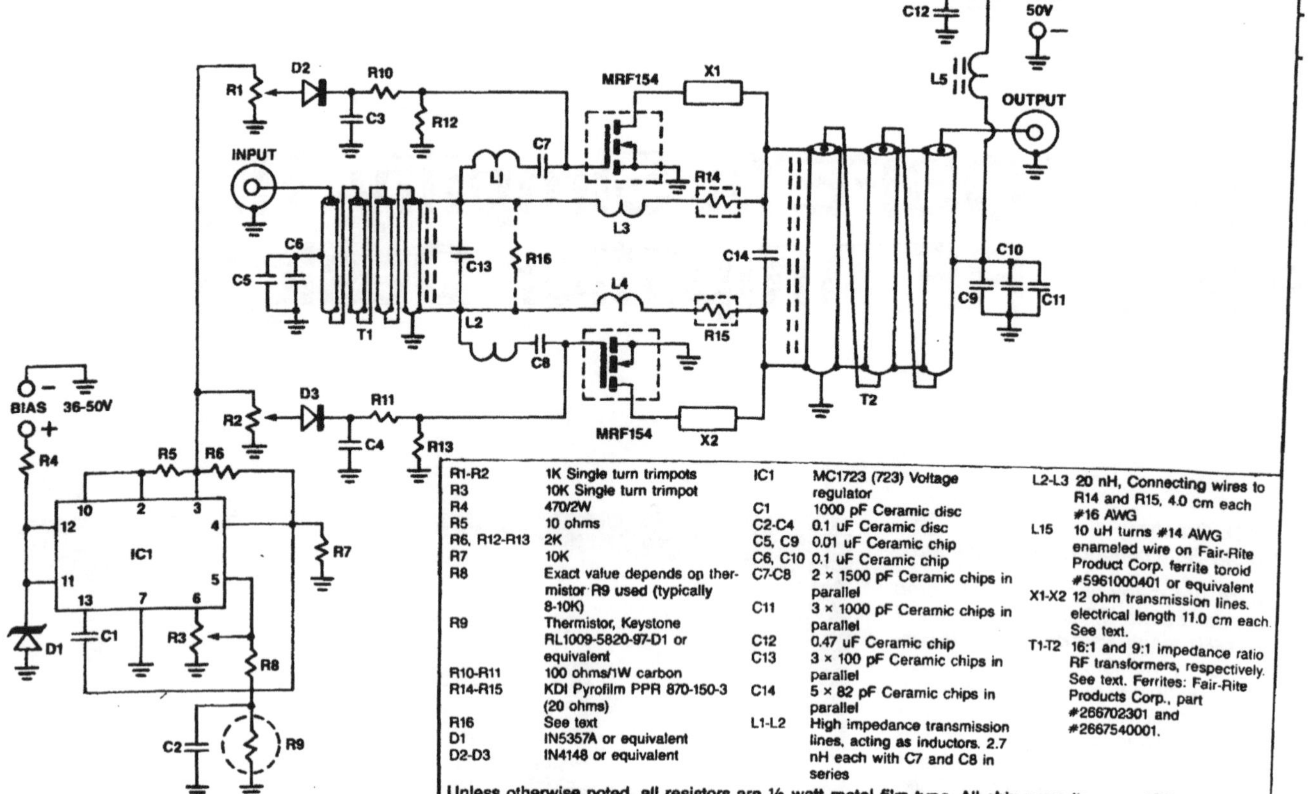

R1-R2	1K Single turn trimpots
R3	10K Single turn trimpot
R4	470/2W
R5	10 ohms
R6, R12-R13	2K
R7	10K
R8	Exact value depends on thermistor R9 used (typically 8-10K)
R9	Thermistor, Keystone RL1009-5820-97-D1 or equivalent
R10-R11	100 ohms/1W carbon
R14-R15	KDI Pyrofilm PPR 870-150-3 (20 ohms)
R16	See text
D1	IN5357A or equivalent
D2-D3	IN4148 or equivalent
IC1	MC1723 (723) Voltage regulator
C1	1000 pF Ceramic disc
C2-C4	0.1 uF Ceramic disc
C5, C9	0.01 uF Ceramic chip
C6, C10	0.1 uF Ceramic chip
C7-C8	2 × 1500 pF Ceramic chips in parallel
C11	3 × 1000 pF Ceramic chips in parallel
C12	0.47 uF Ceramic chip
C13	3 × 100 pF Ceramic chips in parallel
C14	5 × 82 pF Ceramic chips in parallel
L1-L2	High impedance transmission lines, acting as inductors. 2.7 nH each with C7 and C8 in series
L2-L3	20 nH, Connecting wires to R14 and R15. 4.0 cm each #16 AWG
L15	10 uH turns #14 AWG enameled wire on Fair-Rite Product Corp. ferrite toroid #5961000401 or equivalent
X1-X2	12 ohm transmission lines. electrical length 11.0 cm each. See text.
T1-T2	16:1 and 9:1 impedance ratio RF transformers, respectively. See text. Ferrites: Fair-Rite Products Corp., part #266702301 and #2667540001.

Unless otherwise noted, all resistors are ½ watt metal film type. All chip capacitors are ATC type 100/200B or Tansitor MPR 2/7, case 2 series or equivalent.

Figure 2. Schematic of the 1 kW FET amplifier. The MRF154 is supplied in matched pairs for g_{FS} and gain. It is necessary to have gate bias voltages individually adjustable.

mal time constant desired. The slope can be adjusted with R8, for which the exact value is determined by the FET g_{FS}. The value shown typically results in bias current tracking of less than 20 percent for 25° to 75°C. The bias can be turned completely off by grounding R8. However, this cannot be used for high speed switching of the amplifier due to the limiting time constant of the FET input capacitances and the bias voltage source path. Since the bias voltages are individually adjustable with R1 and R2 in addition to a common adjustment R3, the FET gate threshold voltages do not need to be matched. The power gain of an FET is mainly dictated by the g_{FS} and not by the V_g(th).

The bias setting procedure is as follows: 1) adjust R1 and R2 to minimum; 2) adjust R3 for a voltage higher than the device V_g(th) at pin 3 of IC1. This should be typically 7-9 volts in order to place the R1 and R2 settings in the middle of the tuning range; 3) measure current at 50 volt supply point; 4) with power supply connected, advance R1 for desired current reading; 5) advance R2 until the current reading is doubled; 6) R1 and R2 need no adjustment after this, and the bias currents of both FETs can now be set with R3. During this operation (1-6) the input and output should be both terminated into 50 ohms with no RF drive applied.

On the output side of the circuit board design, the DC paths must be able to handle current levels of *50 amperes* and more, and the maximum RF currents are in the order of 15 amperes RMS at the low impedance points. The DC current would require almost 10^4 mils² for the conductor cross sectional area in free air, but since the conductor will be heat sunk to the circuit board surface, a number about one fourth of this is adequate. Even then, circuit board material with at least 2 ounces of copper is required, and should be solder plated for added conductor thickness. In regards to the skin effect, a certain foil thickness is also necessary for the conductors carrying the RF currents. The skin depth at the high frequency end (90 MHz) is about 0.40 mils, and the foil thickness in the RF conducting paths should be at least five times that, or 2.0 mils, according to a rule of thumb. Since the skin depth varies as an inverse function of the frequency, it is really only meaningful at high frequencies, where the dimensional conditions can be met. It is then desirable to have a conductor with a large surface area and a thickness that meets the minimum requirement. Normally this will also be sufficient for low frequencies, where the conductor losses become nearly purely resistive.

Input-Output Impedance Matching

Since the output impedance matching and transformer design are far more critical than the input side, they will be discussed first. According to (1) and (2) dips in transmission line transformer response will occur when the physical line length reaches ¼ wavelength, if the Z_o differs from the optimum required value or if the terminating impedances are incorrect. These dips, which actually are changes in the transformer impedance characteristics, have been noticed with ⅛ and 1/16 wavelength increments as well. Their magnitude also strongly relates to the amount of leakage inductance present, which has the same effect as an incorrect line impedance. Standard practice is to keep the line lengths as short as possible to reduce the IR losses, preferably shorter than ⅛ wavelength at the highest operating frequency. Although operation between the incremental frequencies is possible, the bandwidth would be limited to less than one octave.

The effective line length varies with the transformer configuration. In a balanced

4:1 as shown in Figure 3B, the two lines are electrically in series, making the effective line length twice the actual. Cascading these for 16:1 impedance ratio further doubles the effective line length, making the total four times the length of one line. In a 9:1 impedance ratio transformer (Figure 3A) the two lines *a* and *b* are electrically in parallel, making the effective line length equal to the length of one line. However, since a balun is required for the balanced to unbalanced function, its length must be added to the total.

In high power solid state RF amplifiers it is desirable to eliminate the need for output DC blocking capacitors. Even if they are located at the 50 ohm points, they must be able to handle large RF currents (4.5A at 1 kW) and should be chip type to minimize the series inductance. This can be done either by replacing the typical autotransformer configuration with a design as in Figure 3C, or replacing the balun with a 1:1 isolating transformer (Figure 3Ad and 3Dh). A disadvantage with 3D is that an impractically low characteristic impedance may be required for line *h*, but its high frequency performance is excellent due to the equal delay unbalanced 9:1 section.

The design shown in Figure 3C is probably the most practical one for its simplicity and ease of manufacture. It lends itself to high impedance ratios such as 16:1 and up, which would be difficult to implement with other types. The impedance transformation is achieved by parallel connection of one conductor of the lines and series connection of the other. In principle it resembles the multifilar type transformers described in references (3) and (4), and must be considered a conventional (non-transmission line) transformer, although the low impedance line provides most of the coupling between the primary and secondary at high frequencies. The line impedance is not defined in the same manner as in transmission line transformers, and is not as critical. For increased coupling and low IR losses it can be lowered to a point where the resonance of the line capacitance and the leakage inductance falls outside the highest frequency of operation.

As discussed earlier, the high frequency limit of an RF transformer is set by the physical length of the line or the winding on the high impedance side. Considering the ⅛ wavelength rule, mentioned in several of the references, the maximum total line length at 90 MHz would be:

$$\frac{\lambda}{8} \times V_p = \frac{333}{8} \times 0.63 = 26.2 \text{ cm.}$$

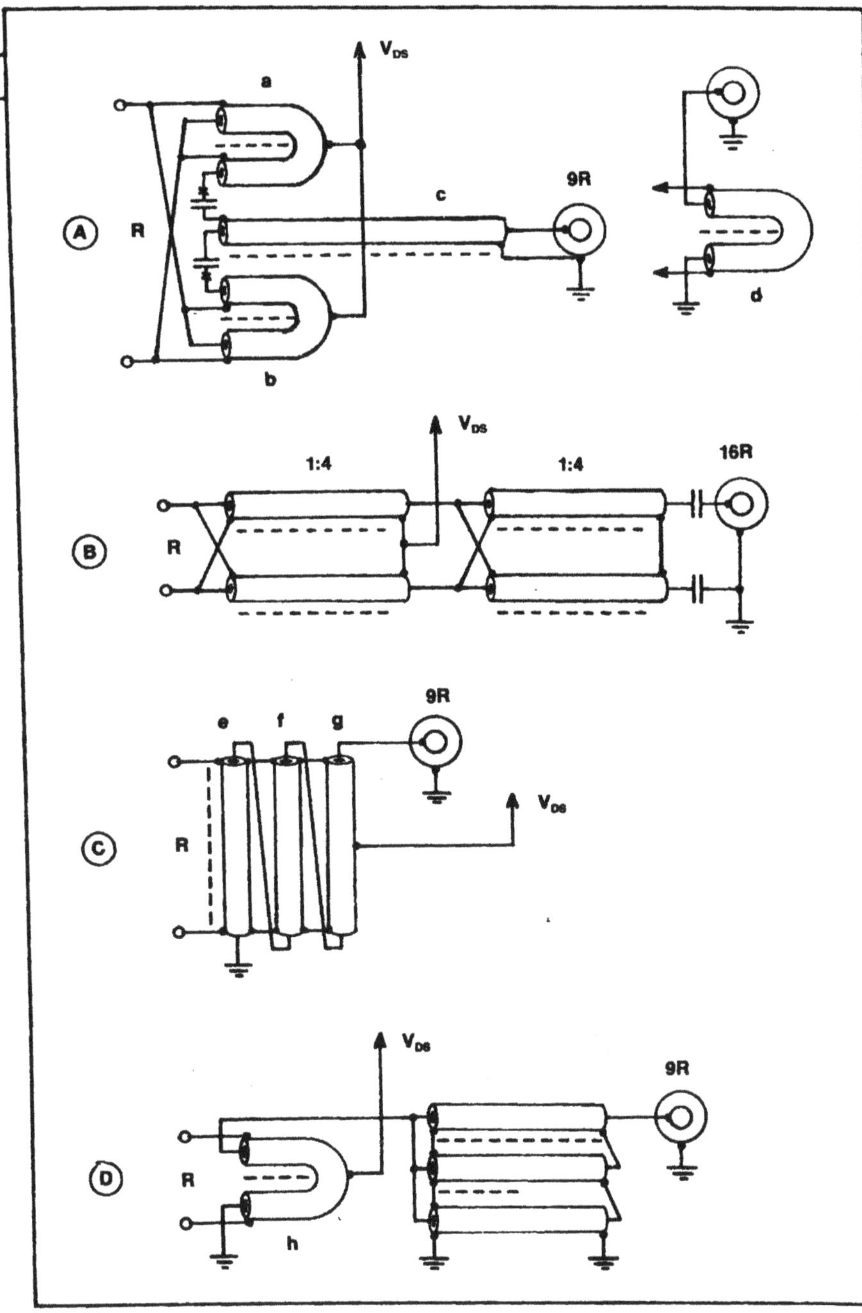

Figure 3. Wideband RF transformer configurations for balanced to unbalanced impedance matching at high power levels. Coaxial transmission lines are most convenient for compact physical designs.

Where: V_p = velocity factor (0.63 for low impedance TFE co-axial cable).

Figure 4 shows the amplifier frequency versus power gain characteristics, *A* with a 9:1 transmission line output transformer, and *C* with a transformer shown in Figure 3C. Both were designed to have equal line or winding lengths of 16 ohm TFE insulated coax cable (Type CXN 1848 W.L. Gore Co.), except for the added 50 ohm balun in *A*. The balun and the higher leakage inductance resulting from the interconnections and the blocking capacitors in *A* possibly cause the slight roll-off at the high end. The type *C* transformer was also tried with 50 ohm coax line, which resulted in a 0.8 dB gain reduction across the band, plus an additional 0.3 dB at the high end compared to the unit made with 16 ohm cable. This indicates that the IR losses in the cable with a smaller center conductor are dominant over the line impedance.

Both transformer types reached temperatures of 80°C in a five minute CW test at the full power output, although type *A* had twice the ferrite cross sectional area. This leads us to determine that the ferrite dielectric losses are a problem in high power and high frequency applications such as this. Lower permeability material could be a solution, but larger cross sectional area would be required, making it more difficult to meet the maximum line length criteria. This could be a major problem and limiting factor in designing wide band amplifiers of this type, unless ferrite materials with lower dielectric losses can be developed. It would also be worth investigating how powdered iron material would behave in broadband power transformers, although suitable core shapes have not been available thus far.

The input transformer used in this design is of similar type and design as the output unit, except having a smaller physical size. The primary winding is made of 25 ohm miniature coax cable (Microdot 260-4118). It must be able to handle less than 100 watts of power, and its losses only affect the power gain, whereas the quality of the output transformer determines the overall system efficiency as well.

The high values of the gate and drain capacitances of the FETs make both the input and output matching difficult for large bandwidths. The effect of the drain capacitance can especially be noticed at frequencies above 50 to 60 MHz in reduced efficiency. Part of this capacitance can be compensated for with small values of series inductance or stripline. The stripline would be of extremely low characteristic impedance, and probably practical only in the output where high RF currents are also involved. Using calculated and measured drain parameters to be matched into the 9:1 transformer and computer optimization, values as shown for X1 and X2 were obtained. As an etched line on a 62 mil G10 substrate, the line width is 0.7 inches. In practice the lines had to be folded into a form of U, but this allows part of the line to be conveniently shorted for adjustment purposes. The effect of the output lines can be noticed in increased efficiency at 90 MHz (6% to 8%), but at a cost of reduced power gain by approximately 0.5 dB.

Similarly, using the data sheet numbers converted to parallel form, indicated values for input lines L1 and L2 were obtained. They are high impedance etched lines on similar G10 substrate. They act as inductors and their values are also adjustable by shorting part of the line or by moving the input transformer connection points. The values of L1 and L2 will finally depend on the amount of negative feedback necessary for the desired gain slope. Although not adapted to this design, a dummy resistor (R16) can be used to make the input look more resistive and improve the input VSWR. A suitable amount of L in series with it, combined with negative feedback, is a common technique in applications requiring extremely large bandwidths.

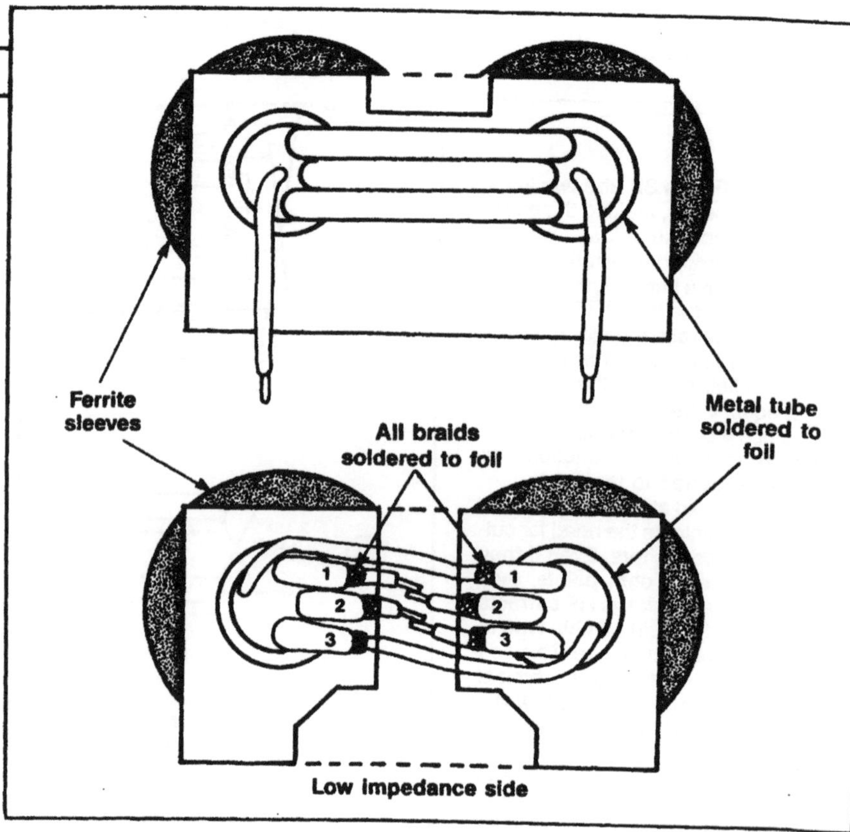

Figure 4. Physical construction of RF transformers used in the amplifier. The electrical details are shown in Figure 3C.

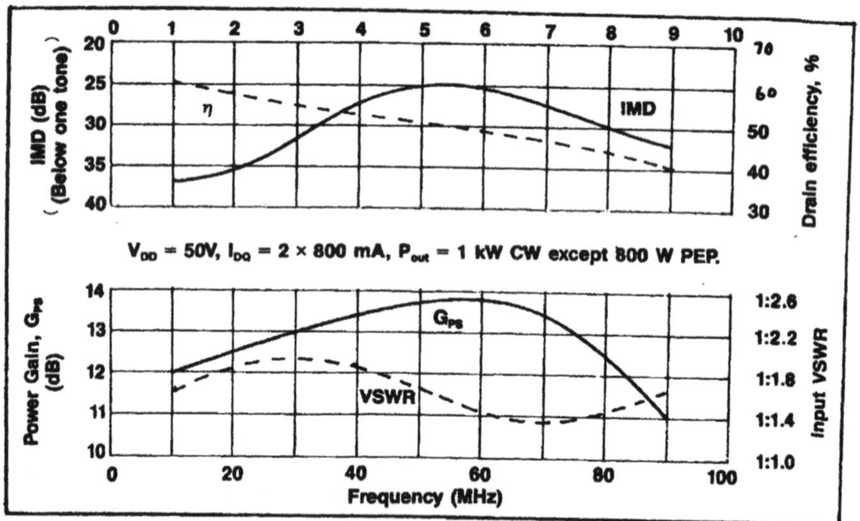

Figure 5. 1 kW, 10-90 MHz amplifier typical performance. The numbers may vary slightly depending on the exact device parameters within specified limits.

Gain Leveling with Negative Feedback

The negative feedback term usually refers to a condition where part of the output power (voltage) is fed back to the input out of phase. Out of phase generally means 180° phase difference, although in practical systems the voltage fed back usually lags the input voltage due to delays in the circuitry. Any inductance in the feedback path causes a delay and a phase error, but seldom has effects large enough to cause instability. Sometimes an amount of inductance is intentionally included in the feedback circuit in order to prevent the feedback from affecting the high end. In most circuit configurations the phase shift is close to 180° between the input and the output, in which case the feedback is easy to implement. Otherwise a phase reversing component, such as a transformer must be employed.

Here the series inductances L3 and L4 are limited to their minimum values of 20-25 nH by the physical distance between the input and the output, although they can be controlled to a degree by varying the conductor diameter. Their reactances are about 8 ohms at the midband, where lower values would result in increased feedback and a flatter gain response than shown in Figure 5. It must be noted that the reactances at 10 MHz may be also significant, and should be deducted from the values of R_{fb} to be calculated.

In addition to gain reduction, negative feedback lowers the effective input impedance. Ideally the amount of feedback voltage should be inversely proportional to the frequency in such amplitude that the gain would be reduced just the correct amount at all frequencies below the high end. This is not possible with simple feedback networks consisting only of L and R. Even with more sophisticated networks the feedback voltage source should be adjustable in some manner. Such a system is described in (6), but due to the high frequencies and higher power level involved it would be difficult to implement in this design. Here the feedback voltage is derived directly from the FET drains, which will limit the optimization of the system in this respect.

The MRF 154 data sheet shows a power gain of 22 dB for the device at 10 MHz, and a one to four difference in the gate input impedance from 10 to 90 MHz, or 12 ohms and 3 ohms composite parallel, respectively, from gate to gate. The 22 dB can be considered typical, and can vary as much as 3 dB at 30 MHz. However, the devices are supplied in matched pairs for operation in push-pull systems. The 22 dB translates to a power input of 6.5 watts for

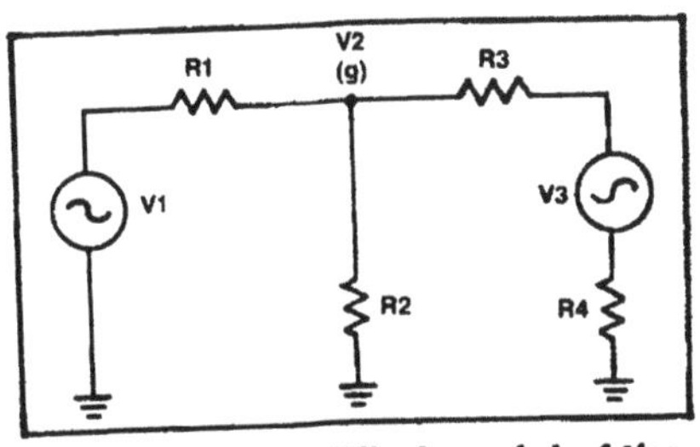

Figure 6. A simplified model of the negative feedback network can be used to figure the loop parameters with sufficient accuracy.

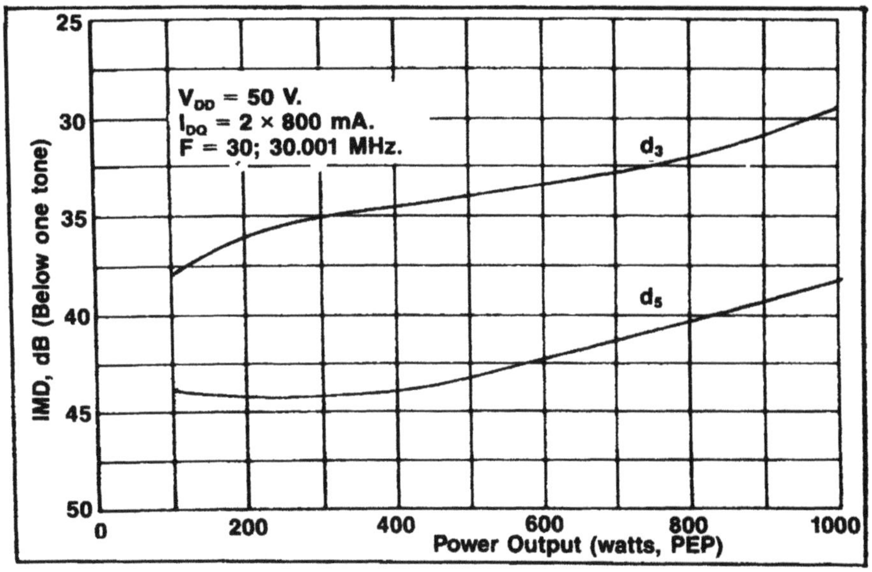

Figure 7. Intermodulation distortion versus power output of 1 kW amplifier. Output impedance matching is optimum only at 800 watts with the transformer impedance ratio employed.

two devices at 1 kW output. Assuming a 10 dB gain reduction, the power input then would be 65 watts. Since the device power gain is not affected by the feedback and part of the input power is cancelled by the feedback voltage, the difference power must be dissipated somewhere. Most of this occurs in the feedback resistors, which also control the amount of feedback.

A 16:1 impedance ratio input transformer was selected with an idea in mind that the feedback would bring the low frequency gate to gate impedance down to the 90 MHz value (3 ohms). Because of the reasons discussed ahead, the gain slope cannot be controlled, and since no loss of gain in the high end can be afforded the feedback is limited to an amount that results in a low VSWR at 10 MHz. The computer analysis of the input matching mentioned earlier assumed a constant 3 ohm impedance, but this will deviate considerably at the mid-band. A computer program with all these variables would be very complex, and since the system could not be totally optimized anyway it was decided to discard the effort at this point.

Since an FET is a voltage-controlled device, the feedback loop can be modeled at low frequencies or DC as shown in Figure 6, where:

R1 = Transformer source impedance
R2 = FET gate to gate impedance
R3 = Feedback resistor. (Divided into two equal values in the amplifier.)
V1 = input voltage
V2 = FET gate to gate voltage

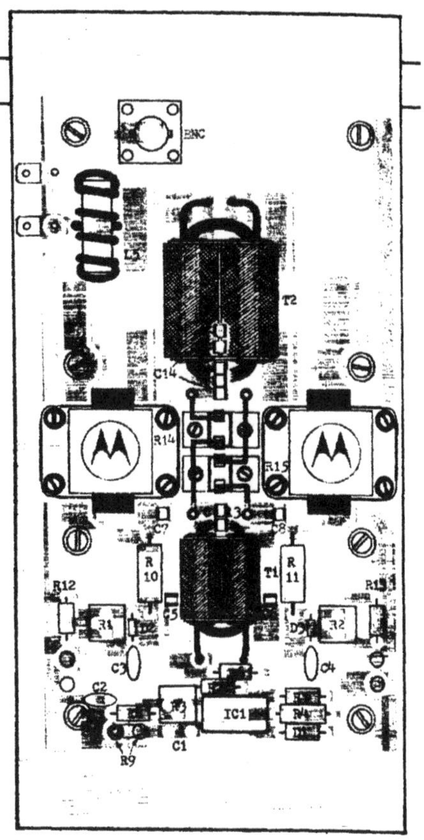

Figure 8a. Component layout.

V3 = Drain to drain output voltage across R4

For a given level of power output, the values of R1, V2 and V3 will remain virtually unchanged regardless of the amount of feedback.

It is assumed for simplification, that the value of R2 will be reduced by the feedback from 12 ohms to equal R1. Then:

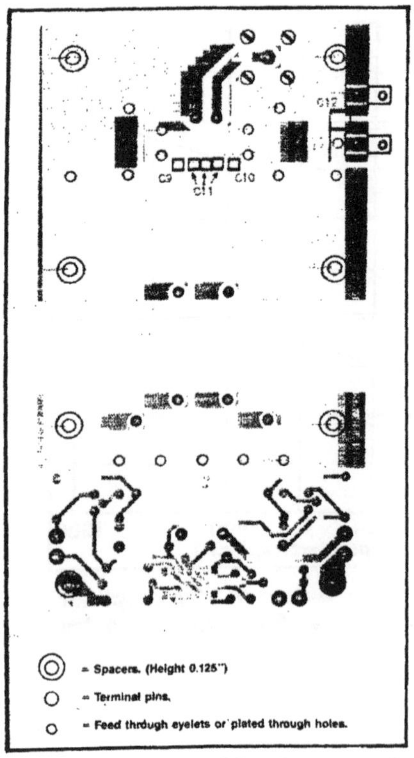

Figure 8(b). Underside of p.c. board.

R1, R2 = 3.12 ohms.
V1 = 14.23 volts (−10 dB)
V2 = 4.5 volts (22 dB)
V3 = −74.2 volts (R4 = 5.5 ohms)

$$R3 = \frac{(V2 + V3)}{\left(\frac{V1 - V2}{R1}\right) - \left(\frac{V2}{R2}\right)} - R4 =$$

$$\frac{(4.5 + 74.2)}{\left(\frac{14.23 - 4.5}{3.12}\right) - \left(\frac{4.5}{3.12}\right)} - 5.5 =$$

$$\frac{78.7}{1.70} - 5.5 = 40.8 \text{ ohms}$$

or 20.4 ohms each.

The feedback resistors then must be able to dissipate 65-6.5 watts plus any excess power resulting from the non-optimum voltage source. In this case the dissipation is 78.7 × 1.70 = 133.8 watts, or 66.9 watts per resistor. It is obvious that they must be of a type with a provision for heat sinking, and with low parasitic inductance (8).

Design and Construction Summary

The amplifier performance, shown in Figures 5 and 7 can be affected by circuit parameters such as type of components, their values and exact locations. In this respect, it can be compared to a low power UHF design if the impedance levels are scaled down with frequency. Although the layout resembles that of a typical 2 to 30 MHz amplifier, considerable differences in the construction techniques are essential to ensure proper operation, especially at the high frequencies.

Due to the high gate capacitance (C_{iss}) of the devices, the input matching is critical for good broadband performance. The values of L1 and L2, as well as the physical locations of T1 and C13, have a dominant effect in the input VSWR. The above is also true concerning the output matching, where these variables affect the power gain, efficiency, saturated power, and the IM distortion. Special attention must be paid to the location and quality of C14, which is essential to the amplifier operation above 50 MHz. Once all these criteria have been established, duplication of the system should not be a problem, although it is not possible to give physical details with sufficient accuracy in an article of this proportion to guarantee the exact results without minor adjustments.

The RF currents associated with the high power level and low impedances also introduce new problems to the designer in the form of passive components. The weakest link probably is the capacitors, which in certain locations must be composed of several paralleled smaller values in order to achieve the current carrying capability required and reduce the series inductance. Other limitations associated with the passive components have been discussed earlier. One of these is the circuit board itself, where the DC currents and the skin effect place a minimum limit to the foil thickness.

The low impedance levels are not new to a solid state power designer, but their association with a kW power level in a single amplifier is unique at these frequencies. This places new requirements on all passive components and presents challenges in thermal design.

Finally, it must be pointed out that the component values given or the mechanical design may not be exactly optimum for the specific goals described. The intent was to make the circuit board layout, including the output section, as universal as possible to allow its use for designs with other devices and frequency ranges. rf

References

1. Pitzalis-Couse, "Broadband Transformer Design for RF Transistor Amplifiers," ECOM-2989, U.S. Army Electronics Command, Fort Monmouth, N.J., July 1968.
2. Roderick K. Blocksome, "Practical Wideband RF Power Transformers, Combiners and Splitters," Proceedings of RF Expo, February 1986.
3. Krauss-Allen, "Designing Toroidal Transformers to Optimize Wideband Performance," Electronics, August 16, 1973.
4. T.A.O. Gross, "Broaden Transformer Frequency Response," Electronics, August 24, 1982.
5. A.H. Hilbers, "Design of HF Wideband Power Transformers," Amperex (Philips) Application Laboratory Report ECO6907 and ECO7213.
6. Mueller-Vatis-Edelstein-Bottomley, "RF Technology for NMR Imaging," Proceedings of RF Expo, February 1986.
7. H.O. Granberg, "A Two-Stage 1 kW Solid-State Linear Amplifier," Motorola Application Note AN-758.
8. KDI Electronics, "RF/Microwave Components & Sub-Assemblies," Catalog 1985.

About the Author

Helge H. Granberg is a member of the technical staff at Motorola Semiconductor Products, Inc., Box 20912, Phoenix, AZ 85036. Mr. Granberg is a leading authority on HF and VHF solid state power amplifier design.

October 1986

Chapter 5

HYBRID POWER COMBINERS AND SPLITTERS

Power splitters and combiners are used when more power is needed than can practically be provided by a single power amplifier (PA). The use of multiple power-combined PAs also increases system reliability by allowing a transmitter to stay on the air in spite of the failure of one or more individual PAs. Combiners and splitters are used over a wide range of frequencies and power levels. At HF, VHF, and UHF, power-combining techniques are used to provide kilowatts from multiple PAs with outputs of hundreds of watts. In millimeter-wave integrated circuits (MMICs), multiple PAs whose outputs are a fraction of a watt are combined to produce MMIC outputs of several watts.

This chapter discusses "hybrid" power splitters and combiners. The term hybrid comes from the early days of long-distance telegraphy and telephony; the isolation afforded by the hybrid circuit allowed transmission and reception over the same line. In modern transmitters, the isolation allows two or more PAs to deliver power to the same load without interacting. This chapter first presents basic terminology, then develops the theory upon which hybrid operation is based, and finally reviews applicable literature. Selected reprints appear at the end of the chapter.

5.1 TERMINOLOGY AND CLASSIFICATION

Most power splitters and combiners can be classified in three basic ways: corporate vs. serial configuration, hybrid vs. nonhybrid operation, and transformer vs. transmission-line topology.

Some combiners and splitters can also be classified as having inputs or outputs that are of the same phase (0°), of opposite phases (180°), or in phase quadrature (90°).

The majority of power combiners and splitters are based upon the corporate configuration shown in Figure 5.1. In this configuration, the drive (input) signal is split (possibly in multiple steps) into (ideally) identical drive signals for each of the individual power amplifiers. The (ideally) identical outputs of the amplifiers are then combined (again, possibly in multiple steps) to produce the system output. Multiple binary splitters and combiners can be used as shown in Figure 5.1, or N-way splitters can be used. Similar designs can be used for all splitters and all combiners in a system.

The serial (also called "traveling-wave" or "chain") structure shown in Figure 5.2 is an alternative to the corporate structure as shown in Figure 5.1. In the serial approach, each splitter routes a fraction of its input to one of the power amplifiers. The remainder of the power is passed to the next splitter. Each combiner is analogously designed to combine unequal input signals. Its output is then passed on to become the input of the next combiner. Each splitter and combiner must be designed for a specific splitting or combining ratio. For example, ratios of 1/4, 1/3, and 1/2 are required for the splitters and combiners in Figure 5.2.

Power splitters and combiners are also classified by operation as hybrids or nonhybrids. Amplifier outputs can be combined by summing voltages or currents as shown in Figure 5.3. These

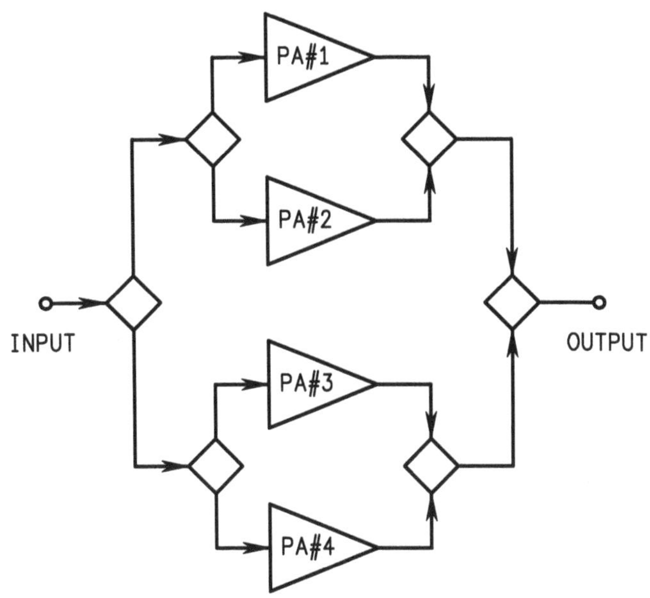

Figure 5.1 *Corporate structure for power splitting and combining.*

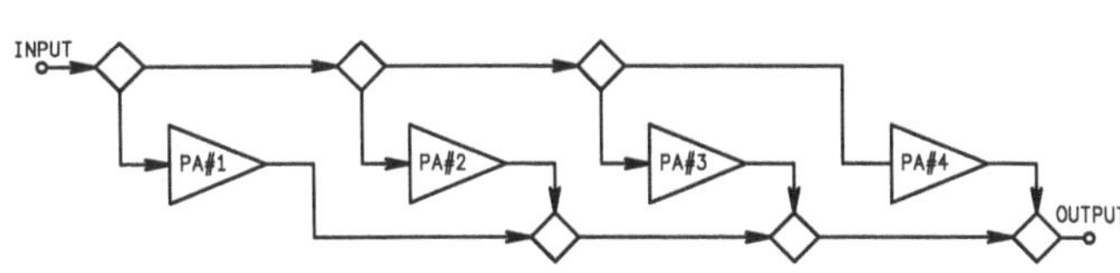

Figure 5.2 *Serial structure for power splitting and combining.*

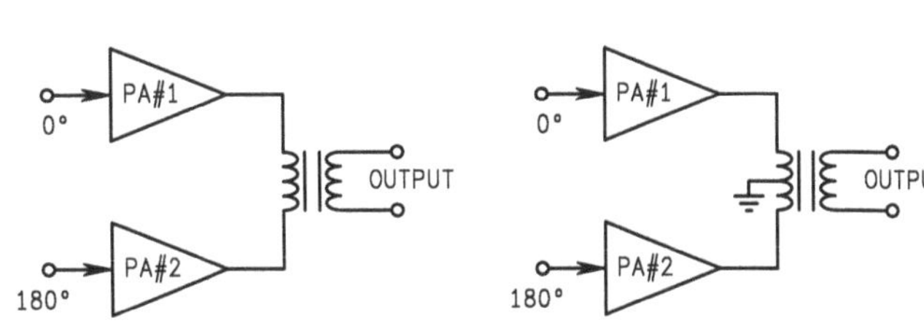

Figure 5.3 *Nonhybrid power combiners. (left) Voltage-summing combiner. (right) Current-summing combiner.*

simple combiners are effective and result in an efficient system as long as all PAs remain operational and their outputs do not deviate significantly from their nominal values. Since the load impedance presented to a given PA depends upon the voltage or current produced by all PAs, variations in the output of one PA result in changes in the load impedances presented to the other PAs. This in turn can lead to erratic operation, delivery of more or less power than desired, and excessive stress. Failure of one PA generally shuts down the output of the whole system.

The basic hybrid combiner shown in Figure 5.4 adds a resistor at the center tap of the transformer. The presence of this resistor allows each PA to see the same load resistance regardless of the output of the other PA. If the outputs of the two PAs are of the same amplitude and opposite phases, the voltage produced on hybrid resistor R_h is zero and all power is delivered to the load. Any difference in the outputs is dissipated in the hybrid resistor. If one PA fails, half of the power from the operational PA is dissipated in the hybrid resistor, but the other half is delivered to the load, allowing the system to remain "on the air," albeit at a lower output power.

A broadband hybrid combiner implemented with transmission-line transformers is shown in Figure 5.5. The combination of a Ruthroff transformer and a balun, or Guanella transformer, (see Section 4.3) produces the equivalent of a transformer with a center-tapped input. The currents or voltages produced by the two PAs are summed in the load. The difference between the PA outputs flows through hybrid resistor R_h.

The third major classification of power combiners is whether they are based upon transformers (including transmission-line transformers) or transmission lines. The classic Wilkinson topology (Figure 5.6) is a hybrid combiner based upon the action of two quarter-wavelength transmission lines. In contrast to the previously shown transformer-based combiners, all inputs to a Wilkinson combiner must be of the same phase. The Wilkinson combiner can be adapted to three or more inputs by connecting multiple hybrid resistors in a "Y" or "Δ" configuration. Wilkinson combiners can also be implemented using lumped-element equivalents of transmission lines as shown in Figure 5.7.

As shown in the preceeding discussion, two-input combiners can be of the 0° type (similarly phased signals) or the 180° type (oppositely phased signals). An analogous classification applies to the outputs of power splitters. Generally, in combiners and splitters with more than two outputs or inputs, all signals are of the same phase. Quadrature splitters and combiners (discussed in Chapter 6) are based upon signals with 90° phase differences.

5.2 BASIC THEORY OF THE HYBRID SPLITTER

A simplified circuit of the hybrid splitter is shown in Figure 5.8. The complete circuit includes a balun T1 (analogous to the balun in Figure 5.5) that applies input signal to Ruthroff transformer T2. Balun T1 is not needed for the subsequent analysis and therefore omitted for simplicity. The drive from a class-A, -B, or -F PA is represented by current source I_i.

Splitting of the input current into transformer and output currents yields

$$I_i = I_1 + I_T \tag{5.1}$$

and

$$I_i = -I_2 - I_T \tag{5.2}$$

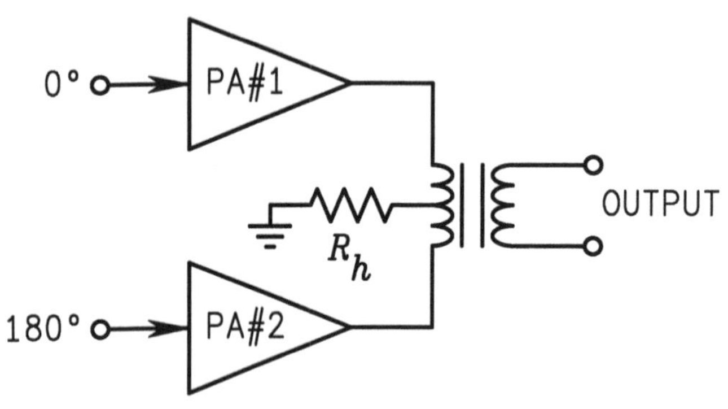

Figure 5.4 *Basic hybrid combiner.*

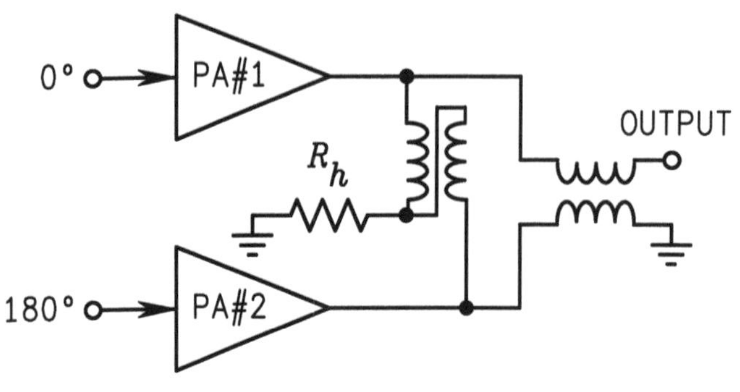

Figure 5.5 *Hybrid combiner based upon transmission-line transformers.*

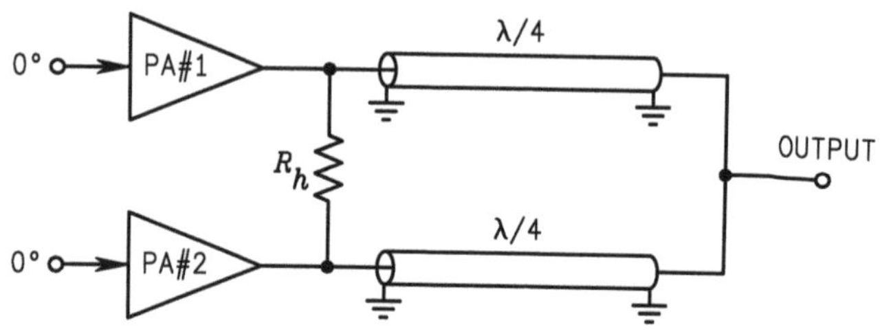

Figure 5.6 *Classic Wilkinson combiner.*

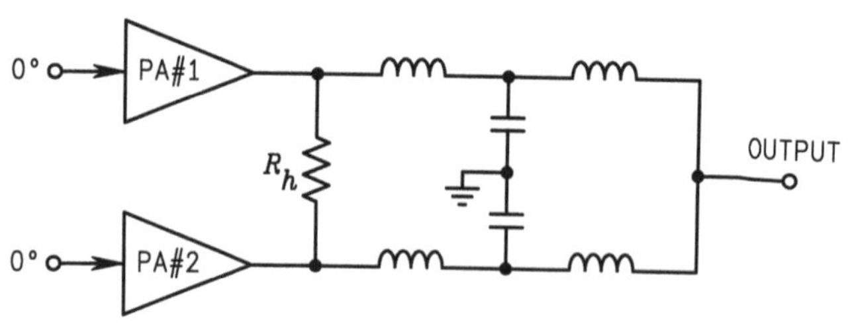

Figure 5.7 *Lumped-element Wilkinson combiner.*

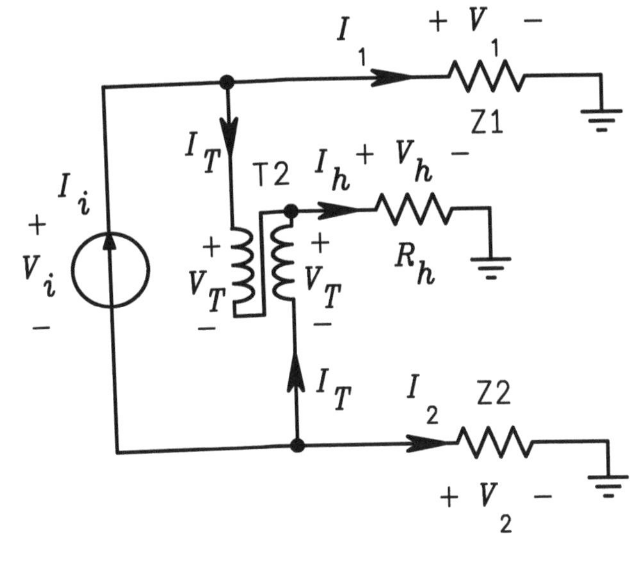

Figure 5.8 *Simplified circuit of hybrid splitter.*

from which

$$I_i = \frac{1}{2}(I_1 - I_2) \tag{5.3}$$

and

$$I_h = 2I_T = -(I_1 + I_2) \tag{5.4}$$

Following the method in [1],

$$V_T = V_1 - V_h = V_h - V_2 \tag{5.5}$$

hence

$$V_h = \frac{1}{2}(V_1 + V_2) = \frac{1}{2}(I_1Z_1 + I_2Z_2) \tag{5.6}$$

and

$$I_h = \frac{V_h}{R_h} = \frac{I_1Z_1 + I_2Z_2}{2R_h} \tag{5.7}$$

Comparison of (5.4) and (5.7) yields

$$-2I_1R_h - 2I_2R_h = I_1Z_1 + I_2Z_2 \tag{5.8}$$

and collection of terms then yields

$$\frac{I_1}{I_2} = -\frac{Z_2 + 2R_h}{Z_1 + 2R_h} \tag{5.9}$$

It is apparent that current splits evenly only if $Z_1 = Z_2$ or $R_h = \infty$. If $R_h = 0$, T2 forces equal voltages on the two loads.

Insertion of (5.9) into (5.3) and minor rearrangement yields

$$I_i = I_1\frac{Z_1 + Z_2 + 4R_h}{2(Z_2 + 2R_h)} \tag{5.10}$$

and

$$I_1 = I_i\frac{2(Z_2 + 2R_h)}{Z_1 + Z_2 + 4R_h} \tag{5.11}$$

The voltage produced at the input to PA #1 is then

$$V_1 = I_1Z_1 = I_iZ_1\frac{2(Z_2 + 2R_h)}{Z_1 + Z_2 + 4R_h} \tag{5.12}$$

Combining (5.12) and an analogous expression for V_2 yields

$$V_i = I_i\frac{4[Z_1Z_2 + (Z_1 + Z_2)R_h]}{Z_1 + Z_2 + 4R_h} \tag{5.13}$$

Division by I_i yields the input impedance presented to the driver as

$$Z_i = \frac{4[Z_1Z_2 + (Z_1 + Z_2)R_h]}{Z_1 + Z_2 + 4R_h} \tag{5.14}$$

The nominal impedance of $Z_i = 4R_h$ occurs when $Z_1 = Z_2 = 2R_h$. Other load impedances produce nonnominal input impedances.

5.3 BASIC THEORY OF HYBRID COMBINER

A simplified version of the hybrid combiner of Figure 5.5 is shown in Figure 5.9. Again, balun T4 is omitted for simplicity as it does not enter the analysis. The outputs from PA#1 and PA#2 are represented by current sources I_1 and I_2. The response to output from PA #2 is obtained by connecting a current source to input (B) and leaving input (A) open. The response to drive from PA #1 is analogous, and the complete response is obtained by superpositioning the two individual responses.

Transformer T3 forces equal currents in each winding, thus

$$I_o = -I_2/2 \tag{5.15}$$

and

$$I_h = I_2 \tag{5.16}$$

The voltages on the two windings are also equal, so

$$V_o = I_oZ_o = -I_2Z_o/2 \tag{5.17}$$

$$V_h = I_hR_h \tag{5.18}$$

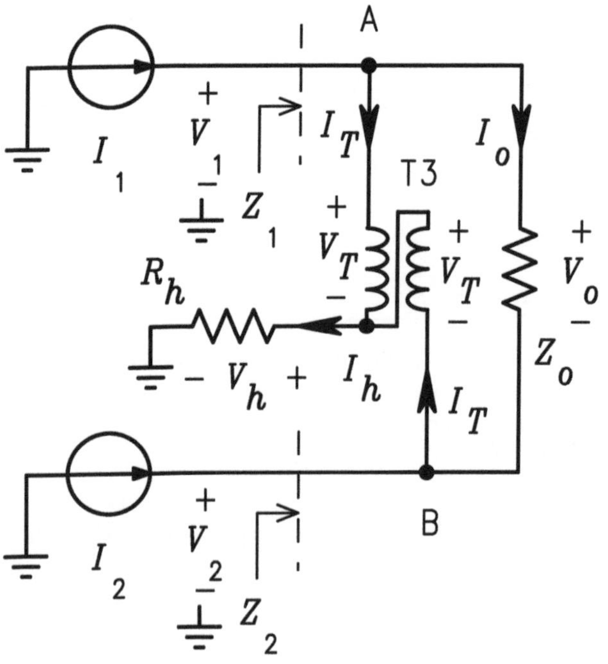

Figure 5.9 *Simplified circuit of hybrid combiner.*

and

$$V_2 = V_h - V_T = V_h - V_o/2 \tag{5.19}$$

Addition of these responses to the analogous responses for PA #1 driving I_1 into input A of T3 then yields

$$I_o = \frac{1}{2}(I_1 - I_2) \tag{5.20}$$

$$V_o = \frac{1}{2}(I_1 - I_2)Z_o \tag{5.21}$$

$$I_h = I_1 + I_2 \tag{5.22}$$

$$V_h = (I_1 + I_2)R_h \tag{5.23}$$

and

$$V_2 = (I_1 + I_2)R_h - \frac{1}{4}(I_1 - I_2)Z_o \tag{5.24}$$

Division of V_2 by I_2 then yields the load resistance seen by PA #2 as

$$Z_2 = \frac{V_2}{I_2} = \frac{I_1}{I_2}\left(R_h - \frac{Z_o}{4}\right) + \left(R_h + \frac{Z_o}{4}\right) \tag{5.25}$$

Equal PA Currents

If the two PAs produce equal output currents, then $I_2 = -I_1$ and (5.25) yields

$$Z_2 = Z_o/2 \tag{5.26}$$

regardless of R_h. In this case, $I_h = 0$ and current circulates from PA #1 through R_o to PA #2. The output power is then

$$P_o = \frac{I_1^2 R_o}{2} \tag{5.27}$$

This is the case when the outputs of two voltage-source PAs are combined. The load impedance (of whatever value) is simply split between the two PAs and impedance mismatch if present is transferred directly to the power amplifiers.

Unequal PA Currents

When current-source PAs are combined, differences in gain cause I_1 and I_2 to differ. Inspection of (5.25) suggests setting

$$R_h = R_o/4 \tag{5.28}$$

which causes (for a resistive load)

$$R_2 = R_o/2 \tag{5.29}$$

regardless of the relative amplitudes of I_1 and I_2. This maintenance of constant PA-load impedance is the principle feature of the hybrid powercombiner.

If PA #2 fails (open), then

$$I_h = I_1 \tag{5.30}$$

$$P_h = \frac{I_1^2 R_h}{2} = \frac{I_1^2 R_o}{8} \tag{5.31}$$

$$I_o = I_1/2 \tag{5.32}$$

$$P_o = \frac{I_1^2 R_o}{8} \tag{5.33}$$

Thus a hybrid-combined system with one failed power amplifier delivers one-fourth of the normal output power and dissipates an equal amount of power in the hybrid resistor R_h. Similar results are obtained if PA #2 fails shorted. The hybrid resistor must therefore be capable of dissipating up to one fourth of the system output power. This generally requires selection of an appropriate RF-power resistor and mounting it on a suitable heat sink. It is important to note that the "hybrid" action that maintains constant PA-input impedances (isolation) occurs only when $R_h = R_o/4$. If R_o varies from its nominal value, the PA-load impedances also vary from their nominal value.

5.4 LITERATURE

Early Work

Pitzalis and Couse [2] make an early introduction of hybrid power splitters and combiners to the production of RF power at HF and VHF. They utilize both Ruthroff and Guanella transformers to implement a broadband hybrid transformer, and show the effect of line length upon bandwidth. The basic theory for a four-port hybrid based upon ideal transformers is developed by Sartori [3]. The Communication Transistor Corporation applications note Solid Circuits [4] was one of the first to explain in some detail how to make and to use combiners and splitters with the newly developed solid-state power amplifiers.

Topologies and Construction

A general introduction to the principles and uses of splitters and combiners appears in [1]. Dye and Granberg in [5] give an excellent summary of various approaches to implementing broadband ferrite-loaded transformers, splitters, and combiners in Chapter 11 of their book [5]. Joe Johnson also gives a good overview of techniques in [6].

Rod Blocksome presents a comprehensive review of various types of RF transformers, combiners, and splitters in [7], which is included in the reprints section. His paper covers a variety of topologies, including both 0° and 180° types and combiners and splitters with two and four inputs. Illustrations show how the combiners are built from transmission lines and ferrite beads or blocks.

Applications

Granberg [8] describes a solid-state HF linear power amplifier that uses a four-way splitter and a four-way combiner to obtain a 1-kW combined output from four 300-W PAs. Gibbons [9] combines sixteen 1.2-kW PAs to obtain 15-kW pulses for a magnetic-resonance imaging transmitter for HF and VHF. The 1.2-kW modules are in turn based upon four-way splitting and combining of PA outputs [10]. All of the preceding are based upon corporate splitting and combining. An example of the use of serial combiners and splitters is given by Burdick [11] and additional discussion is given by Walker [12].

Transmission-Line Combiners and Splitters

The quarter-wavelength transmission-line combiner is introduced by Wilkinson in his classic paper [13], which is included in the reprints section. Various topologies are shown, including more than two inputs. Replacing the quarter-wavelength lines with lumped-element equivalents allows the Wilkinson combiner to be used at lower frequencies (e.g., VHF) for which quarter-wavelength transmission lines would be inconveniently long. Dufour [14] and Noriega [15] develop the lumped-element equivalents of the quarter-wavelength lines, and a number of examples of lumped-element designs are given by Shor [16].

A number of variants of the Wilkinson combiner have been developed. Noriega [15] further shows how to implement versions for unequal power division. London [17] describes some multi-line variations upon the Wilkinson combiner. The Gizel combiner described by Gonzales [18] can be regarded as a variation with a different connection to the hybrid resistors.

An excellent example of the use of the Wilkinson combiner at VHF is given by Brounley [19], who combines the outputs from seven PAs to produce a 1-kW total output.

Effects of Losses, Unequal Inputs, and Failures

An ideal combiner with perfectly balanced input signals delivers all of the input power to its load. However, real splitters and combiners are made from real RF transformers (Chapters 3 and 4) and are therefore subject to the same sorts of losses due to the core and windings.

The hybrid combiner delivers all of its power to its load only when the input signals from the two PAs are of exactly the same amplitude and exactly in the proper phase (0° or 180°, depending upon the type of combiner). Maintaining constant PA-load impedance (isolation between the PAs) happens because the difference is dissipated in the hybrid resistor instead of generating a load voltage or current. The losses associated with amplitude and phase imbalances are presented by Blocksome [20].

Failure of one of the PAs in a two-input combiner causes half of the power from the PA remaining active to be dissipated in the hybrid resistor. The system output is thus reduced to one fourth of its normal output. The effects of multiple PA failures in a corporate hybrid-combining system are quantified by Ernst [21], Saleh [22], and Walker [12]. Saleh also develops schemes to improve efficiency by switching-out failed amplifiers.

REFERENCES

[1] Krauss, H.L., C.W. Bostian, and F.H. Raab, *Solid State Radio Engineering*, New York: Wiley, 1980.

[2] Pitzalis, O., and T.P. Couse, "Broadband Transformer Design for RF Transistor Power Amplifiers," *Proc. IEEE Electronic Components Conf.*, 1968, pp. 207–216. Also Report AD-676 816, U.S. Army Electronic Command, Ft. Mommouth, NJ.

[3] Sartori, E.F., "Hybrid Transformers," *IEEE Trans. Parts, Materials, and Packaging*, Vol. PMP-4, No. 3, September 1968, pp. 59–66.

[4] Johnson, J.H. (ed.), *Solid Circuits: Circuit Techniques*. Applications Book 2.2.8.0A, Communications Transistor Corp., San Carlos, CA, March 1973.

[5] Dye, N., and H. Granberg, *Radio Frequency Transistors*, Boston, MA: Butterworth-Heinemann, 1993.

[6] Johnson, J.H., "RF Combining and Protection Techniques," *Proc. RF Expo East '87*, Boston, MA, November 11–13, 1987, pp. 197–215.

[7] Blocksome, R.K., "Practical Wideband RF Power Transformers, Combiners, and Splitters," *Proc. RF Tech. Expo '86*, Anaheim, CA, January 30– February 1, 1986, pp. 207–227. Also *Proc. RF Expo East*, Boston, MA, November 10–12, 1986, pp. 1–21.

[8] Granberg, H.O., "A Two-Stage 1 kW Solid-State Linear Amplifier," Appl. Note AN-758, Motorola Semiconductor Products, Phoenix, AZ, 1976. Copyright of Freescale Semiconductor Inc. 2005. Used by permission.

[9] Gibbons, H., "High-Power VHF Power Dividing and Combining Techniques," *Proc. RF Expo West*, San Diego, CA, March 18–20, 1992, pp. 181–187.

[10] Haupt, D., "The Design of RF Modules Intended for Combining to High Power," *Proc. RF Expo West*, San Diego, CA, March 18–20, 1992, pp. 173–179.

[11] Burdick, M., G. Kinnetz, J-M. Boulay, and J-J. Bouny, "Splitters/Combiners Team with RF LDMOS in UMTS Amp," *Microwaves & RF*, Vol. 38, No. 7, July 1999, pp. 80–92.

[12] Walker, J.L.B. (ed.), *High-Power GaAs FET Amplifiers*, Norwood, MA: Artech House, 1993.

[13] Wilkinson, E.J., "An N-Way Hybrid Power Divider," *IEEE Trans. Microwave Theory Techniques*, Vol. MTT-8, No. 1, January 1960, pp. 116–118.

[14] Dufour, K., "60 Watt VHF Amplifier Using Splitting/Combining Techniques," *RF Design*, Vol. 4, No. 4, September/October 1981, pp. 19–23. Also *Engineering Bulletin EB93*, Motorola (Freescale) Semiconductor, Phoenix, AZ, circa 1981. Copyright of Freescale Semiconductor Inc. 2005. Used by permission.

[15] Noriega, F., and P.J. Gonzalez, "Designing LC Wilkinson Power Splitters," *RF Design*, August 2002, pp. 18–24.

[16] Shor, A., "Broadband Lumped Quadrature Coupler," *Proc. RF Technology Expo*, Anaheim, CA, March 27–29, 1990 [P-168], pp. 139–143.

[17] London, S. Y., "A New Broadband Coupled-Line N-Way Power Combiner/Splitter," *Applied Microwave & Wireless*, Vol. 13, No. 5, May 2001, pp. 28–38.

[18] Gonzales, F. J. O., and J. L. J. Martin, "High-Power Amps Blend Low Cost, High Performance," *Microwaves & RF*, Vol. 34, No. 5, May 1995, pp. 87–103.

[19] Brounley, R. W., "A 1000 Watt VHF Linear Amplifier," *RF Design*, Vol. 4, No. 4, September/October 1981, pp. 12–17.

[20] Blocksome, R. K., "Predicting RF Output Power from Combined Power Amplifier Modules," *RF Design*, Vol. 11, No. 2, February 1988, pp. 40–49.

[21] Ernst, R. E., "Graceful Degradation Properties of a Matched N-Port Power Amplifier Combiners," *1977 MTT-S Int. Microwave Symp. Dig.*, June 1977, pp. 174–177.

[22] Saleh, A. A. M., "Improving the Graceful-Degradation Performance of Combined Power Amplifiers," *IEEE Trans. Microwave Theory and Techniques*, Vol. MTT-28, No. 10, October 1980, pp. 1068–1070.

PRACTICAL WIDEBAND RF POWER TRANSFORMERS, COMBINERS, AND SPLITTERS

by

Roderick K. Blocksome
Manager, HF PA/PS Design Group
Rockwell International
High Frequency Communications Division
855 35th Street, NE
Cedar Rapids, Iowa 52498

INTRODUCTION

This paper will deal with the practical aspects of designing and building wideband RF power transformers, combiners hybrids), and splitters. Emphasis will be on topology. A consistent approach to represent these transformers pictorially and schematically with equivalent circuits showing source and load connections will be developed to help provide an intuitive understanding of the devices. Laboratory test data comparing various designs and topologies is included.

Modern solid state HF power amplifiers are required to operate over increasingly wider bandwidths and at higher power levels for applications in communications as well as electronic countermeasures. Wideband RF power transformers are required for coupling into and out of the solid state devices. conventional or so-called "wire-wound" transformer and two topologies of the transmission line transformer conventional and equal delay are presented.

A wideband RF power combiner (or hybrid) is required to achieve output levels above the capabilities of a single solid state amplifier stage. The RF outputs of two or more identical amplifier modules can be combined to reach these higher powers. Design examples of in-phase, 180-degree, and quadrature combiners are detailed. Two basic topologies for in-phase 180-degree combiners are presented.

A wideband RF power splitter (or divider) is simply a combiner or hybrid used in reverse. The splitter topology is the same as a combiner, however splitters are usually operated at lower power levels. The discussion centers around combiners but is equally applicable to power splitter applications.

TRANSFORMERS

The bandwidth of rf transformers does not refer to the usual -3 dB points since in power applications this represents an unacceptable loss. Typical HF amplifier designs require operation from 2 to 30 MHz and sometimes lower to 1.6 MHz. transformer losses must be as low as possible over this operational bandwidth. Transformer losses translate to heat that must be removed as well as extra power that must be supplied by the transistors (at the collector/drain efficiency and ultimately by the power supply (at its conversion efficiency). A few tenths of a dB of unnecessary loss in output transformers or combiners can mean significant increases in primary power consumption.

RF power FET devices have operational bandwidths of 1 to 175 MHz making possible extended range amplifiers covering HF and the lower VHF frequencies. Transformer designs covering over six octaves of bandwidth are required.

A wideband RF power transformer performs one or more of any combination of three basic functions:

(a Impedance transformation

(b) Balanced to unbalanced transformation

Phase inversion

Transformation of a secondary load to a desired load impedance at the primary of the transformer is the most common function. RF transformers are often referred to by their impedance transformation ratio rather than primary to secondary turns ratio. The former is simply the turns ratio squared. In this application, we are most often interested in manipulating impedances rather than voltages or currents with the transformers. Balanced-to-unbalanced transformers, commonly termed "Baluns" are extremely useful in wideband amplifier designs. A single-ended load can be driven by a push-pull (balanced) source or vice-versa by using a balun transformer A wideband transformer can also perform a phase reversal from primary to secondary by proper winding connections.

Transformer connections between a source and a load may be either balanced or unbalanced. Additionally, the balanced source or load may be either entirely floating or with center grounded such as two single ended sources phased 180-degrees apart or a load resistor with grounded center tap. The distinction between "balanced, floating" and "balanced, center grounded" may seem unimportant for wideband transformer design, but it is not. A proposed balun transformer equivalent circuit with source and load connected should be drawn showing the magnetization current path.

Figure 1 (a) is an example of a simple 1:1 balun with a floating balanced load. The magnetization current, i_m, flows through the load resistor as shown. Figure 1 (b) illustrates what happens to the magnetization current path if the balanced load is changed to a balanced, center tap grounded load. The magnetization current flows through only one winding and only one-half of the load resistance. This causes undesirable phase and amplitude imbalance in the balun restricting the bandwidth. The balance can be restored by using a third or tertiary winding, as shown in figure 1 (c), to shunt the magnetization current around the load. This illustrates the necessity of considering the type of source and load connections when selecting wideband transformer topologies.

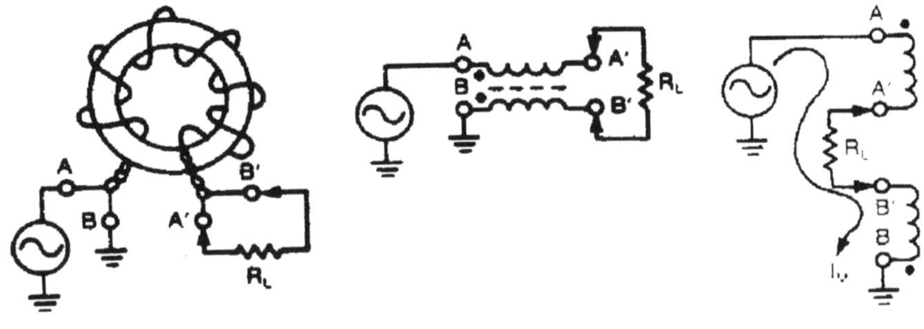

(A) BALANCED LOAD, FLOATING

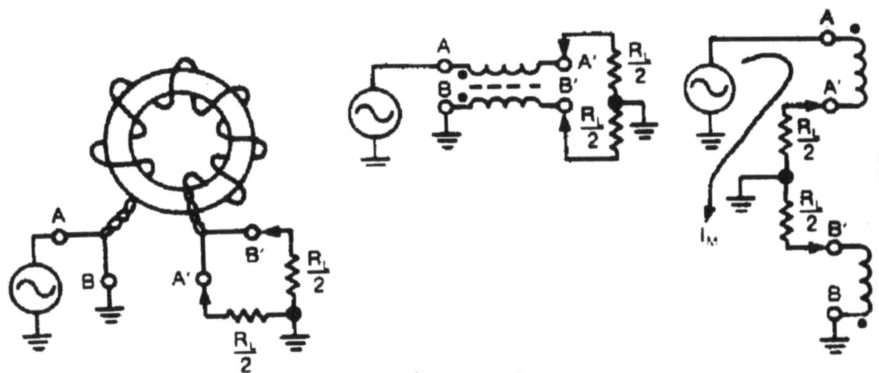

(B) BALANCED LOAD, CENTER GROUNDED

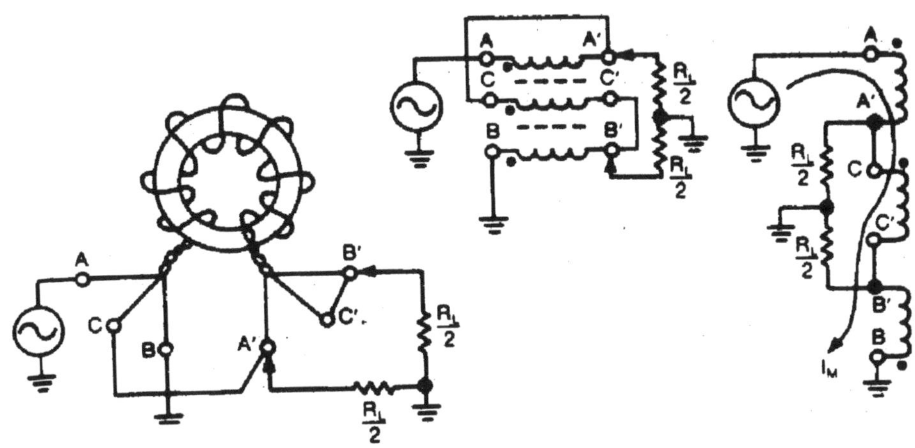

(C) BALANCED LOAD, CENTER GROUNDED WITH TERTIARY WINDING (C–C')

FIGURE 1. 1:1 TRANSMISSION LINE BALUN TRANSFORMERS

The nine possible transformer connections are given below:

SOURCE	LOAD
Unbalanced	Unbalanced
Unbalanced	Balanced, floating
Unbalanced	Balanced, center grounded
Balanced, floating	Unbalanced
Balanced, floating	Balanced, floating
Balanced, floating	Balanced, center grounded
Balanced, center grounded	Unbalanced
Balanced, center grounded	Balanced, floating
Balanced, center grounded	Balanced, center grounded

Wideband RF transformers and combiners typically use a magnetic core. The magnetic cores used in wideband RF transformers are available in a wide variety of shapes and sizes. Balun core, toroidal, sleeves, tubes, beads, and cup cores are the common names for the various shapes. The earliest material used was powdered iron followed by modern ferrites. Ferrite is composed of iron oxide in combination with various proportions of oxides of manganese, magnesiun, nickel, and zinc. In general the ferrites composed of iron, nickel, and zinc are applicable for the HF/VHF frequencies. Various mixes of ferrites are available. A high permeability and moderately low loss material is used for HF/VHF power transformers. Operational flux densities must be kept well within the linear portion of the B-H curve of the material. The area inside the

B-H curve represents the relative loss, therefore the narrow curves are preferred for low loss designs. Detailed information is available from the various ferrite manufacturers.

Core losses and winding dielectric losses heat the core. The core temperature must be held well below the Curie temperature of the ferrite, otherwise the magnetic properties of ferrite will be permanently altered. Operation near the Curie temperature is not recommended as some materials can go into thermal runaway. The high temperature increases the core loss which in turn further increases the core temperature until the core is ruined.

CONVENTIONAL OR "WIRE-WOUND TRANSFORMERS"

The conventional broadband RF transformer is characterized by a power transfer from the primary to secondary windings via magnetic coupling through the ferrite core. The transmission line transformer, by contrast, is characterized by the use of a transmission line of characteristic impedance, Z_0, and a ferrite core. The core suppresses common mode or non-transmission line currents which would otherwise flow due to the transmission line interconnections. A core wound with wire or may not be a conventional transformer, depending upon how source and load are connected. Figure 2 illustrates this distinction.

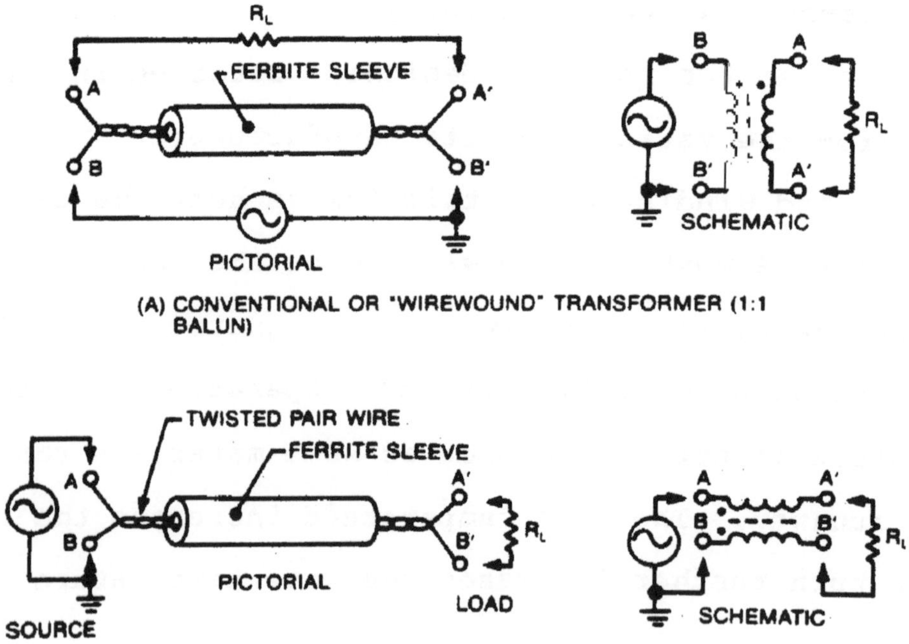

(A) CONVENTIONAL OR "WIREWOUND" TRANSFORMER (1:1 BALUN)

(B) TRANSMISSION LINE TRANSFORMER (1:1 BALUN)

FIGURE 2. COMPARISON OF CONVENTIONAL AND TRANSMISSION LINE TRANSFORMERS

In general, the conventional transformer is inferior to the transmission line transformer for the combination of high power capability, low loss, and wide bandwidth. The conventional transformer can be constructed for a wider range of impedance transformation ratios than the transmission line type. Some ratios will have wider bandwidths than others due to the number of turns to achieve the desired turns ratio. There are no fractional turns. If the wire or line passes through the core, it is one turn

Figure 3 is a conventional transformer that finds wide usage at low impedances (3 to 20 ohms). The core is commonly referred to as a balun core, yet the transformer may or may not be connected to perform as a balun. Metal sleeves of copper or brass are inserted into the core and connected together at end to form a primary winding. Connections to the circuit made at each of the two sleeves at the opposite end. Two pieces of copper clad G-10 circuit board work nicely at each end secondary winding is constructed by winding the required turns of insulated wire through the primary tubes

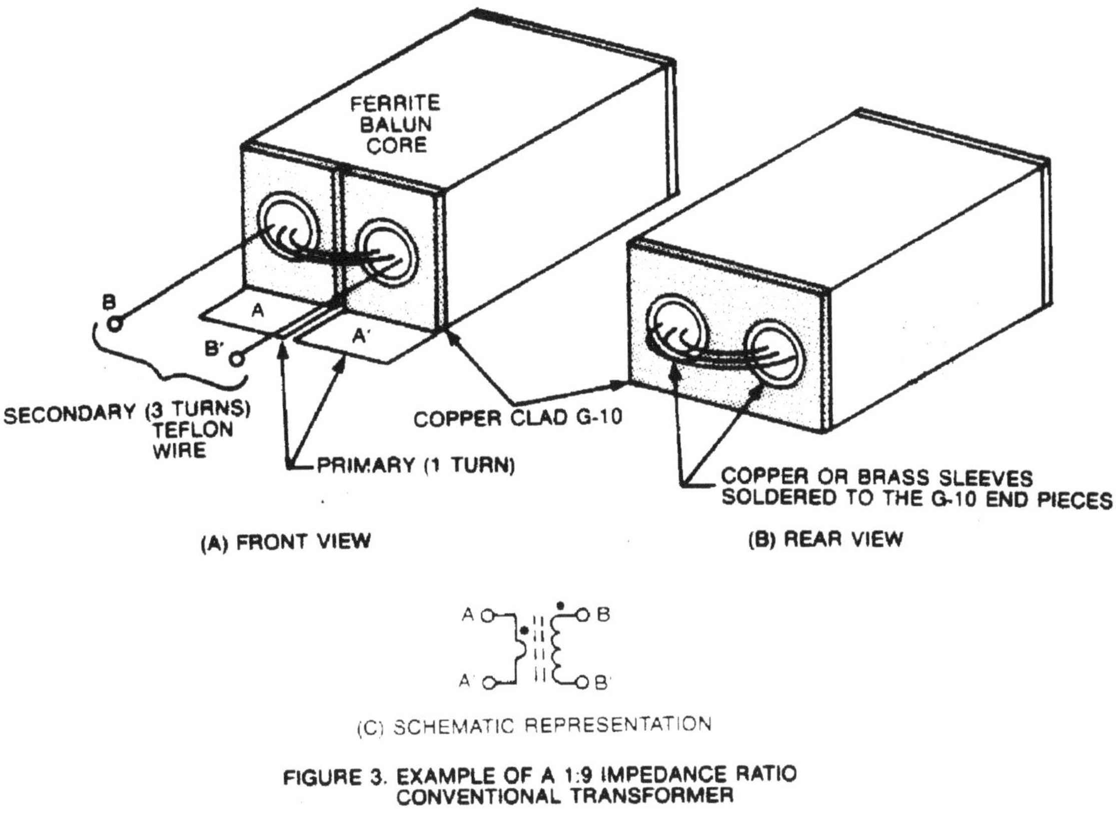

FIGURE 3. EXAMPLE OF A 1:9 IMPEDANCE RATIO CONVENTIONAL TRANSFORMER

One of the factors limiting the high frequency response of the transformer is leakage inductance. Leakage inductance is due to any flux lines that do not link the primary and the secondary. To minimize the leakage inductance, the primary copper tubes should fit quite close in the core holes. They should not be so tight that thermal expansion will cause the core to break. The lead inductances of the primary and secondary windings from the point they exit the core to the circuit connection will also limit performance at the high frequency end, especially on low impedance applications. Shunt capacitance on either the primary or secondary or both will compensate the leakage reactance and extend the useful frequency limit.

TRANSMISSION LINE TRANSFORMERS

The simplest transmission line transformer is a quarter-wavelength line whose characteristic impedance, Z_O, is chosen to give the correct impedance transformation. This relationship is illustrated in figure 4. Note that this transformer is a narrowband device valid only at frequencies for which the line is odd multiples of a quarter wavelength. The transformation ratio is given by the square of the ratio of the line impedance to the load connected to the line.

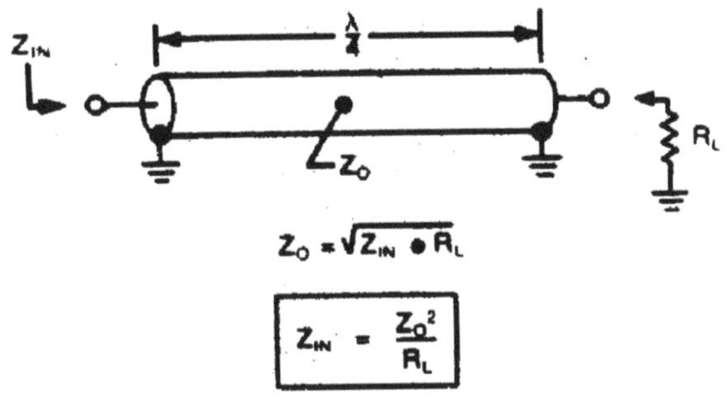

FIGURE 4. SIMPLE QUARTER WAVE TRANSMISSION LINE TRANSFORMER

If a ferrite sleeve is added to the transmission line (see figure 5), common mode currents (currents flowing in both transmission line conductors in phase and in the same direction) are suppressed and the load may be balanced and floating above ground. The line can now be any length with characteristic impedance equal to the balanced load impedance. The result is a 1:1 balun. Low frequency operation is limited by the amount of impedance offered to common mode currents. A good rule-of-thumb requires the impedance presented to common mode currents be not less than five times the load impedance. The line length limits the high frequency response of transmission line transformers.

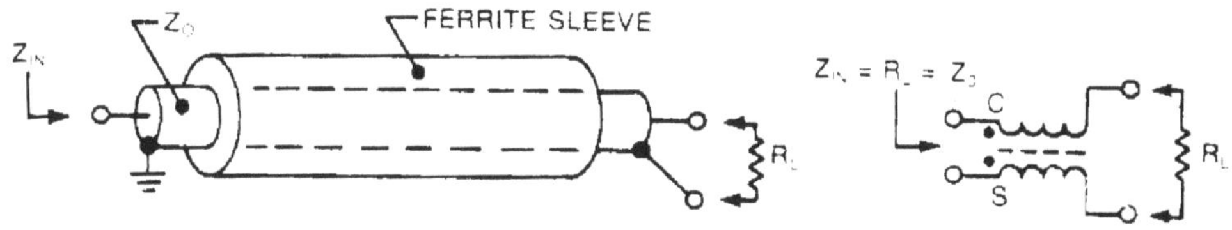

FIGURE 5. 1:1 TRANSMISSION LINE BALUN

If the ferrite loaded length of transmission line in figure 5 is folded back so that the two ends may be interconnected, a impedance transformer is formed. A load resistance, R_L, connected as shown in figure 6 is reflected to the input of the transformer as $R_L/4$. The line Z_o should be equal to the geometric mean of R_L and Z_{in} for maximum bandwidth. The length must be as short as possible for extended high frequency operation. The practical high frequency limit for this type of transformer is reached when the line length approaches 1/8 wavelength and appreciable phase error difference occurs at the interconnection of the lines.

A 1:4 transmission line balun transformer may be constructed as shown in figure 7. Two cores are required and may be either balun cores (as shown) or toroids or sleeve cores. The transmission line Z_o should be the geometric mean of the input and load impedances. This transformer may also be used for balanced-to-balanced source and load connections. Transmission line baluns for 1:9 and 1:16 impedance ratios are constructed similarly as shown in figures 8 and 9. The limitation of squared integer transformation ratios is the biggest disadvantage of this type of transmission line transformer. The availability of coaxial cable in a variety of impedances is another limitation. 50 and 75-ohm cables are by far the most common but impedances of 25, 35, 60, 95, and 125-ohms are available.

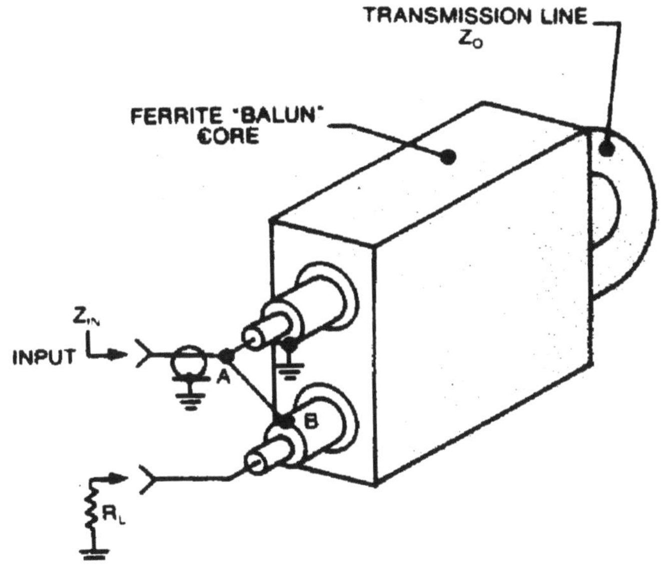

(A) PICTORIAL VIEW

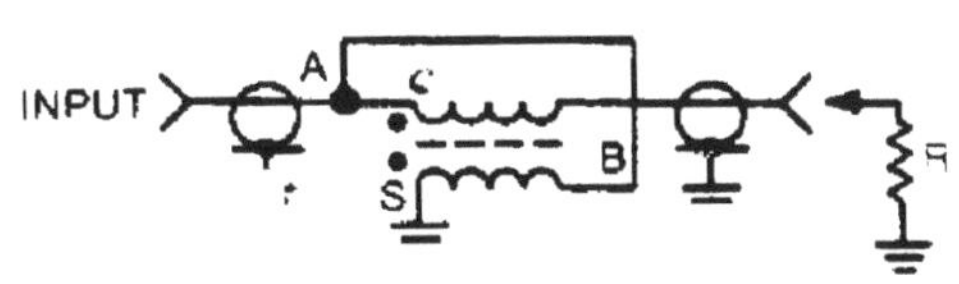

(B) SCHEMATIC

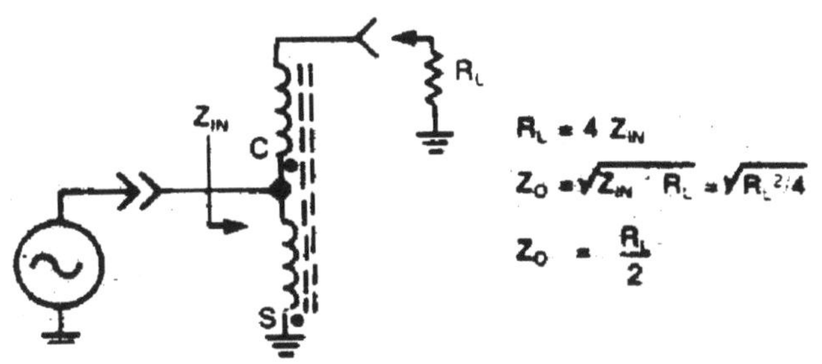

(C) EQUIVALENT CIRCUIT

FIGURE 6. TRANSMISSION LINE TRANSFORMER; 4:1 UNBALANCED-TO-UNBALANCED

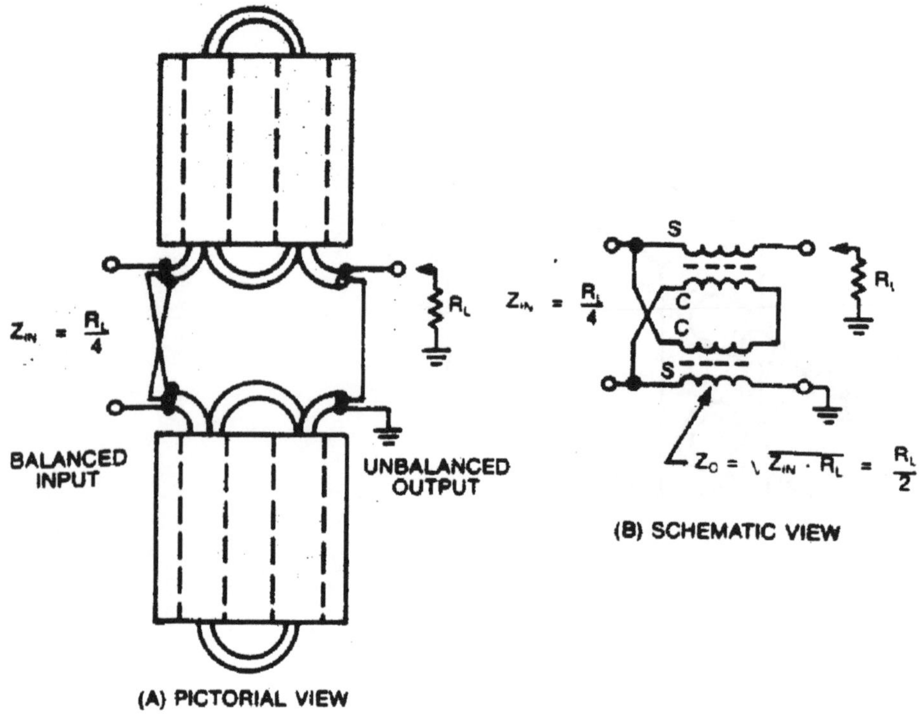

FIGURE 7. 1:4 TRANSMISSION LINE BALUN TRANFORMER

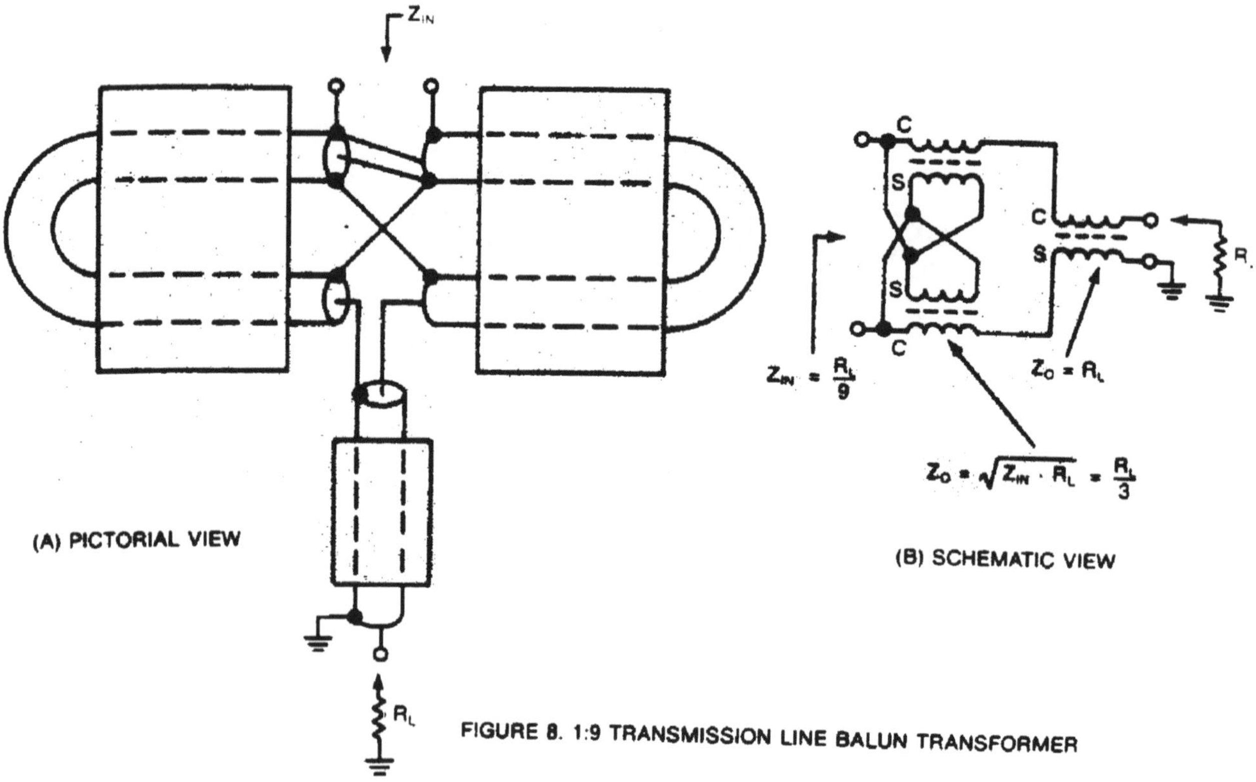

FIGURE 8. 1:9 TRANSMISSION LINE BALUN TRANSFORMER

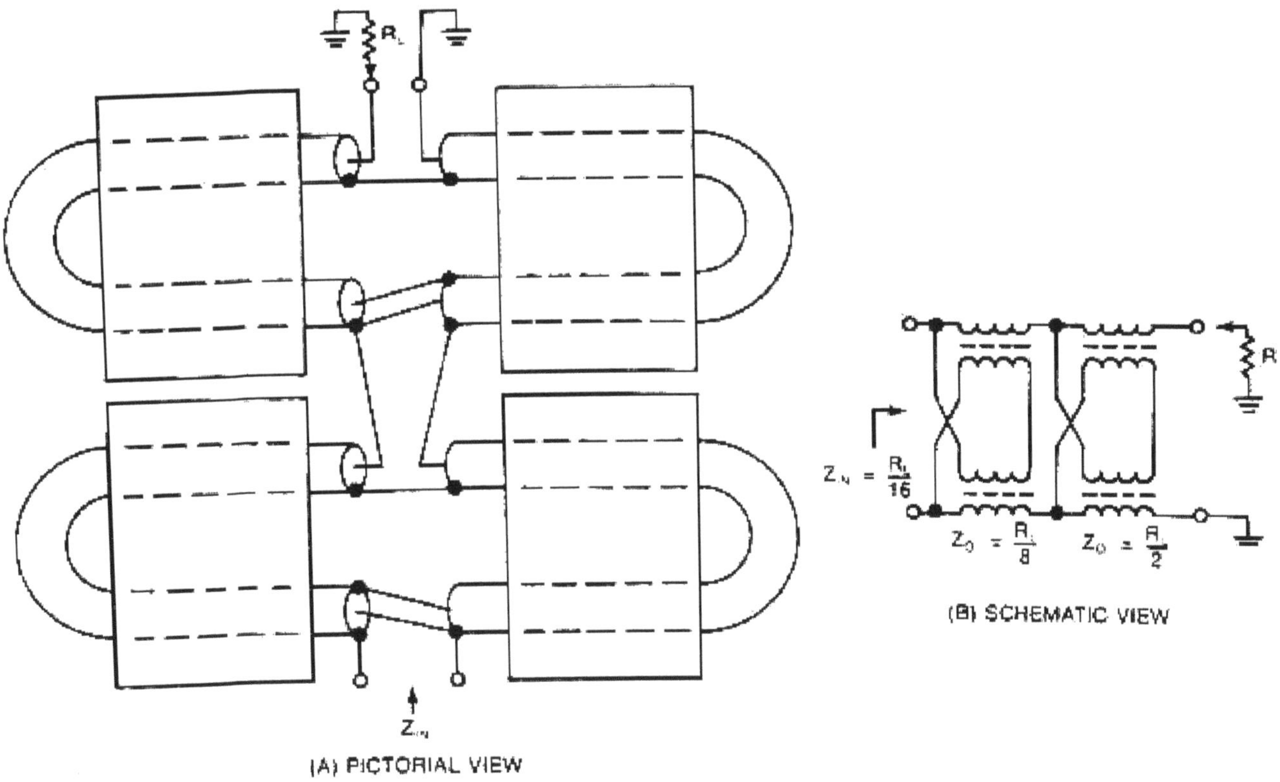

FIGURE 9. 1:16 TRANSMISSION LINE BALUN TRANSFORMER

Several techniques to achieve nonstandard impedance lines include simply parallel connecting two or more lines. For example, two parallel 50-ohm lines provide an effective 25-ohm line. The parallel lines do not have to be the same impedance either. Bifilar or twisted enameled wire can easily be constructed for odd characteristic impedances also. The impedance depends upon the wire diameter, insulation dielectric spacing, and number of twists per unit length. Multiples of even numbers of wire may be twisted together and then parallel connected to achieve low characteristic impedances. The characteristic impedance of experimentally constructed bifilar or twisted pair transmission lines may be determined by measuring the reactance of an open circuit, 1/8-wavelength, sample. The magnitude of the reactance is equal to the line impedance at the frequency for which the line is 45-degrees in electrical length. Remember to account for the velocity of propagation when determining the frequency of 1/8 wavelength.

Micro-strip transmission lines on printed circuit boards is another technique for achieving virtually any desired line impedance. Mechanical problems with the strip lines in ferrite cores may be more difficult but interconnections with the amplifier circuit may be improved.

The bandwidth degradation experienced by not using the correct value of line impedance may be acceptable in some applications. Figure 10 is a comparison of two identical 1:4 balun transformers; one wound with the proper 25-ohm line, the other wound with 50-ohm line. The measurement was made by connecting two identical transformers back-to-back to provide matched 50-ohm impedance ports to interface with the network analyzer. The indicated loss of one transformer is half of the measured value. This technique is valuable for evaluating various transformer designs and initially chooseing values of compensation capacitors for leakage reactance.

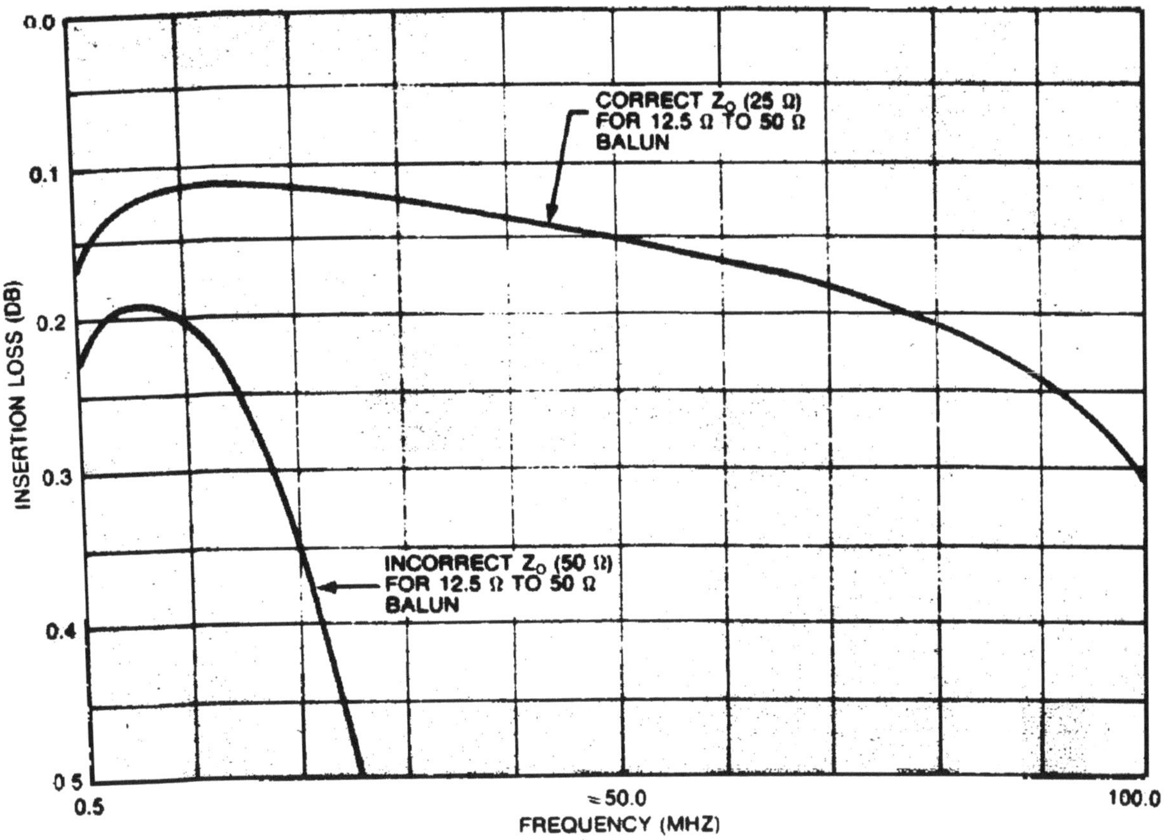

FIGURE 10. COMPARISON OF EFFECT OF Z_0 ON 1:4 BALUN TRANSFORMER

As pointed out earlier, the 1:4 transmission line transformers' high frequency response is limited when appreciable phase error is introduced at the interconnection point a-b shown in figure 11. If the connection a-b were made with a transmission line of equal impedance and length as the ferrite loaded line, the phase difference between input and output is eliminated. The transformer topology remains the same, except the a-b connection has the same phase delay as the main transformer line. For this reason this subclass of transmission line transformers are called "Equal Delay

Transmission Line Transformers". The transformer input and output connections can be physically separated which is advantageous in some applications.

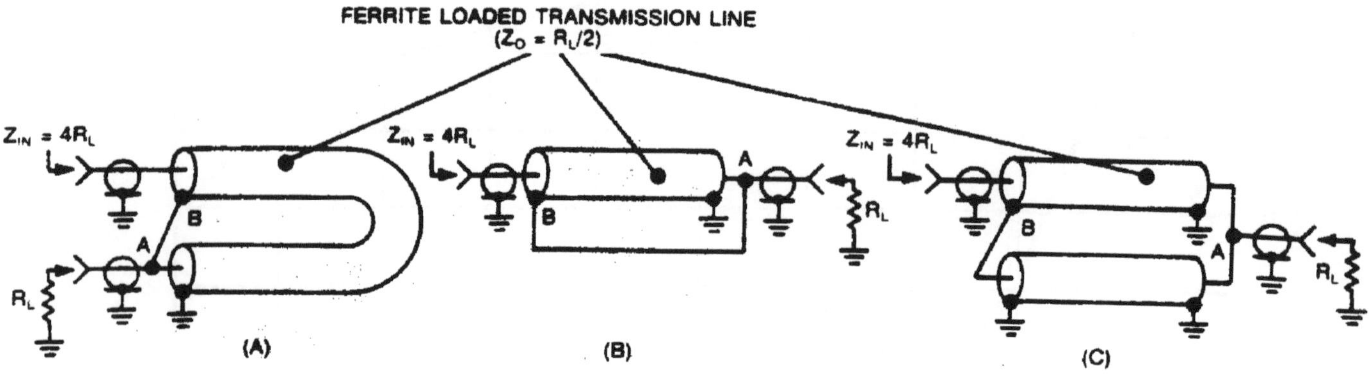

FIGURE 11. DERIVATION OF THE EQUAL DELAY TRANSFORMER

Figure 12 (a is the usual pictorial and schematic representation of a 1:4 equal delay transformer. If a third line is stacked on the 1:4 design, a 1:9 impedance transformer results. In like manner, four lines produce a 1:16 transformer and so on. Figure 12 (b) and (c) illustrates these ratios. For comparison, if one unit of ferrite is required on the 1:4 transformer for a given bandwidth, then two units will be required for the third line on the 1:9 transformer. In like manner, the fourth line requires three units of ferrite for the same bandwidth. Notice that these designs are all unbalanced-to-unbalanced transformers. Suppose we add ferrite to the bottom line on the 1:4 transformer. Now we can lift the grounds on the parallel connected end (still keeping the shields connected) and connect a balanced, floating load between the center conductors and the shields to form a 1:4 balun. The stray capacitance to ground can be balanced better by interconnecting the center conductor of one coax to the shield of the other coax. The result is the balun transformer described earlier in figure 7

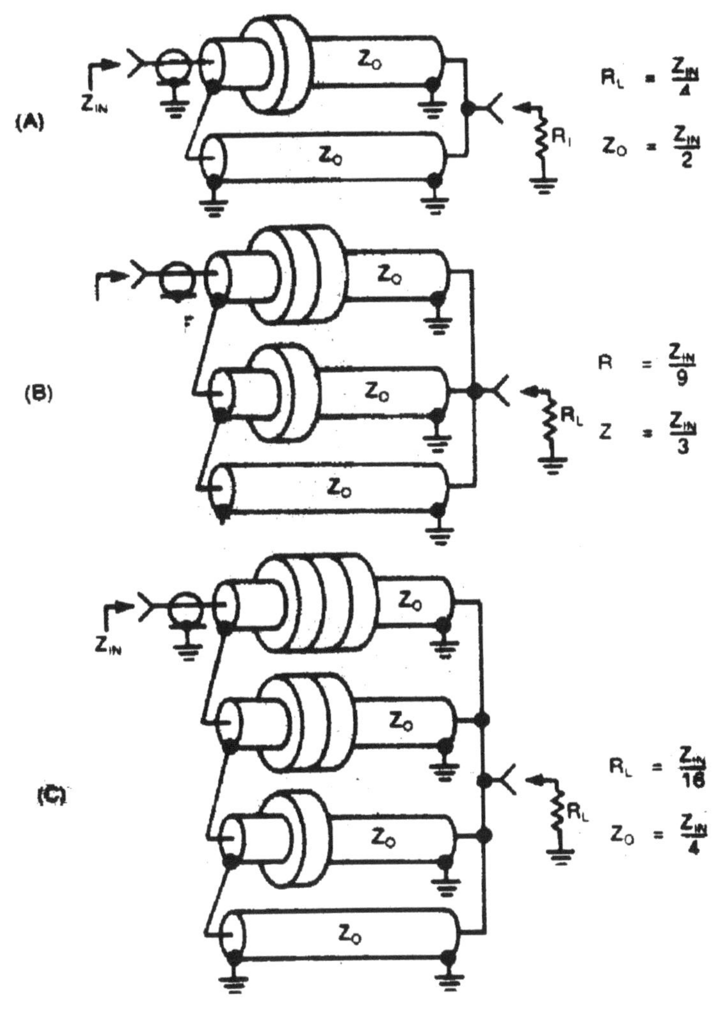

FIGURE 12. EQUAL DELAY TRANSFORMER CONFIGURATIONS

How much improvement in bandwidth does the equal delay transformer give compared to the conventional transmission line transformer? Figure 13 is a plot of insertion loss versus frequency for the two types constructed on identical cores. Again, the test consisted of measuring two identical transformers connected back-to-back, so the actual loss for one transformer is one-half the measured value.

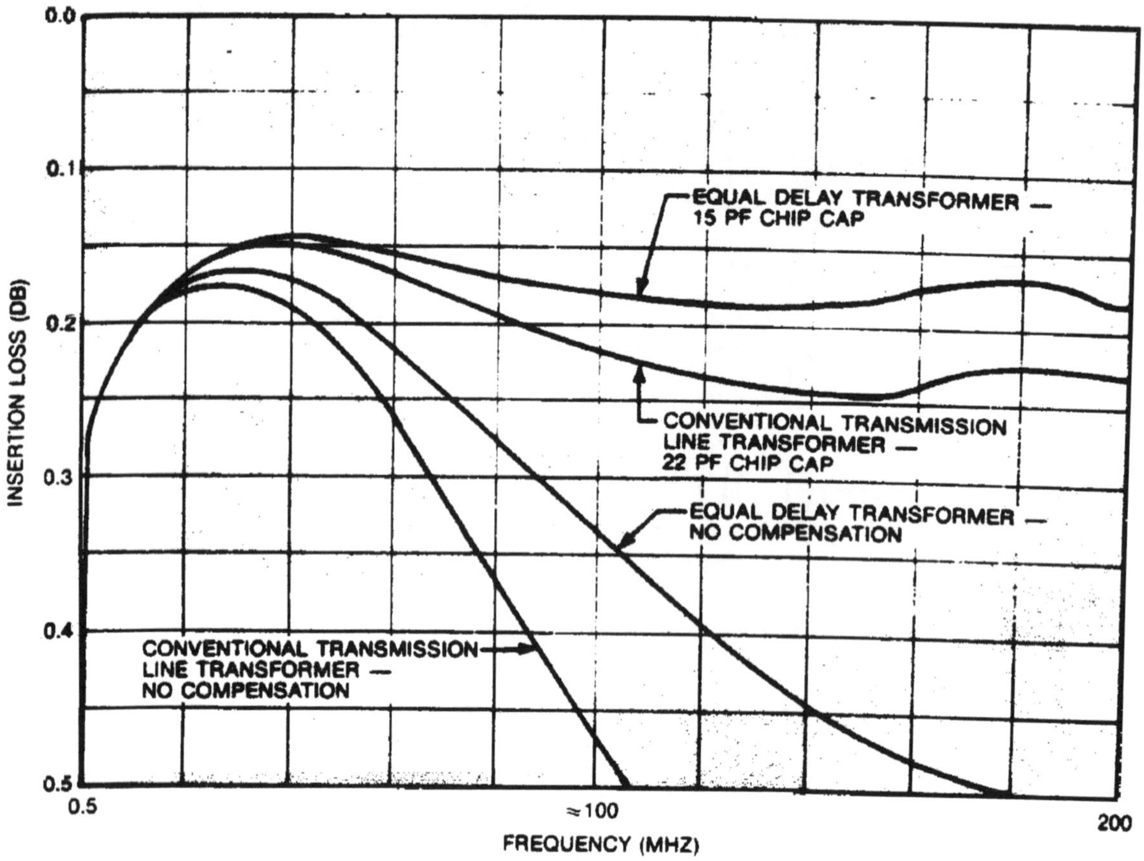

FIGURE 13. BANDWIDTH COMPARISONS OF CONVENTIONAL AND EQUAL DELAY 1:4 TRANSMISSION LINE TRANSFORMERS

COMBINERS AND SPLITTERS

When required output power levels exceed the capabilities of a single power amplifier stage, two or more stages or modules are combined to produce the required output. The combiner is closely related to wideband transformers in design and techniques. A power splitter is simply a lower powered version of the combiner used in reverse. The splitter divides the drive signal into multiple equal amplitude outputs to be applied to the amplifier inputs. The power combiner then recombines the amplified outputs into a single signal. Since the splitter is the same as a combiner, the following discussion will mention only combiners.

A wideband power combiner must perform the following basic functions:

a. Provide low insertion loss over the required bandwidth.

b. Provide isolation (minimum coupling) between the input ports.

c. Provide a low VSWR load at the input ports over the required bandwidth

The operating bandwidth of combiners must be as wide or wider than the amplifiers to not restrict the overall bandwidth of the transmitter. Transmission line techniques are used for lowest loss and widest bandwidth. The primary function of the combiner is to maintain port-to-port isolation. By isolating the output of one amplifier from the others, multiple failures as a result of a single amplifier failure are avoided. For example, in a two-input-port combiner, if one amplifier is disabled the output power drops by 6 dB. The output drops 3 dB due to lack of power from the disabled module and an additional 3 dB is due to the power from the remaining module dividing equally between the bridging resistor and the output load.

The bridging resistor must dissipate -6 dB of the maximum combiner output power. The bridging resistor value is prescribed by the type and configuration of the combiner as detailed later. Some topologies require either single-ended or balanced, floating bridging resistors. Sometimes the bridging resistor is referred to as the "dump" load or "dump" port since power due to phase or amplitude imbalance is dumped to this load.

The bridging resistor dissipates power due to any slight differences in either the phase or amplitude of the input signals. This relationship is given in figure 14

There are three basic types of combiners:

a. In-phase combiner or hybrid (two or more input ports)

b. 180-degree combiner or hybrid (two input ports)

c. 90-degree combiner or quadrature hybrid (two input ports)

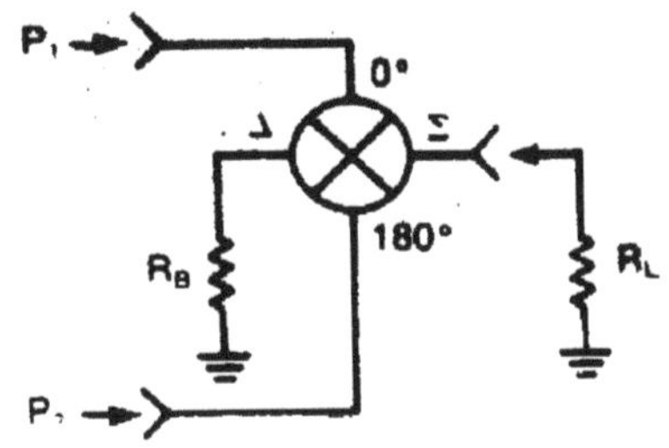

$P_\Sigma \quad \frac{P_1 + P_2}{2} \quad \sqrt{P_1 \cdot P_2}\ \cos\theta$: WHERE θ IS DIFFERENCE IN PHASE BETWEEN INPUT PORT SIGNALS

$P_{RB} = P_1 + P_2 - P_\Sigma$

(A) 180-DEGREE HYBRID COMBINER

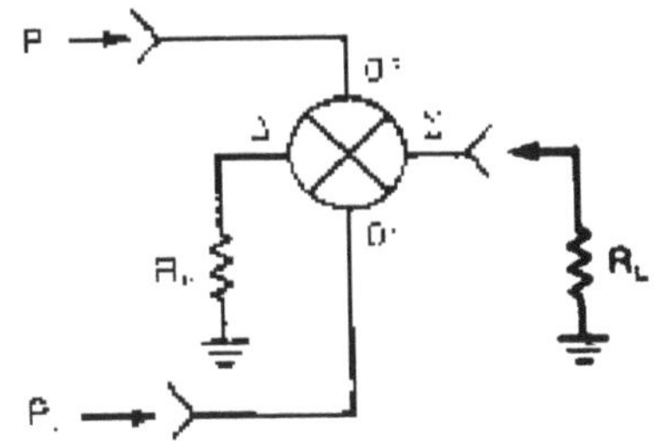

$P_{RL} = \frac{P_1 + P_2}{2} - \sqrt{P_1 \cdot P_2}\ \cos\theta$ WHERE θ IS DIFFERENCE IN PHASE BETWEEN INPUT PORT SIGNALS

$P_{RB} = P_1 + P_2 - P_{RL}$

(B) IN-PHASE HYBRID COMBINER

FIGURE 14 HYBRID COMBINER POWR RELATIONSHIPS

If more than two signals are combined in-phase, the term "combiner" is used since the term "hybrid" refers to a device with two input ports. The topologies of each of the three basic configurations will be examined.

The following definitions apply:

output load resistance

R_B= bridging resistor

transmission line characteristic impedance

input impedance (with output port terminated)

shield connection of coaxial cable

C = center connection of coaxial cable

IN-PHASE COMBINERS

In-phase combiners operate with two or more inputs of equal phase and amplitude to combine into a single output. There are two basic topologies for in-phase combiners, examples of which are shown in Figures 15 and 16. The differences are in the number and configurations of the ferrite cores and the value of the bridging resistor. The type-I configuration has a single balun core or toroidal core and a bridging resistor equal to four times the output load. The type-II combiner has two separate cores; either sleeves or toroidal. The bridging resistor is equal to the load resistance. Consideration of physical layout, practical transmission line impedances (Z_O), and bridging resistance (R_B) will determine the best type of combiner for a particular design.

A comparison of input impedance and port-to-port isolation between typical type-I and type-II combiners yields interesting results as shown in Figure 17. Both combiners were constructed with a single turn of 50-ohm coax in the cores. Core material was Stackpole 7D for both types. The test data indicates superior port-to-port isolation with a type-II combiner while the type-I combiner exhibited lower input VSWR.

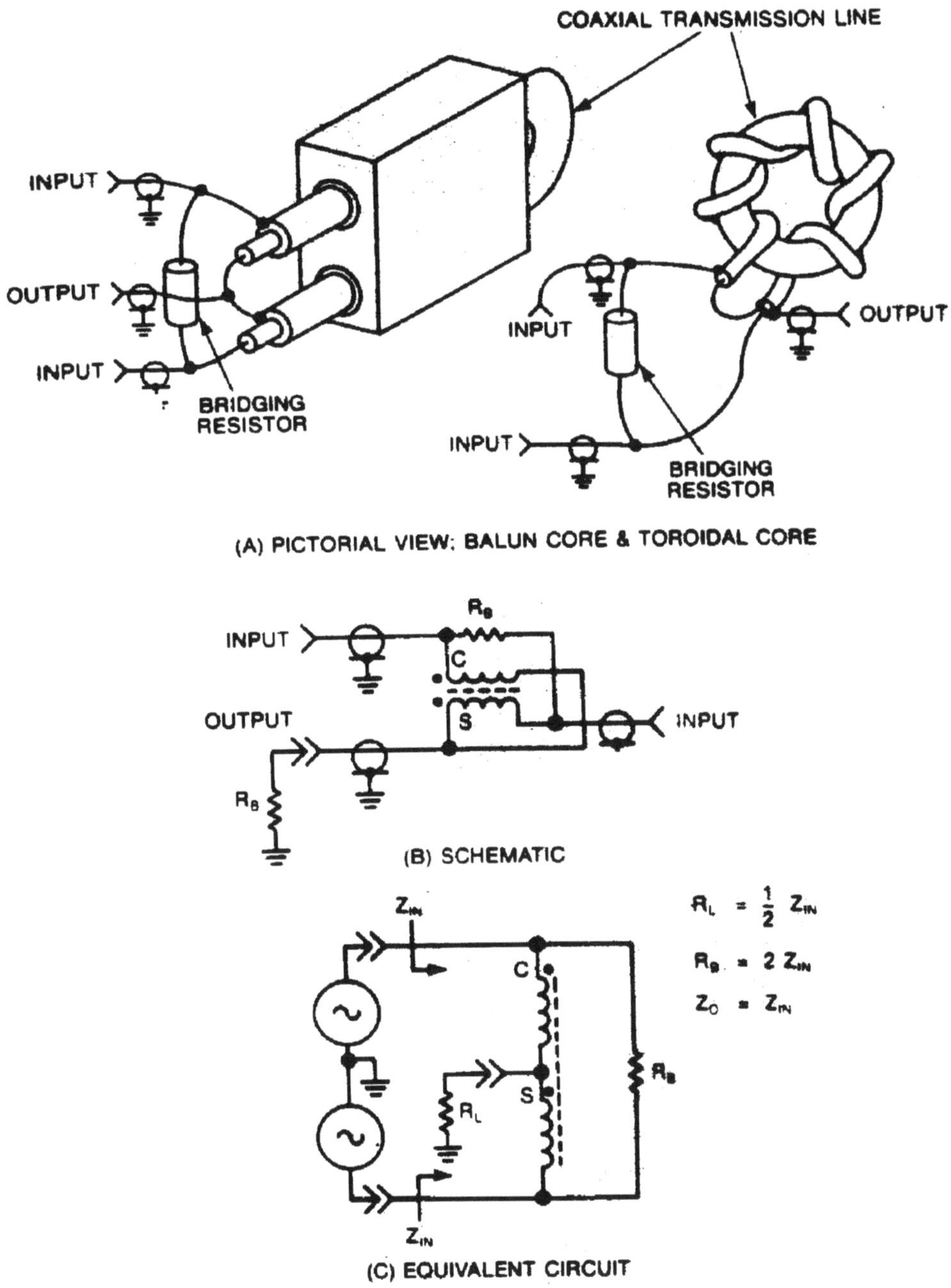

FIGURE 15 TYPE I IN-PHASE TWO PORT COMBINER

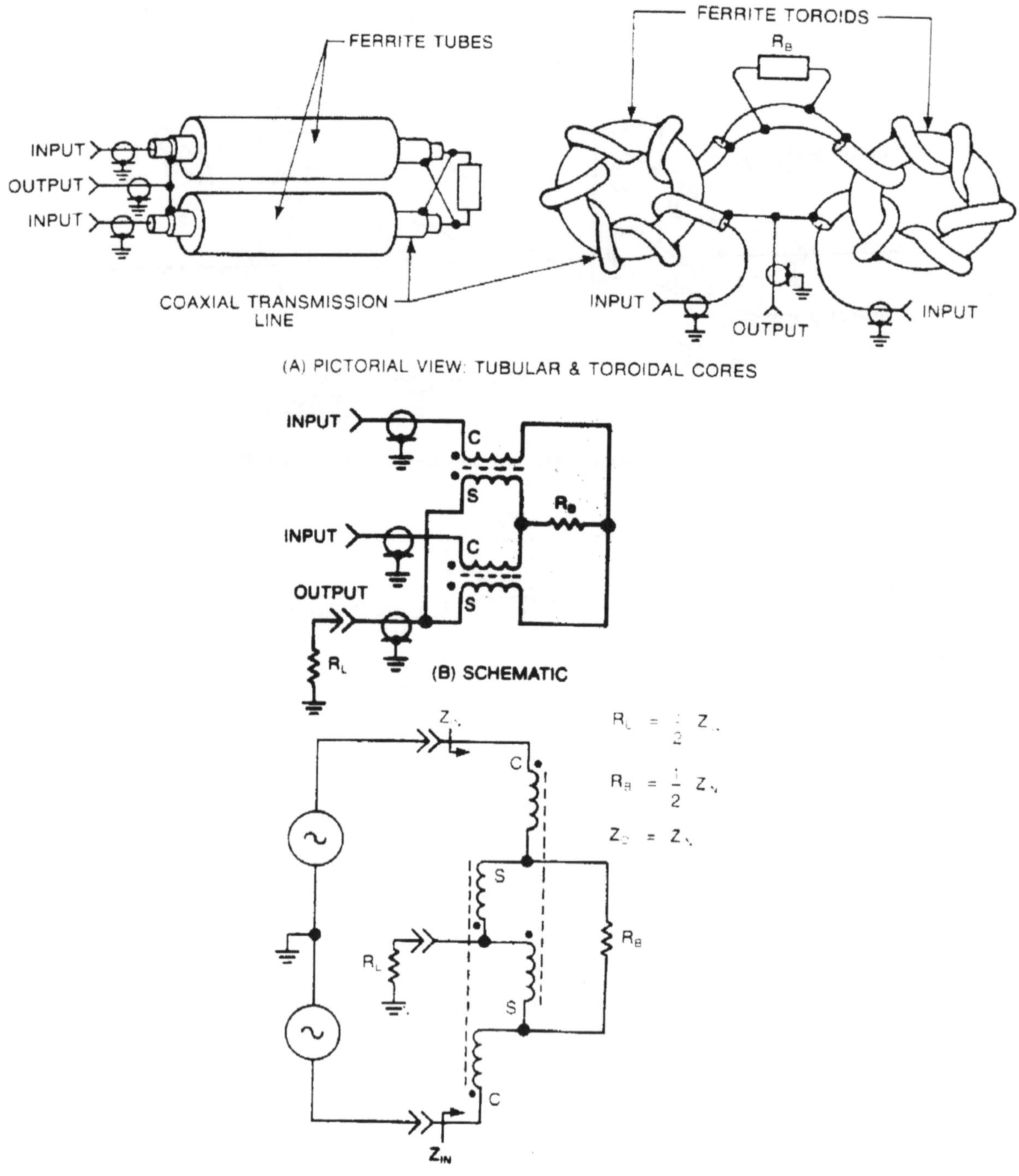

FIGURE 16 TYPE II IN-PHASE TWO-PART COMBINER

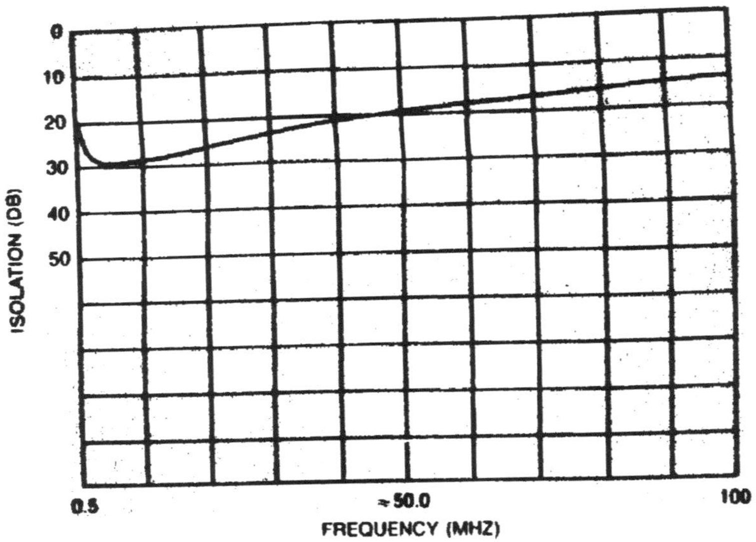

(A) PORT-TO-PORT ISOLATION, TYPE I IN-PHASE COMBINER

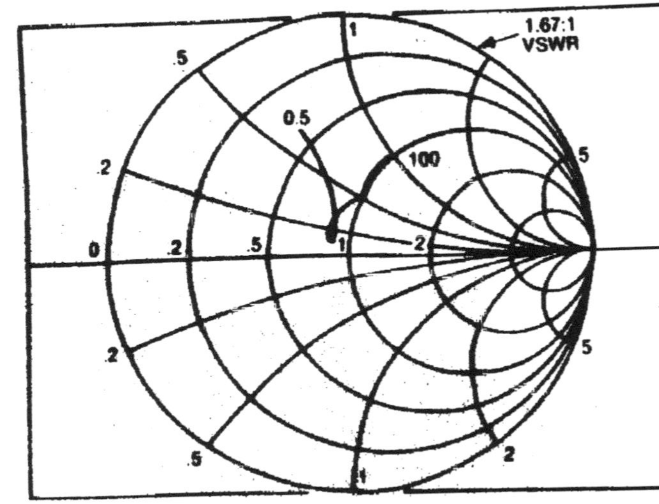

(B) INPUT IMPEDANCE, TYPE I IN-PHASE COMBINER

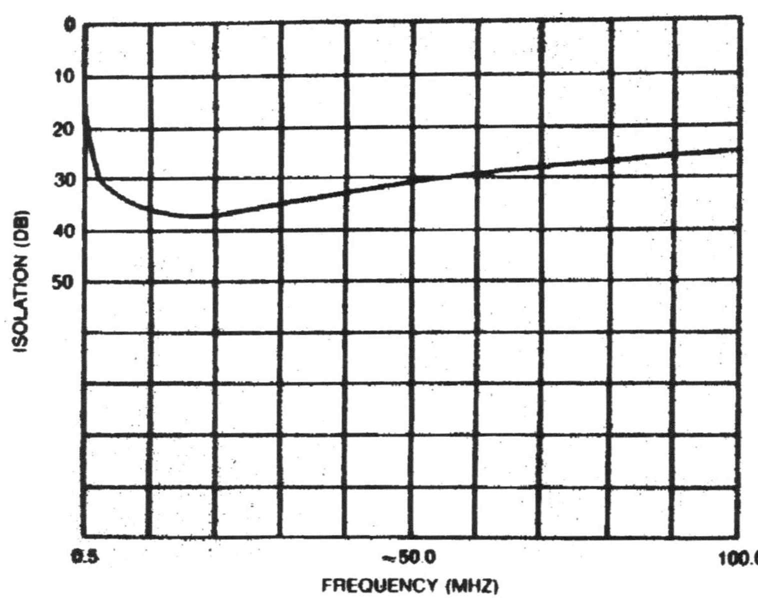

(C) PORT-TO-PORT ISOLATION, TYPE II IN-PHASE COMBINER

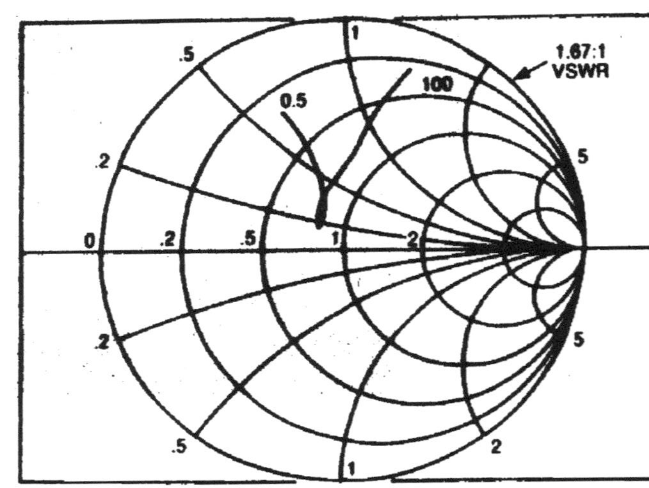

(D) INPUT IMPEDANCE, TYPE II IN-PHASE COMBINER

The combiner output load impedance is usually transformed to another desired value such as 50 or 75 ohms. This is readily accomplished by one of the wideband transformers described earlier. Usually the output impedance transformer is physically integrated into the combiner assembly. The interconnection between combiner and transformer can be made using micro-strip line techniques if it is not a standard coax impedance

Theoretically any number of inputs may be combined with an in-phase combiner, but a practical limit is reached when the output impedance becomes too low to allow efficient wideband transformation back to the desired load impedance. An example of a type-II four port in-phase combiner is given in figure 18.

Four-port combiners may also be implemented by cascading two port combiners. This technique is illustrated in figure 19 for both types of two port combiners.

In-phase combiners all use a floating bridging resistor. This may be difficult to implement, especially in combiners handling high power. A wideband balun transformer allows using a single-ended or unbalanced load. The balun could also transform the balanced impedance to 50 or 75 ohms. Standard coaxial dummy loads, connected to the combiner with coax cable, may then be used as bridging resistors.

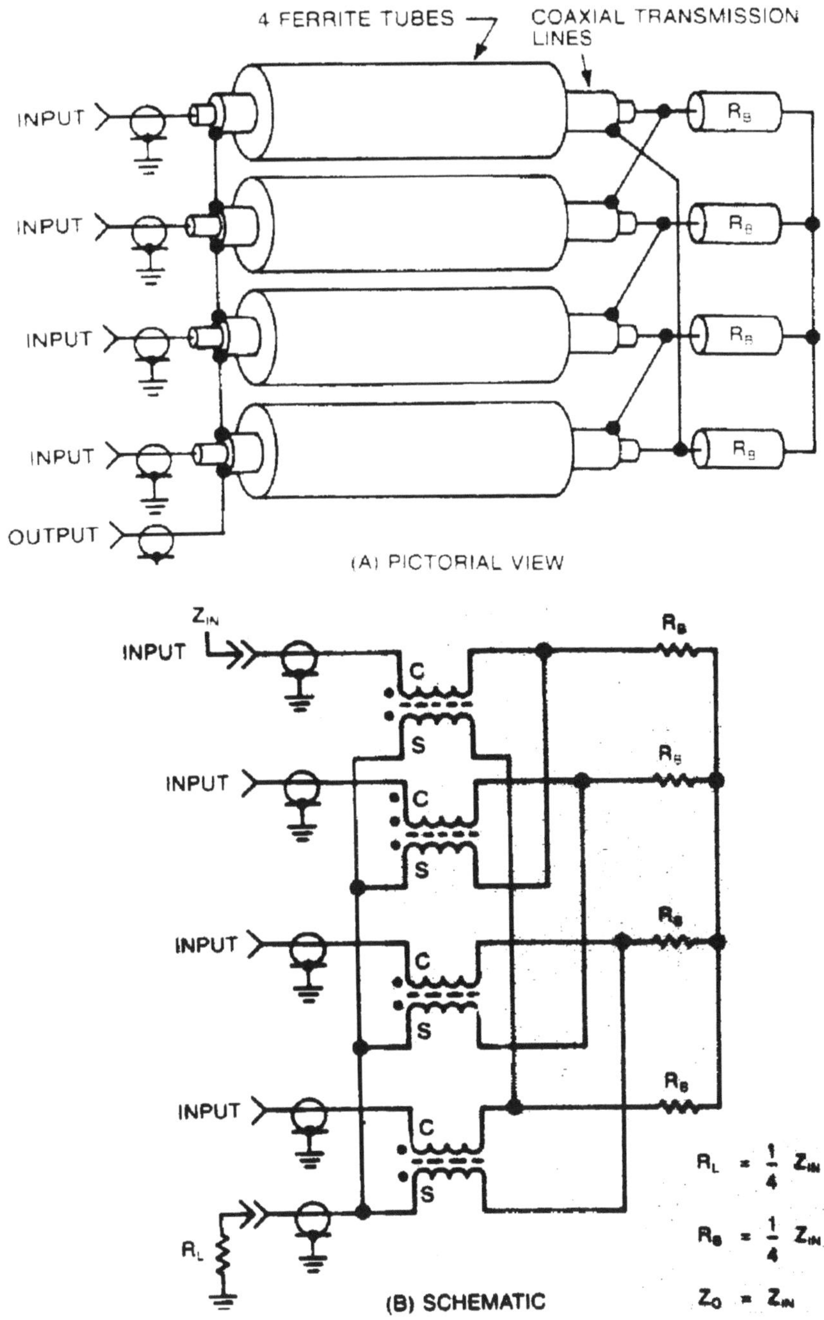

FIGURE 18 TYPE II FOUR PORT IN-PHASE COMBINER

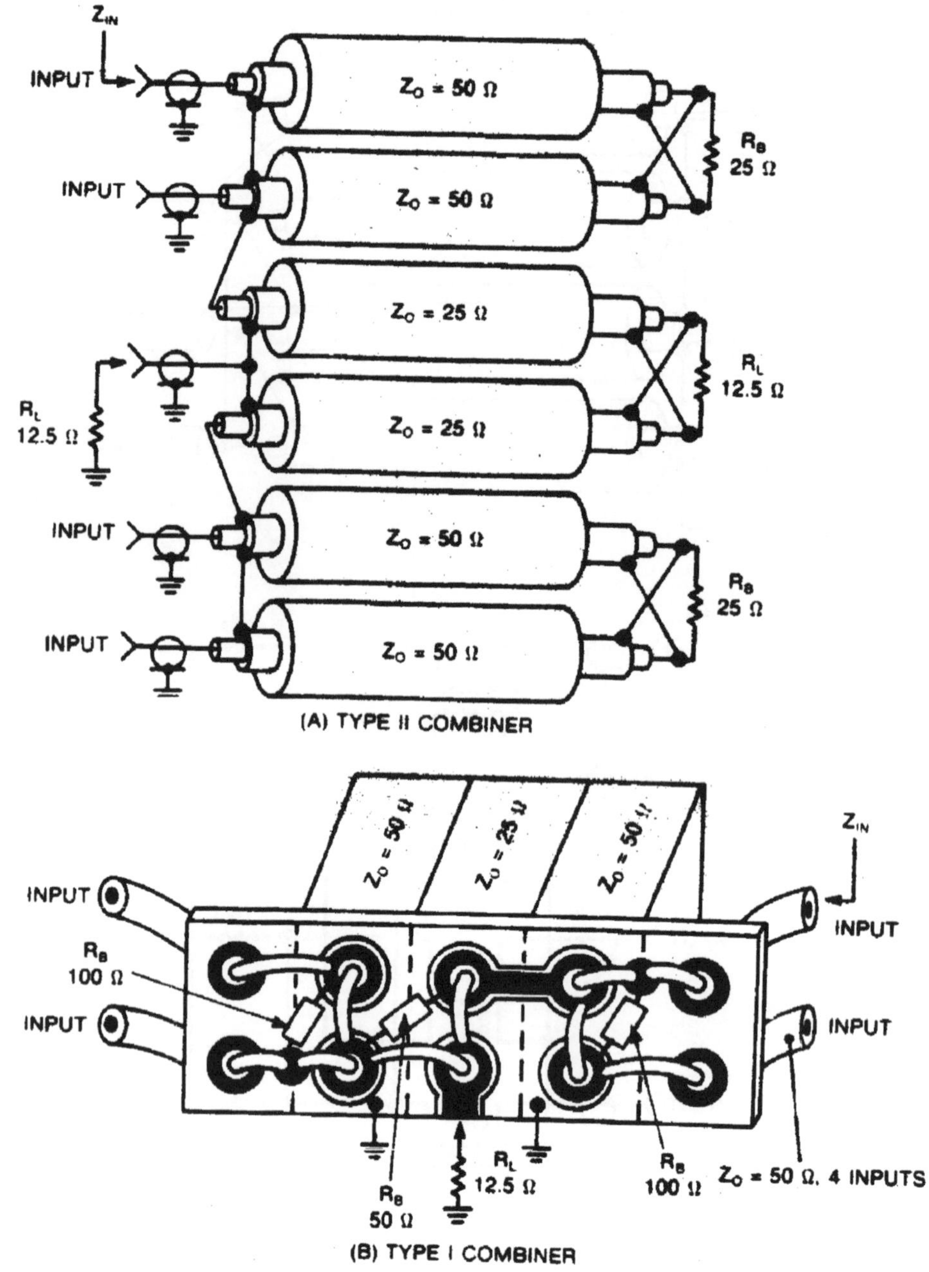

FIGURE 19 FOUR PORT COMBINERS IMPLEMENTED WITH TWO PORT IN-PHASE HYBRIDS

180-DEGREE HYBRIDS

If the roles of the bridging resistor and the load are interchanged, the result is a 180-degree hybrid combiner. The two input signals must be 180-degrees out of phase and of equal amplitude. The output is balanced to ground unless the usual balun is used. Examples of type-I and type-II 180-degree hybrid combiners with output baluns are shown in figures 20 and 21.

Many unique combiner designs are possible by using various combinations of basic combiner types and balun transformers. The combiner described in figures 22 (pictorial) and 23 schematic) is an example of a four-port combiner using two each type-I in-phase combiners (cores A and F) and two each, parallel connected type-II 180-degree hybrid combiners (cores D and C) and a 4:1 balun transformer (cores B and E) to couple the combined output to a 50-ohm load. Connecting two 180-degree hybrids in parallel avoids using 25-ohm coax cable and provides the extra core material to handle the higher rf power.

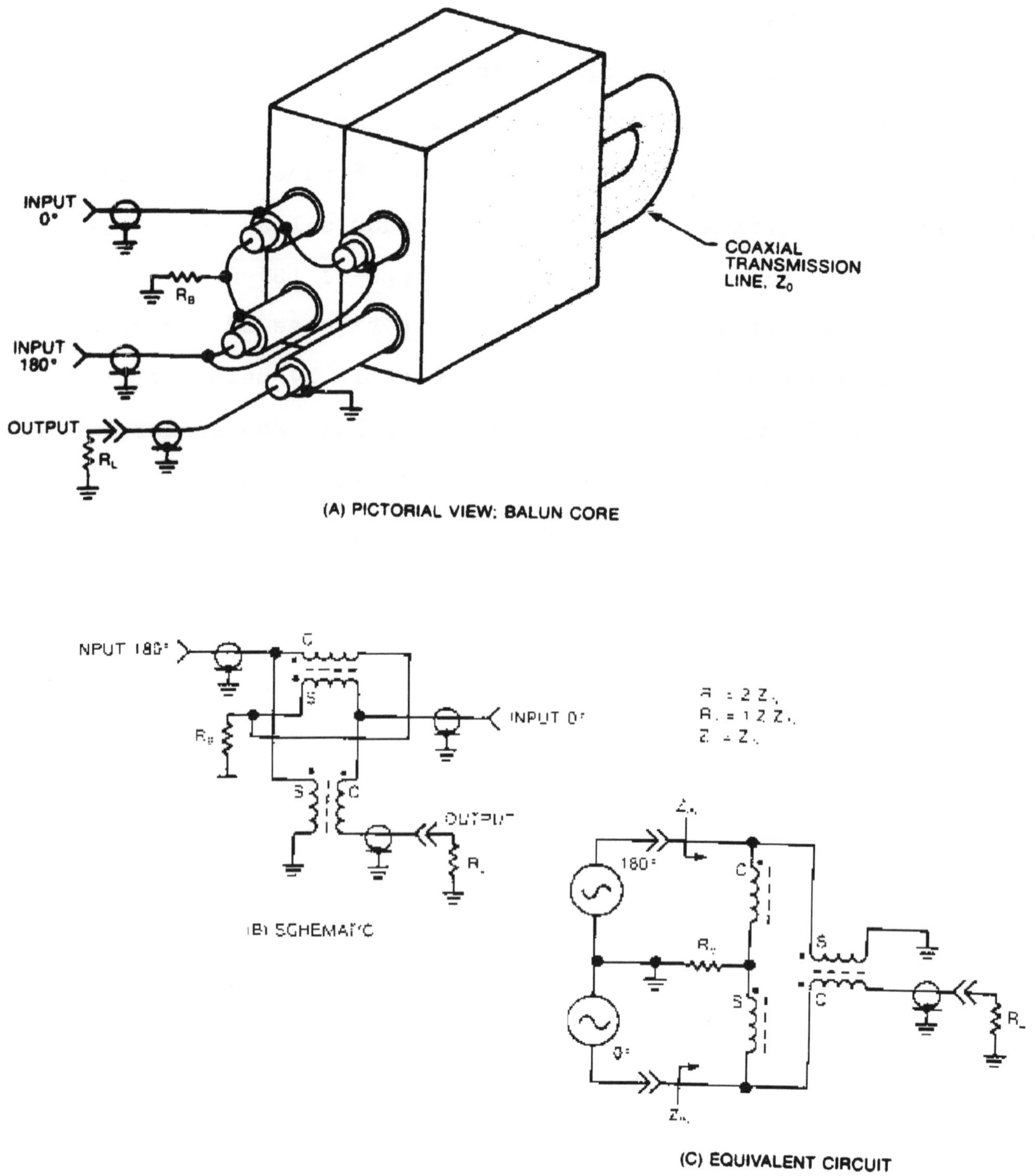

FIGURE 20. TYPE I 180-DEGREE HYBRID COMBINER

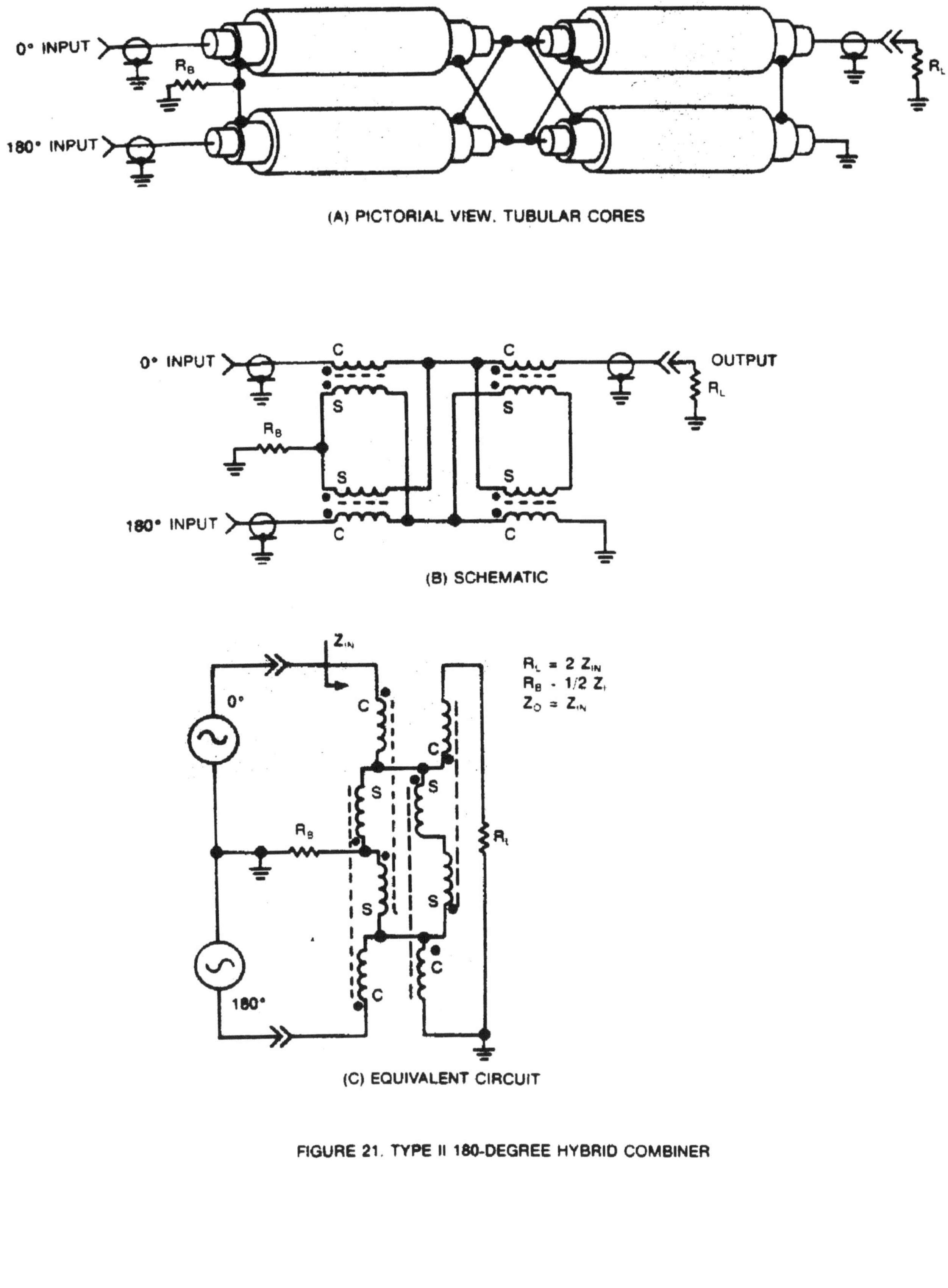

FIGURE 21. TYPE II 180-DEGREE HYBRID COMBINER

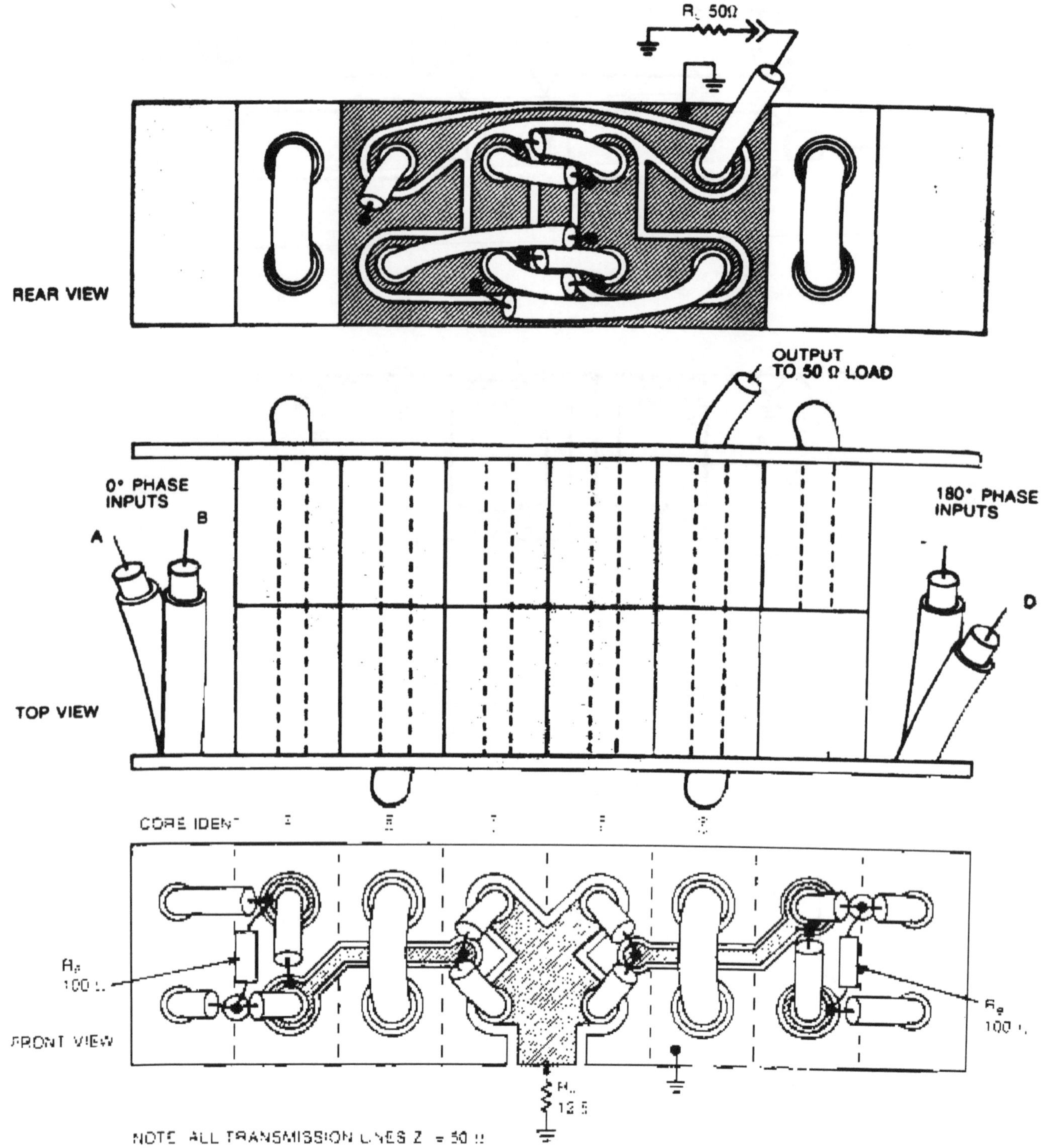

FIGURE 22. FOUR PORT, TWO STAGE COMBINER USING TYPE I AND II HYBRIDS; PICTORIAL

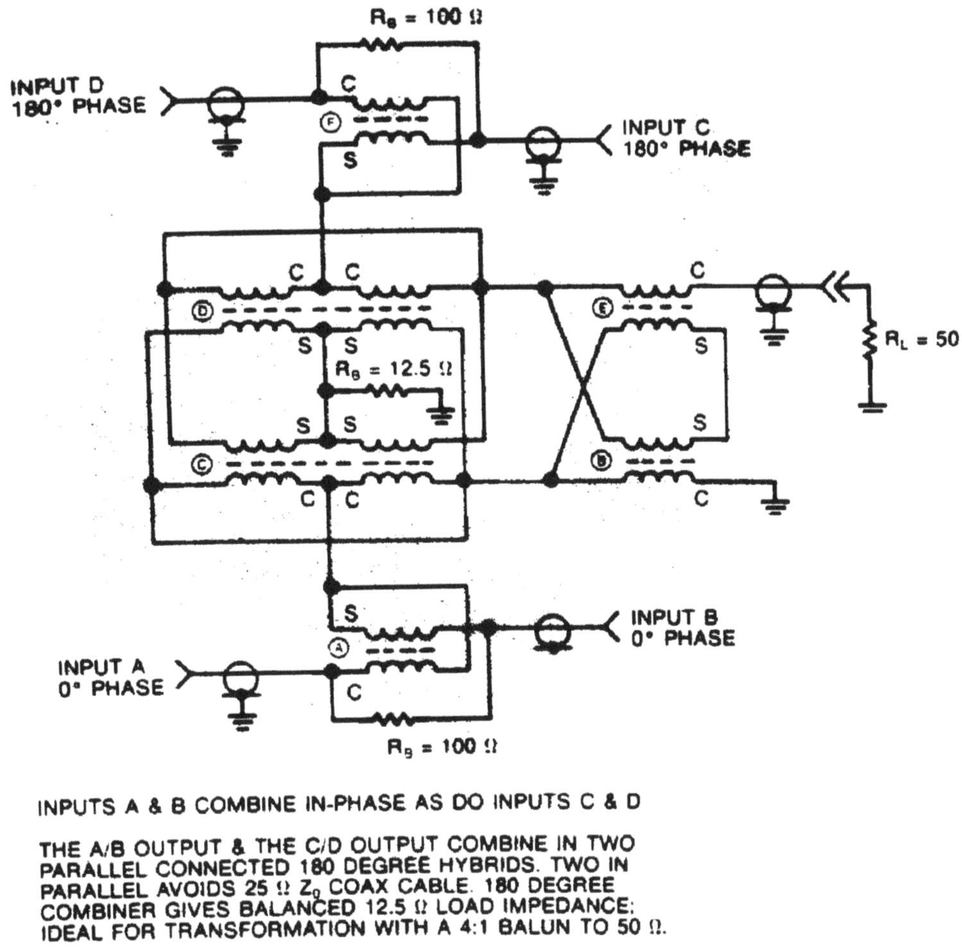

FIGURE 23. FOUR PORT, TWO STAGE COMBINER USING TYPE I & II HYBRIDS: SCHEMATIC

QUADRATURE HYBRIDS

The quadrature hybrid has two input ports, each of equal amplitude but one is 90-degrees out of phase relative to the other. Four-phase combining of four amplifier modules is feasible using two quadrature hybrids and a 180-degree combiner

The quadrature hybrid is constructed by using "all-pass"

networks and a wideband hybrid as shown in the block diagram of figure 24. Two all-pass networks are required; one for 0-degree (reference) phase shift and the other for 90-degree phase shift relative to the reference output. The absolute phase shift across an all-pass network changes with frequency, however, the two networks are designed to maintain a constant 90-degree phase difference between their outputs as the input frequency to both networks is varied.

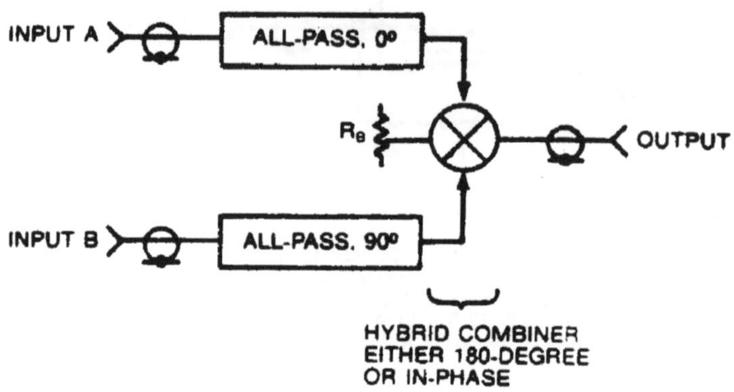

FIGURE 24. BLOCK DIAGRAM OF A QUADRATURE COMBINER

The all-pass networks may either be balanced or unbalanced circuits. Typical circuit topologies for both are shown in figure 25. Note that the mutual coupling in the unbalanced network must be negative and of a prescribed value

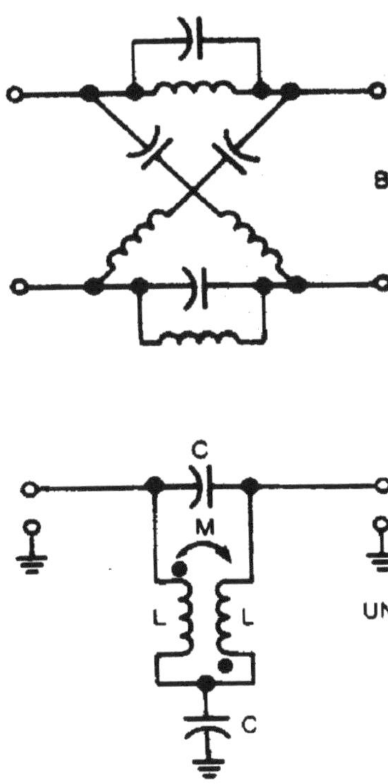

BALANCED ALL-PASS NETWORK

UNBALANCED ALL-PASS NETWORK

FIGURE 25. TYPICAL ALLPASS NETWORK TOPOLOGIES

Both circuits exhibit difficulties in practical implementation. The balanced lattice network may require leadlengths and possibly a balun transformer for interface to an unbalanced hybrid. It requires more components than an equivalent unbalanced network and the component values must be closely matched to achieve low VSWR across the design bandwidth.

Implementation of an unbalanced all pass network allows shorter lead lengths and eliminates the balun transformer. It requires fewer components than an equivalent balanced all network. No closely matched component values are required, however, an exact amount of mutual coupling is required between two inductors.

The quadrature hybrid offers three advantages over the in-phase and 180-degree hybrids when used as an output combiner in solid state wideband power amplifiers. The third harmonic and certain other odd order harmonics cancel in the output port and add in the bridging resistor. The all pass phase shift networks and the basic combiner specifications must hold up to the frequency of the highest harmonic of concern to achieve this in practice. For example, the all-pass networks must provide the 90-degree phase difference up to at least 90 MHz in order to cancel the third harmonic of a 30 MHz fundamental signal in the hybrid's output.

RF power flowing into the output port of a quadrature hybrid will split, go through the all pass networks, partially reflect at the signal source impedance, go back through the all pass networks, and cancel in the output port and add across the bridging resistor. This is happens whether the power flowing into the output port is the result of a mismatched load or coupling from an adjacent transmitting antenna. The result is the combined power amplifier output behaves as though it has a matched source impedance. The situation of reverse power flow from adjacent transmitter coupling is especially important since the two signals cross modulate each other in the active devices. The intermodulation products caused would be radiated along with the desired signal. The quadrature hybrid will cancel <u>some</u> of these "backdoor" IMD products in it's output port by terminating the energy in the bridging resistor

Two wideband linear power amplifiers, combined with a quadrature hybrid, will exhibit nearly constant gain under varying load impedances (varying VSWR). The ratio of input power into the quadrature hybrid splitter to the <u>forward</u> power out of the quadrature hybrid combiner will be nearly constant in contrast to the same situation using in-phase or 180-degree hybrids.

Techniques for wideband RF power transformers, combiners, and splitters have been presented with emphasis on topology. Various types of transformers and combiners were examined and classified. Examples of each were presented in pictorial form, schematic, and equivalent circuit in an effort to bridge the gap from theory to working hardware. A rich variety of combinations of the basic transformers and combiners are possible though not covered here. It is hoped that a more complete understanding of the basic types presented here will enable the reader to produce more sophisticated designs.

BIBLIOGRAPHY

Badger, G., "A New Class of Coaxial-line Transformers", Ham Radio, Part 1 Feb. 1980, pp. 12-18; Part 2 Mar. 1980, pp. 18-29

2. Benjamin, J.A., "Use Hybrid Junctions for more VHF Power" Electronic Design, Aug. 1, 1968, pp. 54-59.

3. Bruene, W.B., "Quadrature Hybrid Behavior When Used as a Power Amplifier Combiner", Rockwell International Working Paper, Feb. 27, 1980.

4. Burwasser, A.J., "Wideband Monofilar Autotransformers", R.F. Design, Part 1 Jan./Feb. 1981, pp. 38-44, Part 2 Mar./Apr. 1981, pp. 20-29.

5. Burwasser, A.J., "Taking the Magic Out of the "Magic Tee"" R.F. Design, May/June 1983, pp. 44-60.

6. Granberg, H.O., "Broadband Transformers and Power Combining Techniques for RF", Motorola Application Note AN-749, 1975.

7 Granberg, H.O., "Combine Power Without Compromising Performance", Electronic Design, July 19, 1980, pp. 181-187

8. Granberg, H.O., "Four MOSFETs Deliver 600 W of RF Power Microwaves & RF, Jan. 1983, pp. 89-91 & 120

9. Krauss, H.L. & Allen, C.W., "Designing Toroidal transformers to Optimize Wideband Performance", Electronics, Aug. 16, 1973, pp. 113-116.

10. Lefferson, P., "Twisted Magnet Wire Transmission Line", IEEE Transactions on Parts, Hybrids, and Packaging, Vol. PHP-7 No. 4, Dec. 1971, pp. 148-154.

11. Lewis. W.A., "Low-Impedance Broadband Transformer Techniques in the HF and VHF Range", Collins Radio Co. Working Paper, July 1, 1965.

12. Munk, P.R. & Sartori, E.F., "A Theoretical and Experimental Study of Transformer Balance", IEEE Transactions on Parts, Materials and Packaging, Vol. PMP-4, No.1, Mar. 1968, pp. 12-21.

13. Nagle, J.J., "High-performance Broadband Balun", Ham Radio, Feb. 1980, pp. 28-34.

14. Nagle, J.J., "Testing Baluns", Ham Radio, Aug. 1983, pp. 30-39.

15. Pitzalis. O. & Couse, T.P.M., "Broadband Transformer Design for RF Transistor Power Amplifiers", US Army Electronics Command, ECOM-2989, July 1968.

16. Ruthroff, C.L., "Some Broadband Transformers", Proceedings, Vol. 47, August 1959, pp. 1337-1342.

17. Sevick, J., "Broadband Matching Transformers Can Handle Many Kilowatts", Electronics, Nov. 25, 1976, pp. 123-128

18. Waight, M., "Report on Quadrature Hybrid Investigation Aug.1985", Unpublished.

AN-758
Application Note

A TWO-STAGE 1 kW SOLID-STATE LINEAR AMPLIFIER

Prepared by:
Helge O. Granberg
RF Circuits Engineering

Two basic push-pull amplifiers operating from 50 V supplies are discussed in this note. Both amplifiers cover a 2.0 – 30 MHz frequency band. The high power 300 W amplifier is employed as a building block to form a 1 kW system, in which four of these units are combined. The low-power driver amplifier drives the four modules, resulting in a total power gain of over 30 dB.

MOTOROLA *Semiconductor Products Inc.*

A TWO-STAGE 1 kW SOLID-STATE LINEAR AMPLIFIER

INTRODUCTION

This application note discusses the design of 50 W and 300 W linear amplifiers for the 1.6 to 30 MHz frequency band. Both amplifiers employ push-pull design for low, even harmonic distortion. This harmonic distortion and the 50 Vdc supply voltage make the output impedance matching easier for 50-Ohm interface, and permits the use of efficient 1:1 and 4:1 broadband transformers.

Modern design includes integrated circuit bias regulators and the use of ceramic chip capacitors throughout the RF section, making the units easily mass producible.

Also, four 300 W modules are combined to provide a 1 to 1.2 kW PEP or CW output capability. The driver amplifier increases the total power gain of the system to approximately 34 dB.

Although the transistors employed (MRF427 and MRF428) are 100% tested against 30:1 load mismatches, in case of a slight unbalance, the total dissipation ratings may be well exceeded in a multi-device design. With high drive power available, and the power supply current limit set at much higher levels, it is always possible to have a failure in one of the push-pull modules under certain load mismatch conditions. It is recommended that some type of VSWR based protective circuitry be adapted in the equipment design, and separate dc regulators with appropriate current limits provided for each module.

The MRF428 is a single chip transistor with the die size of 0.140 x 0.248", and rated for a power output of 150 W PEP or CW. The single chip design eliminates the problem of selecting two matched die for balanced power distribution and dissipation. The high total power dissipation rating (320 W) has been achieved by decreasing the thermal resistance between the die and the mount by reducing the thickness of the BeO insulator to 0.04" from the standard 0.062", resulting in $R_{\theta JC}$ as low as 0.5°C/W.

The MRF427 is also a single chip device. Its die size is 0.118 x 0.066", and is rated at 25 W PEP or CW. This being a high voltage unit, the package is larger than normally seen with a transistor of this power level to prevent arcing between the package terminals.

The MRF427 and MRF428 are both emitter-ballasted, which insures an even current sharing between each cell, and thus improving the device ruggedness against load mismatches.

The recommended collector idling currents are 40 mA and 150 mA respectively. Both devices can be operated in Class A, although not specified in the data sheet, providing the power dissipation ratings are not exceeded.

GENERAL DESIGN CONSIDERATIONS

Similar circuit board layouts are employed for the four 300 W building block modules and the preamplifier. A compact design is achieved by using ceramic chip capacitors, of which most can be located on the lower side of the board. The lead lengths are also minimized resulting in smaller parasitic inductances and smaller variations from unit-to-unit.

Loops are provided in the collector current paths to allow monitoring of the individual collector currents with a clip-on current meter, such as the HP-428B. This is the easiest way to check the device balance in a push-pull circuit, and the balance between each module in a system such as this.

The power gain of each module should be within not more than 0.25 dB from each other, with a provision made for an input Pi attenuator to accommodate device pairs with larger gain spreads. The attenuators are not used in this device however, due to selection of eight closely matched devices.

In regards to the performance specifications, the following design goals were set:

Devices: 8 x MRF428 + 2 x MRF427A

Supply Voltage: 40 – 50 V

η, Worst Case: 45% on CW and 35% under two-tone conditions

IMD, d_3: -30 dB Maximum (1 kW PEP, 50 V and 800 W PEP, 40 V)

Power Gain, Total: 30 dB Minimum

Gain Variation: 2.0 – 30 MHz: ±1.5 dB Maximum

Input VSWR: 2.0:1 Maximum

Continuous CW Operation, 1 kW: 50% Duty Cycle, 30-minute periods, with heatsink temperature <75°C.

Load Mismatch Susceptibility: 10:1, any phase angle

Determining the figures above is based on previous performance data obtained in test circuits and broadband amplifiers. Some margin was left for losses and phase errors occurring in the power splitter and combiner.

THE BIAS VOLTAGE SOURCE

Figure 1 shows the bias voltage source employed with each of the 300 W modules and the preamplifier. Its basic components are the integrated circuit voltage regulator MC1723C, the current boost transistor Q3 and the temperature sensing diode D1.

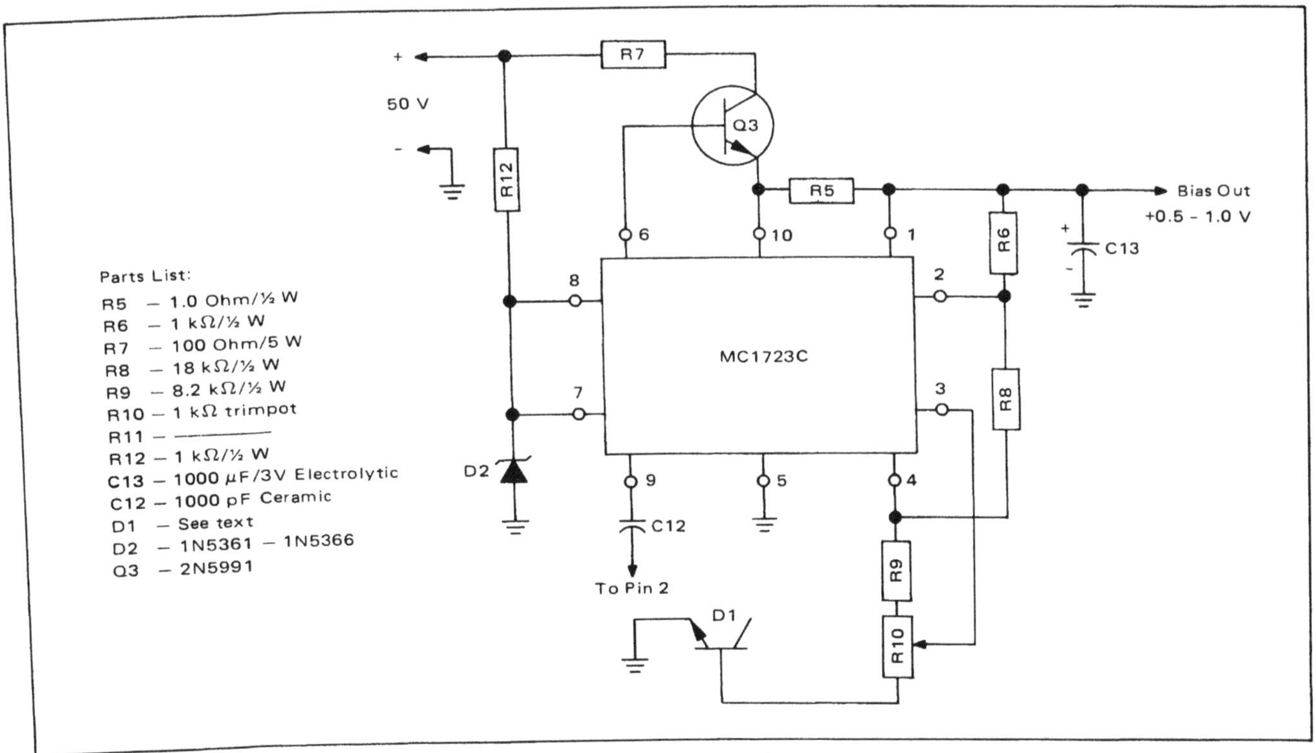

FIGURE 1 – Bias Voltage Source

Although the MC1723C is specified for a minimum V_O of 2 Volts, it can be used at lower levels with relaxed specifications, which are sufficient for this application. Advantages of this type bias source are:

1. Line voltage regulation, which is important if the amplifier is to be operated from various supply voltages.
2. Adjustable current limit.
3. Very low stand-by current drain.

Figure 1 is modified from the circuit shown on the MC1723 data sheet by adding the temperature sensing diode D1 and the voltage adjust element R10. D2 and R12 reduce the supply voltage to a level below 40 V, which is the maximum input voltage of the regulator.

D1 is the base-emitter junction of a 2N5190, in a Case 77 plastic package. The outline dimensions allow its use for one of the circuit board stand-offs, attaching it automatically to the heatsink for temperature tracking.

The temperature compensation has a slight negative coefficient. When the collector idling current is adjusted to 300 mA at 25°C, it will be reduced to 240 – 260 mA at a 60°C heatsink temperature. (-1.15 to -1.7 mA/C°.)

The current limiting resistor R5 sets the limiting to approximately 0.65 A, which is sufficient for devices with a minimum h_{FE} of 17, ($I_B = \frac{I_C}{h_{FE}}$) when the maximum average I_C is 10.9 A. (2 MHz, 50 V, 250 CW.) Typically, the MRF428 h_{FE}'s are in the 30's.

The measured output voltage variations of the bias source (0 – 600 mA) are ±5 to 7 mV, which amounts to a source impedance of approximately 20 milliohms.

THE 300 W AMPLIFIER MODULE

Input Matching

Due to the large emitter periphery of the MRF428, the series base impedance is as low as 0.88, -J.80 Ohm at 30 MHz. In a push-pull circuit a 16:1 input transformer would provide the best impedance match from a 50-Ohm source. This would however, result in a high VSWR at 2 MHz, and would make it difficult to implement the gain correction network design. For this reason a 9:1 transformer, which is more ideal at the lower frequencies, was chosen. This represents a 5.55 Ohm base-to-base soure impedance.

In a Class C push-pull circuit, where the conduction angle is less than 180°, the base-to-base impedance would be about four times the base-to-emitter impedance of one device. In Class A where the collector idling current is approximately half the peak collector current, the conduction angle is 360°, and the base-to-base impedance is twice the input impedance of one transistor. When the forward base bias is applied, the conduction angle increases and the base-to-base impedance decreases rapidly, approaching that of Class A in Class AB.

A center tap, common in push-pull circuits, is not necessary in the input transformer secondary, if the transistors are balanced. (C_{ib}, h_{FE}, V_{BEf}.) The base current return path is through the forward biased base-emitter

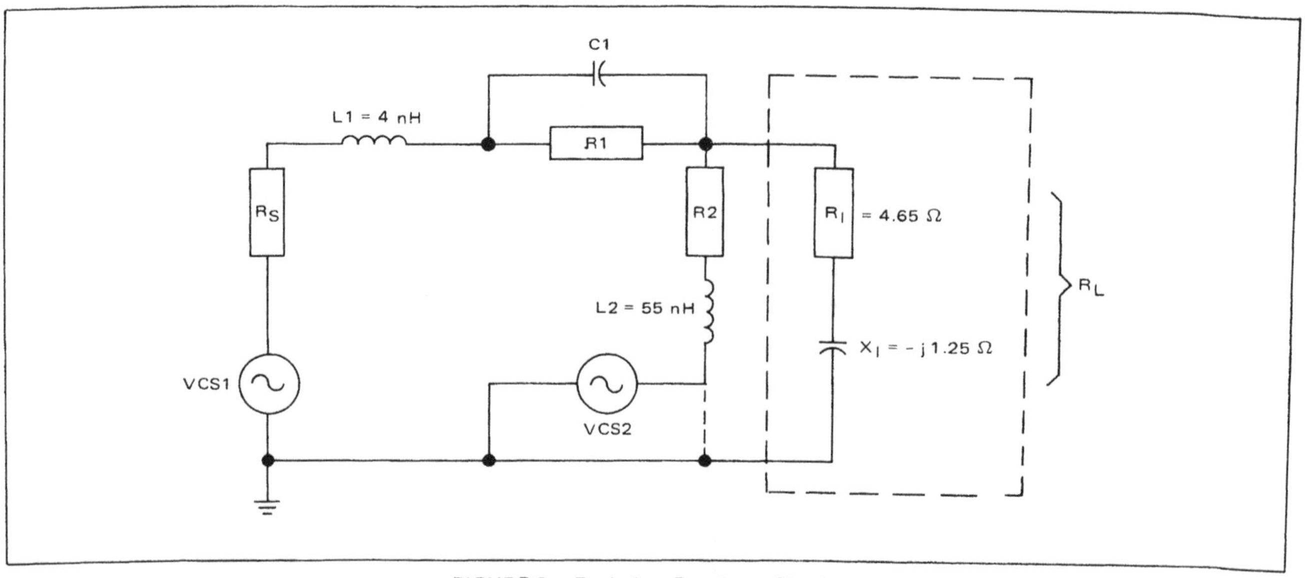

FIGURE 2 – Equivalent Base Input Circuit

junction of the off transistor. This junction acts as a clamping diode, and the power gain is somewhat dependent upon the amount of the bias current. The equivalent input circuit (Figure 2) represents one half of the push-pull circuit, and for calculations R_S equals the total source impedance (R_S') divided by two.

Since a junction transistor is a current amplifier, it should ideally be driven from a current source. In RF applications this would result in excessive loss of power gain. However, input networks can be designed with frequency slopes having some of the current source characteristics at low frequencies, where excess gain is available.

The complex base input characteristics of a transistor would place requirements for a very sophisticated input compensation network for optimum overall performance. The design goal here was to maintain an input VSWR of 2:1 or less and a maximum gain variation of ±1.5 dB from 2 to 30 MHz. Initial calculations indicated that these requirements can be met with a simple RC network in conjunction with negative collector-to-base feedback. Figure 2 shows this network for one device. L1 and L2 represent lead lengths, and their values are fixed. The feedback is provided through R2 and L2. Because the calculations were done without the feedback, this branch is grounded to simulate the operating conditions.

The average power gain variation of the MRF428 from 2 to 30 MHz is 13 dB. Due to phase errors, a large amount of negative feedback in an RF amplifier decreases the linearity, or may result in instabilities. Experience has shown that approximately 5 – 6 dB of feedback can be tolerated without noticeable effects in linearity or stability, depending upon circuit layout. If the amount of feedback is 5 dB, 8 dB will have to be absorbed by the input network at 2 MHz.

Omitting the reactive components, L1, L2, C1, and the phase angle of X_I which have a negligible effect at 2 MHz, a simple L-pad was calculated with $R_S = 2.77\ \Omega$, and $R_L = \sqrt{4.65^2 + 1.25^2} = 4.81\ \Omega$.

From the device data sheet we find the G_{PE} at 2 MHz is about 28 dB, indicating 0.24 W at R_L will produce an output power of 150 W, and the required power at R_S = 0.24 W + 8 dB = 1.51 W.

Figuring out currents and voltages in various branches, results in: R1 = 1.67 Ω and R2 = 1.44 Ω.

The calculated values of R1 and R2 along with other known values and the device input data at four frequencies were used to simulate the network in a computer program. An estimated arbitrary value of 4000 pF for C1 was chosen, and VCS2 represents the negative feedback voltage (Figure 2.) The optimization was done in two separate programs for R1, R2, C1 and VCS2 and in several steps. The goals were: a) VCS and R2 for a transducer loss of 13 dB at 2 MHz and minimum loss at 30 MHz. b) R1 and C1 for input VSWR of <1.1:1 and <2:1 respectively. The optimized values were obtained as:

C1 = 5850 pF	R2 = 1.3 Ω
R1 = 2.1 Ω	VCS2 = 1.5 V

The minimum obtainable transducer loss at 30 MHz was 2.3 dB, which is partly caused by the highest reflected power at this frequency, and can be reduced by "over-compensation" of the input transformer. This indicates that at the higher frequencies, the source impedance (R_S) is effectively decreased, which leaves the input VSWR highest at 15 MHz.

In the practical circuit the value of C1 (and C2) was rounded to the nearest standard, or 5600 pF. For each half cycle of operation R2 and R4 are in series and the value of each should be $\frac{1.3\ \Omega}{2}$ for VCS2 = 1.5 V. Since the voltage across ac and bd = V_{CE}, a turns ratio of 32:1 would be required. It appears that if the feedback voltage

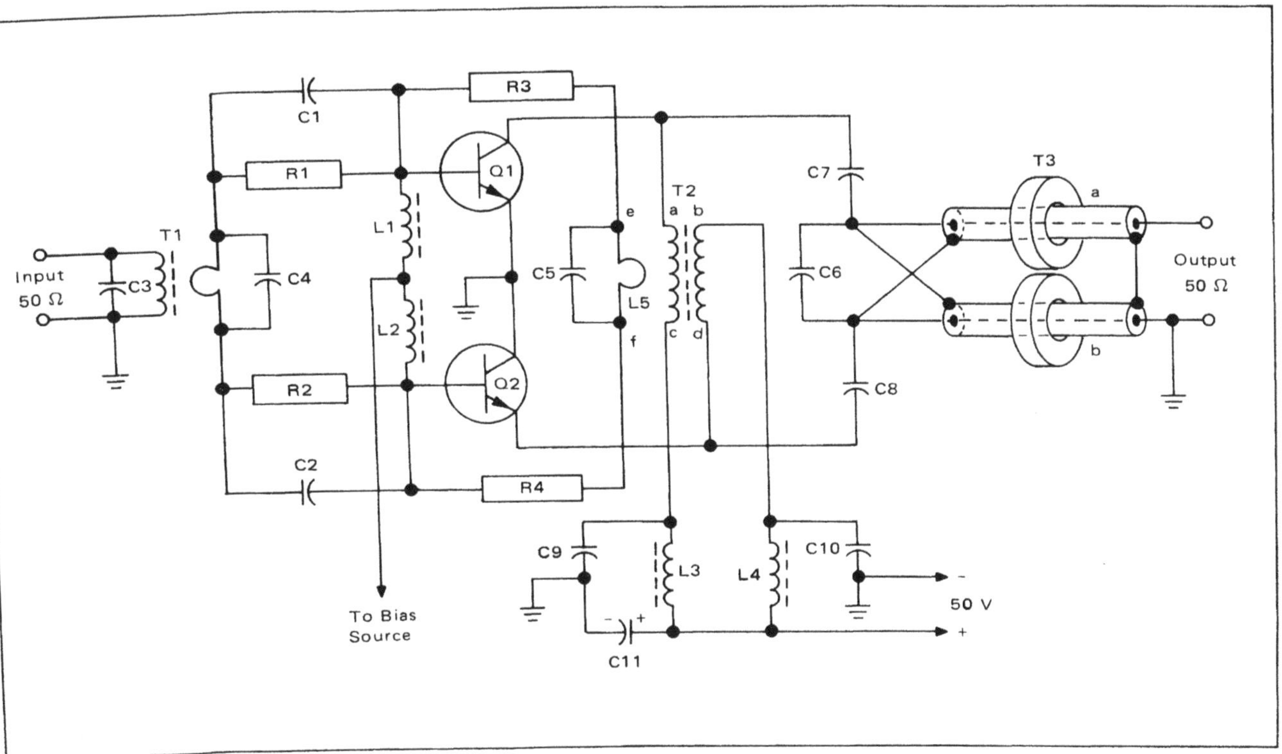

FIGURE 3

on the bases remains unchanged, the ratio of the voltage across L5 (VCS2) and R2R4 can be varied with only a small effect to the overall input VSWR. To minimize the resistive losses in the bifilar winding of T2 (Figure 3), the highest practical turns ratio should not be much higher than required for the minimum inductance, which is

$$\frac{4R}{2\pi f} = \frac{50}{12.5} = 4.0\ \mu H.$$

R = Collector-to-Collector Impedance = 12.5 Ω

f = 2 MHz

ac or bd will then be 1.0 μH, which amounts to 5 turns. (See details on T2.) 25% over this represents a 7:1 ratio setting VCS2 to 6.9 V.

In addition to providing a source for the negative feedback, T2 supplies the dc voltage to the collectors as well as functions as a center tap for the output transformer T3.

The currents for each half cycle are in opposite phase in ac and bd, and depending on the coupling factor between the windings, the even harmonic components will see a much lower impedance than the fundamental. The optimum line inpedance for ac, bd would equal the collector-to-collector impedance, but experiments have shown that increasing this number by a factor of 2-3 affects the 2nd and 4th harmonic amplitudes by only 1 to 2 dB.

Since the minimum gain loss obtainable at 30 MHz with network as in Figure 2, and the modified VCS2 source was about 3.8 dB at 30 MHz, C5 was added with the following in mind: C5 and L5 form a parallel resonant circuit with a Q of approximately 1.5. Its purpose is to increase the shunting impedance across the bases, and to disturb the 180° phase difference between the input signal and the feedback voltage at the higher frequencies. This reduces the gain loss of 3.8 dB, of which 1.4 dB is caused by the feedback at 30 MHz. The amount depends upon the resonant frequency of C5 L5, which should be above the highest operating frequency, to avoid possible instabilities.

When L5 is 45 nH, and the resonance is calculated for 35 MHz, the value of C5 becomes 460 pF, which can be rounded to the closest standard, or 470 pF. The phase shift at 30 MHz is:

$$\tan^{-1}\left[\frac{2\pi f L}{R\left(1 - \frac{f^2}{f_0^2}\right)}\right] = \tan^{-1}\left[\frac{6.28 \times 30 \times 0.045}{6.8\left(1 - \frac{900}{1225}\right)}\right]$$

$$= \tan^{-1}\left(\frac{8.48}{1.80}\right) = 78.0^\circ$$

$$\text{The impedance is: } \frac{R}{\cos\theta} = \frac{6.8}{\cos 78^\circ} = 32.7\ \Omega$$

At 2 MHz the numbers are respectively 4.76° and 6.83 Ω.

The 1.4 dB feedback means that the feedback voltage is 16% of the input voltage at the bases. By the aid of vectors, we can calculate that the 78° phase shift and the increased impedance reduces this to 4%, which amounts to 0.35 dB. These numbers were verified in another computer program with VCS2 = 6.9 V, and including C5. New values for R1 and R2 were obtained as 1.95 Ω and 6.8 Ω respectively, and other data as shown in Table 1.

The VSWR was calculated as

$$\frac{Z1 - Z2}{Z1 + Z2} \qquad \text{where:}$$

Z1 = Impedance at transformer secondary.

TABLE1:

Frequency MHz	Input VSWR	Input Impedance Real	Input Impedance Reactive	Attenuation dB
2.0	1.07	2.79	-0.201	13.00
4.0	1.16	2.66	-0.393	12.07
7.5	1.33	2.35	-0.615	10.42
15	1.68	1.77	-0.611	7.40
20	1.82	1.57	-0.431	5.90
30	1.74	1.62	-0.21	2.70

Although omitted from the preliminary calculations, the 2 x 5 nH inductances, comprising of lead length, were included in this program.

The input transformer is a 9:1 type, and uses a television antenna balun type ferrite core, made of high permeability material. The low impedance winding consists of one turn of 1/8" copper braid. The sections going through the openings in the ferrite core are rounded to resemble two pieces of tubing electrically. The primary consists of AWG #22 TFE insulated wire, threaded through the rounded sections of braid, placing the primary and secondary leads in opposite ends of the core. (4) (5). The saturation flux density is about 60 gauss which is well below the limits for this core. For calculation procedures, see discussion about the output transformer.

This type physical arrangement provides a tight coupling, reducing the amount of leakage flux at high frequencies. The wire gauge, insulation thickness, and number of strands have a minimal effect in the performance except at very high impedance ratios, such as 25:1 and up. The transformer configuration is shown in Figure 4. By using a vector impedance meter, the values for C3 and C4 were measured to give a reasonable input match at 30 MHz, (Z_{in} = 1.62 - j 0.21 x 2 = 3.24 - j 0.42) with the smallest possible phase angle.

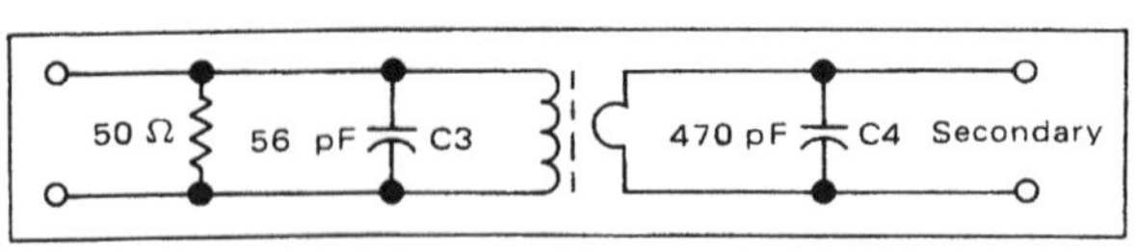

FIGURE 4 – Transformer Configuration

When the high impedance side was terminated into 50 Ω, the following readings were obtained at the secondary:

Z2 = Input impedance of compensation network x 2 (R_S in Figures 2 and 3) as in computer data presented ahead.

The effect of the lower VSWR to the power loss in the input network can be calculated as follows:

$$10 \text{ Log} \left[\frac{\left(1 - \left(\frac{S1 - 1}{S1 + 1}\right)^2\right)}{\left(1 - \left(\frac{S2 - 1}{S2 + 1}\right)^2\right)} \right]$$

where:
S1 = VSWR 1 (Lower)
S2 = VSWR 2 (Higher)

$$\text{which at 30 MHz} = 10 \text{ Log} \frac{\left(1 - \left(\frac{1.11 - 1}{1.11 + 1}\right)^2\right)}{\left(1 - \left(\frac{1.74 - 1}{1.74 + 1}\right)^2\right)}$$

$$= 10 \text{ Log} \left(\frac{0.997}{0.927}\right) = 0.32 \text{ dB}, 2.7 - 0.32 = 2.38 \text{ dB}$$

These figures for other frequencies are presented with the data below. Later, some practical experiments were done with moving the resonance of C5 L5 lower, to find out if instabilities would occur in a practical circuit. When the resonance was equal to the test frequency, slight break-up was noticed in the peaks of a two-tone pattern. It was then decided to adjust the resonance to 31 MHz, where C5 = 560 pF, and the phase angle at 30 MHz increases to 87°. The transducer loss is further reduced by about 0.2 dB.

Several types of output transformer configurations were considered. The 12.5 Ω collector-to-collector im-

TABLE 2:

Frequency MHz	R_S Ohms	X_S Ohms	VSWR	Attenuation dB
2.0	5.59	+0.095	1.05	12.99
4.0	5.55	+0.057	1.15	12.06
7.5	5.50	+0.046	1.32	10.40
15	4.90	+0.25	1.48	7.28
20	4.32	+0.55	1.38	5.63
30	3.43	+0.73	1.11	2.38

(Above readings with transformer and compensation network.)

A.

B.

FIGURE 5 – Bottom and Top of the 300 W Module Circuit Board

pedance estimated earlier, would require a 4:1 transformer for a 50 Ω output. The type used here as the input transformer exhibits good broad band characteristics with a convenient physical design. However, according to the low frequency minimum inductance formula presented earlier in connection with T2, the initial permeability required would be nearly 3000, with the largest standard core size available. High permeability ferrites are almost exclusively of Nickel-Manganese composition, and are lossy at radio frequencies. Although their Curie points are higher than those of lower permeability Nickel-Zinc ferrites, the core losses would degrade the amplifier performance. With the core losses being a function of the power level, these rules can sometimes be disregarded in low power applications.

A coaxial cable version was adapted for this design, since the transmission line type transformers are theoretically ideal for RF applications, especially in the 1:4 impedance ratio. A balanced to unbalanced function would normally require three separate transmission lines including a balun (5) (6). It appears that the third line can be omitted, if lines a and b (Figure 3) are wound on separate magnetic cores, and the physical length of the lines is sufficient to provide the necessary isolation between the collectors and the load. In accordance to formulas in (7), the minimum line length required at 2 MHz, employing Stackpole 57-9074 or equivalent ferrite toroids is 4.2", and the maximum permissible line length at 30 MHz would be approximately 20". The 4.2" amounts to four turns on the toroid, and measures 1.0 μH, which in series with the second line is sufficient for 2 MHz. Increasing the minimum required line length by a factor of 4 is still within the calculated limits, and in practical measurements the isolation has been found to be over 30 dB across the band. The main advantage with this arrangement is a simplified electrical and physical lay-out.

The maximum flux density of the toroids is approximately 200 gauss (3), and the number of turns has been increased beyond the point where the flux density of the magnetic core is the power limiting factor.

The 1:4 output transformer is not the optimum in this case, but it is the closest practical at these power levels. The optimum power output at 50 V supply voltage and 50 Ω load is:

$V_{RMS} = 4 \times (V_{CC} - V_{CE(sat)} \times 0.707) = 135.75$ V, when $V_{CE(sat)} = 2$ V

$$I = \frac{135.75}{50} = 2.715 \text{ A}, P_{out} = 2.715 \times 135.75 = 368.5 \text{ W}$$

The optimum V_{CC} at $P_{out} = 300$ W would be:

$$V_{CC} = V_{CE(sat)} + (\sqrt{R_{in} \times 2 P_{out}}) = 2 + (\sqrt{6.25 \times 300}) = 45.3 \text{ V}$$

The above indicates that the amplifier sees a lower load line, and the collector efficiency will be lowered by 1-2%. The linearity at high power levels is not affected, if the device h_{FE} is maintained at the increased collector currents. The linearity at low power levels may be slightly decreased due to the larger mismatch of the output circuit.

The required characteristic line impedance (a and b, Figure 3) for a 1:4 impedance transformer is: $\sqrt{R_{in} R_L} = \sqrt{12.5 \times 50} = 25$ Ω, enables the use of standard miniature 25 Ω coaxial cable (i.e., Microdot 260-4118-000) for the transmission lines. The losses in this particular cable at 30 MHz are 0.03 dB/ft. With a total line length of 2 x 16.8" (2 x 4 x 4.2"), the loss becomes 0.084 dB, or

$$300 - \left(\frac{300}{10 \text{ antilog } 0.084 \text{ dB}}\right) = 5.74 \text{ W}.$$

For the ferrite material employed, Stackpole grade #11 (or equivalent Indiana General Q1) the manufacturers data is insufficient for accurate core loss calculations (6). The B_H curves indicate that 100-150 gauss is well in the linear region.

The toroids measure 0.87" x 0.54" x 0.25", and the 16.8" line length figured above, totals to 16 turns if tightly wound, or 12-14 turns if loosely wound. The flux density can then be calculated as:

$$B_{max} = \frac{V_{max} \times 102}{2 \pi f n A}$$

where:
- f = Frequency in MHz
- n = Total number of turns.
- A = Cross sectional area of the toroid in cm^2.
- V = Peak voltage across the 50 Ω load, $\sqrt{\left(\frac{300}{50}\right)} \left(\frac{50}{0.707}\right) = 173$ V

$$B_{max} \text{ (for each toroid)} = \frac{86.5 \times 10^2}{6.28 \times 2 \times 28 \times .25} = 98.3 \text{ gauss}$$

Practical measurements showed the core losses to be negligible compared to the line losses at 2 MHz and 30 MHz. However, the losses increase as the square of B_{max} at low frequencies.

With the amount of HF compensation dependent upon circuit layout and the exact transformer construction, no calculations were made on this aspect for the input (or output) transformers. C3, C4, and C6 were selected by employing adjustable capacitors on a prototype whose values were then measured.

A photo of the circuit board is shown in Figure 5, A-bottom and B-top. The performance data of the 300 W module can be seen in Figure 6.

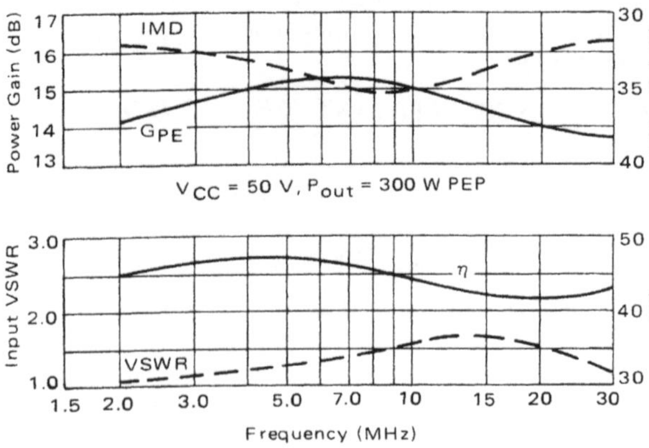

FIGURE 6 – IMD, Power Gain, Input VSWR and Efficiency versus Frequency of a 300 W Module

THE DRIVER AMPLIFIER

The driver uses a pair of MRF427 devices, and the same circuit board layout as the power amplifier, with the exception of the type of the output transformer.

The input transformer is equal to what is used with the power amplifier, but has a 4:1 impedance ratio. The required minimum inductance in the one turn secondary (Figures 3 and 4) being considerably higher in this case,

$$\frac{4R}{2\pi f} = \frac{4 \times 12.5}{12.5} = 4\ \mu H$$

the A_L product of the core is barely sufficient. The measured inductances between a number of cores range 3.8 - 4.1 μH.

This formula also applies to the output transformer, which is a 1:1 balun. The required minimum inductance at 2 MHz is 16 μH, amounting to 11 turns on a Ferroxcube 2616P-A100-4C4 pot core, which was preferred over a toroid because of ease of mounting and other physical features. Although twisted wire line would be good at this power level, the transformer was wound with RG-196 coaxial cable, which is also used later for module-driver interconnections.

The required worst case driver output is 4 x 12 W = 48 W. The optimum P_{out} with the 1:1 output transformer is

$$\frac{V_{RMS}}{50} \times V_{RMS} = \frac{67.7}{50} \times 67.7 = 92\ W.$$

The MRF427 is specified for a 25 W power output. Having a good h_{FE} versus I_C linearity, the 1 to 2 load mismatch has an effect of 2-3 dB in the IMD at the 10% power level, and the reduced efficiency in the driver is insignificant regarding the total supply current in the system.

The component values for the base input network and the feedback were established with the aid of a computer, and information on the device data sheet, as described earlier with the 300 W module. The HF compensation was done in a similar manner as well. Neither amplifier employs LF compensation. C7 and C8 are dc blocking capacitors, and their value is not critical.

In T2 (Figure 7), b and c represent the RF center tap, but are separated in both designs – partly because

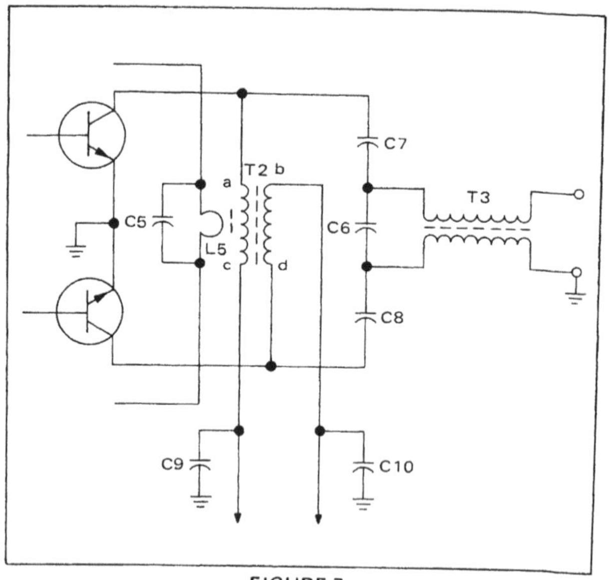

FIGURE 7

of circuit lay-out convenience and partly for stabilization purposes.

The test data of the driver is presented later along with the final test results.

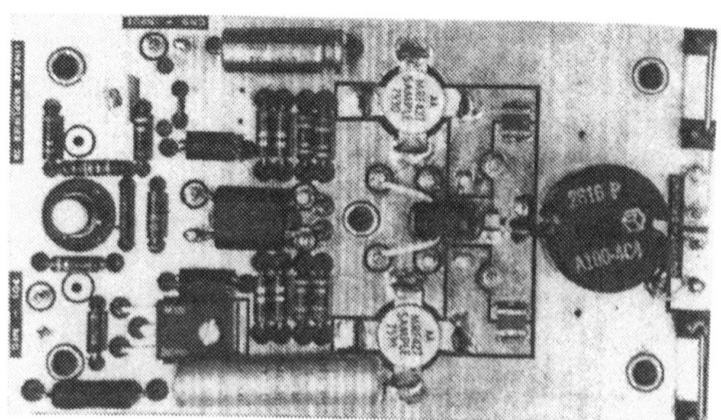

FIGURE 8 – Driver Amplifier Board Layout

COMBINING FOUR 300 W POWER MODULES

The Input Power Divider

The purpose of the power divider is to divide the input power into four equal sources, providing an amount of isolation between each. The outputs are designed for 50 Ω impedance, which sets the common input at 12.5 Ω. This requires an additional 4:1 step down transformer to provide a 50 Ω load for the driver amplifier. Another requirement is a 0° phase shift between the input and the 50 Ω outputs, which can be accomplished with 1:1 balun transformers. (a, b, c and d in Figure 10.) For im-

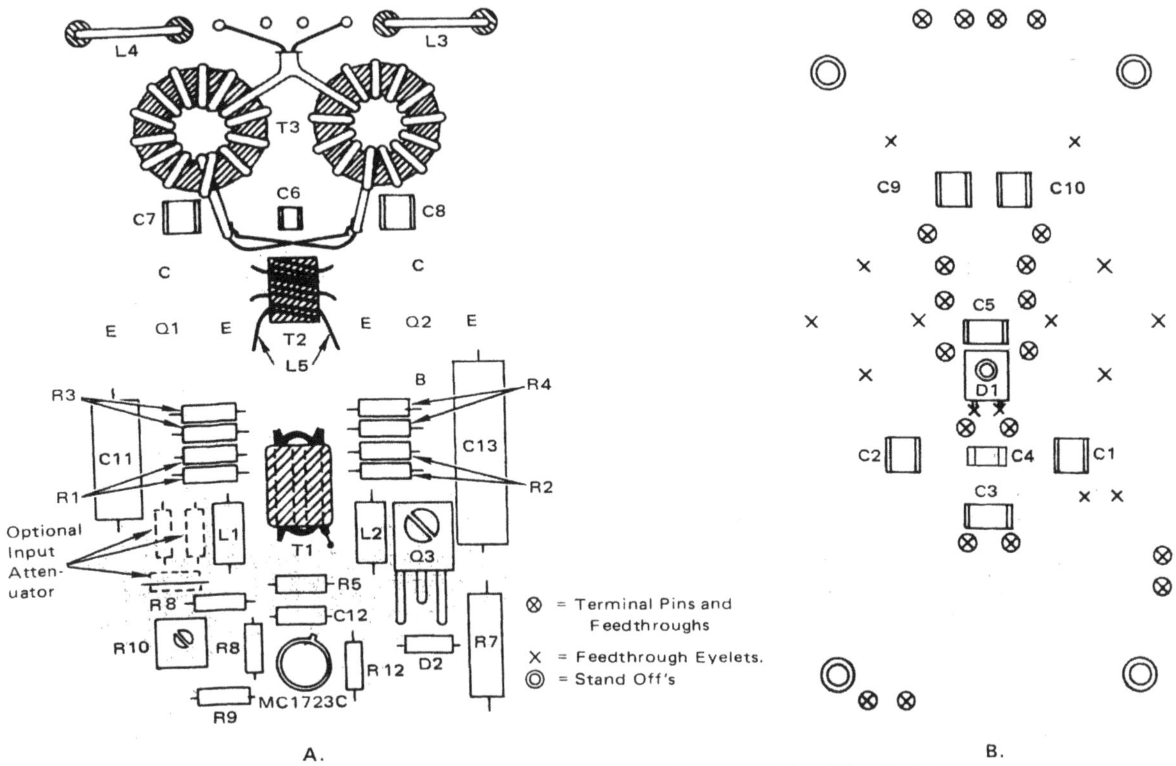

FIGURE 9 – Component Layout of the 300 W Amplifier Module

PARTS LIST
(Power Module and Driver Amplifier)

	Power Module	Driver Amplifier
C1, C2	5600 pF	3300 pF
C3	56 pF	39 pF
C4	470 pF	Not Used
C5	560 pF	470 pF
C6	75 pF	51 pF
C7, C8	0.1 μF	0.1 μF
C9, C10	0.33 μF	0.33 μF
C11	10 μF/150 V	10 μF/150 V
R1, R2	2 x 3.9 Ω/½ W in parallel	2 x 7.5 Ω/½ W in parallel
R3, R4	2 x 6.8 Ω/½ W in parallel	2 x 18 Ω/½ W in parallel
L1, L2	Ferroxcube VK200 19/4B ferrite choke	Ferroxcube VK 200 19/4B ferrite choke
L3, L4	6 ferrite beads each, Ferroxcube 56 590 65/3B	6 ferrite beads each, Ferroxcube 56 590 65/3B
	All capacitors, except C11, are ceramic chips. Values over 100 pF are Union Carbide type 1225 or 1813 or Varadyne size 18 or 14. Others ATC Type B.	
T1	9:1 type, see text.	4:1 type, see text.
	(Ferrite cores for both: Stackpole 57-1845-24B or Fair-Rite Products 287300201 or equivalent.)	
T2	7 turns of bifilar or loosely twisted wires. (AWG #20.) Ferrite cores for both: Stackpole 57-9322, Indiana General F627-8Q1 or equivalent.	
T3	**14 turns of Microdot 260-4118-00*** 25 Ω miniature coaxial cable wound on each toroid. (Stackpole 57-9074, Indiana General F624-19Q1 or equivalent.)	11 turns of RG-196, 50 Ω miniature coaxial cable wound on a bobbin of a Ferroxcube 2616P-A100-4C4 pot core.

*Equivalent product is available from: W. L. Gore, Inc., Newark, Del. Part number CXN1286

proved low frequency isolation characteristics the line impedance must be increased for the parallel currents. This can be done, without affecting the physical length of the line, by loading the line with magnetic material. In this type transformer, the currents cancel, making it possible to employ high permeability ferrite and a relatively short physical length for the transmission lines. In an absolutely balanced condition, no power will be dissipated in the magnetic cores, and the line losses are reduced. The minimum required inductance for each line can be calculated as shown for the driver amplifier output transformer, which gives a number of 16 μH minimum at 2 MHz. A low inductance value degrades the isolation characteristics between the 50 Ω output ports, to maintain a low VSWR in case of a change in the input impedance of one or more of the power modules. However, because of the base compensation networks, the power splitter will never be subjected to a completely open or shorted load.

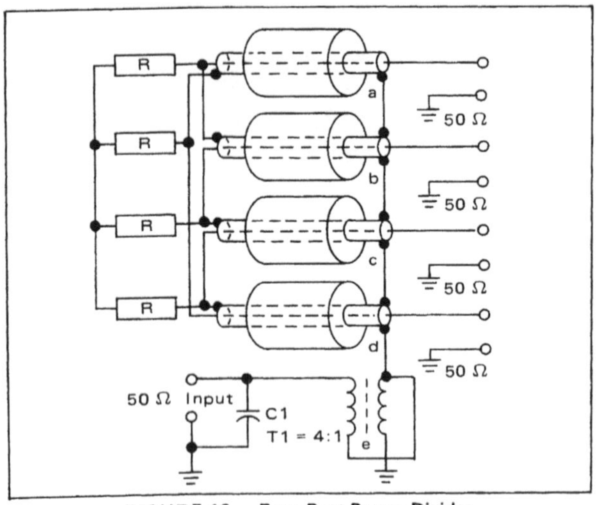

FIGURE 10 – Four Port Power Divider

The purpose of the balancing resistors (R) is to dissipate any excess power, if the VSWR increases. Their optimum values, which are equal, are determined by the number of 50 Ω sources assumed unbalanced at one time, and the resistor values are calculated accordingly.

Examining the currents with one load open, it can be seen that the excess power is dissipated in one resistor in series with three parallel resistors. Their total value is 50 - 12.5 = 37.5 Ω. Similarly, if two loads are open, the current flows through one resistor in series with two parallel resistors, totaling 37.5 Ω again. This situation is illustrated in Figure 11.

Except for a two port power divider [5], the resistor values can be calculated for odd or even number systems as:

$$R = \left(\frac{R_L - R_{in}}{n+1}\right) n \text{ where:}$$

R_L = Impedance of the output ports, 50 Ω.
R_{in} = Impedance of the input port, 12.5 Ω.
n = Number of output ports properly terminated.

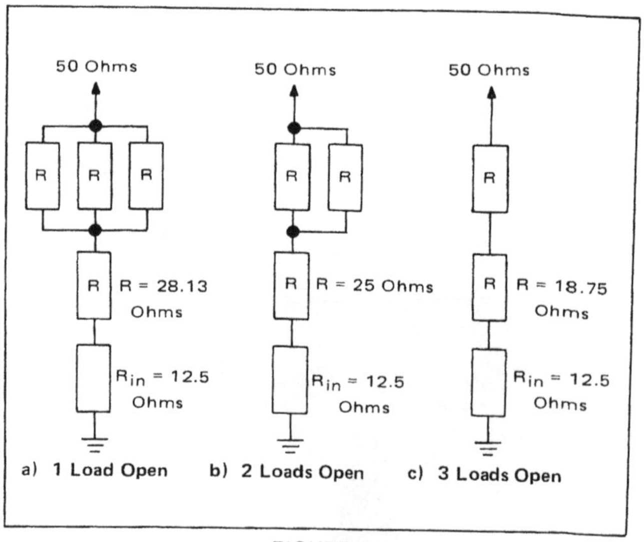

FIGURE 11

Although these resistor values are not critical in the input divider, the formula also applies to the output power combiner, where mismatches have a larger effect in the total power output and linearity.

The practical power divider employs large ferrite beads (Fair-Rite Products 2673000801 or Stackpole 57-1511-24B or equivalent) over a 1.2 inch piece of RG-196 coaxial cable. The arrangement is shown in Figure 10. Both above ferrite materials have a μr of about 2500, and the inductance for one turn is in excess of 10 μH.

The step-down transformer (T1, Figure 10) is wound on a Stackpole 57-9322-11 toroid with 25 Ω miniature coaxial cable. (Microdot 260-4118-000 or equivalent.) Seven turns will give a minimum inductance of 4/16 μH, required at 2 MHz.

For the preamplifier interface, C1 could be omitted in order to achieve the lowest input VSWR.

The structure is mounted between two phenolic terminal strips as can be seen in the foreground of Figure 14, providing a sufficient number of tie points for the coaxial cable connections.

THE OUTPUT COMBINER

The operation of the output combiner is reversed from that of the input power divider. In this application we have four -50 Ω inputs and one 12.5 Ω output, which is transformed to 50 Ω by a 1:4 impedance ratio transformer.

An arrangement similar to the input power divider is employed in the combiner. The baluns consist of straight pieces of coaxial cable loaded by a sleeve of magnetic material (ferrite). The line length is determined by the physical dimensions of the ferrite sleeves. The μr versus cross sectional area should be calculated or measured to give sufficient loading inductance.

Straight line baluns as these have the advantage over multiturn toroidal types in introducing a smaller possibility for phase errors, due to the smaller length of the line. The largest possible phase errors occur in the input and output connecting cables, whose lengths are 18" and 10" respectively. All four input and output cables must be of equal length within approximately ¼", and the excess in some, caused by the asymmetrical system layout, can be coiled or formed into loops.

The output connecting cables between the power amplifier outputs and the combiner are made of low loss RG-142B/U coaxial cable, that can adequately handle the 300 W power with the average current of 2.45 A.

The balun transmission lines are also made of RG-142B/U coaxial cable, with an outer diameter of 0.20". The line length is not critical as it is well below the maximum length permitted for 30 MHz (7). The minimum inductance, as in the input divider, is 16 μH per line. Measurements were made between two port combiners, one having the line inductance of 17 μH (7 Ferroxcube 768 series 3E2A toroids) and the other 4.2 μH (one Stackpole 57-0572-27A ferrite sleeve). The results are shown in Table 3.

The power output with various numbers of disabled sources, referring to Figures 11 and 12 can be calculated as:

$$P_n - P_R + \frac{P_R}{n}$$

where: n = Number of Operative Sources.
P_n = Total Power of Operative Sources.
P_R = Power Dissipated in Balancing Resistors.

For one disabled source:

$$P_R = 250\left(\frac{28.13}{50}\right) = 140.65,$$

$$P_{out} = (250 \times 3) - \left(140.65 + \frac{140.65}{3}\right) =$$

$$750 - 187.5 = 562.5\ \text{W}$$

This is assuming that the phase errors between the active sources are negligible. Otherwise the formula in (7) can

f MHz	Isolation dB (Line Inductance 17 μH)	Isolation dB Line Inductance 4.2 μH)
2.0	40.2	29.1
4.0	40.0	38.3
7.5	39.6	39.1
15	37.5	37.8
20	35.8	36.2
30	33.4	33.5

TABLE 3:

The main difference is at 2 MHz – and it was decided that the 29 dB of isolation is sufficient, as the high frequency isolation in either case is not much better. The 3E2A and other similar materials are rather lossy at RF, and with their low Curie points, would present a danger of overheating in case of a source unbalance.

Figure 12 shows the electrical design of the four-port power combiner.

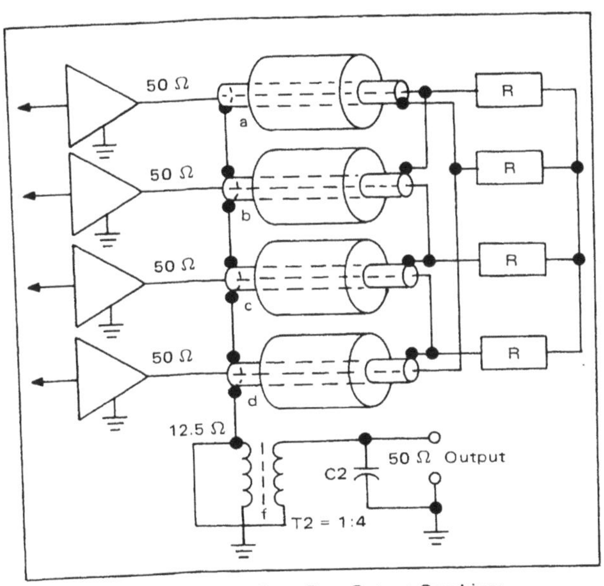

FIGURE 12 – Four Port Output Combiner

be adapted, but if the errors between the active sources are unequal, the situation will get rather complex.

From above we see that 140.65 W will be dissipated by one of the balancing resistors and only 15.6 W by the other three. For this high power dissipation the resistors must be the type which can be mounted to a heat sink, and noninductive. After experiments with the "noninductive" wirewound resistors which exhibited excess inductance at 30 MHz and were bulky with 50 and 100 W ratings, thin film terminations were specially fabricated in-house for this application.* These terminations are deposited on a BeO wafer, which is attached to a copper flange. They are rated for 50 W continuous power, but can be operated at 100 or even 150 W for nonextended periods if the flange temperature is kept moderately low. The balancing resistors can be seen on the upper side of the combiner, which is shown in the foreground of Figure 15.

The purpose of the step-up transformer T2, (Figure 12) is to transform the 12.5 Ω impedance from the combiner up to 50 Ω. It is a standard 1:4 unbalanced-to-unbalanced transmission line type transformer, (6, 7, 8) in which the line is made of two RG-188 coaxial cables connected in parallel in the manner as shown in Figure 13.

Normally the loss in RG-188 at 30 MHz is 0.08 dB/foot. In this connection arrangement, the currents in both directions are carried by the braid in parallel with the

*Similar attenuators and terminations are available from Solitron, EMC Technology, Inc., and other manufacturers of microwave components.

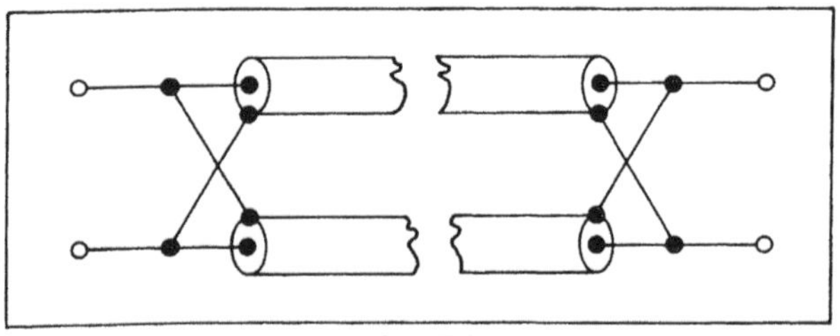

FIGURE 13

inner conductor and the power loss is reduced to approximately 0.025 dB/foot. The impedance becomes 25 Ω, and depending on how close the cables are to each other physically, it can be as low as 22 Ω. The minimum line inductance can be calculated as shown before, and is 16 μH for the 50 Ω side. This inductance is achieved by winding several turns of the dual cable line on a magnetic core. In contrast to the balun transformers in the combiner, the line currents do not cancel and the magnetic core must handle the full power, and must be made of lower loss material. The form of a toroid was figured to require the shortest line length for a specific inductance, and out of the standard sizes, two stacked units resulted in a shorter line length than a single larger one with similar cross sectional area.

Six turns on two Indiana General F626-12-Q1 toroids give 4.8 and 23 μH for the secondary; the line length being 16 inches.

In continuous operation the core temperature was measured as 95-90°C. This resulted in a decision to change the core material to Q2, which exhibits about 70% lower losses at 30 MHz. The permeability is also lower (35), and with the same number of turns gives only 13 μH.

The line length could not be increased according to (7), and the measurements indicated no difference in operation at 2 MHz, so the Q2 toroids with the low inductance were considered permanent.

The maximum flux density of the toroids is calculated as shown before:

$$B_{max} = \frac{V_{max} \times 10^2}{2\pi f \eta A} \text{ gauss, where:}$$

V = Peak voltage across the secondary, (50 Ω point) |316.2 V

f = Frequency in MHz (2.0)

η = Number of turns at the 50 Ω point. (12)

A = Core cross sectional area (1.21 cm^2)

$$B_{max} = \frac{316.2 \times 10^2}{6.28 \times 2 \times 12 \times 1.21} = 260 \text{ gauss}$$

From the BH curves we can see that the linear portion extends to 800-1000 gauss, and the saturation occurs at over 3000 gauss. Comparable materials are Stackpole grade 14 and Fair-rite products 63.

The core losses are minimal compared to the line losses, which for the 16" length amount to 0.035 dB or 0.81%.

As in the input transformer, the HF compensation (C2) was not required. The lay-out of the combiner and T2 is such that minimum lead lengths are obtained, and the structure is mounted on a PC board having feedthrough eyelets to a continuous ground plane on its lower side.

FIGURE 14 – 1 kW Linear Amplifier showing the input power divider in the foreground, to the right is the preamplifier. Two of the four 300 W modules can be seen on the upper side of the structure. The other two modules are shown in Figure 15.

FIGURE 15 – 1 kW Linear Amplifier showing the output combiner in the foreground, to the right is the 1:4 stepup transformer. The four balancing resistors, mounted to the heat sink, can be seen directly above the combining network.

MEASUREMENTS

Six 300 W modules were built using matched pair production MRF428's. The maximum gain distribution was 0.9 dB, and in the four units selected for the amplifier, the gain varied from 13.7 to 14.1 dB at 30 MHz, so it was not necessary to utilize the option of the input attenuators.

Figure 16 shows the test set-up arrangement employed for testing the modules and the combined amplifier.

The heatsink design was not optimized as it was felt to be outside the scope of this report; concentration was made in the electrical design. However, it was calculated to be sufficient for short period testing under two-tone or CW conditions at full power. The heatsink consists of

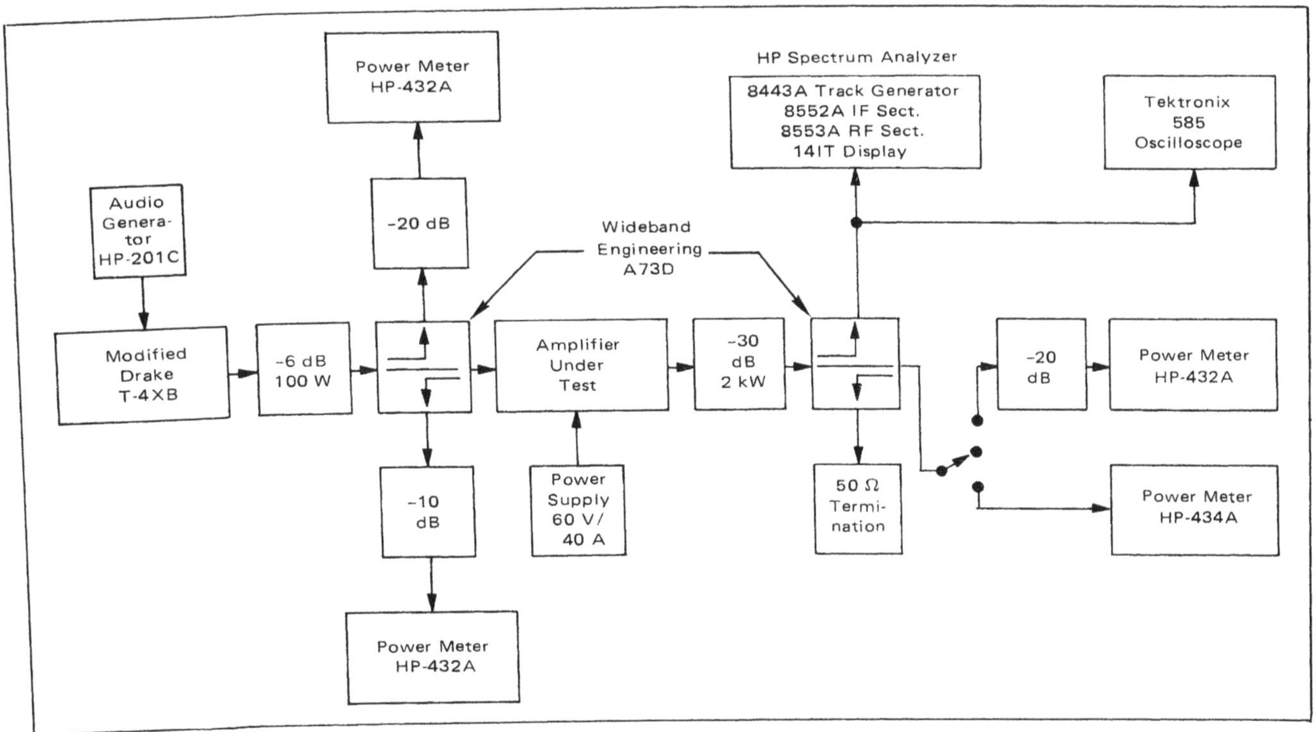

FIGURE 16 – For two tone operation, a signal from an external audio oscillator is added to a signal from the T-4XB built-in oscillator, which has been adjusted to 800 Hz.

During single tone testing, the external oscillator (1200 Hz) is switched off. A calorimeter wattmeter in the output can be used to calibrate the HP-432A's at frequencies below ≈10 MHz, where their response roll-off begins.

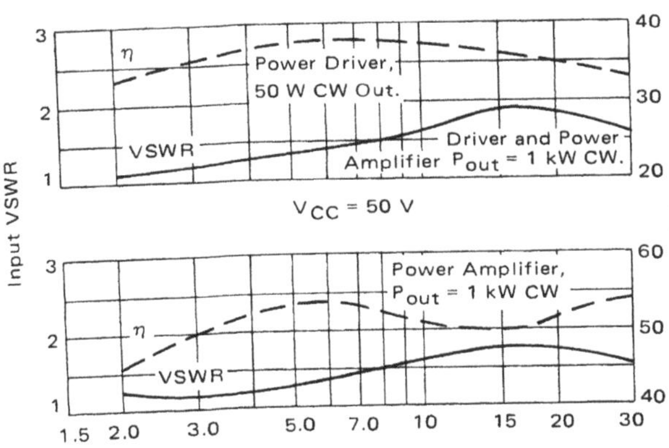

FIGURE 17 – VSWR and Efficiency versus Frequency

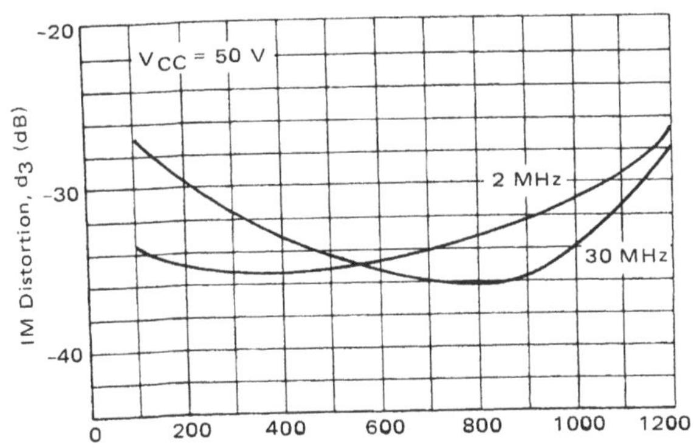

FIGURE 18 – IMD versus Power Output

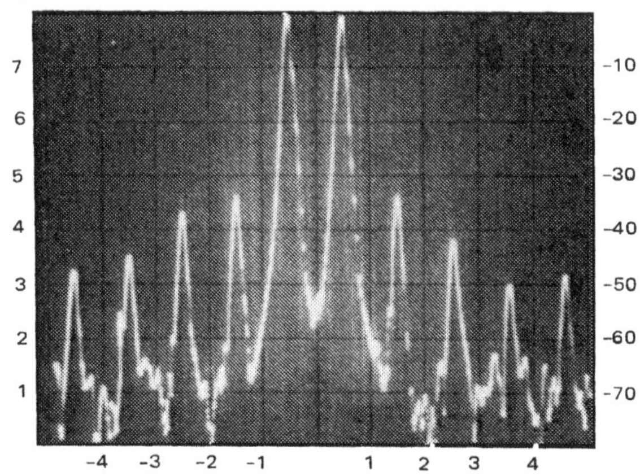

FIGURE 19 – Photo of Spectrum Analyzer Display Showing the IMD Products to the 9th Order. Power Output = 1 kW at 30 MHz (50 V).

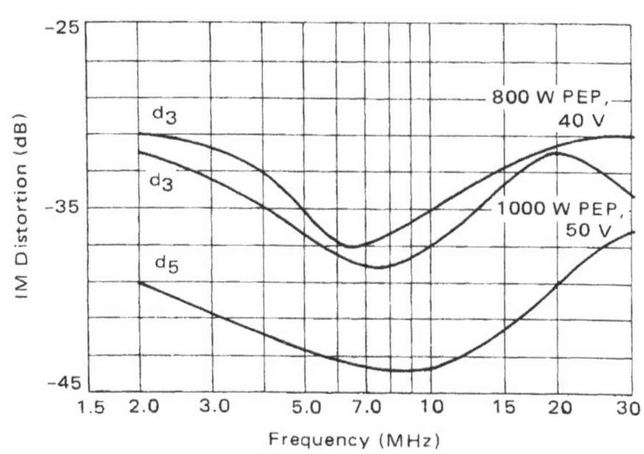

FIGURE 20 – IMD versus Frequency

four 9" lengths of Thermalloy 6151 extrusion, each having a free air thermal resistance of 0.7°C/W. They are bolted in pairs to two 9" x 8½" x 3/8" copper plates, to which the four power modules are mounted. Assuming a coefficient of 0.85 between two parallel extrusions, a total thermal resistance of 0.4°C/W is realized. Two of these dual extrusions are mounted back-to-back to provide

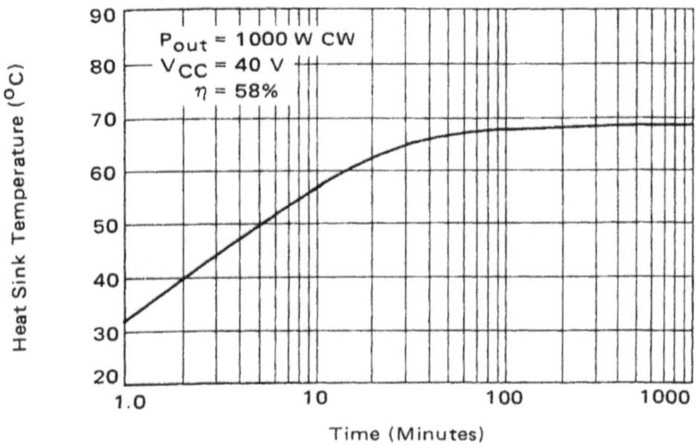

FIGURE 21 – Heat Sink Temperature versus Time

a channel for the air flow from four Rotron SP2A2 4" fans. Two are mounted in each end of the heatsink, and the four fans operating in the same direction provide an air flow of approximately 150 CFM.

The third order harmonic is 14 dB below the fundamental at certain frequencies, as can be seen in Figure 22. This number is typical in a four octave amplifier, and it is obvious that some type of output filter is required when it is used for communications purposes.

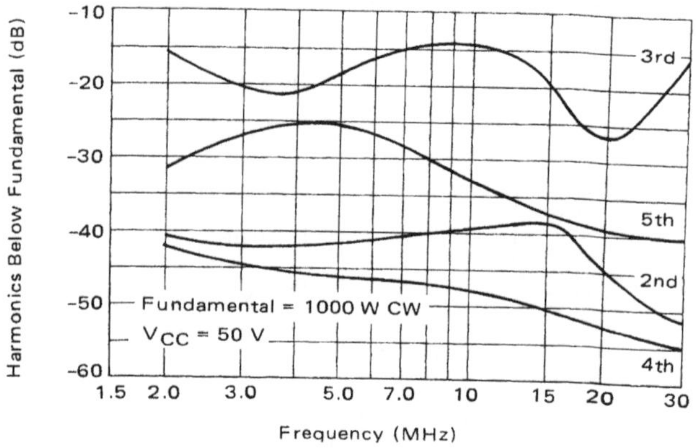

FIGURE 22 – Output Harmonic Contents versus Frequency

The 10:1 load mismatch was simulated with 34 feet of RG-58 coaxial cable, which has an attenuation of approximately 0.9 dB at 30 MHz, representing 1.8 dB return loss. The coaxial was terminated into an LC network consisting of a 2 x 15 – 125 pF variable capacitor and two inductors as shown in Figure 23.

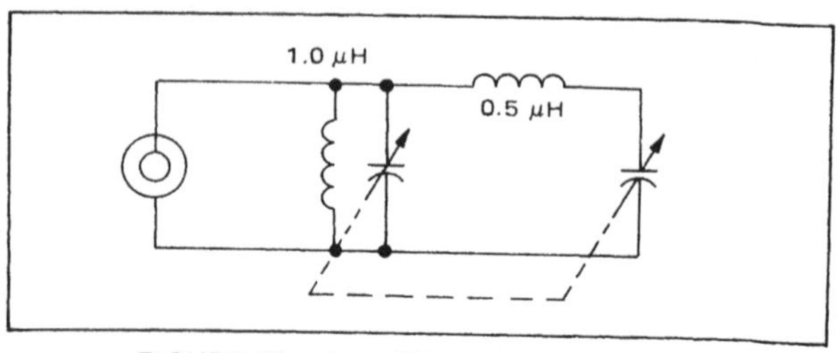

FIGURE 23 – Load Mismatch Test Circuit

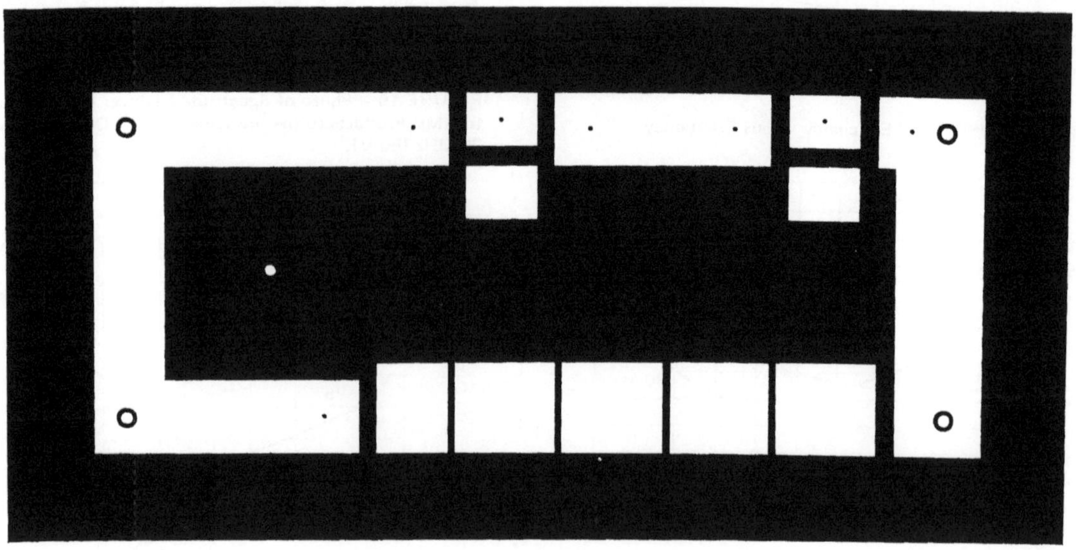

FIGURE 24 – Circuit Board Layout of the Power Combiner Assembly

Actual Size

The high current mode appears at a phase angle of -90° and 20 Ω, where the monitored individual collector currents increased to 6.8 A. At 50 V this amounts to 340 W, which almost entirely represents device dissipation.

At 20:1 load mismatch an equal power dissipation is reached at a power output of approximately 650 W CW.

Since the collector voltages remain below the device breakdowns at the high impedance mode ($+90^{\circ}$, 150 Ω), it may be concluded, that the load mismatch susceptability is limited by overdissipation of the transistors.

Actual Size

FIGURE 25 – Board Layout of the Power Combiner Transmission Line Assembly

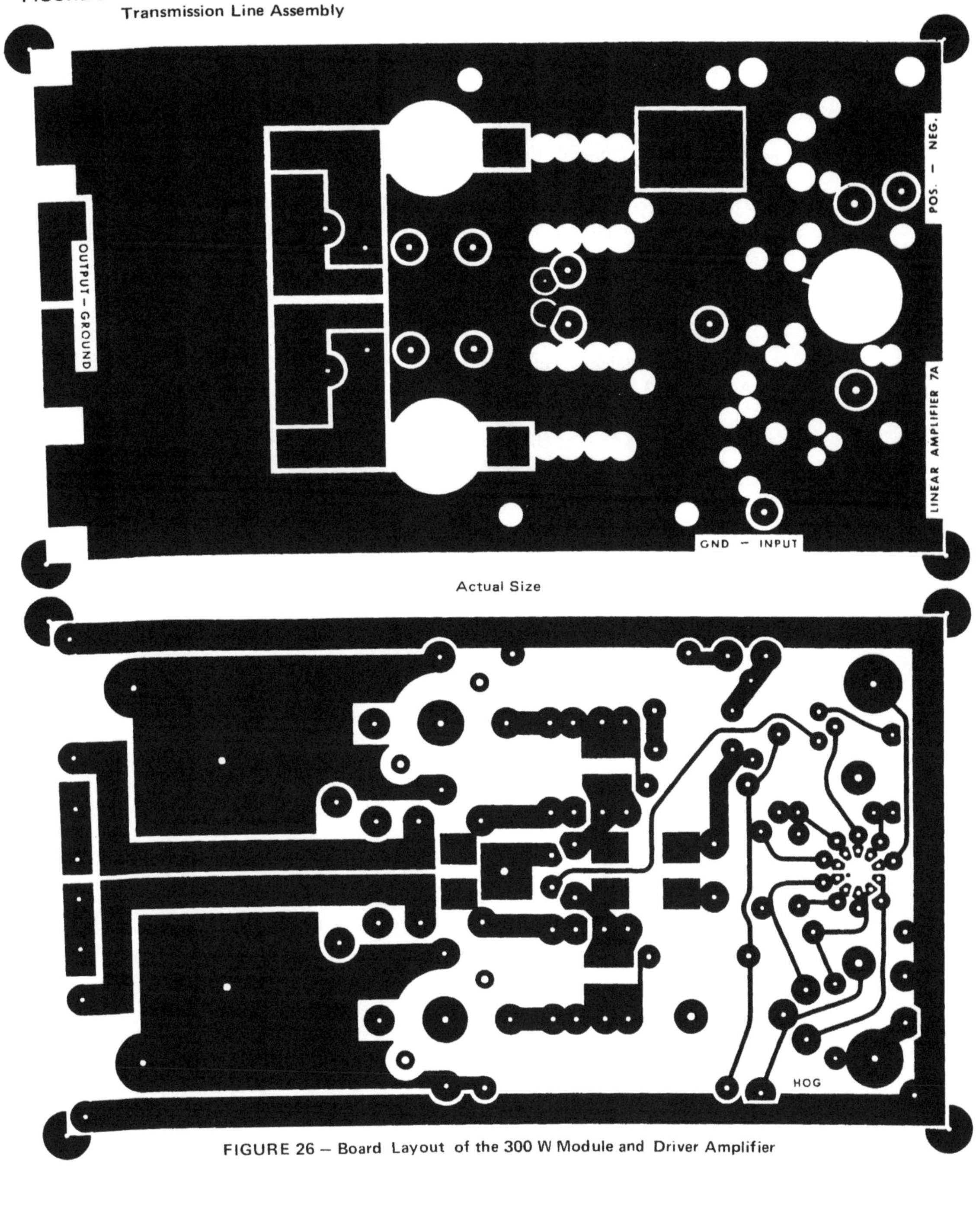

Actual Size

FIGURE 26 – Board Layout of the 300 W Module and Driver Amplifier

REFERENCES

1. Ruthroff: Some Broad Band Transformers, *IRE, Volume 47,* August 1975.
2. Lewis: *Notes on Low Impedance H.F. Broad Band Transformer Techniques,* Collins Radio Company, November 1964.
3. Granberg, H.: *Broadband Linear Power Amplifiers Using Push-Pull Transistors,* AN-593, Motorola Semiconductor Products Inc.
4. Granberg, H.: *Get 300 Watts PEP Linear Across 2 to 30 MHz From This Push-Pull Amplifier,* EB-27, Motorola Semiconductor Products Inc.
5. Granberg, H.: *Broadband Transformers and Power Combining Techniques for RF,* AN-749, Motorola Semiconductor Products Inc.
6. Hilbers: Design of H.F. Wideband Power Transformer Techniques, *Phillips Application Information #530.*
7. Pizalis-Couse: Broadband Transformer Design for RF Transistor Amplifiers, *ECOM-2989,* U.S. Army Electronics Command, Fort Monmouth, New Jersey, July 1968.
8. *Philips Telecommunication Review,* Volume 30, No. 4, pp. 137-146, November 1972.
9. Lefferson: Twisted Wire Transmission Line, *IEEE Transactions on Parts, Hybrids and Packaging,* Volume PHP-7, No. 4, December 1971.
10. Krauss-Allen: Designing Toroidal Transformers to Optimize Wideband Performance, *ELECTRONICS,* August 1973.

MOTOROLA Semiconductor Products Inc.

BOX 20912 • PHOENIX, ARIZONA 85036 • A SUBSIDIARY OF MOTOROLA INC

10149-1 PRINTED IN USA 10-77 IMPERIAL LITHO B67349 2H

AN-758

An *N*-Way Hybrid Power Divider*

ERNEST J. WILKINSON†

Summary—A circularly symmetric power divider is described which splits a signal into *n* equiphase equiamplitude parts where *n* can be odd or even. The power divider provides isolation between output terminals and approximately matched terminal impedances over about a 20 per cent band. A theory of operation is given which yields the necessary design parameters, and an experimental model is described which has a minimum isolation of −27 db between output terminals, an output VSWR of 1.6, and an input VSWR of 1.2.

Introduction

THE problem of dividing up an input signal into a number of equiphase equiamplitude output signals is a familiar one to RF engineers, particularly those working in the field of phased arrays. It is often desired that the equiphase equiamplitude condition be obtained in a manner which is fundamentally independent of frequency, and, in addition, that a fairly high degree of isolation exist between output terminals over some specified frequency band. Two of the more commonly used power dividers are illustrated in Fig. 1. In Fig. 1(a), a corporate feed structure of *T* junctions provides the symmetry necessary to preserve the equiphase equiamplitude condition independent of frequency. In Fig. 1(b), circular symmetry is employed where all the output terminals are connected to the common center conductor of a coaxial line, and the combined load is matched by means of a quarter-wave transformer. To provide isolation between output terminals, the *T* junctions of Fig. 1(a) may be replaced by hybrid junctions. In addition to the cost and complexity of providing a number of separate, yet identical, hybrids, the corporate feed structure has the limitation of providing only a binary number of outputs. If nine outputs were required, for example, a 16-to-1 power divider would have to be used with $\frac{7}{16}$ of the power output being dissipated in matched loads. Fig. 1(b), on the other hand, because of the circular symmetry, provides any number of outputs. It has, however, the serious disadvantage of having no isolation between outputs, as the output terminals are badly mismatched because all loads appear in parallel across any given output terminal. This paper describes a device which possesses the circular symmetry of Fig. 1(b), so that it maintains phase and amplitude equality between any number of outputs independent of frequency, but, in addition, provides isolated and matched outputs.

Theory of Operation

The power divider as sketched in Fig. 2 consists of a coaxial line in which the hollow inner conductor has been split into *n* splines of length $\lambda/4$. A shorting plate

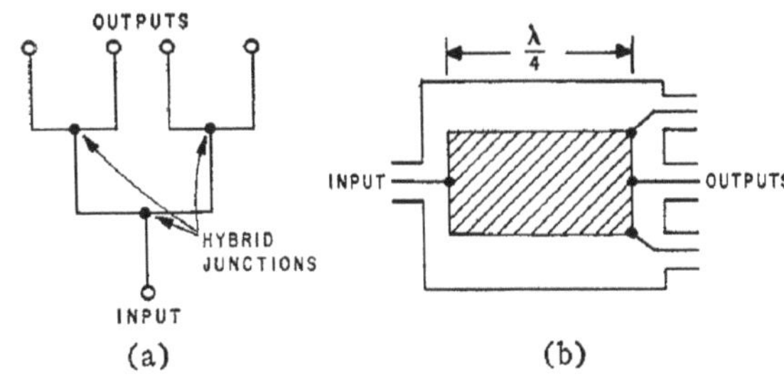

Fig. 1.

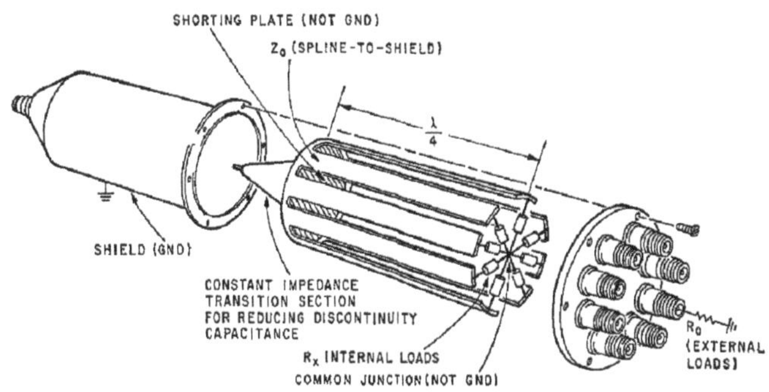

Fig. 2.

connects the splines at the input end,[1] and resistors are connected in a radial manner between each spline at the output end and a common junction. Output connectors are shown connected to the splines in an "in line" manner, although they may also be connected in radial fashion, *i.e.*, at right angles to the splines. When a signal is fed into the power divider, it divides by virtue of symmetry into *n* equiphase equiamplitude parts. No power is dissipated by the resistances when matched loads are connected to the outputs, since all splines will be at the same potential. However, if a reflection occurs at one of the output terminals, the reflected signal will split; part of it will travel directly to the remaining output terminals via the resistors, and the rest of it will travel back to the input, splitting again at the junction of the splines and then returning to the remaining output terminals. Thus, the reflected wave arrives at the remaining output terminals in two parts, and the path length difference between the two paths of travel will be 180° when the splines are $\lambda/4$ in length. It will be shown that when the value of the resistors and the characteristic impedance of the spline transmission lines are properly chosen, the two parts of the reflected wave are also equal in amplitude; hence, complete cancellation occurs.

* Manuscript received by the PGMTT, June 3, 1959; revised manuscript received, September 10, 1959.
† Sylvania Electronic Systems, Waltham, Mass.

[1] A tapered section is shown at the input end for reducing the discontinuity caused by the abrupt change in line sizes when a standard type *N* connector is used. A more convenient form of this device, which eliminates the need for a separate tapered section, can be realized by uniformly tapering the splined coaxial section itself.

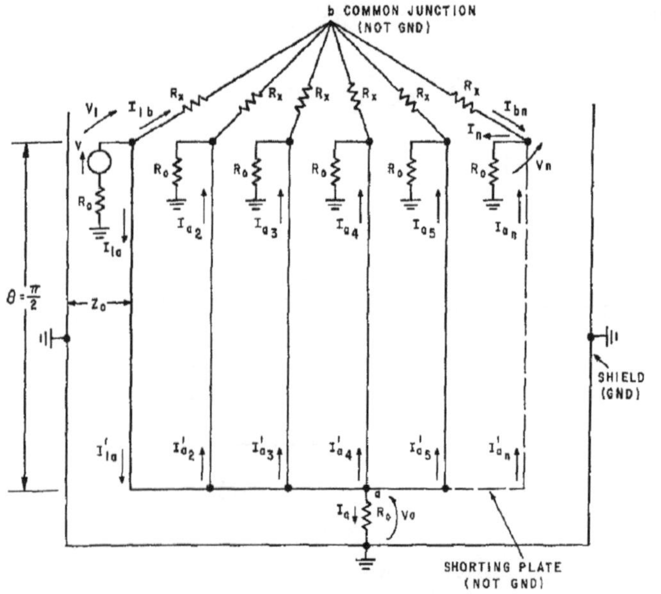

Fig. 3—R_0=external loads connected between splines and shield (GND); R_x=internal loads connected between splines and common junction point; Z_0=characteristic impedance of each spline-to-shield transmission line; V=voltage applied to a single output terminal; V_n=voltage appearing at nth output terminal; V_a=voltage appearing at input terminal.

Referring to Fig. 3, if a voltage V is applied to output terminal 1 by a generator of internal resistance R_0, then the voltages V_n appearing at the other output terminals must all be equal because of symmetry. Applying the transmission line equations when each spline transmission line is a quarter wavelength long ($\theta = \pi/2$), we have

$$\text{spline 1:}\quad V_1 = V_a \cos\theta + jI_{1a}'Z_0 \sin\theta = jI_{1a}'Z_0$$

$$I_{1a} = I_{1a}' \cos\theta + j\frac{V_a}{Z_0}\sin\theta = j\frac{V_a}{Z_0}$$

$$\text{spline } n\text{:}\quad V_a = V_n \cos\theta + jI_{an}Z_0 \sin\theta = jI_{an}Z_0 \tag{1}$$

$$I_{an}' = I_{an}\cos\theta + j\frac{V_n}{Z_0}\sin\theta = j\frac{V_n}{Z_0}.$$

It is also true that

$$I_{1a}' = \frac{V_a}{R_0} + (n-1)I_{an}'$$

$$I_{an} = \frac{V_n}{R_0} - \frac{I_{1b}}{n-1}$$

$$(I_{1b} + I_{1a})R_0 = V - V_1$$

$$\left(I_{1b} + \frac{I_{1b}}{n-1}\right)R_x = V_1 - V_n. \tag{2}$$

Combining (1) with (2) we obtain the following set of simultaneous equations for the unknown voltages V_1, V_a, V_n:

$$V_a + j(n-1)\frac{R_0}{Z_0}V_n + jV_1\frac{R_0}{Z_0} = 0$$

$$jV_a\frac{R_0}{Z_0} + \left(1 + \frac{R_0}{nR_x}\right)V_n - V_1\frac{R_0}{nR_x} = 0$$

$$jV_a\frac{R_0}{Z_0} - \left(\frac{n-1}{n}\right)\frac{R_0}{R_x}V_n + V_1\left(1 + \frac{R_0}{R_x}\left(\frac{n-1}{n}\right)\right) = V. \tag{3}$$

For perfect isolation $V_n = 0$, (3) may be combined to yield

$$\frac{V_1}{V} = \frac{1}{1 + \dfrac{R_0}{R_x}}$$

$$\left(\frac{R_0}{Z_0}\right)^2 = \frac{R_0}{nR_x}, \tag{4}$$

and for matched output admittances, we obtain

$$Y_1 = \frac{I_{1b} + I_{1a}}{V_1} = \frac{V - V_1}{R_0V_1} = \frac{1}{R_0}. \tag{5}$$

Combining (4) and (5) yields

$$R = R_0$$

$$Z_0 = \sqrt{n}\,R_0. \tag{6}$$

Thus, if the internal loads R and the characteristic impedance of the spline-to-shield transmission lines are adjusted according to (6), the outputs will be completely isolated and matched. The input impedance under these conditions will be the parallel combination of the n output loads R_0, after each has been transformed through a quarter wavelength of line Z_0. Hence,

$$Z_{\text{input}} = \frac{\dfrac{Z_0^2}{R_0}}{n} = R_0$$

or, in other words, the input of the power divider is also matched when the conditions for isolation between outputs are satisfied.

Experimental Results

In order to determine whether or not the isolation and matched output features could be realized in a power divider of practical design, a power divider of the type illustrated in Fig. 1(b), was modified according to the following procedure. Narrow slots were milled in the hollow center conductor of an 8-to-1 power divider of this type to create eight spline-to-shield transmission lines having a characteristic inpedance of approximately eight times that of the unslotted coaxial line whose characteristic impedance was $R_0/\sqrt{8}$. Hence, the spline-to-shield lines had

$$Z_0 = 8\frac{R_0}{\sqrt{8}} = \sqrt{8}R_0$$

Fig. 4.

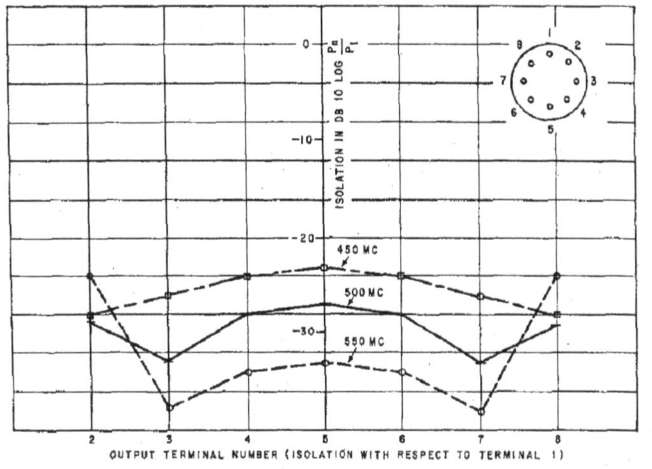

Fig. 5—Isolation between output terminals for N-way hybrid power divider.

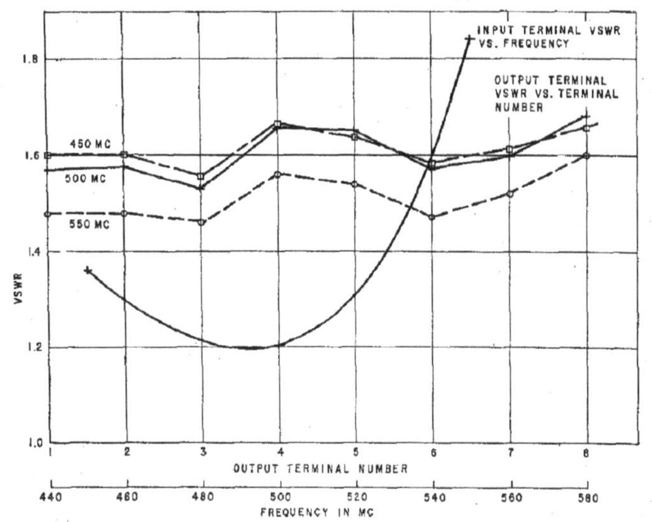

Fig. 6—VSWR characteristics for N-way hybrid power divider.

in accordance with (6). Eight resistors equal in value to R_0 (50 ohms) were then soldered to the ends of the splines, joining in a radial manner to a short circuiting ring (see Fig. 4). Isolation and output VSWR values were obtained using a parallel combination of three 150-ohm $\frac{1}{2}$-watt composition resistors for the 50-ohm internal loads. This combination was found to yield a very low reactance-to-resistance ratio at the operating frequencies. Standard General Radio 50-ohm loads were used to terminate the unused terminals when measuring VSWR and isolation characteristics. The measured isolation between outputs is plotted in Fig. 5 over the frequency range 450 to 550 mc. It will be noted, that at the center frequency of 500 mc the minimum isolation is 27 db, although it is not constant for all output terminals—a fact contrary to the theory based on the model of Fig. 3. The output VSWR is plotted in Fig. 6 and is about 1.6 at 500 mc, where the input VSWR is about 1.2. The output-to-input ratio was measured and found to be -9 db at all frequencies, indicating that the device was behaving as an 8-to-1 lossless power divider with negligible insertion loss.

The experimental model described above differs in several respects from the theoretical model. First, the theoretical model assumes a diameter which is negligible in wavelengths. In the experimental model it was 0.1λ. The principle effect of this appears to be that it shifts the center frequency for best performance to a lower value than the design frequency, and, perhaps, changes the effective resistance of the internal loads, since the current throughout the length of the resistors may no longer be uniform. Second, measurement of the phase between outputs showed several degrees of phase inequality, indicating that the perfect symmetry assumed in the theory had not been achieved in the splined structure and internal loads. Finally, the exact value of the characteristic impedance of the spline-to-shield transmission lines was probably not exactly $\sqrt{8}R_0$. Although the effects mentioned above can be overcome with further development work, the inequality of isolation between output terminals is believed to be due to a more fundamental reason. It will be noted that, shunting any pair of output terminals, there is a short circuited two-wire line formed by the pair of splines leading to these terminals. When the length of the splines is $\lambda/4$, an infinite impedance is shunted between output terminals; but at frequencies off the center frequency, a finite reactance appears between output terminals which is directly proportional to the characteristic impedance of the two-wire line formed by the splines. Because of the circular arrangement of output terminals, diametrically opposite splines (terminals 1–5 in Fig. 5), for example, would form a transmission line of higher characteristic impedance than adjacent splines (terminals 1–2 in Fig. 5). Since, the shunting effect is difference for each pair of output terminals, it is to be expected that the isolation will not be uniform for all terminals, except at the center frequency where there is no shunting effect for any terminal.

60 WATT VHF AMPLIFIER USES SPLITTING/COMBINING TECHNIQUES

Prepared by
Ken Dufour
RF Product Group

Using proven combining techniques to obtain higher output power or added reliability at VHF can be accomplished with excellent results. Simple matching networks and power transistors featuring moderate gain capability can produce a level of performance comparable to that of a single-stage amplifier using a larger, more expensive device. Though not the ultimate answer in VHF amplifier design, the splitter/combiner method does have distinct advantages over designs that brute force the transistors into a parallel configuration. Current hogging and reduced impedance level problems associated with that technique are minimized. The exotic materials or expensive board layout required to produce a true push-pull design operating at VHF again makes combining techniques more appealing.

This 60 W amplifier operates from 150 to 175 MHz and features two, low-cost Motorola MRF264 transistors. These devices are designed for operation at VHF and individually produce 30 watts of rated output power and 6.0 dB of gain with a 12.5 volt supply. The amplifier design makes use of a modified Wilkinson combiner technique to produce 60 watts output with a drive level of 15 watts.

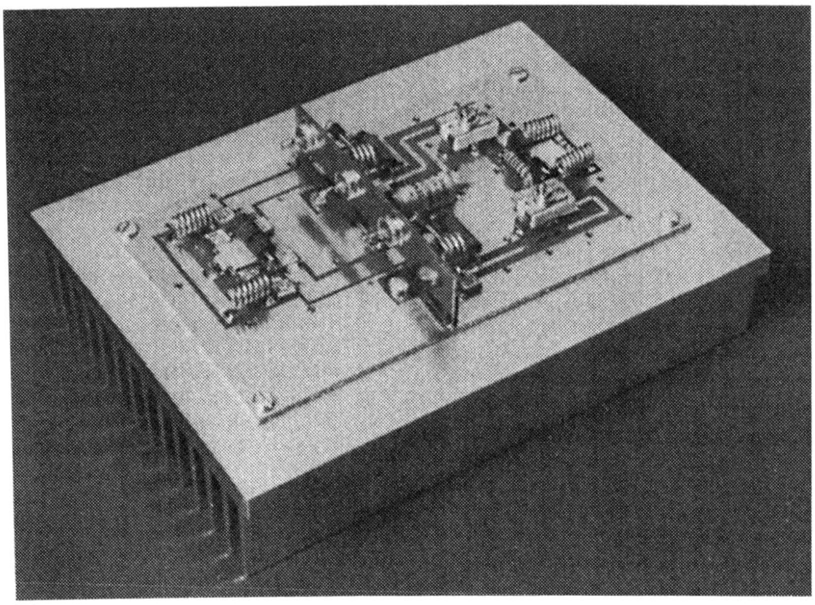

FIGURE 1 — Engineering Model

Design Considerations

Experimental work with 90° (quadrature) couplers proved unsuitable for this application. Generally, they are sensitive to mismatch and tend to create instability and loss of power when used in an amplifier. In-phase (Wilkinson) couplers provide an adequate solution to this problem. (Ref: 1) They are relatively insensitive to phase changes and offer good bandwidth characteristics.

Printed transmission lines for the frequency of interest can become somewhat cumbersome on standard circuit board material. Therefore, lumped reactances (L1, 2, 9, 10 and C1, 2, 3, 14, 15, 16, Figure 5) are used to simulate 70.7 ohm 1/4 wave transmission lines, the main element in the couplers. This approach not only conserves board space, but provides a means to compensate for small variations in associated component values.

Microstrip techniques are incorporated in the amplifier networks to balance RF performance and promote reproducibility. Because of the lower circulating currents and reduced component heating in the collector circuitry of low-powered stages, smaller capacitors can be used in the networks at that point than would be required for a single-ended 60 watt design. Separating the major heat producing devices to two areas on the heatsink produces a more even heat transfer to the ambient air. The combined amplifier presented here has good harmonic suppression (Figure 8). A low-pass filtering effect is noticeable with the Wilkinson combiners.

Construction and Alignment

A 1:1 photomask of the circuit is provided in Figure 9 and double-sided G-10 fiberglass board with two-ounce copper cladding is recommended for construction. The ground points are indicated on the PCB photomask.

The inductors required for the splitter/combiner are constructed by winding the appropriate number of turns (closewound) on a temporary 1/8 inch form and then separating the individual turns by 0.020 inch. An Xacto number 11 knife blade was used for this purpose and provides the correct turns spacing. The 100-ohm isolation resistors, R1 and R2, must be noninductive and carbon composition resistors proved to be entirely adequate. In a properly tuned and balanced amplifier these resistors should remain fairly cool to the touch during normal operation. Each amplifier and coupler input and output port is designed to be terminated into 50-ohms to facilitate testing into a 50-ohm system.

A PCB bridge (Figures 3 and 9) is used to carry all of the dc feed circuitry. It acts as a continuation of the ground plane and enhances circuit stability. Solid copper (0.027 inch) and double-sided circuit board were used as a construction medium and no difference in performance was noted with either material.

Initial alignment is accomplished by driving the amplifier with a 5 watt CW source at approximately 160 MHz. The applied voltage is set at 12.5 volts and the variable capacitors, C4 and C5, are adjusted in an alternating manner to provide maximum output power. Full drive (15 watts) is then applied and the capacitor adjustments are repeated. At this point, the circuitry should be delivering 60 watts or more to the 50-ohm load with the 15 watts input. After the final adjustments are made, the isolation resistor temperature in either coupler should be relatively cool to the touch and the input VSWR should be at a minimum. Best results will be obtained if the transistors are beta matched (±10%) prior to installing them in the circuit.

Additional Comments

This amplifier has been extensively tested for ruggedness and reproducibility. The 15 watt input level makes it compatible with the EB-90 two-stage VHF amplifier as a driver. Together they form a chain requiring 200 mW of input power for a 60 watt or more output.

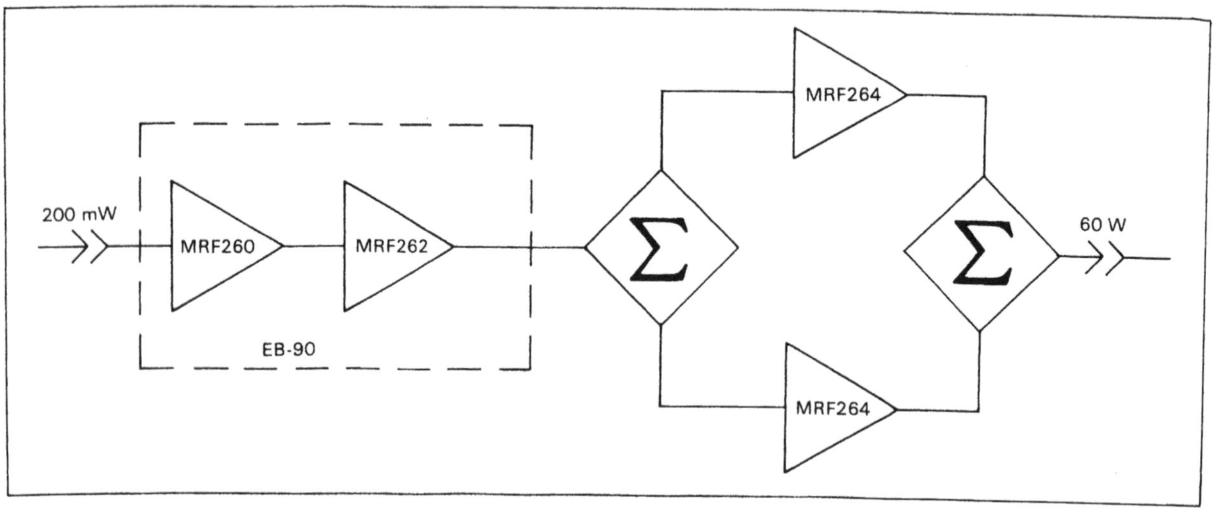

References

1. Lawrence R. Laveller; "Two Phased Transistors Shortchange Class C Amps," *Microwaves*, Pg. 48-54, February, 1978.

2. Ernest J. Wilkinson; "An N-Way Hybrid Power Divider," *PGM TT Transactions*, pg. 116-118, January, 1960.

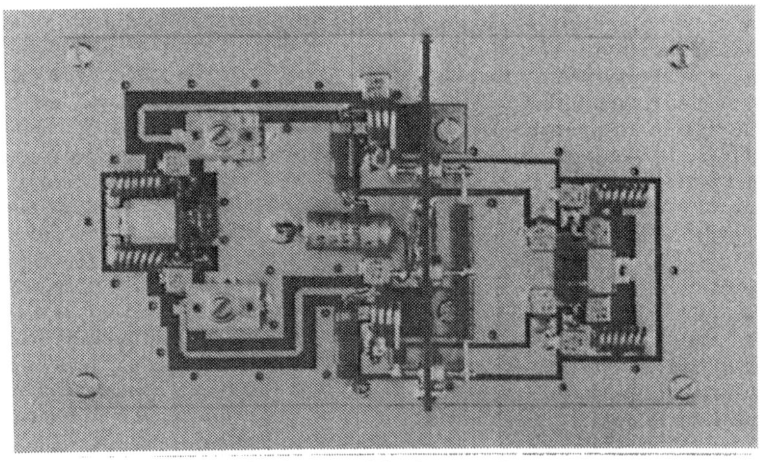

FIGURE 2 — Amplifier Layout - Top View

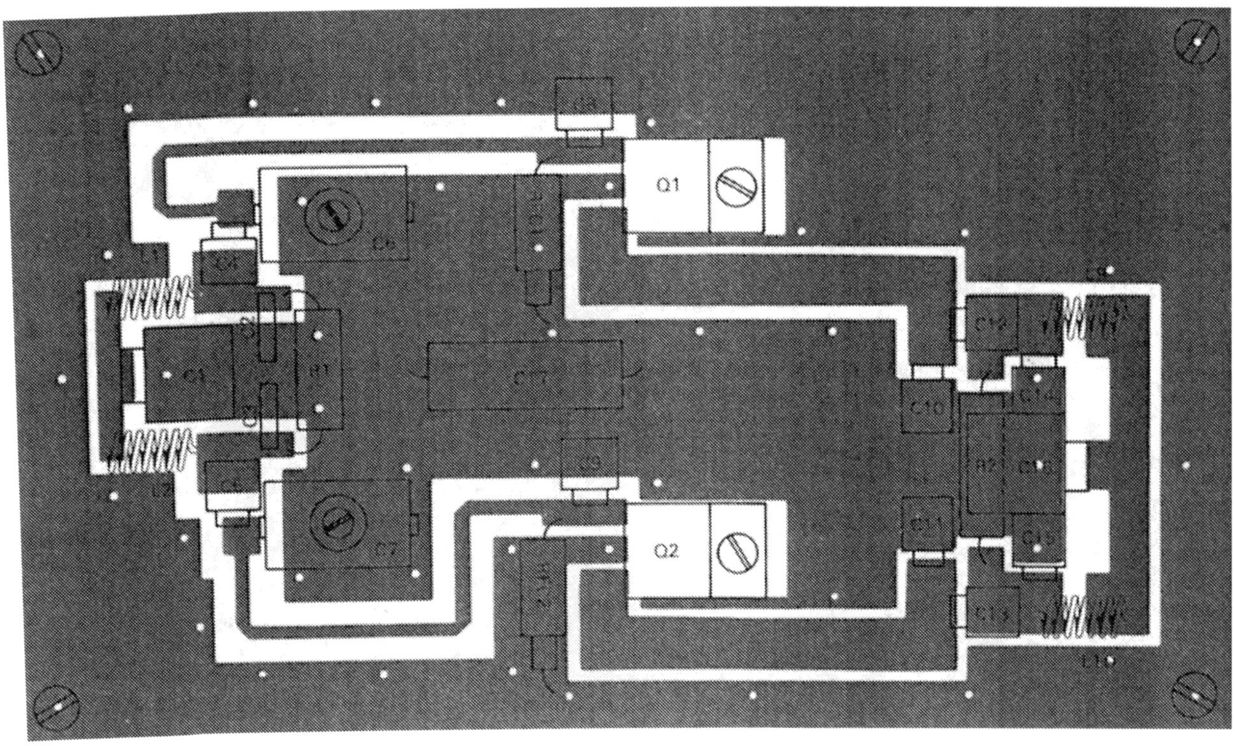

FIGURE 3 — Component Placement

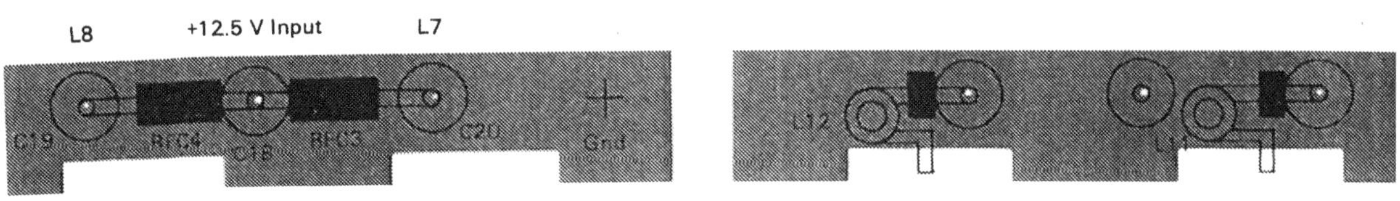

a. Output Side

b. Base Side

FIGURE 4 — PCB Bridge Details

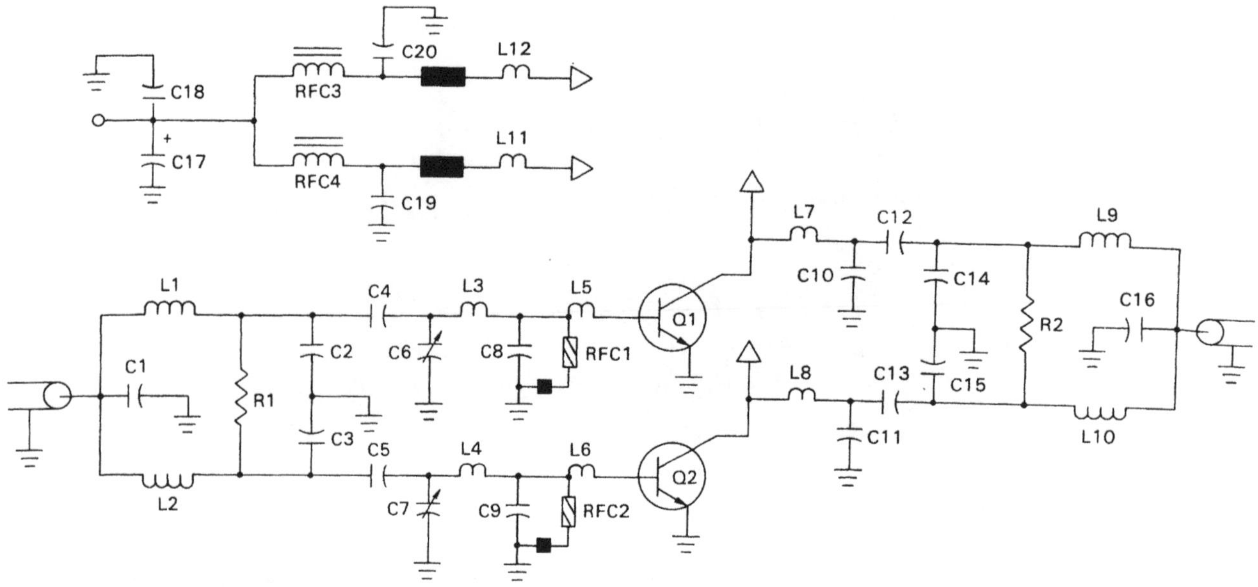

C1, C16 — 25 pF Unelco (J101)
C2, C3 — 15 pF CM04 Mica
C4, C5 — 68 pF Standex
C6, C7 — Arco 404 Variable
C8, C9 — 150 pF Standex
C10, C11 — 56 pF Standex
C12, C13 — 39 pF Standex
C14, C15 — 15 pF Standex
C17 — 100 μF @ 16 V Electrolytic
C18, C19, C20 — 680 pF Allen Bradley Feedthru

L1, L2 — 7 Turns #18, 0.125" ID
L3, L4, L5, L6 — Printed Inductors
L7, L8 — Printed Inductors
L9, L10 — 7 Turns #18 AWG, 0.125 ID
L11, L12 — 4 Turns #18 AWG, 0.250 ID w/Bead
Q1, Q2 — MRF264
RFC1, RFC2 — 0.15 μH Molded Choke w/Bead, Ferroxcube 56-590 65/3B
RFC3, RFC4 — 4 Ferrite Beads each on #18 AWG
R1 — 100 Ω 1/2 W Carbon
R2 — 100 Ω 2.0 W Carbon

FIGURE 5 — Schematic - 60 W Amplifier

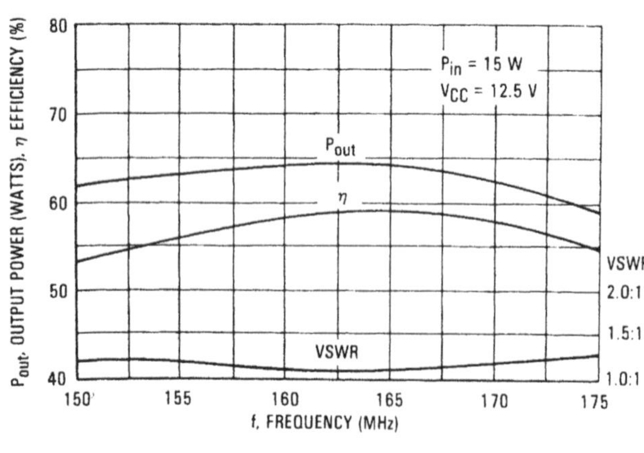

FIGURE 6 — Output Power, Efficiency, and Input VSWR versus Frequency

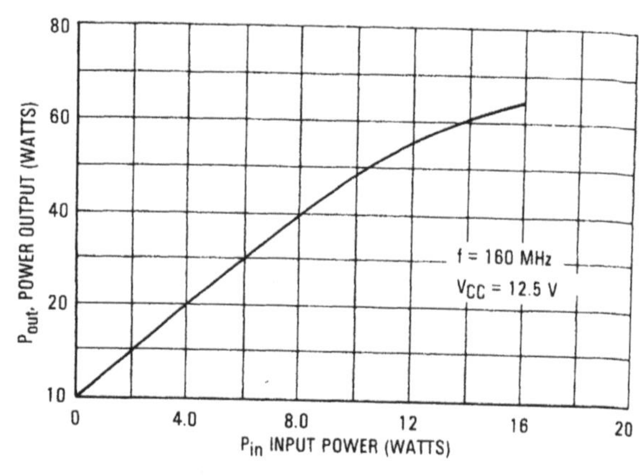

FIGURE 7 — Output Power versus Input Power

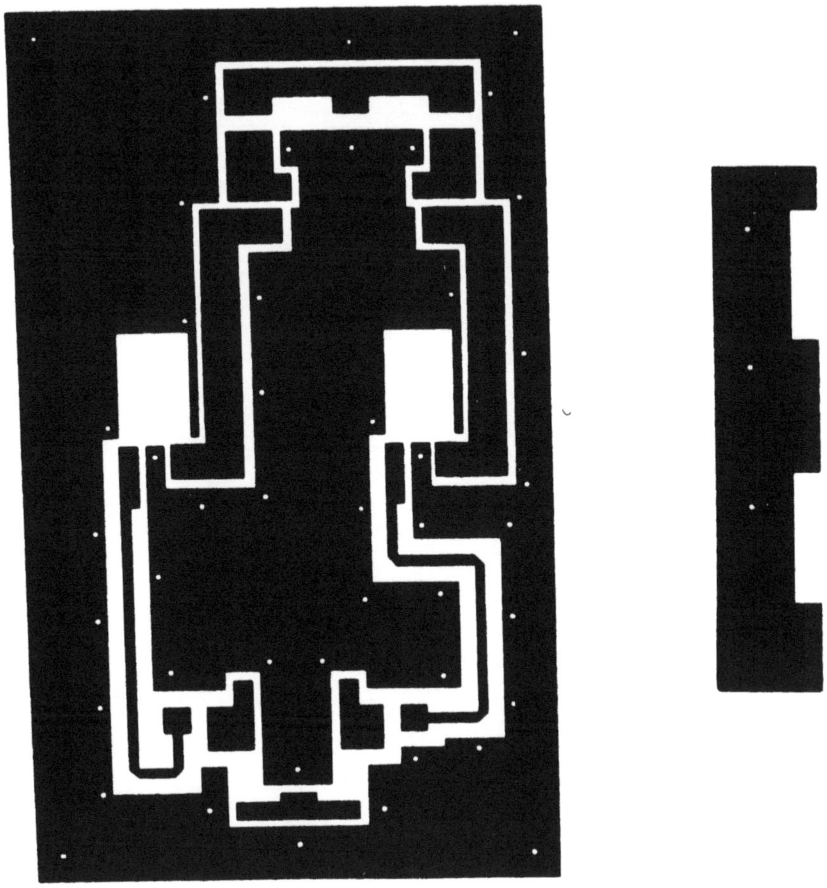

NOTE: The Printed Circuit Board shown is 75% of the original.

FIGURE 8 — PCB Photomaster

Predicting RF Output Power From Combined Power Amplifier Modules

By Roderick K. Blocksome
Rockwell International Corporation
Collins Defense Communications

A power level of a few hundred watts is presently the practical limit for a push-pull pair of RF transistors in a broadband linear power amplifier application. This type of amplifier is referred to as a module. When greater power levels are required, the outputs from two identical modules may be combined into a single output. For even greater power levels, many modules and multiple stages of power combining are required. For example, two groups of two modules each may be combined and the two combiner outputs combined in a third combiner for a single output. In similar manner, eight modules may be combined into a single output by using four combiners feeding two combiners feeding one combiner. This is referred to as three stages of combining.

Losses in wideband combiners are due to normal insertion loss and excess insertion loss. Normal insertion loss of combiners is characterized as mismatch loss and dissipation loss (heat loss in the magnetic core, winding dielectric, and winding resistance). These are not treated in this paper. Excess insertion loss is an additional loss to the output due to amplitude and or phase imbalance of the two input signals. Imbalance in amplitude and phase may result from two causes: (1) variations in the applied input signals and (2) less than perfect combiner construction.

A two input port combiner actually has two output ports. The output we are interested in is the *sum* port. The other port is the *difference* port and is terminated in the bridging resistor. Power lost to the output (sum port) because of excess insertion loss is delivered to the difference port and dissipated in the bridging resistor.

If two input sine wave signals were precisely in phase and of exactly the same amplitude, the combiner output (assuming an in-phase perfect combiner) would equal the sum of the two inputs less the combiner normal insertion loss. The bridging resistor would dissipate zero power and could be removed without any effect on the operation of the combiner in this example.

Combining of perfectly matched input signals is never achieved. Any difference in amplitude or phase or combination of both results in less than perfect combining. The question naturally becomes "How much difference in phase and amplitude can be tolerated for acceptable combining in a particular application?"

Reference 1 addresses this subject for narrow band applications. The expression for calculating the excess insertion loss and combiner output phase is applicable to wideband (multi-octave) hybrid combiners. First derive the expressions to find the excess insertion loss and the phase shift of the output relative to a reference input. The derivation of these expressions uses the same symbols as the referenced article but fills in missing details and adds clarifications so that the reader may readily understand the expressions and apply the technique to combiners with any number of inputs. The phase shift is necessary for calculating excess insertion loss in successive stages of combining.

Derivation of the expressions

A detailed derivation of the expressions for excess insertion loss and phase shift for a two-port combiner is given. Figure 1 is a schematic diagram of a two-port in-phase combiner. A and B designate the two input ports. The output port of summing port is designated by Σ and is connected to the load R_L. The difference port is designated by Δ and is terminated by R_B, the bridging resistor.

With only one input to the combiner, the input power divided equally between the R_B and R_L, thus:

$$P_{out} = \frac{P_{in}}{2}$$

$$\frac{\left(\frac{V_{out}}{\sqrt{2}}\right)^2}{R_L} = \frac{\left(\frac{V_{in}}{\sqrt{2}}\right)^2}{2R_L}$$

$$V_{out}^2 = \frac{V_{in}^2}{2}$$

$$V_{out} = \frac{V_{in}}{\sqrt{2}}$$

Note that V_{out} and V_{in} are peak voltages.

For an in-phase two-port combiner, the relationship can be

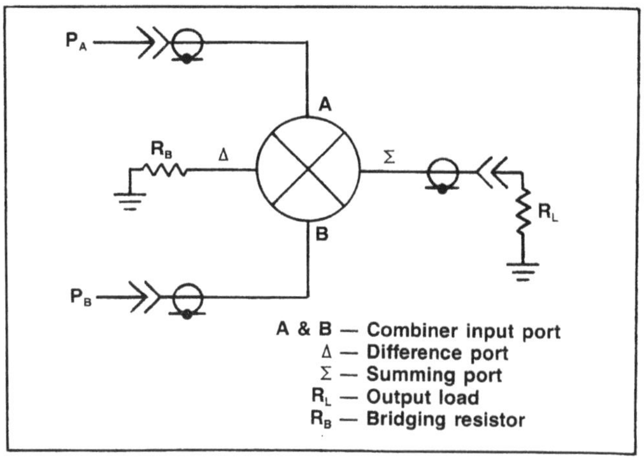

Figure 1. Two-port, in-phase combiner schematic.

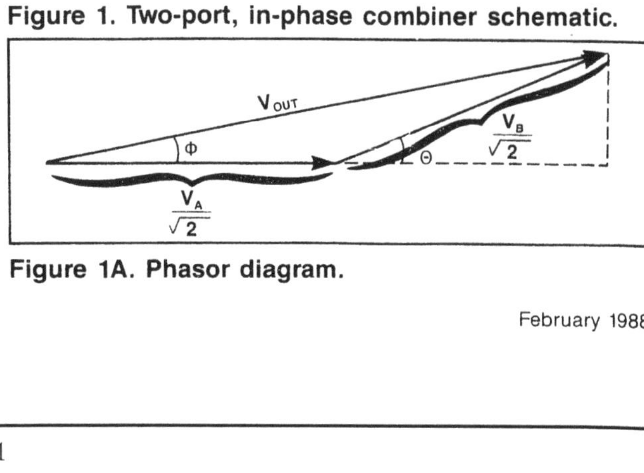

Figure 1A. Phasor diagram.

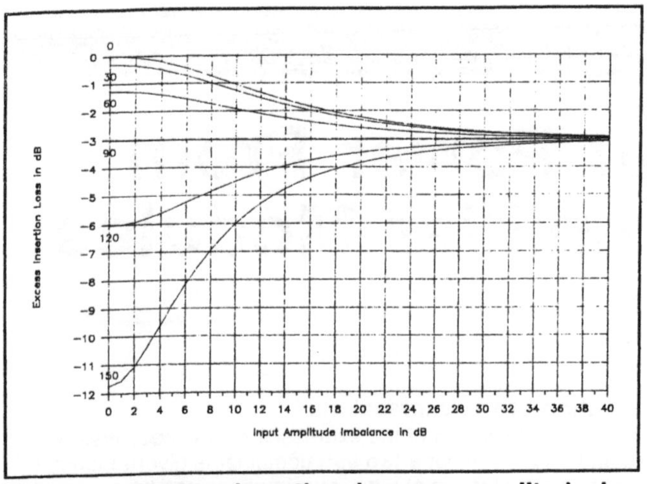

Figure 2. Excess insertion loss vs. amplitude imbalance for various phase differences.

represented between the input voltages and the output voltage (across the load R_L) by the phasor diagram in Figure 1A.

A general expression can now be derived for V_{out} in terms of V_A, V_B, and Θ, the phase difference between the two input signals. For multi-stage combining Φ, the difference between the combiner output signal and one of the input signals chosen as a reference (in this case V_A) is of interest. This expression requires trigonometric manipulation as follows:

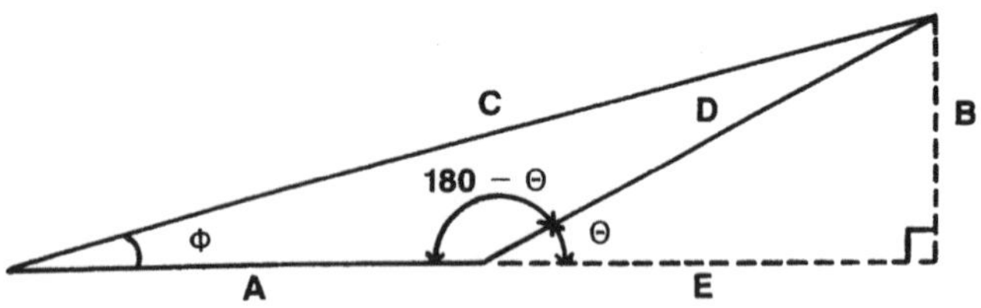

$$C^2 = (A + E)^2 + B^2$$

$$C^2 = A^2 + 2AE + E^2 + B^2$$

Substituting: $B = D \sin \Theta$ and $E = D \cos \Theta$

$$C^2 = A^2 + 2AD \cos \Theta + D^2 \cos^2 \Theta + D^2 \sin^2 \Theta$$

$$C^2 = A^2 + 2AD \cos \Theta + D^2 (\cos^2 \Theta + \sin^2 \Theta)$$

Since $\cos^2 \Theta + \sin^2 \Theta = 1$

$$C^2 = A^2 + D^2 + 2AD \cos \Theta$$

Substituting the combiner voltages:

$$V_{out}^2 = \left(\frac{V_A}{\sqrt{2}}\right)^2 + \left(\frac{V_B}{\sqrt{2}}\right)^2 + 2\left(\frac{V_A}{\sqrt{2}}\right)\left(\frac{V_B}{\sqrt{2}}\right) \cos \Theta$$

$$V_{out}^2 = \frac{V_A^2}{2} + \frac{V_B^2}{2} + 2\frac{V_A V_B}{2} \cos\Theta$$

$$V_{out}^2 = \frac{1}{2}\left(V_A^2 + V_B^2 + 2 V_A V_B \cos \Theta\right)$$

The combiner output power is:

$$P_{out} = \frac{\left(\frac{V_{out}}{\sqrt{2}}\right)^2}{R_L} = \frac{V_{out}^2}{2 R_L}$$

$$P_{out} = \frac{\frac{1}{2}(V_A^2 + V_B^2 + 2V_A V_B \cos \Theta)}{2 R_L} \qquad (1)$$

The total combiner input power is:

$$P_{in} = P_A + P_B = \frac{\left(\frac{V_A}{\sqrt{2}}\right)^2}{R_L} + \frac{\left(\frac{V_B}{\sqrt{2}}\right)^2}{R_L}$$

$$P_{in} = \frac{V_A^2 + V_B^2}{2 R_L} \qquad (2)$$

The excess insertion loss is defined as:

$$I.L. = 10 \operatorname{Log}\left(\frac{P_{out}}{P_{in}}\right)$$

Substituting equation 1 and 2, we obtain:

$$I.L. = 10 \operatorname{Log}\left(\frac{\frac{V_A^2 + V_B^2 + 2 V_A V_B \cos \Theta}{4 R_L}}{\frac{V_A^2 + V_B^2}{2 R_L}}\right)$$

$$I.L. = 10 \operatorname{Log}\left(\frac{V_A^2 + V_B^2 + 2 V_A V_B \cos \Theta}{2 (V_A^2 + V_B^2)}\right)$$

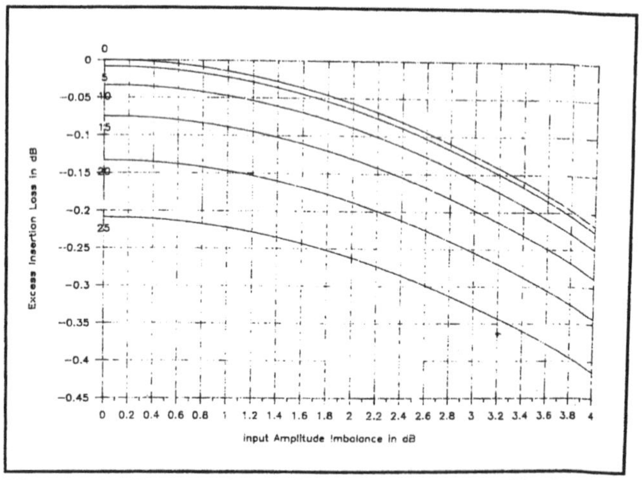

Figure 3. Excess insertion loss vs. amplitude imbalance for various input phase differences.

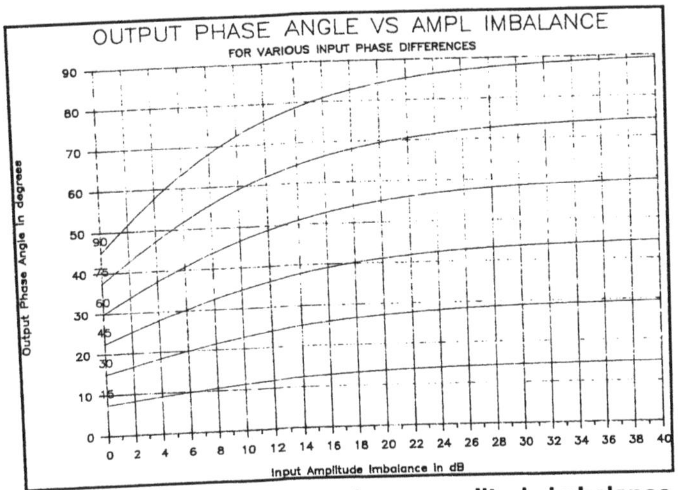

Figure 4. Output phase angle vs. amplitude imbalance for various input phase differences.

$$\text{I.L.} = 10 \text{ Log}\left[\tfrac{1}{2}\left(\frac{V_A^2 + V_B^2}{V_A^2 + V_B^2} + \frac{2\,V_A V_B \cos\Theta}{V_A^2 + V_B^2}\right)\right]$$

$$\text{I.L.} = 10 \text{ Log}\left[\tfrac{1}{2}\left(1 + \frac{2\,V_A V_B \cos\Theta}{V_A^2 + V_B^2}\right)\right] \qquad (3)$$

We now define the ratio of the individual input powers P_A and P_B as:

$$P_r = \frac{P_B}{P_A} = \frac{\left(\frac{V_B}{\sqrt{2}}\right)^2 / R_L}{\left(\frac{V_A}{\sqrt{2}}\right)^2 / R_L} = \frac{V_B^2}{V_A^2} \qquad (4)$$

We arbitrarily chose P_A as the largest amplitude input and designate it as the reference input. We now find V_B in terms of V_A and P_r from equation 4:

$$V_B = V_A \sqrt{P_r}$$

Substituting into equation 3, we obtain:

$$\text{I.L.} = 10 \text{ Log}\left[\tfrac{1}{2}\left(1 + \frac{2\,V_A \cdot V_A \cdot \sqrt{P_r} \cos\Theta}{V_A^2 + V_A^2 P_r}\right)\right]$$

$$\text{I.L.} = 10 \text{ Log}\left[\tfrac{1}{2} + \tfrac{1}{2}\left(\frac{2\,V_A^2 \sqrt{P_r} \cos\Theta}{V_A^2 (1 + P_r)}\right)\right]$$

$$\text{I.L.} = 10 \text{ Log}\left[0.5 + \frac{\sqrt{P_r} \cos\Theta}{1 + P_r}\right] \qquad (5)$$

The phase Φ of the combiner output voltage with respect to V_A is found from the phasor diagram, but first the necessary trigonometry:

$$B = D \sin\Theta$$
$$E = D \cos\Theta$$

$$\tan\Phi = \frac{B}{A + E}$$

$$\tan\Phi = \frac{D \sin\Theta}{A + D \cos\Theta}$$

Substituting the combiner voltages:

$$\tan\Phi = \frac{\frac{V_B}{\sqrt{2}} \sin\Theta}{\frac{V_A}{\sqrt{2}} + \frac{V_B}{\sqrt{2}} \cos\Theta} = \frac{V_B \sin\Theta}{V_A + V_B \cos\Theta}$$

Next divide both numerator and denominator by V_A and substituting the ratio of the individual input powers:

$$\tan\Phi = \frac{\frac{V_B}{V_A} \sin\Theta}{\frac{V_A}{V_A} + \frac{V_B}{V_A} \cos\Theta} = \frac{\sqrt{P_r} \sin\Theta}{1 + \sqrt{P_r} \cos\Theta}$$

$$\tan\Phi = \left(\frac{\sin\Theta}{\frac{1}{\sqrt{P_r}} + \cos\Theta}\right)$$

$$\Phi = \tan^{-1} \frac{\sin\Theta}{\frac{1}{\sqrt{P_r}} + \cos\Theta} \qquad (6)$$

Sometimes it is convenient to work with the ratio of the input powers expressed in dB:

$$P_r\,(\text{dB}) = 10 \text{ Log } P_r$$

$$P_r = \text{Log}^{-1}\left(\frac{P_r(\text{dB})}{10}\right) = 10^{\frac{P_r(\text{dB})}{10}}$$

$$\sqrt{P_r} = 10^{\frac{P_r(\text{dB})}{20}}$$

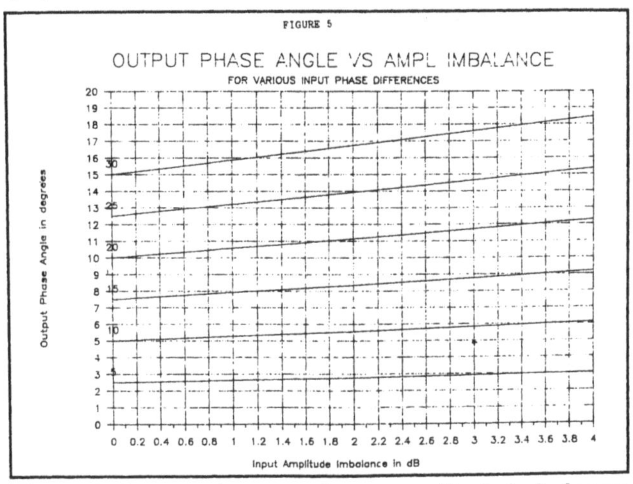

Figure 5. Output phase angle vs. amplitude imbalance for various input phase differences.

$$\frac{1}{\sqrt{P_r}} = 10^{\frac{-P_r(dB)}{20}}$$

Substituting into the expressions for excess insertion loss (equation 5) and the output phase angle (equation 6):

$$I.L. = 10 \operatorname{Log}\left[0.5 + \frac{10^{\frac{P_r(dB)}{20}} \operatorname{Cos}\Theta}{1 + 10^{\frac{P_r(dB)}{10}}}\right] \tag{7}$$

$$\Phi = \operatorname{Tan}^{-1}\left[\frac{\operatorname{Sin}\Theta}{10^{\frac{-P_r(dB)}{20}} + \operatorname{Cos}\Theta}\right] \tag{8}$$

The total loss in a hybrid combiner is the sum of the excess insertion loss and the normal insertion loss which is composed of mismatch loss at the inputs and heat loss in the combiner.

A candidate combiner's insertion loss can be readily measured in the laboratory. The excess insertion loss can be calculated with a knowledge of the amplitude and phase imbalances to obtain the actual combined output power. If more than one stage of combining is used, the relative phase of the previous combiners output voltage is required to calculate the phase imbalance of the succeeding combiners.

Results

Figures 2 and 3 are plots of the excess insertion loss (equation 7) on two different scales. Figures 4 and 5 are plots of the combiner phase shift (equation 8) on two different scales. Figures 2 and 4 provide an overall understanding of the combiner behavior while Figures 3 and 5 are useful for designing combiner systems of power amplifier modules. These plots are valuable aids for determining acceptable amounts of phase and amplitude imbalance among multiple combined amplifier modules.

CAD of Multiple Combined Amplifiers

As the number of combined modules gets larger than four, the amount of calculation work becomes quite tedious. A computer spreadsheet program can be used to rapidly analyze large trees of combiners.

Such a program using Lotus 1-2-3-® and the equations derived earlier has been derived. The spreadsheet is most valuable when evaluating different amounts of phase and amplitude imbalance and their affect upon combiner output power and bridging resistor power. The amplitude imbalance can be taken to the extreme such as a module failure with the combiner results immediately available from the spreadsheet model.

An example of spreadsheet analysis is shown in Figure 6. It is set up for combining up to 16 modules using four stages of 2- port combiners. The matched loss in dB of each combiner is input as well as the individual module power output in watts and its phase relative to a chosen reference (module #1). The Lotus 1-2-3® spreadsheet then calculates four parameters at each two-port combiner which are:

1. Phase in degrees relative to a reference (module #1 in this case) of the combiner output signal.
2. The excess insertion loss in dB of the combiner due to phase and amplitude imbalance between the input signals.
3. The combiner output power in watts.
4. The power dissipated in the combiner bridging resistor.

The equations used in the spreadsheet are derived as follows (referring back to figure 1):

$$P_A + P_B = P_O + P_{BR} + P_L$$

Where: P_A & P_B = input power to the combiner ports A & B P_{BR} = power dissipated in the bridging resistor due to the excess insertion loss of the combiner.

P_O = combiner output power delivered to a matched load.

ANALYSIS OF MULTI-STAGE COMBINING OF AMPLIFIER MODULES	Input Variable Column	FIRST STAGE				SECOND STAGE				THIRD STAGE				FOURTH STAGE			
		PHASE DEGREES	EXCESS I.L.(DB)	OUTPUT POWER	BRIDGING POWER	PHASE DEGREES	EXCESS I.L.(DB)	OUTPUT POWER	BRIDGING POWER	PHASE DEGREES	EXCESS I.L.(DB)	OUTPUT POWER	BRIDGING POWER	PHASE DEGREES	EXCESS I.L.(DB)	OUTPUT POWER	BRIDGING POWER
#1 Power, Watts	280.00																
#1 Phase, Degrees REF	0.00																
1ST STAGE COMB. I.L., DB	-0.20	-1.52	-0.00	522.91	0.42												
#2 Power, Watts	268.00																
#2 Phase, Degrees	-3.00																
2ND STAGE COMB. I.L., DB	-0.20					-3.15	-0.43	572.82	58.99								
#3 Power, Watts	0.01																
#3 Phase, Degrees	-4.00																
1ST STAGE COMB. I.L., DB	-0.20	-3.99	-2.96	138.66	135.43												
#4 Power, Watts	287.00																
#4 Phase, Degrees	-3.00																
3RD STAGE COMB. I.L., DB	-0.20									-0.94	-0.10	1487.19	33.49				
#5 Power, Watts	287.00																
#5 Phase, Degrees	2.00																
1ST STAGE COMB. I.L., DB	-0.20	0.99	-0.00	534.55	0.25												
#6 Power, Watts	273.00																
#6 Phase, Degrees	0.00																
2ND STAGE COMB. I.L., DB	-0.20					2.01	-0.00	1019.53	0.32								
#7 Power, Watts	266.00																
#7 Phase, Degrees	4.00																
1ST STAGE COMB. I.L., DB	-0.20	3.02	-0.00	533.37	0.47												
#8 Power, Watts	293.00																
#8 Phase, Degrees	2.00																
4TH STAGE COMB. I.L., DB	-0.25													-0.67	-0.02	3218.28	14.07
#9 Power, Watts	272.00																
#9 Phase, Degrees	-5.00																
1ST STAGE COMB. I.L., DB	-0.20	-1.53	-0.02	525.16	1.99												
#10 Power, Watts	280.00																
#10 Phase, Degrees	2.00																
2ND STAGE COMB. I.L., DB	-0.20					-1.01	-0.00	1010.16	0.10								
#11 Power, Watts	270.00																
#11 Phase, Degrees	0.00																
1ST STAGE COMB. I.L., DB	-0.20	-0.49	-0.00	532.71	0.18												
#12 Power, Watts	288.00																
#12 Phase, Degrees	-1.00																
3RD STAGE COMB. I.L., DB	-0.20									-0.36	-0.00	1936.69	0.26				
#13 Power, Watts	268.00																
#13 Phase, Degrees	1.00																
1ST STAGE COMB. I.L., DB	-0.20	-0.96	-0.01	533.90	0.90												
#14 Power, Watts	292.00																
#14 Phase, Degrees	-3.00																
2ND STAGE COMB. I.L., DB	-0.20					0.30	-0.00	1018.07	0.49								
#15 Power, Watts	271.00																
#15 Phase, Degrees	5.00																
1ST STAGE COMB. I.L., DB	-0.20	1.56	-0.02	532.67	2.13												
#16 Power, Watts	289.00																
#16 Phase, Degrees	-2.00																

FIGURE 6 EXAMPLE OF MULTI-STAGE COMBINER ANALYSIS ON A SPREADSHEET

Figure 6. Example of multi-stage combiner analysis on a spreadsheet.

P_L = power lost as heat in the combiner as a result of the matched insertion loss.

I.L. = insertion loss (dB) of the combiner under matched conditions and no phase or amplitude imbalance on the two input signals.

X.I.L. = excess insertion loss (dB) due to amplitude and/or phase imbalance between the two input signals.

The power lost as heat in the combiner comes off the top:

$$P_L = P_A + P_B - \text{Log}^{-1}\left[\frac{\text{I.L.}}{10} + \text{Log}\,(P_A + P_B)\right]$$

The power out of the combiner is found from the following:

$$P_O = \text{Log}^{-1}\left\{\frac{\text{X.I.L.}}{10} + \text{Log}\left[\text{Log}^{-1}\left(\frac{\text{I.L.}}{10} + \text{Log}\,(P_A + P_B)\right)\right]\right\}$$

Likewise the power in the bridging resistor:

$$P_{BR} = \text{Log}^{-1}\left[\frac{\text{I.L.}}{10} + \text{Log}\,(P_A - P_B) \; -P_O\right]$$

The phase of the various modules and combiner outputs are referenced to module #1, thus 0 is entered for it and all other signals are referenced to it. The phase angle of a combiner output is:

$$\Phi = \Theta_A - \text{Tan}^{-1}\left[\frac{\text{Sin}\,(\Theta_A - \Theta_B)}{\frac{1}{\sqrt{P_r}} + \text{Cos}\,(\Theta_A - \Theta_B)}\right]$$

Where: Θ_A and Θ_B are the phase angles of the signals at port A & B and $P_R = P_A/P_B$

Remember that Lotus 1-2-3® requires all angles of trig functions to be in radians. Multiply an angle in degrees by Pi/180 to convert to radians. Divide an angle in radians by Pi/180 to convert back to degrees.

Figure 7 is an example of the Lotus 1-2-3® combiner three model. It is interesting to note the affect through the combiner tree of module #3 failure.

Summary

The equations modeling the behavior of two-port power combiners have been derived. The derivations are presented in detail so the reader may be assured of their accuracy and to enable one to similarly analyze combiners with an unusual number of input ports (i.e. 3-port or 7-port). Results are presented in graphical form to provide an intuitive understanding of combiner behavior. Finally a unique approach to computer aided design of large combining trees is presented using Lotus 1-2-3 spreadsheet software. rf

References

1. Earnest A. Franke, "Excess Insertion Loss at the Input Ports of a Combiner Hybrid", *RF Design,* Nov 1985, p. 43-48.

2. Roderick K. Blocksome, "Practical Wideband RF Power Transformers, Combiners, and Splitters", *Proceedings of RF Technology Expo 86,* p. 207-227.

About the Author

Roderick Blocksome is Manager of HF Power Amplifier Group at Rockwell International Corp., Collins Defense Communications, 855 35th St., NE, Cedar Rapids, IA 52498.

Improving the Graceful-Degradation Performance of Combined Power Amplifiers

ADEL A. M. SALEH, SENIOR MEMBER, IEEE

Abstract—The standard method of employing n-way hybrid power divider/combiner to combine n amplifiers offers some sort of graceful-degradation performance when one or more of the amplifiers fail. Two schemes are discussed that improve that performance significantly. The first, which was proposed previously, involves replacing each failed amplifier by an equal-delay "through" connection. The second involves using a resistor-free combiner, and separating each failed amplifier from the combiner by an appropriately placed short or open circuit.

I. INTRODUCTION

CONSIDER n identical amplifiers whose output powers are being combined by means of an n-way power divider and an n-way power combiner as shown in Fig. 1. A basic requirement of such an arrangement, beside the obvious one of obtaining n times the output power of a single amplifier, is for the total output power to fail gracefully as one or more of the component amplifiers suffer arbitrary modes of failure. The use of n-way hybrid power dividers/combiners [1]–[6] (i.e., in Fig. 1, ports $1',2',\cdots,n'$ are reasonably matched and isolated from one another, and so are ports $1,2,\cdots,n$) is the usually accepted way of accomplishing this requirement [7]. In that case, when one or more amplifiers fail, the input and output impedances of each of the remaining amplifiers are unchanged. Thus, if m amplifiers fail, the output voltage at port 0 becomes simply $(n-m)/n$ times the corresponding no-failure voltage; and hence [7]

$$P_{\text{out}}/P_{\text{out,max}} = (1-m/n)^2. \tag{1}$$

That quantity, in decibels, is plotted versus m/n by the equal-segment dashed line in Fig. 2, which, for example, shows about a 2-dB drop in the output power when one fifth of the amplifiers fail completely, or a 6-dB drop when half of the amplifiers fail. While this represents some sort of graceful-degradation performance, the power output is still significantly smaller than the sum of the output powers of the remaining amplifiers, which is given by

$$P_{\text{remaining}}/P_{\text{out, max}} = 1-m/n. \tag{2}$$

That quantity, in decibels, is plotted versus m/n by the solid line Fig. 2, which, for example, shows only about a 1-dB drop in the output power when one fifth of the amplifiers fail completely, or only a 3-dB drop when half of the amplifiers fail.

Manuscript received May 7, 1980; revised June 24, 1980.
The author is with Bell Laboratories, Crawford Hill Laboratory, Holmdel, NJ 07733.

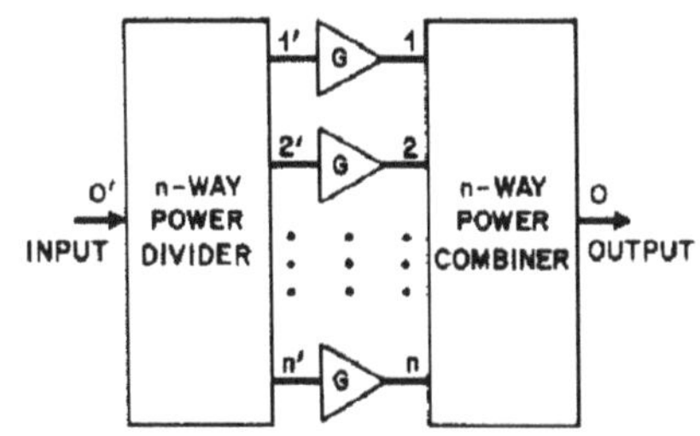

Fig. 1. Combining n identical amplifiers by means of an n-way power divider and an n-way power combiner.

The purpose of this paper is to discuss two combining schemes that give significantly better graceful-degradation performance than that represented by (1).

II. THE FIRST SCHEME

It was suggested by Ernst, Camisa, and Presser [7] to replace each failed amplifier by a "through" connection having the same phase delay as a working amplifier. In that case, it can be shown that if the gain of each amplifier is G, the total output power when m out of the n amplifiers fail is given by

$$P_{\text{out}}/P_{\text{out, max}} = \left[1-(1-1/\sqrt{G})m/n\right]^2 \tag{3}$$

which approaches (1) as G approaches ∞. This quantity, in decibels, is plotted versus m/n by the dotted lines in Fig. 2 for $G=3$, 6, and 10 dB. The figure shows a marked improvement over the performance of (1) (equal-segment dashed line). When $G \leqslant 6$ dB, one even gets better performance than that of (2) (solid line). This is possible since a part of the input power to the failed amplifiers is recovered in the output.

The problem with this scheme is in the practical difficulty of replacing a failed amplifier with an equal-phase "through" connection. A more practical scheme is described next.

III. THE SECOND SCHEME

As previously, this scheme requires a *hybrid* power divider in the input side of the amplifiers, but a *nonhybrid*, i.e., lossless, power combiner at the output side. For example, assuming that the amplifiers are driven in phase, the power combiner could be of the Wilkinson [1], radial [2]–[4] or fork [4]–[6] type without any isolation resistors, as shown schematically in Fig. 3. In addition, each of ports 1–n of the combiner is equipped with a shorting device which is activated when the amplifier connected to it fails. (The failed amplifier can be identified, for exam-

0018-9480/80/1000-1068$00.75 ©1980 IEEE

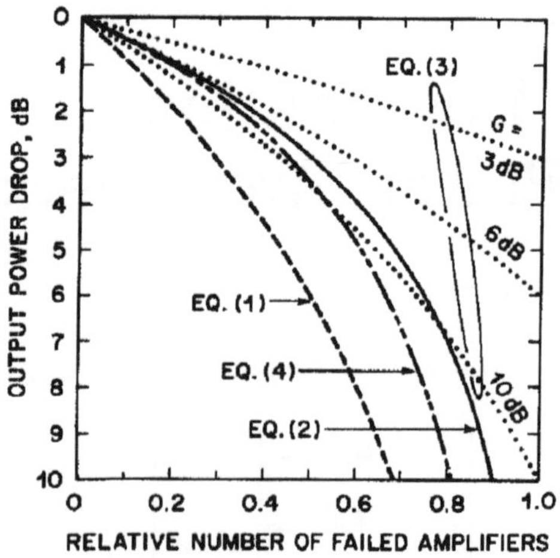

Fig. 2. Degradation of the output power of the combined amplifiers of Fig. 1 versus the relative number of failed amplifiers. The solid line represents the total output power of the remaining amplifiers, the equal-segment dashed line represents the standard scheme of using hybrid power divider/combiner, the dotted lines represent the first improved scheme, and the unequal-segment dashed line represents the second improved scheme.

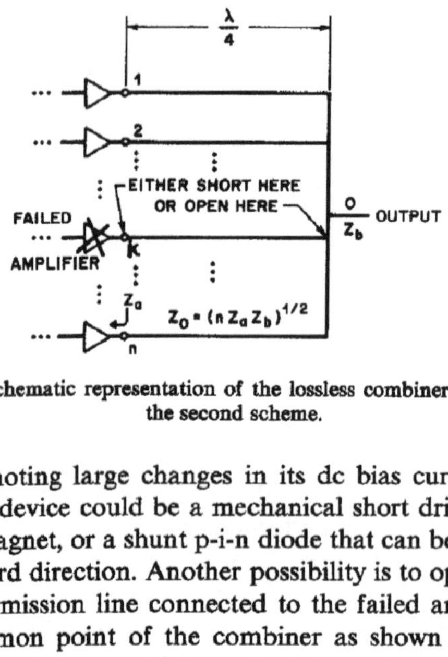

Fig. 3. Schematic representation of the lossless combiner discussed in the second scheme.

ple, by noting large changes in its dc bias current.) The shorting device could be a mechanical short driven by an electromagnet, or a shunt p-i-n diode that can be biased in its forward direction. Another possibility is to open circuit the transmission line connected to the failed amplifier at the common point of the combiner as shown in Fig. 3. The open circuit could be activated mechanically by an electromagnet, or by using a bonded wire that can be burned out by passing through it a large current pulse as in a fuse. A third possibility exists if the devices being combined give a repeatable *reactive* output impedance when they fail. In this case, they could be placed at a distance from the common point of the combiner such that they present an open circuit at that point in the event of failure. This would eliminate the need for external shorting or opening devices.

With any of the above provisions, it can be shown that if m out of the n amplifiers fail, the total output power is

$$P_{\text{out}}/P_{\text{out, max}} = [1 - m/(2n-m)]^2. \qquad (4)$$

This quantity, in decibels, is plotted versus m/n by the unequal-segment dashed line in Fig. 2, which, for example, shows only about 1-dB drop in the output power when one fifth of the amplifiers fail completely, or only a 3.5-dB drop when half of the amplifiers fail. It is observed that (4) approaches (2) for small values of m/n.

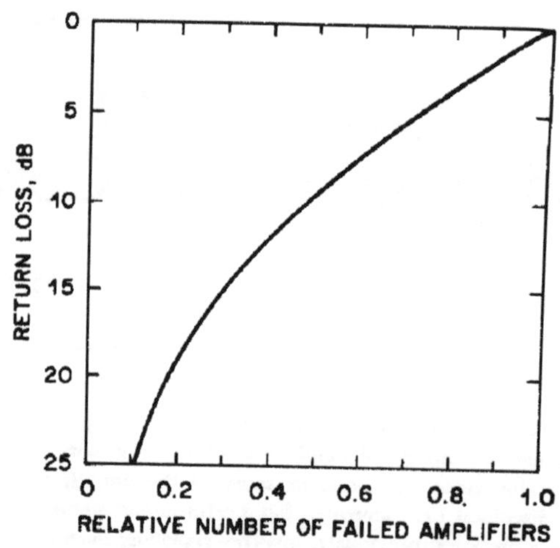

Fig. 4. The return loss presented to the working amplifiers in the second scheme versus the relative number of failed amplifiers.

It can be shown that the return loss presented to each of the working amplifiers is given by

$$\text{Return Loss} = -20 \log [m/(2n-m)] \text{ dB}. \qquad (5)$$

That quantity is plotted in Fig. 4, which, for example, shows a return loss of 19.1 dB when one fifth of the amplifiers fail, and 9.5 dB when half of the amplifers fail. That return loss was taken into account in deriving (4) under the assumption that the *available* power from each of the working amplifiers is constant. That would be the case, for example, if these amplifiers are linear. On the other hand, if they are nonlinear, the so called "load-pull effect" could reduce the available power from the working amplifiers, thus introducing an additional degradation in (4). However, because of the rather slow degradation of the return loss indicated by (5) and Fig. 4, the load-pull effect is expected to be negligible perhaps up to the point where half of the amplifiers fail.

It is worth noting that in the other schemes, as well as in the input side of the present scheme, the working amplifiers remain matched independently of the number of failures.

IV. Conclusions

The standard method of employing hybrid n-way power divider/combiner to combine n amplifiers offers some sort of graceful-degradation performance when one or more of the amplifiers fail. Two schemes have been discussed that improve that performance significantly. The first, which has been previously suggested in the literature [7], involves replacing each failed amplifier by an equal-phase "through" connection. While that scheme gives good performance especially when the amplifiers have a small gain, it is hard to implement in practice. The second scheme, which gives good performance independently of the gain of the amplifiers, involves the use of a lossless combiner equipped with short-circuiting or open-circuiting devices. That scheme is more practially realizable than the first scheme, and offers the added advantage of not needing isolation resistors in the power combiner, and hence, it has no problems of power dissipation that would occur when such resistors are used.

Acknowledgment

The author thanks Dr. James W. Gewartowski of Bell Laboratories, and all of the MTT reviewers for useful comments and suggestions.

References

[1] E. J. Wilkinson, "An *N*-Way Hybrid Power Divider," *IRE Trans. Microwave Theory Tech.*, vol. MTT-8, pp. 116–118, Jan., 1960.

[2] J. M. Schellenberg and M. Cohn, "A Wideband Radial Power Combiner for FET Amplifiers," *1978 IEEE Int. Solid-State Circuit Conf. Dig.*, Feb., 1978, pp. 164–165, 273.

[3] M. Cohn, B. D. Geller, and J. M. Schllenberg, "A 10-Watt Broadband FET Combiner/Amplifier," *1979 IEEE MTT-S Int. Microwave Symp. Dig.*, April, 1979 pp. 292–297.

[4] A. A. M. Saleh, "Planar, Electrically-Symmetric, N-Way, Hybrid Power Dividers/Combiners," *IEEE Trans. Microwave Theory Tech.*, vol. MTT-28, pp. 555–563, June 1980.

[5] Z. Galani and S. J. Temple, "A Broadband *N*-Way Combiner/Divider," *1977 IEEE MTT-S Int. Microwave Symp. Dig.*," June, 1977, pp. 499–502.

[6] N. Nagai, E. Maekaura, and K. Ono, "New *n*-Way Hybrid Power Dividers," *1977 IEEE MTT-S Int. Microwave Symp. Dig.*, June 1977, pp. 503–504.

[7] R. L. Ernst, R. L. Camisa, and A. Presser, "Graceful Degradation Properties of Matched *n*-Port Power Amplifier Combiners," *1977 IEEE MTT-S Int. Microwave Symp. Dig.*, June 1977, pp. 174–177.

Chapter 6

QUADRATURE COMBINERS AND SPLITTERS

Quadrature splitters and combiners are a type of hybrid splitter and combiner that employ 90° phase differences between the two outputs or inputs. A power-amplifier (PA) system based upon quadrature couplers is shown in Figure 6.1. The splitter produces drive signals that are of equal amplitudes, but introduces a 90° phase shift in the drive signal to PA#2. The combiner introduces a 90° phase shift in its input from PA#1 so that the two signals are combined without loss. Quadrature-combined power amplifiers are also known as "balanced power amplifiers."

Quadrature combining of power amplifiers offers the following:

- Matched input impedance, given equal PA-input impedances;
- PA load impedance independent of the output of the other PA;
- System power output independent upon load angle;
- System power output decreases slowly as SWR increases;
- Cancellation of third harmonics;
- Cancellation of back-IMD.

It is often thought by design engineers that all power reflected by a mismatched load is dumped into the hybrid resistor and the drain impedances are therefore constant for all loads. This is shown subsequently not to be true and it is in fact shown that variations in the PA drain impedances are greater than the variations in the system load impedance.

A basic implementation of a quadrature-combined PA system (Figure 6.2) uses Ruthroff 180° hybrid transformers, baluns, and quarter-wavelength transmission lines. The Ruthroff transformers T2 and T3 and dump resistors R1 and R2 comprise the hybrid splitter and combiner. Baluns T1 and T4 allow one side

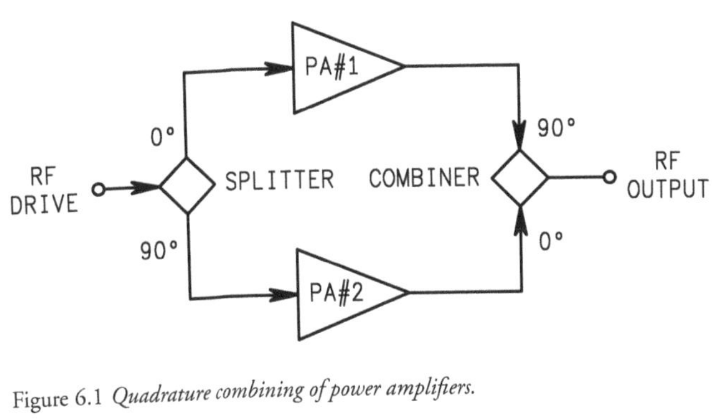

Figure 6.1 *Quadrature combining of power amplifiers.*

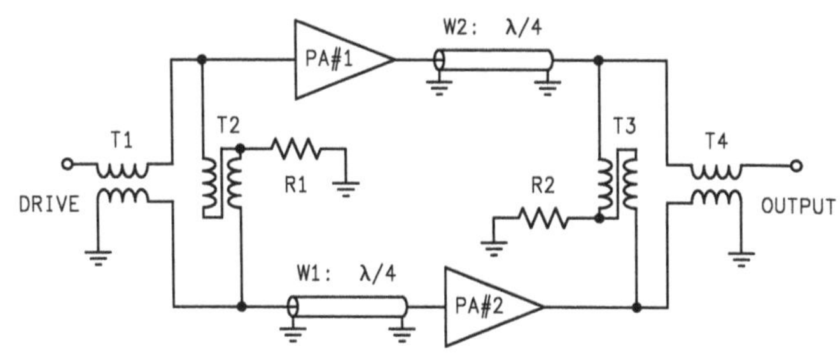

Figure 6.2 *Implementation of quadrature-combined power amplifiers.*

of the input or output to be grounded. Quarter-wavelength transmission lines W1 and W2 are simple means of obtaining the required 90° phase shifts and (as shown subsequently) impedance inversions. Most HF and VHF implementations use lumped-element T or Π networks that provide the 90° phase shift and impedance inversion over a relatively large bandwidth.

This chapter first presents a detailed analysis of the quadrature splitter and quadrature combiner. The performance of a quadrature-combined PA system is then determined, based upon ideal class-B linear RF power amplifiers. Also analyzed is cancellation of the third harmonic and backward intermodulation distortion (IMD). Literature applicable to both theory and implementation is then reviewed. Selected reprints appear at the end of the chapter.

6.1 BASIC THEORY OF QUADRATURE SPLITTER

The quadrature splitter (Figure 6.3) adds a 90° phase shift and impedance inversion (represented by a quarter-wavelength line of characteristic impedance R_L) to one output of the hybrid splitter (Figure 5.8). The impedance presented to the lower terminal of T2 by line W1 and load Z_2 is

$$Z_3 = \frac{R_L^2}{Z_2} \tag{6.1}$$

As discussed in Section 5.3, proper operation of the hybrid requires nominal loads of $Z_1 = Z_3 = 2R_h$. This implies that $R_L = 2R_h$ and the nominal value of Z_2 is also R_h. Consequently, for an arbitrary value of Z_2,

$$Z_3 = \frac{4R_h^2}{Z_2} \tag{6.2}$$

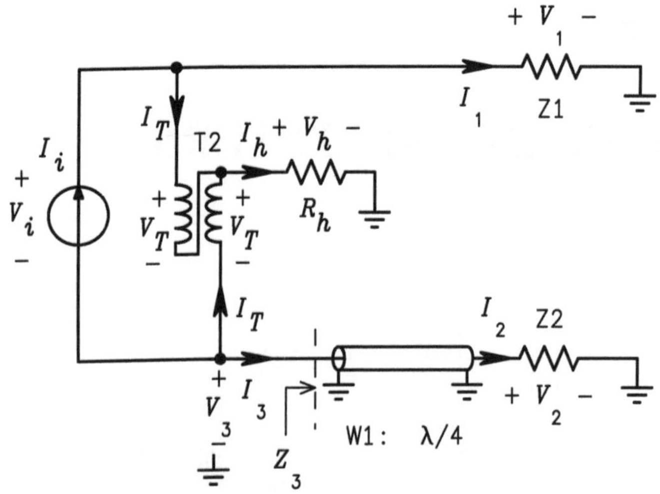

Figure 6.3 *Simplified circuit of quadrature splitter.*

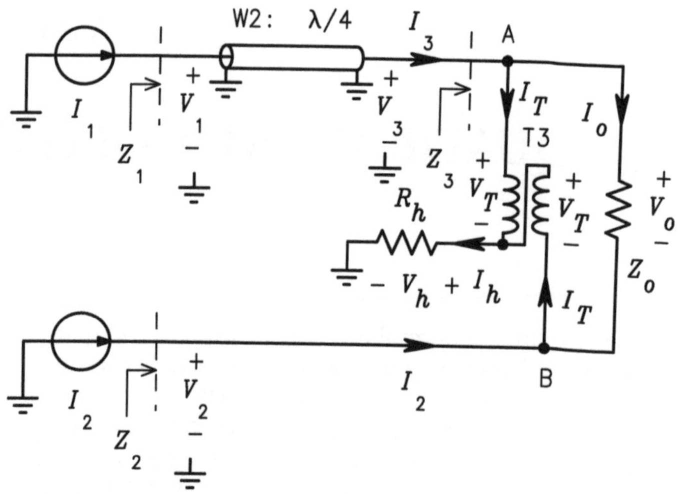

Figure 6.4 *Simplified circuit of quadrature combiner.*

The input impedance and other parameters of the quadrature-hybrid splitter are found by replacing Z_2 in the equations of Section 5.3 with Z_3 above.

Under normal operating conditions, the input impedances of the two PAs are expected to be similar, although not necessarily the desired nominal resistance. If $Z_2 = Z_1$, then insertion of Z_3 into (5.14) produces

$$Z_i = \frac{4[4R_h^2 + (Z_1 + 4R_h^2/Z_1)R_h]}{Z_1 + 4R_h^2/Z_1 + 4R_h} = 4R_h \tag{6.3}$$

The quadrature splitter thus produces the nominal, resistive input impedance as long as its two load impedances are equal. Power that is not absorbed by the two loads is dumped into R_h.

6.2 QUADRATURE COMBINER WITH NOMINAL LOAD

The quadrature power combiner (Figure 6.4) is modeled as a 180° hybrid combiner with a quarter-wavelength transmission line W2 between point A on T3 and PA #1 to produce the 90° phase shift and impedance inversion. Derivation of the basic equations for the combiner is achieved by superpositioning the responses to PAs #1 and #2.

Quarter-Wavelength Transmission Line

Power transmitted through a lossless line must be conserved, thus

$$\frac{|V_3|^2}{2R_3} = \frac{|I_1|^2 R_1}{2} \tag{6.4}$$

and

$$|V_3|^2 = |I_1|^2 R_1 R_3 \tag{6.5}$$

For a quarter-wavelength line, $R_L^2 = R_1 R_3$, thus

$$V_3 = -jI_1 R_L \tag{6.6}$$

where $-j$ indicates a phase lag by 90°. Similarly,

$$V_1 = +jI_3 R_L \tag{6.7}$$

Current-source excitation of the left side of W2 is therefore turned into voltage-source excitation of the upper connection of Ruthroff transformer T3.

The nominal input impedance at A on T3 is $R_o/2$, hence the value of R_L is set (ideally) at

$$R_L = R_o/2 = 2R_h \tag{6.8}$$

which causes

$$V_3 = -j2I_1 R_h \tag{6.9}$$

Similarly, (6.7) becomes

$$V_1 = +j2I_3 R_h \tag{6.10}$$

Current-Source Excitation

From the viewpoint of the current-source excitation I_2 at B, the voltage source at A is equivalent to a ground connection. The first problem is, therefore, to determine how much of the output current flows into the ground connection. Since the voltages on the two windings of T2 must be equal,

$$V_h = -\frac{1}{2}V_o = -\frac{1}{2}I_o Z_o \tag{6.11}$$

Since the voltage on R_h also depends upon the current flowing through it,

$$V_b = I_b R_b = 2 I_T R_b \tag{6.12}$$

Comparison of these yields

$$I_T = -I_o \frac{Z_o}{4R_b} \tag{6.13}$$

The drive current is split between the transformer and load, thus

$$I_2 = I_T - I_o = -I_o\left(1 + \frac{Z_o}{4R_b}\right) \tag{6.14}$$

and

$$I_o = -I_2 \frac{4R_b}{Z_o + 4R_b} \tag{6.15}$$

From this,

$$V_o = I_o Z_o = -I_2 Z_o \frac{4R_b}{Z_o + 4R_b} \tag{6.16}$$

The current in R_b is (via rearrangement of (6.15)

$$I_b = 2I_T = 2(I_2 + I_o) = I_2 \frac{2Z_o}{Z_o + 4R_b} \tag{6.17}$$

so

$$V_b = I_b R_b = I_2 R_b \frac{2Z_o}{Z_o + 4R_b} \tag{6.18}$$

Finally,

$$V_2 = -V_o = I_2 Z_o \left(\frac{4R_b}{Z_o + 4R_b}\right) \tag{6.19}$$

The current drawn from transmission line W2 is obtained from (6.15) and (6.13):

$$I_3 = I_o + I_T = -I_2 \frac{Z_o - 4R_b}{Z_o + 4R_b} \tag{6.20}$$

The voltage induced on PA #1 is obtained by inserting (6.20) into (6.10), which yields

$$V_1 = jI_2 \frac{2R_b(Z_o - 4R_b)}{Z_o + 4R_b} \tag{6.21}$$

Voltage-Source Excitation
From the viewpoint of the voltage-source excitation V_3 at A, the current source at B is equivalent to an open circuit. Transformer current I_T is therefore equal to output current I_o and

$$I_b = 2I_T = 2I_o = I_3 \tag{6.22}$$

Since the voltages on the two windings of T3 must be equal,

$$V_T = \frac{1}{2} V_o = \frac{1}{2} I_o Z_o = \frac{1}{4} I_3 Z_o \tag{6.23}$$

Since $V_b = I_b R_b$,

$$V_3 = V_T + V_b = I_3\left(R_b + \frac{Z_o}{4}\right) \tag{6.24}$$

and

$$I_3 = V_3 \frac{4}{Z_o + 4R_b} \tag{6.25}$$

Since $I_b = I_3$,

$$V_b = V_3 \frac{4R_b}{Z_o + 4R_b} \tag{6.26}$$

Since $I_o = I_3/2$,

$$I_o = V_3 \frac{2}{Z_o + 4R_b} \tag{6.27}$$

and

$$V_o = V_3 \frac{2Z_o}{Z_o + 4R_b} \tag{6.28}$$

Finally, combining (6.23) and (6.28) yields

$$V_2 = V_b - V_T = I_3(R_b - Z_o/4) = -V_3 \frac{Z_o - 4R_b}{Z_o + 4R_b} \tag{6.29}$$

Excitation by Current Source Through Transmission Line
The partial responses due to excitation by PA #1 are obtained by insertion of (6.9) into (6.25)–(6.29), which produces

$$I_3 = -jI_1 \frac{8R_b}{Z_o + 4R_b} \tag{6.30}$$

$$V_b = -jI_1 \frac{8R_b^2}{Z_o + 4R_b} \tag{6.31}$$

$$I_o = -jI_1 \frac{4R_b}{Z_o + 4R_b} \tag{6.32}$$

$$V_o = -jI_1 \frac{4Z_o R_b}{Z_o + 4R_b} \tag{6.33}$$

and

$$V_2 = +jI_1 \frac{2R_b(Z_o - 4R_b)}{Z_o + 4R_b} \tag{6.34}$$

Applying (6.10) to (6.30) gives

$$V_1 = I_1 \frac{16R_b^2}{Z_o + 4R_b} \tag{6.35}$$

Combined Response

The output current is obtained by summing (6.32) and (6.15), which produces

$$I_o = (-jI_1 - I_2)\frac{4R_b}{Z_o + 4R_b} \tag{6.36}$$

hence

$$V_o = (-jI_1 - I_2)Z_o\frac{4R_b}{Z_o + 4R_b} \tag{6.37}$$

The current in R_b is similarly determined by summing (6.31) and (6.17), thus

$$I_b = -jI_1\frac{8R_b}{Z_o + 4R_b} + I_2\frac{2Z_o}{Z_o + 4R_b}$$

$$= \frac{2(-j4R_bI_1 + Z_oI_2)}{Z_o + 4R_b} \tag{6.38}$$

$$V_b = \frac{2R_b(-j4R_bI_1 + Z_oI_2)}{Z_o + 4R_b} \tag{6.39}$$

The voltage induced at the output of PA #1 is obtained by summing (6.35) and (6.21), which produces

$$V_1 = I_1\frac{16R_b^2}{Z_o + 4R_b} + jI_2\frac{2R_b(Z_o - 4R_b)}{Z_o + 4R_b} \tag{6.40}$$

The load impedance of PA #1 is therefore

$$Z_1 = \frac{16R_b^2}{Z_o + 4R_b} + \frac{jI_2}{I_1}\frac{2R_b(Z_o - 4R_b)}{Z_o + 4R_b} \tag{6.41}$$

It is apparent that Z_1 depends upon Z_o, R_b, and I_2/I_1.

The voltage produced on PA #2 is obtained by summing (6.34) and (6.19), which yields

$$V_2 = jI_1\frac{2R_b(Z_o - 4R_b)}{Z_o + 4R_b} + I_2Z_o\frac{4R_b}{Z_o + 4R_b} \tag{6.42}$$

The load impedance for PA #2 is therefore

$$Z_2 = \frac{V_2}{I_2} = \frac{jI_1}{I_2}\frac{2R_b(Z_o - 4R_b)}{Z_o + 4R_b} + \frac{4Z_oR_b}{Z_o + 4R_b} \tag{6.43}$$

It is apparent that, in general, Z_2 also depends upon Z_o, R_b, and I_1/I_2 and has a form analogous to that of Z_1.

Nominal Load

The PA-load impedances in general depend upon Z_o, R_b, and the relative amplitudes of I_1 and I_2. If, however, the load resistance is its nominal value and

$$R_b = R_o/4 \tag{6.44}$$

as in an ordinary hybrid combiner, the dependence upon the opposite PA is eliminated. The PA load impedances are then

$$Z_1 = Z_2 = R_o/2 \tag{6.45}$$

which are the same as those of an ordinary hybrid combiner.

If the two PAs produce equal currents, then (6.36) gives the system output current as

$$I_o = I_1 = -jI_2 \tag{6.46}$$

and (6.38) gives

$$I_b = -jI_1 + I_2 = 0 \tag{6.47}$$

With a nominal load, the quadrature combiner behaves just like the ordinary hybrid combiner with the exception of a 90° phase shift of the output.

6.3 QUADRATURE COMBINER WITH ARBITRARY LOAD

The behavior of a quadrature-combined PA system with an arbitrary load is most readily analyzed by assuming perfect matching of the output currents from the two PAs. The equations for currents, voltages, impedances, powers, and the like are first simplified for the case of equal PA outputs and then used to derive equations for output power and efficiency. Behavior and performance characteristics are first reviewed for a given SWR and then as functions of SWR.

Basic Equations

For equal output currents from the two PAs, $I_2 = jI_1$.

Insertion of this into the equations of the previous section yields

$$I_o = -jI_1\frac{8R_b}{Z_o + 4R_b} \tag{6.48}$$

$$V_o = -jI_1Z_o\frac{8R_b}{Z_o + 4R_b} \tag{6.49}$$

$$I_b = jI_1\frac{2Z_o - 8R_b}{Z_o + 4R_b} \tag{6.50}$$

$$V_b = jI_1R_b\frac{2R_o - 8R_b}{Z_o + 4R_b} \tag{6.51}$$

$$V_1 = I_1\frac{24R_b^2 - 2R_bZ_o}{Z_o + 4R_b} \tag{6.52}$$

$$Z_1 = \frac{24R_b^2 - 2R_bZ_o}{Z_o + 4R_b} \tag{6.53}$$

$$V_2 = jI_1\frac{6Z_oR_b - 8R_b^2}{Z_o + 4R_b} \tag{6.54}$$

$$Z_2 = \frac{6Z_oR_b - 8R_b^2}{Z_o + 4R_b} \tag{6.55}$$

Power Output of System

The power output is given by

$$P_o = \frac{1}{2}\mathrm{Re}[V_oI_o^*] \tag{6.56}$$

Insertion of (6.48) and (6.49) yields

$$P_o = \frac{|I_1|^2}{2}\mathrm{Re}\left[\frac{64Z_oR_b^2}{(Z_o + 4R_b)(Z_o + 4R_b)^*}\right] = \frac{32R_b^2R_o|I_1|^2}{(Z_o + 4R_b)(Z_o^* + 4R_b)} \tag{6.57}$$

Since the nominal load impedance is $4R_b$, the load impedance corresponding to a given reflection coefficient Γ is

$$Z_o = \frac{1 + \Gamma}{1 - \Gamma}(4R_b) = \frac{1 - |\Gamma|^2 + 2j\mathrm{Im}[\Gamma]}{(1 - \Gamma)(1 - \Gamma^*)}(4R_b) \tag{6.58}$$

Expansion of the denominator of (6.57) by use of (6.58) similarly yields

$$(Z_o + 4R_b)(Z_o^* + 4R_b) = \frac{64R_b^2}{(1 - \Gamma)(1 - \Gamma^*)} \tag{6.59}$$

Insertion of (6.59) and the resistance from (6.58) into (6.57) now produces

$$P_o = \frac{32R_b^2 \cdot 4R_b(1 - |\Gamma|^2)|I_1|^2}{64R_b^2} = 2|I_1|^2R_b(1 - |\Gamma|^2) \tag{6.60}$$

Since $|\Gamma| = (S - 1)/(S + 1)$, it is apparent that the system output power is a function of SWR and constant around a given SWR circle.

Power Delivered by PAs

Summation of (6.53) and (6.55) yields

$$Z_1 + Z_2 = \frac{4Z_oR_b + 16R_b^2}{Z_o + 4R_b} = 4R_b. \tag{6.61}$$

Consequently, $R_1 + R_2$ is also constant as is $P_1 + P_2$, which is the total power delivered by the PAs to the combiner. Since the input and output RF powers of the combiner are constant for a given SWR, the combiner efficiency is also constant for a given SWR.

Supply Voltage

The supply voltage required to ensure linear operation at full output is determined by the maximum PA drain-load impedance and the current required to deliver the desired full output power to that load impedance. On a given SWR circle, the magnitude of the impedance is highest at the right-side pure-resistance point. At this point,

$$Z_o = 4SR_b \tag{6.62}$$

Insertion of this into (6.55) yields

$$Z_{\mathrm{MAX}} = \frac{6S - 2}{S + 1}R_b = (1 + 2|\Gamma|)(2R_b) \tag{6.63}$$

The minimum supply voltage is then

$$V_{DD} = |I_1|Z_{\mathrm{MAX}} \tag{6.64}$$

This value can be scaled upward to allow for the voltage drop across the on-state resistance of the transistors. Allowance for the effects of load reactance is included in (6.64). For example, a higher supply voltage is required to drive the same power into a load with a higher impedance [1].

PA Performance

For two ideal class-B PAs [1] with equal currents,

$$I_{\mathrm{dc}} = 2(2/\pi)|I_1| = (4/\pi)|I_1| \tag{6.65}$$

the total dc current is the same whether the PAs are single ended (total of two FETs) or push-pull (total of four FETs). The resultant dc-input power is

$$P_i = V_{DD}I_{\mathrm{dc}} \tag{6.66}$$

The system efficiency is determined by combining (6.66), (6.65), (6.64), (6.63), and (6.60), which produces

$$\eta_S = \frac{P_o}{P_i} = \frac{\pi}{4}\frac{1 - |\Gamma|^2}{1 + 2|\Gamma|} = \frac{\pi}{4}\frac{4S}{3S^2 + 2S - 1} \tag{6.67}$$

For class-A PAs [1], the factor $4/\pi$ is replaced by 2, corresponding to a PEP efficiency of 50%. For class-F PAs, a factor corresponding to the harmonics in the current waveform is substituted [2]. A measured factor for a real PA can also be used.

Operation Around SWR Circle

It is first instructive to examine the behavior of a quadrature-combined PA system on the 2:1-SWR circle, and to compare their behavior to that of a hybrid-combined system.

Linear class-B PAs behave as fixed sources of RF current. The dc-input current (hence dc-input power) is proportional to the amplitude of the RF current and therefore fixed, given a fixed supply voltage.

With equal PA-output currents, the current in the hybrid resistor of a hybrid combiner is zero, hence no power is dissipated in the hybrid resistor. The combiner output power is therefore directly proportional to the series resistance of the load. In an ordinary hybrid combiner with equal PA-output currents, the PA-load impedance is the transformed combiner-load impedance.

The variations of output power and efficiency of hybrid-coupled linear class-B PAs on the 2:1 SWR circle are shown in Figure 6.5. In this and subsequent figures, the load angle γ is defined as 0 at $Z = \infty$ and increases positively with counterclockwise rotation. The nominal load impedance is $R_o = 2\Omega$ so that the nominal values of R_1 and R_2 are 1Ω. This normalization and the associated normalization of the nominal supply voltage to 1V are also used in subsequent graphs. With the nominal load, a supply voltage of V_{DD} = 1V is sufficient for linear operation of the PAs and a maximum combiner output of 1 W. For linear operation at all impedances on the 2:1-SWR circle, however, V_{DD} must be increased to 2V. The output power varies over a 6 dB range from 2W at $Z_o = 2\Omega$ to 0.5W at $Z_o = 1/2\Omega$. The efficiency similarly varies by a factor of 4 from 0.785 to 0.195.

In contrast, the output power and efficiency of a quadrature-combined pair of linear class-B PAs remain constant for all impedances on the $S = 2$ circle (Figure 6.5). The nominal load impedance is $R_o = 1\Omega$ as for the hybrid combiner, but the supply voltage is reduced to 1.67V. The power output is slightly lower (0.889 or −0.51 dB) than the 1-W output for a nominal load. However, the efficiency is reduced from 0.785 for the nominal load to 0.419.

The variations of the drain-load impedances around the 2:1-SWR circle are shown in Figure 6.6. The magnitudes of the impedances range from 0.333 to 1.667Ω, with $|Z_1|$ peaking when $|Z_2|$ dips and vica versa. The impedance of PA #2 varies roughly with the load impedance (since it is connected directly), while that of PA #1 varies inversely (since it is connected to the load

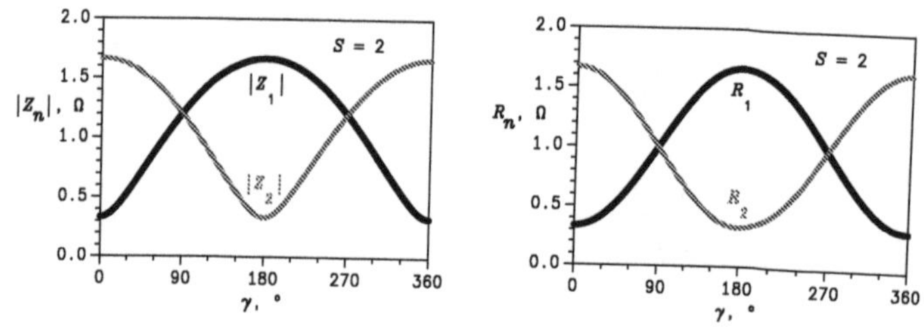

Figure 6.6 *Drain-load impedances on SWR circle for quadrature combiner.*

through an impedance-inverting network). The series resistances vary sinusoidally with load angle, and it is apparent that their sum is a constant value of $4R_h$, which is twice the nominal load resistance. If the PAs are designed for a nominal drain load of 1Ω (corresponding to a nominal combiner load of 2Ω), the SWR at the drain varies from a minimum of 1.667 to a maximum of 3.

With a maximum drain impedance of 1.67V and the RF drain current of 1A, linear operation at all load impedances on the 2:1 SWR circle requires that $V_{DD} \geq 1.667$V. The resultant PA power outputs and dissipations are shown in Figure 6.7. The power output of each PA varies from 0.1667 to 0.8333W and its dissipation varies from 0.8944 to 0.2277. PA #2 delivers the majority of the system output power and dissipates least when $Z_o = 2\Omega$, but delivers only a small fraction of the system output and dissipates most when $Z_o = 1/2\Omega$. Since the sum of R_1 and R_2 is constant, the combined output of the two PAs is also constant. Consequently, the total dissipation is constant as is the combiner efficiency.

The RF drain voltage varies from 0.333 to 1.667V, hence the peak drain voltage varies from 2 to 3.333V. The per-device power-output capability [1] is thus decreased by 60 percent from 0.125 to 0.067. In contrast, operation of a class-B PA directly into a load with $S = 2$ requires a factor of two increase in ratings, with a corresponding decrease in power-output capability to 0.0625.

Stabilization of output power and efficiency by a quadrature combiner are thus obtained at the cost of requiring PAs with higher ratings than are required for operation into a nominal load. The increase in ratings is, however, slightly less than that required for direct delivery of the same power into the combiner load impedance. The individual PAs must operate into load im-

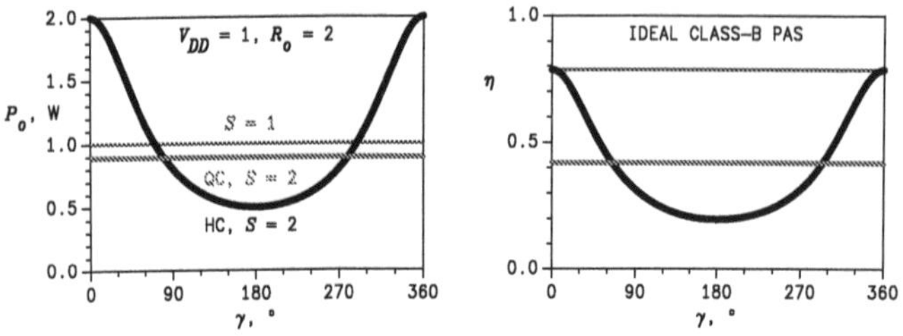

Figure 6.5 *System output power and efficiency on SWR circle.*

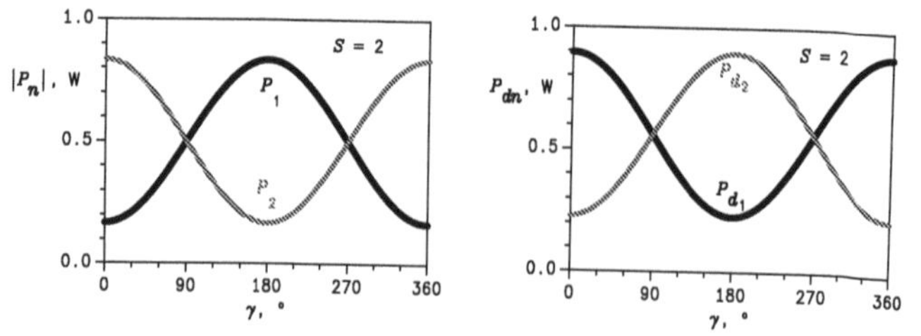

Figure 6.7 *PA output and efficiency around SWR circle.*

pedances that vary considerably and have higher SWRs than does the combiner load itself.

Effect of SWR

The variations of system power output and efficiency with SWR are shown in Figure 6.8 and values are given in Table 6.1. Supply voltage V_{DD} is set to just large enough to ensure linear operation for all impedances on the circle for the specified maximum SWR S_{max}. System power output decreases relatively slowly with SWR, dropping from 1.0 at $S = 1$ to 0.750 (−1.25 dB) at $S = 3$ and 0.556 (−2.54 dB) at $S = 5$.

The system efficiency at the maximum design SWR S_{max} decreases significantly with SWR, dropping from 0.785 at $S = 1$ to 0.294 at 3 and 0.184 at 5. Efficiency increases slightly for loads inside the maximum-SWR circle, as shown by the light lines in Figure 6.8. For example, with a design for S_{max}=3, the efficiency varies from 0.294 at $S = 3$ (outer circle) to 0.393 at $S = 1$ (nominal load).

The load presented to the PAs is variable and generally reactive. The ranges of drain-load impedances are shown as functions of SWR in Figure 6.9. The maximum impedance increases progressively with SWR. The minimum impedance drops to zero (short circuit) at $S = 3$. For $S < 3$, negative drain resistances occur for some system-load impedances (Figure 6.10). When the drain resistance is negative, power is delivered from one PA to the load, hybrid resistor, and the other PA. A negative load resistance does not necessarily preclude proper operation of the system. However, its effect upon PA operation must be considered. Negative resistance also causes the drain voltage to peak at the peak drain current, resulting in increased second-breakdown stress.

The supply voltage (Figure 6.11) must increase in proportion to the maximum value of $|Z|$ that occurs at the design SWR S_{max}. The maximum drain voltage is $2V_{DD}$. The supply voltage must be 2 for operation at $S = 3$ and 2.33 for $S = 5$; the corresponding peak drain voltages are 4 and 4.67, respectively (Table 6.1).

The ranges of power output and dissipation for an individual PA are shown in Figure 6.12. For $S > 3$, the negative resistance causes one PA to sink power from the other, resulting in increased dissipation.

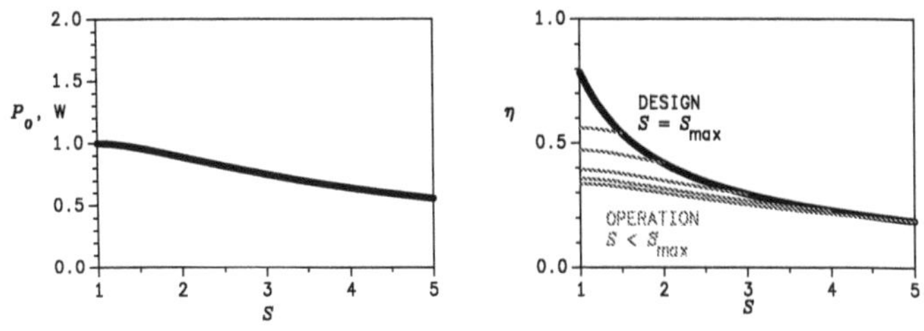

Figure 6.8 *System output and efficiency as functions of SWR.*

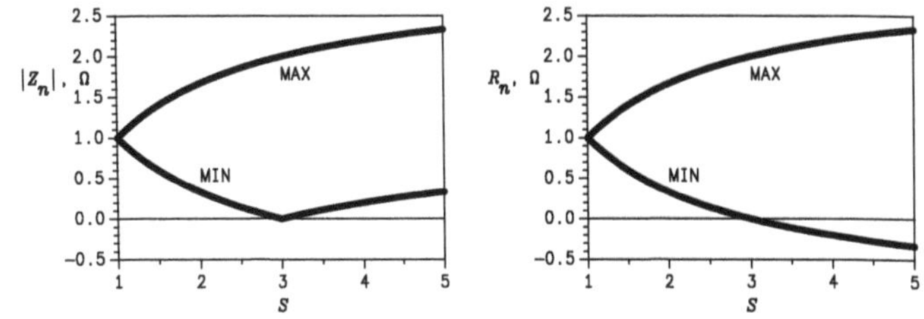

Figure 6.9 *Range of drain-load impedances as functions of SWR.*

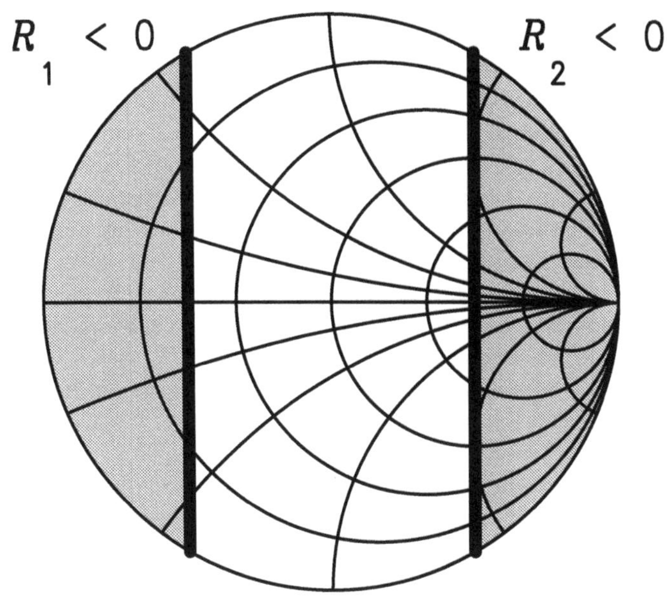

Figure 6.10 *Boundary of negative-resistance region.*

Table 6.1 *Performance at Maximum Design SWR*

	P_o							
S_{max}	W	dB	η_S	V_{DDmax} V	v_{Dmax} V	P_d W	P_{max}	γ
1.0	1.0000	0.000	0.7854	1.0000	2.0000	0.1366	0.1250	1.000
1.5	0.9600	−0.177	0.5386	1.4000	2.8000	0.5913	0.0857	1.458
2.0	0.8889	−0.511	0.4189	1.6667	3.3334	0.8944	0.0667	1.874
2.5	0.8163	−0.882	0.3452	1.8572	3.7143	1.1109	0.0549	2.277
3.0	0.7500	−1.249	0.2945	2.0000	4.0000	1.2733	0.0469	2.665
3.5	0.6914	−1.603	0.2572	2.1111	4.2222	1.3995	0.0409	3.056
4.0	0.6400	−1.938	0.2285	2.2000	4.4000	1.5006	0.0364	3.434
4.5	0.5950	−2.255	0.2056	2.2727	4.5455	1.5832	0.0327	3.823
5.0	0.5556	−2.552	0.1870	2.3334	4.6667	1.6521	0.0298	4.195

NOTE: Normalized for P_o = 1W at S = 1 and $I_1 = I_2$ = 1A.

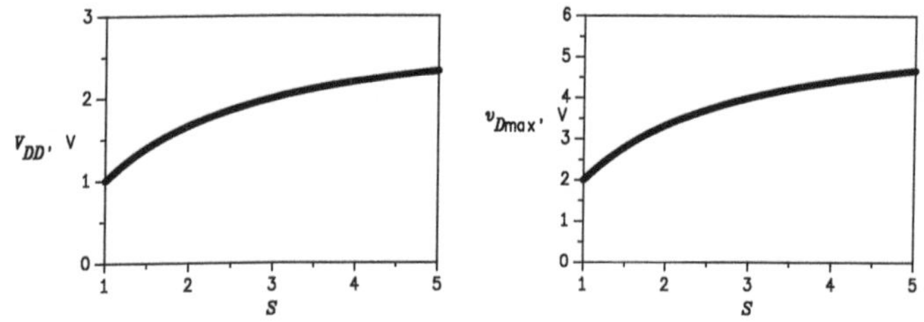

Figure 6.11 *Supply voltage and maximum drain voltage as functions of SWR.*

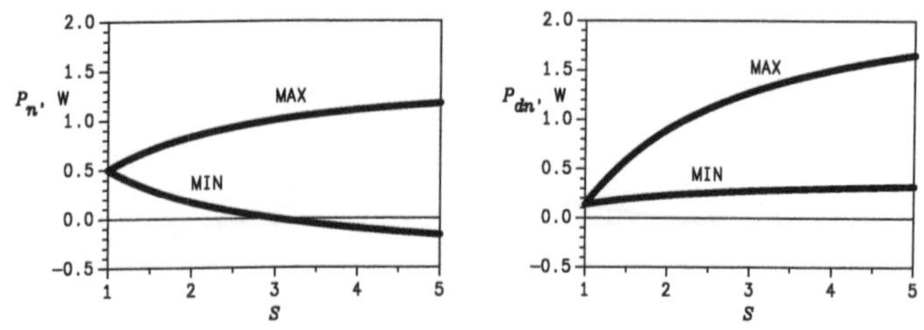

Figure 6.12 *Ranges of PA output and dissipation as functions of SWR.*

The normalized per-device power-output capability U or $P_{\max}$ is defined [1] as the actual power ouptut divided by the peak voltage, peak current, and number of transistors ($P_o/Nv_{D\max}\ i_{D\max}$). It drops from 0.125 of the ideal class-B PA to 0.0469 at $S = 3$ and 0.0298 at $S = 5$, as shown in Figure 6.13. To maintain the same output power, device voltage ratings must be increased accordingly. For example, as $S = 3$, the voltage rating (hence power-output capability) of the transistor must be 2.67 times as large as is required for operation at the same power into the nominal load. The rating factor Y shown in Table 6.1 is the ratio of the products of the peak voltage and current for the actual load to those of the nominal load for delivery of the same power ($v_{D\max}(Z)i_{D\max}(Z)/v_{D\max}(R)i_{D\max}(R)$).

The increase in device ratings is slightly less than the maximum SWR for full-power operation. In contrast, the rating factor of a class-B PA connected directly to the load increases directly with SWR because both voltage and current ratings must increase by $S^{1/2}$.

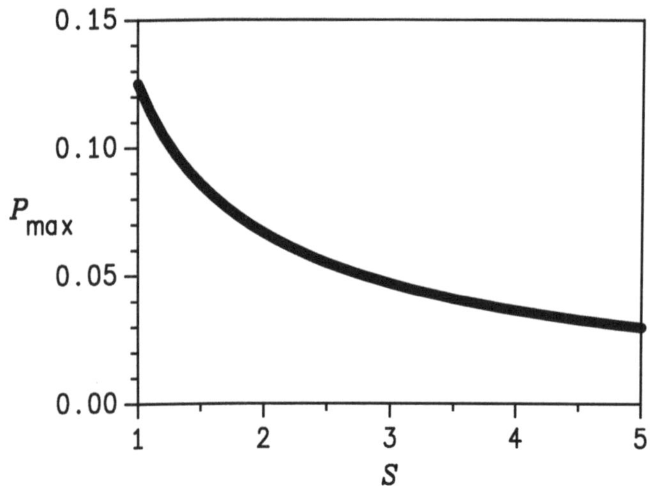

Figure 6.13 *Power-output capability as a function of SWR.*

6.4 CANCELLATION OF HARMONICS

The subsequent analysis of harmonic cancellation is based upon the assumption of a nominal load impedance and equal output currents from the two PAs.

The input splitter and line W1 in Figure 6.4 introduce 180° and –90° phase shifts into the drive to PA #2. The total 90° phase shift in the output of PA #2 causes $I_2 = jI_1$ at the fundamental frequency. Insertion of this into (6.36) causes the PA-induced load currents to reinforce each other. The contributions of the two PAs to hybrid current I_h are opposite, causing I_h to be zero. Thus, ideally all fundamental-frequency power is delivered to the load and none is expended in the hybrid resistor.

The third harmonic is produced by the cubic term of the nonlinearity in the PA. The resultant triple-mixing process sums three contributions of the phase of the signal. The phase of the drive to PA #1 is 0°, hence, the phase of the third harmonic in its output is also 0° (complex amplitude +1). Propagation through W1 results in a phase of –90° (complex amplitude $-j$) at input A to T3.

The phase of the drive to PA #2 is 180° – 90° = 90°, hence the phase of the third harmonic at the output of PA #2 (and input B to T3) is 3 × 90° = 270° = –90° (complex amplitude $-j$). Consequently, for the third harmonic,

$$I_2 = -jI_1 \tag{6.68}$$

Substitution of this into (6.36) yields

$$I_o = 0 \tag{6.69}$$

and substitution into (6.47) yields

$$I_h = -jI_1 + I_2 = -2jI_1 \tag{6.70}$$

All third harmonic power is therefore (ideally) dumped into R_h and none reaches the load.

The phases of all harmonics at input A of T3 are affected only by W2 and are therefore –90°. The production of higher-order harmonics by nonlinearities in PA #2 causes proportionally larger phase shifts, thus the fifth and seventh harmonic are shifted by 90 × 5 = +90° and 90 × 7 = –90°, respectively. Harmonics of orders 5, 9, 13, and so forth therefore add in phase much as the desired signal, while those of orders 3, 7, 11, etc. are cancelled.

The phases of the even harmonics in the output of PA #2 are either 180° (2 × 90°, 6 × 90°, etc.) or 0° (4 × 90°, etc.). This causes them to add to the harmonics in the output of PA #1 in phase quadrature. The resultant harmonic in the system output is therefore 3 dB below that in either PA. Elimination of the even harmonics is best accomplished by using push-pull PAs for PA #1 and PA #2.

6.5 CANCELLATION OF BACKWARD IMD

Operation of multiple transmitters with colocated antennas causes signals from one to enter the antenna of the other. Large-signal, nonlinear operation of the transistors in the PA intermodulates the forward- and reverse-direction signals, producing "backward intermodulation distortion" or simply "back IMD."

A signal entering the antenna is modeled by replacing load impedance Z_o in Figure 6.4 with a voltage source V_B as shown in Figure 6.14. Since PA #1 and its quarter-wavelength line cause T3 to be grounded at A and the voltages on both windings are V_T,

$$V_T = V_B/2 \tag{6.71}$$

and

$$V_h = -V_T = -V_B/2 \tag{6.72}$$

Consequently,

$$I_h = \frac{V_h}{R_h} = -\frac{V_B}{2R_h} \tag{6.73}$$

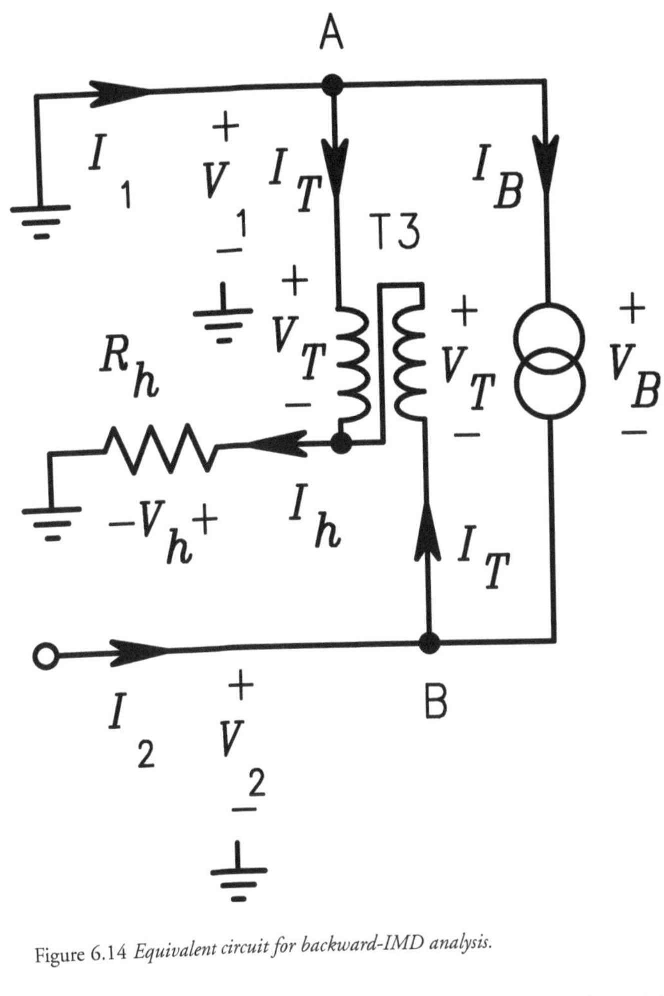

Figure 6.14 *Equivalent circuit for backward-IMD analysis.*

Since PA #2 appears as an open circuit and the currents in the two windings are equal,

$$I_B = \frac{I_h}{2} = -\frac{V_B}{4R_h} \tag{6.74}$$

The input impedance to the reverse-direction signal is therefore

$$R_B = \frac{V_B}{-I_B} = 4R_h \tag{6.75}$$

If $R_h = R_o/4$, then $R_B = R_o$. Given a nominal load impedance, the reverse-direction signal is therefore dissipated completely in hybrid resistor R_h. This process is analogous to the quadrature splitter.

Dumping of the reverse direction signal into R_h does not by itself eliminate production of back IMD. A nonzero voltage is, for example, produced on the output of PA #2, and slight imperfections in the PA and coupler can cause nonzero current to enter W1 at A. A reverse-direction signal reaches PA #1 with a phase of –90° and there meets a forward-direction signal with a phase of 0°. In traveling back to the combiner, another –90° is added by W1, resulting in an IMD phase of –180°. The reverse-direction signal has a phase of 180° and meets a forward-direction signal of –90°, resulting in IMD with a phase of +90°. The two IMDs are thus dumped into R_h in much the same manner as the third harmonics.

6.6 LITERATURE

The classic paper by Seidel [3] is generally regarded as the first well-published use of quadrature combiners. Seidel applies both quadrature combiners and feedforward error cancellation to power amplifiers operating at VHF and UHF. One of the first appearances of the "balanced amplifier" is in the classic CTC Solid Circuits application note [4]. The use of quadrature combiners and balanced power amplifiers is discussed in a number of papers and texts [5–9]. In many cases, however, construction details are considered proprietary and not discussed.

Design

Fisher [10] introduces the concept of using a transmission-line transformer to make a quadrature-hybrid coupler. Capacitors shunting both ends of the transmission line work in combination with the inductance of the winding to produce a balanced Π network. Design equations and examples for 7 and 200 to 400 MHz are given.

Ho and Chen [11, 12] develop a relatively simple design proceedure for a broadband quadrature. Their coupler is based upon three narrowband quadrature couplers separated by transmission lines of specific lengths and impedances. The narrowband couplers are based upon twisted-wire transformers with shunt capacitors as described by Fisher. The narrowband couplers are designed first, after which the transmission lines are chosen for broadband operation.

Boulouard [13] develops a single-stage lumped-element quadrature coupler for balanced input and output. One version is based upon inductors from both inputs to both outputs and capacitors connected across both inputs and from both sides of both inputs to ground. Shor [14] analyzes siimilar types of couplers. He also uses an even/odd-mode analysis to develop an *S*-parameter analogy between the lumped-element quadrature couplers and quarter-wavelength transmission lines.

Andrews and Aitchison provide the most comprehensive treatment of the theory and design procedures for quadrature couplers. They begin by developing the transfer functions and approximations for them [15, 16]. The theory is then used [17] to develop a technique for synthesis. Two classes of lumped-element quadrature hybrids are defined. In one class, the power split is exactly the correct power split and the phase shift is approximately correct. In the other class, the phase shift is correct and the power split is approximate. Explicit formulas are derived for circuits of first through fourth order using inductors, transformers, and capacitors. The formulas are followed by design examples including a 50 to 150 MHz, second-order coupler with 0.2° phase error and 0.8-dB amplitude imbalance and a 2–30 MHz, third-order coupler with a 1.3-dB amplitude imbalance and 2° phase error. Finally, the theory is extended to cascaded (multisection) quadrature couplers in [18]. Andrews and Aitchison illustrate their theory with a design that uses transmission lines, toroidal transformers, and capacitors to operate from 10 to 100 MHz with a 2.3° maximum phase error.

Applications

One coupler designed by Andrews and Aitchison [19] achieves a 20% bandwidth with a 1° phase error at 1.2 GHz. Another design achieves a 20% bandwidth with a 2° phase error at 6.4 GHz.

Raskmark [20] uses quadrature power combining to produce a 150W, 70-MHz power amplifier. Max [21] describes a 4-kW amplifier subsystem for FM broadcast that uses a variety of combining techniques. Four push-pull PAs are first combined in pairs. The outputs of the two pairs are then quadrature-combined to produce a module with a 600-W ouptut. Finally, the outputs of eight 600-W modules are combined through a nonhybrid Wilkinson combiner to produce the 4-kW subsystem output.

Podcameni and Riback [22] use a quadrature-combined structure to cancel third-order intermodulation products in a 90-MHz PA system. The third-order IM product is reduced by 17 dB. Shumaker [23] compares hybrid-combined and quadrature-combined PAs based upon GaAs MESFETs in a 150W S-band amplifier system. He shows that quadrature combining offers not only isolation but also improves input match and reduces IMD. Ku [24] uses quadrature combining to cancel back IMD in co-located transmitters. He also provides further analysis of this process, using both *S* parameters and Volterra series.

REFERENCES

[1] Krauss, H.L., C.W. Bostian, and F.H. Raab, *Solid State Radio Engineering*, New York: Wiley, 1980.

[2] Raab, F.H., "Maximum Efficiency and Output of Class-F Power Amplifiers," *IEEE Trans. Microwave Theory and Techniques*, Vol. 49, No. 6, June 2001, pp. 1162–1166.

[3] Seidel, H., H.R. Beurrier, and A.N. Friedman, "Error-Controlled High Power Linear Amplifiers at VHF," *Bell System Technical Journal*, Vol. 47, No. 3, May–June 1968, pp. 651–722.

[4] Johnson, J.H. (ed.), *Solid Circuits: Circuit Techniques. Applications*, Book 2.2.8.0A, Communications Transistor Corp., San Carlos, CA, March 1973.

[5] Johnson, J.H., "RF Combining and Protection Techniques," *Proc. RF Expo East '87*, Boston, MA, Nov. 11–13, 1987, pp. 197–215.

[6] Blocksome, R.K., "Practical Wideband RF Power Transformers, Combiners, and Splitters," *Proc. RF Tech. Expo '86*, Anaheim, CA, January 30–February 1, 1986, pp. 207–227. Also *Proc. RF Expo East*, November 10–12, 1986, Boston, MA, pp. 1–21.

[7] Sabin, W.E., and E.O. Shoenike, (eds.), *Single-Sideband Systems and Circuits*, Second Edition, New York: McGraw-Hill, 1995.

[8] Vizmuller, P., *RF Design Guide: Systems, Circuits, and Equations*, Norwood, MA: Artech House, 1995.

[9] Walker, J.L.B., (ed.), *High-Power GaAs FET Amplifiers*, Norwood, MA: Artech House, 1993.

[10] Fisher, R.E., "Broad-Band Twisted-Wire Quadrature Hybrids," *IEEE Trans. Microwave Theory and Techniques*, Vol. MTT-21, No. 5, May 1973, pp. 355–357.

[11] Ho, C.Y., and G-L. Chen, "Design of Wideband Quadrature Couplers for UHF/VHF: Part I," *RF Design*, Vol. 12, No. 11, November 1989, pp. 58–61.

[12] Ho, C.Y., and G-L. Chen, "Design of Wideband Quadrature Couplers for UHF/VHF: Part II," *RF Design*, Vol. 12, No. 12, December 1989, pp. 51–55.

[13] Boulouard, A., "Lumped-Element Quadrature Couplers," *RF Design*, Vol. 12, No. 7, July 1989, pp. 56–59.

[14] Shor, A., "Broadband Lumped Quadrature Coupler," *Proc. RF Technology-Expo*, Anaheim, CA, March 27–29, 1990, pp. 139–143.

[15] Andrews, D.P., and C.S. Aitchison, "Optimum Approximation Functions for Lumped Element Quadrature Hybrids," *IEE Proc. Circuits, Devices and Systems*, Vol. 148, No. 1, February 2001, pp. 5–9.

[16] Andrews, D.P., and C.S. Aitchison, "Transfer Functions of Optimum Lumped Element Quadrature Hybrids," *IEE Proc. Circuits, Devices and Systems*, Vol. 148, No. 3, June 2001, pp. 135–139.

[17] Andrews, D.P., and C.S. Aitchison, "Synthesis of Symmetric Lumped Element Quadrature Hybrids," *IEE Proc. Microwaves, Antennas and Propagation*, Vol. 148, No. 5, October 2001, pp. 269–274.

[18] Andrews, D.P., and C.S. Aitchison, "Design of Cascaded Lumped Element Quadrature Hybrids," *IEE Proc. Microwaves, Antennas and Propagation*, Vol. 148, No. 5, October 2001, pp. 275–279.

[19] Andrews, D.P., and C.S. Aitchison, "Microstrip Lumped Element Quadrature Couplers for Use at Microwave Frequencies," *IEE Proc. Microwaves, Antennas and Propagation*, Vol. 147, No. 4, August 2000, pp. 267–271.

[20] Raskmark, P., "Power Amplifier Design Using Quadrature Hybrids," *RF Design*, Vol. 17, No. 6, June 1994, pp. 70–74.

[21] Max, L.B., "FM Broadcast 88 to 108 MHz 4,000 Watt Amplifier Subsystem," *RF Design*, Vol. 9, No. 2, February 1986, pp. 36–38.

[22] Podcameni, A., and C.R. Riback, "An Amplifier Linearization Method Based on a Quadrature Balanced Structure," *IEEE Trans. Broadcasting*, Vol. 48, No. 2, June 2002, pp. 158–162.

[23] Shumaker, J., R. Basset, and A. Skuratov, "High-Power GaAs FET Amplifiers: Push-Pull Versus Balanced Configurations," The AMW Archive 1989–2002 CD-ROM, Noble Publishing, http://www.noblepub.com, 2003.

[24] Ku, W.H., J.E. Erikson, R.E. Rabe, and G.L. Seasholz, "Design Techniques and Intermodulation Analysis of Broad-Band Solid-State Power Amplifiers," *IEEE Trans. Electromagnetic Compatibility*, Vol. EMC-19, No. 2, May 1977, pp. 57–65.

Reprinted from IEEE TRANSACTIONS
ON *MICROWAVE THEORY AND TECHNIQUES*
Volume MTT-21, Number 5, May, 1973
pp. 355-357

Broad-Band Twisted-Wire Quadrature Hybrids

R. E. FISHER

***Abstract*—A symmetrical 3-dB quadrature hybrid, consisting chiefly of a bifilar pair of twisted wires, is described. A cascade of two such hybrids can achieve an octave bandwidth with a 0.7-dB coupling error. Since this class of hybrid is simple, compact, and low in cost, its use may be preferred over the more common coaxial line or printed-circuit types in the frequency region below 1 GHz.**

I. INTRODUCTION

The concept of using twisted wires wrapped upon ferrite toroids to form compact, asymmetric, 180° hybrids was first introduced by Ruthroff [1]. It has also been found by several other investigators [2]–[4] that twisted-wire structures can be made to function as *symmetric*, 3-dB quadrature (90°) hybrids, thus permitting the construction of compact directional couplers at arbitrarily low frequencies.

Examples of twisted-wire quadrature hybrids centered at approximately 7 and 300 MHz are shown schematically and pictorially in Figs. 1, 2, and 5. For both hybrids, the coupling section consists of two strands of insulated copper magnet wire tightly twisted together to form a bifilar pair. For the 7-MHz hybrid shown in Fig. 2, the pair is wrapped upon a small ferrite toroid which is then soldered to four BNC along with two mica capacitors. For the 225–400-MHz two-stage hybrid shown in Fig. 5, where much less inductance is required, the wire pairs are bent into loops, and then soldered onto an epoxy fiberglass circuit board along with four omni-spectra-miniature (OSM) connectors.

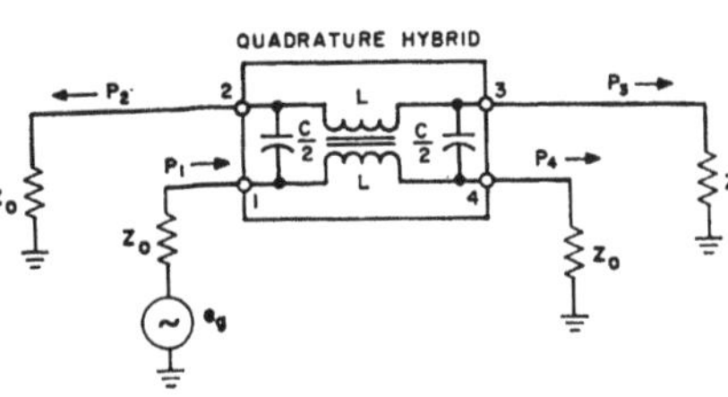

Fig. 1. Quadrature-hybrid functional diagram.

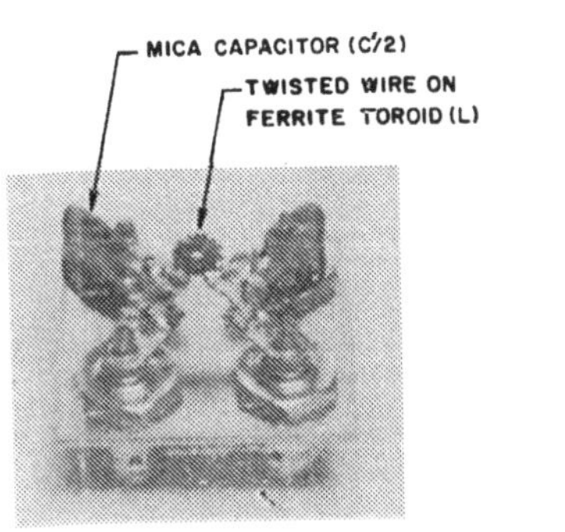

Fig. 2. Photograph of 7-MHz quadrature hybrid.

Manuscript received May 1, 1972; revised November 8, 1972.
The author is with Bell Laboratories, Whippany, N. J. 07981.

II. SINGLE-STAGE HYBRID DESIGN

Consider Fig. 1. If terminal 1 is connected to terminal 2, and terminal 3 is connected to terminal 4, the resulting two-terminal network is now simply a lumped inductance L, since the magnetic coefficient of coupling between the twisted wires is nearly unity.

If terminal 1 is connected to terminal 4, and terminal 2 is connected to terminal 3, this different two-terminal network can be approximated by a lumped capacitor C, which is the sum of the interwinding capacitance and the external capacitors. At UHF, the external capacitors may not be required.

The 4-port will display hybrid properties when

$$Z_0 = \sqrt{\frac{L}{C}}. \tag{1}$$

Equal power division between ports 2 and 4 will occur at a frequency f_0 where

$$\omega_0 L = \frac{1}{\omega_0 C} = Z_0$$
$$\omega_0 = 2\pi f_0. \tag{2}$$

The transducer loss between ports 1 and 2 is

$$\frac{P_1}{P_2} = 1 + \left(\frac{\omega_0}{\omega}\right)^2 \tag{3}$$

and between ports 1 and 4 is

$$\frac{P_1}{P_4} = 1 + \left(\frac{\omega}{\omega_0}\right). \tag{4}$$

The port 3 isolation and return loss (at all ports) is theoretically infinite at all frequencies. Also, the voltages at ports 2 and 4 should differ in phase by 90° over an infinite frequency band.

The 7-MHz hybrid shown in Fig. 2 exhibited a measured insertion loss of 0.1 dB, and the isolation and return loss both exceeded 30 dB. One stage of the 300-MHz dual hybrid shown in Fig. 5 had a measured insertion loss of 0.2 dB, while the isolation and return loss were about 20 dB.

III. BROAD-BANDING THE HYBRID

Equations (3) and (4) indicate that a single-stage twisted-wire quadrature hybrid is not a very broad-band network, since equal power division occurs only at one frequency.

However, the hybrid can be broad-banded by cascading two sections, as shown in Fig. 3. Ports 3 and 4 of the first hybrid are con-

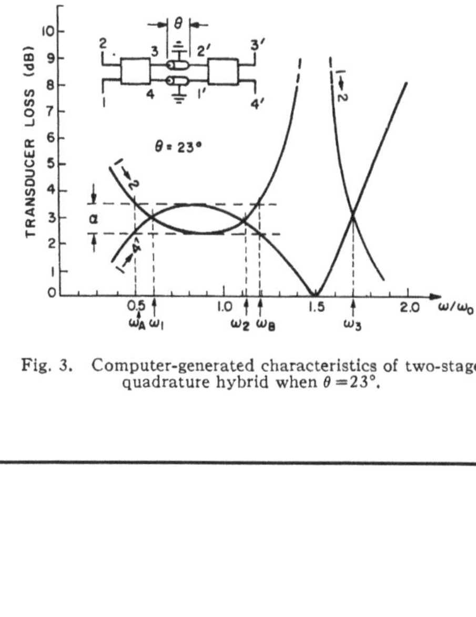

Fig. 3. Computer-generated characteristics of two-stage quadrature hybrid when $\theta = 23°$.

nected, respectively, to ports 1′ and 2′ of the second identical hybrid via sections of Z_0 coaxial cable each having an electrical length θ at ω_0.

When this two-stage cascade was computer analyzed,[1] it exhibited the transmission characteristics shown in Fig. 3. All frequencies are normalized to ω_0, which is the 3-dB frequency of each hybrid used in the cascade. Note that there exist three frequencies ω_1, ω_2, and ω_3 where equal division of generator power occurs. The frequencies

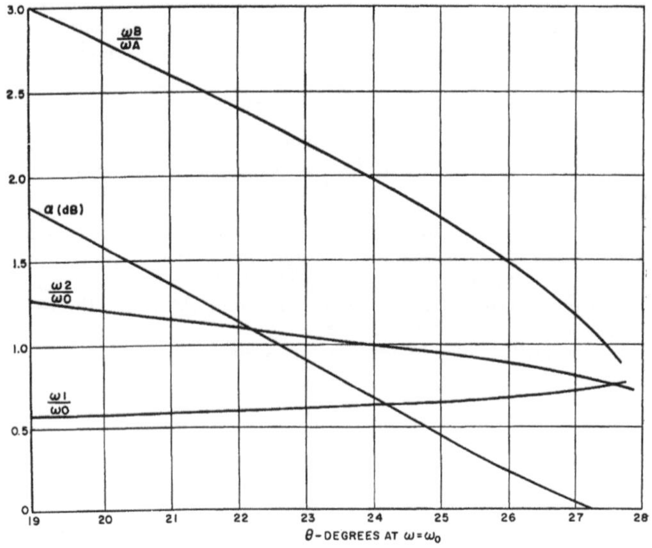

Fig. 4. Two-stage quadrature-hybrid transmission parameters versus connecting cable length.

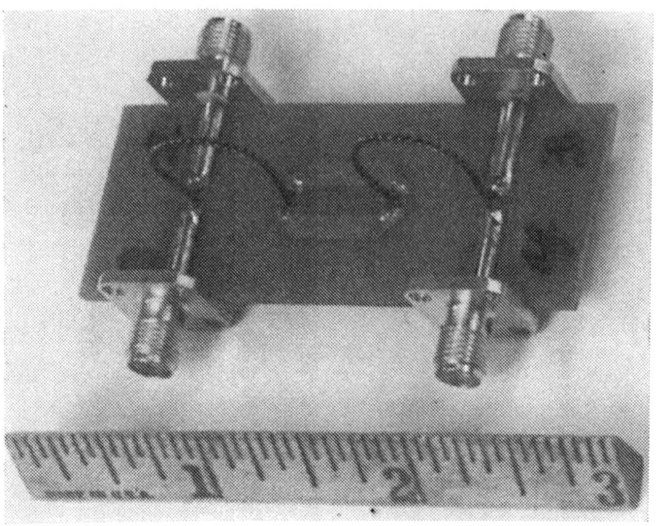

Fig. 5. Photograph of two-stage 225–400-MHz quadrature hybrid.

[1] The computer program used was BELNAP-II, a Bell Laboratories linear circuit-analysis program.

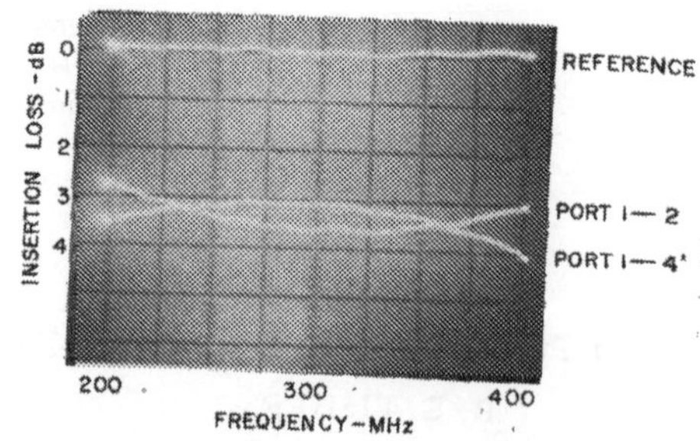

Fig. 6. Measured characteristics of two-stage 225–400-MHz quadrature hybrid.

ω_A and ω_B define that frequency band over which the difference in coupling never exceeds α.

Fig. 4 presents some graphs which can be used to design a broadband two-stage hybrid. These show how some of the transmission characteristics given by Fig. 3 are related to the electrical length of the connecting cables. An octave bandwidth $(\omega_B/\omega_A=2)$ results when $\theta\cong 24°$. Over this band the coupling difference α will never exceed 0.7 dB.

Figs. 5 and 6 show, respectively, the photograph and transmission characteristics of a two-stage hybrid designed to cover the 225–400-MHz communications band. Note that α is about 0.5 dB, which implies from Fig. 4 that the coupling lines are about 24.7° long. The isolation and return loss exceeded 20 dB over the measured band. The insertion loss never exceeded 0.5 dB.

An apparent disadvantage of this cascade technique is that the coaxial connecting cables become prohibitively long as the hybrid operating frequency is lowered. This problem can be bypassed by substituting lumped-element low-pass filters for the coaxial cables.

REFERENCES

[1] C. L. Ruthroff, "Some broad-band transformers," *Proc. IRE*, vol. 47, pp. 1337–1342, Aug. 1959.
[2] W. L. Firestone, U.S. Patent 2 972 121, Feb. 1961.
[3] K. Kurokawa, "Design theory of balanced transistor amplifiers," *Bell Syst. Tech. J.*, vol. XLIV, no. 8, pp. 1675–1698, Oct. 1965.
[4] J. D. Cappucci and H. Seidel, U.S. Patent 3 452 300 and 3 452,301, June 1969.

Design of Wideband Quadrature Couplers for UHF/VHF: Part I

By Chen Y. Ho,
M/A-COM Active Assemblies Div.
and Ge-Lih Chen, Chung Shan Institute of Science and Technology

This article presents design information for a class of wideband quadrature couplers consisting of three narrowband quadrature couplers separated by two pairs of transmission lines of adequate length and characteristic impedance Z_o. Due to the use of lumped elements, the coupler is particularly useful in the UHF/VHF band. Experimental results agree well with the theoretical computations.

The configuration of the narrowband quadrature coupler is shown in Figure 1. On a toroid, two strands of insulated wire are tightly twisted together to form a bifilar pair. The self inductance and mutual inductance of the transformer toroid are each equal to L. (This assumes unity magnetic coupling between the inductors.) Two lumped capacitors of value C/2 are shunted across the primary and secondary windings of the transformer. The design equations for this coupler can be found in References 1 and 2 and are summarized below. Let Z_0 be the input and output impedance of the coupler. L and C are chosen such that:

$$Z_o^2 = L/C \quad (1)$$

The coupler is then perfectly matched for all frequencies, and port 1 (or 2) and port 3 (or 4) are perfectly isolated for all frequencies. The voltages at ports 2 and 4 with 1 volt applied at port 1 are:

$$V_{21} = \frac{j\omega L}{Z_o + j\omega L} \quad (2a)$$

and

$$V_{41} = \frac{Z_o}{Z_o + j\omega L} \quad (2b)$$

The phase of V_{21} leads that of V_{41} by 90 degrees for all frequencies. The amplitude responses of V_{21}^2 and V_{41}^2 are plotted in Figure 2, showing the 0.6 dB amplitude ripple bandwidth to be about 10 percent.

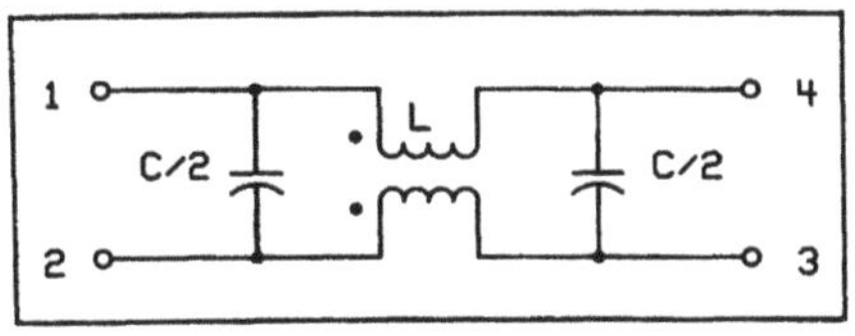

Figure 1. A narrowband quadrature coupler.

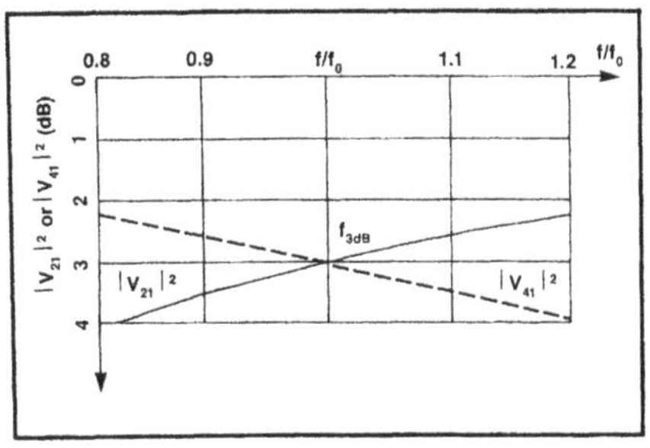

Figure 2. Amplitude response of the narrowband quadrature coupler.

The frequency at which $V_{21} = V_{41}$ is designated as f_{3dB}. At f_{3dB}:

$$Z_o = \omega L = 2\pi f(3\text{ dB})L \quad (3a)$$

or,

$$L = \frac{Z_o}{2\pi f(3\text{ dB})} \quad (3b)$$

The capacitor which is related to L by equation 1 can be expressed as:

$$C = \frac{1}{2\pi f(3\text{ dB})Z_o} \quad (4a)$$

or

$$\frac{C}{2} = \frac{1}{4\pi f(3\text{ dB})Z_o} \quad (4b)$$

If the f_{3dB} of the narrowband quadrature hybrid is known, the capacitor C and inductor L can be computed by equations 3 and 4. The amplitude response

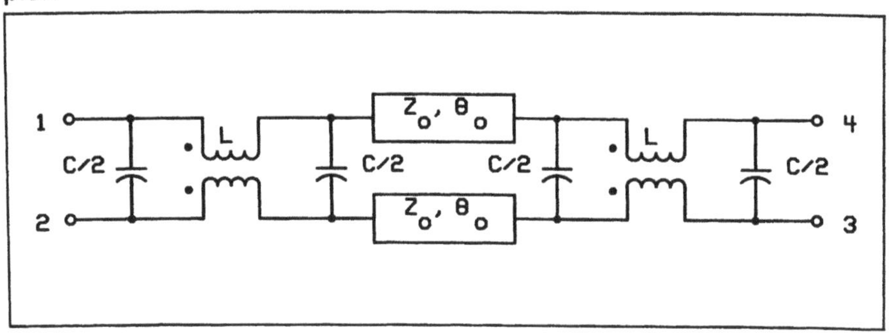

Figure 3. Two identical narrowband quadrature couplers.

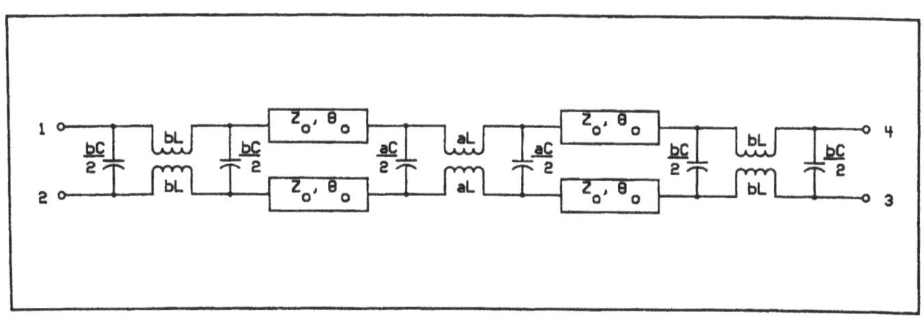

Figure 4. Configuration of the broadband quadrature coupler.

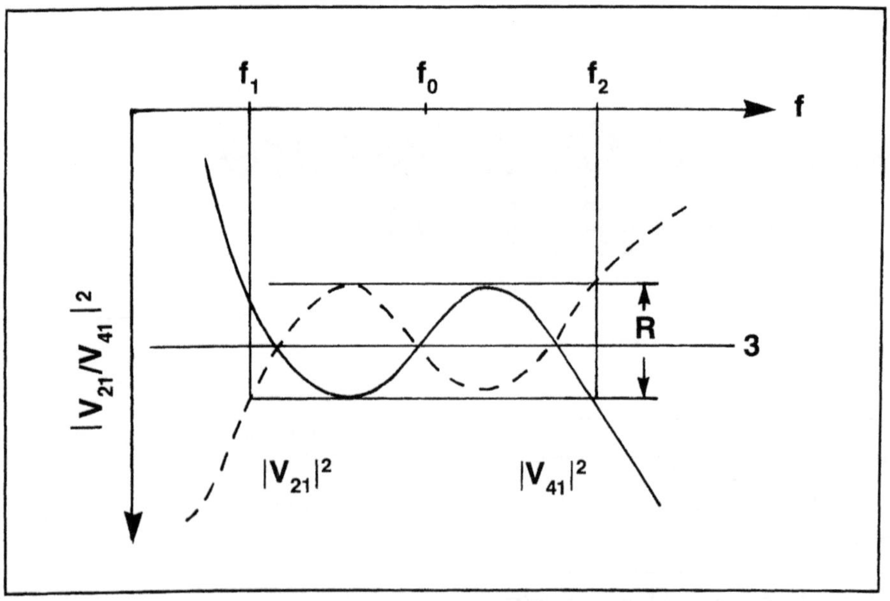

Figure 5. A typical frequency response of the broadband quadrature coupler.

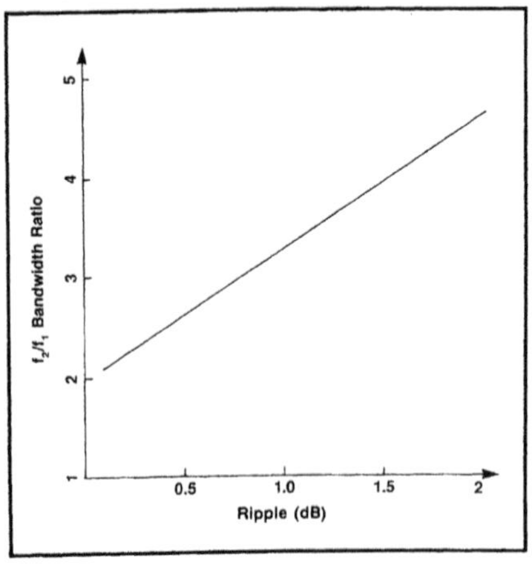

Figure 6. f_2/f_1 vs. Ripple R(dB).

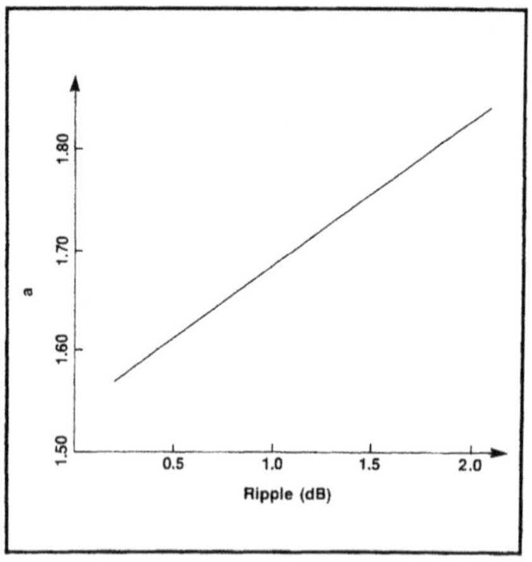

Figure 7. a vs. Ripple (dB).

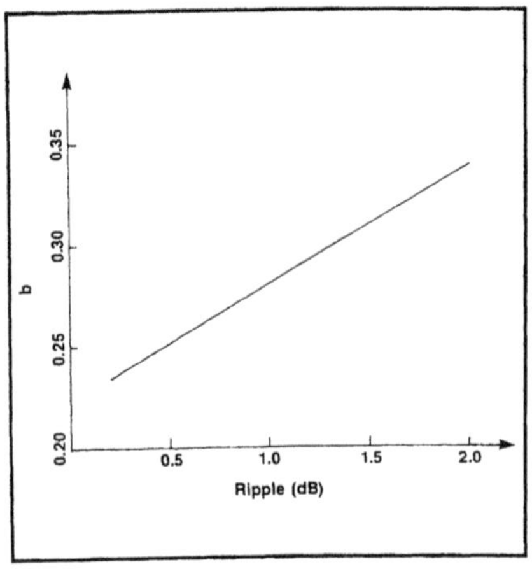

Figure 8. b vs. Ripple (dB).

of the coupler can be seen in Figure 1. Therefore, f_{3dB} completely represents the characteristics and the design of this narrowband quadrature hybrid.

The operating frequency bandwidth of the coupler can be broadened by using two identical narrow couplers of this type interconnected by a pair of transmission lines of characteristic impedance Z_o and of length θ, as shown in Figure 3.

The design parameters of this coupler can be found in References 1 and 3. Several design curves are available to facilitate the design. It has been reported that the 0.6 dB amplitude ripple bandwidth of this coupler is about 67 percent.

Design of Broad Bandwidth Quadrature Couplers

The broadband quadrature coupler under consideration has the configuration shown in Figure 4. It consists of two identical narrowband couplers (described in the previous section) at both ends and another narrowband coupler of different f_{3dB} in between.

The couplers are separated by two pairs of transmission lines of characteristic impedance Z_o and electrical length θ_o at f_o, the center frequency of interest. Note that the narrowband couplers at both ends have f_{3dB} equal to f_o/b, and the middle coupler has its f_{3dB} at f_o/a. Design parameters for this broadband coupler can then be simplified to the three parameters a, b and θ_o. The inductor L and capacitor C are related to the center frequency f_o and source/load impedance Z_o by:

$$L = \frac{Z_o}{2\pi f_o} \tag{5a}$$

$$C = \frac{1}{2\pi f_o Z_o} \tag{5b}$$

Analysis of the broadband coupler of Figure 4 can be carried out either by an even-mode and odd-mode approach of analysis, or by cascading techniques which use a transmission matrix and scattering parameter matrix (4). Both approaches arrive at identical results. Let R be the output power ratio between port 2 and port 4 for input power at port 1. Then:

$$R = \frac{|V_{21}|^2}{|V_{41}|^2} \tag{6}$$

$$= \left[\frac{a\omega L}{Z_o} + \frac{2b\omega L \cos(2\theta)}{Z_o} - \frac{(2a+b)b\omega^2 L^2 \sin(2\theta)}{Z_o^{\,2}} + \frac{4ab^2\omega^3 L^3 \sin^2(\theta)}{Z_o^{\,3}}\right]^2$$

$$= \left[a\left(\frac{f}{f_o}\right) + 2b\left(\frac{f}{f_o}\right)\cos\left(\frac{2\theta f}{f_o}\right) - (2a+b)b\left(\frac{f}{f_o}\right)^2 \sin\left(\frac{2\theta_o f}{f_o}\right) + 4ab^2\left(\frac{f}{f_o}\right)^3 \sin^2\left(\frac{\theta_o f}{f_o}\right)\right]$$

To find the design parameters a, b and θ_o for a given R, a computer program has been written to compute the response of the broadband quadrature coupler. Given the desired amplitude ripple R in a nominal 3 dB design and center frequency f_o, the program (after some optimization) will provide these design parameters. Figure 5 shows a typical frequency response of the broadband quadrature coupler after parameter optimization.

The optimization criteria are established to adjust a, b and θ_o such that equal ripple response of the amount R dB is achieved with the resulting bandwidth ratio, f_2/f_1. The relationship between the amplitude ripple R and the bandwidth ratio f_2/f_1 is shown in Figure 6 where higher passband ripple results in wider bandwidth ratio. Figures 7, 8 and 9 present the design information for a, b and θ_o, also as a function of the passband amplitude ripple.

As a design example, f_o = 80 MHz, Z_o = 50 ohms and R = 0.6 dB; f_2/f_1 = 3.0 with f_2 = 120 MHz and f_1 = 40 MHz; a and b are equal to 1.63 and 0.26, respectively; and θ_o = 35 degrees at 80 MHz. The narrowband coupler at the ends has an f_{3dB} of 307.7 MHz and the middle coupler has an f_{3dB} of 49.08 MHz.

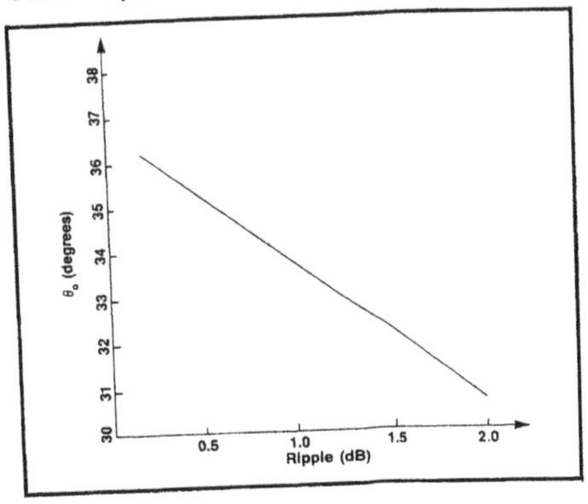

Figure 9. θ_o vs. Ripple (dB).

$$L = \frac{50}{(2\pi)(80 \times 10^6)} = 99.47 \text{ nH} \qquad (7a)$$

$$\frac{C}{2} = \frac{1}{2(2\pi)(80 \times 10^6)(50)} = 19.89 \text{ pF} \qquad (7b)$$

The element values of the broadband quadrature coupler for this design are summarized in Figure 10. The computer simulation of the frequency response of the design is shown in Figure 11.

Experimental Results

A broadband quadrature coupler was fabricated and tested based on the design information provided in the previous section. The circuit was fabricated using 25 mil thick Epsilom-10 substrate. The circuit was constructed using ATC chip capacitors and Johanson variable capacitors. Eight turns of 34 bifilar wires are used on a T20-10 toroid core for the realization of 162.1 nH, and five turns of 34 bifilar are used on a T20-0 toroid core to obtain 25 nH.

A photo of this coupler is shown in Figure 12 and the measured frequency responses in Figure 13. The measured passband amplitude ripple is about 0.8 dB and the bandwidth ratio is 143/45 = 3.18. The center frequency is slightly shifted to 84 MHz. The agreement between the theoretical and the measured responses is excellent.

Part II will be published in an upcomming issue of *RF Design*.

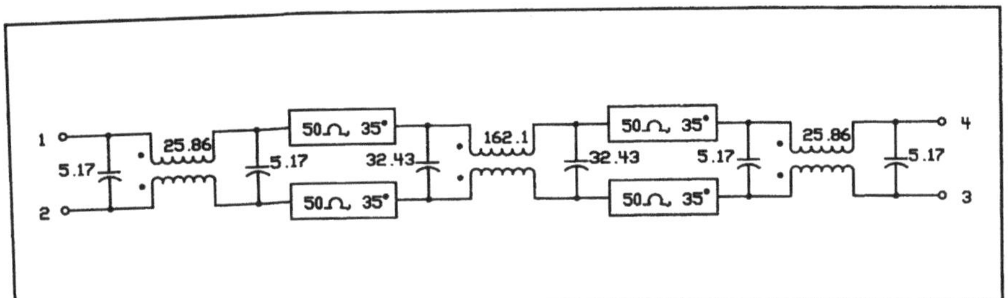

Figure 10. Experimental circuit.

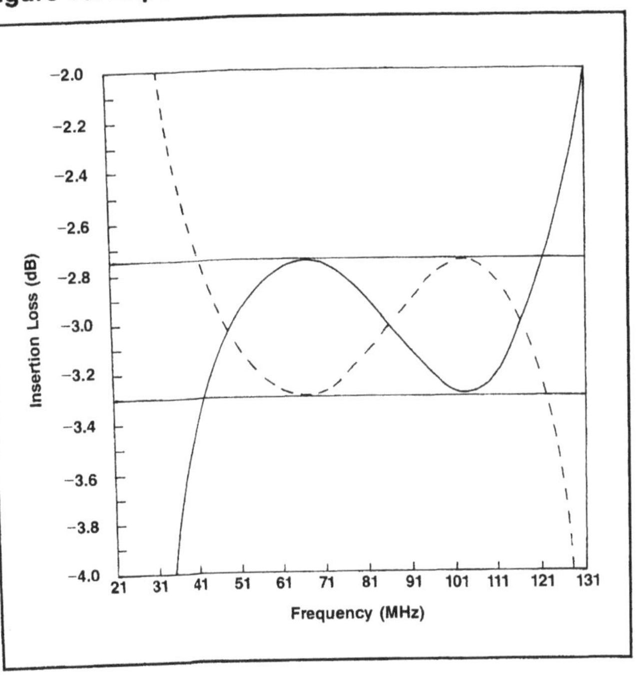

Figure 11. Computer simulation frequency response.

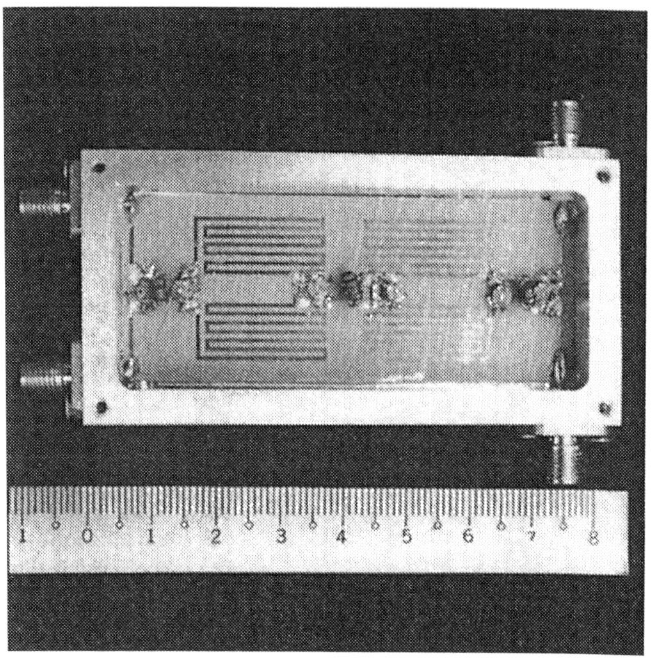

Figure 12. Photo of the coupler.

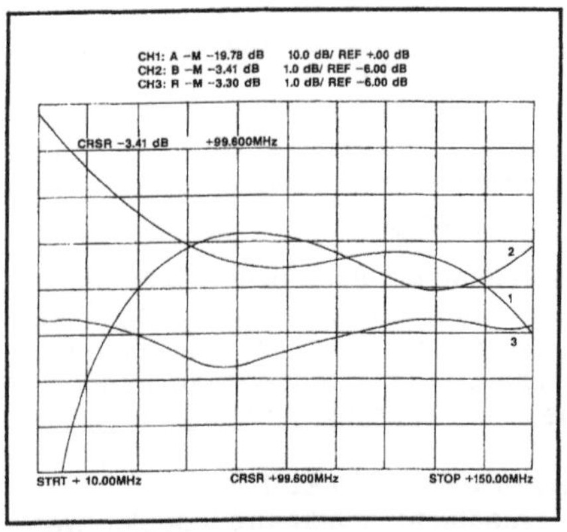

Figure 13a. Measured frequency response.

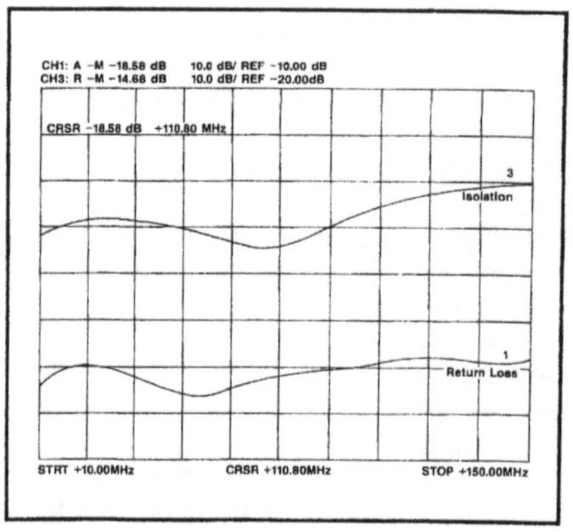

Figure 13b. Return loss and isolation.

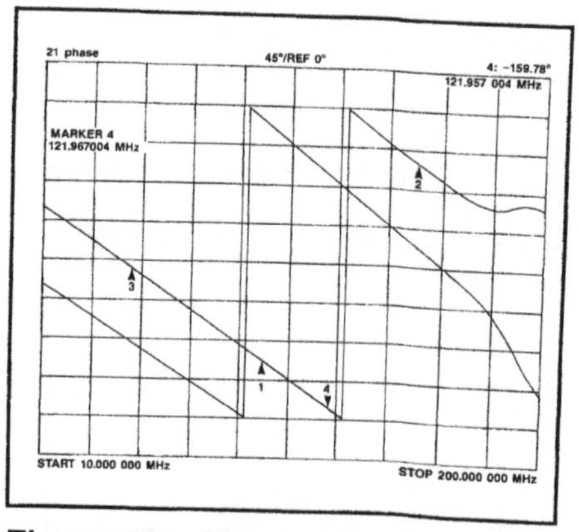

Figure 13c. Measured phase relation between T2I and T4I.

References

1. R.E. Fisher, "Broadband Twisted Wire Quadrature Hybrids," *IEEE Transactions on MTT*, Vol. 21, May 1973, pp. 355-357.
2. C.Y. Ho, R.Furrow, "Analysis and Design of VHF Quadrature Couplers: Part 1," *RF Design*, July/August 1979, pp. 48-50.
3. C.Y. Ho, R. Furrow, "Analysis and Design of VHF Quadrature Couplers: Part 2," *RF Design*, September/October, 1979, pp. 32-39.
4. G.L. Chen, Theory and Design of a Class of Wideband Quadrature Hybrids for UHF/VHF, M.S. Thesis, 1988, National Chiao Tung University, Hsin Chu, Taiwan.

About the Authors

Dr. Chen Y. Ho is vice-president—engineering, at M/A-COM, Active Assemblies Division, 340 North Roosevelt Ave., Chandler, AZ 85226. Tel: (602) 966-1471. Ge-Lih Chen is a research associate at the Chung Shan Institute of Science and Technology in Taiwan.

Design of Wideband Quadrature Couplers for UHF/VHF: Part II

By Chen Y. Ho,
M/A-COM Active Assemblies Div.
and Ge-Lih Chen, Chung Shan Institute of Science andTechnology

This article presents the same design approach as the wideband quadrature coupler discussed in Part I (November 1989 RF Design*). It uses a lowpass filter of proper phase characteristics to replace the transmission lines which interconnect the narrowband quadrature coupler. The results are particularly useful for the lower frequency spectrum where transmission lines become excessive in length. Experimental results show good agreement with the theoretical computations.*

At the lower end of the VHF band, the physical length of the interconnecting transmission lines used in the wideband quadrature coupler becomes excessively long and in many cases they are impractical to realize. Since the main function of the transmission lines is to provide a proper phase relationship between the narrowband couplers, it can be replaced by lumped element lowpass filters with proper phase characteristics. Any circuit which has minimum insertion loss and mismatch and possesses proper phase characteristics over the frequency band of interest can be used to replace the transmission lines. Some degradation in performance can be expected due to imperfect match and the non-ideal linear phase characteristics of the filter. An advantage is that the size of the coupler can be considerably reduced.

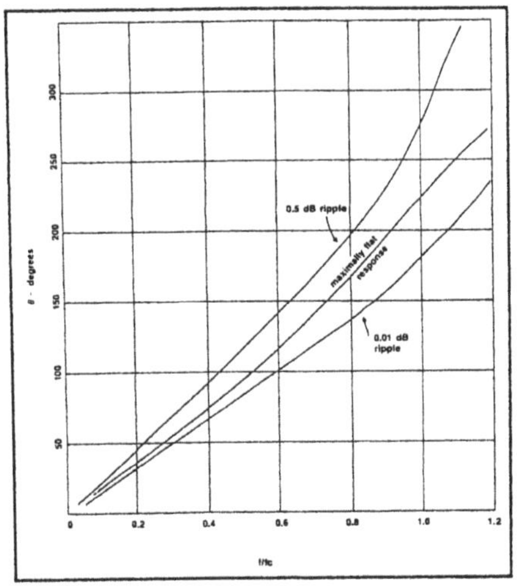

Figure 1. Phase shift characteristics with maximally flat or Chebyshev attenuation responses and n=5.

Phase Characteristics of a Lowpass Filter

The phase characteristics of a lowpass filter prototype (n = 5) are shown in Figure 1. It can be seen from the figure that the phase characteristic of a low ripple Chebyshev response filter design is almost the same as that of a maximally flat response design for $f \leqslant 0.6f_c$, where f_c is the cut-off frequency of the filter. It is known that filters with maximally flat responses have poorer

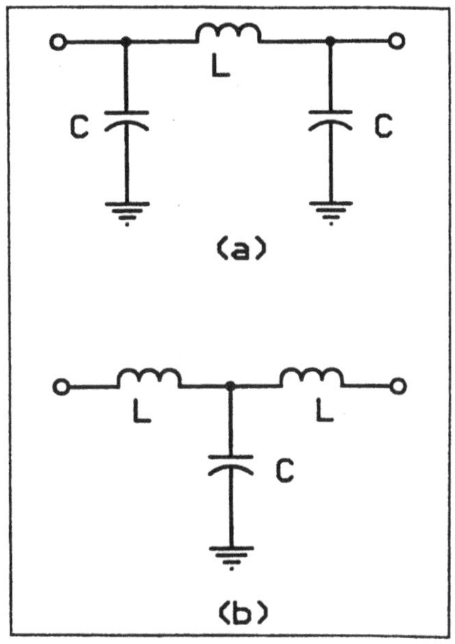

Figure 2. Lowpass filter configurations.

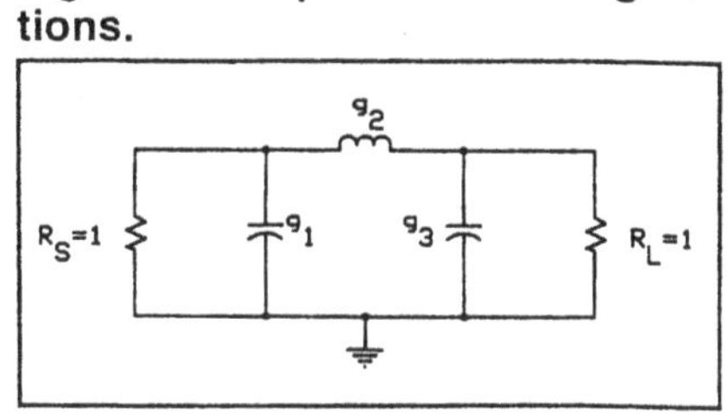

Figure 3. Lowpass filter prototype.

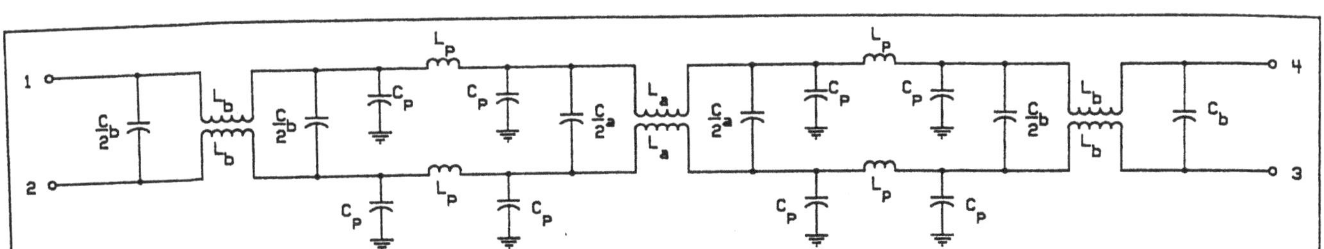

Figure 4. The wideband quadrature coupler with lowpass filters.

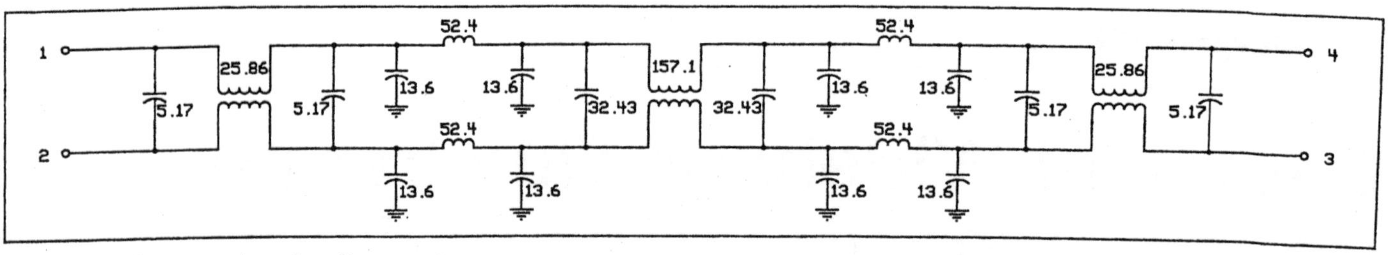

Figure 5. The coupler circuit.

return loss $(0, f_c)$ for the same amount of ripple. Only filters with Chebyshev frequency response will be considered.

The total phase shift of an nth order filter is equal to $n\pi/2$ for frequencies from DC to infinite. However, most of the phase shift occurs from DC to f_c. For a filter with a Chebyshev frequency response, n should be an odd number in order to avoid an unequal termination for source and load. For $n = 1$, the amount of phase shift may not be sufficient to achieve the phase shift required to replace the transmission lines. For $n \geqslant 5$, the number of elements is large and it becomes difficult to build two or more lowpass filters of similar phase characteristics.

With the above consideration in mind, only filter designs with low ripple Chebyshev response and $n = 3$ are used for the replacement of the transmission lines. The passband ripple of a filter is directly related to the return loss in the same band.

The analysis for the filters in Figure 2 is similar. Hence, only the circuit in Figure 2(a) will be discussed. For the lumped element lowpass filter prototype shown in Figure 3, where $g_1 = g_3$ for symmetry, θ can be expressed as:

$$\theta = \tan^{-1}\left[\frac{g_2\left(\frac{f}{f_c}\right)}{2\left(1 - g_1g_2\left(\frac{f}{f_c}\right)^2\right)} + \frac{g_1\left(\frac{f}{f_c}\right)\left(2 - g_1g_2\left(\frac{f}{f_c}\right)^2\right)}{2\left(1 - g_1g_2\left(\frac{f}{f_c}\right)^2\right)}\right] \quad (1)$$

where f_c is the cut-off frequency of the filter and g_1 and g_2 are the element values of the lowpass prototype. The required length of the transmission lines for the wideband quadrature coupler is θ_o at f_o. At $f = f_o$, equation 1 becomes:

$$\theta_o = \tan^{-1}\left[\frac{g_2\left(\frac{f_o}{f_c}\right)}{2\left(1 - g_1g_2\left(\frac{f_o}{f_c}\right)^2\right)} + \frac{g_1\left(\frac{f_o}{f_c}\right)\left(2 - g_1g_2\left(\frac{f_o}{f_c}\right)^2\right)}{2\left(1 - g_1g_2\left(\frac{f_o}{f_c}\right)^2\right)}\right] \quad (2)$$

If g_1 and g_2 are known for a particular design, f_c can be computed by solving equation 2. L_p and C_p are related to f_c by:

$$C_p = \frac{g_1}{2\pi\, f_c Z_o} \quad 3(a)$$

$$C_p = \frac{g_1}{Z_o(2\pi f_c)} = 13.6\ \text{pF} \quad 3(b)$$

By replacing the transmission lines with the lowpass filters, the configuration of the wideband quadrature coupler shown in Figure 4 is obtained.

Experimental Results

The same design examples used in Part I are used to demonstrate the design approach described here. For a wideband quadrature coupler centered at 80 MHz, $L = 99.47$ nH, $C/2 = 19.89$ pF, $a = 1.63$, $b = 0.26$ and $\theta = 35$ degrees. The lowpass filter chosen has 0.01 dB ripple or 26 dB return loss for $f \leqslant f_c$. This results in element values of $g_1 = g_3 = 0.6291$ and $g_2 = 0.9702$.

For $\theta_o = 35$ degrees at 80 MHz, using equation 2, f_c is 147.34 MHz. The element values for L_p and C_p can be computed from equation 3(a) and 3(b):

$$L_p = \frac{g_2}{2\pi f_c}(Z_o) \quad 4(a)$$

$$L_p = \frac{g_2 Z_o}{2\pi f_c} = 52.4\ \text{nH} \quad 4(b)$$

A wideband coupler has been fabricated and tested. The circuit is shown in Figure 5. The coupler is fabricated using the same techniques as those used in Part I, except that the transmission lines are replaced by lumped element lowpass filters. The capacitors used in the lowpass filter are ATC chip capacitors and variable capacitors from Johanson. The inductors are air-core winding, 5 turns, 0.250 in. I.D. The photo of the coupler is shown in Figure 6. The measured response (after some adjustments) of the wideband quadrature coupler is shown in Figure 7 (amplitude ripple response). Figure 8 illustrates the return loss and isolation, and Figure 9

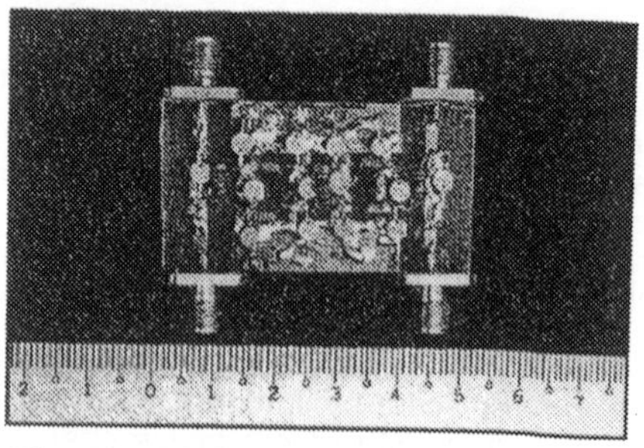

Figure 6. The coupler prototype.

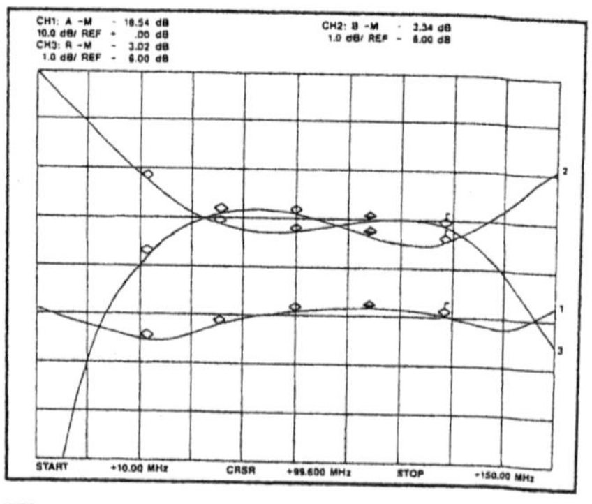

Figure 7. Measured frequency response.

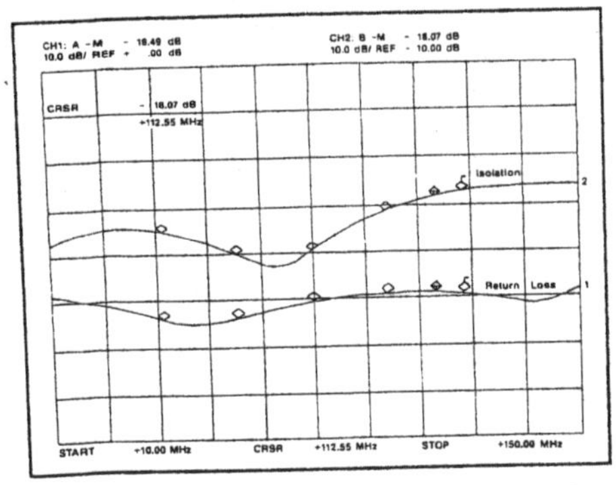

Figure 8. Return loss and isolation. of the coupler.

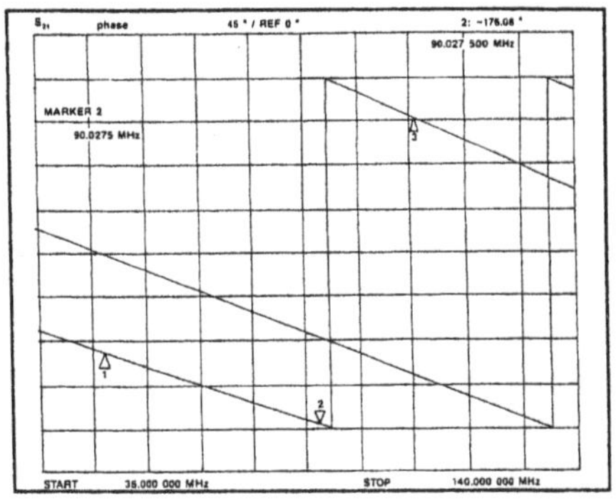

Figure 9. Quadrature phase relationships.

shows the quadrature phase relationships. The measured ripple is about 0.5 dB from 47 MHz to 130 MHz or bandwidth ratio of 2.75. The return loss is better than 18 dB and isolation is better than 16 dB. rf

Reference

1. Matthaei, Young and Jones, *Microwave Filters, Impedance Matching Networks, and Coupling Structures,* McGraw-Hill Book Co., 1964.

About the Authors

Dr. Chen Y. Ho is vice-president—engineering, at M/A-COM, Active Assemblies Division, 340 North Roosevelt Ave., Chandler, AZ 85226. Tel: (602) 966-1471. Ge-Lih Chen is a research associate at the Chung Shan Institute of Science and Technology in Taiwan.

Synthesis of symmetric lumped element quadrature hybrids

D.P.Andrews and C.S.Aitchison

Abstract: The synthesis of a class of lumped element quadrature hybrids is described. The circuits are applicable to RF systems, and have wide operating band potential. The circuits possess symmetry in both horizontal and vertical planes. Two examples have been designed and constructed; one, a second-order example, operates over a 50–150 MHz band, exhibiting 0.3 dB loss, 23 dB return loss and isolation and 0.2° phase accuracy, the other, a third-order example operates over a 2–30 MHz band exhibiting 0.3 dB loss, 25 dB return loss, 30 dB isolation and 2° phase accuracy.

1 Introduction

There are two distinct categories of lumped element quadrature hybrid in use in RF systems; those which exhibit exact quadrature properties while approximating an equal power division [1–6], and those which are exact in their power division but approximate quadrature performance [7, 8]. Of the former category, only the circuits invented by Cappucci [6] are capable of achieving bandwidths substantially greater than an octave. However, as design information is not available in the literature for these circuits, the latter category of circuits are usually favoured for wideband operation, for which design information is readily available.

Recent developments by the authors in the theory of optimum approximation functions [9] and hence transfer functions [10], have shown the potential for exact quadrature hybrids to operate over wide bandwidths with a predictable performance. Although it can be shown that hybrids of the Cappucci-type implement these transfer functions, this will be left for another paper. The purpose of this paper is to demonstrate the synthesis procedure of a class of quadrature hybrids, which implement the transfer functions of [10] and are also symmetric in form.

2 Outline of the synthesis problem

A generic quadrature hybrid is as shown in Fig. 1. It will be assumed that the circuit is lossless and perfectly matched on all ports. These two constraints already imply the further condition that power applied to any particular port will be transmitted to two further ports, with the remaining port isolated [11]. Furthermore, the circuit is also constrained to exhibit perfect quadrature response to the two outputs. This last condition imposes a symmetry constraint on the circuit [12] (except for an arbitrary assignment of the polarity of the terminations), and this is shown in Fig. 1.

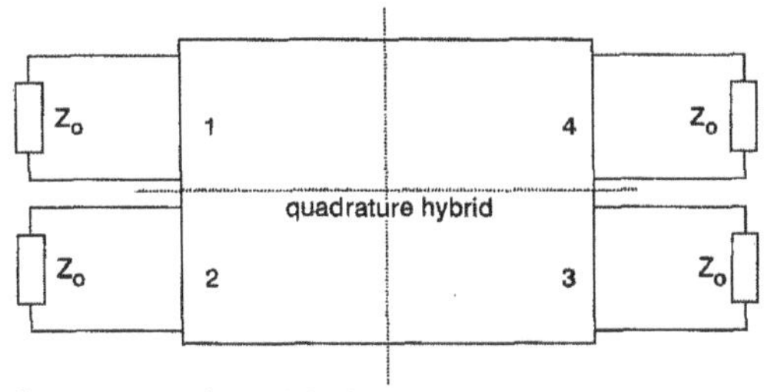

Fig. 1 *Generic quadrature hybrid*
.... Line of symmetry

It has been shown [10] that for a hybrid with transfer functions in terms of s-parameters defined as

$$s_{41}(s) = \frac{N_4(s)}{D(s)} \tag{1}$$

$$s_{21}(s) = \frac{N_2(s)}{D(s)} \tag{2}$$

the numerator and denominator expressions are given by

$$D(s) = \prod_{r=1}^{n} (\sigma_r + s) \tag{3}$$

$$N_4(s) = \text{even part of } N(s) \tag{4}$$

$$N_2(s) = \text{odd part of } N(s) \tag{5}$$

where

$$N(s) = \prod_{r=1}^{n} (\sigma_r + (-1)^{r-1}.s) \tag{6}$$

$$\sigma_r = \frac{1}{\omega_U} sc\left(\left(r - \frac{1}{2}\right).\frac{K_1'}{n}, k_1'\right) \quad r = 1 \ldots n \tag{7}$$

and where ω_u is the upper radian frequency normalised to a geometric mean frequency of unity, $k_1 = 1/\omega_U^2$ the modulus for the elliptic function system, $k_1' = \sqrt{(1 - k_1^2)}$ is the complementary modulus, and K_1' is the complete elliptic integral of the complementary modulus.

IEE Proceedings online no. 20010670
DOI: 10.1049/ip-map:20010670
Paper first received 13th December 2000 and in revised form 9th August 2001

D.P. Andrews is with Vectawave Technology Ltd., Unit 1, 17C Riverway, Newport, Isle of Wight PO30 5UX, UK
C.S. Aitchison is with the School of Electronic Engineering, University of Surrey, Guildford, Surrey GU2 5XH, UK

Reduction of the problem of determining elliptic functions can be effected by use of the equation

$$\sigma_r = \frac{1}{\sigma_{n+1-r}} \tag{8}$$

The problem of synthesis is much more complicated than in the case of lowpass prototype filters. It is necessary to devise circuits which provide a perfect match on all ports and simultaneously realise two transfer functions. One approach described by the literature is to excite the input and coupled ports in even and odd modes [2]. The apparent circuits in the two modes are duals of each other. The through response is the same as the transfer function of either the even or the odd mode equivalent circuit, and the coupled response is the same as the reflection coefficient of either the even or odd mode. The requirement of quadrature response will be assured by the nature of the transfer function synthesised [12]. The resulting even and odd mode equivalent circuits will automatically be symmetrical (electrically, though not necessarily physically). This approach is familiar in the design of transmission line couplers, where the duality arises from arranging them such that the even and odd mode characteristic impedances of each section are geometrically related. Even and odd mode analysis is a sound approach for lumped element quadrature hybrids as well, though the combination of the even and odd mode equivalent circuits into one circuit may not be obvious and may not be unique. A further complication arises when considering eqns. 1 and 2. [10] showed that the numerator of eqn. 1 is even and that of eqn. 2 is odd. It follows that for all but the most trivial case of $n = 1$, the transfer functions possess right half-plane zeros. This is a particular problem when considering the through response, as it means the even and odd mode equivalent circuits cannot be synthesised using a ladder circuit, but must employ bridging elements or coupled inductors.

It is also desirable to synthesise a circuit where all the ports have one of their terminals connected to a common ground, so that they are all available for measurement without additional isolation.

As the final circuits are not unique, and a common ground is not guaranteed, it is useful to approach the synthesis problem in a way which guarantees both vertical and horizontal symmetry, and also assumes a common ground. A way of achieving these aims is shown in Section 3.

3 Analysis of the hybrid by four modes

In a similar manner to analysis by even and odd modes, consider the circuit of Fig. 1 to be excited at all four ports. The first mode, which we shall call mode 1, places an equal excitation at all four ports. Consider the apparent circuit from the viewpoint of port 1. As the hybrid is perfectly matched, the generator at port 1 will receive no reflection from this port. Furthermore, the isolation qualities of the hybrid will ensure that the generator at port 3 will not transmit a signal to port 1. Only the generators at ports 2 and 4 will transmit signals to port 1, and these will equate to $s_{21}(s)$ and $s_{41}(s)$, respectively, so that the apparent reflection coefficient seen at port 1 will be the sum of these two quantities. We use the well known equation for impedance in terms of reflection coefficient

$$Z(s) = \frac{1+\Gamma(s)}{1-\Gamma(s)} \tag{9}$$

where Z_0, the reference impedance, is normalised to unity. The expressions for $s_{21}(s)$ and $s_{41}(s)$ from eqns. 1 and 2 can be substituted for $\Gamma(s)$ in eqn. 9, and the apparent impedance seen by port 1 will be given by

$$Z_1(s) = \frac{1+\left(\frac{N_4(s)}{D(s)}+\frac{N_2(s)}{D(s)}\right)}{1-\left(\frac{N_4(s)}{D(s)}+\frac{N_2(s)}{D(s)}\right)} = \frac{D(s)+N(s)}{D(s)-N(s)} \tag{10}$$

Consider a second type of excitation whereby the generators at ports 1 and 2 are identical, and the generators at ports 3 and 4 are identical to each other, and equal but opposite in sign to those at ports 1 and 2. We shall designate this as mode 2. Now, the contribution of the generator at port 2 to the signal emanating at port 1 will be the same as it was in mode 1. However, the contribution of the generator at port 4 to the signal emanating from port 1 will be opposite in polarity, and so the resultant apparent reflection coefficient at port 1 will be $s_{21}(s) - s_{41}(s)$. Substituting this into eqn. 9 gives after simplification

$$Z_2(s) = \frac{D(s)-N(-s)}{D(s)+N(-s)} \tag{11}$$

The third excitation, which we shall designate mode 3, is defined such that the generators at ports 1 and 3 are identical, but opposite in sign to those at ports 2 and 4. Proceeding similarly gives the apparent impedance at port 1

$$Z_3(s) = \frac{D(s)-N(s)}{D(s)+N(s)} \tag{12}$$

Finally, the fourth excitation, which we shall designate mode 4, is defined such that the generators at ports 1 and 4 are identical, but opposite in sign to those at ports 2 and 3. The apparent impedance at port 1 then becomes

$$Z_4(s) = \frac{D(s)+N(-s)}{D(s)-N(-s)} \tag{13}$$

These four modes are sufficient to define any excitation to the circuit, for example an excitation at port 1 only can be simulated by a quarter of the sum of modes 1–4 inclusive. The mode numbers have been chosen to correspond with the port which is identical to port 1, except for mode 1, where all the generators are identical.

Each of the four modes expresses a different aspect of the circuit symmetry. Mode 1 places virtual open circuits along the vertical and horizontal planes of symmetry. Mode 2 places a virtual short circuit along the vertical plane and an open circuit along the horizontal plane. Mode 3 places a short circuit along both planes. Mode 4 places an open circuit along the vertical plane and a short circuit along the horizontal plane.

In practice, the expressions on the right-hand side of eqns. 10–13 readily simplify, as eqns. 3 and 6 show there to be common factors. Note that the apparent impedance in modes 1 and 3 are duals, as are those in modes 2 and 4. Equivalent circuits with element values for the apparent impedances in all four modes are given in Fig. 2 for orders up to the fourth. Use is made of eqn. 8.

Designing networks which exhibit the equivalent circuits shown in Fig. 2 still requires a degree of innovation. Configuring the capacitive part is most difficult, as all values must be positive. There is more flexibility in the inductive part, as mutual inductance element values can be positive or negative. However, once the network has been determined, the reduction into the four modes makes component value determination a simple task.

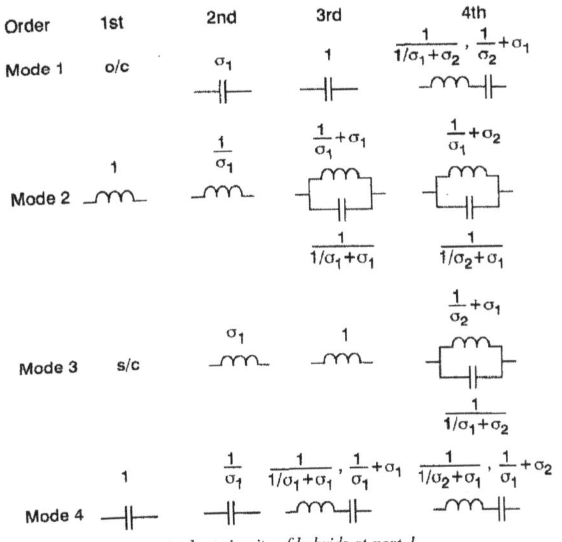

Fig. 2 *Apparent equivalent circuits of hybrids at port 1*

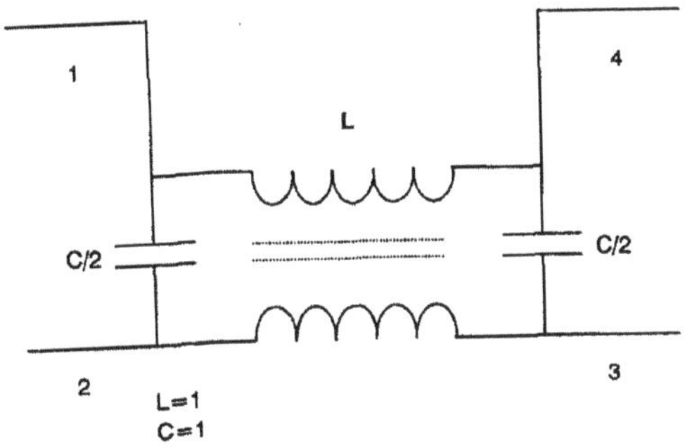

Fig. 3 *First-order hybrid solution*

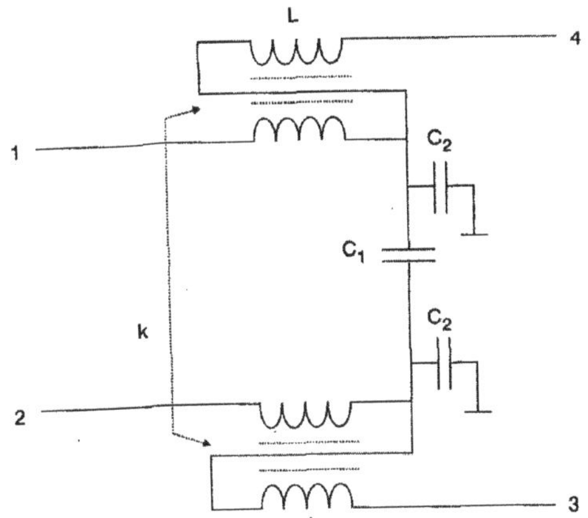

$$L=\frac{1}{4}\left(\frac{1}{\sigma_1}+\sigma_1\right)$$

$$M=\frac{1}{4}\left(\frac{1}{\sigma_1}-\sigma_1\right)$$

$$k=\frac{1/\sigma_1-\sigma_1}{1/\sigma_1+\sigma_1}$$

$$C_1=\frac{1}{\sigma_1}-\sigma_1$$

$$C_2=2\sigma_1$$

Fig. 4 *Second-order hybrid solution*

4 Network designs

Networks that successfully implement the terminal requirements are shown in Figs. 3–6, for solutions up to the fourth order. The first-order circuit of Fig. 3 is well known [2–4], and has been included for the sake of thoroughness. The second-order circuit solution of Fig. 4 appears to be absent from the literature, but is still simple enough to be synthesised by even and odd mode analysis. The third- and fourth-order circuit solutions of Figs. 5 and 6, also absent from the literature, were determined by inspection of the equivalent circuit requirements of Fig. 2. Note that for simplicity, all inductors wound on the same core are assumed to be perfectly coupled. Where imperfect coupling is required, a coupling factor is indicated.

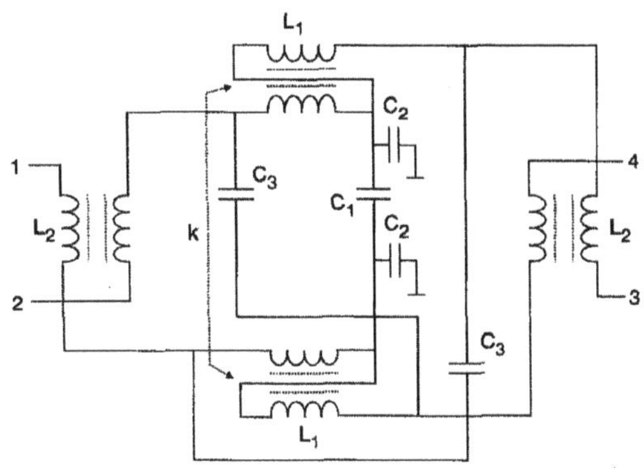

$$L_1=\frac{1}{4}\left(\frac{1/\sigma_1^2+1/\sigma_1+1+\sigma_1+\sigma_1^2}{1/\sigma_1+\sigma_1}\right)$$

$$M_1=\frac{1}{4}\left(\frac{1/\sigma_1^2-1/\sigma_1+3-\sigma_1+\sigma_1^2}{1/\sigma_1+\sigma_1}\right)$$

$$k=\frac{1/\sigma_1^2-1/\sigma_1+3-\sigma_1+\sigma_1^2}{1/\sigma_1^2+1/\sigma_1+1+\sigma_1+\sigma_1^2}$$

$$L_2=\frac{1}{2}\left(\frac{1}{1/\sigma_1+\sigma_1}\right)$$

$$C_1=\frac{1/\sigma_1^2-1/\sigma_1+1-\sigma_1+\sigma_1^2}{1/\sigma_1+\sigma_1}$$

$$C_2=2,\quad C_3=\frac{1}{2}\left(\frac{1}{1/\sigma_1+\sigma_1}\right)$$

Fig. 5 *Third-order hybrid solution*

To illustrate the procedure for determining element values, consider the fourth-order circuit of Fig. 6. First, apply mode 1 excitation to the circuit. Certain elements can be identified as redundant in this mode, owing to the effects of symmetry. Opposing currents in each L_1 ensure zero flux generated by these elements, which appear as short circuits. C_1 and each C_3 and C_4 have zero potential across them, and so appear as open circuits. Only each L_2 and C_2 appear in circuit. The polarity of coupling between each L_2 is defined such that mutual coupling will subtract from the inductance in this mode. From the point of view of port 1, the circuit appears as an inductor of value twice the difference of L_2 and M_2 in series with a capacitor of value half C_2. Comparing the circuit given in Fig. 2 for mode 1 fourth-order gives two equations for the circuit values

$$2(L_2-M_2)=\frac{1}{1/\sigma_1+\sigma_2} \tag{14}$$

$$\frac{C_2}{2}=\frac{1}{\sigma_2}+\sigma_1 \tag{15}$$

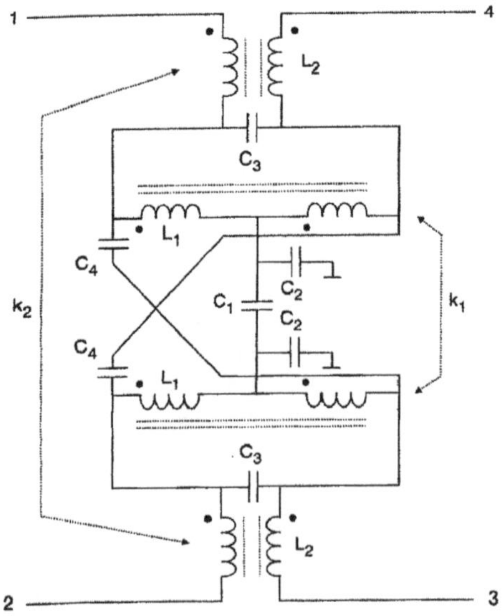

$$L_1=\frac{1}{4}\left(\frac{1}{\sigma_1}+\frac{1}{\sigma_2}+\sigma_2+\sigma_1\right) \qquad M_1=\frac{1}{4}\left(\frac{1}{\sigma_1}-\frac{1}{\sigma_2}+\sigma_2-\sigma_1\right)$$

$$L_2=\frac{\sigma_2+\sigma_1}{4(1+\sigma_1\sigma_2)} \qquad M_2=\frac{\sigma_2-\sigma_1}{4(1+\sigma_1\sigma_2)}$$

$$k_1=k_2=\frac{\sigma_2-\sigma_1}{\sigma_2+\sigma_1} \qquad C_1=(\sigma_2-\sigma_1)\left(1+\frac{1}{\sigma_1\sigma_2(1+\sigma_1\sigma_2)}\right)$$

$$C_2=2\left(\frac{1}{\sigma_2}+\sigma_1\right) \qquad C_3=\frac{1}{2(1/\sigma_1+\sigma_2)} \qquad C_4=\frac{\sigma_2-\sigma_1}{2(1+\sigma_1\sigma_2)}$$

Fig. 6 *Fourth-order hybrid solution*

Turning our attention now to mode 2, symmetry will dictate that there is zero flux generated in each L_2. There will also be no potential difference across C_1 and each C_2. Only each L_1, C_3 and C_4 will contribute. The polarity of coupling between each L_1 is defined such that mutual coupling will add to the inductance in this mode. From the point of view of port 1, the circuit will appear as the parallel combination of an inductor of value twice the sum of L_1 and M_1 and a capacitor of value twice the sum of C_3 and C_4. Two more equations can then be identified from comparison with Fig. 2. These are

$$2(L_1+M_1)=\frac{1}{\sigma_1}+\sigma_2 \tag{16}$$

$$2(C_3+C_4)=\frac{1}{1/\sigma_2+\sigma_1} \tag{17}$$

Consider now the circuit excited in mode 3. Flux cancellation will occur in each L_2. In addition, no voltage will appear across C_1 and each C_2 and C_4. The polarity of coupling between each L_1 will be such as to reduce inductance by mutual coupling. From the point of view of port 1, the circuit will appear as a parallel combination of an inductor of value twice the difference of L_1 and M_1 and a capacitor of value twice C_3. Comparison with Fig. 2 gives the two equations

$$2(L_1-M_1)=\frac{1}{\sigma_2}+\sigma_1 \tag{18}$$

$$2C_3=\frac{1}{1/\sigma_1+\sigma_2} \tag{19}$$

Finally, consider the circuit excited in mode 4. This time flux cancellation will occur in each L_1, and no potential will appear across each C_3. The polarity of coupling between each L_2 will be such as to increase inductance by mutual coupling. From the point of view of port 1, the circuit will appear as the series combination of an inductor whose value is twice the sum of L_2 and M_2 and a capacitor of value C_1 plus half C_2 plus twice C_4. Comparison with Fig. 2 then gives two more equations

$$2(L_2+M_2)=\frac{1}{1/\sigma_2+\sigma_1} \tag{20}$$

$$C_1+\frac{1}{2}C_2+2C_4=\frac{1}{\sigma_1}+\sigma_2 \tag{21}$$

Manipulation of eqns. 14–21 gives the element values in Fig. 6.

5 Design examples

5.1 Second-order hybrid

To demonstrate the application of the second-order solution, it was decided to apply the circuit of Fig. 4 to the problem of designing a hybrid operating over a 50–150 MHz frequency band, for which the geometric centre frequency is 86.6 MHz. Clearly, $\omega_u=\sqrt{3}$. Application of eqn. 7 gives $\sigma_1=0.39332$. The formulae in Fig. 4 will then give the element values shown in Table 1.

Table 1: Element values

Element	Prototype value	Scaled value
L	0.7339	67.4 nH
M	0.5373	49.4 nH
(k	0.7321	0.7321)
C_1	2.1491	79 pF
C_2	0.7866	28.9 pF

The capacitor values are easily implemented using fixed values. The inductor element has to be custom wound. A suitable implementation for L was to wind a bifilar pair of 0.2 mm diameter enamelled copper wire five times on a Micrometals T16-2 carbonyl iron core. Both instances of L were wound on the same core, with the required coupling between the two windings being achieved by overlapping the end turns of each winding. Optimum match and isolation were achieved with a fixed value of 82 pF for C_1 and a parallel combination of 27 pF and 2.7 pF for C_2.

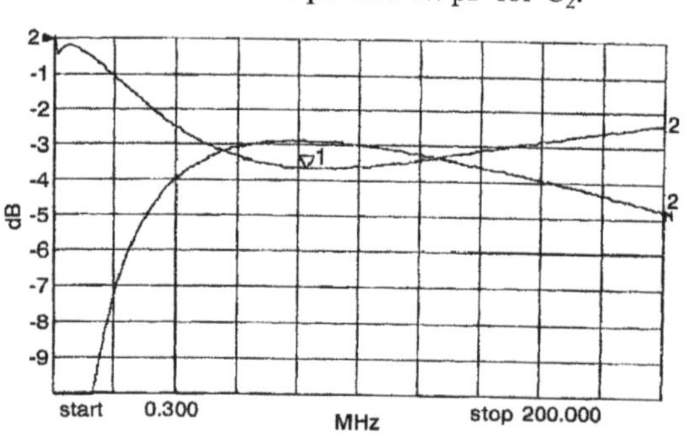

Fig. 7 *Measured through and coupling response of the 50–150 MHz hybrid*

The measured insertion loss of the hybrid is shown in Fig. 7. It can be seen that there are two places where the traces cross, with an overcoupled portion in between, as expected for a second-order hybrid. Over the design frequency band, the amplitude imbalance is about 0.8 dB for the worst case. This compares favourably with the theoretical value of 0.62 dB (determined using the techniques presented in [9]). The average excess loss over 3dB is about 0.3dB at the upper frequency limit.

The measured isolation and return loss is shown in Fig. 8. Both have a worst case value of 23dB at the upper frequency limit.

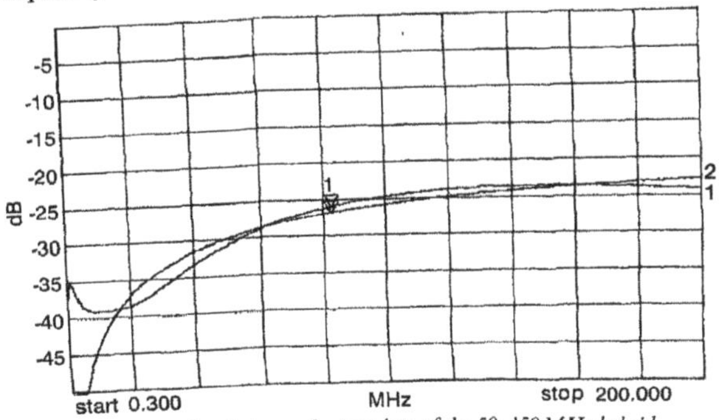

Fig. 8 *Measured isolation and return loss of the 50–150 MHz hybrid*

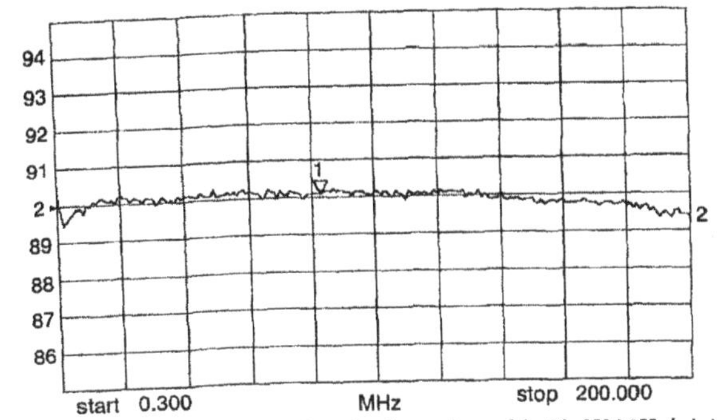

Fig. 9 *Measured coupled to through relative phase of the 50–150 MHz hybrid*

The measured phase of the coupled output with respect to the through output is shown in Fig. 9. Deviation from quadrature over the 50–150MHz design band is less than 0.2°.

5.2 Third-order hybrid

To demonstrate the third-order circuit, a design frequency of 2–30MHz was chosen, for which the geometric centre frequency is 7.75MHz and ω_u =3.873. Application of eqn. 7 then gives σ_1 = 0.19046. Substitution into the formulae in Fig. 5 then gives the component values shown in Table 2.

Table 2: Component values

Element	Prototype value	Scaled value
L_1	1.5643	1.307 μH
M_1	1.156	966 nH
(k	0.739	0.739)
L_2	0.0919	94.4 nH
C_1	4.2571	1.749 nF
C_2	2	822 pF
C_3	0.0919	37.8 pF

As was the case for the second-order coupler, the capacitors are easily implemented using fixed values, with the inductor elements having to be custom wound. L_1 was wound using 10 turns of bifilar 0.2mm enamelled copper wire on a Micrometals T30-3 carbonyl iron core. Both instances of L_1 were wound on opposite sides of the same core, with coupling adjusted through positioning of the windings. L_2 was wound using five turns of bifilar 0.2mm enamelled copper wire on a Micrometals T16-2 carbonyl iron core.

The measured insertion loss of the 2–30MHz hybrid is shown in Fig. 10. This time there are three points where the traces cross over, as would be expected with a third-order hybrid. Over the design frequency, the amplitude imbalance is about 1.3dB maximum, a little worse than the theoretical maximum of 0.93dB (determined using the techniques presented in [9]). The average excess loss over 3dB is about 0.3dB at the upper frequency limit.

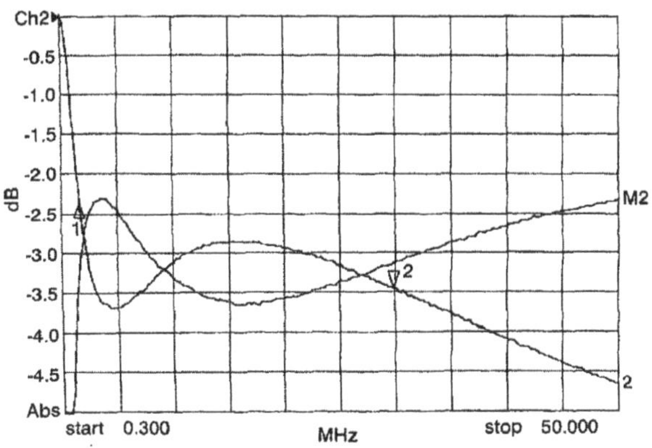

Fig. 10 *Measured through and coupling response of the 2–30 MHz hybrid*

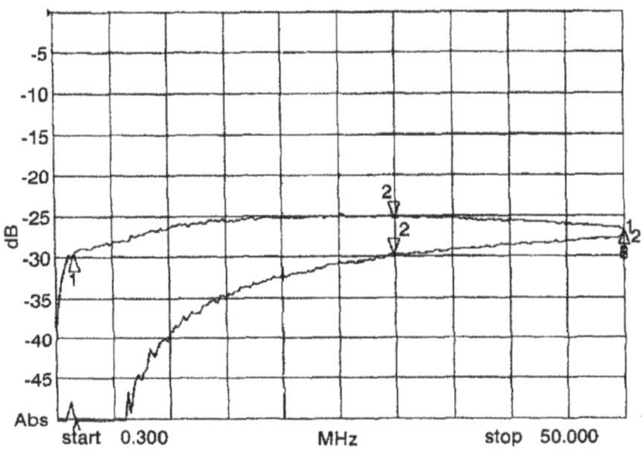

Fig. 11 *Measured isolation and return loss of the 2–30 MHz hybrid*

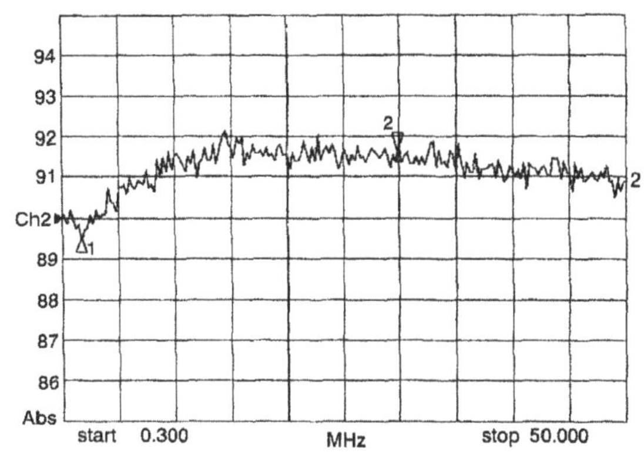

Fig. 12 *Measured through to coupled relative phase of the 2-30 MHz hybrid*

The measured isolation and return loss are shown in Fig. 11. The isolation achieves 30dB over almost the entire frequency range, while the return loss is better than 25dB. The measured phase of the coupled output with respect to the through output is shown in Fig. 12. Maximum deviation from quadrature is less than 2°. The greater deviation, as compared with the second-order example, can be attributed to the more complicated circuit and the limits to inductor adjustment using fixed cores.

6 Conclusions

The circuits of Figs. 3–6 show how it is possible to realise quadrature hybrids of up to fourth-order. It is anticipated that the publication of the second-order circuit of Fig. 4 will be of significant value, as many applications are of modest bandwidth, and the circuit is not difficult to construct. The measured performance of the 50–150 MHz example shows excellent agreement with theory, as does the 2–30 MHz example, albeit with a less accurate phase performance.

7 References

1 FIRESTONE, W.L.: 'Coupling system'. US Patent 2 972 121, February 1961

2 CAPPUCCI, J.D., and SEIDEL, H.: 'Four port directive coupler having electrical symmetry with respect to both axes'. US Patent 3 452 300, June 1969

3 FISHER, R.E.: 'Broad-band twisted-wire quadrature hybrids', *IEEE Trans.*, 1973, **MTT-**, pp. 355–357

4 CHEN, Y.H., and FURLOW, B.: 'Design of VHF quadrature hybrids - Part 1', *R.F. Des.*, 1979, pp. 49–54

5 CHEN, Y.H., and CHEN, G.-L.: 'Design of wideband quadrature couplers for UHF/VHF: Part 1', *R.F. Des.*, 1989, pp. 58–61

6 CAPPUCCI, J.D.: 'Networks using cascaded quadrature couplers, each coupler having a different center operating frequency'. US Patent 3 514 722, May 1970

7 DARLINGTON, S.: 'Realization of constant phase differences', *Bell Syst. Tech. J.*, 1950, **29**, pp. 94–104

8 KEELY, T.A.: 'Design of constant phase difference networks', *R.F. Des.*, 1989, pp. 32–42

9 ANDREWS, D.P., and AITCHISON, C.S.: 'Optimum approximation functions for lumped element quadrature hybrids', *IEE Proc., Circuits, Devices Syst.*, 2001, **148**, (1), pp. 5–9

10 ANDREWS, D.P., and AITCHISON, C.S.: 'Transfer functions of optimum lumped element quadrature hybrids', *IEE Proc., Circuits, Devices Syst.*, 2001, **148**, (3), pp. 135–139

11 MONTGOMERY, C.G., DICKE, R.H., and PURCELL, E.M.: 'Principles of microwave circuits' (Dover, 1965)

12 CRISTAL, E.O., and YOUNG, L.: 'Theory and tables of optimum symmetrical TEM-mode coupled transmission line directional couplers', *IEEE Trans.*, 1965, **MTT-13**, (5), pp. 544–558

Design of cascaded lumped element quadrature hybrids

D.P.Andrews and C.S.Aitchison

Abstract: Methods to implement lumped element quadrature hybrids using a cascade of simpler hybrids are described. Both first and second order cascades are considered in detail, the former requiring an ideal inverting transformer to give an optimum equi-ripple response. Both techniques are employed to design a fourth order hybrid operating over a 10–100 MHz band, and details of fabrication and test results are given, showing consistency with theory.

1 Introduction

In 1968, Cappucci filed a patent describing the design of a class of lumped element quadrature hybrids, using a cascade of first order hybrids together with an ideal inverting transformer [1]. Although Cappucci gave some insight into the mathematics of the hybrids, a complete design procedure was not given, so it remains to be determined how to calculate the centre frequency of each of the first order hybrids, and how they should be configured. Significantly, however, it was noted that the hybrids had wide bandwidth potential, and their frequency responses were described as rational Chebychev.

Recent work by the authors [2–4] has investigated the use of rational Chebychev functions in the design of optimum quadrature hybrids; optimum in the sense of giving the best approximation to an equal power division over a specified frequency range, for a given order. This work predicts the frequency response of an optimum quadrature hybrid [2], and then proceeds to give the transfer functions of such hybrids [3]. Finally, some practical circuits are described, whose performance gives a realisation of the optimum transfer functions [4].

This paper considers two new aspects of quadrature hybrid design. First, it shows how the hybrids of the Cappucci class are a realisation of the optimum transfer functions (described in eqns. 1–8 of [3]), and thus how they can be designed. Secondly, a new class of quadrature hybrid is described, in which second order hybrids are cascaded to form higher order quadrature hybrids. Finally, the design of two circuits is described, with an identical specification, one of the Cappucci class, and the other of the new class. However, before these objectives can be dealt with, it is necessary to consider some aspects of cascaded quadrature hybrids, to derive common characteristics, and also to define some notation.

IEE Proceedings online no. 20010703

DOI: 10.1049/ip-map:20010703

Paper first received 14th March and in revised form 5th September 2001

D.P. Andrews is with Vectawave Technology Ltd., 2 Winston Close, Ryde, Isle of Wight PO33 3QF, UK

C.S. Aitchison is with the School of Electronic Engineering, University of Surrey, Guildford, Surrey GU2 5XH, UK

2 Theory

2.1 Properties of cascaded symmetric quadrature hybrids

Consider the block diagram shown in Fig. 1, an electrically symmetric (though not necessarily physically symmetric) quadrature hybrid. The ports are defined such that the input port is at upper left, the isolated port is at lower left, the through port is at upper right and the coupled port is at lower right. The quadrature hybrid is assumed to be a lossless passive network (and also lumped, linear, finite and reciprocal), with all ports matched to the same characteristic impedance Z_0. The transfer function from the input to the through port is defined to be:

$$through \equiv \frac{N_e}{D}(s) \quad (1)$$

The transfer function from the input to the coupled port is defined to be:

$$coupled \equiv \frac{N_o}{D}(s) \quad (2)$$

where $N_e(s)$, $N_o(s)$ and $D(s)$ are all polynomials of s. In [3], it was determined that for a quadrature hybrid, $N_e(s)$ is an even function of s and $N_o(s)$ is an odd function of s, hence the choice of subscripts. $D(s)$ is a Hurwitz polynomial, and will subsequently be referred to as the denominator polynomial. The symmetry constraint means that the output ports are isolated from each other, and the transfer functions from the isolated to the output ports are the same as eqns. 1 and 2, except that the definition of through and coupled ports is reversed.

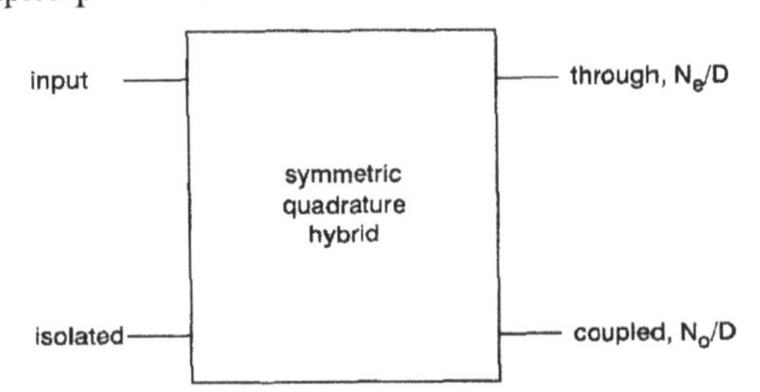

Fig. 1 *Symmetric quadrature hybrid port definitions*

Furthermore, the sum of the numerator expressions is defined by the equation:

$$N(s) = N_e(s) + N_o(s) \quad (3)$$

where $N(s)$ is subsequently referred to as the numerator polynomial.

Consider now a cascade of two quadrature hybrids of the type shown in Fig. 1, but of different and arbitrary specification, as shown in Fig. 2. Analysing the transfer function from the input to the through ports gives the equation:

$$\frac{N_e}{D}(s) = \frac{N_{ae}N_{be} + N_{ao}N_{bo}}{D_a D_b}(s) \tag{4}$$

where the left-hand side denotes the response of the total network, and the right-hand side the functions of the individual components, the subscripts a and b referring to the networks designated in Fig. 2.

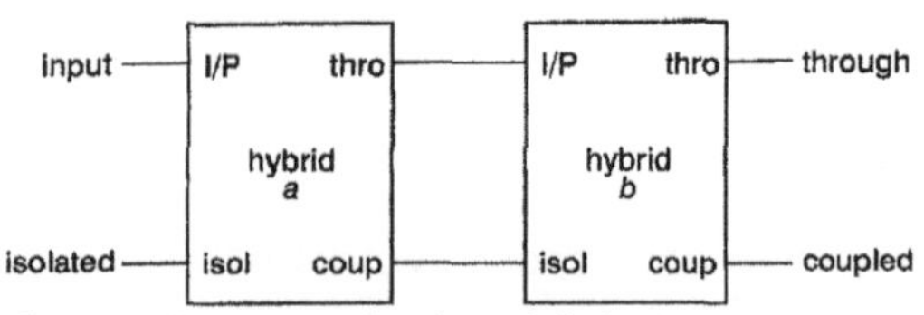

Fig. 2 *Cascade connection of quadrature hybrids*

Similarly, the transfer function from the input to the coupled port is given by:

$$\frac{N_o}{D}(s) = \frac{N_{ao}N_{be} + N_{ae}N_{bo}}{D_a D_b}(s) \tag{5}$$

Some important results can be deduced from eqns. 4 and 5. First, the numerator of the through transfer function remains even and that of the coupled transfer function remains odd, so that the subscripts on the left-hand side continue to be meaningful. This shows that the total network is also a quadrature hybrid (by substituting $s = j\omega$ into both equations). Secondly, examination of the subscripts indicates that the equations would be exactly the same if the two constituent hybrids were reversed. The resultant hybrid is therefore electrically symmetric in the vertical plane. Symmetry in the horizontal plane has of course not been affected. The resultant cascade continues to be perfectly matched on all ports, and appropriately isolated. It is therefore a symmetric quadrature hybrid.

Now, it can be appreciated that in Fig. 2, each hybrid itself may be composed of a cascade of lower order hybrids, which have already been combined by the argument above. This leads to the following theorem.

Theorem 2.1: The cascade connection of symmetric quadrature hybrids is also a symmetric quadrature hybrid. In theorem 2.1 it is understood that the individual quadrature hybrids are arranged as in Fig. 1 and assembled as in Fig. 2.

Furthermore, from eqn. 3 (dropping the variable s for clarity):

$$\begin{aligned} N &= N_e + N_o \\ &= N_{ae}N_{be} + N_{ao}N_{bo} + N_{ao}N_{be} + N_{ae}N_{bo} \\ &= (N_{ae} + N_{ao})(N_{be} + N_{bo}) \\ &= N_a N_b \end{aligned} \tag{6}$$

The result of eqn. 6 suggests a quick method of determining the transfer functions of a quadrature hybrid composed of an arbitrary number of cascaded sections. It is evident from eqn. 6 that the numerator function $N(s)$ of a quadrature hybrid, composed of a cascade of quadrature hybrids, can be determined by multiplying the numerator functions of the individual components; that is to say, the equation can be used recursively. Clearly, from eqns. 4 and 5, the same is true of the denominator function $D(s)$. Assigning the even and odd parts of $N(s)$, respectively, and dividing by $D(s)$ can then determine the through and coupled outputs' transfer functions.

2.2 Hybrids composed of cascade of first order sections

Consider the first order quadrature hybrid as in Fig. 3, which has been orientated in accordance with Fig. 1 (this way of drawing the hybrid is contrary to convention, but is chosen so that the position of the through and coupled outputs is consistent in a cascade). For element values chosen as in Fig. 3, the through transfer function is given by:

$$\frac{N_e}{D}(s) = \frac{\sigma}{\sigma + s} \tag{7}$$

and the coupled transfer function is given by:

$$\frac{N_o}{D}(s) = \frac{s}{\sigma + s} \tag{8}$$

where σ is a parameter defining the centre frequency of operation. Clearly, in this case

$$N(s) = D(s) = \sigma + s \tag{9}$$

To implement an optimum quadrature hybrid, the numerator and denominator functions should be [3]:

$$N(s) = \prod_{r=1}^{n} (\sigma_r + (-1)^{r-1} s) \tag{10}$$

$$D(s) = \prod_{r=1}^{n} (\sigma_r + s) \tag{11}$$

where

$$\sigma_r = \frac{1}{\omega_U} \mathrm{sc}\left(\left(r - \frac{1}{2}\right)\frac{K_1'}{n}, k_1'\right), \quad r = 1, \ldots, n \tag{12}$$

ω_u is the normalised upper radian frequency, $k_1 = 1/\omega_U^2$ is the modulus for the elliptic function, $k_1' = \sqrt{(1 - k_1^2)}$ is the complementary modulus, K' is the complete elliptic integral of the complementary modulus, n is the order of the circuit, and $\mathrm{sc}(x) = \mathrm{sn}(x)/\mathrm{cn}(x)$.

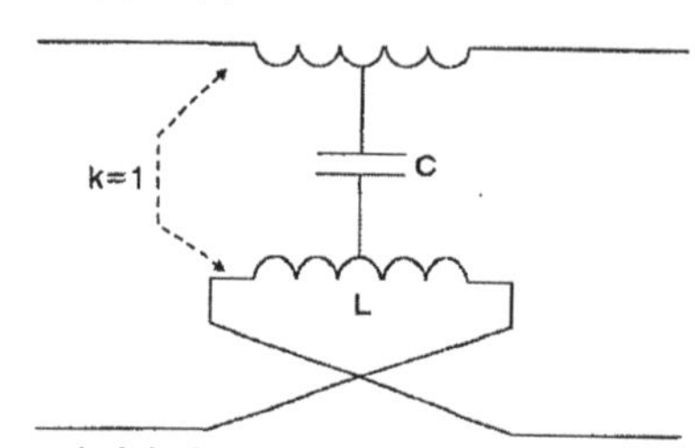

Fig. 3 *First order hybrid section*
$C = 1/(\sigma Z_0)$, $L = Z_0/\sigma$

It is evident from eqn. 6 that a cascade of first order hybrids will yield a numerator function consisting of the product of terms as in eqn. 9. Comparing this with eqn. 10, it is seen that such a cascade cannot yield an optimum hybrid.

2.3 Cappucci's solution to hybrid problem

To circumvent the problem that a first order cascade cannot implement an optimum hybrid, Cappucci [1] invented a solution using an ideal inverting transformer in the hybrid cascade. To see how this operates, consider the block diagram of Fig. 4. An inverting transformer has been placed in the coupled line between two symmetric quadrature hybrids, each of which is a cascade of first order hybrids. Clearly, no change in the denominator function results from the inclusion of the inverter; however, the numerator function is now given by:

$$\begin{aligned} N &= N_{ae}N_{be} - N_{ao}N_{bo} + N_{ae}N_{bo} - N_{ao}N_{be} \\ &= (N_{ae} - N_{ao})(N_{be} + N_{bo}) \\ &= N_a(-s)N_b(s) \end{aligned} \quad (13)$$

A method of implementing the optimum quadrature hybrid can now be seen. If in Fig. 4, cascade a is composed of hybrids with characteristic frequencies given by the terms in eqn. 10 for r even and cascade b those for which r is odd, then the right-hand sides of eqns. 10 and 13 will be equal, and Fig. 4 will be an optimum quadrature hybrid.

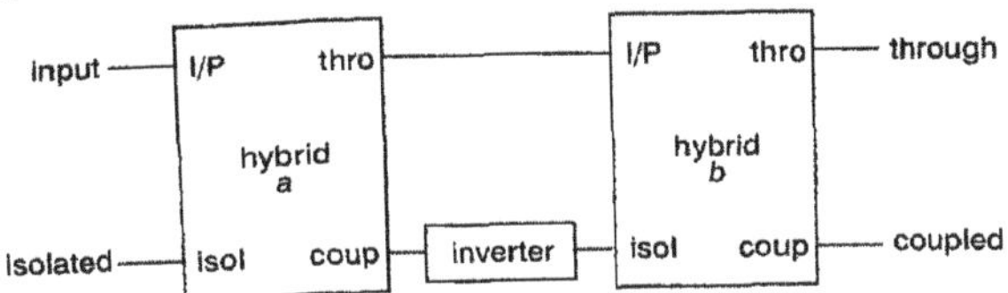

Fig. 4 *Hybrid cascade with inverter in coupled line*

It should be noted that the resulting quadrature hybrid is not symmetric. Reflection in the horizontal plane will change the sign of both outputs. Also, reflecting in the vertical plane will change the sign of the coupled output, while not affecting the through output. The hybrid can be made electrically symmetric by placing a second inverting transformer in the isolated output. However, practical applications will seldom require this.

2.4 Hybrids composed of cascade of second order sections

Having shown how first order sections can be interconnected to produce an optimum hybrid, one might enquire as to whether higher order sections can also be cascaded. A second order section is as shown in Fig. 5, together with expressions for the element values. This circuit was presented by the authors in a previous paper [4]. The element values have been recalculated for an arbitrary frequency band so that the numerator function is:

$$N(s) = (\sigma_1 + s)(\sigma_2 - s) \quad (14)$$

and the corresponding denominator function is:

$$D(s) = (\sigma_1 + s)(\sigma_2 + s) \quad (15)$$

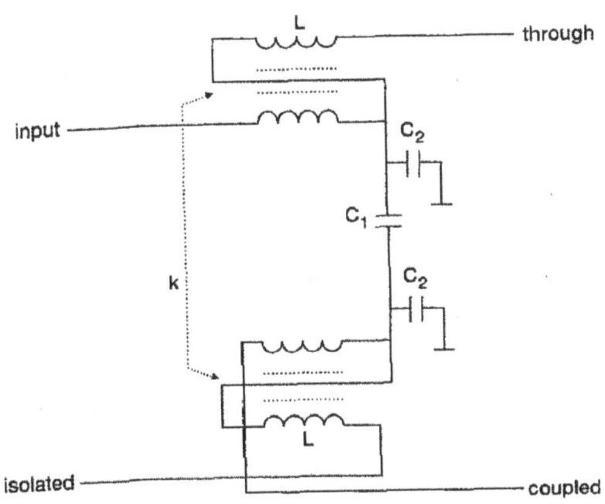

Fig. 5 *Second order hybrid section*
$L = (Z_0/4)(1/\sigma_1 + 1/\sigma_2)$, $M = (Z_0/4)(1/\sigma_1 - 1/\sigma_2)$, $k = (\sigma_2 - \sigma_1)/(\sigma_2 + \sigma_1)$, $C_1 = (1/Z_0)(1/\sigma_1 - 1/\sigma_2)$, $C_2 = 2/(Z_0\sigma_2)$

Now, applying eqn. 6 recursively to the whole cascade, it can be seen from eqn. 14 that the numerator function of a cascade of second order hybrids will have an equal number of positive and negative zeros. Eqn. 10 shows this to be the desired result. The question now is how the factors in eqn. 10 should be distributed in the cascaded second order sections. Eqn. 14 constrains σ_2 to be greater than σ_1 for the circuit to be realisable, otherwise a negative value for C_1 will result (see Fig. 5). Applying this consideration to eqn. 10, it is seen that the factor containing σ_1 can be paired with any factor with a negative sign in it. However, σ_2 can only be paired with σ_1, so one second order section in the cascade must implement these two values. A similar argument will follow, continuing with σ_3. The end result will be the cascade of second order sections with adjacent characteristic frequencies taken from the list in eqn. 10.

So far this formula will only give an even order hybrid. However, as eqn. 10 gives a final factor of the form $(\sigma_n + s)$ when n is odd, the final section can be implemented as a first order circuit, as in Fig. 3. As the individual sections are all symmetric, they can be assembled in any order, according to theorem 2.1 above.

3 Practical circuits

To demonstrate the application of the preceding analysis, it was decided to use both techniques to implement a fourth order quadrature hybrid, operating over a 10–100MHz frequency band, and with a minimum return loss of 20dB.

Using the techniques of [2], it can be verified that the ideal solution will exhibit a maximum amplitude imbalance of about 0.16dB. Normalising to a centre radian frequency of unity, eqn. 12 will give the characteristic frequencies as follows: $\sigma_1 = 0.15139$, $\sigma_2 = 0.59702$, $\sigma_3 = 1.67498$, $\sigma_4 = 6.60526$.

To afford a fair comparison between the first and second order cascade techniques, it was decided to use similar construction standards in their assembly.

3.1 Implementation using Cappucci class hybrid

Considering first a Cappucci implementation, element values from Fig. 3 (after scaling for frequency and a 50Ω characteristic impedance) become: L_a = 1.662μH, C_a = 665pF; L_b = 421.5nH, C_b = 168.6pF; L_c = 150.2nH, C_c = 60.1pF; L_d = 38.1nH, C_d = 15.2pF.

In addition, an inverting transformer will also be necessary. To achieve the return loss specification, it was decided to specify this component with a magnetising inductance of 4μH minimum.

All the inductors were implemented using a twisted pair of 0.2mm diameter enamelled copper wire, wound on suitable ring cores. The capacitances were implemented using fixed values, placed as close as possible to the mid-point of each inductor winding. The inverting transformer was implemented using a winding of the twisted pair on a dual hole bead. A second bead was wound with the same number of terms, but connected as a non-inverting transmission line. This was placed on the opposite side to the inverting transformer, in order to equalise the phase length that component introduces. To give the correct transfer functions, the lowest and second highest frequency hybrids were placed on one side of the transformers, while the remaining two hybrids were placed on the other side, as discussed in Section 4. A photograph of the completed hybrid is shown in Fig. 6. It was found that extra tuning capacitors were desirable in order to compensate for stray inductance, and these can be seen in Fig. 6.

Fig. 6 *Photograph of 10–100 MHz, fourth order Cappucci hybrid*

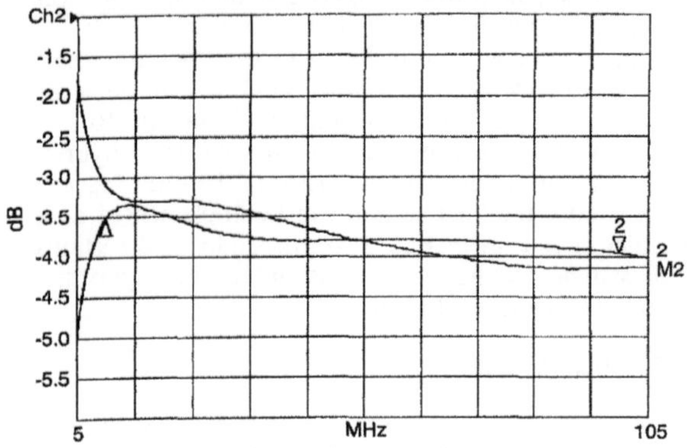

Fig. 7 *Through and coupled loss of Cappucci hybrid*
Ch2: Mkr2: 100 MHz, –3.95 dB

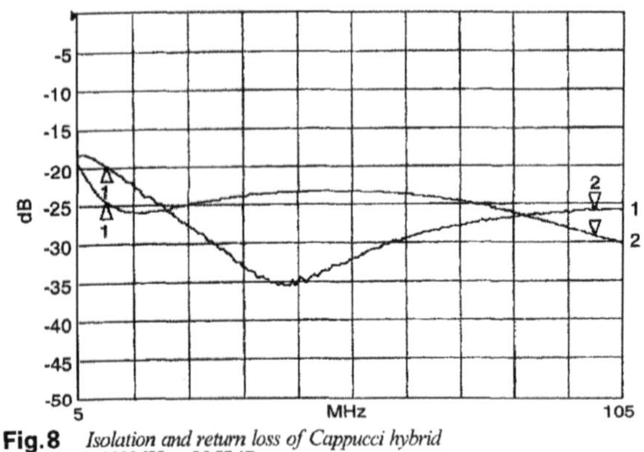

Fig. 8 *Isolation and return loss of Cappucci hybrid*
Ch1: Mkr2: 100 MHz, –25.77 dB
Ch2: Mkr2: 100 MHz, –29.19 dB

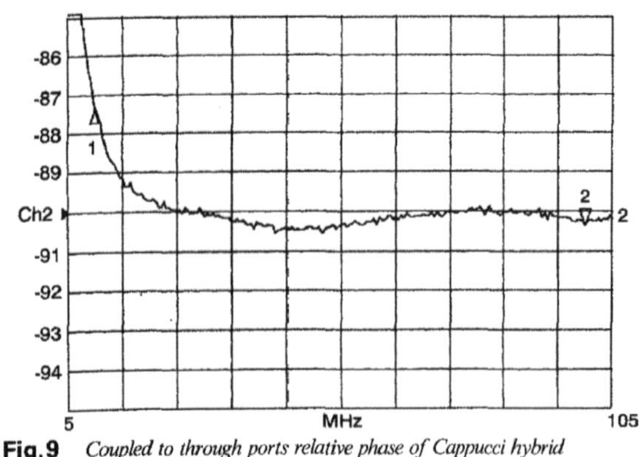

Fig. 9 *Coupled to through ports relative phase of Cappucci hybrid*
Ch2: Mkr2: 100 MHz, –90.37°

Fig. 7 shows the loss to the through and coupled ports. The maximum imbalance is 0.3 dB over most of the band, but degrades to 0.5 dB at the lower band edge, owing to insufficient coupling. The worst case insertion loss is at the upper frequency limit, where the average value is about 1 dB in excess of the ideal 3 dB.

Fig. 8 shows the input return loss and isolation. The specification limit of 20 dB for the return loss is marginally achieved at the lower frequency limit, but is comfortably achieved thereafter. The worst case isolation is 23 dB.

Fig. 9 shows the relative phase of the coupled output compared with the through output. Quadrature accuracy is better than 1° over most of the band, only degrading to 2.6° at the lower band edge as a result of the shunt susceptance of the inverting transformer's magnetising inductance.

3.2 Implementation using second order cascade

Proceeding now to the second order cascade implementation, the element value expressions in Fig. 5 give the following component values for each of the two second order sections (after scaling for frequency and impedance): L_a = 520.9 nH, M_a = 310.2 nH, C_{a1} = 496.3 pF, C_{a2} = 337.2 pF, L_b = 47.1 nH, M_b = 28 nH, C_{b1} = 44.9 pF, C_{b2} = 30.5 pF.

It can be verified that the coupling factor in each individual section is 0.595, which corresponds to a –4.5 dB maximum coupling.

Once again, 0.2 mm enamelled copper wire twisted pairs were used to implement the inductors on suitable ring cores. The capacitors were implemented as fixed values. A photograph of the completed hybrid is shown in Fig. 10. The capacitors were fitted underneath for the sake of convenience. Tuning capacitors were again found to be desirable to compensate for stray inductance.

Fig. 10 *Photograph of 10–100 MHz second order cascade hybrid*

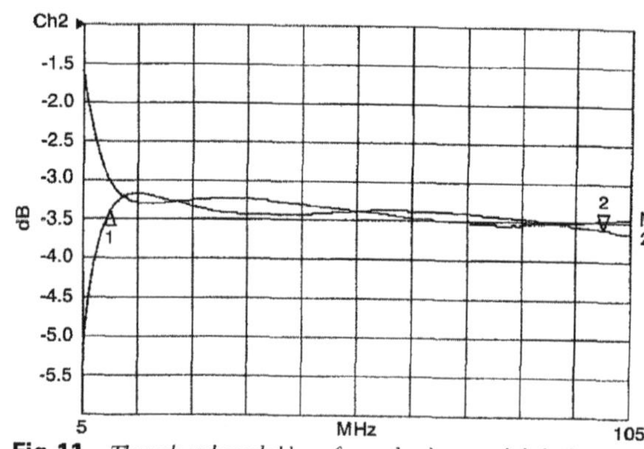

Fig. 11 *Through and coupled loss of second order cascade hybrid*
Ch2: Mkr2: 100 MHz, –3.59 dB

Fig. 11 shows the loss to the through and coupled ports. The maximum imbalance is 0.3 dB over most of the band, but degrades to 0.4 dB at the lower band edge, owing to insufficient coupling. The worst case excess loss is about 0.5 dB average at the upper frequency limit.

Fig. 12 shows the input return loss and isolation. The input return loss is extremely good at 30dB over the band, while isolation is an acceptable 22dB.

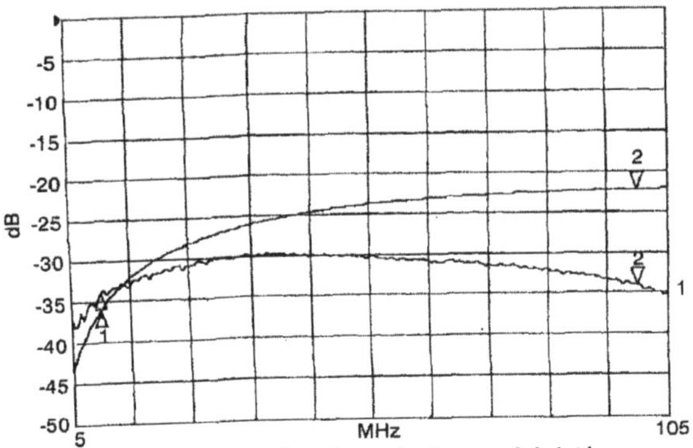

Fig. 12 *Isolation and return loss of second order cascade hybrid*
Ch1: Mkr2: 100MHz, −34.03dB
Ch2: Mkr2: 100MHz, −22.15dB

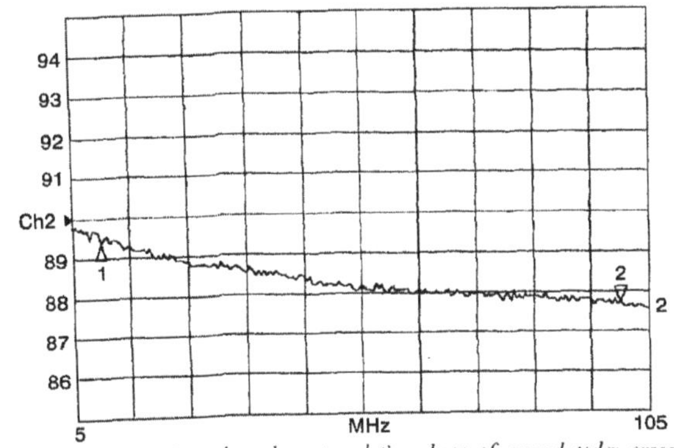

Fig. 13 *Coupled to through ports relative phase of second order cascade hybrid*
Ch2: Mkr2: 100MHz, 87.71°

Fig. 13 shows the relative phase of the coupled output compared with the through output. The performance shows an increasing departure from quadrature, with the maximum deviation being 2.3° at the upper frequency limit.

3.3 Comparison of results for two implementations

Both hybrids achieved the specification requirements in terms of frequency range and return loss. The other quantities measured may or may not be acceptable, depending on the particular application. Where the performance of the hybrids departs from the ideal (that is to say, perfect match, return loss, isolation and quadrature accuracy), the reason can be attributed to component losses, stray effects and element tolerances.

Comparing the performance of the two implementations, the insertion loss is better for the second order cascade quadrature hybrid. Part of the reason for this is the fewer hybrids required; each individual section contributes a little to the loss. The Cappucci hybrid requires more hybrid sections, and an inverter as well. The inverter has to operate satisfactorily over the entire operating frequency, and its magnetic component must exhibit high impedance. This can only be achieved through the use of a relatively high-permeability core material, which in turn implies significant loss. The amplitude balance is comparable.

There is little to choose between the two hybrids in terms of return loss and isolation. With design iteration, both quantities might be optimised in both hybrids.

Both hybrids also exhibited similar worst case departure from quadrature. However, the Cappucci hybrid was better over most of the band, and the poorer performance at the lower frequency limit would likely be improved by using a different inverter design.

Both hybrids were also similar with regard to ease of construction. Although the Cappucci hybrid has more parts, each is easier to construct.

For applications where small size is required, the second order cascade hybrid has the advantage. However, it is anticipated that the second order version will become more difficult to construct as the frequency increases. The twisted pairs in each section have a finite electrical length, which causes their performance to deteriorate at higher frequencies. The same is not true of the Cappucci hybrid, where the first order sections can be made to operate even up to microwave frequencies, if suitably designed [5]. A common advantage to both types of hybrid over the use of the symmetric circuits described in [4], is the ability to distribute the circuit, as the length of connecting lines between sections does not affect quadrature performance.

4 Conclusions

It has been shown that the Cappucci class of lumped element quadrature hybrids gives designs which realise the same optimum approximation functions as derived in recent work by the authors. In addition, the cascade principle has been extended to include second order sections, and so avoid the necessity for an inverting transformer.

A practical example of each realisation technique has been presented, operating over the same frequency band of 10–100MHz. Both achieve better than 20dB return loss and isolation, and departure from quadrature of less than 3°. The Cappucci implementation gives an excess loss of 1dB maximum, while the second order cascade implementation gives an excess loss of 0.5dB maximum.

5 References

1 CAPPUCCI, J.D.: 'Networks using cascaded quadrature couplers, each coupler having a different center operating frequency'. US patent 3,514,722, 1970

2 ANDREWS, D.P., and AITCHISON, C.S.: 'Optimum approximation functions for lumped element quadrature hybrids', *IEE Proc., Circuits Devices Syst.*, 2001, **148**, (1), pp. 5–9

3 ANDREWS, D.P., and AITCHISON, C.S.: 'Transfer functions of optimum lumped element quadrature hybrids', *IEE Proc., Circuits Devices Syst.*, 2001, **148**, (3), pp. 135–139

4 ANDREWS, D.P., and AITCHISON, C.S.: 'The synthesis of symmetric lumped element quadrature hybrids', *IEE Proc., Microw. Antennas Propag.*, 2001, **148**, (5), pp. 269–274

5 ANDREWS, D.P., and AITCHISON, C.S.: 'Microstrip lumped element quadrature couplers for use at microwave frequencies', *IEE Proc., Microw. Antennas Propag.*, 2000, **147**, (4), pp. 267–271

FM Broadcast 88 to 108 MHz 4,000 Watt Amplifier Subsystem

By Lee B. Max
Microwave Modules and Devices

During the last five years, ultra high power solid-state transmitters have truly come of age. A variety of multikilowatt, all solid-state systems have been developed. Initially, military systems, where more dollars were available, were developed using ultra high power solid-state transmitters. The more cost conscious commercial applications remained tube systems. Today, however, the technology and economics have reached the level where one of the most cost sensitive commercial applications is being converted to solid-state. The FM broadcast transmitter (88-108 MHz) market has already converted to wide band, solid-state, driver amplifiers for power levels through 1,000 watts. But even beyond this, the performance, initial cost, operational cost, and reliability considerations have made solid-state transmitters the choice of a new FM broadcast market.

As a result of an interest in the FM radio station business, the Federal Communications Commission (FCC) has taken the following actions:

1. Relative to the presently licensed high power FM radio stations, the FCC, has adopted to "use it or lose it" policy. Meaning that a time limit has been set when a radio station whose license allows for more "effective radiated power" (ERP) than it is presently delivering must either increase its ERP level or be relicensed down to its present ERP level.
2. After the "use it or lose it" cut-off date, the FCC will begin issuing new licenses for FM radio stations to fill in the open areas.
3. Approximately 700 new licenses for FM radio stations will be issued for 3,500 watt transmitters.

In response to the creation of this entirely new 3,500 watt "low power" transmitter market opportunity, complete 3.5 kW solid-state transmitters are already being built.

Microwave Modules and Devices (MMD) has developed a 4,000 watt subsystem for use in these low power FM broadcast transmitters. The major features of this subsystem are:
- 4,000 watt output
- electrically self-protected
- mechanical rugged
- 50 watt input
- low cost
- high efficiency
- rack mounted

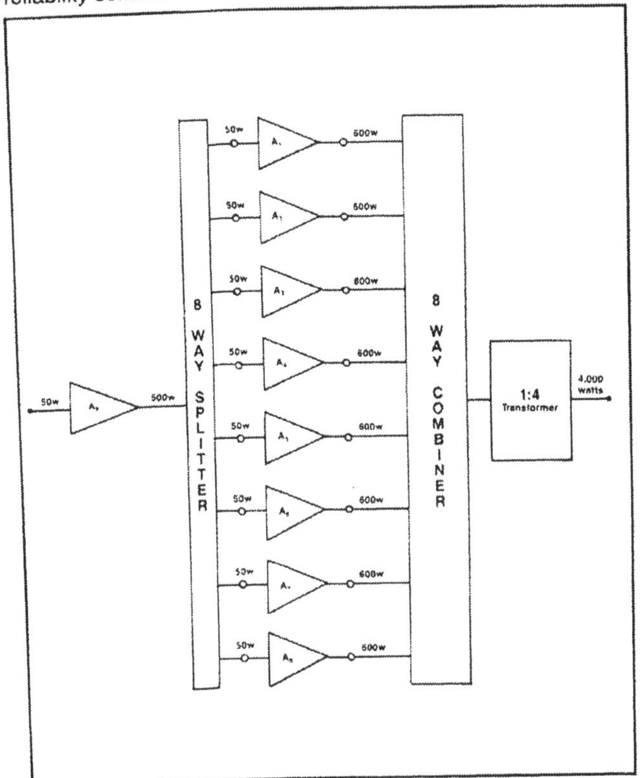

Figure 1. 4,000 watt amplifier chain

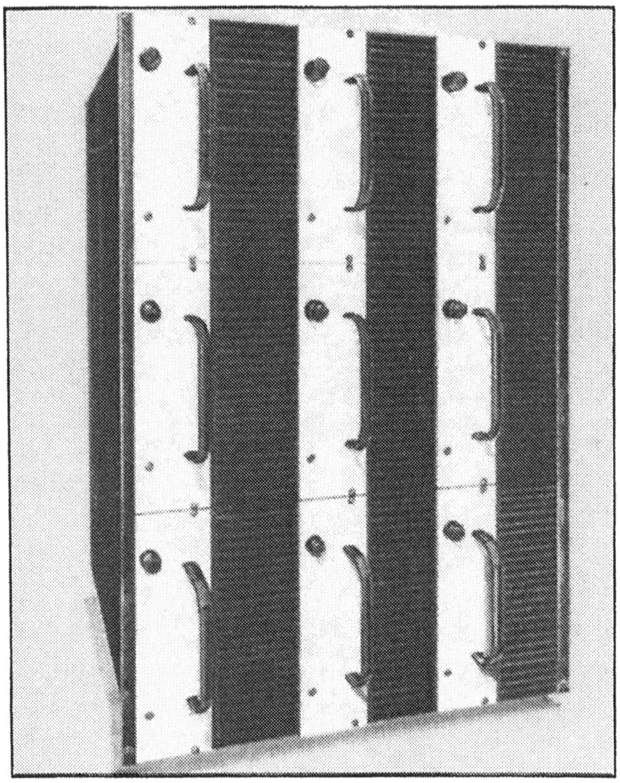
Figure 2. 19" rack-mounted subsystem

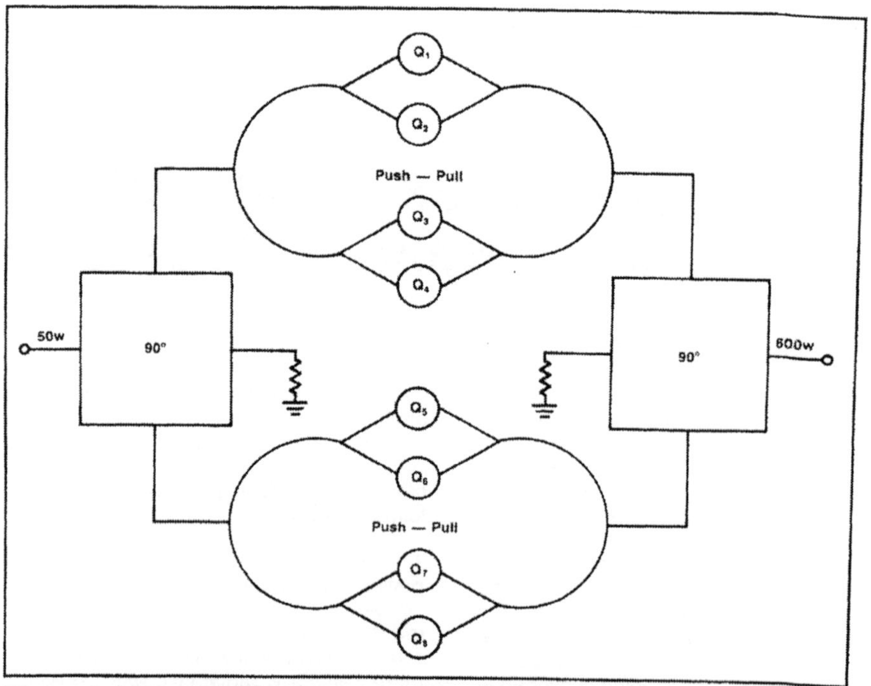

Figure 3. 600 watt FET module amplifier architecture

Figure 4. 600 watt building block amplifier module

The most crucial requirement is that the unit be totally self-protected and virtually indestructable.

Subsystem Architecture

The 4,000 watt subsystem (Figure 1) is modular in construction. It uses nine identical 600 watt amplifier modules in a one driving eight configuration. The mechanical structure consists of three rows of three modules per row (Figure 2). All nine modules are plug-in replaceable from the front of the cabinet. The subsystem is also designed to allow for "hot replacement" (i.e., a module can be removed and replaced without turning the transmitter off). This helps maximize the stations "up" time. The subsystem is cooled using forced air with each 600 watt module having its own self-contained blower assembly.

600 Watt Self-Protected FET Amplified Module

The heart of the 600 watt module is the RF amplifier section (Figure 3). The unit uses eight 125 watt silicon FETs. The die are similar to those used in the Motorola MRF-174 transistor. These die are mounted in special high performance MMD packages. Since high temperature reliability was critical and small size was not a driving consideration, MMD's module was targeted for minimum thermal density (Figure 4). The unit easily delivers the rated 600 watts of power with eight FETs operating at well under 140°C even during worst case environmental conditions.

The eight thermally individually packaged transistors are binarily interconnected through three levels of combining. To end up with the maximum combination of benefits, each level of combining uses a different technique. At the first level, a pair of transistors is combined in the simplest possible way, i.e., in phase or parallel combined. Isolation resistors are used across the pair to smooth out slight amplitude or phase imbalances. The next level of interconnecting uses 180°, i.e., push-pull combining techniques. An optimally designed transmission line trans-

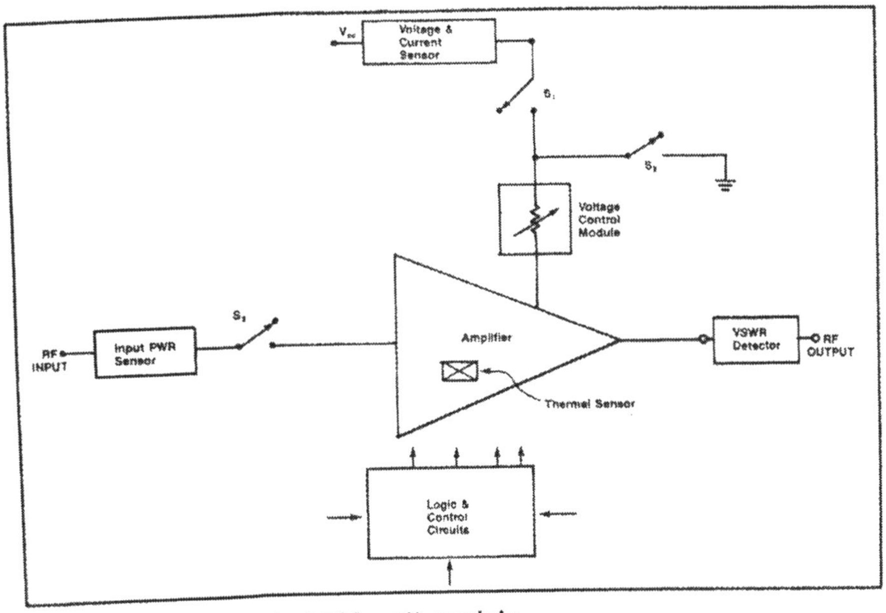

Figure 5. Self-protected 600 watt module

former structure accomplishes this. The push-pull combiner is an ultra-wide band-width structure that provides very predictable impedance transformation, even harmonic attenuation, as well as efficient power combining. The third and final level of interconnection uses a special 90° combiner. This combiner is the key to making the 600 watt module a true building block unit. Use of the 90° (or Quad) combiner produces a module with the following advantages:

- 1.5:1 VSWR input and output
- Improved forward power delivery into a mismatched load
- Improved back IMD
- Reduced 3rd harmonic levels
- Sub-module isolation
- Improved efficiency into a mismatched load

Since it is crucial that the 600 watt building block module be indestructable, sophisticated protection circuitry is built in. As seen in Fig. 5, the module contains both an input power sensor and a load VSWR detector. The critical parts of these sensors are the special, low-loss, bi-directional couplers used to pick off forward and reflected power samples that are then rectified and converted into DC reference levels. The 600 watt module also contains DC voltage, DC current, and baseplate temperature monitoring circuits. If abnormal conditions appear, the unit can automatically reduce the DC voltage to the module, or crowbar the supply to the ground, depending on what is required. The entire 600 watt module can be shut down in less than 2 milliseconds.

When any module shuts itself off after sensing an abnormality, it must wait a preset time (not more than 2 seconds) before resetting itself and trying to come back on line. Each 600 watt module has a slightly different preset time so that if the entire 4 kW subsystem shuts down, all nine modules will not try to come up at the same instant and overload the power supply.

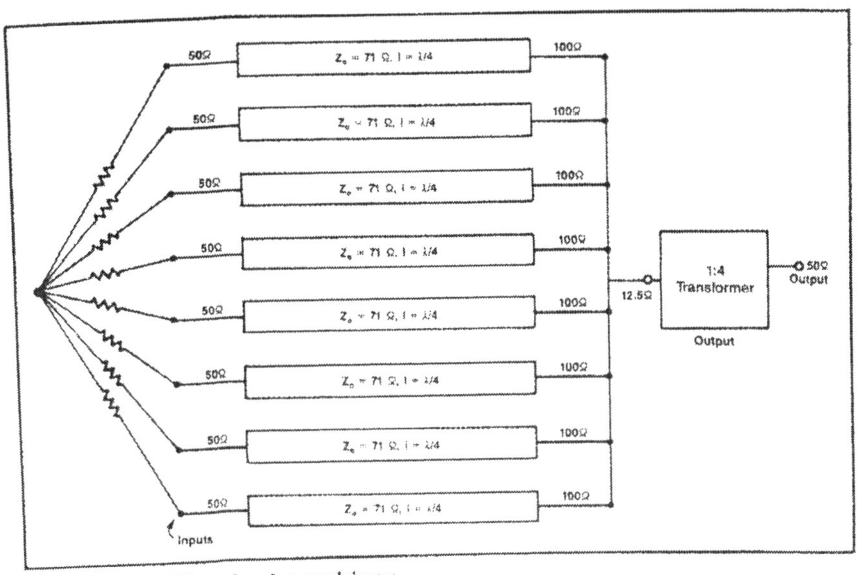

Figure 6. 4 kW output combiner

4,000 Watt Combiner

The low-loss, 4,000 watt, 8-way combiner is critical to the system performance and reliability. As seen in Fig. 6 it is constructed using eight 71 ohm, 1¼ inch cables assembled in a ring configuration on both the input and output of the combiner. The circular structure allows for the balancing terminations to be conveniently mounted on the input, and the high power combining to occur with the minimum loss to the output. It should be noted that the output impedance of the combined eight large coaxial cables is 12.5 ohm. To transform from here to 50 ohm, a low loss 1:4 transformer made from a 25 ohm, quarter wave length transmission line is used. The entire combining system has less than 0.2 dB insertion loss. Furthermore, the eight isolation resistors, rated at 500 watts each, provide a minimum of 20 dB isolation between any one module and its neighbors. Measurements have shown that if a module fails or is removed, none of the remaining seven modules sees a load VSWR any worse than 1.5:1. Thus, if one module fails or is removed, the subsystem can still deliver over 80 percent of its full power capability.

The eight way low power splitter used on the input (not described in detail) is a conventional in-phase Wilkinson structure.

About the Author

Lee Max is vice president of Microwave Modules and Devices, 550 Ellis St., Mountain View, CA 94043. His telephone number is (415) 961-1473.

High-Power GaAs FET Amplifiers: Push-Pull Versus Balanced Configurations

By Jonathan Shumaker, Raymond Basset and Alex Skuratov
Fujitsu Compound Semiconductor, Inc.

Most high-power microwave gallium arsenide (GaAs) field-effect transistor (FET) modules [2] up to L- and S-band (and soon up to C-band) consist of two independent devices without any internal transversal connection between them. Though often called push-pull amplifiers, the two devices can be combined in a variety of configurations created by external components, such as 180-degree splitter/combiners (ba-luns [3]); 3 dB quadrature couplers, such as branch line or Lange couplers; or in-phase couplers, such as Wilkinsons.

Push-pull [2] and balanced [4] configurations are both used for high-power GaAs FET amplifier designs for relatively narrow band commercial applications from UHF to S-band. In the near future, these devices will be available at higher frequencies.

The question is whether push-pull amplifiers are better suited for these applications than balanced ones. This article compares the push-pull and balanced configurations for amplifiers using dual GaAs push-pull devices for commercial applications with less than 1 octave bandwidth.

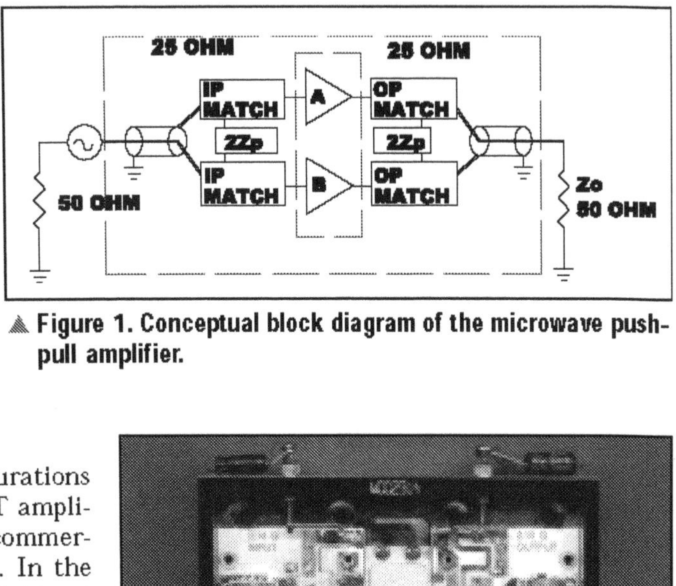

▲ Figure 1. Conceptual block diagram of the microwave push-pull amplifier.

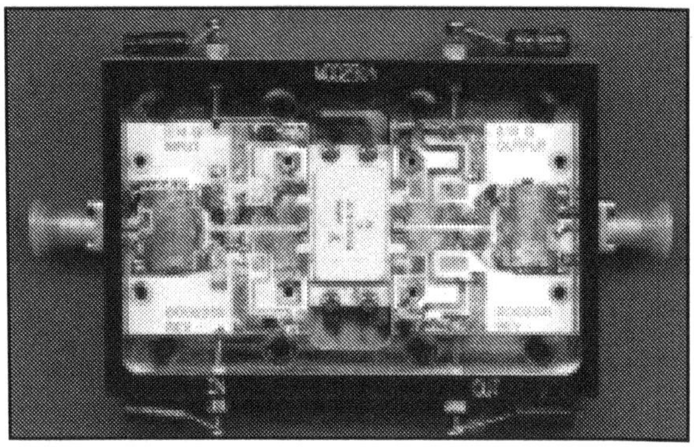

▲ Figure 2. Push-pull amplifier as built and tuned.

Push-pull amplifier

A push-pull amplifier (see Figures 1 and 2) consists of an input 0 to 180-degree power splitter driving two identical devices in antiphase and a 0 to 180-degree output power combiner adding the output power of the two devices in the amplifier load. The splitter and combiner are baluns [3] (BALanced UNbalanced). Baluns transform a balanced system that is symmetrical with respect to ground to an unbalanced system with one side grounded.

Note that the push-pull amplifier contains two independent devices each amplifying an individual signal of half the total power.

Advantages of the push-pull amplifier:

- Four times higher device impedance [5] (Z_{in} Gate-to-Gate and Z_{out} Drain-to-Drain) compared to single-ended device impedances with

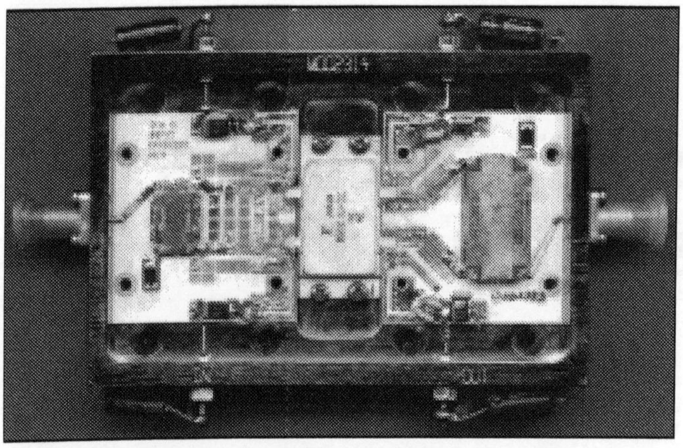

▲ **Figure 3. Balanced amplifier as built and tuned.**

the same output power. Thus it is easier to match.

- Virtual ground [5], which can be used for more compact and simpler matching structures.
- Cancellation of even products and harmonics, such as F2-F1, 2F1, 2F2, F1 + F2...

Disadvantages of the push-pull amplifier:

- Poor input and output external match because the baluns used for push-pull amplifiers do not eliminate the input and output power reflected by the device.
- With conventional baluns, isolation between the two sides of the part is theoretically only 6 dB; this poor interdevice isolation can cause instability problems.
- Use of baluns: coaxial baluns are simple to make for laboratory use but in production they have high labor cost. Surface-mount technology (SMT) baluns are available but add cost and tend to be larger than equivalent quadrature couplers.

Balanced amplifier

The balanced amplifier [3, 6] (see Figures 3 and 4) uses a splitter at the input and a combiner at the output, with a 90-degree phase difference between the coupled and through ports. The fourth port of the splitter/combiner must be terminated with a good 50-ohm load (for a 50-ohm system impedance). This resistor must reliably dissipate the reflected power by the input circuit of the devices, the reflected power from the amplifier load and the power due to the unbalance between the two sides of the amplifier for the combiner. As power goes up, the power rating and size of these resistors must increase. The load of the splitter can be relatively small compared to the load of the combiner.

The input signal is split at 0 and 90 degrees, amplified, then added in the load by the 90-degree combiner. Due to the phase shift, the output voltage of the two signals in the load of the isolated port of the combiner are cancelled and the load connected at the load combiner port sees the sum of these two signals.

Advantages of the balanced amplifier configuration:

- Good isolation between the two halves of the device. This improves amplifier stability within the bandwidth of the coupler.
- Good input and output external match, since the reflected power is absorbed by the 50-ohm load in the isolated coupler port. This gives a constant well-defined load to the driver stage, improving amplifier stability and driver circuit power flatness versus frequency.
- Cancellation in the load of third-order products and harmonics such as 2F1+F2, 2F2+F1, 3F1, 3F2.... and attenuation by 3 dB of second-order products such as F1-F2, F1+f2, 2F1, 2F2...
- Easy to design and integrate within a printed or SMT 3 dB quadrature coupler.

Disadvantages of the balanced amplifier configuration:

- Requires the use of a 50-ohm load in the decoupled port of the input and output couplers. This is an extra part that must be purchased and installed. High-power resistors can be expensive and require proper heat sinking.
- Couplers must be used that require either design effort (this type of coupler is well documented and easy to design) for printed quadrature couplers or purchased and installed.
- No virtual ground. This leads to a generally less compact tuning structure.

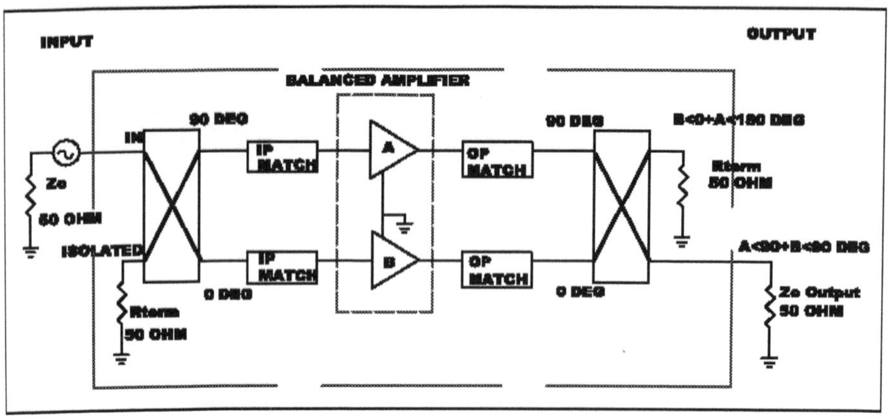

▲ **Figure 4. Conceptual block diagram of the microwave balanced amplifier.**

Push-pull versus balanced:

- Single tone performance should be equivalent. The external elements matching circuits and the splitters and combiners have similar loss. Note that both styles have different advantages in multi-octave amplifiers that do

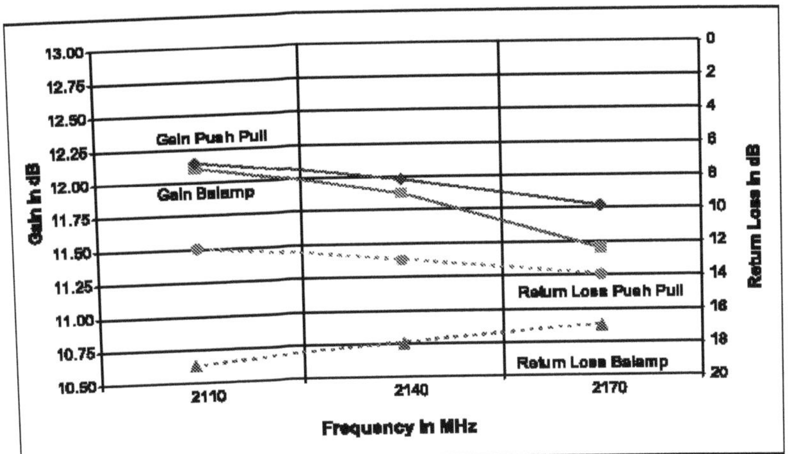

▲ **Figure 5. Gain and return loss versus frequency for average of five devices in each configuration.**

not come into play in this discussion of narrow-band (5 to 10 percent) commercial amplifiers. Because even products and harmonics are out of the passband of the matching circuits and the baluns, the cancellation of these products does not exist. Consequently, for narrow-band applications, this advantage of the push-pull configuration is not realized.

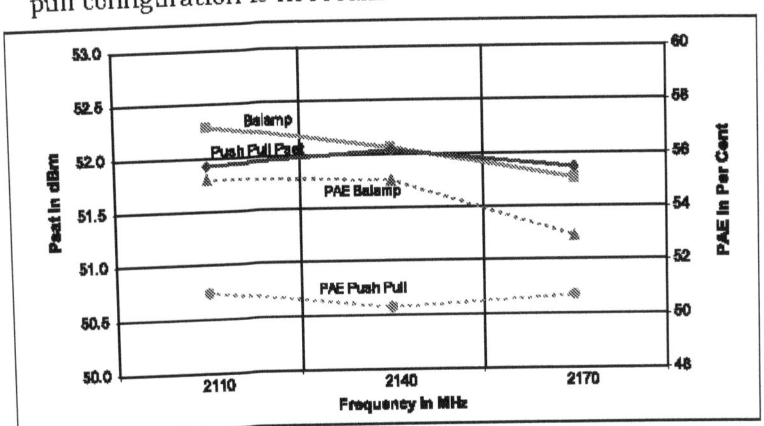

▲ **Figure 6. Saturated power and PAE versus frequency for balanced and push-pull amplifier configurations.**

- Linearity is the same for amplifiers with less than one octave in bandwidth. Matching circuits filter the even products and harmonics for the push-pull configuration and the products and harmonics are attenuated or cancelled by the balanced configuration before they reach the output combiner. The output balun and quadrature coupler have limited bandwidth and cannot cancel products and harmonics that are out of the passband.
- The push-pull configuration has an impedance transformation ratio advantage of two for conventional baluns. This can make the design easier, depending on the impedance to be matched.
- Balanced amplifiers have a significant external match advantage.
- Balanced amplifiers are more stable due to good isolation between the two device sides.
- The virtual ground present for the push-pull configuration can be used with lumped tuning capacitors between the two sides for simpler tuning.
- Both configurations can be tuned using open stubs, preferred in production to lumped capacitors for their lower cost, lower loss, easy modeling and better power handling capability.

After reviewing the advantages and disadvantages of push-pull versus balanced configurations, the amplifier designer may select the best configuration for his application.

To illustrate, two amplifier designs were created at 2.11 to 2.17 GHz, using the GaAs FET device FLL1500IU-2C(1), a Fujitsu 150-watt internally partially matched device. Both balanced and push-pull designs were realized. Five devices were selected from three lots and tested in each amplifier type without changing the amplifier tuning. The measured results from testing the same five devices in each amplifier are summarized in Table 1.

Test data analysis

Figures 5, 6 and 7 show that the two configu-

Parameter/Condition for Vds = 12 volts, Idsq = 4A	Push-Pull Data	Balanced Data
Linear Gain (GL)	12 dB	11.8 dB
Input Return Loss (RL)	13.4 dB	20.2 dB
Saturated Output Power (Psat)	51.9 dBm	52.0 dBm
Power-Added Efficiency at Psat (PAE or Nadd)	51 percent	54 percent
Third-Order Intermodulation Ratio for 43 dBm Total Power out (IMD3)	–36.4 dBc	–38.5 dBc
Adjacent Channel Power Ratio (ACPR) for 3GPP (3.8 MHz) Test Model 1, 64 DPCH CDMA modulation at 43 dBm Pout	–40.5 dBc	–41.8 dBc

▲ **Table 1. Push-pull versus balanced amplifier data. The data for GL, RL, Psat, PAE and ACPR are average for five samples from three lots tested at three frequencies in each amplifier without retuning. IMD3 data is the average of five samples at one frequency in each fixture.**

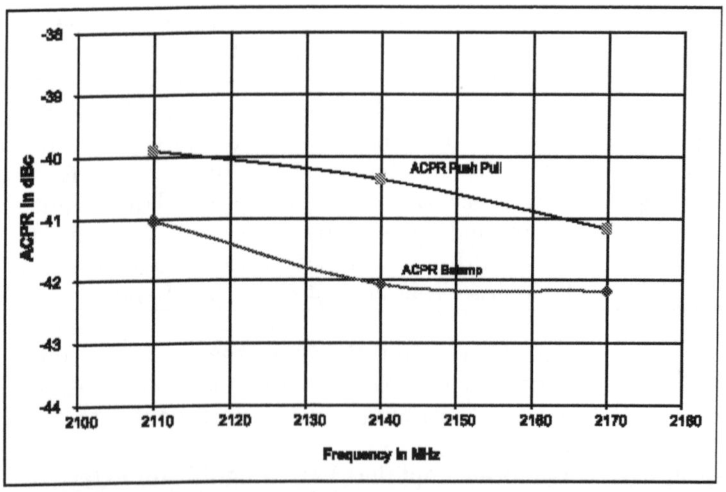

▲ **Figure 7. Adjacent channel power ratio versus frequency for P_{out} = 43 dBm for both the balanced and push-pull amplifier configurations.**

rations have similar (within 0.1 to 0.2 dB) linear gain (GL) and Psat. The linearity performance is slightly better (2.1 dBc for IMD3 and 1.3 dBc for adjacent-channel power ratio (ACPR)) for the balanced configuration. The balanced amplifier exhibits a better power-added efficiency (PAE) by 3.7 percentage points.

This unexpected difference can be misleading. Theoretically, the two types should have the same PAE or linearity. However, this may be explained by the fact that two external matching circuits present to the device the same impedances at the fundamental frequency and different ones for the harmonics. This difference of impedances for harmonics may affect the linearity and efficiency.

The input return loss data shows the superiority of the balanced configuration (6.8 dB better input return loss) concerning the external match.

Conclusion

This article discussed the relative performance of push-pull and balanced amplifiers. A sample 150-watt amplifier of each type was constructed and tested. The performance verified what was anticipated in theory: the basic performance parameters are similar, with the exception of the external match, which is better for the balanced configuration. ■

References

1. FLL1500IU-2C Device Data Sheet, available at http://www.fcsi.fujitsu.com.

2. K. Inoue, et al, "A 240 W Push-Pull GaAs Power FET for W-CDMA Base Stations," *2000 IEEE MTT-S Digest.*

3. R. Basset, "Three Balun Designs For Push-Pull Amplifier," *Microwaves*, July 1980.

4. S. Song and R. Basset, "S-Band Amplifier Modeled For Wireless Data," *Microwaves & RF*, November 2000.

5. L. Max, "Balanced Transistors: A New Option For RF Design," *Microwaves*, June 1977.

6. S. Cripps, *RF Power Amplifiers For Wireless Communications*, Boston: Artech House, 1999.

Author information

Jonathan Shumaker received a bachelor of science degree in electrical engineering from California Poly, SLO, in 1977. Between 1976 and 2000, he worked for Hughes Aircraft, Avantek, Inc., and Litton Solid State. Since 2000, he has been employed with Fujitsu Compound Semiconductors, where he designs high-power amplifiers. He may be reached via E-mail: jshumaker@fcsi.fujitsu.com.

Raymond Bassett has more than 30 years of experience in microwave circuits and active devices. He joined Fujitsu Compound Semiconductor in 2000 as director of Power Product Engineering Design and Application Engineering. He previously worked for CEL/NEC, San Jose, CA. He holds patents on the Gunn oscillator and push-pull package and has authored or co-authored more than 15 technical articles.

Alex Skuratov received a bachelor of science degree in electrical engineering from the Polytechnic University, Kiev, Ukraine. He has more than 20 years of experience in the design and development of RF/microwave components. In 1998, he joined Fujitsu Compound Semiconductor, Inc., where he is manager of the Microwave and Millimeterwave Application Laboratory He previously worked with Mini-Circuits, Inc., Brooklyn, NY.

ABOUT THE EDITORS

John L. B. Walker obtained a B.Sc. in electronic engineering, an M.Sc. in microwaves and communications, and a Ph.D. in microwave directional filters, all from the University of Leeds, in 1970, 1971, and 1976, respectively. In 1974 he joined GEC Hirst Research Centre to work on the design of microwave transistor amplifiers before becoming the group leader of IMPATT diodes and, subsequently, the chief engineer of the Compound Semiconductor Laboratory. In 1984 he joined Thorn-EMI Electronics as the manager of the Microwave Hybrids activity, and in 1995 took up his present position as the RF division manager at Semelab plc. He joined the Board of Directors at Semelab in 2002. Dr. Walker is a Fellow of the IEE and a Senior Member of the IEEE, the chairman of the IEEE MTT-S Technical Program Committee on HF/VHF/UHF Technology, and the editor of the book *High Power GaAs FET Amplifiers* (Artech House, 1993).

Daniel P. Myer is the president and founder of Communication Power Corporation (CPC, http://www.cpcamps.com), a manufacturer of high-power, broadband, RF/microwave amplifiers for the medical, industrial, scientific, and defense industries. CPC's Multi-Nuclear Spectroscopic amplifiers drive 9 out of 10 of the most powerful MRI scanners in the United States, Europe, and Asia, including the world's highest field strength MRI scanner operating at 9.4 tesla at the University of Minnesota's Center for Magnetic Resonance Research Center (CMRR) in Minneapolis, Minnesota. CPC is ISO-9001-2000 registered and a Deloitte & Touche "Fast 50" company for five consecutive years and a North America "Fast 500" company in 2005.

Mr. Myer received a B.S. in electrical engineering from Polytechnic University in Brooklyn, New York, and an A.A.S. from S.U.N.Y. Farmingdale, New York. He is the author of more than 15 technical papers and has been issued two U.S. patents. In 1990 and 1992, he introduced a new topological network synthesis procedure for equal delay transmission line transformers along with the fundamental realization criteria.

Prior to founding CPC, Mr. Myer was an RF design engineer at M/A-COM Microwave Power Devices (MPD), where he codesigned high-power solid-state RF amplifiers for high-profile military programs such as Rivet Rider (Commando Solo), employed in the USAF's Airborne EC-130E-based, psychological warfare programs. In addition, he codeveloped RF amplifiers for use in marine search radar, missile command/destruct, L-band airborne radar tracking, submarine position emergency radio beacon, and multicarrier feedforward networks for cellular phone/wireless base stations.

Mr. Myer is a member of the Eta Kappa Nu Honor Society, the IEEE, the Association of Old Crows (AOC) and the IEEE MTT-S Subcommittee 17 on HF/VHF/UHF Engineering.

Frederick H. Raab is the chief engineer and owner of Green Mountain Radio Research, a consulting firm he founded in 1980. He received a B.S., an M.S., and a Ph.D. in electrical engineering from Iowa State University in 1968, 1970, and 1972, respectively. Dr. Raab is the coauthor of the classic textbook *Solid State Radio Engineering*, is the author of more than 90 technical papers, and has been issued 11 patents. He was the program chairman of RF Expo East '90 and the cofounder of the IEEE MTT-S Subcommittee 17 on HF/VHF/UHF Technology. In 1995 he received ISU's Professional Achievement Citation in Engineering. His professional activities include research, development, and consulting on RF power amplifiers, radio transmitters, and radio-communication/navigation systems. Dr. Raab is a member of the IEEE, HKN (Eta Kappa Nu), ΣΞ, AOC, AFCEA, and RCA.

Chris Trask is the principal engineer of Sonoran Radio Research, where he actively pursues methods for linearizing amplifiers and mixers. He has published numerous articles and papers in hobby, trade, and professional publications on the subjects of amplifier and mixer linearization, transductors, active loop antennas, and wideband transformers. Mr. Trask presently holds six patents in the application of feedback in mixers and the linearization technique of augmentation. He received a B.S.E.E. and an M.S.E.E. from Pennsylvania State University and is a Senior Member of the IEEE.

Recent Titles in the Artech House Microwave Library

Active Filters for Integrated-Circuit Applications, Fred H. Irons

Advanced Techniques in RF Power Amplifier Design, Steve C. Cripps

Automated Smith Chart, Version 4.0: Software and User's Manual, Leonard M. Schwab

Behavioral Modeling of Nonlinear RF and Microwave Devices, Thomas R. Turlington

Broadband Microwave Amplifiers, Bal S. Virdee, Avtar S. Virdee, and Ben Y. Banyamin

Classic Works in RF Engineering: Combiners, Couplers, Transformers, and Magnetic Materials, John L. B. Walker, Daniel P. Myer, Frederick H. Raab, and Chris Trask, editors

Computer-Aided Analysis of Nonlinear Microwave Circuits, Paulo J. C. Rodrigues

Design of FET Frequency Multipliers and Harmonic Oscillators, Edmar Camargo

Design of Linear RF Outphasing Power Amplifiers, Xuejun Zhang, Lawrence E. Larson, and Peter M. Asbeck

Design of RF and Microwave Amplifiers and Oscillators, Pieter L. D. Abrie

Distortion in RF Power Amplifiers, Joel Vuolevi and Timo Rahkonen

EMPLAN: Electromagnetic Analysis of Printed Structures in Planarly Layered Media, Software and User's Manual, Noyan Kinayman and M. I. Aksun

FAST: Fast Amplifier Synthesis Tool—Software and User's Guide, Dale D. Henkes

Feedforward Linear Power Amplifiers, Nick Pothecary

Generalized Filter Design by Computer Optimization, Djuradj Budimir

High-Linearity RF Amplifier Design, Peter B. Kenington

High-Speed Circuit Board Signal Integrity, Stephen C. Thierauf

Integrated Circuit Design for High-Speed Frequency Synthesis, John Rogers, Calvin Plett, and Foster Dai

Intermodulation Distortion in Microwave and Wireless Circuits, José Carlos Pedro and Nuno Borges Carvalho

Lumped Elements for RF and Microwave Circuits, Inder Bahl

Microwave Circuit Modeling Using Electromagnetic Field Simulation, Daniel G. Swanson, Jr. and Wolfgang J. R. Hoefer

Microwave Component Mechanics, Harri Eskelinen and Pekka Eskelinen

Microwave Engineers' Handbook, Two Volumes, Theodore Saad, editor

Microwave Filters, Impedance-Matching Networks, and Coupling Structures, George L. Matthaei, Leo Young, and E.M.T. Jones

Microwave Materials and Fabrication Techniques, Second Edition, Thomas S. Laverghetta

Microwave Mixers, Second Edition, Stephen A. Maas

Microwave Radio Transmission Design Guide, Trevor Manning

Microwaves and Wireless Simplified, Thomas S. Laverghetta

Modern Microwave Circuits, Noyan Kinayman and M. I. Aksun

Neural Networks for RF and Microwave Design, Q. J. Zhang and K. C. Gupta

Nonlinear Microwave and RF Circuits, Second Edition, Stephen A. Maas

QMATCH: Lumped-Element Impedance Matching, Software and User's Guide, Pieter L. D. Abrie

Practical Analog and Digital Filter Design, Les Thede

Practical RF Circuit Design for Modern Wireless Systems, Volume I: Passive Circuits and Systems, Les Besser and Rowan Gilmore

Practical RF Circuit Design for Modern Wireless Systems, Volume II: Active Circuits and Systems, Rowan Gilmore and Les Besser

Production Testing of RF and System-on-a-Chip Devices for Wireless Communications, Keith B. Schaub and Joe Kelly

Radio Frequency Integrated Circuit Design, John Rogers and Calvin Plett

RF Design Guide: Systems, Circuits, and Equations, Peter Vizmuller

RF Measurements of Die and Packages, Scott A. Wartenberg

The RF and Microwave Circuit Design Handbook, Stephen A. Maas

RF and Microwave Coupled-Line Circuits, Rajesh Mongia, Inder Bahl, and Prakash Bhartia

RF and Microwave Oscillator Design, Michal Odyniec, editor

RF Power Amplifiers for Wireless Communications, Steve C. Cripps

RF Systems, Components, and Circuits Handbook, Second Edition, Ferril A. Losee

Stability Analysis of Nonlinear Microwave Circuits, Almudena Suárez and Raymond Quéré

TRAVIS 2.0: Transmission Line Visualization Software and User's Guide, Version 2.0, Robert G. Kaires and Barton T. Hickman

Understanding Microwave Heating Cavities, Tse V. Chow Ting Chan and Howard C. Reader

For further information on these and other Artech House titles, including previously considered out-of-print books now available through our In-Print-Forever® (IPF®) program, contact:

Artech House
685 Canton Street
Norwood, MA 02062
Phone: 781-769-9750
Fax: 781-769-6334
e-mail: artech@artechhouse.com

Artech House
46 Gillingham Street
London SW1V 1AH UK
Phone: +44 (0)20 7596-8750
Fax: +44 (0)20 7630 0166
e-mail: artech-uk@artechhouse.com

Find us on the World Wide Web at: www.artechhouse.com

www.ingramcontent.com/pod-product-compliance
Ingram Content Group UK Ltd.
Pitfield, Milton Keynes, MK11 3LW, UK
UKHW052301280926
15471UKWH00006B/1050

9 781596 933767